BURGER'S MEDICINAL CHEMISTRY, DRUG DISCOVERY AND DEVELOPMENT

BURGER'S MEDICINAL CHEMISTRY, DRUG DISCOVERY AND DEVELOPMENT

Editors-in-Chief
Donald J. Abraham
Virginia Commonwealth University

David P. Rotella
Wyeth Research

Consulting Editor
Al Leo
BioByte Corp

Editorial Board
John H. Block
Oregon State University

Robert H. Bradbury
AstraZeneca

Robert W. Brueggemeier
Ohio State University

John W. Ellingboe
Wyeth Research

William R. Ewing
Bristol-Myers Squibb Pharmaceutical Research Institute

Richard A. Gibbs
Purdue University

Richard A. Glennon
Virginia Commonwealth University

Barry Gold
University of Pittsburgh

William K. Hagmann
Merck Research Laboratories

Glen E. Kellogg
Virginia Commonwealth University

Christopher A. Lipinski
Melior Discovery

John A. Lowe III
JL3Pharma LLC

Jonathan S. Mason
Lundbeck Research

Andrea Mozzarelli
University of Parma

Bryan H. Norman
Eli Lilly and Company

John L. Primeau
AstraZeneca

Paul J. Reider
Princeton University

Albert J. Robichaud
Lundbeck Research

Alexander Tropsha
University of North Carolina

Patrick M. Woster
Wayne State University

Jeff Zablocki
CV Therapeutics

Editorial Staff
VP & Director, STMS Book Publishing: **Janet Bailey**
Editor: **Jonathan Rose**
Production Manager: **Shirley Thomas**
Production Editor: **Kris Parrish**
Illustration Manager: **Dean Gonzalez**
Editorial Program Coordinator: **Surlan Alexander**

BURGER'S MEDICINAL CHEMISTRY, DRUG DISCOVERY AND DEVELOPMENT

Seventh Edition

Volume 3: Drug Development

Edited by

Donald J. Abraham
Virginia Commonwealth University

David P. Rotella
Wyeth Research

Burger's Medicinal Chemistry, Drug Discovery and Development
is available Online in full color at
http://mrw.interscience.wiley.com/emrw/9780471266945/home/

A JOHN WILEY & SONS, INC., PUBLICATION

Copyright © 2010 by John Wiley & Sons, Inc. All rights reserved

Published by John Wiley & Sons, Inc., Hoboken, New Jersey
Published simultaneously in Canada

No part of this publication may be reproduced, stored in a retrieval system, or transmitted in any form or by any means, electronic, mechanical, photocopying, recording, scanning, or otherwise, except as permitted under Section 107 or 108 of the 1976 United States Copyright Act, without either the prior written permission of the Publisher, or authorization through payment of the appropriate per-copy fee to the Copyright Clearance Center, Inc., 222 Rosewood Drive, Danvers, MA 01923, (978) 750-8400, fax (978) 750-4470, or on the web at www.copyright.com. Requests to the Publisher for permission should be addressed to the Permissions Department, John Wiley & Sons, Inc., 111 River Street, Hoboken, NJ 07030, (201) 748-6011, fax (201) 748-6008, or online at http://www.wiley.com/go/permission.

Limit of Liability/Disclaimer of Warranty; While the publisher and author have used their best efforts in preparing this book, they make no representations or warranties with respect to the accuracy or completeness of the contents of this book and specifically disclaim any implied warranties of merchantability or fitness for a particular purpose. No warranty may be created or extended by sales representatives or written sales materials. The advice and strategies contained herein may not be suitable for your situation. You should consult with a professional where appropriate. Neither the publisher nor author shall be liable for any loss of profit or any other commercial damages, including but not limited to special, incidental, consequential, or other damages.

For general information on our other products and services or for technical support, please contact our Customer Care Department within the United States at (800) 762-2974, outside the United States at (317) 572-3993 or fax (317) 572-4002.

Wiley also publishes its books in a variety of electronic formats. Some content that appears in print may not be available in electronic formats. For more information about Wiley products, visit our web site at www.wiley.com.

Library of Congress Cataloging-in-Publication Data:

Abraham, Donald J., 1936-
  Burger's medicinal chemistry, drug discovery, and development/Donald J. Abraham, David P. Rotella. – 7th ed.
       p. ; cm.
  Other title: Medicinal chemistry, drug discovery, and development
  Rev. ed. of: Burger's medicinal chemistry and drug discovery. 6th ed. / edited by Donald J. Abraham. c2003.
    Includes bibliographical references and index.
    ISBN 978-0-470-27815-4 (cloth)
 1. Pharmaceutical chemistry. 2. Drug development. I. Rotella, David P. II. Burger, Alfred, 1905-2000. III. Burger's medicinal chemistry and drug discovery. IV. Title. V. Title: Medicinal chemistry, drug discovery, and development.
    [DNLM: 1. Chemistry, Pharmaceutical–methods. 2. Biopharmaceutics–methods. 3. Drug Compounding–methods.   QV 744 A105b 2010]
    RS403.B8 2010
    615'.19–dc22
                                                            2010010779

Printed in Singapore

10 9 8 7 6 5 4 3 2

CONTENTS

PREFACE		vii
CONTRIBUTORS		ix
1	Large-Scale Synthesis	1
2	Physicochemical Characterization and Oral Dosage Form Selection Based on the Biopharmaceutics Classification System	25
3	The FDA and Drug Regulatory Issues	63
4	Intellectual Property in Drug Discovery and Biotechnology	101
5	Polymorphic Crystal Forms and Cocrystals in Drug Delivery (Crystal Engineering)	187
6	Prodrugs: Strategic Deployment, Metabolic Considerations, and Chemical Design Principles	219
7	Process Development of Protein Therapeutics	289
8	Cost-Effectiveness Analyses Throughout the Drug Development Life Cycle	345
9	Provisional BCS Classification of the Leading Oral Drugs on the Global Market	353
10	The Role of Permeability in Drug ADME/PK, Interactions and Toxicity, and the Permeability-Based Classification System (PCS)	367
11	Salt Screening and Selection	381
12	Enzymatic Assays for High-Throughput Screening	401
13	Crystallographic Survey of Albumin Drug Interaction and Preliminary Applications in Cancer Chemotherapy	437
14	Nanotechnology in Drug Delivery	469
INDEX		493

PREFACE

The seventh edition of Burger's Medicinal Chemistry resulted from a collaboration established between John Wiley & Sons, the editorial board, authors, and coeditors over the last 3 years. The editorial board for the seventh edition provided important advice to the editors on topics and contributors. Wiley staff effectively handled the complex tasks of manuscript production and editing and effectively tracked the process from beginning to end. Authors provided well-written, comprehensive summaries of their topics and responded to editorial requests in a timely manner. This edition, with 8 volumes and 116 chapters, like the previous editions, is a reflection of the expanding complexity of medicinal chemistry and associated disciplines. Separate volumes have been added on anti-infectives, cancer, and the process of drug development. In addition, the coeditors elected to expand coverage of cardiovascular and metabolic disorders, aspects of CNS-related medicinal chemistry, and computational drug discovery. This provided the opportunity to delve into many subjects in greater detail and resulted in specific chapters on important subjects such as biologics and protein drug discovery, HIV, new diabetes drug targets, amyloid-based targets for treatment of Alzheimer's disease, high-throughput and other screening methods, and the key role played by metabolism and other pharmacokinetic properties in drug development.

The following individuals merit special thanks for their contributions to this complex endeavor: Surlan Alexander of John Wiley & Sons for her organizational skills and attention to detail, Sanchari Sil of Thomson Digital for processing the galley proofs, Jonathan Mason of Lundbeck, Andrea Mozzarelli of the University of Parma, Alex Tropsha of the University of North Carolina, John Block of Oregon State University, Paul Reider of Princeton University, William (Rick) Ewing of Bristol-Myers Squibb, William Hagmann of Merck, John Primeau and Rob Bradbury of AstraZeneca, Bryan Norman of Eli Lilly, Al Robichaud of Wyeth, and John Lowe for their input on topics and potential authors. The many reviewers for these chapters deserve special thanks for the constructive comments they provided to authors. Finally, we must express gratitude to our lovely, devoted wives, Nancy and Mary Beth, for their tolerance as we spent time with this task, rather than with them.

As coeditors, we sincerely hope that this edition meets the high expectations of the scientific community. We assembled this edition with the guiding vision of its namesake in mind and would like to dedicate it to Professor H.C. Brown and Professor Donald T. Witiak. Don collaborated with Dr. Witiak in the early days of his research in sickle cell drug discovery. Professor Witiak was Dave's doctoral advisor at Ohio State University and provided essential guidance to a young scientist. Professor Brown, whose love for chemistry infected all organic graduate students at Purdue University, arranged for Don to become a medicinal chemist by securing a postdoctoral position for him with Professor Alfred Burger.

It has been a real pleasure to work with all concerned to assemble an outstanding and up-to-date edition in this series.

DONALD J. ABRAHAM
DAVID P. ROTELLA

March 2010

CONTRIBUTORS

Renée J.G. Arnold, Arnold Consultancy & Technology LLC, New York, NY; Mount Sinai School of Medicine, New York, NY; Arnold and Marie Schwartz College of Pharmacy, Brooklyn, NY

Gregory E. Amidon, University of Michigan, Ann Arbor, MI

Gordon L. Amidon, University of Michigan, Ann Arbor, MI

Daniele Carettoni, Axxam, Milano, Italy

Daniel C. Carter, New Century Pharmaceuticals, Inc., Huntsville, AL

Arik Dahan, Ben-Gurion University of the Negev, Beer-Sheva, Israel

Paul W. Erhardt, University of Toledo, Toledo, OH

Urban Fagerholm, AstraZeneca R&D Södertälje, Södertälje, Sweden

S. Sam Guhan, Amgen, Inc., Seattle, WA

Frank Gupton, Virginia Commonwealth University, Richmond, VA

Michael J. Hageman, Bristol-Myers Squibb Company, Princeton, NJ

Xiaorong He, Boehringer-Ingelheim, Danbury, CT

Janan Jona, Amgen, Inc., Thousand Oaks, CA

Richard A. Kaba, Fitch, Even, Tabin & Flannery, Chicago, IL

Rahul Khupse, University of Toledo, Toledo, OH

Rudy Kratz, Fitch, Even, Tabin & Flannery, Chicago, IL

James P. Krueger, Fitch, Even, Tabin & Flannery, Chicago, IL

Timothy P. Maloney, Fitch, Even, Tabin & Flannery, Chicago, IL

Calista J. Mitchell, Fitch, Even, Tabin & Flannery, Chicago, IL

Matthew L. Peterson, Amgen, Inc., Cambridge, MA

Dean K. Pettit, Amgen, Inc., Seattle, WA

Ramnarayan S. Randad, U.S. Food and Drug Administration, Rockville, MD

Jeffrey G. Sarver, University of Toledo, Toledo, OH

Andreas G. Schatzlein, University of London, London, UK

David J. Semin, Amgen, Inc., Thousand Oaks, CA

Ning Shan, Thar Pharmaceuticals, Inc., Tampa, FL

James N. Thomas, Amgen, Inc., Seattle, WA

Jill A. Trendel, University of Toledo, Toledo, OH

Ijeoma F. Uchegbu, University of London, London, UK

Philippe Verwaerde, iNovacia, Stockholm, Sweden

Roger Zanon, Amgen, Inc., Thousand Oaks, CA

Michael J. Zaworotko, University of South Florida, Tampa, FL

LARGE-SCALE SYNTHESIS

Frank Gupton
Department of Chemistry,
Department of Chemical and Life
Science Engineering,
Virginia Commonwealth University,
Richmond, VA

1. INTRODUCTION

The ability to produce active pharmaceutical ingredients (APIs) to support the various disciplines of the drug development process is an enabling element of pharmaceutical product development. In the initial stages of drug development, bulk active materials are typically supplied from bench-scale laboratory synthesis. However, API requirements can quickly exceed the capacity of normal laboratory operations, thus making it necessary to carry out the synthesis of the drug candidate on a larger scale. Section 32.2 is devoted to providing a general overview of the issues and requirements associated with the scale-up of chemical processes from the laboratory to pilot and commercial-scale operations.

Section 3 describes the process development of nevirapine, a novel nonnucleoside reverse transcriptase (NNRT) inhibitor used in the treatment of AIDS. This case study details the evolution of the nevirapine process from conception in medicinal chemistry through process development, pilot plant scale-up, and commercial launch of the bulk active drug substance. Restricting the case study to nevirapine allows the process and rationale to be described in more detail. The author is aware of the vast amount of excellent process development that has been performed in the commercialization of other drug products. The processes described herein are not necessarily unique solutions to this particular synthesis. To some extent, they reflect the culture, philosophy, raw materials, equipment, and synthetic tools available during the period 1990–1996, as well as the ingenuity of the process chemists.

2. SCALE-UP

2.1. General

The process development and scale-up of APIs require a multidisciplinary cooperation between organic chemists, analytical chemists, and engineers, quality control, quality assurance, and plant operations. Furthermore, the development of a drug candidate requires collaboration with pharmaceutics for formulation studies, drug metabolism and pharmacokinetics, toxicology, clinical studies, purchasing, and marketing. Outsourcing specialists, working in concert with purchasing, also play a key role in identification, coordination, and procurement of key raw materials in support of the scale-up effort. This particular function has gained greater importance in recent years as a result of the increasing emphasis in the pharmaceutical industry to improve the overall efficiency of the drug development process.

In the early stages of process development, the chemist must often balance the need to optimize each synthetic step with the API delivery requirements for toxicology, formulation, and clinical trials. To fulfill these requirements, the process chemist may often scale-up a process in the pilot plant with less than optimal process conditions. As a result, the first quantities of API produced in the pilot plant can be the most challenging to prepare. However, as the drug candidate passes through the various stages of drug development, the probability of commercialization increases and the need to address the commercial viability of the process becomes more important. This section presents an overview of the issues associated with the preparation of multi-kilogram quantities of APIs throughout the drug development process.

2.2. Synthetic Strategy

The types of development activities that are associated with the large-scale synthesis of a drug candidate can be divided into a series of discrete functions. Although the terminology used to describe these activities may vary, for the purpose of these discussions the specific functions of the drug development process

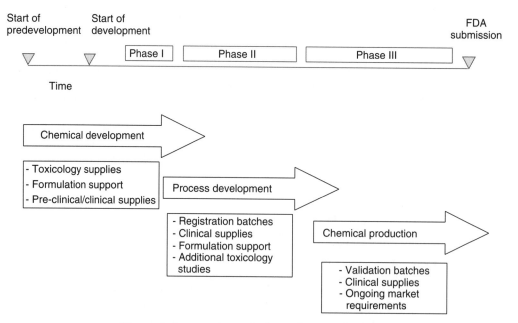

Figure 1. Large-scale synthesis requirements for drug.

related to chemical synthesis will be divided into the following three categories: (1) chemical development, (2) process development, and (3) commercial production.

Figure 1 indicates the specific areas of the drug development process where each of these activities occurs. Although each function has specific requirements and outputs from its respective activities, the overlap that is indicated between these activities is critical to the successful implementation of the project.

In the initial stages of *chemical development*, the focus of the effort is to supply materials to assess the viability of the drug candidate. The emphasis of this effort is on the expeditious supply of these materials rather than the commercial viability of the process used to produce the compound. Unique raw materials, reagents, solvents, reaction conditions, and purification techniques can and will be employed in this phase of the process to produce the desired compound in a timely fashion. The initial transition from laboratory to pilot-scale operations typically takes place during this portion of the drug development process to supply larger quantities of the bulk active material for toxicology, formulation, and preclinical evaluations. As the project proceeds through drug development, chemical development personnel continue to evaluate potential improvements to the synthesis. The insights obtained from these efforts provide the platform for future process development investigation.

The role of *process development* is to balance the timeline and material requirements of the project with the need to develop a commercially viable method for the preparation of the drug candidate. This stage of the drug development process will concentrate on such issues as (1) synthetic strategy, (2) improvement of individual reaction yields, (3) identification and use of commercially available raw materials and reagents, (4) evaluation of alternative solvent systems, (5) compatibility of process conditions with existing manufacturing assets, (6) identification and quantification of potential process safety hazards, (7) simplification of purification methods, (8) evaluation of process waste streams, and (9) the improvement of the overall process economics.

Both chemical and process development activities typically require that the drug candidate be prepared on a pilot plant scale. Although the batch size may vary depending on the drug substance requirements, these operations are usually conducted in 100–2000-L reactors. The scale-up factor from the laboratory to the pilot plant is quite large (1–200 or more), and particular emphasis is placed on detailed safety analysis of this scale-up. The outcome of these efforts is a documented process that is included in the drug submission package to the U.S. Food and Drug Administration.

The overall objective of *chemical production* activities is to reproduce the process that has been transferred from process development to meet the current and future market requirements for the drug product. Particular emphasis is placed on issues related to process safety, environmental issues, equipment requirements, and production economics. The scale-up factor from the pilot plant to commercial production is usually rather small (approximately 1–20). As a result, the information obtained from the process development efforts can be quite valuable in the successful implementation of the commercial process. The reproducibility of the process is confirmed and documented as part of the process validation package, which in turn is part of the transfer process.

2.2.1. Route Selection When considering the merits of alternative synthetic pathways to produce a specific molecule, the route that incorporates the most convergent subroutes is generally the most advantageous option, provided yields for the individual steps are essentially equivalent [1]. For example, an eight-step linear synthesis (Fig. 2), in which each step has an 85% yield, results in a 27% overall yield (Case I). However, if the eight steps can be divided into two three-step converging pathways leading to two final steps, as in Case II, the overall process yield is increased to 44%, which is a 63% improvement over the Case I scenario. Furthermore, if the process is broken down to even shorter converging pathways, as in Case III, the overall yield improves by 25%, from Case I, to 61%.

In addition to the obvious yield advantages, an important benefit of a convergent approach is the proximity of the starting materials to

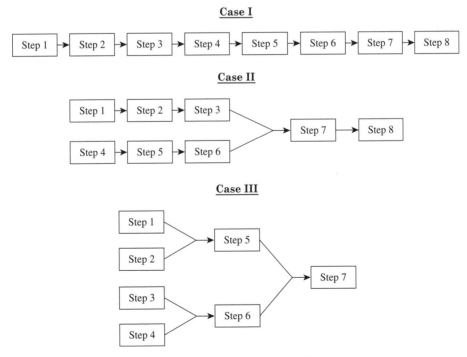

Figure 2. Convergent synthesis.

the product. In Case II, the raw materials are only five steps away from the product and only three steps away in Case III. This can significantly reduce the time required to respond to an unexpected need for additional product. Also, the value of each intermediate in a linear synthesis becomes greater with each additional step as a result of the resources required to produce material from that step. In a convergent synthesis, the cost is spread over two or more intermediates, thus reducing the overall risk in the event of material losses. Convergence may also provide an advantage in developing the regulatory filing strategy. In this manner, greater chemical complexity can be built into the registered starting materials that can greatly simplify the ability to implement process improvements following the product launch.

2.2.2. Chiral Requirements Over the last several decades, drug development efforts have placed increasing emphasis on the development of the biologically active stereoisomers of drug products. Chiral APIs offer the opportunity to provide higher drug potency while reducing the metabolic burden and risk of undesirable side effects to the patient [2]. It has been estimated that over half the best-selling drugs worldwide are single enantiomers [3]. As a result, the process chemist is presented with the challenge of developing commercially viable processes for the production and isolation of these chiral compounds. Several approaches can be used to produce enantiomerically enriched bulk active pharmaceutical products. The resolution of racemic mixtures with chiral adjuvants has been a common approach in the past to isolate the desired optical isomer of drug products. Chiral amines and acids are typically used to isolate an enantiomer by crystallization of the diastereomeric salt. The major drawback with this approach is the significant loss of material as the undesired enantiomer. This can be mitigated by racemization of the off isomer followed by recycling of the racemate back into the resolution. However, the equipment requirements to execute this procedure can be significant and must be economically justified.

An alternative approach for the preparation of chiral APIs is the use of chiral raw materials. The increased availability of functionalized chiral raw materials from both synthetic and natural sources has made this a more viable option in recent years. When desired chiral precursors are not commercially available, asymmetric synthetic techniques may be employed to introduce one or more stereogenic centers into the molecule. Many elegant techniques have been developed using chiral induction [4], chiral templates [5], and chiral catalysts [6] to produce enantiomerically enriched drug substances, and this area of research continues to be at the forefront of organic chemistry.

Regardless of the approach used to introduce the stereogenic center(s) into the molecule, a significant cost is incurred in achieving this objective. For this reason, it is important to introduce the chiral component later in the synthesis and employ the principles of convergent synthesis (Section 2.2.1) to effectively minimize the impact of this cost to the overall process economics.

2.3. Bench-Scale Experimentation

A significant laboratory effort is required to define the operating ranges of the critical process parameters in order to scale up a chemical process into pilot or commercial-scale operations. A critical process parameter is any process variable that may potentially affect the product quality and/or yield. This information is required to prepare a process risk analysis, which is an FDA prerequisite for process validation. Process parameters that are often evaluated as part of the risk analysis include reaction temperature, solvent systems, reaction time, raw material and reagent ratios, rate and orders of addition, agitation, and reaction concentration. If catalysts are employed as part of the process, additional laboratory evaluation may also be required to further define the process limits. Process recycling of solvents and other materials must require data that define the impact on product quality as well as the limits and specifications of recyclability.

Experimental design is often used for the evaluation of critical process parameters to minimize the total laboratory effort [7]. This technique is equally important in identifying interdependent process parameters that can have a synergistic impact on product yield and quality. In-process control (IPC) requirements are also defined during this phase of the development process. All these bench-scale activities help provide a better understanding of the capabilities and limitations of the process and are discussed in further detail in this section.

2.3.1. Selection of Reaction Solvents Solvents are generally used to promote the solubility of reagents and starting materials in a reaction mixture. Reactants in solution typically undergo conversion to product at a higher rate of reaction and are generally easier to scale up because of the elimination of mass transfer issues. For this reason, the solubility properties of the reagents and raw materials are a major consideration in the solvent selection process for scale-up. In addition, the solvent must be chemically compatible with the reagents and raw materials to avoid adverse side reactions. For example, an alcohol solvent would be a poor choice for a reaction when a strong base such as butyl lithium is being employed as a reagent. Information pertaining to the physical properties of solvents is available to assist in the solvent selection process [8].

Solvents can also be used to promote product isolation and purification. An ideal solvent system is one that exhibits high solubility with the reagents and starting materials but only limited solubility with the reaction product. Precipitation of the reaction product from the mixture can increase the reaction rate, drive reactions in equilibrium to completion, and isolate the product in the solid state to minimize the risk of undesirable side reactions. Solvents can also aid in the regiocontrol of the reaction pathway. It was found in the preparation of nevirapine (**3**) that when diglyme was used as the reaction solvent with sodium hydride, the ring closure of (**1**) (Scheme 1) proceeded by the desired reaction pathway [9]. However, when dimethyl formamide was used for this reaction, the exclusive product was the oxazolopyridine (**2**). In this particular case, the solvation effects may have helped stabilize the transition state of the desired product.

One of the most challenging aspects of solvent selection is the avoidance of certain classes of solvent that are routinely used in laboratory operations but are inappropriate for pilot and commercial-scale applications. Solvents such as benzene and 1,4-dioxane can present significant health risks to employees

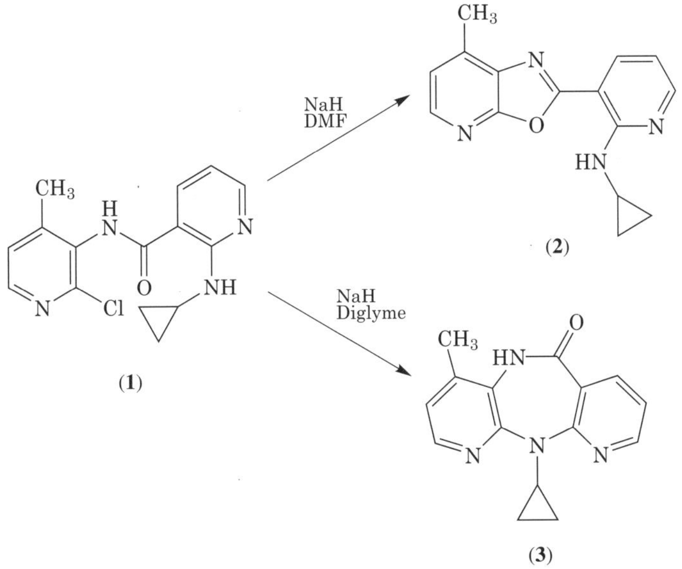

Scheme 1.

handling large quantities of these materials [10]. Toluene is routinely used as a commercial substitute for benzene and other aromatic solvents. Likewise, solvents that promote peroxide formation such as diethyl ether and tetrahydrofuran present significant safety hazards in scale-up operations [11]. Methyl *tert*-butyl ether is a good commercial substitute for these materials. The autoignition temperature of the solvent should also be considered against the process operating conditions and electrical classifications of the equipment being used. With regard to environmental issues, several chlorinated solvents have been identified as priority pollutants [12] and can present permitting issues if adequate environmental containment capabilities are not incorporated into the scale-up facility. Although specific health, safety, and environmental issues for a given solvent can usually be addressed, it is important to evaluate the advantages of using an undesirable solvent against the additional cost and operational constraints that are imposed on the process.

2.3.2. Reaction Temperature Before conducting a reaction temperature profile experiment, it is important to understand the temperature limitations of the specific scale-up equipment that is to be used. For example, the typical operating temperature for a pilot or production facility employing a silicone-based heat transfer system ranges from -20 to $180°C$. It is also important to understand the capabilities of the temperature control system used in the scale-up facility. The selected reaction temperature range must also be consistent with the accuracy and precision limits of the equipment. Given these constraints, the objective of this effort is to identify the optimal temperature range that gives the maximum conversion of starting materials to product in the shortest period of time and with the minimum amount of impurity formation. A general rule for the evaluation of reaction temperature is that increasing the reaction temperature by $10°C$ will double the reaction rate. However, this will also increase the potential for by-product formation, which could adversely impact both product yield and quality. The optimal temperature range is typically a balance between these three dependent variables.

2.3.3. Reaction Time In a laboratory environment, reactions are often run overnight with limited concern for the actual time requirements to complete the reaction. When selecting the reaction time for a specific process step to be scaled up, consideration should be given both to the potential reaction yield improvement and to the equipment utilization requirements. In many cases, doubling the reaction time will result in only a small percentage increase in yield. The cost of the additional equipment time can more than offset the potential yield benefit for cases in which the raw material costs are low. However, in cases where raw materials of high cost and greater chemical complexity are employed, the additional reaction time may be easily justified on an economic basis.

Consideration should also be given to the quantification of potential adverse effects from extending the reaction time beyond the optimum condition. Product decomposition and by-product formation are often observed under these circumstances. This information can be beneficial in scale-up operations when reaction times are extended beyond the specified period due to unforeseen circumstances. This information is also important in evaluation of this variable as a potential critical process parameter for the process risk assessment.

2.3.4. Reaction Stoichiometry and Order of Addition Reaction rates, product yields, and by-product formation can often be effectively managed by the selection of appropriate ratios of reactants and raw materials as well as by the rate and order of addition of these materials. A fundamental mechanistic understanding of the process is essential for the effective evaluation of these parameters. Reaction kinetic information can be beneficial in defining the limiting reagent for the reaction under evaluation. More often, the financial impact of specific raw materials will be a key driver of the overall process economics, and as a result, optimization efforts will focus on the minimization of these materials. This issue has gained increasing importance because of the chemical complexity of advanced starting materials in bulk pharmaceutical production. Likewise, a statistical design of experiments can assist in the evaluation of multiple process parameters and also identify interactions between multiple process variables.

The minimization of by-product formation can be a particularly difficult task because of the high degree of chemical functionality in bulk pharmaceutical intermediates and products. Oligomerization reactions are a major mode of impurity formation in these types of chemical processes and can often be effectively minimized by the control of addition rates. Characterization of these impurities can also provide valuable insights into the control of these side reactions. The order and rate of addition are also frequently used to control extremely exothermic reactions. Chlorinating reagents such as thionyl chloride and phosphorous oxychloride, as well as strong bases such as butyl lithium, lithium diisopropylamide, and sodium hydride are usually added in a controlled manner to limit both heat and by-product formation in these reactions.

2.3.5. Solid-State Requirements The solid-state properties of active pharmaceutical ingredients can have a dramatic impact on critical dosage form parameters such as bioavailability and product stability. For this reason, FDA filing requirements include the definitive characterization of drug substance physical properties as part of the NDA information package. Formulation activities during the drug development process are directly linked to these parameters, and control of these physical properties during laboratory, pilot, and commercial-scale operations can be challenging.

The particle size distribution of the API can affect the dissolution rate of the formulated product and thus the drug bioavailability. Once particle size requirements have been defined from formulation studies, the process must be capable of routinely meeting these requirements. One of the ways that particle size distribution can be controlled is by the conditions under which the product is crystallized. Typically for cooling crystallizations, the particle size distribution depends on the rate of cooling. In general, smaller size particles are formed under rapid cooling conditions, whereas larger crystal growth is experienced with slower cooling rates. Milling and grinding techniques can also control particle size. However, these methods exclusively result in particle size reduction. Both the milling conditions and the solid-state characteristics of the bulk active material being charged to the mill thus determine the particle size distribution of the API. Milling parameters are discussed in more detail in Section 2.4.4.

Bulk drug products often exist in different crystalline or polymorphic forms. Because the polymorphs of a specific API can exhibit distinguishably different bulk stability properties and bioavailability characteristics as a result of the differences in surface area between the different crystalline forms, specification of the polymorphic form is typically required for FDA submission. Products such as ranitidine [13], lorazepam [14], and natamycin [15] serve as examples of APIs that exist in several different polymorphic forms. The solvent system and the crystallization conditions generally determine the specific crystallization form that is isolated. Polymorph selection for regulatory submission is usually based on the ability to reliably produce and process the material in the same crystalline form. In many cases, this is the thermodynamically most stable polymorphic form. In the event that a less stable polymorphic form is desired, because of the stability or bioavailability issues, seeding techniques can be used to control the crystallization selectivity of a specific polymorph. However, when seeding is required a seeding strategy must be developed including specifications and documentation of the seed history or genealogy.

2.4. Scale-Up from Bench to Pilot Plant

Bulk active pharmaceutical ingredients are most often produced at the pilot scale under batch-mode operations with multipurpose equipment. In contrast, continuous operations are typically reserved for high volume products that can be produced in dedicated facilities. Although there have been significant advances in the development of continuous microreactor technology in recent years (see Section 2.6) the focus of scale-up discussions in this section will be restricted to batch-scale operations. From a procedural perspective, batch operations more closely resemble conventional bench-scale operations. However, the successful transformation of bench-scale experiments in laboratory glassware to pilot and commercial-scale operations requires a more detailed understanding of the physical issues related to scale-up, such as heating and cooling requirements, agitation, liquid–solid separation techniques, and solids handling requirements. Particular emphasis is placed on understanding the thermal requirements because this can often be the area of greatest perceived risk. This can influence the rate of by-product formation, which has an impact on both the impurity profile and the yield. Fortunately, reactions proceed by the same mechanism regardless of the scale, and problems in scale-up are typically restricted to physical parameters.

2.4.1. Heating and Cooling A pilot plant is generally outfitted with multipurpose vessels that can obtain an operating temperature range of -20 to $+150°C$. Broader temperature ranges can be obtained with silicone-based heat-transfer fluids. Temperatures lower than $-20°C$ can be required in API production and can be achieved with liquid nitrogen cooling systems.

The heating and cooling capabilities of a reactor system are determined by several factors. Variables such as reactor surface area, materials of construction, the temperature of the heating and cooling media, and the heat capacity of the reactor contents contribute to the thermal properties of the reactor system. The effects of these parameters on heating and cooling are greatly magnified upon scale-up from the bench to the pilot plant. For example, a 250-mL round-bottom flask in the laboratory has a large surface area to volume ratio. As a result, the flask can be heated and cooled quickly. In comparison, the surface area to volume ratio of a 100-L glass-lined steel reactor is drastically reduced and may influence the ability to effectively control the reactor contents. In general, in transitioning from a 250-mL flask to a 100-L reactor, the surface area versus volume is reduced by a factor of 10. Likewise, the surface heat constant (k) of a stainless steel reactor is much greater than that of a laboratory reaction flask, which could result in a thermal transfer that is much more rapid than that of the laboratory experience.

This effect of heating and cooling can be calculated as follows [16]:

$$T = t_s - (t_s - t^0)e^{-kF/C} \qquad (1)$$

where T is the temperature of the vessel in degree-centigrade, t^0 is t at the beginning of the heating, t_s is the temperature of the heat-exchange fluid, F is the reactor surface, k is the heat constant on the surface (kcal/m^2/h/C), F is the heat surface, and C is the heat capacity of the reaction vessel with contents.

2.4.2. Agitation The key function of agitation is to ensure homogeneity of the reactor contents. The major factors that affect reactant homogeneity are both the reactor-agitator configuration and the physical properties of the reactor contents. Miscible liquids of low viscosity, such as ethanol and water, represent mixtures with which one can easily attain homogeneity with minimal agitation. As one might expect, biphasic mixtures require more vigorous agitation than miscible solutions. The extent of the additional agitation requirement depends on the viscosities of the individual phases. Liquid–solid mixtures also require greater agitation to increase the uniform dispersion of reactor contents. In many cases, the solid is formed later in the process, resulting in different agitation requirements over the duration of the reaction.

Catalytic hydrogenations can represent some of the most challenging agitation issues. A typical hydrogenation reaction will require the dispersion of a heterogeneous catalyst and hydrogen gas throughout a specific solution containing the material that is to undergo the reduction. Hydrogenation agitators are often specifically designed to maximize the dispersion of the hydrogen gas throughout the liquid phase.

The ability to transfer heat to the reaction mixture is also a function of agitation. A typical agitation heat-transfer correlation is as follows:

$$k \propto \frac{L^{4/3}N^{2/3}}{D} \qquad (2)$$

where k is the surface heat constant, L is the agitator impeller length, N is the agitator speed, and D is the vessel diameter.

2.4.3. Liquid–Solid Separations In the majority of drug syntheses, the reaction product is a solid. The isolation of the solid product from the reaction mixture is often accomplished in bench-scale operations by rotary evaporation of the volatile components of the reaction mixture, leaving a solid residue that is easily recovered. This technique is clearly not amenable to scale-up, and therefore alternative methods of solids isolation are required. Crystallization of the desired product from the reaction mixture is the most desirable approach as the first step to product isolation. Laboratory, pilot, and commercial-scale crystallizations are typically carried out by cooling, evaporative concentration, or by pH adjustment to precipitate the salt form of the product. However, the use of cosolvents to reduce the product solubility can also be effective in promoting dissolution. Typical liquid–solid slurries are manageable in the 20–30% solids range in a pilot plant or commercial operation. At higher solids concentrations transfers can become significantly more difficult.

Separation of the solid product from the liquid phase is usually accomplished at the bench scale by vacuum filtration through a single-stage filter such as a Buchner funnel. Although pilot and commercial-scale facilities are equipped with similar types of equipment, centrifugation is commonly used for liquid–solid separations. This is particularly true for commercial-scale operations. One of the major advantages of centrifuge systems is their ability to effectively remove liquid from a product cake. This can result in a significant reduction in both the product drying time requirements and the impurity content. For example, the residual solvent content of solids isolated by centrifugation is typically in the 5–10% range, whereas solids isolated by vacuum filtration can be in the 20–30% range. Measurement of filtration rates and cake compressibility at the bench scale can provide valuable insights into the commercial feasibility of the isolation conditions and the selection of appropriate equipment.

2.4.4. Drying and Solid Handling Drying operations under laboratory conditions are typically restricted to the use of vacuum ovens. Similar types of equipment are often used in pilot operations and are commonly referred to as tray dryers. These types of dryers fall into a specific FDA class of dryer systems referred to as indirect conduction heating static solid-bed dryers and are very versatile when processing wet solids that are difficult to dry. One of the drawbacks of these systems is the static nature of the drying operation that limits the ability for heat transfer to occur across the solid mass. In addition, these units are very labor intensive and can present significant industrial hygiene and validation challenges on a commercial scale. For these reasons, pilot plants are often equipped with a variety of types of dryers to make an effective transition between the laboratory and the commercial-scale operations.

The most commonly used commercial drying systems are rotary tumble dryers. This type of dryer falls into the FDA classification of indirect conduction, moving solids bed dryers. These units work well for free-flowing solids that have high volume requirements but are less effective with solids that have a tendency to agglomerate and cake while drying. Agitated drying systems such as paddle and spherical dryers are another type of solids drying system that are of the same FDA dryer class as the rotary tumble dryers. These units typically have a fixed heated surface and internal agitation to maximize heat transfer while breaking up any agglomerated solids. Agitated dryers are often outfitted with chopper attachments to the agitation system that can also affect particle size reduction and potentially avoid an additional milling step. As a result, these units can provide high-throughput drying of a variety of difficult-to-handle materials, are applicable for both pilot and commercial applications, and are commonly found in more modern installations. Fluidized bed dryers represent a second FDA classification of drying system. These units use a hot inert gas flowing at a high velocity to suspend and dry the solid in a finely divided state. This type of dryer equipment falls into the FDA classification of direct heating, dilute solids bed, and flash dryers and has been used for both batch and continuous drying operations on a more limited basis.

Whenever possible, particle size distribution is controlled by crystallization parameters such as agitation and cooling rate. Once the solid material has been isolated and dried, particle size reduction can be achieved by various milling techniques. Both the particle size requirements and the physical properties of the solid dictate the type of milling equipment used for a specific application. The particle size requirements are usually defined during drug formulation development and impact the bioavailability of the drug candidate. Some of the physical properties of the solid that can affect the selection of milling equipment and conditions include hardness, crystal morphology, and thermal stability. The stability of the solid is a critical issue with regard to milling operations because of the energy applied by the milling equipment. Fluid impact mills, such as jet mills, are one type of milling equipment that can be used in both development and commercial applications. These mills promote particle size reduction through high-speed particle-to-particle collisions. In contrast, impact mills, such as hammer and pin mills, impart particle size reduction both by particle-to-mill surface and by particle-to-particle collisions. These units are also used routinely in pilot and production environments. Other milling techniques such as compression milling and particle size classification can also be applied, depending on the particle size specifications and the physical properties of the solid.

2.4.5. Safety When transferring processes from the laboratory to the pilot plant, it is of utmost importance to identify and address potential safety issues as early as possible in the transfer process. Typically, calorimetry studies and process hazard reviews are carried out to meet this requirement. Calorimetry experiments can assist in the identification and quantification of reaction exotherms associated with the process. This information can then be used to determine the capability of the pilot equipment to control the reaction.

Process hazard reviews are conducted to identify potential hazards that could occur as a result of operational failures such as loss of power or cooling capacity. The process hazard review is conducted subsequent to the calorimetry studies to avail the benefit of this additional information in risk assessment. The compatibility of the pilot plant electrical classification with the process solvents, reagents, and solids is also evaluated as part of this process. Predictive evaluations are often made during the process hazard review by using information obtained from reactions carried out using similar reaction conditions and raw materials. In addition, the pilot plant materials of construction should also be evaluated against the reaction conditions that are to be employed, as part of the scope of the review process.

Because these drug candidates have potential biological activity, precautions should be taken to limit worker exposure during scale-up operations. Personal protective equipment requirements and adequate containment and ventilation provisions should also be defined as part of the safety review process. Often, this assessment can be difficult because the material produced from the pilot plant will be used for toxicology evaluation purposes. In these cases, structure–activity relationship evaluations with regard to the relative toxicity of the compound may be appropriate to estimate the extent of risk.

2.5. Commercial-Scale Operations

The commercial implementation of a new process primarily depends on three factors: (1) the quality of the information obtained from laboratory and piloting efforts, (2) the effective transfer of the knowledge gained from these efforts, and (3) the ability to match the process requirements with the production capabilities. The production capabilities may be new and/or existing but in all cases should incorporate the effective utilization of existing assets while meeting the process requirements. Fortunately, the scale-up factor from the pilot plant to commercial operations is usually 1–20 or less, so that pilot information can be easily transferred to commercial practice. Likewise, the information transfer can be facilitated by the participation of production personnel in process development scale-up operations. The issues of equipment requirements, implementation of in-process controls, and validation requirements, as well as safety and environmental issues related to commercial production, are addressed in this section.

2.5.1. Identification of Processing Equipment Requirements When transferring processes from pilot to commercial-scale operations, a comparative analysis is usually made between the equipment used in the pilot operation with the proposed commercial facility. Process flow diagrams (PFD's) that include material balances from pilot plant experiments can facilitate this analysis. Specifications and requirements for agitation, filtration, drying, and milling devices are established based on experimental results that support these specifications and are documented.

Vent treatment requirements are also established during the evaluation of the process equipment requirements. The compatibility of the existing vent gas treatment system is evaluated against the process information obtained from the pilot runs and the existing environmental permit constraints. Permissible levels of venting are then established on the basis of this assessment and the design requirements are documented.

Process streams from pilot plant experiments are usually retained to evaluate the compatibility of various types of materials of construction. Mass balance information is often sufficient to determine these requirements based on pH, halide content, solvents used, and process temperatures. Corrosion testing of pilot plant process streams through the use of coupon testing and/or electrochemical techniques is often recommended. These results are compiled and documented along with the specifications for the materials of construction.

The process requirements for both temperature and pressure should also be evaluated against the production equipment capabilities as part of the production equipment assessment. Normal operating conditions are used as a base case, but upset conditions should also be included as part of the evaluation. If venting is chosen to control unintended reactions, vent-sizing calculations must be performed and peripheral equipment selected as needed. Experiments and simulations to determine consequences of unintended reactions and the interpretation of these experiments need to be documented as part of the production process safety review.

The ignition prevention requirements for the process must also be defined on the basis of both the electrical classification of solvents used in the process and the ignition characteristics of dry powders used in the process. Preferably, dry powder characteristics such as minimum ignition energy and temperature can be established on the basis of testing of the solid materials. However, dry powder characteristics may also be estimated using experience with similar materials. Experiments to establish ignition characteristics and the interpretation of experimental results should also be documented as part of the production process safety review.

2.5.2. In-Process Controls In-process controls are used in both pilot and commercial operations to confirm that the process is in a state of control and that the reactions and unit operations have been carried out to their expected completion point. Other process control points such as pH measurement, reactor content volume, distillation endpoints, filter cake washings, and drying endpoints are often considered to be critical process parameters that are also incorporated into IPCs. In pilot operations, numerous IPCs are taken to establish benchmarks for various process parameters, such as reaction time, drying time, distillation endpoints, and many other process variables. In a production environment, these benchmarks have been established and fewer IPCs are typically used to control the process. It is important when transferring a process from the pilot plant to review, identify, and separate the IPCs that were used only to establish process benchmarks in the pilot operations from those IPCs that would be appropriate for routine commercial operations. The appropriate commercial IPCs should then be documented and incorporated into the production procedure. This is important because the New Drug Application (NDA) will define the IPC testing requirements in the Chemistry, Manufacturing, and Control (CMC) section of the submission, and the elimination of an unnecessary control beyond this point can be a significant undertaking.

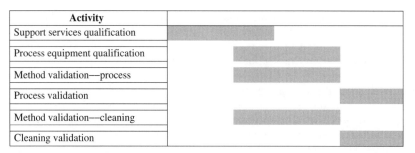

Figure 3. Validation activity timeline.

2.5.3. Validation The purpose of a validation program is to establish documented evidence that provides a high degree of assurance that specified processes consistently produce product that meet predetermined specifications and quality attributes. The validation program attempts to ensure that all systems, instruments, and equipment impacting the quality or integrity of the product have been validated.

The validation program is composed of several different elements and is designed to ensure that all validation requirements are addressed. General validation requirements for each of these elements are outlined in a master plan. Descriptions of the various validation elements are as follows:

- *Equipment/Systems Qualification*: Qualification of the equipment in which the product is manufactured, the support services, and computer systems supporting the process.
- *Process Validation*: Validation of the manufacturing process through the execution of production batches to establish that all product performance criteria have been met.
- *Cleaning Validation*: Validation of the cleaning procedures used to clean the product from production equipment.
- *Method Validation*: Validation of the analytical methods used to support the process validation, cleaning validation, in-process testing, and release testing of the product.

All FDA-regulated products should be validated. Validation of each product is performed in a phased approach, encompassing all the elements mentioned. The typical sequence for these activities is shown in Fig. 3.

In addition to validating these elements, several support programs must be in place to ensure that, once validated, manufacturing processes will be maintained in a validated state on an ongoing basis. These programs include calibration, preventive maintenance (PM), personnel training, and change control. The *personnel training* program should be designed so that all operational procedures and requirements are defined and communicated. It is also important that documentation of the training activities is completed and readily retrievable.

The *calibration* program ensures that instruments associated with manufacturing processes are calibrated and maintained. Instruments in the facility are typically classified as critical, noncritical, or reference. Those instruments deemed either critical or noncritical are typically calibrated using NIST or other applicable standards on a routine basis.

A *preventive maintenance* program should be in place to support the ongoing qualification requirements for all production and support facilities. The objective of the program is to ensure that preventive maintenance requirements for the equipment are carried out throughout the operational life of the equipment. The PM requirements are established using the equipment manufacturer's recommendations and any additional requirements established by the operation site.

A *change control* program is used to regulate the alteration of systems and changes to processes. The program should outline the methods to be followed when a system, process, or equipment change is proposed. The change control program ensures that proposed changes are reviewed and approved by the quality unit and other appropriate departmental representatives before initiating changes. The program should also ensure implemented changes are reviewed before use in manufacturing. It is through this review that any required testing and documentation is defined to verify that the proposed change is acceptable and that the equipment remains in a validated state. The review also ensures that governing regulatory issues for the affected process/operation are addressed.

Validations should be performed in accordance with preapproved written protocols. The execution of the approved validation protocol will generate documentation that supports the intended use of the equipment and demonstrates compliance with current good manufacturing practices (cGMP's). A validation package that contains all documentation relevant to the validation study should be prepared at the completion of each protocol. The validation package should include the summary report that documents the results, observations, and conclusions from the implementation of the protocol, a copy of the protocol, and completed data sheets corresponding to each section of the protocol. The combination of the summary report and the protocol with completed attachments serves as a permanent document of the validation study.

2.5.4. Chemical Safety in Production Before scaling the process from the pilot plant to commercial scale, a process hazard review is performed based on the additional data obtained during the pilot campaign. This review should include the evaluation of reaction calorimetry data, powder ignitability, and the results of the acute and chronic toxicity testing of all raw materials, intermediates, and the product. The battery of toxicity testing can include mutagenicity, teratogenicity, and carcinogenicity; acute dermal and ocular irritation testing results; absorption routes for raw materials, intermediates, and products; and potential sensitizers in the process. Antidotes to acutely toxic materials should also be identified as part of this evaluation. Personnel protective equipment such as gloves, goggles, and protective garments should also be reevaluated at this point based on experience gained from pilot plant operations. Any industrial hygiene sampling data obtained from various operations during the pilot studies should also be included in this evaluation.

2.5.5. Environmental Controls in Production

Environmental permit requirements should be evaluated based on the commercial-scale material balance and new equipment specifications. Testing requirements for environmental evaluation should include acute fish and invertebrate toxicity for raw materials, intermediates, and products; biodegradation of raw materials, intermediates, and products; microbial growth inhibition of raw materials, intermediates, and products; water coefficients (KOW) and water solubility for raw materials, intermediates, and products; and waste treatability test results. Particular emphasis should be placed on the evaluation of the compatibility of the new process waste streams with the existing waste-treatment systems. If any process waste streams require off-site disposal into regulated hazardous waste landfills, additional leaching experiments may also be required.

2.6. Process Trends and Technologies

The ability to minimize scale-up variables such as reactor configuration, agitation, and temperature control has been one of the major challenges of process development. Likewise, the capability to obtain real-time analytical feedback of reaction composition is a strategic scale-up capability that has been evolving over the past several years. The advancements in microfluidic reactor design and development provide the promise of continuous chemical processing at all phases of process development and scale-up. Similarly, the promotion of process analytical technology (PAT) has been used to monitor product quality parameters for both batch and continuous applications. These two related technology developments are discussed in this section.

2.6.1. Continuous Microfluidic Reactors

Batch processing is the current industry standard for API process development and manufacturing. Likewise, the issues associated with

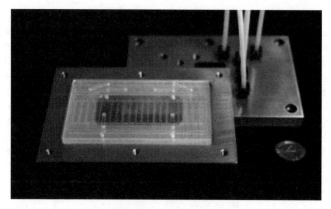

Figure 4. Microreactor.

large-scale synthesis in this chapter have centered on batch unit operations. However, recent advances in microfluidic technology [17] have provided a viable alternative to batch process development and manufacturing with the development of continuous microreactor systems (Fig. 4). The channels within the microreactors are configured to promote mixing through turbulent flow and the capacity of the systems can be incrementally expanded to meet the API requirements at all stages of process development. Scaling out rather than up greatly simplifies the process evolution from the laboratory to production by minimizing the impact of variables such as reactor configuration, temperature control, and agitation. Furthermore, the capital investment and operating cost for these continuous flow systems can offer significant advantages over conventional batch systems. The barriers to implementation of this technology include the significant investments that exist in batch operations as well as the prevailing culture of process chemists to work in batch glassware at the laboratory scale. Continuous operations can also require a greater level of control and feedback through in-process control that will be discussed in the following section.

2.6.2. Process Analytical Technology The ability to monitor reactions with in-line probes is a desirable capability and a strategic initiative of the Food and Drug Administration, commonly referred to as PAT [18]. The driving force behind this effort is the ability to obtain continuous control of product quality rather than "after the fact" release testing in the laboratory. In-line probes have been developed using infrared and Raman spectroscopic methodologies to monitor reaction profiles. Likewise, in-line laser diffraction probes have been developed to monitor particle size distribution during crystallization. As already stated, the development of the continuous monitoring capabilities will also be required to advance the development of continuous microreactor systems.

3. NEVIRAPINE

3.1. Background

In 1986, Boehringer Ingelheim Pharmaceuticals initiated an antiviral HIV research program that focused on the identification of potential nonnucleosidic reverse transcriptase inhibitors with the specific intent to develop an AIDS drug with reduced adverse side effects. A high-throughput screening method was established using AZT as a standard. Promising candidates were screened for mammalian DNA polymerase as well as other enzymes and receptors. Nine months after the first lead compound was identified, nevirapine (3) was approved as a development candidate.

The nevirapine clinical program focused on both single and combination drug therapies. Clinical results indicated that this material not only was effective in the treatment of HIV-related illness but also was found to be well tolerated and safe. Boehringer Ingelheim submitted an NDA in February 1996 and the new AIDS drug was approved in July 1996.

3.2. Evolution of the Nevirapine Synthesis

3.2.1. Medicinal Chemistry Synthetic Route

The initial nevirapine synthesis developed by the Medicinal Chemistry Group entailed the condensation of 2-chloro-3-amino-4-methylpyridine (CAPIC, **4**) with 2-chloronicotinoylchloride (**5**), to give the 2,2′-dihaloamide (**6**). Treatment of (**6**) with four equivalents of cyclopropylamine (CPA) in xylene at 120–140°C under autogenous pressure produced the 2′-alkylamino adduct (**1**) followed by ring closure with sodium hydride in pyridine at 80–100°C, to give nevirapine (**3**), as shown in Scheme 2.

The basic synthetic strategy developed during this phase of drug development was quite sound and provided an excellent starting point for future process development efforts. The synthesis exhibited significant elements of convergence, starting with two functionalized pyridine precursors, which were only three chemical steps away from the target molecule. However, technical barriers existed impeding the ability to meet the short-term API requirements for toxicology and clinical supplies, and additional process issues would need to be addressed to obtain a commercially viable synthesis from this point in the nevirapine process development. The most critical short-term issue was with the ability to obtain significant quantities of raw materials to meet the bulk active needs to support the drug evaluation efforts.

The 2-chloronicotinoylchloride (**5**) was easily prepared from 2-chloronicotinic acid, which was commercially available in multiton quantities. The most significant initial concern was with the ability to obtain pilot-scale quantities of CAPIC (**4**). Gram quantities of this material were initially obtained by the reduction of 2-chloro-4-methyl-3-nitropyridine.

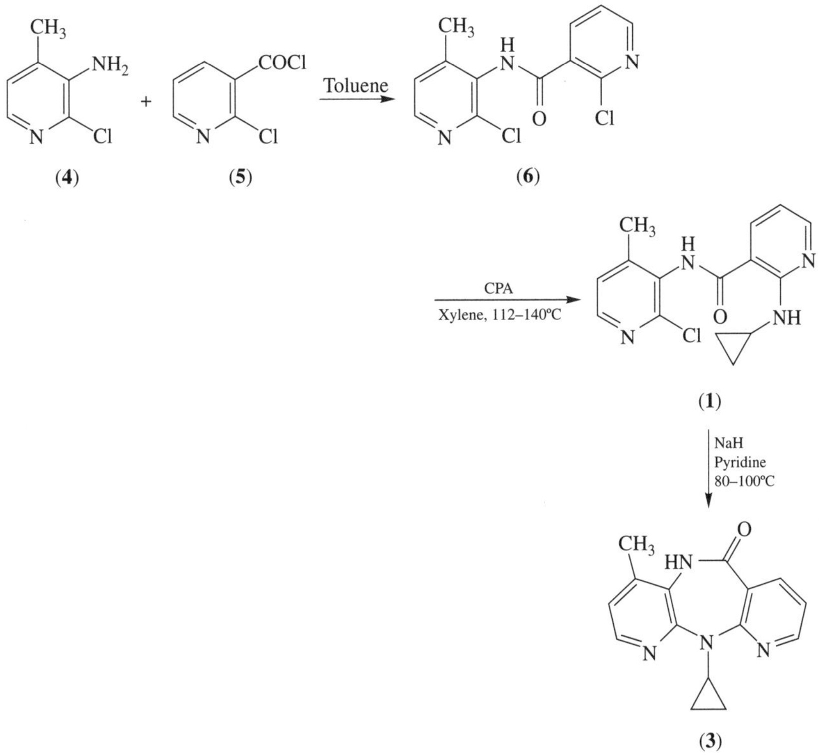

Scheme 2.

Small quantities of this material were initially obtained from laboratory supply houses, but significant scale-up quantities were not commercially available.

Attempts were made to nitrate both 2-hydroxy-4-methylpyridine and 2-amino-4-methylpyridine, which are commercially available, by conventional synthetic methods [19]. However, the major product from both of these reactions was the 5-nitro adduct, with less than 30% of the desired 3-nitro isomer present in the reaction mixture. The yield of the desired isomer was improved to 82% by adding the nitric acid/sulfuric acid premix to the respective substrates at 0–5°C, followed by heating at 60–80°C for 1 h [20]. However, the physical separation of the products proved to be industrially impractical because of the similarity in physical properties of the respective isomers.

Based on the information obtained from these experiences, the initial nitration approach was abandoned. The thrust of chemical and process development activities was redirected toward the development of a CAPIC process that could, as a minimum requirement, be scaled up to produce pilot plant quantities of this raw material to support toxicology, formulation, and clinical studies.

3.2.2. Chemical Development and Pilot Plant Scale-Up
Having benefited from the experience gained during the development of the nevirapine medicinal chemistry route, technical efforts directed at chemical development initially shifted to the identification and evaluation of synthetic alternatives for the preparation of CAPIC. As previously stated, 2-chloro-3-nitro-4-methylpyridine could be readily converted to CAPIC by catalytic reduction. However, obtaining significant quantities of this material was problematic because of the lack of reaction selectivity. Alternative approaches to the introduction of the 3-amino group were examined and found to be quite promising. Functionalized nicotinonitriles have been produced in high yield from readily available acyclic precursors by the Guareschi–Thorpe condensation [21]. The 3-cyano substituent could be readily converted to the corresponding amine by hydrolysis to amide followed by a Hofmann rearrangement [22]. A process for the preparation of 2,6-dihydroxy-4-methyl-3-cyanopyridine (**9**) using this method was identified, which employed ethyl acetoacetate (**7**) and cyanoacetamide (**8**) as relatively inexpensive and readily available raw materials [23] (Scheme 3). Further investigation revealed that (**9**) is commercially available in multiton quantities.

Conditions were established to chlorinate this intermediate by using phosphorous oxychloride, to give 2,6-dichloro-4-methyl-3-cyanopyridine (**10**), followed by acid hydrolysis of the 3-cyano substituent and conversion to the amine under Hofmann rearrangement conditions (Scheme 4).

Efforts to selectively remove the 6-chloro substituent from either (**10**), (**11**), or (**12**) were unsuccessful. However, removal of both chlorine atoms by catalytic dechlorination [24] followed by selective rechlorination in the 2-position gave the desired product, as shown in Scheme 5 [25].

This method was used to produce the nevirapine API requirements through phase III clinical trials as well as for commercial launch and production. Although this synthetic

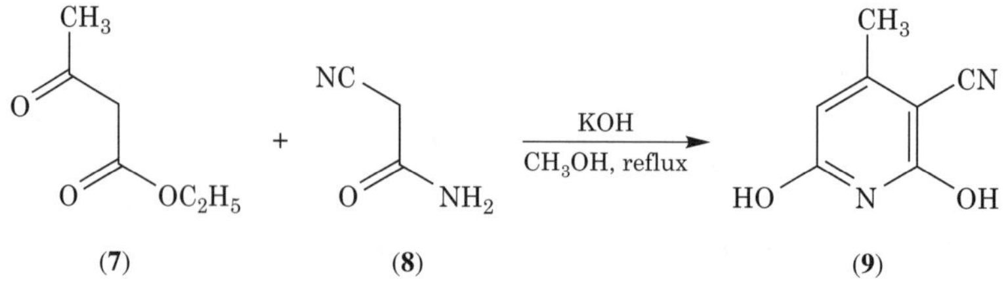

92% Yield

Scheme 3.

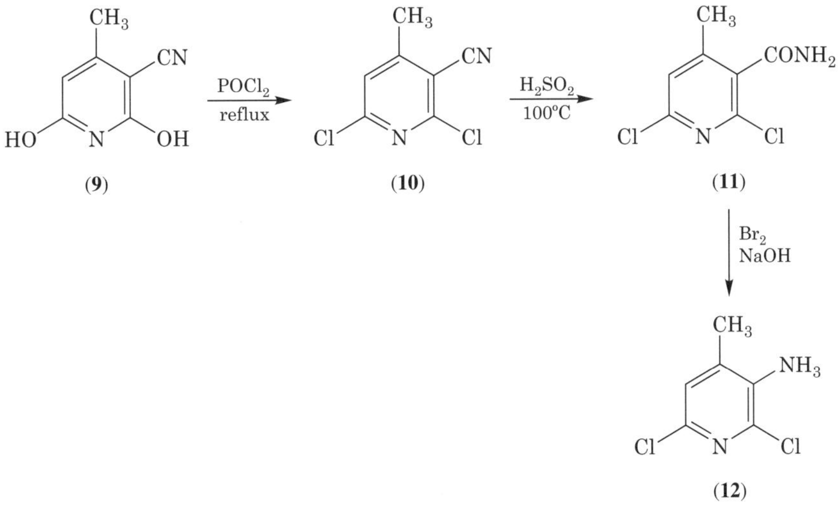

Scheme 4.

approach lacked atom economy with respect to the removal and addition of chlorine atoms, it provided the opportunity to meet the short-term API supply needs and established a synthetic strategy from which further development activities would benefit. It should also be noted that all process steps from the commercially available raw material 2,6-dihydroxy-4-methyl-3-cyanopyridine (9) are carried out in aqueous media, making this option also environmentally attractive.

An alternative synthetic option was also examined that would eliminate the dechlorination/rechlorination process steps for the preparation of CAPIC, which is shown in Scheme 6. In this approach the chloride was removed in the last chemical step from (17) eliminating the dechlorination of (12) and selective rechlorination of (13). Although this option provided significant synthesis advantages over the existing process, this alternative method produced a different impurity profile from that of the original process (Scheme 2). For this reason, revalidation of the API impurity profile, toxicology, and other pharmacological and regulatory issues would be required. Because this option was identified late in the chemical development process, it was decided that the potential process benefits would be more than offset by the additional efforts required to requalify the alternative process, and this option was eliminated from further commercial consideration.

3.2.3. Process Development and Pilot Plant Scale-Up

The nevirapine process scheme used during chemical development (Scheme 2) provided the basis on which to begin process development studies, with the objective of defining reaction conditions that would allow production to be carried out on a routine commercial basis. In this process, the basic

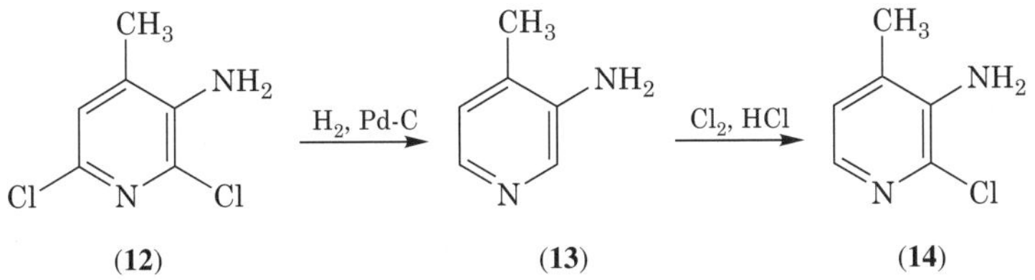

Scheme 5.

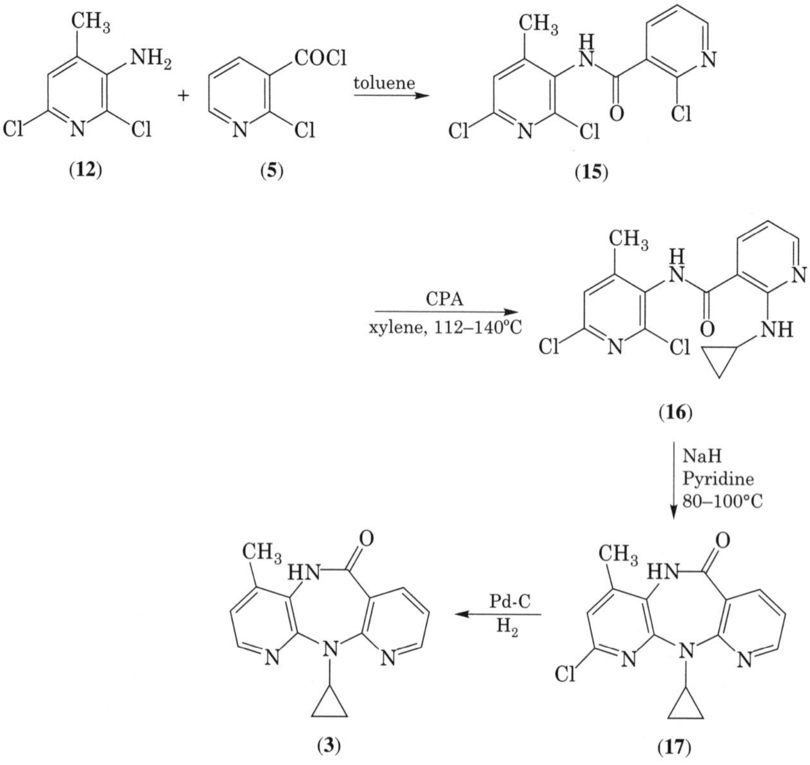

Scheme 6.

elements of the molecule are introduced with the condensation of CAPIC (**4**) and 2-chloroniconinoylchloride (**5**). Using FDA guidelines [26] for defining the starting point in the synthesis for regulatory purposes, CAPIC was considered a raw material in the synthesis. This provided the opportunity to implement further CAPIC process improvements in the preparation of CAPIC after the product launch with limited regulatory impact. With this in mind, priority was given to the condensation ring closure and purification steps that were filed in the NDA.

In the manufacture of CAPIC, the reaction conditions were found to be quite acceptable for commercial operations, with minimal process modifications required to maximize the reactor utilization. However, reaction selectivity problems were encountered with Step 4 during pilot runs that had not been observed in previous work. An alternative set of reaction conditions was established that employed hydrochloric acid and hydrogen peroxide to selectively chlorinate the 2-position of (**13**). Process research established that a very narrow temperature range is required for this step, and the reaction temperature is controlled by the rate of addition of the hydrogen peroxide.

The condensation reaction required significant process development modifications from the procedure used to produce the initial drug development requirements. 2-Chloronicotinoylchloride (**5**) was prepared *in situ* during medicinal and chemical development runs by adding 2-chloronicotinic acid to a 5 molar excess of thionyl chloride as a neat reaction mixture. Upon completion of the reaction, excess thionyl chloride was removed by distillation. The residue was then redissolved in toluene, followed by the addition of CAPIC in toluene and sodium carbonate to neutralize the excess HCl liberated from the condensation. An alternative procedure was developed with the use of a 10% molar excess of thionyl chloride in toluene to produce (**5**). The excess thionyl chloride was removed by distillation and to the resulting mixture was added the CAPIC/toluene solution, to give (**14**).

The work-up conditions for the condensation step (Scheme 6) were also modified to accommodate commercial operations. Sodium carbonate was used in the initial chemical development pilot plant batches to absorb the by-product HCl from the reaction. The quantities of carbon dioxide produced from the neutralization made this approach impractical in a commercial plant. To complicate matters, the amide bond formed during the condensation was subject to hydrolysis under strongly acidic conditions. Solid sodium acetate was added to the reaction mixture as a buffer to address this issue. A significant quantity of the diacetylation product (**18**) was also detected in the reaction mixture before work-up. However, this material rapidly hydrolyzes to the condensation product (**6**) and 2-chloronicotinic acid upon exposure to water (Scheme 7).

The use of cyclopropylamine for the preparation of (**1**) (Scheme 6) also presented a significant process optimization opportunity. In the initial pilot studies, four molar equivalents of CPA were used in the reaction medium. Although CPA appears to be a simple building block, it is rather expensive on a per kilogram basis and represents a significant cost contribution to the overall drug substance. One mole of CPA was initially used to absorb the by-product HCl from the reaction. Calcium oxide was found to be a much more cost-effective substitution as a neutralizing agent. To efficiently remove the calcium salts before further processing, a centrifugation step was added to the work-up. However, even with calcium oxide present, a 2.5 molar excess of cyclopropylamine was still required to carry the reaction to completion. Efforts to telescope this operation into the ring-closure step were successful, and (**1**) was treated as a nonisolated intermediate in process development pilot runs as well as on a commercial scale after removal of calcium salts. Reaction temperature profile studies of this step indicated that the cyclopropylamine addition reaction occurred between 125°C and 145°C. However, a significant exothermic side reaction was observed above 145°C. Although this side reaction was not observed in any pilot trials, redundant cooling and ventilation systems were installed to pilot and commercial equipment to ensure safe operation of this process step.

One of the most critical issues to be addressed from these development activities concerned the specific reaction conditions employed in the final cyclization step (Scheme 8). The medicinal chemistry route used pyridine as the reaction solvent medium and sodium hydride as the base. It was later recognized from solvent screening studies that the reaction pathway for the ring closure was solvent dependent. When dimethylformamide (DMF) was used as the solvent, an alternative cyclization pathway was observed (Scheme 8). The oxazolo[5,4-]pyridine (**2**) is the exclusive product under these conditions. This product arises from the displacement of the chlorine atom by the amide carbonyl oxygen. A 2.8 molar excess of sodium hydride is required to carry the reaction to completion. The first mole of sodium hydride is consumed with the deprotonation of the more acidic amide proton.

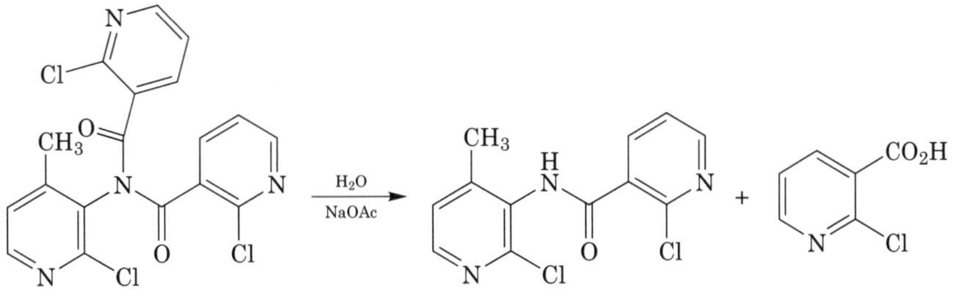

Scheme 7.

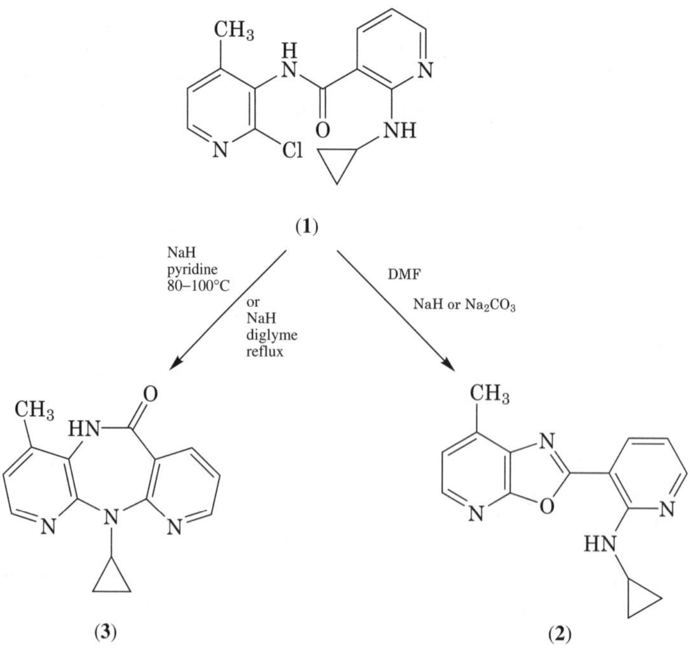

Scheme 8.

If a base of insufficient strength is used in this reaction, the ring closure to the oxazol (12) primarily occurs. Although no industrially practical substitute could be found for sodium hydride as a reagent base, diglyme was found to be an effective alternative solvent to promote the conversion of (1) to nevirapine (3). This was accomplished through an effective collaboration between members of the respective chemical and process development teams. Because of the low autoignition temperature of diglyme, significant equipment modifications were required to upgrade the pilot and production facilities to meet the more stringent electrical code requirements.

One issue with the use of sodium hydride as a reagent in pilot and commercial operations is the storage and handling requirements for this material. Sodium hydride is typically obtained commercially as a 60% amalgam in mineral oil to stabilize the reagent. In the nevirapine process, the mineral oil tends to agglomerate with the product upon precipitation from the reaction mixture. An intermediate purification step was developed using DMF as a crystallization medium. The crude product was dissolved in hot DMF followed by charcoal treatment to absorb the residual mineral oil associated with the product. The charcoal was then removed by filtration followed by evaporative crystallization of the product. A final aqueous crystallization was carried out to remove residual quantities of DMF from the product by acidification with hydrochloric acid followed by treatment with caustic to precipitate the product.

3.2.4. Commercial Production and Process Optimization

On February 23, 1996, Boehringer Ingelheim Pharmaceuticals submitted the NDA for nevirapine to the FDA. Production of the nevirapine API launch batches began within weeks after the submission. The company received regulatory approval for the product in July that year. A priority review of the NDA was initiated by the FDA based on the nature of the drug indication. Because of the accelerated drug development timeline, the procedure used in the final process development piloting campaign was transferred to the production unit virtually unchanged. Only minor modifications to the existing production equipment were made to address electrical code requirements.

As previously noted, having developed a relatively converging synthesis for nevirapine provided the opportunity to define CAPIC as a raw material rather than a registered intermediate in the process. This in turn provided Boehringer Ingelheim with the flexibility to manage the CAPIC manufacturing requirements more effectively. This can be a particularly important issue with new product launches in general, given the high level of uncertainty in initial market forecasts.

As it turned out, nevirapine was well received in the marketplace as an effective AIDS treatment and postlaunch sales consistently exceeded the market projections. With this rapid growth came an increasing awareness of the need to improve the synthesis of CAPIC (4) to meet the growing drug substance demands. Both the linear nature of the CAPIC synthesis and the lack of atom economy in the method were recognized as the major process deficiencies. A retrosynthetic analysis of (4) (Scheme 9) was carried out to evaluate alternative options for the preparation of this material. The goal of this effort was to limit the number of chemical transformations in the synthesis by constructing a pyridine ring with the optimal functionalization from acyclic precursors. The conditions used in the existing commercial process to introduce the amino group in the 4-position by the hydrolysis and Hofmann rearrangement appeared to be an effective approach. The 2-chloro-3-cyano-4-picoline (21) could be readily obtained from 3-cyano-4-methyl-2-pyridone (22) by chlorination with phosphorous oxychloride.

Research efforts were directed toward the evaluation of options for the preparation of (22) from commercially available starting materials. Several approaches were examined, all with the common feature to use a Knovenagel condensation reaction to establish the desired regiochemistry for the target molecule.

Option 1 [27], as shown in Scheme 10, employs acetone and 2-cyanoethyl acetate (23) in the initial Knovenagel condensation. The resulting α-β-unsaturated cyanoacetate (24) is reacted with DMF-acetal to produce (25). The ring-closure step was conducted under Pinner reaction conditions to give ethyl 2-chloro-4-methylnicotinate (26) that is converted to (4) in three steps. However, low yields were observed in the ring-closure steps in this route. An alternative approach [28] was examined using an alternative synthetic approach (Option 2). In this procedure, acetone was reacted with malononitrile (27), to produce (28) followed by reaction with trimethylorthoformate, to give a mixture of (29) and (30). Although (30) is the predominant product from this reaction, both compounds are readily converted to (22) upon treatment with sulfuric acid. Low yields observed in the formylation step led to the development of Option 3. In this procedure, the formylation step is avoided by using a protected β-ketoaldehyde (31) (Scheme 10) in the Knovenagel condensation with malononitrile (27). The protected β-ketoaldehyde (31) is prepared from acetone, methylformate, and sodium methoxide, and is readily available in commercial quantities. The Knovenagel intermediate (30) was converted into (22) under acidic reaction conditions. Upon completion of an economic evaluation of these procedures, Option 3 [29] was selected for commercialization.

More recently, a second generation process [30] was commercialized that included changes in chemistry of the registered process in order to address safety issues that had been identified in the launch process. By

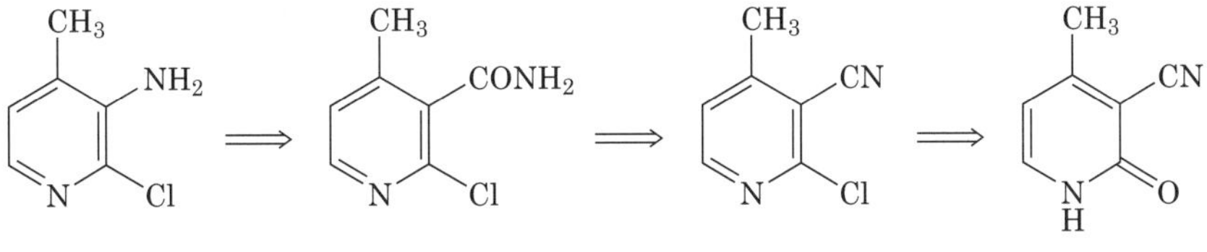

Scheme 9.

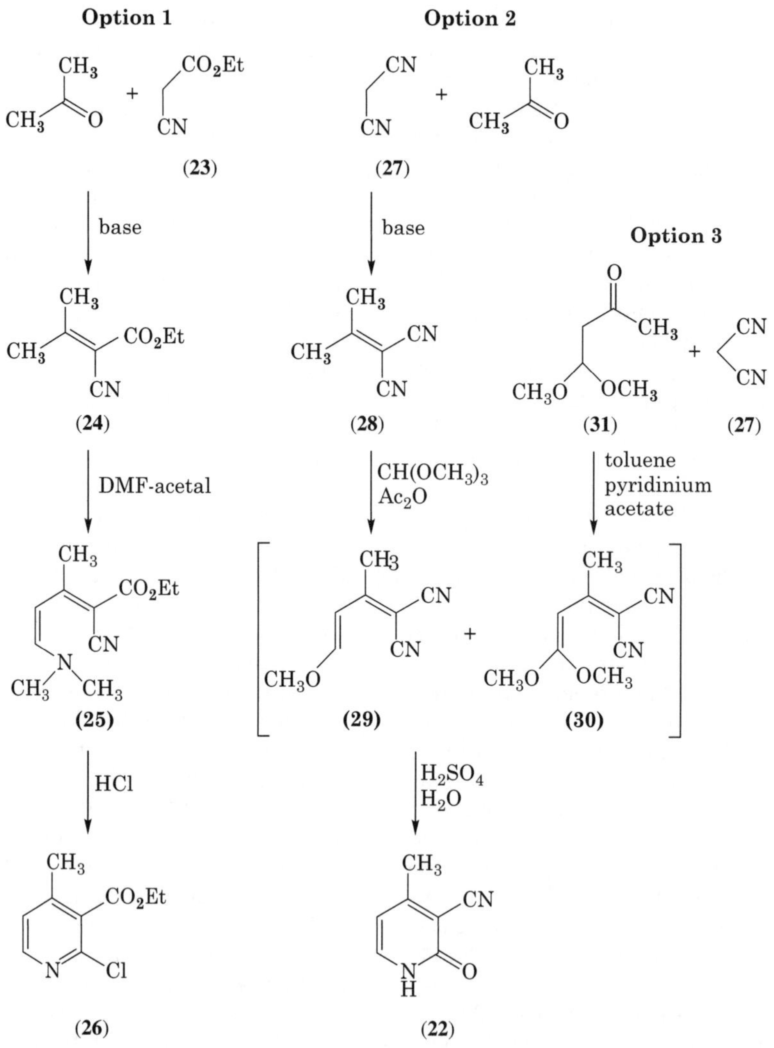

Scheme 10.

incorporating the cyclopropylamino moiety into a more advanced starting material (**31**), the more convergent process led to the elimination of one synthetic step and an intermediate purification resulting in significant improvements in yield, throughput, and production capacity.

4. SUMMARY

The FDA approval of nevirapine for the treatment of AIDS was granted less than 7 years after the submission of the IND. During this period, many technical and regulatory barriers were overcome to bring this product to the marketplace. From a process development perspective, the challenge, as always, is in ensuring the uninterrupted supply of bulk active drug substance in support of the overall drug development effort without sacrificing the ability to deliver a commercially viable chemical process. Although these issues represent a common theme in most drug development case studies, the accelerated pace of the nevirapine project significantly magnified the complexity of the drug development effort. Fortunately, the major elements of the original synthesis remained intact throughout the various phases of process development and provided the opportunity to conduct these activities in parallel with minimal regulatory impact.

REFERENCES

1. Warren S. Organic Synthesis: The Disconnection Approach. New York: John Wiley & Sons, Inc.; 1982.
2. Collins AN, Sheldrahe GN, Crosby J. Chirality in Industry. New York: John Wiley & Sons, Inc.; 1982.
3. Stinson SC. Chem Eng News 1998;76:83.
4. Seyden-Penne J. Chiral Auxiliaries and Ligands in Asymmetric Synthesis. New York: John Wiley & Sons, Inc.; 1995.
5. Ahuja S. Chiral Separation: Applications and Technology. Washington, DC: American Chemical Society; 1997.
6. Noyori R. Asymmetric Catalysis in Organic Synthesis. New York: John Wiley & Sons, Inc.; 1994.
7. Hicks C. Fundamental Concepts in the Design of Experiment. New York: Oxford; 1999.
8. Reichardt C. Solvents and Solvent Effects in Organic Chemistry. 2nd ed. Weinheim, Germany: Wiley-VCH GmbH Verlag; 1990.
9. Norman MH, Minick DJ, Martin GE. J Heterocycl Chem 1993;30:771.
10. Anderson NG. Practical Process Research. San Diego, CA: Academic Press; 2000.
11. Kelly RJ. Chem. Health Saf 1996;3(5): 28.
12. Federal Water Pollution Control Act (Clean Water Act), 33, U.S.C.A. Section 1251 et seq.
13. Murthy K, Radatus B, Kanwarpal S. US patent 5,523,423. 1996.
14. Flory K. Analytical Profiles of Drug Substances. Vol. 23. New York: Academic Press; 1988.
15. Chemburkar SR, Bauer J, Deming K, Spi H, Spanton S, Dzjlki W, Porter W, Quick J, Soldani IM, Riley D, McFarland K. Org Process Res Dev 2000;4:413–417.
16. Perry RH, Green DW. Perry's Chemical Engineers' Handbook. 7th ed. New York: McGraw-Hill; 1997.
17. Ley SV, Baxendale IR. Chimia 2008;62:162.
18. Hinz DC. Anal Bioanal Chem 2006;384:1036.
19. Grozinger K, Fuchs V, Hargrave K, Mauldin S, Vitous J, Campbell SA. J Heterocycl Chem 1995;32:259.
20. Burton AG, Halls PJ, Katritzky AR. Tetrahedron Lett 1971;24:20211.
21. Bobbitt M, Scala DA. J Org Chem 1960;25:560.
22. Wallis R, Lane G. Org React 1946;3:267–306.
23. Grozinger KG, Hargrove KD.US patent 5,200,522. 1993.
24. Grozinger KG, Hargrave KD, Adams J. US patent 5,668,287. 1997.
25. Grozinger KG, Hargrave KD, Adams J. US patent 5,571,912. 1996.
26. Guidance for Industry BACPAC I: Intermediates in Drug Substance Synthesis Bulk Active Post Approval Changes: Chemistry, Manufacturing, and Controls Documentation. U.S. Department of Health and Human Services Food and Drug Administration Center for Drug Evaluation and Research (CDER), Center for Veterinary Medicine (CVM). Feb 2001.
27. Grozinger KG.US patent 6,136,982. 2000.
28. Grozinger KG.US patent 6,111,112. 2000.
29. Gupton BF.US patent 6,399,781. 2002.
30. Boswell RF, Gupton BF, Lo YS. US patent 6,680,383. 2004.

PHYSICOCHEMICAL CHARACTERIZATION AND ORAL DOSAGE FORM SELECTION BASED ON THE BIOPHARMACEUTICS CLASSIFICATION SYSTEM

Gregory E. Amidon[1]
Xiaorong He[2]
Michael J. Hageman[3]

[1] University of Michigan, College of Pharmacy, Ann Arbor, MI
[2] Boehringer-Ingelheim, Danbury, CT
[3] Bristol-Myers Squibb Company, Princeton, NJ

1. INTRODUCTION

While thousands of compounds, really hundreds of thousands in today's pharmaceutical industry, are synthesized and evaluated every year, very few make it to clinical testing and fewer still make it to the market. The reasons for failure are many. One thing is always true though—a marketable dosage form with the desired drug delivery properties must be developed to commercialize a product! Because of the challenges associated with drug discovery and development, the opportunity to identify and develop a safe and effective product benefits greatly from the integration of pharmacology, chemistry, toxicology, metabolism, clinical research, and formulation development. Every discovery team will benefit by keeping this in mind. The ability to identify a suitable dosage form can make or break a product. The dosage form must achieve the desired concentration at the desired site (often considered the blood) for the desired duration. Furthermore, the dosage form must be robust and manufacturable!

In order to initiate formulation development activities—that is the identification of a "marketable" product—important physical and chemical properties (physicochemical properties) as well as permeability properties need to be determined. Evaluation of these properties early in the discovery process can help discovery teams select which templates to pursue as well as identify the most promising leads. This information is also valuable "downstream" as decisions regarding dosage form design are being made. The focus of this chapter will be primarily on those physicochemical properties that are most important in the evaluation of discovery leads and that provide the information needed as dosage form development is started and progresses. It will become apparent that the dosage forms that make sense to consider are dictated by both the physicochemical and biopharmaceutical properties of the drug molecule. Throughout this chapter, there will be an emphasis on the tools and principles that will help the medicinal chemist during the drug discovery process with particular emphasis on oral dosage forms.

It must be emphasized that, throughout the discovery and development processes, collaboration between the drug discovery experts, medicinal chemistry scientists, clinical development, and drug delivery scientists is extremely important in identifying lead compounds that have the best chance of surviving the development process and lead to a marketable product. In some ways, drug discovery and development is a bit like a crapshoot; the best we can do is load the dice in our favor to improve our odds of identifying successful compounds. We can best do this by understanding and applying sound scientific principles throughout the drug discovery and development process.

2. SOLID FORM SELECTION

Usually, for oral drug delivery, crystalline solids are preferred. This is especially true for solid dosage forms such as tablets or capsules where the solid form of the active ingredient often is a critical component in determining dosage form manufacturing, performance and stability. Ideally, though not always, the most thermodynamically stable form is used, as it will generally provide the greatest physical and chemical stabilities. In recent years, the utilization of amorphous, noncrystalline solid forms has increased along with our understanding of the physical, chemical, and dosage form considerations needed to develop these materials [1–4]. Therefore, early identification and selection of the solid form to be used in development becomes paramount as it has a direct impact on physicochemical and drug delivery attributes.

Many of the physicochemical properties discussed in this chapter are, in fact, dependent on the solid form. Aqueous solubility, hygroscopicity, and chemical stability are three obvious examples where very large differences may exist between solid forms. It is therefore valuable to begin a rigorous process of identifying solid forms early in the discovery process. This information is valuable as discovery efforts proceed. As lead compounds are identified, a systematic approach to the synthesis of crystalline salts should take place where possible and appropriate using different solvents and counterions [5]. Once salts of interest are identified, a systematic approach to the crystallization, identification, and characterization of all solid forms including salts should be undertaken. This often requires a systematic approach to crystallization, often from a variety of solvents to identify polymorphs as well as alternative pseudopolymorphs such as solvates [6,7]. Selection of the right solid form or salt can allow the formulation scientist to design the dosage form with optimal physicochemical and drug release properties. A thorough understanding of the polymorphs, pseudopolymorphs, hydrates, solvates, salt forms, and amorphous forms maximizes the opportunity to understand, control, and predict the behavior of a compound in the solid state, identify the appropriate dosage forms to consider and develop a marketable product.

This section provides a brief overview of solid form considerations. The physicochemical characterization described in this chapter can, in effect, be applied to each of the forms that have been identified and isolated, as each solid form will have a unique set of physicochemical properties. Careful consideration of these properties will inevitably lead to the identification of better lead compounds and forms with which to enter development.

2.1. Salts

The preparation of salts is frequently undertaken to improve the physicochemical properties of an ionizable compound. Most often improvement of solubility and dissolution rate is desired. However, improvement in crystallinity (e.g., melting point), stability or hygroscopicity may be possible [5–7]. Figure 1 shows the solubility of several salts of terfenadine and the free base as a function of pH. Development of salts, in particular soluble salts of insoluble compounds, is not without challenge. Complete characterization of salt forms is needed [8,9]. If the salt is very soluble, precipitation of the insoluble free base (or free acid) may occur *in vivo* under physiological pH conditions. Systematic screening for salts is advantageous at an early stage to identify the most desirable salts from a solubility, dissolution rate, solid state stability, hygroscopicity, toxicity, and drug delivery perspective. In particular, consideration of the toxicological properties of the counterion is needed. For some counterions, toxicological concerns may exist because of the high doses needed during drug safety evaluation while the quantity of the counterion present in the marketed dosage form may not be problematic. Table 1 is a summary of the most commonly used counterions for pharmaceutical salt formation. A review of FDA approved drugs shows more than 60 different counterions have been used [5,9] and progress in automated screening methodologies have been reported [10,11].

2.2. Polymorphs

Polymorphs differ in solid crystalline phase structure (crystal packing) but are identical in the liquid and vapor states [12]. As such, different polymorphs may exhibit very different properties. In effect, two different polymorphs of the same molecule may be as different in their physicochemical properties as the crystals of two different compounds. Properties such as melting point, solubility, dissolution rate, stability, hygroscopicity, and density can all vary with polymorph. In fact, it is a safe bet that multiple polymorphs will exist of a crystalline form and a large number of pharmaceutical compounds have been shown to crystallize in multiple polymorphic forms. It is therefore appropriate and even "mandatory" to actively pursue polymorph identification [7,13]. The first form isolated in the medicinal chemistry laboratory is not necessarily the most stable, soluble or desirable form for testing or development. Identification of the most stable form, which is generally the

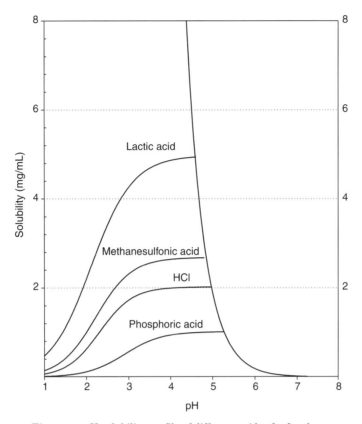

Figure 1. pH-solubility profile of different acids of a free base.

form that is most desirable for development, is critical to the successful development of a stable solid dosage form. It is also valuable to identify and characterize the other metastable polymorphic forms that may have more desirable properties. Even if a metastable crystalline form is ultimately chosen for development, an understanding of the properties of the thermodynamically stable form is critical to the development of a dosage form with maximum stability.

2.3. Pseudopolymorphs

Pseudopolymorphs are not strictly polymorphs since they differ from each other in the solid crystalline phase such as through the incorporation of solvents (solvates) or water (hydrates). Pseudopolymorphs may also be suitable for development and should be identified and evaluated during solid form screening activities.

2.3.1. Hydrates Generally, hydrates are considered appropriate pseudopolymorphs for development. Many drugs are marketed as hydrates, presumably because either they are the most stable form at "typical" relative humidities or they offer other drug delivery advantages. Hydrates often, though not always, are less soluble in water than the corresponding anhydrous form. If the hydrate is less soluble, it often crystallizes when the anhydrous form is suspended in water and allowed to equilibrate.

An important consideration for hydrates is the humidity range in which interconversion of the anhydrous form and hydrate occurs. In theory, exposing anhydrous or hydrous forms to different relative humidities and evaluating the solid form and moisture content present at "equilibrium" can characterize this. However, the interconversion rate between anhydrous and hydrate forms is often slow. A more rapid and efficient method is to slurry

Table 1. Selected Counterions Suitable for Salt Formation

	Anions	Cations
Preferred Universally accepted by regulatory agencies with no significant limit on quantities	Citrate[a] Hydrochloride[a] Phosphate[a]	Calcium[a] Potassium[a] Sodium[a]
Generally accepted Generally accepted but with some limitation on quantities and/or route of administration	Acetate[a] Besylate Gluconate Lactate Malate Mesylate[a] Nitrate[a] Succinate Sulfate[a] Tartrate[a] Tannate Tosylate	Choline Ethylenediamine Tromethamine
Suitable for Use More limited approval with some limitation on quantities which are acceptable in safety studies or human use and/or route of administration.	Adipate Benzoate Cholate Edisylate Hydrobromide Fumarate Maleate[a] Napsylate Pamoate Stearate	Arginine Glycine Lycine Magnesium Meglumine

[a] Commonly used.

anhydrous and hydrate forms in solvents of varying water activity [14,15]. The slurry facilitates and speeds conversion to the most stable form at the specified water activity and temperature. Characterization of the solid form at equilibrium in the slurry allows a more precise determination of the relative humidity (water activity) at which conversion occurs. Appropriate selection of excipients, manufacturing processing conditions, packaging conditions, and storage conditions may minimize changes in form that can occur for such compounds in a dosage form. However, the development needs and regulatory burdens to demonstrate adequate physical stability (e.g., no conversion to another form in the dosage form) may be undesirable [13] and it is often best to avoid these forms if possible.

As a general rule, anhydrous forms that do not convert to the hydrate below 75% RH (at equilibrium) are likely to exhibit adequate physical and chemical stability in oral solid dosage forms. Adequate manufacturing and packaging can be designed to protect most oral solid dosage forms from exposure to >75% RH. Conversely, hydrates that do not convert to the anhydrous form until the relative humidity drops below about 20% RH (at equilibrium) are also likely to exhibit adequate physical stability in solid dosage forms.

2.3.2. Solvates Generally, solvates are undesirable as a final form since the solvent is frequently unacceptable for human use. However, the formation of solvates can frequently occur during drug synthesis and understanding their physicochemical properties is a key to proper control and crystallization of the desired solid form. Desolvated solvates can retain the structure of the solvate even after the solvent is removed by drying or vacuum. Small changes in lattice parameters may occur and the remaining desolvated solvate form tends to be less ordered [7,16]. Under some circumstances, a desolvated solvate may be used as an intermediate in the preparation of the final solid form.

2.4. Amorphous Solids

While development of oral dosage forms most often takes advantage of crystalline solid forms, amorphous solids have occasionally been used and offer some unique opportunities to overcome limitations associated with crystalline forms [1–4]. This is sometimes necessary if no crystalline forms can be identified. Amorphous solids may be prepared by solvent evaporation, freeze drying or coprecipitation processes. Because of their higher energy state, amorphous solids may be used to improve solubility or dissolution rate. Hancock and Parks evaluated the theoretical and experimental differences in solubilities of crystalline and amorphous drugs and concluded that, while amorphous phases offer significant advantages in solubility, experimentally determined values are often not as great as predicted. Experimentally measured solubility enhancement of the amorphous form of the drugs studied varied between 1.1- and $\sim$20-fold more than that of the crystalline form [17]. Therefore, caution and careful formulation is required to assure that no undesirable form changes occur within the dosage form that could compromise physical, chemical or performance properties of the product throughout its shelf life. Amorphous solids are generally quite hygroscopic, less chemically stable than crystalline forms, and more difficult to handle [3,17–19].

2.5. Nonsolid Forms

Nonsolid forms of active pharmaceutical ingredients are rarely used in oral dosage forms primarily because of difficulties in isolation and in carrying out weighing and processing steps commonly used in pharmaceutical manufacturing. It is possible, however, to incorporate liquids directly into liquid formulations such as syrups and in liquid filled capsule preparations. However, this approach rarely offers any advantage over the use of a solid form and should be avoided if possible through the use of appropriate solid form crystallization or salt selection studies.

3. PHYSICOCHEMICAL PROPERTY EVALUATION

Measurement of key physicochemical properties of the most relevant solid forms of lead compounds early in the discovery process can help discovery teams select which templates to pursue as well to identify promising leads. This information is particularly valuable "downstream" as decisions regarding dosage form design are being made. Some of the key physical, chemical and biological properties that should be of interest to the discovery team as well as the formulation scientist are listed in Table 2. Of these, the physicochemical properties that have a clear impact on feasibility of oral formulation development are melting point, partition coefficient, solubility, biological membrane permeability, hygroscopicity, ionization constants, solution stability, and solid state stability. These are discussed in greater detail in the following sections. It should be kept in mind that many of these properties are dependent on the solid form and complete characterization of each of the most relevant solid forms is needed to provide a complete physicochemical picture.

Many of the physicochemical properties of interest are dependent on solid form and, unfortunately, successful prediction of polymorphic forms is inexact. This, in combination with the fact that prediction of physicochemical properties is also very challenging, makes *ab initio* prediction very difficult and imprecise. However, some discussion of predictive tools is included in this chapter. A general comment regarding *ab initio* prediction is that "order of magnitude" predictions are often possible once some basic physicochemical information is available. However, the complexity and diversity of the chemistry space makes reliable predictions across a broad spectrum of chemical structures difficult. Physicochemical predictions across more narrowly defined chemical spaces (e.g., chemical or therapeutic classes) can be more reliable and useful. Drug delivery, formulation and computational chemistry experts are often able to provide a useful perspective on opportunities to take advantage of such *ab initio* predictions within the chemistry space that discovery teams operate.

Table 2. Important Physical, Chemical, and Biological Properties for Oral Drug Delivery

Physical Properties	Chemical Properties	Biological Properties
Polymorphic form(s)	Ionization constant (pK_a)	Membrane permeability
Crystallinity	Solubility product (K_{sp}) of salt forms	Gut metabolism
Melting point	Chemical stability in solution	First-pass metabolism
Particle size, shape, surface area	Chemical stability in solid state	Systemic metabolism
Density	Photolytic stability	
Hygroscopicity	Oxidative stability	
Aqueous solubility as a function of pH	Incompatibility with formulation additives	
Solubility in organic solvents	Complexation with formulation additives	
Solubility in presence of surfactants (e.g., bile acids)		
Dissolution rate		
Wettability		
Partition coefficient (octanol–water)		

3.1. Melting Point

Melting point is defined as the temperature at which the solid phase exists in equilibrium with its liquid phase. As such, the melting point is a measure of the "energy" required to overcome the attractive forces that hold the crystal together. Melting point determination is of great value and can successfully be accomplished by any of several commonly used methods including visual observation of the melting of material packed in a capillary tube (Thiele arrangement), by hot-stage microscopy, or other thermal analysis methods such as differential scanning calorimetry (DSC). Careful characterization of thermal properties such as that possible with DSC provides the investigator with an opportunity to assess and quantify the presence of impurities as well as the presence or interconversion of polymorphs and pseudopolymorphs. Melting points and the energetics of desolvation can also be evaluated, as can the enthalpies of fusion for different solid forms.

As a practical matter, low melting materials tend to be more difficult to handle in conventional solid dosage forms. Melting points below ~60°C are generally considered to be problematic and melting points greater than 100°C are considered desirable. Temperatures in conventional manufacturing equipment such as high shear granulation equipment; fluid bed granulation and drying as well as production tablet machines can exceed 70°C. While amorphous solids do not have a distinct melting point, they undergo softening as temperatures approach the glass transition temperature. Furthermore, common handling procedures (e.g., weighing and processing) can be difficult for low melting materials. Alternative dosage forms (liquid type) may be required for liquid or low melting materials. A comparison of melting points of polymorphs also provides a perspective on the relative stability of polymorphic forms [12]. For monotropic polymorphs, the highest melting polymorph is the most stable at all temperatures while for enantiotropic polymorphs, the highest melting polymorph is not necessarily the most stable at all conditions [12].

Ab initio prediction of melting point is not currently very practical because there is no general relationship yet which relates melting points of compounds to chemical structure. Some success has been achieved for small datasets of hydrocarbons and substituted aromatics [20]. Yalkowsky and coworkers [21–23] have had some success using a group contribution and molecular geometry approach to predict the melting points of aliphatic compounds. To date, melting point predictions for these limited datasets are in the range of ±35°C. The melting point of organic molecules is primarily controlled by the intermolecular forces (van der Waals, dipolar forces, and hydrogen bonds) and molecular symmetry [23].

Use of computational tools to predict polymorphs and melting points have been used to a limited extent thus far [24]. Greater molecular symmetry, which determines how efficiently molecules will pack in a crystal, and the presence of hydrogen donor groups both significantly increase intermolecular interactions in the solid state and increase melting point.

3.2. Partition Coefficient

The partition coefficient is defined for dilute solutions as the molar concentration ratio of a single, neutral species between two phases at equilibrium:

$$p = \frac{[A]_o}{[A]_w} \quad (1)$$

Usually the logarithm (base 10) of the partition coefficient (log p) is used because partition coefficient values may range over 8–10 orders or magnitude. Indeed, the partition coefficient, typically the octanol–water partition coefficient, has become a widely used and studied physicochemical parameter in a variety of fields including medicinal chemistry, physical chemistry, pharmaceutics, environmental science and toxicology. While p is the partition coefficient notation generally used in the pharmaceutical and medicinal chemistry literature, environmental and toxicological sciences have more traditionally used the term K or K_{ow}.

One of the earliest applications of oil/water partitioning to explain pharmacological activity was the work of Overton [25] and Meyer [26] over a century ago that demonstrated that narcotic potency tended to increase with oil/water partition coefficient. The estimation and application of partition coefficient data to drug delivery began to grow rapidly in the 1960s [27,28] to become one of the most widely used and studied physicochemical parameters in medicinal chemistry and pharmaceutics.

Selection of the octanol–water system is often justified in part because, like biological membrane [25] components, octanol is flexible and contains a polar head and a nonpolar tail. Hence, the tendency of a drug molecule to leave the aqueous phase and partition into octanol is a viewed as a measure of how efficiently drug will partition into and diffuse across biological barriers such as the intestinal membrane. While the octanol–water partition coefficient is, by far, most commonly used, other solvent systems such as cyclohexane–water and chloroform–water systems offer additional insight into partitioning phenomena.

Partition coefficients are relatively simple, at least in principle, to measure. However, the devil is in the details and certain aspects demand sufficient attention that rapid throughput methodologies have not yet been successfully developed. Several recent reviews of experimental methods provide an abundance of practical information on the accurate determination of partition coefficients [29]. Indeed, some of the motivation to develop reliable predictions of partition coefficient lies in the fact that measurement is often time-consuming and challenging [28,30].

With the widespread application of lipophilicity and partitioning to biophysical processes, a wide variety of tools are currently available to estimate partition coefficient. Several recent reviews of programs and methods that are commercially available have been published [31–34]. Predictive methods may be broken down into the following basic "approaches" to partition coefficient estimation: (1) group contribution methods using molecular fragments, (2) group contribution methods using atom-based contributions, (3) conformation-dependent or molecular methods, (4) combined fragment and atom-based methods, and (5) other physicochemical methods.

As mentioned above, partition coefficient refers to the distribution of the neutral species. For ionizable drugs where the ionized species does not partition into the organic phase, the apparent partition coefficient, D, can be calculated from Equations 2 and 3.

Acids: $\log D = \log p - \log(1 + 10^{(pH - pK_a)})$
$$(2)$$

Bases: $\log D = \log p - \log(1 + 10^{(pK_a - pH)})$
$$(3)$$

Critical reviews of computational methods are available in the literature [30–33]. Each computational method has strengths and weaknesses but it is important to keep in mind that any computational tool is only as good as its "database" and extrapolation to compound structures which lie outside that dataset is risky. In general, predictive methods can be viewed as providing the best estimates for the chemistry spaces used to develop the model. For medicinal chemists, this means that greatest success is likely to be achieved with the development of specific relationships for the class of compounds of interest. Measurement of representative compounds within a therapeutic class may very likely allow more accurate prediction of the properties of the entire class.

3.3. Aqueous Solubility

At its simplest, the importance of aqueous solubility in determining oral absorption can be seen from the Equation 4 describing the flux of drug across the intestinal membrane.

$$\text{Flux} = P_m \times (C_i - C_b) \quad (4)$$

where P_m is the intestinal membrane permeability, C_i is the aqueous drug concentration (unionized) in the intestine, and C_b is the portal blood concentration.

If solid drug is present in the intestine, the concentration in the intestinal tract may approach or equal its aqueous solubility if dissolution and release of drug from the dosage form is sufficiently rapid. Then, from Equation 4, it is apparent that the flux of drug across the intestine is directly proportional to its aqueous solubility. For drugs that have high intestinal membrane permeability, P_m, aqueous solubility may be the limiting factor to adequate drug absorption. Generally, only the unionized species is absorbed so, for ionizable compounds, the concentration of the unionized form should be considered in Equation 4.

Generally, drug solubility is determined by adding excess drug to well-defined aqueous media and agitating until equilibrium is achieved. Appropriate temperature control, solute purity, agitation rate, and time as well as monitoring of solid phase at equilibration are needed to assure high quality solubility data is obtained [35]. In particular, it is important to evaluate the suspended solid form present at equilibrium as conversion to another solid form (e.g., polymorph, pseudopolymorph, hydrate, and salt) may occur during equilibration. If a form change has occurred, the measured solubility is more likely representative of the solubility of the final form present rather than the starting material. Efforts to develop high-throughput screening methods to measure or classify solubility have recently been undertaken [36] with some success, although they generally suffer from being a dynamic measure (i.e., not equilibrium) and increased variability due to higher throughput.

A wide variety of techniques have been proposed for estimating aqueous solubility. They can broadly be classified as (1) methods based on group contributions, (2) techniques based on experimental or predicted physicochemical properties (e.g., partition coefficient and melting point), (3) methods based on molecular structure (e.g., molar volume, molecular surface area, and topological indices), and (4) methods that use a combination of approaches [35,37,38]. While all of the methods have some theoretical basis, their use in predicting aqueous solubility is largely empirical. Detailed discussions may be found in the literature of the fundamentals of solubility measurement and prediction [36,39]. Each approach has advantages and has been successfully applied to a variety of classes of compounds to develop and test the accuracy of solubility predictions. Usually, approaches that are developed from structurally related analogs yield more accurate predictions [37].

Aqueous solubility is, in a simple sense, determined by the interaction of solute molecules in the crystal lattice, interactions in solution, and the entropy changes as solute passes from the solid phase to the solution phase. Accordingly, the pioneering work of Yalkowsky and Valvani [40–42] successfully estimated the solubility of rigid short-chain nonelectrolytes with Equation 5.

$$\log(S) = -\log(p) - 0.01 \times (\text{MP}) + 1.05 \quad (5)$$

where S is the molar solubility, p is the octanol–water partition coefficient, and MP is the melting point in degree centigrade.

The Yalkowsky–Valvani equation provides insight into the relative importance of crystal energy (melting point) and lipophilicity (partition coefficient). This semiempirical approach has subsequently been applied to and refined for a variety of solutes and classes of compounds [38,41–43]. From Equation 5, one can see that the octanol–water partition coefficient is a significant predictor of aqueous solubility. A one log unit change in aqueous solubility can be expected for each log unit change in partition coefficient. By comparison, a melting point change of 100°C is required to have the same one log unit change on solubility. The Yalkowsy–Valvani and similar equations can be used to predict aqueous solubility often within a factor of 2 using predicted partition coefficients and measured melting points.

Aqueous solubility prediction continues to be an active area of research with a wide variety of approaches being applied to this important and challenging area. To date, group contribution approaches as well as correlation with physicochemical properties (partition coefficient) appear to be the most promising [35,37]. It is important to keep in mind that correlations that are developed from structurally related analogs would consistently yield more accurate predictions.

3.4. Dissolution Rate

Aqueous solubility can also play a critical role in the rate of dissolution of drug and hence release from dosage forms. The dissolution rate of a solute from a solid was shown by Noyes and Whitney [44] as

$$\text{Dissolution rate} = \frac{A}{V} \times \frac{D}{h} \times (S - C) \quad (6)$$

where A is the drug surface area, V is the volume of dissolution medium, D is the aqueous diffusion coefficient, h is the "aqueous diffusion layer thickness" that is dependent on viscosity and agitation rate, S is the aqueous drug solubility at the surface of the dissolving solid, and C is the concentration of drug in the bulk aqueous phase

From the Noyes Whitney equation, dissolution rate is seen to be directly proportional to the aqueous solubility, S, as well as the surface area, A, of drug exposed to the dissolution medium. It is common practice, especially for low solubility drugs, to increase dissolution rate by increasing the surface area of a drug through particle size reduction. If drug surface area is too low, the dissolution rate may be too slow and absorption may become dissolution rate limited. For high solubility drugs, the dissolution rate is generally fast enough that a high drug concentration is achieved in the lumen and extensive particle size reduction is not needed. Synthesis of high solubility salts of weak acids or bases is commonly undertaken to facilitate rapid dissolution in the GI tract. Based on theoretical considerations, as a rough "rule of thumb," if the particle diameter in um is less than the aqueous solubility in microgram per milliliter, further particle size reduction is probably not needed to achieve conventional immediate release dissolution profiles.

3.5. Permeability

Lead compounds generated in today's pharmaceutical research environment, frequently using high-throughput screen programs, often have unfavorable biopharmaceutical properties. These compounds are generally more lipophilic, less soluble and higher molecular weight [36]. Indeed, permeability, solubility and dose have been referred to as the "triad" [36] that determine if a drug molecule can be developed into a commercially viable product with the desired properties. As described by Equation 4, intestinal permeability can be critically important in controlling the rate and extent of absorption and to achieving desired plasma levels.

It is often assumed that the major factors determining transport of drugs across the intestinal membrane are molecular size and hydrophobicity. While true to some extent, this is a simplistic view of drug transport across a complex biological barrier [45]. There are essentially four mechanisms by which drugs may cross the intestinal membrane: passive diffusion, carrier mediated transport, endocytosis, and paracellular transport. Of these, passive diffusion across the intestinal membrane follows Fick's law (Equation 4) and is often the major mechanism for low molecular weight, lipophilic compounds. Transport is proportional to intestinal membrane permeability. Passive transport depends to a large extent on three interdependent physicochemical parameters: lipophilicity, polarity and molecular size [46–48]. Maximizing passive absorption generally involves optimizing these. Molecules, most commonly hydrophilic in nature, may also pass through the tight junctions that exist between adjacent epithelial cells. However, tight junctions are estimated to comprise only 0.1% of the surface area of the intestine [49] and so limit this mechanism. Some compounds are reported to open tight junctions thereby increasing paracellular transport.

Epithelial cells are also known to contain P-glycoprotein (P-gp) efflux pumps that serve to pump drugs out of the cells and back into the lumen against a concentration gradient. The small intestine is particularly rich in P-gp pumps and this mechanism has been shown to limit the oral absorption of a variety of molecules such as cyclosporin, digoxin, ranitidine, and cytotoxic drugs [45,50,51]. The role of efflux pumps in drug absorption is a topic of great interest and research and its importance may be greater than is currently recognized [50].

While physicochemical descriptors of drug molecules are generally not adequate to precisely predict oral bioavailability, there is certainly value to the medicinal chemist to understand some of the basic molecular properties that influence permeability. Of them, passive transport in particular depends to a large extent on three interdependent physicochemical parameters: lipophilicity, polarity and molecular size [47,48,52–55].

Drug lipophilicity is widely used as a predictor of membrane permeability since partitioning of drug into the lipophilic epithelial cells is a necessary step for passive diffusion. Chief among the measures of lipophilicity is the octanol–water partition coefficient discussed in greater detail above. However, lipophilicity alone is inadequate to accurately predict bioavailability and this is no surprise considering the complicated multifaceted nature of transport across the intestinal membrane [56]. Even when limiting predictions to drugs that are absorbed by passive diffusion, only an approximate relationship between lipophilicity and permeability is observed [48]. However, excessively high lipophilicity is clearly a detriment to efficient absorption probably because it is reflected in a very low aqueous solubility. Also, highly lipophilic drugs may become sequestered in the cell with little improvement in permeability across the membrane. In a recent study by Lipinski and coworkers [54], only about 10% of the compounds that entered late stage clinical testing (Phase II) had a $\log p$ greater than 5.

Additional factors that appear to influence permeability are polarity and molecular weight. An excessive number of hydrogen-bond donors and hydrogen bond acceptors have been correlated to decreased permeability. Lipinski and others [54,57,58] also concluded that molecules with more than five hydrogen bond donors (number of NH + number of OH) and more than ten hydrogen-bond acceptors (number of nitrogens + number of oxygens) are not common in compounds that have reached later stage clinical development and this is likely due to decreased absorption. Finally, they conclude that molecular weight also appears to be a factor. Very few compounds with a molecular weight greater than 500 proceed very far in development. Based on these observations, Lipinski and coworkers defined the "Rule of 5" as described in Table 3 as a reasonable rule-based guideline to consider [36]. If molecules exceed two or more of the "limits," the medicinal chemist should be concerned that oral absorption may be a significant problem.

With the difficulties associated with accurate estimation of permeability based only on physicochemical properties, a variety of methods of measuring permeability have been

Table 3. Rule of 5 [53,54]

Molecular weight > 500
$\log p > 5$
Number of H-bond acceptors > 10 (sum of nitrogens and oxygens in molecule)
Number of H-bond donors > 5 (sum of OHs and NHs in molecule)

developed and used. Among them are (1) cultured monolayer cell systems such as Caco-2 or MDCK, (2) diffusion cell systems that utilize small sections of intestinal mucosa between two chambers, (3) *in situ* intestinal perfusion experiments performed in anesthetized animals such as rats, and (4) intestinal perfusion studies performed in humans [45,59–68]. All of these methods offer opportunities to study transport of drug across biological membranes under well-controlled conditions. Caco-2 monolayer systems in particular have become increasingly commonly used in recent years and human intestinal perfusion methods are also becoming more commonly available. Correlations between Caco-2 permeability and absorption in humans have been developed in several laboratories [69–77]. As shown in Fig. 2, a correlation between absorption in humans and Caco-2 permeability is obtained in each laboratory but with a displacement of the curve [78]. A comparison of the variability in Caco-2 permeability values from different laboratories demonstrates that direct comparison of results between laboratories must be done with caution. For this reason, it is necessary to use a set of reference compounds to accurately characterize a drug molecule as poorly or highly permeable. Generally, the best correlation to the *in vivo* situation is obtained for drugs absorbed by passive transport, but transport by different mechanisms may also be characterized using these *in vitro* and *in situ* methods, depending on selected cell lines or *in situ* models.

3.6. Ionization Constant

Knowledge of acid–base ionization properties is essential to an understanding of solubility properties, partitioning, complexation, chemical stability, and drug absorption. The ionized molecule exhibits markedly different properties from the corresponding unionized form.

For weak acids, the equilibrium between the free acid, HA, and its conjugate base, A^- is described by the following equilibrium equation:

$$HA = H^+ + A^- \tag{7}$$

The corresponding acid dissociation constant is given by

$$K_a = \frac{[H+][A^-]}{[HA]} \tag{8}$$

By definition, pK_a is described as

$$pK_a = -\log(K_a) = pH + \log\frac{[HA]}{[A^-]} \tag{9}$$

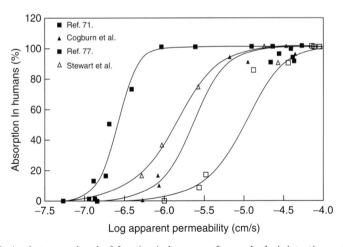

Figure 2. Correlation between absorbed fraction in humans after oral administration and permeability in Caco-2 monolayers obtained from four different laboratories Data obtained from Refs [71,73,75,77,78]. Reprinted with permission from the publisher of Ref. [78].

Corresponding equations for a weak base, B, and its conjugate acid, BH^+, are described by the following equations:

$$BH^+ = B + H^+ \qquad (10)$$

$$K_a = \frac{[B][H^+]}{[BH^+]} \qquad (11)$$

$$pK_a = pH + \log\frac{[BH^+]}{[B]} \qquad (12)$$

Of particular interest to the medicinal chemist and formulation scientist is the impact of pK_a on apparent aqueous solubility and partitioning.

Taking a weak base as an example, the total aqueous solubility, S_T, is equal to the sum of the ionized, $[BH^+]$, and unionized species, $[B]$, concentrations in solution.

$$S_T = [B] + [BH^+] \qquad (13)$$

$$S_T = [B] + \frac{[B][H^+]}{K_a} = [B] \times \left(1 + \frac{[H^+]}{K_a}\right) \qquad (14)$$

Generally, the unionized form, in this case the free base, is the less soluble species in water. The solubility of the unionized free base form is defined as the intrinsic solubility, S_b.

Assuming that the solution is saturated with respect to free base at all pH values, Equation 14 can be written as

$$S_T = S_b\left(1 + \frac{[H^+]}{K_a}\right) \qquad (15)$$

or alternatively

$$S_T = S_b[1 + 10^{(pK_a - pH)}] \qquad (16)$$

While the corresponding solubility equation for a weak acid with an intrinsic solubility of S_a is given by

$$S_T = S_a\left(1 + \frac{K_a}{[H^+]}\right) \qquad (17)$$

or alternatively

$$S_T = S_a[1 + 10^{(pH - pK_a)}] \qquad (18)$$

typical solubility profiles are shown in Fig. 3 for a weak acid and a weak base and several significant conclusions and implications are worth pointing out. Taking the free base as an example once again, at pH values greater than the pK_a, the predominant form present in solution is the unionized form hence the total solubility is essentially equal to the intrinsic solubility. At $pH = pK_a$, the drug is 50% ionized and the total solubility is equal to twice the intrinsic solubility. As the pH drops significantly below the pK_a, a rapid increase in total solubility is observed since the percent ionized is dramatically increasing. In fact, for each unit decrease in pH, the total aqueous solubility will increase 10-fold as seen in Fig. 3. The total solubility will continue to increase in such a manner as long as the ionized form continues to be soluble. Such dramatic increases in solubility as a function of pH demonstrate the importance of controlling solution pH and also offer the formulation scientist a number of possible opportunities to modify dosage form and factors leading to oral absorption properties.

For weak acids, one will observe a rapid increase in total solubility as the pH exceeds the pK_a since, in this case, the ionized conjugate base concentration will increase with increasing pH. Often, for weak acids and bases, the medicinal chemist and formulation scientist must understand the solubility properties of both the unionized species and its corresponding conjugate form since each may limit solubility. In this regard, the work of Kramer and Flynn [79] is particularly instructive. As seen from the figure, the free base solubility curve is as predicted at $pH > 8$. In this pH range, the free base form is the least soluble form and limits the total solubility as predicted by Equation 16. Also shown in Fig. 3 is the solubility curve for the corresponding salt. At $pH < 8$, it is the solubility of the HCl salt that limits the total solubility. From this solubility curve, one can correctly conclude which solid form will exist at equilibrium as a function of pH. This basic principal is of significance *in vivo*, for example, since one

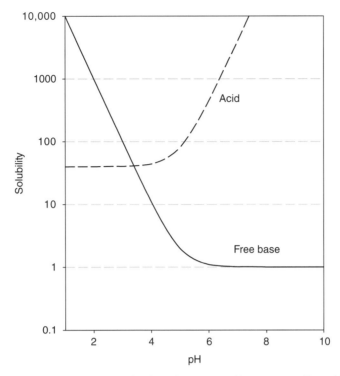

Figure 3. pH-solubility profile for a free base and its corresponding acid.

might imagine dosing patients with a soluble salt, which could rapidly dissolve in the low pH of the stomach, but as drug in the gastric contents entered the intestine where solution pH approaches neutral, precipitation of the free base could occur. Such changes have been proposed as an explanation for the poor bioavailability of highly soluble salts of weak bases.

There are currently a number of software packages that allow for reasonably accurate pK_a estimates based on a variety of approaches including the application of linear free energy relationships based on group contributions, chemical reactivity, and calculated atomic charges. In addition to predictive tools, a variety of reliable methods for measuring pK_a are available including tritrametric and spectroscopic methods as well as aqueous solubility curve measurements as a function of pH.

3.7. Hygroscopicity

Moisture uptake is a significant concern for pharmaceutical powders. Moisture has been shown to have a significant impact, for example, on the physical, chemical, and manufacturing properties of drugs, excipients, and formulations. It is also a key factor in decisions related to packaging, storage, handling and shelf life and successful development requires a sound understanding of hygroscopic properties. Moisture sorption isotherms relate the equilibrium water content of the solid material to the atmospheric relative humidity (i.e., water activity in the vapor phase) to which it is exposed. Isotherms can yield an abundance of information regarding the physical state of the solid and the conditions under which significant changes may occur. Conversion from an anhydrous form to a hydrated form may be observed when the relative humidity exceeds a critical level and moisture content rapidly increases in the solid. Quantitative measurement of moisture content also provides valuable information on the type of hydrate that formed.

Measurement of moisture uptake is typically done by either of the two general methods. The classical approach involves equilibration

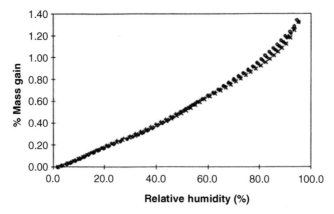

Figure 4. Moisture sorption as a function of relative humidity for an antiarrhythmic compound Reprinted from Ref. [80] with permission from Elsevier Science.

of solid at several different humidities and the subsequent determination of water content either by gravimetric methods or by analytical methods such as Karl Fischer titration or loss on drying. Moisture adsorption or desorption may be measured using this method and the process is effective but tedious and time consuming. A relatively recent development is the use of automated controlled atmosphere systems in conjunction with an electronic microbalance [80–82]. Such systems can generate an atmosphere with well-controlled humidity passing over a sample (often, only a few milligrams are needed) and weight change is monitored. They can be programmed to carry out a series of humidity increments to generate the adsorption curve and/or a series of decrements to generate moisture desorption. In this way, hysteresis may be observed as well as any phase or form changes that are associated with moisture sorption. Examples of moisture sorption curves are shown in Figs 4 and 5.

Dynamic moisture sorption, in particular, provides an excellent opportunity to study solid form conversion and Fig. 6 depicts a typical sorption curve of an antiarrhythmic compound that shows the conversion of an

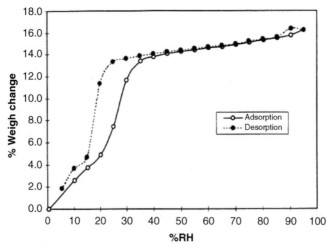

Figure 5. Moisture sorption as a function of relative humidity of a protein kinase inhibitor. Reprinted with permission from the publisher of Ref. [81].

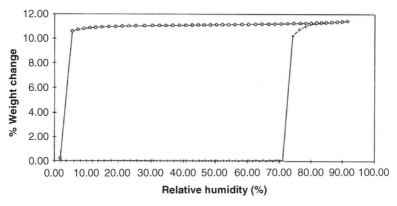

Figure 6. Moisture sorption as a function of relative humidity of an antiarrhythmic compound.

anhydrous to a monohydrate. Moisture uptake by the anhydrous form is very small on the moisture uptake curve until a critical humidity of about 70% is achieved. At this point, rapid moisture uptake occurs and a hydrate form containing 10% moisture is formed. Subsequent reduction in the humidity (desorption) shows the hydrate to remain until approximately 5% RH when it spontaneously converts to the anhydrous form. It is important to recognize, however, that conversion between solid forms is very time dependent. The relative humidities at which conversion was seen in Fig. 6 are very dependent upon the length of time the solid material was equilibrated. For the material shown in Fig. 6, conversion from the anhydrous to the hydrate "at equilibrium" will occur somewhere between 10% and 70% RH. More precise determination of the critical humidity at which conversion occurs may be determined as described in Section 2.3.1.

Prediction of moisture sorption is not possible currently though certainly it is influenced by crystal structure, amorphous components, and solubility. In general, water adsorption to the surface of crystalline materials will result in very limited moisture uptake. Only 0.1% water uptake would be predicted for monolayer coverage of a crystalline material with an average particle size of 1 um [83]. Typically, pharmaceutical powders are in the range of 1–200 μm in diameter so significant moisture uptake by powders is likely due to reasons other than simple surface adsorption. Amorphous regions tend to be much more prone to moisture sorption and high moisture uptake is likely to reflect the presence of amorphous regions or form changes such as the formation of stochiometric hydrates or nonstochiometric hydrates (clathrates). Moisture sorption has, in fact, been used to quantitate the amorphous content of predominantly crystalline materials.

3.8. Stability

Both solution and solid state stability are key considerations for oral delivery. The drug molecule must be adequately stable in the dosage form to assure a satisfactory shelf life. For oral dosage forms, it is generally considered that 2 years is the minimum acceptable shelf life. This allows sufficient time for the manufacture and storage of API, the manufacture of the dosage form, shipping, storage and finally sale to and use by the consumer. Loss of potency is an obvious consideration and generally stability guidelines require that at least 90% of the drug remain at the end of the shelf life. More often, shelf life is determined by the appearance of relatively low levels of degradation products. While, perhaps, 5–10% loss of drug may be considered acceptable, the appearance of a degradation product or impurity of unknown toxicity at a level of 0.1–1% will likely require identification or qualification. Detailed guidance regarding stability has been provided by regulatory agencies such as those in the FDA Guidance for Industry and the International Conference on Harmonization (ICH) [84–86].

3.8.1. Solution Stability
Solution stability is important for oral products because the drug generally has to dissolve in the gastric or intestinal fluids prior to absorption. Residence time in the stomach varies between 15 min and a couple of hours depending on fasting/fed state. In addition, the stomach is generally quite acid for a majority of subjects but may depend on disease state. In this context, stability under acid conditions over a period of a couple of hours at 37°C is satisfactory with no significant appearance of degradation products of unknown toxicity. Residence time in the small intestine is approximately 3 h where the pH may range from 5 to 7 while residence in the large intestine ranges up to 24 h. Stability studies for up to 24 h in the pH range of 5–7 at 37°C with no significant appearance of degradation products of unknown toxicity generally indicates that significant decomposition in the intestine will not occur.

Buffered aqueous solution stability studies are typically done at pH of 1.2–2 and in the range of 5–7. A complete degradation rate profile can provide valuable information regarding the degradation mechanism and degradation products. A complete study and understanding of solution stability is particularly critical for aqueous and cosolvent solution formulations that may be developed for pediatric or geriatric populations. The medicinal chemist will likely have an excellent understanding of the possibility of acid or base catalysis and degradation in aqueous solutions. A close collaboration with the formulation scientist can assure that a careful analysis and study of potential decomposition mechanisms are adequately studied early in development. This will minimize the chances of surprises later in development.

3.8.2. Solid State Stability
Adequate solid state stability is often critical for many drugs since solid dosage forms (tablets and capsules) are generally the preferred delivery system. Stability of the drug in the dosage form for several years at room temperature is generally required. Unstable drugs may be developed, but the time and resources needed are generally much greater and the chances of failure far greater.

Accelerated stability studies are often carried out early in development on pure drug to assess stability and identify degradation products and mechanism. Testing at 50°C, 60°C, or even 70°C under dry and humid conditions (75% RH) for 1 month are often sufficient to provide an initial assessment. More quantitative assessments of drug and formulation stability are carried out to support regulatory filings and generally follow regulatory guidance [84–89].

The field of solution and solid state stability is expansive, varied and beyond the scope of this chapter. Stability studies described above at a variety of conditions provide the perspective and understanding needed to make meaningful predictions of long-term stability and shelf life [90,91]. Typically solid state decomposition occurs either by zero-order or by first-order processes. Arrhenius analysis and extrapolation to room temperature may provide additional confidence that the dosage form will have acceptable stability. Generally though, regulatory guidance allows for New Drug Applications to project shelf life based on accelerated conditions but data at the recommended storage temperature is generally required to support the actual shelf life of marketed products.

4. DOSAGE FORM DEVELOPMENT STRATEGIES

An important aspect of pharmaceutical formulation development is to "facilitate" drug absorption and ensure that an adequate amount of drug reaches the systemic circulation. Most orally administered drugs enter systemic circulation via a passive diffusion process through the small intestine. This is easily seen from Equation 19 [92]:

$$M = P_{\text{eff}} \times A \times C_{\text{app}} \times t_{\text{res}} \qquad (19)$$

where the amount of drug absorbed (M) is proportional to the effective membrane permeability (P_{eff}), the surface area available for absorption (A), the apparent luminal drug concentration (C_{app}) and the residence time (t_{res}). Since it is difficult to alter or control surface area and residence time, formulation strategies often focus on enhancing either drug permeability across the apical membrane or drug concentration at the absorption site. Recently, a Biopharmaceutics Classification System (BCS) has been proposed as a tool to categorize compounds into four classes according to these two key parameters: solubility and permeability [93]. Although the BCS does not address other important factors such as the drug absorption mechanism and presystemic degradation or complexation, it nonetheless provides a useful framework for identifying appropriate dosage forms and strategies to consider which may provide opportunities to overcome physicochemical limitations. It is within the BCS context that this chapter discusses dosage form development strategies that should be considered.

4.1. Biopharmaceutics Classification System

According to the BCS, compounds are grouped into four classes according to their solubility and permeability as shown in Table 4 [93].

4.1.1. Solubility In recent guidelines issued by the U.S. Food and Drug Administration (FDA), solubility within the BCS is defined as the "minimum concentration of drug, milligram/milliliter (mg/mL), in the largest dose strength, determined in the physiological pH range (pH 1–7.5) and temperature ($37 \pm 0.5^\circ$C) in aqueous media" [84,94]. Drugs with a dose to solubility ratio of less than or equal to 250 mL are considered highly soluble. Otherwise, they are considered poorly soluble. In other words, the highest therapeutic dose must dissolve in 250 mL of water at any physiological pH. Since dissolution rate is closely tied to solubility (see Section 3.4), FDA also provides dissolution criteria for immediate release (IR) products. A rapidly dissolving IR drug product should release no less than 85% of the labeled drug content within 30 min, using USP dissolution apparatus I at 100 rpm in each of the following media: (1) 0.1 N HCl or Simulated Gastric Fluid (USP) without enzymes; (2) pH 4.5 buffer; and (3) pH 6.8 buffer or simulated Intestinal Fluid (USP) without enzymes. One needs to realize, however, that these compendial dissolution media may drastically underestimate *in vivo* performance of poorly soluble compounds, especially lipophilic compounds. Even though a lipophilic compound is poorly soluble in an aqueous environment, it may be sufficiently soluble in the presence of bile salts and other native components of the GI tract. To better simulate physiological conditions of the GI tract, Dressman and coworkers recently proposed use of "biorelevant" media to take into account the effect of composition, volume, and hydrodynamics of the luminal contents on drug dissolution and solubility [95].

4.1.2. Permeability In the same guidance as mentioned above, permeability is defined "as the effective human jejunal wall permeability of a drug and includes an apparent resistance to mass transport to the intestinal membrane" [94,96]. High permeability drugs are considered to be those "with greater than 90% oral absorption in the absence of documented instability in the gastrointestinal tract, or whose permeability attributes have been determined experimentally" [94,96]. A list of compounds has been compiled to allow researchers to establish a correlation between *in vitro* permeability measurements and *in vivo* absorption. Accordingly, a drug with a human permeability greater than $2-4 \times 10^{-4}$ cm/s would be expected to have greater than 95% absorption [93]. Permeability measurements in predictor models such as Caco-2 or *in situ* perfusions are generally related back to the reference compounds. A rough guide is that compounds with permeability greater than metoprolol are considered high permeability. Even though this BCS was

Table 4. The Biopharmaceutics Classification System [169]

Class I	Class II	Class III	Class IV
High solubility	Low solubility	High solubility	Low solubility
High permeability	High permeability	Low permeability	Low permeability

designed to guide decisions with respect to *in vivo* and *in vitro* correlations and the need for bioequivalence studies, it can also be used to categorize the types of formulation strategies that might be pursued. Table 5 summarizes dosage form options for each biopharmaceutics class.

4.2. Class I: High Solubility and High Permeability

Compounds belonging to Class I are highly soluble and permeable. When formulated in an immediate release dosage form, a Class I compound should rapidly dissolve and be well absorbed across the gut wall. Because of this, recent developments in regulatory policy have simplified the requirements for introducing Class I compounds into the market by providing the opportunity to file an application with minimal or no human data under some circumstances [94,97]. However, absorption problems may still occur if the compound is unstable, forms an insoluble complex in the lumen, undergoes presystematic metabolism, or is actively secreted from the gut wall. Potential formulation strategies that may overcome such absorption barriers are discussed in details later.

In many cases, the challenge to formulate Class I drugs is not to achieve rapid absorption, but to achieve the target release profile associated with a particular pharmacokinetic and/or pharmacodynamic profile. In such cases, a controlled release dosage form may be more desirable to tailor the blood profile to maintain the plasma concentration at a more sustained level (see Fig. 7). Controlled release technology is well established in the pharmaceutical industry and there are at least 60 commercial available oral controlled release products [98]. Depending on release mechanism, controlled release technology can be classified into four major categories: dissolution controlled, osmotically controlled, diffusion controlled, and chemically controlled dosage forms. Extensive reviews are available in this area [98,99]. It is beyond the scope of this chapter to review all controlled release technology. However, it still worth pointing out the considerations in designing oral controlled release dosage forms, particularly those relevant to Class I compounds. These considerations include the following:

(1) How long is the GI transit time?
(2) Is there any substantial colonic absorption?
(3) What is the dose required?
(4) Is there a safety concern if dose dumping occurs?

Dose dumping of Class I compounds, in particular, may cause more safety concerns than for other classes of compounds because Class I compounds are expected to be absorbed rapidly.

4.3. Class II: Low Solubility and High Permeability

Compounds belonging to Class II have high permeability but low aqueous solubility. In the past decade, an increasing number of Class II compounds have emerged from the discovery pipeline thereby stimulating the development of a variety of dosage forms and drug delivery technologies. Major Class II technologies are designed to deal with poor solubility and dissolution characteristics and include salt formation, size reduction, use of metastable forms, complexation, solid dispersion, and lipid-based formulations. Despite the diversity of these technologies, the central theme of Class II formulation approaches remains the same, that is, to enhance drug dissolution rate and/or solubilize drugs at the absorption site to provide faster and more complete absorption. The following section discusses the advantages and limitations of major technologies for Class II compounds.

4.3.1. Salt Formation
Salt formation is one of the most commonly used approaches to deal with Class II compounds as a way to enhance drug solubility and dissolution rate. Salt selection largely depends on pK_a. It is generally accepted that a minimum difference of 3 units between the pK_a value of the group and that of its counterion is required to form stable salts [100]. Many other factors also influence salt selection such as the physical and chemical characteristics of the salt, safety of the

Table 5. Dosage Form Options Based on Biopharmaceutics Classification System

Class I: High Solubility High Permeability	Class II: Low Solubility High Permeability	Class III: High Solubility Low Permeability	Class IV: Low Solubility Low Permeability	Class V: Metabolically or Chemically Unstable Compounds[a]
• No major challenges for immediate release dosage forms • Controlled release dosage forms may be needed to limit rapid absorption profile	Formulations designed to overcome solubility or dissolution rate problems: • Salt formation • Particle size reduction • Precipitation inhibitors • Metastable forms • Solid dispersion • Complexation • Cocrystallization • Lipid technologies	Approaches to improve permeability: • Prodrugs • Permeation Enhancers • Ion pairing • Bioadhesives	• Formulation would have to use a combination of approaches identified in Class II and Class III to overcome dissolution and permeability problems • Strategies for oral administration are not really viable. Often use alternative delivery methods, such as intravenous administration	Approaches to stabilize or avoid instability: • Prodrugs • Enteric coating (protection in stomach) • Lipid vehicles (micelles or emulsions/microemulsions) • Enzyme inhibitor • Lymphatic delivery (to avoid first-pass metabolism) • Lipid prodrugs • P-gp efflux pump inhibitors

[a] Class V compounds do not belong to Biopharmaceutics Classification System. Compounds in this class may have acceptable solubility and permeability, but can still pose significant absorption challenge if they undergo luminal degradation, presystemic elimination or are effluxed by P-glycoproteins.

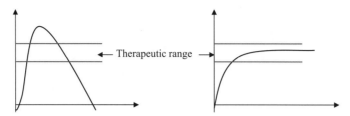

Figure 7. Modification of drug release profile to achieve maximum therapeutic effect. Reprinted with permission from the publisher of Ref. [167].

counterion, therapeutic indications, and route of administration. The main purpose of salt formation is to enhance the rate at which the drug dissolves. For this purpose, sodium and potassium salts are often first considered for weakly acidic drugs whereas hydrochloride salts are often first considered for weakly basic drugs. A wide variety of counterions have been successfully used in pharmaceuticals and many of the most common ones are listed in Table 1.

Salt formation does have its limitations. It is not feasible to form salts of neutral compounds and it may be difficult to form salts of very weak bases or acids. Even if a stable salt can be formed, the salt may be hygroscopic, exhibit complicated polymorphism, or have poor processing characteristics. In addition, formulation of a stable and soluble salt may not be as straightforward as one would expect. Conversion from salt to either free acid or free base has been a common problem both *in vitro* and *in vivo*. Such conversion may cause surface deposition of a less soluble free acid or free base on a dissolving tablet and prevent further drug release. Even if a solid dosage form completely dissolves, the unionized form may still precipitate out in the lumen prior to absorption. In such cases, use of precipitation inhibitors may significantly improve the bioavailability of rapidly dissolving salts.

4.3.2. Particle Size Reduction Particle size reduction is a common method to enhance the dissolution rate of poorly soluble drugs. The underlying principle is that the dissolution rate is directly proportional to the surface area, which increases with size reduction. The most common way to reduce particle size is through milling. There are several types of milling equipment including the cutter mill, revolving mill, hammer mill, roller mill, attrition mill and fluid-energy mill. Equipment selection depends on target particle size distribution as well as characteristics of drugs. For example, the cutter mill is often used for fibrous material and product size is 180–850 μm, whereas the fluid-energy mill (sometimes referred to as micronizing) is often used for moderately hard and friable crystalline materials with typical size distribution created of 1–30 μm [101]. Although size reduction generally enhances the dissolution rate of poorly soluble compounds, there is a critical threshold below which further reduction in particle size will not enhance absorption. For compounds that are extremely insoluble, the critical threshold may be in the nanoparticle range. In such cases, nanoparticle technology may become handy and recent reviews on nanoparticle technology applied to oral, parenteral, and protein delivery are available [102–105]. Refer to Section 3.4 for further details on the importance and impact of particle size on dissolution rate.

Heat and mechanical impact generated during the milling process can cause both physical and chemical instabilities. During the milling process, the localized temperature may rise as high as a hundred degrees and cause chemical degradation or physical conversion. Therefore, it is important to always evaluate the effect of milling on the physical and chemical properties of drugs. In addition to stability problems, very small particle size powder often possesses poor flow properties and wettability. In general, size reduction of hydrophobic material increases the tendency for powder to aggregate in an aqueous environment. Powder aggregation reduces the effective surface area of a drug thereby reducing dissolution rate. Excipients such as surfactants, sugars, polymers or other excipients may be added to a formulation to minimize aggregation. Small amount of surfactants are often added to formulations to significantly improve drug wettability.

4.3.3. Precipitation Inhibition As mentioned earlier, the central theme of Class II technologies is to enhance drug dissolution rate or solubilization to provide more rapid and complete absorption. However, a significant increase in "free drug" concentration above equilibrium solubility results in supersaturation that can lead to drug precipitation. This has been a common problem of many Class II technologies. The supersaturation (σ), which is the driving force for both nucleation and crystal growth, is frequently defined as follows:

$$\sigma = \ln\left(\frac{C}{S}\right) \qquad (20)$$

where C is the solution concentration and S is the solubility of the compound of interest at a given temperature. Crystallization rate generally increases with σ and decreases with viscosity of the crystallization medium. Certain inert polymers, such as hydroxypropyl methylcellulose (HPMC), polyvinylporrolidone (PVP), polyvinyl alcohol (PVA), and polyethylene glycol (PEG) are known to prolong the supersaturation of certain compounds from a few minutes to hours. It is suggested that these polymers increase the viscosity of crystallization medium, thereby reducing the crystallization rate of drugs [106–108]. In addition, these polymers may present a steric barrier to drug molecules and inhibit drug crystallization through specific intermolecular interactions on growing crystal surfaces. Polymers such as acacia, poloxamers, HPMC, and PVP have been shown to be adsorbed onto certain faces of host crystals, reduce the crystal growth rate of the host, and produce smaller crystals [109,110].

In addition to polymers, synthetic impurities may also influence the crystallization rate of drugs. Extensive research has been carried out to examine effect of "tailor-made additives" on crystallization of drugs [111]. These additives are generally structurally tailored to resemble drug molecules and can be incorporated into the lattice of the drug to some extent. Upon incorporation, additives can impede further drug crystallization through specific host-additive interactions. In theory, potent additives can be designed to inhibit drug precipitation more effectively than polymers. However, unless these additives are pharmaceutically acceptable, use of such additives in a formulation can raise safety and regulatory concerns.

4.3.4. Metastable Forms The solid state structure of drugs, such as the state of hydration, polymorphic form and crystallinity have a significant effect on physicochemical properties such as solubility and dissolution rate and this has been discussed earlier in this chapter. In general, anhydrous forms, for example, dissolve faster and have higher solubility than hydrates in an aqueous environment. Although some studies have shown that hydrates of certain drugs dissolve faster than anhydrous forms, such studies may be complicated by phase transition between anhydrous to hydrated forms or differences in particle size and wettability between anhydrous and hydrated materials.

Polymorphic form also influences dissolution rate and solubility. By definition, metastable polymorphs should have higher solubility and faster dissolution rates than their more stable crystalline counterpart because they possess a higher Gibbs free energy. Generally only a moderate enhancement of solubility and dissolution rate can be achieved through polymorphic modification although exceptions do exist [112]. Greater increases can be achieved through the use of amorphous material, which is a noncrystalline solid that is metastable with respect to the crystalline form. Amorphous forms can be viewed as an extension of the liquid state below the melting point of the solid [113]. In some cases, amorphous materials can significantly enhance dissolution rate and solubility and lead to 3–4-fold increase in bioavailability [17,114].

Common processing methods, such as freeze drying, spray drying, and milling may partially or completely transform a crystalline material into amorphous forms. Because the solid state structure can significantly impact bioavailability, it is important to control processing methods so that a pure form (or a mixture of forms with fixed ratio) is produced consistently. Even if a reproducible processing method is available, one still faces the inevitable challenge: metastable forms are destined to convert to thermodynamically more stable polymorphs with time. Polymorphic transformation is a kinetic issue that depends on many factors such as crystal defects, residual solvent, processing, and storage conditions. Amorphous forms are particularly sensitive to moisture level in the product as well as in the atmosphere because water significantly lowers the glass transition temperature of the amorphous form and facilitates recrystallization. For this reason, there is often a reluctance to develop a metastable form of a drug unless there is enough confidence that the metastable form will not transform to the stable form during storage within a desirable shelf life. To overcome this problem, several approaches have been used to prolong the shelf life of metastable forms.

4.3.5. Solid Dispersion Sekiguchi and Obi were the first to develop a solid dispersion method to enhance the bioavailability of a poorly water-soluble drug [115]. Their method involved melting a physical mixture of drug with hydrophilic carriers to form a eutectic mixture in which the drug was present in a microcrystalline state. When drug is homogeneously dispersed throughout the solid matrix, this type of formulation is also termed a "solid dispersion." However, a drug may not always be present in a "microcrystalline state." As later demonstrated by Goldberg et al., a certain fraction of drug might be molecularly dispersed in a carrier, thereby forming a solid solution [116,117]. The key difference between a solid dispersion and a solid solution is that the former is a homogeneous physical mixture of components whereas the latter is a molecular dispersion of one component in another. In the solid dispersion, each component still preserves its own crystal lattice whereas, in solid solution, there are no individual crystals of each component but rather, the molecules are mixed together at the molecular level. A solid dispersion is generally considered to release drug as very fine colloidal particles upon contact with aqueous environment, thereby enhancing the dissolution rate of poorly soluble drugs through increased surface area [118]. Other factors that could lead to enhanced dissolution rate include possible creation of amorphous drug as well as generally increased solubility and wettability of drug in the solid dispersion matrix.

Despite the promises, complicated processing methods have limited commercial viability of solid dispersions. There are two common methods to produce solid dispersions. One is to melt drugs and hydrophilic carriers such as PEG, PVP, PVA, HPMC, or other sugars, followed by cooling and hardening of the melt. The other is to dissolve drug and carrier in a common solvent, followed by solvent evaporation. The melting technique often involves high temperature, which presents a challenge for processing thermal-labile compounds while the cosolvent technique has its own problems. Since solid dispersion often uses hydrophilic carriers and hydrophobic drugs, it is difficult to find a common solvent to dissolve both components. Regardless of processing method, solid dispersion materials are often soft, waxy, and possess poor compressibility and flowability. This presents additional manufacturing challenges, especially during the scale-up process.

Physical instability and the preparation of reproducible material are two significant challenges in developing solid dispersions. It is not uncommon to produce wholly or partially amorphous drug during processing, which will eventually transform to a more stable crystalline form over time. The rate of transformation may be greatly influenced by storage conditions, formulation composition as well as processing methods. So far very few solid dispersion products have been marketed.

4.3.6. Complexation It has been well established in the literature that complexation is an effective way to solubilize hydrophobic compounds. Nicotinimide is known to complex with aromatic drugs through π donor–π acceptor interaction [119]. Similar π–π interaction also occurs between salts of benzoic acid or salicylic acid and drugs containing aromatic rings such as caffeine [120]. Obviously, aromaticity is an important factor in this type of complexation. Unfortunately, from a safety perspective the use of these types of complexing agents for products is not really very viable.

Cyclodextrin (CD) exemplifies another type of complexation, that is, complexation through inclusion. CD are torus shaped "oligosaccharides composed of 6–8 dextrose units (α-, β- and γ-cyclodextrins, respectively) joined through 1–4 bonds" [121]. It has a lipophilic cavity with 6.0–6.5 Å opening, which can form inclusion complexes by taking up a guest molecule into the central cavity. Formation of complexes alters the physicochemical properties such as solubility, dissolution rate, stability, and volatility of both drug molecules and CD molecules. A drug's solubility and dissolution rate usually increases when forming inclusion complex with CDs [122].

Inclusion complexation has also been used to stabilize, decrease the volatility, and ameliorate the irritancy and toxicity of drug molecules. In addition, modified CD such as carboxymethyl derivatives (e.g., CME-β-CD) exhibits pH-dependent solubility, and therefore can be used in enteric formulations.

Many drugs interact most favorably with β-CD. Unfortunately, β-CD has the lowest solubility (1.8% in water at 25°C) among the three CDs, thereby, limiting its solubilization capacity. To enhance solubility of the β-CD as well as to improve its safety, derivatives of β-CD have been developed. Among them, hydroxypropyl-β-cyclodextrin (HP-β-CDs) and sulfobutylether-β-cyclodextrins (SBE-β-CDs, mainly as SBE7-β-CD, while 7 refers to the average degree of substitution) have captured interest in formulating poorly soluble drugs in immediate-released dosage forms. A disadvantage of going to these derivatized cyclodextrins as an excipient is higher cost.

It is well demonstrated in the literature that complexation with CD could significantly enhance bioavailability of poorly soluble compounds. Apart from bioavailability enhancement, other advantages of using cyclodextrin include good stability, ease of manufacturing, and reproducibility as compared to other Class II strategies such as use of "higher energy forms" and solid dispersion. So far, at least 10 oral products containing CD have gained approval from the FDA [121].

A frequent consideration when forming inclusion complexes is how fast drug is released from the complex in vivo. Stella and Rejewski showed that weakly to moderately bound drug, in fact, rapidly dissociates from CD upon dilution [121]. For strongly bound drugs or when dilution is minimal, competitive displacement is important for rapid and complete dissociation. Although rapid reversibility of the complexation process is essential for drug absorption, it also poses the potential to reduce bioavailability. Excess free drug may precipitate upon dilution in GI tract prior to absorption. If such precipitation is significant, it may prove valuable to combine drug/CD complexes with precipitation inhibitors.

Another potential concern with CD is safety since there is only limited experience with marketed products at this point. Although oral administration of CD is generally considered safe, it may cause increased elimination of bile acids and certain nutrients [123]. In addition, CD may also cause membrane destabilization through its ability to extract membrane components such as cholesterol and phospholipids. In this regard, cyclodextrin may act as a permeation enhancer to enhance the mucosal permeation of the drug [124].

4.3.7. Cocrystallization The utilization of cocrystalline forms to modify the physical and chemical properties of solid forms is a recently developing concept within the pharmaceutical industry [125]. While there is disagreement on the exact definition of a cocrystal, most agree that a cocrystal is a crystalline solid containing multiple components that are associated through intermolecular interactions. Typically, the individual components must exist as solid forms—a distinction that separates cocrystals from hydrates and solvates. Furthermore, a distinction between salts and cocrystals can be made based on whether a proton has been transferred from an acid to a base [126,127]. A recent example of the application of cocrystals to pharmaceutical solid design is the enhancement of the physical stability (i.e., decreased moisture adsorption) of theophylline when formed as a cocrystal with oxalic, malonic, maleic, or glutaric acid [128]. The application of cocrystals to enhance bioavailability has also been recently demonstrated, illustrating their potential as an alternative solid form for pharmaceutical product development [129,130].

4.3.8. Lipid Technologies It has been known for a long time that lipid-based formulations can significantly improve the bioavailability of hydrophobic drugs by facilitating drug dissolution, dispersion, and solubilization either directly from administered lipids or through intraluminal lipid processing [131–133]. However, the preference for solid dosage forms usually prevails because of physical and chemical instability associated with lipid formulations. It was not until recently that lipid technologies generated much interest, likely because of the increasing numbers of hydrophobic compounds emerging from discovery programs. Several common types of lipid based dosage forms include lipid suspensions and solutions, micelle solubilization, microemulsions, macroemulsions (or emulsions), and liposomes.

Lipid Suspensions and Solutions A typical lipid solution is composed of triglycerides or mixed glycerides and surfactants [131,132,134]. Although lipid solutions are easy to formulate, they have limited solvent capacities except for very lipophilic drug ($\log p > 5$) [133,135]. Consequently, the design of unit dose lipid solutions is often not a practical approach, especially for high dose compounds. Further, a typical lipid solution may be poorly dispersible in water. In such cases, digestibility of the lipid formulation may be important to achieve good bioavailability because lipolysis is commonly believed to facilitate dispersion of drug into colloidal solution, thereby leading to faster absorption [136]. Nondigestible lipids such as mineral oil (liquid paraffin) and sucrose polyesters "can actually limit/reduce drug absorption by retaining a portion of the coadministered drug" [137]. Studies have also shown that the bioavailability of digestible lipid formulation tends to be higher than nondigestible formulations. However, given the complexity of lipid digestion, it may be difficult to interpret the effects of lipid vehicles. Common digestible lipids are dietary lipids (including glycerides, fatty acids, phospholipids, and cholesterol/cholesterol esters) and their synthetic derivatives. A few excellent reviews on how digestibility may influence bioavailability are available [136,137].

Lipid suspensions are also known to enhance the bioavailability of hydrophobic drugs. Unlike lipid solutions, suspended drug needs to undergo additional dissolution prior to the absorption. Therefore, factors such as drug particle size and amount suspended may also influence the bioavailability.

Micelle Solubilization The early interest in lipid formulations was initiated by the findings that coadministration of drug with food enhanced the bioavailability of many drugs. Intake of food stimulates secretion of bile salts into the duodenum, increasing bile salts concentration in the duodenum from a typical 1–4 mmol/L in the fasted state to 10–20 mmol/L in the fed-state [138]. It is hypothesized that micelles or mixed micelles formed by bile salts and digested lipids could significantly solubilize hydrophobic drug thereby enhancing drug absorption. This prompted investigation of using simple lipid solutions and suspensions for hydrophobic drugs discussed above. As summarized in Table 6, normal micelles are transparent and thermodynamically stable liquid solutions consisting of water and amphiphile. Micelles have low viscosity, long shelf life and are easy to prepare. However, they have limited capacity to solubilize oil and hydrophobic drugs.

Small amount of surfactants are often added to formulations to significantly improve drug wettability. Also, when added below their critical micelle concentration (CMC), surfactant can adhere to the surface of drug and reduce the interfacial tension between the drug and the dissolution medium. Above the CMC, surfactants form micelles that can solubilize drug. Micelle solubilization either may increase drug absorption by increasing the amount of drug which is solubilized and available at the absorption surface [139] or may, in some cases, reduce diffusion of the drug to the absorption surface and reduce absorption [140].

Emulsions Emulsions have much higher solvent capacity for hydrophobic materials than micelles. However, emulsions are metastable

Table 6. Physical Characteristics of Different Lipid Colloidal Systems

	Micelles	Microemulsions	Emulsions	Liposome
Spontaneously obtained	Yes	Yes	No	No
Thermodynamically stable	Yes	Yes	No	No
Turbidity	Transparent	Transparent	Turbid	From transparent to turbid, depending on droplet size
Typical size range	<0.01 µm	~0.1 µm or less	0.5–5 m	0.025–2 µm
Cosurfactant used	No	Yes	No	No
Surfactant concentration	<5%	>10%	1–20%	0.5–20%
Dispersed phase concentration	<5%	1–30%	1–30%	1–30%

Modified from Ref. [168].

colloids that will phase separate over a period of time and form a two-phase system (i.e., oil phase and aqueous phase). Because of its physical instability, large energy input (usually mechanical mixing) is required to form an emulsion.

Microemulsions Unlike emulsions, microemulsions are transparent and thermodynamically stable colloidal systems, formed under certain concentrations of surfactant, water and oil (Fig. 8). The transparency is because the droplet size of the microemulsions is small enough (<100 nm) that they do not reflect light. Because of its thermodynamic stability, microemulsions may have long shelf lives and spontaneously form with gentle agitation. However, microemulsions are not infinitely stable upon dilution because dilution changes the composition of the colloidal system. Microemulsions also have a high capacity for hydrophobic drugs that further adds to their attractiveness as a promising drug delivery system for poorly water-soluble compounds.

Liposomes Liposomes are a metastable colloidal system consisting of natural lipids and cholesterol. Unlike micelles, emulsions, microemulsions, and liposomes use ingredients that are part of biological membranes. Therefore, liposomes have relatively few problems with toxicity unlike the exogenous surfactants present in other colloidal systems. Although liposomes have generated great interest in the past decades, oral administration of liposomes remains highly controversial, thereby, will not be discussed in detail.

Self-Emulsifying Drug Delivery Systems Self-emulsifying drug delivery systems (SEDDS) are closely related to microemulsions and emulsions. SEDDS is an isotropic lipid solution typically consisting of a mixture of surfactant, oil, and drug that rapidly disperses to form fine emulsion droplets upon administration. If the droplet size is comparable to typical microemulsion droplet size, SEDDS become SMEDDS or self-micro-emulsifying drug delivery systems. One apparent advantage of SMEDDS and SEDDS over microemulsions is elimination or reduction of the aqueous phase, thereby significantly reducing the dose volume. SEDDS often have a dose volume that is small enough to allow encapsulation into soft or hard gelatin capsules. In addition, use of SEDDS avoids or partially avoids common physical stability problems associated with emulsions.

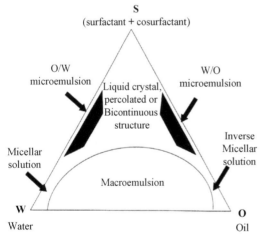

Figure 8. Hypothetical phase regions of microemulsion systems of oil (O), water (W), and surfactant consurfactant (S). Reprinted with permission from the publisher of Ref. [168].

Early work by Pouton demonstrated that a good SEDDS formulation could significantly enhance the dissolution and bioavailability of poorly soluble compounds [133,141]. Pouton proposed two criteria to describe the efficiency of SEDDS formulation: (1) the rate of emulsification and (2) the particle size distribution of the resultant emulsion. An efficient SEDDS should produce fine dispersions (<1 µm) rapidly at reproducible rate. Efficient SEDDS or SMEDDS have demonstrated their potential in delivering hydrophobic compounds. The most notable case is a SMEDDS formulation of cyclosporin A ("Neoral") where the formulation has significantly increased the bioavailability as well as decreased patient variability [142,143]. Recent success with griseofulvin [144], torcetrapib [145], and paclitaxel [146] has also been achieved.

Despite the potential of SEDDS and SMEDDS in oral delivery of poorly soluble drugs, few oral SEDDS formulations have been marketed so far. This is partly due to traditional preference to develop a solid dosage form, partly due to inherent limitations associated with lipid products. One limitation of SEDDS systems in general is physical and chemical instability caused by undesirable interactions between drug–excipient or among excipients. Another major limitation of SEDDS is that many hydrophobic drugs may not have sufficient solubility in pharmaceutically acceptable lipids. It is a common misperception that poor water solubility means good lipid solubility. Although SEDDS may have relatively higher solubilization capacities than simple lipid solutions and micelles, most of hydrophobic drugs ($\log p < 4$) are not very soluble in long-chain hydrocarbon oils. In general, many hydrophobic drugs ($2 < \log p < 4$) are more soluble in small/medium-chained oils such as Miglyol 812 than long-chain oils. However, it is rare that the drug load of SEDDS formula can exceed 30%.

High concentrations of surfactants in SEDDS also raise safety concerns, especially for drugs intended for chronic therapy. In addition, common lipid components such as fatty acids, glycerides, and several surfactants are known to act as absorption enhancers [147]. There are a host of safety issues associated with absorption enhancers (for details, refer to "absorption enhancer" under Class III technology). Furthermore, formulating SEDDS systems is not a trivial exercise. It requires an understanding of the complicated phase-behavior of a system, consisting of at least four basic components: oil, drug, surfactant(s) (Table 7), cosurfactant(s), as well as water for microemulsion formation. Hopefully, research in this area will build a large enough database to help formulate SEDDS in the future.

Table 7. List of Common Surfactants that are Pharmaceutically Acceptable

Nonionic	Polysorbates (Tweens), sorbitan esters (Spans), polyoxyethylene monohexadecyl ether
Anionic	Sodium lauryl sulfate, SLS or SDS, sodium docusate
Cationic	Quaternary ammonium alkyl salts such as hexadecyl trimethyl ammonium bromide (CTAB), didodcecylammonium bromide (DDAB)
Zwitter-ionic	Phospholipids

4.4. Class III: High Solubility and Low Permeability

The limiting factor for Class III compounds is the effective permeability across the GI tract. Given the difficulty of altering membrane permeability, Class II technologies are often used to formulate Class III compounds. The underlying principle for such substitution is that increasing the drug concentration in the GI tract should increase absorption of a drug if it is absorbed through the passive diffusion process (the assumption is valid in most cases). However, Class II solubilization technologies may not significantly enhance drug absorption if solubility and dissolution rate is high. Theoretically, the most effective way to enhance the absorption of Class III compound is to overcome the absorption rate-limiting barrier—permeation. The following sections will review some methods that have shown promise in formations to formulate Class III compounds.

4.4.1. Prodrugs
Poor membrane permeation is most commonly due to either low partitioning into the lipid membrane or low membrane

diffusivity. The most direct solution is to (1) modify a drug's structure to increase lipophilicity, (2) reduce molecular weight or (3) remove hydrogen-bonding groups. Prodrugs are one way to structurally modify the active compound to improve membrane permeability and still maintain activity of the parent drug upon bioreversion. Successful prodrug approaches include an approved antihypertensive agent -Fosinopril (an acyloxyalkyl prodrug of Fosinoprilat), and various angiotensin converting enzyme (ACE) inhibitors [148].

There are five important criteria in prodrug design: (1) adequate stability to variable pH environment of the GI tract, (2) adequate solubility or solubilization mechanisms, (3) enzymatic stability to luminal contents as well as the enzymes found in the brush border membrane, (4) good permeability and adequate $\log p$, and (5) the prodrug should revert to the parent drug either in the enterocyte or once absorbed into systemic circulation [148]. Postenterocyte reversion is more desirable because conversion in the enterocyte would also allow for back diffusion into the GI lumen.

Among these five criteria, knowledge of rate of bioreversion as well biological distribution of reconversion sites is often most critical for prodrug success. If bioreversion is fast and nonspecific, prodrug reversion may take place before the limiting barrier is overcome; if too slow, prodrug may readily reach the site of action but not release enough parent drug to elicit a pharmacological response. Knowledge about the biological distribution of reconversion sites will help predict the location of active drug. Ideally, reconversion sites should coincide with the target site.

Unfortunately, it is often difficult to satisfy all five criteria simultaneously, among which control of reconversion rate proves especially difficult. In addition, increasing lipophilicity often reduces aqueous solubility, which makes it even more difficult to formulate. Perhaps one of the biggest concerns is that prodrugs are considered new chemical entities that require a new set of preclinical studies. Therefore, prodrug approach is often less preferred if simpler formulation approaches are available. The incorporation of a prodrug strategy really should occur in very early preclinical evaluations. Prodrug approaches do offer the opportunity to expand intellectual property through patents.

4.4.2. Permeation Enhancers
Use of permeation enhancers is an alternative way to enhance drug permeation through the biological membrane by transiently altering the integrity of the mucosal membrane. Permeation enhancers may act at either the apical cell membrane (transcellular pathway) or the tight junctions between cells (paracellular pathway). There are several ways that permeation enhancers may interact with the cell membrane. Some fatty acids such as oleic acid have been found to disrupt the configuration of the lipid region [149–151]. Some enhancers such as siacylic acid may interact with membrane protein that carries on important membrane functions. Medium chain monoglycerides may extract cholesterol out of the cell membrane [152]. Chelators such as EDTA and some bile acids could chelate Ca^{2+} in the tight junction that can lead to pore openings from 8–14 Å [153].

Because of its potential damage to the membrane, permeation enhancers elicit great safety concern, especially in chronic therapy. Some of these include (1) potential tissue irritation and damage, (2) effect of the enhancer on structural integrity of the mucosal membrane, (3) reversibility of membrane perturbation, (4) long-term effect of continued exposure to the enhancer and, (5) potential to also enhance absorption of any potential harmful substances that are also present in the intestine. All these issues may have significant toxicity ramifications. Therefore, FDA has not approved any permeation enhancer, although use of some common excipients that are known to enhance absorption may be acceptable (Table 8).

4.4.3. Ion Pairing
Ion pairing has been proposed to enhance effective permeability of polar or hydrophilic drugs that exhibit poor permeability properties. In this approach, an ionizable drug is coadminstered with an excess concentration of a counterion. In theory, the ionized drug will associate with the counterion and partition into the membrane as a more lipophilic ion pair. Although several

Table 8. Compounds Shown to Have Intestinal Absorption Enhancing Effects

Classes	Examples
Bile salts	Sodium deoxycholate, sodium glycocholate, sodium taurocholate and their derivatives
Surfactant	Polyoxyethylene alkyl ethers, polysorbate, sodium lauryl sulfate, dioctyl sodium sulfosuccinate
Fatty acids	Sodium caprate, oleic acid
Glycerides	Natural oils, medium-chain glycerides, phospholipids, polyoxyethylene glyceryl esters
Acyl carnitines and cholines	Palmitoyl carnitine, lauroyl choline
Salicylates	Sodium salicylate, sodium methoxysalicylate
Chelating agents	EDTA
Swellable polymers	Starch, polycarbophil, chitosan
Others	Citric acid, cyclodextrin

Modified from Ref. [147].

animal studies reported moderate success with the ion pairing approach, in most cases the formulation was directly administered onto the absorption surface. Therefore, these studies did not reflect the effect of dilution, dispersion, and other counterions in the GI tract on ion pair dissociation [154]. In fact, although the concept of ion pairing has been around for almost four decades, lack of evidence for commercial feasibility has limited further research and development of this approach.

4.4.4. Improving Residence Time: Bioadhesives

Increasing residence time at the absorption site could also enhance drug absorption. Bioadhesive drug delivery systems have been proposed as a means to increase GI tract residence time. The original concept of bioadhesion is to administer drug in a bioadhesive polymer matrix that adheres to mucosal membranes to prolonge residence time (12–24 h), thereby increasing the contact time between the drug and the absorption site. Several review articles and books have extensively reviewed the concept of bioadhesive polymers [155–158]. This approach seems to have lost its popularity due to disappointing animal and human data.

4.5. Class IV

Class IV compounds exhibit both poor solubility and poor permeability and pose tremendous challenges to formulation development. As a result, a substantial investment in dosage form development with no guarantee of success should be expected. Class IV compounds are rarely developed or reach the market. A combination of Class II and Class III technologies could be used to formulate Class IV compounds. However, redesigning drug molecules to enhance solubility and/or permeability or searching for a nonoral route may be more likely to succeed.

4.6. "Class V": Other Absorption Barriers

Although the BCS provides a useful framework for recognizing solubility and permeability as two key parameters controlling absorption, additional "barriers" that limit drug absorption do exist beyond the scope of the BCS. Luminal complexation can reduce the free drug concentration available for absorption. Luminal degradation further degrades compounds such as proteins and peptides that are susceptible to intestinal enzymes or microorganisms. Presystemic elimination includes both traditional first-pass metabolism and intestinal metabolism. The significance of intestinal drug metabolism is a relatively recent discovery as evidenced by the high intestinal concentration of CYP3A4, which is present in the intestine at approximately 80–100% of the CYP3A4 concentration in the liver [159]. In addition, P-gp, a membrane transporter, further reduces drug absorption by retrograding efflux of drug into the intestinal lumen in the secretory direction (or basolateral-to-apical direction).

This section will review potential formulation strategies addressing the issues mentioned above. However, given the difficult nature of overcoming such absorption barriers and the limited knowledge in this area, most of the formulation strategies reviewed in this section are highly experimental and yet to be proven.

4.6.1. Luminal Degradation

Luminal degradation can be attributed to chemical decomposition in the aqueous intestinal environment or metabolism by luminal digestive enzymes or luminal microorganisms. Degradation in the acidic environment of the stomach is relatively easy to solve either by using an enteric coating dosage form or by formulating with anti-acid agents [160]. Chemical instability in the slightly acidic to neutral pH of the small intestine (pH range from 4 to 7) may be more difficult to solve, especially for molecules that cannot be "protected" from the aqueous environment by complexation or solubilization. Another difficult problem to solve is enzymatic degradation in the GI lumen. Enzymatic degradation, combined with poor permeability, has significantly limited the oral absorption of proteins and peptides. Prodrugs are one approach to protect parent compound from enzymatic degradation. Lipid vesicles and micelles may also be able to shield their encapsulated contents from luminal degradation. Micelle formation, for example, has been shown to slow ester hydrolysis of benzoylthiamine disulfide resulting in increased *in situ* and *in vivo* absorption [161]. Water-in-oil microemulsions have also demonstrated some potential in delivering peptides and proteins, although at a very low capacity or efficiency [162]. Upon aqueous dilution in the GI tract, water in oil (w/o) microemulsions can undergo phase separation or inversion that can cause dose dumping and expose the encapsulated water-soluble drug to luminal degradation. Therefore, in cases where drug absorption is significantly enhanced in w/o microemulsions, it may not be simply because the w/o microemulsion protects the drug from luminal degradation. One cannot rule out the possibility that certain lipid excipients may act as an absorption enhancer and increase absorption of proteins and peptides.

Coadministration of the drug with an enzymatic inhibitor may also protect drug from luminal enzymatic degradation. The key to this approach is that the inhibitor needs to effectively protect the drug until the drug is dissolved and absorbed. This would require large amounts of inhibitor to overcome dilution in the lumen. For the best protection, drug may be encapsulated with the inhibitor in lipid vesicles or polymeric membranes. However, this approach may raise serious safety concerns and would not be recommended as a general approach to overcome enzymatic degradation.

4.6.2. Presystemic Elimination (First-Pass Metabolism and Intestinal Metabolism)

Presystemic elimination includes both intestinal metabolism and hepatic first-pass metabolism. The significance of the former is a relatively recent discovery. Among enzymes discovered in the human intestine, CYP3A4 is by far the most important enzyme to drug metabolism. The CYP3A4 intestinal concentrations are approximately 80–100% of that in the liver. Other enzymes, such as CYP3A5, CYP1A1, CYP2C8-10, CYP2D6, CYP2E1 have also been identified in small intestine but their level is significantly lower than that of the liver [159].

Few oral formulation approaches are available to overcome presystemic elimination. Although coadministration of drug with an enzymatic inhibitor could boost bioavailability, it is not recommended as a viable approach since it raises serious safety concerns. Intestinal lymphatic transport offers the possibility of avoiding hepatic first-pass metabolism. The intestinal lymphatics are the major absorption gateway for natural lipids, lipid derivatives, and cholesterol. However, only highly lipophilic compounds ($\log p > 5$–6), such as lipid-soluble vitamins or xenobiotics can gain significant access to the systemic circulation via the lymphatics. The vast majority of pharmaceutical compounds are not lipophilic enough and, when they are, their solubilities are extremely poor. Lipophilic prodrugs may be designed for the purpose of enhancing intestinal lymphatic drug delivery. The design sophistication varies from simple chemical modification (e.g., derivative compounds via simple ester or ether linkages) to sophisticated functional design where a "functionally based" promoiety is added to facilitate compound incorporation into the normal lipid processing pathways. Comprehensive reviews in this area are available [163,164]. In general, the use of lipid prodrugs to target lymphatic approach is a wide-open research area. Much needs to be done before one can assess the practicality of this approach.

4.6.3. P-gp Efflux Mechanism

Enterocyte P-gp is an apically polarized efflux transporter that was first identified in multidrug-resistant cancer cells but later also found to be present in the intestinal brush border region. P-gp reduces drug absorption by actively transporting drug in the secretory direction back into the intestinal lumen. Interestingly, P-gp appears to share a large number of substrates and inhibitors with CYP3A [159]. Little is known about how to overcome the absorption barrier posed by the P-gp efflux pump. Certain nonionic surfactants such as Cremophor have been shown to inhibit the P-gp efflux pump *in vitro* [165]. However, to counter G.I. dilution, a much higher amount of surfactant may be required to achieve a similar effect *in vivo*. Another approach for overcoming P-gp efflux is to coadminister the drug with a P-gp inhibitor. For example, docetaxel has very poor oral bioavailability partly due to its affinity for the intestinal P-gp efflux pump, partly may due to metabolism of docetaxel by cytochrome P450 in gut and liver. In a recent clinical study with 14 patients, the mean oral bioavailability in patients taking docetaxel is only $8 \pm 6\%$, whereas the bioavailability of docetaxel in patients receiving both the drug and cyclosporine A (both a P-gp substrate and an inhibitor) is $90 \pm 44\%$ [166]. Although effective, this approach does raise a series of safety concerns. Some important questions include (1) since the P-gp efflux pump and CYP3A share a large number of substrates, what is the effect of administrating a P-gp inhibitor on liver and gut metabolism, (2) if both P-gp and CYP3A are inhibited, what are the potential implications caused by potential toxic substances that are usually metabolized by CYP3A, (3) is the inhibitory transient or long-lasting, reversible or irreversible, and (4) what is the effect of inhibitors on intersubject variability? These issues may have serious toxicity implications. Therefore, without fully understanding the mechanism of P-gp inhibitors or the natural role of P-gp transportors, this approach is too risky to be considered as a routine method to improve oral bioavailability.

Overall, many "Class V" compounds face significant delivery challenges that cannot be easily overcome by traditional methods. Although some emerging "Class V" technologies may be able to overcome such challenges to a certain extent, these strategies are highly exploratory and yet to be proven. One may be better off seeking a nonoral delivery route to overcome the absorption barriers posed by Class V compounds. Such delivery routes may include nasal, oral mucosal, or intravenous administration. Details regarding these delivery routes are beyond the scope of this chapter.

4.7. Excipient and Process Selections in Dosage Form Design

The above sections classify the formulation strategies based on the solubility and permeability of drugs. Once a formulation strategy is identified, it is important to choose suitable excipients and processing methods to achieve the objective of a selected dosage form in terms of both dosage form performance and manufacturability. While it is beyond the scope of this chapter to delve into the details of formulation development, a brief overview of some considerations may provide a useful insight into the factors considered by the formulation scientist in developing formulations. Typically, excipients are added to drug to allow for the manufacture of dosage forms that meet performance (e.g., drug release, stability) and manufacturing requirements. Common excipients for tablet formulations, for example, include tablet fillers, binders, disintegrants, wetting agents, glidants, and lubricants. In addition to selecting appropriate excipients, suitable processing steps and processing conditions must be identified. For tablet dosage forms, typical processing steps include mixing, granulation, sizing, and compression. The granulation process can produce product with improved performance and manufacturing properties and either dry granulation or wet granulation can be done depending on whether a granulating fluid is used. For materials with appropriate physical, chemical, and mechanical properties, direct compression without the granulation step may be most appropriate and desirable.

Proper selection of excipients and processes will impact the performance of a dosage form. For example, a poorly soluble drug often tends to be poorly wettable too. If the objective is to obtain a fast dissolving and dispersing dosage form, inclusion of a wetting agent such as sodium lauryl sulfate or polysorbate 80 may be appropriate or even necessary. Processing methods may also significantly impact dosage form performance. For example, it may not be appropriate to wet granulate amorphous drug because water may lower the glass transition temperature and facilitate recrystallization during or after processing. In other situations, wet granulation can be used to avoid potential segregation and content uniformity problems where there is a significant difference in particle size or bulk density between the drug and the excipients. Overall, a wise selection of excipients and processes relies on a sound understanding of the physical, chemical, and mechanical properties of the drug and excipients. A formulation be successfully scaled up and consistently meet performance and manufacturing requirements only when one fully understands the complex relationship between the drug, excipients, processing, and the desired dosage form performance criteria.

5. CONCLUSION

This chapter is composed of two parts. Part I provides an overview of physicochemical characterization and its relevance to drug delivery. Part II categorizes various dosage form options according to the well-established BCS. Physicochemical properties are closely linked to biopharmaceutical properties of drug candidates. The BCS captures this link by highlighting the important effects of solubility and permeability on drug absorption. Therefore, a sound understanding of how physicochemical properties may affect absorption is essential to make a smart choices of which drug candidate(s) to select and which of the wide variety of oral dosage formulation options to pursue.

REFERENCES

1. Craig DQ, et al. Relevance of the amorphous state to pharmaceutical dosage forms: glassy drugs and freeze dried systems. Int J Pharm 1999;179:179–207.
2. Cui Y. A material science perspective of pharmaceutical solids. Int J Pharm 2007;339(1–2):3–18.
3. Blagden N, et al. Crystal engineering of active pharmaceutical ingredients to improve solubility and dissolution rates. Adv Drug Deliv Rev 2007;59(7):617–630.
4. Hilden LR, Morris KR. Physics of amorphous solids. J Pharm Sci 2004;93(1):3–12.
5. Berge SM, Bighley D, Monkhouse DC. Pharmaceutical salts. J Pharm Sci 1977;66(1):1.
6. Byrn SR, et al. Solid-State Pharmaceutical Chemistry. Chem Mater 1994;6:1148.
7. Byrn S.et al Pharmaceutical solids: a strategic approach to regulatory considerations. Pharm Res 1995;12(7):945–950.
8. Morris KR, et al. Integrated approach to the selection of optimal salt form for a new drug candidate. Int J Pharm 1994;105:209–217.
9. Gould PL. Salt selection for basic drugs. Int J Pharm 1986;33:201.
10. Hammond RB, et al. Grid-based molecular modeling for pharmaceutical salt screening: Case example of 3,4,6,7,8,9-hexahydro-2H-pyrimido (1,2-a) pyrimidinium acetate. J Pharm Sci 2006;95(11):2361–2372.
11. Kumar L, Amin A, Bansal AK. An overview of automated systems relevant in pharmaceutical salt screening. Drug Discov Today 2007;12(23–24):1046–1053.
12. Haleblian J, McCrone W. Pharmaceutical applications of polymorphism. J Pharm Sci 1969;58(8):911.
13. International Conference on Harmonization. Draft guidance on specifications. Test procedures and acceptance criteria for new drug substances and drug products: Chemical substances. Washington, DC: Federal Register; 1997. p 62889–62910.
14. Zhu H, Grant D. Influence of water activity in organic solvent plus water mixtures on the nature of the crystallizing drug. Phase 2. Ampicillin. Int J Pharm 1996;139:33–43.
15. Zhu H, Yuuen C, Grant D. Influence of water activity in organic solvent plus water mixtures on the nature of the crystallizing drug. Phase 1. Theophylline. Int J Pharm 1996;135:151–160.
16. Byrn SR, Pfeiffer RR, Stowell JG Solid State Chemistry of Drugs. 2nd ed. West Lafayette, IN: SSCI; 2000.
17. Hancock BC, Parks M. What is the true solubility advantage for amorphous pharmaceuticals?. Pharma Res 2000;17:397–404.
18. Hancock BC. Disordered drug delivery: destiny, dynamics and the Deborah number. J Pharm Pharmacol 2002;54(6):737–746.

19. Hancock BC, Dalton CR. Effect of temperature on water vapor sorption by some amorphous pharmaceutical sugars. Pharma Dev Technol 1999;4(1):125–131.
20. Katritzky AR, et al. Prediction of melting points for the substituted benzenes: a QSPR approach. J Chem Inf Comput Sci 1997;37:913–919.
21. Krzyzaniak JF, et al. Boiling-point and melting-point prediction for aliphatic, non- hydrogen-bonding compounds. Ind Eng Chem Res 1995;34:2530–2535.
22. Simamora P, Miller AH, Yalkowsky SH. Melting-point and normal boiling-point correlations: applications to rigid aromatic compounds. J Chem Inf Comput Sci 1993;33:437–440.
23. Simamora P, Yalkowsky SH. Group-contribution methods for predicting the melting-points and boiling points of aromatic-compounds. Ind Eng Chem Res 1994;33:1405–1409.
24. Verwer P, Leusen FJJ. Computer simulation to predict possible crystal polymorphs. In: Lipkowitz KB, Boyd DB, editors. Reviews in Computational Chemistry. Vol. 12, New York: Wiley-VCH; 1998. p 327–365.
25. Overton E, Phys Chem 1891;8:189–209.
26. Meyer H, Arch Exp Pathol Pharmakol 1899;42:109–118.
27. Buchwald P, Bodor N. Octanol–water partition: searching for predictive models. Curr Med Chem 1998;5:353–380.
28. Leo AL, Hansch C, Elkins D. *Partition coefficients and their uses*. Chemical Reviews 1971;71 IS(6):525.
29. Taylor P, In: Hansch C, Sammes PG, Taylor JB, editors. Comprehensive Medicinal Chemistry. Vol. 4: New York: Pergamon Press; 1990. p 241–294.
30. Leo AJ. Some advantages of calculating octanol–water partition coefficients. J Pharm Sci 1987;76:166–168.
31. Mannhold R, Dross K. Calculation procedures for molecular lipophilicity: a comparative study. Quant Struct Act Relat 1996;15: 403–409.
32. Mannhold R, et al. Comparative-evaluation of the predictive power of calculation procedures for molecular lipophilicity. J Pharm Sci 1995;84:1410–1419.
33. Vandewaterbeemd H, Mannhold R. Programs and methods for calculation of log p-values. Quant Struct Act Relat 1996;15:410–412.
34. Leo AJ. Critique of recent comparison of log p calculation methods. Chem Pharma Bull 1995;43:512–513.
35. Yalkowsky S, Banerjee S, editors. Aqueous solubility: Methods of estimation for organic compounds. New York: Marcel Dekker, Inc; 1992.
36. Lipinski CA, et al. Experimental and computational approaches to estimate solubility and permeability in drug discovery and development settings. Adv Drug Deliv Rev 1997;23:3–25.
37. Huuskonen J. Estimation of aqueous solubility in drug design. Comb Chem High Throughput Screen 2001;4:311–316.
38. Yalkowsky SH, Flynn GL, Amidon GL. Solubility of nonelectrolytes in polar solvents. J Pharm Sci 1972;61:983–984.
39. Yalkowsky SH, Rinal R, Banerjee S. Water solubility, a critique of the solvatochromic approach. J Pharm Sci 1988;77(1):74.
40. Yalkowsky SH, Valvani SC. Solubility and partitioning I: solubility of nonelectrolytes in water. J Pharm Sci 1980;69(8):912.
41. Valvani SC, Yalkowsky SH, Roseman TJ. Solubility and partitoning IV: aqueous solubility and octanol water partition coefficients of liquid nonelectrolytes. J Pharm Sci 1981;70(5):502.
42. Yalkowsky SH, Valvani SC, Roseman TJ. Solubility and partitioning VI: octanol solubility and octanol water partition coefficient. J Pharm Sci 1983;72(8):866.
43. Yalkowsky S, Morozowich W. A physical chemical basis for the design of orally active prodrugs. In: Ariens EJ, editor. Drug Design. Academic Press; 1980. p 121.
44. Noyes AA, Whitney WR. The rate of solution of solid substances in their own solutions. J Am Chem Soc 1897;19:930.
45. Barthe L, Woodley J, Houin G. Gastrointestinal absorption of drugs: methods and studies. Fundam Clin Pharmacol 1999;13:154–168.
46. Vandewaterbeemd H, et al. Estimation of Caco-2 cell permeability using calculated molecular descriptors. Quant Struct Act Relat 1996;15:480–490.
47. Camenisch G, Folkers G, Vandewaterbeemd H. Comparison of passive drug transport through Caco-2 cells and artificial membranes. Int J Pharm 1997;147:61–70.
48. Camenisch G, et al. Estimation of permeability by passive diffusion through Caco-2 cell monolayers using the drugs' lipophilicity and molecular weight. Eur J Pharm Sci 1998;6:313–319.
49. Nellans HN. Mechanisms of peptide and protein absorption. 1. Paracellular intestinal

transport: modulation of absorption. Adv Drug Deliv Rev 1991;7:339–364.
50. Benet LZ, et al. Intestinal drug metabolism and antitransport processes: a potential paradigm shift in oral drug delivery. J Control Rel 1996;39:139–143.
51. Hunter J, Hirst BH. Intestinal secretion of drugs. the role of P-glycoprotein and related drug efflux systems in limiting oral drug absorption. Adv Drug Deliv Rev 1997;25: 129–157.
52. Livingstone DJ, et al. Simultaneous prediction of aqueous solubility and octanol/water partition coefficient based on descriptors derived from molecular structure. J Comput Aided Mol Des 2001;15:741–752.
53. Lipinski CA. Drug-like properties and the causes of poor solubility and poor permeability. J Pharmacol Toxicol Methods 2000;44: 235–249.
54. Lipinski CA, et al. Experimental and computational approaches to estimate solubility and permeability in drug discovery and development settings. Adv Drug Deliv Rev 2001;46:3–26.
55. Desai MC, et al. Physical parameters for brain uptake: optimizing log-p, log D and pK_a. Bioorg Med Chemi Lett 1991;1:411–414.
56. Lee CP, Devrueh RLA, Smith PL. Selection of development candidates based on *in vitro* permeability measurements. Adv Drug Deliv Rev 1997;23:47–62.
57. Paterson DA, et al. A nonaqueous partitioning system for predicting the oral absorption potential of peptides. Quant Struct Act Relat 1994;13:4–10.
58. Abraham MH, et al. Hydrogen-bonding. 32. An analysis of water–octanol and water– alkane partitioning and the delta-log-p parameter of Seiler. J Pharm Sci 1994;83:1085–1100.
59. Fagerholm U, Johansson M, Lennernas H. Comparison between permeability coefficients in rat and human jejunum. Pharm Res 1996;13:1336–1342.
60. Lennernas H. Human jejunal effective permeability and its correlation with preclinical drug absorption models. J Pharm Pharmacol 1997;49:627–638.
61. Lennernas H. Gastrointestinal absorption mechanisms: a comparison between animal and human models. Eur J Pharm Sci 1994;2:39–43.
62. Lennernas H. Human intestinal permeability. J Pharm Sci 1998;87:403–410.
63. Lennernäs H. Does fluid across the intestinal mucosa affect quantitative oral drug absorption? Is it time for a reevaluation. Pharm Res 1995;12:1573.
64. Lennernas H, Crison JR, Amidon GL. Permeability and clearance views of drug absorption: a commentary. J Pharmacokinet Biopharm 1995;23:333–337.
65. Lennernas H, et al. Comparison between active and passive drug transport in human intestinal epithelial (Caco-2) cells *in vitro* and human jejunum *in vivo*. Int J Pharm 1996;127:103–107.
66. Lindahl A, et al. Jejunal permeability and hepatic extraction of fluvastatin in humans. Clin Pharmacol Ther 1996;60:493.
67. Ungell AL, et al. Membrane transport of drugs in different regions of the intestinal tract of the rat. J Pharm Sci 1998;87:360–366.
68. Winiwarter S, et al. Correlation of human jejunal permeability (*in vivo*) of drugs with experimentally and theoretically derived parameters. A multivariate data analysis approach. J Med Chem 1998;41:4939–4949.
69. Artursson P. Cell-cultures as models for drug absorption across the intestinal-mucosa. Crit Rev Ther Drug Carrier Syst 1991;8:305–330.
70. Artursson P, Borchardt RT. Intestinal drug absorption and metabolism in cell cultures: Caco-2 and beyond. Pharm Res 1997;14: 1655–1658.
71. Artursson P, Karlsson J. Correlation Between oral-drug absorption in humans and apparent drug permeability coefficients in human intestinal epithelial (Caco-2) cells. Biochem Biophys Res Commun 1991;175:880–885.
72. Stewart BH, Chan OH, Jezyk N, Fleisher D. Discrimination between drug candidates using models for evaluation of intestinal absorption. Adv Drug Deliv Rev 1997;23:27–45.
73. Stewart BH, Chan OH, Lu RH, Reyner EL, Schmid HL, Hamilton NW, Steinbaugh B. Comparison of intestinal permeabilities determined in multiple *in-vitro* and *in-situ* models: relationship to absorption in humans. Pharm Res 1995;12:693–699.
74. Stewart BH, Wang Y, Surendran N. *Ex vivo* approaches to predicting oral pharmacokinetics in humans. Annual Reports in Medicinal Chemistry 2000;35:299–307.
75. Grass GM, Rubas W, Jezyk N. Evaluation of Caco-2 monolayers as a predictor of drug permeability in colonic tissues. FASEB J 1992;6: A1002.

76. Jezyk N, Rubas W, Grass GM. Permeability characteristics of various intestinal regions of rabbit, dog, and monkey. Pharm Res 1992;9:1580–1586.
77. Rubas W, Jezyk N, Grass GM. Comparison of the permeability characteristics of a human colonic epithelial (Caco-2) cell-line to colon of rabbit, monkey, and dog intestine and human drug absorption. Pharm Res 1993;10:113–118.
78. Artursson P, Palm K, Luthman K. Caco-2 monolayers in experimental and theoretical predictions of drug transport. Adv Drug Deliv Rev 2001;46:27–43.
79. Kramer SF, Flynn GL. Solubility of organic hydrochlorides. J Pharm Sci 1972;61(12):1986.
80. Bergren MS. Automated controlled atmosphere microbalence for the measurement of moisture sorption. Int J Pharm 1994;103: 103–114.
81. Engel G, et al. Salt form selection and characterization of LY-333531 mesylate monohydrate. Int J Pharm 2000;198:239–247.
82. Morris K, et al. Characterization of humidity-dependent changes in crystal properties of a new HMG-CoA reductase inhibitor in support of its dosage form development. Int J Pharm 1994;108:195–206.
83. Ahlneck C, Zografi G. The molecular basis of moisture effects on the physical and chemical stability of drugs in the solid state. Int J Pharm 1990;62:87.
84. U.S. Department of Health and Human Services, U.S. Food and Drug Administration. Guidance for Industry. Q8: Pharmaceutical Development. 2006.
85. International Conference on Harmonization. ICH harmonized tripartite guideline. Specifications: test procedures and acceptance criteria for new drug substances and new drug products: chemical substances Q6A. 1999; p 1–21.
86. ICH Harmonized Tripartite Guideline. Q8: Pharmaceutical Development. 2005; p 1–11.
87. Connors K, Amidon G, Stella V, editors. Draft guidance for industry. Stability testing of drug substances and drug products. In: Chemical Stability of Pharmaceuticals: A Handbook for Pharmacists. New York: Wiley-Interscience; 1986.
88. U.S. Department of Health and Human Services, U.S. Food and Drug Administration, Center for Drug Evaluation and Research (CDER), Center for Biologics Evaluation and Research (CBER). Guidance for industry. Q1A (R2): Stability testing of new drug substances and products. 2003.
89. F.a.D.A.F, U.S. Department of Health and Human Services, Center for Drug Evaluation and Research (CDER), Center for Biologics Evaluation and Research (CBER). Guidance for industry. Q3A: Impurities in New Drug Substances. Washington, DC; 2008.
90. Connors K, Amidon G, Stella V. Chemical Stability of Pharmaceuticals: A Handbook for Pharmacists. New York: Wiley-Interscience; 1986.
91. Carstensen J, Rhodes C. Drug Stability: Principles and Practices. New York: Marcel Dekker, Inc; 2000.
92. Lennernas H. Does fluid across the intestinal mucosa affect quantitative oral drug absorption? Is it time for a reevaluation. Pharm Res. 1995;12:1573.
93. Amidon GL, et al. A theoretical basis for a biopharmaceutical drug classification: the correlation of in vivo drug product dissolution and in vivo bioavailablity. Pharm Res 1995;12:413.
94. U.S. Department of Health and Human, U.S. Food and Drug Administration (FDA), Center for Drug Evaluation and Research. Guidance for industry. Waiver of the in vivo bioavailability and bioequivalence studies for immediate-release solid oral dosage forms based on a biopharmaceutics classification system. Washington, DC; 2000.
95. Dressman JB, Reppas C. In vitro–in vivo correlations for lipophilic, poorly water soluble drugs. Eur J Pharm Sci 2000;11(Suppl 2):S73.
96. U.S. Food and Drug Administration (CDER). Immediate release solid oral dosage forms-scale-up and postapproval changes: chemistry, manufacturing, and controls, in vitro dissolution testing, and in vivo bioequivalence documentation. Washington, DC; 1995.
97. Chen ML, et al. Bioavailability and bioequivalence: an FDA regulatory overview. Pharm Res 2001;18(12):1645.
98. Ranade VV, Hollinger MA. In: Oral Drug Delivery in Drug Delivery Systems. New York: CRC Press; 1996. p 166–167.
99. Kydonieus AF. Treatise on Controlled Drug Delivery: Fundamentals, Optimization Applications. New York: Marcel Dekker; 1992.
100. Bastin RJ, Bowker MJ, Slater BJ. Salt selection and optimization procedures for pharmaceutical new chemical entities. Org Proc Res Dev 2000;4:427.
101. Parrott EL, Milling. In: The Theory and Practice of Industrial Pharmacy, Lachman L, Lieberman HA, Kanig JL, editors. Philadelphia, PA: Lea & Febiger; 1976; 466–485.

102. Sanvicens N, Marco M. Multifunctional nanoparticles: properties and prospects for their use in human medicine. Trends Biotechnol 2008;26(8):425–433.
103. Constantinides P, Chaubal M, Shorr R. Advances in lipid nanodispersions for parenteral drug delivery and targeting. Adv Drug Delivery Rev 2008;60(6):757–767.
104. Peek L, Middaugh C, Berkland C. Nanotechnology in vaccine delivery. Adv Drug Deliv Rev 2008;60(8):915–928.
105. Singh R, Singh S, Lillard J. Past, present, and future technologies for oral delivery of therapeutic proteins. J Pharm Sci 2008;97(7):2497–2523.
106. Suzuki H, Sunada H. Influence of water-soluble polymers on the dissolution of nifedipine solid dispersion with combined carriers. Chem Pharm Bull 1998;46:482.
107. Suzuki H, Sunada H. Comparison of nicotinamide, ethylurea and polyethylene glycol as carriers for nifedipine solid dispersion systems. Chem Pharm Bull 1997;45:1688.
108. Loftsson T. Increasing the cyclodextrin complexation of drugs and drug bioavailability through addition of water-soluble polymer. Pharmazie 1998;53:733.
109. Shefter E. Solubilization by solid-state manipulation. In: Yalkowsky SH, editor. Techniques of Solubilization of Drugs. New York: Marcel Dekker; 1981. p 159–182.
110. Hasegawa A, et al. Supersaturation mechanism of drugs from solid dispersions with enteric coating agents. Chem Pharm Bull 1988;36:4941.
111. Weissbuch I, et al. Understanding and control of nucleation, growth, habit, dissolution, and structure of two- and three-dimensional crystals using 'tailor-made' auxiliaries. Acta Cryst 1995;B51:115.
112. Hancock BC, Parks M. What is the true solubility advantage for amorphous pharmaceuticals?. Pharm Res 2000;17(4):397.
113. Byrn SR, Pfeiffer RP, Stowell JG. Amorphous solids. In: Byrn SR, Pfeiffer RR, Stowell JG, editors. Solid-State Chemistry of Drugs. West Lafayette, IN: SSCI; 1999. p 249.
114. Abdou HM. Effect of the physicochemical properties of the drug on dissolution rate. In: Dissolution Bioavailability and Bioequivalence. Easton, PA: Mack Publishing; 1989. p 56–72.
115. Sekiguchi K, Obi N. Studies on absorption of eutectic mixture. I. A comparison of the behavior of eutectic mixture of sulfathiazole and that of ordinary sulfathiazole in man. Chem Pharm Bull 1961;9:866.
116. Goldberg AH, Gibaldi M, Kanig JL. Increasing dissolution rates and gastrointestinal absorption of drugs via solid solutions and eutectic mixtures. II. Experimental evaluation of eutectic mixture: urea-acetominophen system. J Pharm Sci 1966;55:482.
117. Goldberg AH, Gibaldi M, Kanig JL. Increasing dissolution rates and gastrointestinal absorption of drugs via solid solutions and eutectic mixtures. III. Experimental evaluation of griseofulvin-succinic acid solution. J Pharm Sci 1966;55:487.
118. Serajuddin ATM. Solid dispersion of poorly water-soluble drugs: early promises, subsequent problems, and recent breakthroughs. J Pharm Sci 1999;88:1058.
119. Rasool AA, Hussain AA, Dittert LW. Solubility enhancement of some water-insoluble drugs in the presence of nicotinamide and related compounds. J Pharm Sci 1991;80:387.
120. Hőrter D, Dressman JB. Influence of physicochemical properties on dissolution of drugs in the gastrointestinal tract. Adv Drug Deliv Rev 2001;46:75–87.
121. Stella VJ, Rajewski RA. Cyclodextrins: their future in drug formulation and delivery. Pharm Res 1997;14:556.
122. Uekama K. Pharmaceutical application of cyclodextrins as multi-functional drug carriers. Yakugaku Zasshi 2004;124(12):909–935.
123. Thompson DO. Cyclodextrins-enabling excipients: their present and future use in pharmaceuticals. Crit Rev Ther Drug Carrier Syst 1997;14(1):1–104.
124. Rajewski RA, Stella VJ. Pharmaceutical applications of cyclodextrin. 2. *In vivo* drug delivery. J Pharm Sci 1996;85:1142.
125. Shan N, Zaworotko MJ. The role of cocrystals in pharmaceutical science. Drug Discov Today 2008;13(9–10):440–446.
126. Childs SL, Stahly GP, Park A. The salts-cocrystal continuum: the influence of crystal structure on ionization state. Mol Pharma 2007;4(3):323–338.
127. Etter MC. Hydrogen bonds as design elements in organic chemistry. J Phys Chem 1991;95:4601–4610.
128. Trask AV, Motherwell WD, Jones W. Physical stability enhancement of theophylline via cocrystallization. Int J Pharm 2006;320(1–2):114–123.

129. Basavoju S, Bostroem D, Velaga SP. Indomethacin-saccharin cocrystal: design, synthesis and preliminary pharmaceutical characterization. Pharm Res 2008;25(3):530–541.
130. McNamara DP, et al. Use of a glutaric acid cocrystal to improve oral bioavailability of a low solubility API. Pharm Res 2006;23(8):1888–1897.
131. Hauss DJ. Oral lipid-based formulations. Adv Drug Deliv Rev 2007;59:667–676.
132. Pouton CW. Lipid formulations for oral administration of drugs: non-emulsifying, self-emulsifying, and 'self-microemulsifying' drug delivery systems. Eur J Pharm Sci 2000;11:S93.
133. Pouton CW, Porter CJ. Formulation of lipid-based delivery systems for oral administration: materials, methods and strategies. Adv Drug Deliv Rev 2008;60:625–637.
134. Porter CJ, et al. Enhancing intestinal drug solubilization using lipid-based delivery systems. Adv Drug Deliv Rev 2008;60:673–691.
135. Pouton CW. Formulation of self-emulsifying drug delivery systems. Adv Drug Deliv Rev 1997;25:47.
136. MacGregor KJ, et al. Influence of lipolysis on drug absorption from the gastro-intestinal tract. Adv Drug Del Rev 1997;25:33.
137. Humberstone AJ, Charman WN. Lipid-based vehicles for the oral delivery of poorly water soluble drugs. Adv Drug Deliv Rev 1997;25:103.
138. Sigh BN. Effects of food on clinical pharmacokinetics. Clin Pharmacokinet 1999;37:213.
139. Amidon GE, Higuchi WI, Ho NF. Theoretical and experimental studies of transport of micelle-solubilized solutes. J Pharm Sci 1982;71(1):77.
140. Mall S, Buckton G, Rawlins DA. Slower dissolution rates of supphamerazine in aqueous sodium dodecyl sulphate solutions than in water. Int J Pharm 1994;131:41.
141. Pouton CW. Self-emulsifying drug delivery systems: assessment of the efficiency of emulsification. Int J Pharm 1985;27:335.
142. Mueller EA, et al. Improved dose linearity of cyclosporin pharmackokinetics from a microemulsion formulation. Pharm Res 1994;11:301–304.
143. Kovarik JM, et al. Reduced inter- and intra-individual variability in cyclosporin pharmacokinetics from a microemulsion formulation. J Pharm Sci 1994;83:444–446.
144. Arida AI, et al. Improving the high variable bioavailability of griseofulvin by SEDDS. Chem Pharm Bull 2007;55(12):1713–1719.
145. Perlman ME, et al. Development of a self-emulsifying formulation that reduces the food effect for torcetrapib. Int J Pharm 2008;351(1–2):15–22.
146. Gao P, et al. Development of a supersaturable SEDDS (S-SEDDS) formulation of paclitaxel with improved oral bioavailability. J Pharm Sci 2003;92(12):2386–2398.
147. Aungst BJ, et al. Enhancement of the intestinal absorption of peptides and nonpeptides. J Control Rel 1996;41:19–31.
148. Krise JP, Stella VJ. Prodrugs of phosphate, phosphonates, and phosphinates. Adv Drug Del Rev 1996;19:287.
149. Aungst BJ, et al. Enhancement of the intestinal absorption of peptides and non-peptides. J Control Rel 1996;41:19.
150. Muranishi S. Absorption enhancers. Crit Rev Ther Drug Carrier Syst 1990;7:1.
151. Ganem-Quintanar A, et al. Mechanisms of oral permeation enhancement. Int J Pharm 1997;156:127–142.
152. Watanabe Y, van Hoogdalem EJ, de Boer AG. Absorption enhancement of rectally infused cefoxitin by medium chain monoglycerides in conscious rats. J Pharm Sci 1988;77:847.
153. Tomita M, et al. Enhancement of colonic drug absorption by the paracellular permeation route. Pharm Res 1988;5:341.
154. Aungst BJ. Novel formulation strategies for improving oral bioavailability of drugs with poor membrane permeation or presystemic metabolism. J Pharm Sci 1993;82:979.
155. Duchene D, Touchard F, Peppas NA. Pharmaceutical and medical aspects of bioadhesive systems for drug administration. Drug Dev Ind Pharm 1988;14:283.
156. Moës AJ. Gastroretentive dosage form. Crit Rev Ther Drug Carrier Syst 1993;8:143.
157. Peppas NA, Buri PA. Surface, interfacial, and molecular aspects of polymer bioadhesion on soft tissues. J Control Rel 1985;2:257.
158. Lenaerts VM, Gurny R. Bioadhesive Drug Delivery Systems. Boca Raton, FL: CRC Press; 1990.
159. Wacher VJ, Salphati L, Benet LZ. Active secretion and enterocytic drug metabolism barriers to drug absorption. Adv Drug Del Rev 2001;46:89.
160. Aungst BJ. Novel formulation strategies for improving oral bioavailability of drugs with

poor membrane permeation or presystemic metabolism. J Pharm Sci 1993;82:979–987.
161. Utsumi I, Kohno K, Takeuchi Y. Surfactant effects on drug absorption. III. Effects of sodium glycocholate and its mixtures with synthetic surfactants on absorption of thiamine disulfide compounds in rat. J Pharm Sci 1974;63:676–681.
162. Constantinides PP. Lipid microemulsions for improving drug dissolution and oral absorption: physical and biopharmaceutical aspects. Pharm Res 1995;12:1561.
163. Charman WN, Porter CJH. Lipophilic prodrugs designed for intestinal lymphatic transport. Adv Drug Del Rev 1996;19:149.
164. Charman WN, Lipids Lipophilic drugs, and oral drug deliverysome emerging concepts. J Pharm Sci 2000;89:967.
165. Nerurkar M, Burton P, Borchardt R. The use of surfactants to enhance the permeability of peptides through Caco-2 cells by inhibition of an apically polarized efflux system. Pharm Res 1996;13:528–534.
166. Malingré MM, et al. Coadministration of cyclosporine strongly enhances the oral bioavailability of docetaxel. J Clin Oncol 2001;19:1160.
167. Devane J. Oral drug delivery technology: addressing the solubility/permeability paradigm. Pharm Tech 1998;22:68–80.
168. Bagwe RP, et al. Improved drug delivery using microemulsions: rationale, recent progress, and new horizons. Critical Rev Ther Drug Carrier Syst 2001;18:77.
169. Amidon GL, et al. A theoretical basis for a biopharmaceutic drug classification: the correlation of *in vitro* drug product dissolution and *in vivo* bioavailability. Pharm Res 1995;12(3):413.

THE FDA AND DRUG REGULATORY ISSUES

Ramnarayan S. Randad
Office of Generic Drugs, U.S. Food and Drug Administration, Rockville, MD

Objective: The FDA regulates products accounting for roughly 25% of the United States gross national product. Thus, it is not possible to cover all facets of the FDA's regulation. The goal of this review is to present an overview of the history of drug regulation in the United States, the structure of the FDA, and the approval process for human drug. No attempt has been made to cover animal drugs and devices; readers are advised to consult FDA's Web site for further information.

Caveat: This chapter was written by Dr. Randad in his private capacity and it neither represents the views of the FDA or the U.S. Government nor implies endorsement from the FDA or the U.S. Government. No official support or endorsement by FDA is intended nor should be inferred.

Method: The information provided in this chapter is freely and readily available on the Web from the government and other sources and was identified through main search terms such as FDA, history of the FDA, drug approval process, drug regulation, FDA legislation, ICH, USP, biologics, biological products, generic medications, generic drugs, orange book, and bioequivalence. Additional data sources included the Code of Federal Regulations and regulatory guidance from the FDA Center for Drug Evaluation and Research. The process of new drug approval and its continuous modification reflects both the progress of science and the globalization of the world's economy. The reader is encouraged to follow recent developments by monitoring the FDA Web site as well as industry Web sites.

Definition of Drug (http://www.cfsan.fda.gov/~dms/cos-218.html): The Federal Food, Drug, and Cosmetic Act (FD&C) defines drugs as "articles intended for use in the diagnosis, cure, mitigation, treatment, or prevention of disease" and "articles (other than food) intended to affect the structure or any function of the body of man or other animals" [FD&C Act, sec. 201(g)(1)]. The term *human drug application* means an application for approval for marketing of the drug in the United States and falls under one of the following sections (Table 1).

1. INTRODUCTION: HISTORY OF FOOD AND DRUG REGULATION [1]

To understand the drug review and approval process, one should be familiar with FDA's history (Table 2), regulatory authority, and organizational structure. The history and regulation of pharmaceuticals in the United States has evolved over time, and has been shaped by historical events, various perceptions of medical needs, concerns about cost and economy, and other political and societal pressures (Table 3).

Until the twentieth century, in the United States, drugs were regulated by the state governments like any other "consumer goods." This situation resulted in the widespread distribution of "health" drugs and devices that included totally ineffective, dangerously habit forming, and sometimes deadly products. Regulation became necessary with the emergence of mass production, mass transportation, wide distribution involving distance and anonymous organization, improved medical technology, and high-profile public disasters.

Reports that adulterated, decayed, and/or otherwise ineffective imported drugs (e.g., quinine and others) were major factors in the high mortality of soldiers in war, convinced Congress to enact a general Federal law, the "Drug Importation Act of 1848," at the end of the US–Mexican War (1846–1848). The act provided for inspection, detention, destruction, and reexport of drug shipments that failed to meet prescribed standards.

Federal interest in science originated with the Patent Office. The Commissioner for Patents repeatedly asked Congress to appropriate funds to support scientific investigations, particularly in the food industry. At that time, very little distinction was made between food and drugs. Commissioner Ewbank established an Agricultural Division within the Patent Office sometime between 1858 and 1860, and the first analytical chemist was

Table 1. Type of Drug Applications

505(b)(1)	Application that contains full reports of investigations of safety and effectiveness.
505(b)(2)	Application for which one or more of the investigations relied upon previous finding of safety and effectiveness by the applicant for approval. This provision permits FDA to rely on a previous finding of safety and effectiveness, for example approval for new indication of approved drug product, different dosage form, strength, combination, route of administration, a different salt or polymorph, or new indication for previously approved ingredient. The final approval under 505(b)(2) is contingent upon patent certification. It is also an opportunity to switch to OTC.
505(j)	An application that contains information to show that the proposed product is identical in active ingredient, dosage form, strength, route of administration, labeling, quality, performance characteristics and intended use, among other things to a previously approved application (the RLD). These applications, also known as ANDAs, do not contain clinical studies as required in NDAs but are required to contain information establishing bioequivalence to the RLD.
351 of the Public Health Service Act	Certain biological products are licensed under section 351 of the PHSA, for example polio and varicella vaccines.

hired. That chemist was the first employee of what would later become the FDA. The Agricultural Division set up a chemical laboratory to support their newly hired chemist. In 1862, the Agricultural Division and its chemistry laboratory were moved from the Patent Office to the newly formed Department of Agriculture. The chemical laboratory was initially named the Chemical Division; then the Bureau of Chemistry in 1902; the Food, Drug, and Insecticide Administration in 1927; and the Food and Drug Administration in 1930. The agency has grown from a single chemist in the U.S. Department of Agriculture in 1862 to a staff of approximately 9100 employees and a budget of $2.4 billion in 2008, comprising chemists, pharmacologists, physicians, microbiologists, veterinarians, pharmacists, lawyers, and many others.

Key turning points in the history of FDA regulation have often come in response to outcries over food and drug-related tragedies. Prior to 1906, it was generally believed that the use of food preservatives and other ingredients were harmless but they were considered a dishonest practice. Due to heightened public concern regarding the safety of food products and fraudulent patent medicines, Department of Agriculture employees, popularly known as the "poison squad," volunteered to test various food and drug adulterants. The study was restricted to young volunteers. Although these experiments are considered flawed by today's standards due to the lack of a control group, they established the important step of introducing science into policy making. Findings of severe problems with the nation's food and drug industry as well as extensive lobbying by Dr. Harvey Wiley spurred Congress to pass in 1906 the Pure Food and Drug Act, also known as the Wiley Act. The act prohibited interstate commerce of misbranded and adulterated foods and drugs. The act did not require premarket inspections and approval but it did require "truth-in-labeling." The act was a milestone in our nation's history, representing an unprecedented federal effort to protect consumers and it changed the American food and drug supply forever.

The Pure Food and Drug Act did not limit claims a manufacturer could make with respect to their product as long as the product was correctly identified. The Sherley Amendment of 1912 prohibited drug labels with false therapeutic claims intended to defraud the purchaser. However, establishing fraud required the Bureau of Chemistry to show that the manufacturer knew the product was worthless and that there was intent to defraud the consumer.

The guiding principle behind the Pure Food and Drug Act and the Sherley Amendment was that the consumer was the primary deci-

Table 2. Significant Dates in U.S. Food and Drug Law History (http://www.fda.gov/AboutFDA/WhatWeDo/History/Milestones/ucm128305.htm)

1820	Establishment of the U.S. Pharmacopeia, the first compendium of standard drugs for the United States.
1848	Drug Importation Act, U.S. Customs Service authorized to stop entry of adulterated drugs from overseas.
1906	The original Food and Drugs Act is passed on June 30 and signed by President Theodore Roosevelt.
1911	In U.S. v. Johnson, the Supreme Court rules that the 1906 Food and Drugs Act does not prohibit false therapeutic claims but only false and misleading statements about the ingredients or identity of a drug.
1912	Congress enacts the Sherley Amendment. It prohibits labeling medicines with false therapeutic claims intended to defraud the purchaser, a standard difficult to prove.
1937	Elixir Sulfanilamide, containing the poisonous solvent diethylene glycol, kills 107 persons, many of whom are children, dramatizing the need to establish drug safety before marketing and to enact the pending food and drug law.
1938	The Federal Food, Drug, and Cosmetic Act of 1938 is passed by Congress, requiring new drugs to be shown safe before marketing.
1941	Insulin Amendment requires FDA to test and certify purity and potency of this life-saving drug for diabetes.
1945	Penicillin Amendment requires FDA testing and certification of safety and effectiveness of all penicillin products. Later amendments would extend this requirement to all antibiotics. In 1983, such control would be found no longer needed and abolished.
1951	Durham–Humphrey Amendment defines the kinds of drugs that cannot be used safely without medical supervision and restricts their sale to prescription by a licensed practitioner.
1962	Kefauver–Harris Drug Amendments are passed. For the first time, drug manufacturers are required to prove to FDA the effectiveness of their products before marketing them.
1966	FDA contracts with the National Academy of Sciences/National Research Council to evaluate the effectiveness of 4000 drugs approved on the basis of safety alone between 1938 and 1962.
1972	Over-the-Counter Drug Review initiated to enhance the safety, effectiveness and appropriate labeling of drugs sold without prescription.
1976	Vitamins and Minerals Amendments ("Proxmire Amendments") stop FDA from establishing standards limiting potency of vitamins and minerals in food supplements or regulating them as drugs based solely on potency.
1983	Orphan Drug Act passed, enabling FDA to promote research and marketing of drugs needed for treating rare diseases. The Federal Anti-Tampering Act passed in 1983 makes it a crime to tamper with packaged consumer products.
1984	Drug Price Competition and Patent Term Restoration expedites the availability of less costly generic drugs and the brand name companies can apply for up to five years additional patent protection.
1987	FDA revises investigational drug regulations to expand access to experimental drugs for patients with serious diseases with no alternative therapies.
1991	FDA publishes regulations to accelerate reviews of drugs for life-threatening diseases.
1992	Generic Drug Enforcement Act imposes debarment and other penalties for illegal acts involving abbreviated drug applications. Prescription Drug User Fee requires drug and biologics manufacturers to pay fees for product.
1997	Food and Drug Administration Modernization Act reauthorizes the Prescription Drug User Fee Act of 1992 and mandates the most wide-ranging reforms in agency practices since 1938. Provisions include measures to accelerate review of devices, advertising unapproved uses of approved drugs and devices, health claims for foods in agreement with published data by a reputable public health source, and development of good guidance practices for agency decision-making.
1998	Pediatric Rule, a regulation that requires manufacturers to assess of new drugs for safety and efficacy in children. The Best Pharmaceuticals for Children Act improves safety and efficacy of patented and off-patent medicines for children.
2002	Public Health Security and Bioterrorism Preparedness and Response Act of 2002 is designed to improve the country's ability to prevent and respond to public health emergencies.
2004	Project BioShield Act of 2004 authorizes FDA to expedite its review procedures to enable rapid distribution of treatments as countermeasures to chemical, biological, and nuclear agents.

Table 3. Legislation Spurred by Health and Economic Concerns

Health Concerns	Economic Concerns
1906: Pure Food and Drug Act/Wiley Act	1984: Drug Price Competition and Patent Term Restoration Act
1912: Sherley Amendment prohibition false therapeutic claims	1992: PDUFA
1938: Safety Amendment	1997: FDAMA
1962: Efficacy Amendment	2002: Best Pharmaceuticals for Children Act (BPCA) and PHSBPRA
2003: Pediatric Rule, PREA	2003: MEDIMA, PREA

sion maker regarding medical care and the government's role was in making sure that the consumer received sufficient and accurate information so that he or she could make an informed decision.

The next turning point in food and drug regulation was followed by the tragedy of "Elixir of Sulfonamide." Sulfonamide was one of the first of the miracle drugs that showed great promise as an all-purpose anti-infective agent. In 1937, Messengill Company (Messengill) began investigating a liquid dosage form of sulfonamide, which would be especially useful for children. Sulfonamide was known to be stubbornly insoluble but Messengill was able to create a liquid formulation in diethylene glycol. Diehtylene glycol has a slightly sweet taste and a pleasant pink color, however, it is also highly toxic. Messengill did not test its product prior to marketing because neither clinical tests were required nor marketing approval was needed. The diethylene glycol used in the new liquid formulation of sulfonamide was responsible for the death of 107 patients. (It is interesting to note that the FDA was only able to seize the elixir because of a minor technical detail that rendered the product misbranded: by definition, an "elixir" is supposed to employ alcohol, and Massengill's product did not contain alcohol.) Within 1 year of this disaster, the Federal Food, Drug, and Cosmetic Act of 1938 was passed and signed into law by President Franklin Roosevelt. Scientific testing of drugs to establish safety became a major factor shaping modern medicine. Under the FD&C Act of 1938, drug manufacturers were required to submit a new drug application (NDA) to the FDA before introducing a compound into interstate commerce. It is important to note that a positive response from government on the NDA was not required—the application became effective 60 days after filing.

The next decade saw remarkable growth and change in the pharmaceutical industry (manufacture of insulin (1941); development of miracle drug penicillin (1944)). New production techniques forced FDA to continually adjust its regulatory role (for example, batch certification of penicillin drugs and other antibiotics). With the proliferation of new drugs, it became impossible for the average consumer to make informed choices, even if the product label and insert did contain adequate directions for use. The passage of the Durham–Humphrey Amendment of 1951, which is also known as the "Prescription Drug Amendment," added a statutory declaration of which drugs were required to be labeled for prescription use. The Durham–Humphrey Amendment codified drugs into two categories: (1) over the counter (OTC) medication that could be self-administered by a layperson with adequate warnings and directions, and (2) prescription or ethical drugs that require expert supervision.

The thalidomide tragedy of the early 1960s spurred new regulations for the drug industry. Thalidomide, a sedative mostly used for reducing nausea associated with pregnancy, caused severe limb deformities, not to the user, but to the unborn child. The US consumers were narrowly spared by the efforts of FDA medical officer Frances O. Kelsey, who determined that information in the application by Merrell Company for Kevadron to be inadequate to support safety of the drug and the labeling is unsuitable. In 1962, Merrell reported that thalidomide was withdrawn from sale in Germany because the drug was correlated with severe congenital abnormalities. This disaster exposed regulatory shortcomings regarding the testing of drugs, and in response Congress enacted the Kefauver–Harris Drug Amendments that were signed into law by President Kennedy in October

1962. The Kefauver–Harris Drug Amendments were a landmark in the history of drug regulation and no other law has since been as sweeping in its coverage of the pharmaceutical industry.

The 1962 Kefauver–Harris Drug Amendments introduced number of new requirements to market pharmaceutical products in the United States. The most important change was the requirement that all new drug applications demonstrate "substantial evidence" of the drug's efficacy for a labeled indication, in addition to the existing requirement for premarket demonstration of safety. "Substantial evidence" is described in the Amendments as adequate and well-controlled investigations, including clinical investigations, by the experts qualified to evaluate effectiveness of the drug involved. The Amendments provided FDA with greater control over drug trials, including the requirement for informed consent by the patient involved in the drug trial.

The Kefauver–Harris Drug Amendments also required FDA to approve, rather than just review, an NDA before a drug could be marketed. The legal concept of current good manufacturing practices (cGMPs) was also established. The year 1962 marked the start of the FDA approval process in its modern form.

The 1962 Amendments also required the FDA to review the efficacy of approximately 4000 drugs approved between 1938 and 1962 that were approved based only on safety. Two years later, the FDA initiated the Drug Efficacy and Safety Implementation (DESI) program. Final actions were taken on 3500 drugs with over 1000 classified as ineffective for all labeled claims. Other important provisions of the 1962 Amendments included the requirement that drug companies use the "established" (or what we now might call the "generic") name of a drug along with the trade name, the restriction of drug advertising to FDA-approved indications, and expansion of FDA powers to inspect drug manufacturing facilities.

One important procedural change from the DESI program was a process for approval of generic drugs by means of "paper" NDA or an Abbreviated New Drug Application (ANDA). Prior to 1984, however, this process was only available for generic versions of pre-1962 drugs.

In 1984, the Drug Price Competition and Patent Term Restoration Act, more commonly known as the "Hatch–Waxman Act" was enacted for among other reasons to correct two unfortunate interactions between the new regulations mandated by the 1962 Amendments and existing patent law. The efficacy requirement mandated by the 1962 Amendments significantly delayed the marketing of new drugs and the new regulations were interpreted to require complete safety and efficacy testing for generic copies of approved drugs. The "pioneer" manufacturers obtained court decisions that prevented generic manufacturers from beginning the clinical trial process while a drug was still protected by patent.

The Hatch–Waxman Act extended the patent terms of new drugs, and tied those extensions, in large part, to the length of the FDA approval process for each individual drug. For generic manufacturers, it created a new approval mechanism under the statute—the ANDA. The generic drug manufacturer was only required to demonstrate that their generic formulation has the same active ingredient, route of administration, dosage form, strength, and "bioequivalence" as the corresponding brand name drug. The act specifically prohibited FDA from requiring clinical studies other than bioequivalence studies in the ANDA. This act has been credited with creating the modern generic drug industry.

Substantial other amendments to the law include the Pesticide Amendment (1954), Food Additives Amendment (1958), Color Additive Amendment (1960), Animal Drug Amendment (1968), The Safe Medical Device Act (1976), Orphan Drug Act (1983), Prescription Drug User Fee Act (PDUFA, 1992), Food and Drug Administration Modernization Act (FDAMA, 1997), Pediatric rule (1998), Best Pharmaceuticals for Children Act (2002), and Project BioShield Act (2007).

1.1. Orphan Drug Act

Recognizing that certain diseases affect only a small number of patients (less than 200,000 persons in the United States) and are therefore largely ignored by pharmaceutical companies, the Orphan Drug Act of 1983 provided federal financial incentives to develop these products.

1.2. PDUFA of 1992

Before 1992, all FDA funds were appropriated through the federal budget, except for antibiotics and food additive certification. In response to the pharmaceutical industry's dissatisfaction of drug lag time (time from application filing to approval) and FDA's ever increasing workload and authority, Congress imposed "user fees," under the PDUFA of 1992, which has been reauthorized ever since. User fees (application fees, establishment fees, and product fees) are payment for FDA services, in return for the additional resources for the agency to function effectively. PDUFA introduced nonbinding performance goals: it mandated FDA to complete initial review actions on 90% of "standard" applications within 12 months and within 6 months on "priority" applications and eliminate the application backlog within 2 years. Other actions involve requests for meetings, issuance of meeting minutes, responding to sponsors—clinical hold, dispute resolution, and action letter simplification. The dynamics of user fee legislation seems positive. The PDUFA (1992) was followed by user fee laws for medical devices (2002) and Animal Drug User Fees (2003). FDA is currently developing plans for user fees for generic drugs and for food applications.

1.3. Pediatric Rule

Many drugs used in children are approved based on clinical studies in adults. The Pediatric Rule (1998) and the Best Pharmaceuticals for Children Act (2002) require companies introducing new active ingredients for drugs and biologics, and new dosage forms or indications to assess the effect in children. The laws also require FDA to work with companies to ensure that ethnic and racial groups are adequately represented in clinical studies in children. A provision in the Food and Drug Modernization Act of 1997 (FDAMA) provides 6 months of pediatric exclusivity to a company that conducts pediatric studies.

1.4. FDAMA

The FDAMA, enacted in November 1997, is a complex piece of legislation that introduced several changes to the FD&C Act. Some important provisions of the FDAMA are

1. Elimination of batch certification for insulin and antibiotics.
2. Streamlining the approval process for drugs and biologics (expedited/accelerated approval for serious conditions and unmet needs).
3. Increasing patient access to experimental drugs and devices.
4. Permitting dissemination of information about unapproved uses (off-label use) under certain circumstances.
5. Pediatric studies and exclusivity.

1.5. FDA Reforms in the AIDS Era

Partly in response to the criticisms concerning the length of the drug approval process, the FDA issued new rules to expedite approval of drugs for life threatening diseases, and expanded preapproval access to drugs for patients with limited treatment options. The first of these new rules was the "IND exemption" or "treatment IND" rule, which allowed expanded access to a drug undergoing Phase II or III trials (or in extraordinary cases even earlier) if the drug potentially represented a safer or better alternative to currently available treatments for terminal or serious illness. A second new rule, the "parallel track policy," allowed a drug company to set up a mechanism for access to a new potentially lifesaving drug by patients who for various reasons would be unable to participate in ongoing clinical trials. The "parallel track" designation could be made at the time of IND submission. The accelerated approval rules were further expanded and codified in 1992. All of the initial drugs approved for the treatment of HIV/AIDS were approved through accelerated approval mechanisms.

1.6. Twenty-First Century Initiative

In August 2002, the FDA announced a significant new initiative, Pharmaceutical cGMPs for the Twenty-first century, to enhance and modernize the regulation of pharmaceutical manufacturing and product quality. Twenty-first century initiative:

- Encourage the early adoption of new technological advances.
- Facilitate industry application of modern quality management techniques.
- Encourage implementation of risk-based approaches.
- Ensure that regulatory review, compliance, and inspection policies are based on state-of-the-art pharmaceutical science.
- Enhance the consistency and coordination of FDA's drug quality regulatory programs.

1.7. President's Emergency Plan for AIDS Relief (PEPFAR)

In 2003, President Bush launched the PEPFAR to combat global HIV/AIDS. PEPFAR is the largest commitment ever by a single nation toward an international health initiative. In 2004, FDA initiated an expedited review process to ensure that those being served by the Presidents' Plan would receive safe, effective, and quality antiretroviral drugs.

1.8. Food and Drug Administration Amendments Act (FDAAA)

FDAAA (2007) reauthorized user fees for FDA approval of drugs (PDUFA IV), biologics, and medical devices. Additionally, FDAAA provides for food safety provisions, advisory committee provisions, clinical trial registries, and provisions intended to enhance drug safety.

1.9. Beyond our Borders Initiative

To improve safety and quality of imports entering the United States, in 2008, FDA implemented Beyond Our Borders Initiative (BOB). The BOB involves establishment of permanent FDA offices in the foreign country. FDA is presently considering offices in China, India, Europe, Latin America, and Middle East. The locations were selected by analyzing import and program data, considering the relationship that FDA has with its counterpart regulatory authorities, and the impact that an FDA presence will have on joint cooperative activities.

From this brief history of FDA, it is evident that at every turning point the agency has been guided by the concern for our health and well-being and much of the protection we take for granted today can be traced back to the 1906 act.

2. STATUTE

There are 50 titles in the Code of Federal Regulations (CFR), Title 21 (21 CFR) that provide interpretation of the Federal Food, Drug, and Cosmetic Act.

2.1. Regulations, Guidance, and Adaptations by the FDA

Regulations are typically brief and often difficult to interpret for actual implementation. In an effort to streamline and help establish a useful and effective regulatory scheme, FDA establishes rules and issues Trade Correspondence Letters, Guidance, and Point to Consider (PTC), which provide interpretation of regulations. *Rules* interpret or implement a statute and are legally binding. For more information on rule making, visit http://www.fda.gov/opacom/backgrounders/.

Guidance is less formal documents that represent the Agency's current thinking on a particular subject. These documents are prepared for FDA review staff and applicants/sponsors to provide guidelines on design, production, manufacturing, and testing of regulated products, and content and format for submission and evaluation of applications. Draft guidance is published in the Federal Register for comment. They establish policies intended to achieve consistency in the Agency's regulatory approach and establish inspection and enforcement procedures. Because guidance is not regulations or laws, they are not enforceable through administrative actions or the courts. Guidance is intended to assist industry in fulfilling regulatory requirements and provide greater flexibility with shorter preparation time compared to rules. For the complete list of CDER guidance, please visit http://www.fda.gov/cder/guidance/index.htm.

FDA also publishes *Manuals of Practice and Procedures* (MaPPs), which are detailed instructions for reviewers to standardize reviews of submissions and are available to the public at http://www.fda.gov/cder/mapp.htm.

Readers are encouraged to utilize FDA guidance, MaPPs, and International Conference on Harmonization (ICH) guidance to assist in preparation of scientifically and regulatorily sound submissions.

3. ORGANIZATION

The FDA is organized into eight centers/offices to efficiently implement the many changes to the Federal Food, Drug, and Cosmetic Act (Table 4).

Of the five major centers, CDER and CFSAN have existed in one form or another since 1906. The FDA's federal budget request for 2009 is $2.4 billion. FDA is funded by appropriations from the federal government, user fees submitted with New Drug Applications under PDUFA, and medical devices user fees under the Medical Device User Fee and Modernization Act (MDUFMA). These fees may be waived or reduced for small businesses. Prior to the 1990s, the FDA was solely funded by appropriations from the federal government. Currently, FDA has approximately 10,000 employees.

4. THE FDA'S DRUG REVIEW PROCESS [2]

The path a drug travels from laboratory to medicine cabinet is usually long and every

Table 4. FDA Organization and Responsibilities

Center	Responsibilities
Center for Drug Evaluation and Research (CDER) http://www.fda.gov/cder/	Safety and effectiveness of prescription (Rx) drugs, OTC drugs, and some biologics
Center for Biologics Evaluation and Research (CBER); established in 1972. http://www.fda.gov/cber/	Blood and blood products Protein-based products Vaccines and allergic products Human gene therapy Xenotransplantation Cellular and tissue transplant Transgenic plants and animals Genomics, proteomics, and bioinformatics
Center for Food Safety and Applied Nutrition (CFSAN) http://www.cfsan.fda.gov/	Dietary supplements and safety of 80% of all food consumed in the US (exceptions are meat, poultry, and some egg products, which are regulated by the US Department of Agriculture).
Center for Devices and Radiological Health (CDRH); established in 1974 http://www.fda.gov/CDRH/	Medical devices Radiation-emitting products
Center for Veterinary Medicine (CVM); established in 1965 http://www.fda.gov/cvm/	Safe and effectiveness of drugs to treat animals. Safety of animal feed.
National Center for Toxicological Research (NCTR) http://www.fda.gov/NCTR/	Chemical and pharmaceutical drug toxicity Food contamination risk Terrorism biomarkers
Office of the Commissioner (OC) http://www.fda.gov/oc/	Agency management and operation
Office of Regulatory Affairs (ORA) http://www.fda.gov/ora/	Compliance and review

Adopted from Ref. [1d].

drug takes a unique route. While it is the drug sponsor's obligation to perform research to identify and test the clinical safety and efficacy of a new compound, it is CDER's obligation to evaluate the drug and determine whether the benefits of the drug outweigh its risks.

While the basic framework of the drug approval process is set forth in regulations, it is FDA that has shaped and molded the approval process. For example, there is legal requirement that applicants proposing the use of a new drug present "substantive evidence of effectiveness." "Substantive evidence of effectiveness" is defined as "evidence consisting of adequate and well-controlled investigations, including clinical investigations..." The use of the word "clinical" in the definition has been interpreted as "human." The word "investigations" has also been interpreted routinely as requiring more than one such investigation. These specific words have served as the basis for Agency requirements and decisions during the drug approval process [2d].

The new drug approval process consists of three principle elements:

1. Testing for safety and efficacy through preclinical and clinical studies.
2. Submission of data and other drug-related information through notice of claimed Investigation Exemption for a New Drug (IND) and NDA.
3. FDA review of these premarketing applications. The evaluation of submitted data allows agency reviewers to (1) establish whether the drug submitted for approval works for the proposed use, (2) assess the benefit-to-risk relationship, and (3) determine if the drug will be approved.

5. NEW DRUG APPROVAL PROCESS

The approval process consists of several phases that begin with the sponsor identifying a compound that has therapeutic potential and travels through the general path outlined in Fig. 1.

5.1. Sponsor

By definition, a sponsor is a person or entity who assumes responsibility for compliance with applicable provisions of the Federal Food, Drug, and Cosmetic Act and related regulations and initiates a clinical investigation. A sponsor could be an individual, partnership, corporation, government agency, manufacturer, or scientific institution.

5.2. Pre-IND Phase

This is an early drug development phase that involves (a) structure elucidation and physicochemical characterization of the active pharmaceutical ingredient (API), which becomes the basis for future development of new molecular entity and (b) preclinical testing in animals.

5.3. Preclinical Testing

FDA requires that a new pharmaceutical substance is shown to be safe for use before testing is performed in human subjects to determine therapeutic or diagnostic value. The goal during the preclinical phase is to conduct animal testing and collect data necessary to prove that the investigational compound will not expose human subjects to unreasonable risks during clinical trials. The sponsor can also prove the safety of investigational drugs through existing data obtained from previous testing of the compound in the United States or other countries. The FDA will generally require sponsors to (a) develop pharmacological profiles of the subject drug and (b) determine acute toxicity in at least two animal species using the route of administration proposed for clinical use. Information must be gathered on the absorption, distribution, metabolism, and excretion (ADME) properties of the drug. Animal studies continue after preclinical studies to gain chronic use, carcinogenicity, teratogenicity, and lethality (LD50) information to provide new data and insight on long-term effects. At the pre-IND stage, the sponsor is under no obligation to inform or update the agency of its development plan. However, once a firm decides that there is a benefit in conducting

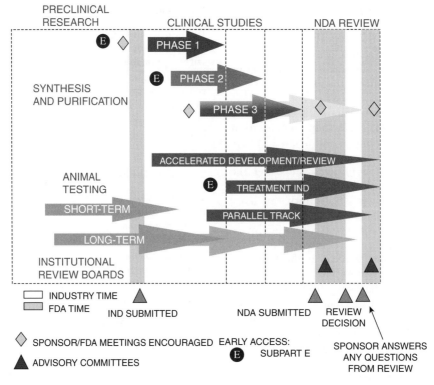

Figure 1. The new drug development process: steps from test tube to new drug application review. Adopted from http://inside.fda.gov:9003/downloads/CFSAN/OfficeofFoodDefenseCommunicationdEmergencyResponse/CAERS/UCM030387.pdf.

human testing, it becomes necessary for the firm to submit an IND application.

5.4. IND Application (http://www.fda.gov/cder/regulatory/applications)

Before clinical testing begins, sponsor must submit an IND which provides results of preclinical testing that has been performed in laboratory animals and outlines a proposal for human testing. An IND application must include (1) results from animal studies and other data to indicate that the drug is reasonably safe for human use, (2) detailed information of the drug composition, source, and manufacture (Chemistry, Manufacturing, and Control (CMC)), (3) the rationale for conducting the study and the protocol that will delineate all aspects of the clinical phase of the investigation, and (4) documentation showing that the clinical investigators are adequately qualified and a commitment to follow the approved protocol and comply with legal and regulatory standards. The notice of claimed Investigation Exemption for a New Drug is an exemption from the legal requirements to transport or distribute a drug across state lines. The IND is often FDA's first exposure to a new drug.

At this stage, the FDA determines whether it is reasonably safe for the company to move forward with testing the drug in humans (Fig. 2). The IND becomes effective within 1 month unless the FDA places a clinical hold on the application due to deficiencies such as concerns about the qualifications of the researchers, whether preclinical studies were conducted according to good laboratory practice (GLP), or about the clinical protocol is unsafe or incomplete. The sponsor may also withdraw an IND, thereby prohibiting further use of the drug.

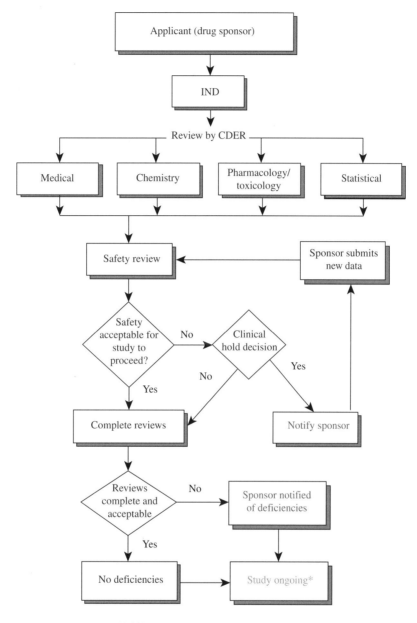

*While sponsor answers any deficiencies

Figure 2. IND review process. Adopted from http://inside.fda.gov:9003/downloads/CFSAN/OfficeofFoodDefenseCommunicationEmergencyResponse/CAERS/UCM030387.pdf.

A Phase I IND Application includes the following (adapted from *The Burger's Medicinal Chemistry and Drug Discovery*, 6th ed) [2e]:

A. FDA forms 1571 (IND application) and 1572 (statement of investigator)
B. Table of contents
C. Introductory statement and general investigational plan
D. Investigator's brochure
E. Protocols

F. CMC information

G. Pharmacology and toxicology information

H. Previous human experience with the investigational drug

I. Additional and relevant information

5.5. Introductory Statement and General Investigational Plan

It should succinctly describe what the sponsor attempts to determine by the first human studies. All previous human experience with the drug, other INDs, previous attempts to investigate followed by withdrawal, and foreign marketing experience relevant to the safety of the proposed investigation should be described. This section should be kept as brief as possible because the detailed development plans are contingent on the results of the initial studies and are limited in scope and subject to change.

5.6. Investigator's Brochure

The sponsor should provide the investigator with an investigator's brochure before the investigation of a drug by participating clinical investigators begins. The recommended elements of an investigator's brochure are set forth in Guidance for Industry E6 Good Clinical Practice: Consolidated Guidance (ICH E6) and should provide a compilation of the clinical and nonclinical data relevant to the study in human subjects. The brochure should include a brief description of the drug substance, summaries of pharmacological and toxicological effects, pharmacokinetics and biological disposition in animals, and if known, in humans from previous clinical studies. Reprints of published studies may be attached. Based on prior experience with the proposed drug or other related drugs, the brochure should describe possible risks and side effects, precautions, and information related to special monitoring.

The investigator's brochure is not intended as a statement of proven claims; the purpose is to provide as much information as possible to the investigators.

5.7. Protocols

Protocols for Phase I studies need not be detailed and may be quite flexible compared with later phases. They should provide the following: (1) an outline of the investigation, (2) the estimated number of patients involved, (3) a description of safety and exclusions, (4) a description of the dosing plan, duration, and dose or method of determining the dose, and (5) specific detail elements critical to safety. Monitoring of vital signs and blood chemistry and toxicity-based stopping or dose adjustment rules should be specified in detail. In this section, the sponsor must list a set of qualifications that would enable an investigator to be qualified to perform high-quality research on the investigational substance. The sponsor must list the names, training, and experience of each investigator and individual charged with monitoring the progress of investigation and evaluating the evidence of safety and effectiveness of the investigational drug received from the investigators. The investigators must complete and forward to the sponsor either Form FD-1572 (pharmacology) or Form FD-1573 (clinical trial), depending on their role in the clinical study. The sponsor then files these forms with the FDA.

5.8. CMC Information

It is recognized that the synthetic methods may change and that additional information may be accumulated as the studies and development progress. Nevertheless, the application should provide CMC information sufficient to evaluate the safety of the drug substance. The governing principle is that the sponsor should be able to relate the drug product proposed for human studies to the drug product used in animal toxicology studies. At issue is the comparability of the (im)purity profiles. Also the issues of stability and the polymorphic form of the drug substance as they might change with the change of synthetic method should be addressed. The Phase I application should include the following CMC information:

1. *CMC Introduction*: This section should address any potential risks and the proposed steps to monitor such risks. It should also describe the chemical and manufacturing differences between batches used in animal studies

and proposed batches for human studies.
2. *Drug Substance*: Since drug substance or API is the compound that provides the therapeutic effect of the drug, it is subject to significant review during the approval processes. This section should include the source and preparation of a new drug substance including the name and address of the manufacturer, physicochemical characterization, elucidation or proof of structure, manufacturing process with a flow chart, certificates of analysis for batches used in animal toxicological studies, stability studies, and batches destined for clinical studies.
3. *Drug Product*: This section should include a statement of the intended route of administration, a list of the drug product components (including quantitative composition), manufacturing, processing, packaging, and proposed labeling of the drug product. Detailed information should be provided on the methods, facilities, and controls used to establish and maintain appropriate standards of identity, strength, quality, and purity as needed for safety and to give significance to clinical investigations conducted with the drug. An environmental assessment should be submitted unless a claim for categorical exclusion is provided. The sponsor should also provide quantitative component information for a placebo or reference listed drug that will be used during clinical investigations.

5.9. Pharmacology and Toxicology Information

The pharmacology and toxicology information is usually divided into the following sections:

1. *Pharmacology and Drug Distribution*: A description of drug pharmacological effects and mechanisms of action in animals and its absorption, distribution, metabolism, and excretion.
2. *Toxicology*: Integrated summary of toxicological effects in animals and *in vitro*. In cases where species specificity may be of concern, the sponsor is encouraged to discuss the issue with the Agency. If the integrated summary is based on unaudited draft reports, the sponsor is required to submit an update by 120 days after the start of the human studies and identify the differences. Any new findings discovered in preparation of the final document that affect patient safety must be reported to FDA in IND safety reports. To support the safety of human investigation the integrated summary should include

   a. Design of toxicological studies and deviations from the design, dates of trials, references to protocols, and protocol amendments.
   b. Systematic presentation of findings highlighting findings that may be considered by an expert as possible risk signals.
   c. Qualifications of individuals who evaluated the animal safety data. The individuals should sign the summary attesting that the summary accurately reflects the data.
   d. Location of animal studies and where records of the studies are located, in case of an inspection.
   e. Declaration of compliance to GLP or explanation why compliance was impossible and how it may affect interpretation of the findings.

3. *Toxicology—Full Data Tabulation*: Each animal toxicology study intended to support the safety of the proposed clinical study should be supported by a full tabulation of data suitable for detailed review. A technical report on methods used and a copy of the study protocol should be attached.

5.10. Previous Human Experience with the Investigational Drug

Previous human experience with the investigational drug may be presented in an integrated summary report. If the drug is marketed commercially or investigated outside of the United

States, an adequate summary of information from preclinical and clinical investigations and experiences should be presented. The absence of previous human experience should also be stated.

5.11. Additional and Relevant Information

Additional and relevant information may be needed if the drug has a dependence or abuse potential, is radioactive, or if a pediatric safety and effectiveness assessment is planned. Any information previously submitted need not to be resubmitted but may be referenced.

Once the IND is submitted to FDA, an IND number is assigned, and the application is forwarded to the appropriate reviewing division. The reviewing division sends a letter to the sponsor investigator providing the IND number assigned, date of receipt of the original application, address where future submissions to the IND should be sent, and the name and telephone number of the FDA contact person to whom questions about the application should be directed. The IND studies shall not be initiated until 30 days after the date of receipt of the IND by the FDA. The sponsor may receive an earlier notification from the FDA that studies may begin or are on "hold" due to safety reasons.

5.12. Clinical Trials

An application for a new chemical entity (NCE), novel dosage form, or major labeling change under 505(b)(1) will require preclinical and clinical trials. Drug studies in humans can begin after an IND is reviewed by the FDA and a local institutional review board (IRB). Clinical trial design may differ greatly from one drug to another. To assist drug sponsors design clinical trials, FDA has published numerous guidelines on the clinical evaluation of specific types of drugs.

Clinical studies must be conducted strictly in accordance to the protocol approved by the FDA. The study protocol will be prepared by the sponsor or clinical investigator, discussed with FDA and agreed upon. It is the responsibility of the clinical investigator to conduct the study, obtain IRB approval before beginning study, complying with the investigational plan, informing patients of their rights, obtaining informed consent, controlling distribution of the study drug, maintaining records, and reporting all observations and adverse events. It is also the responsibility of the clinical investigator to inform the IRB of all changes to the research activity, adverse experience, and to obtain approval for changes to the study protocol.

The clinical studies can be carried out anywhere in the world and submitted for approval in the United States based on a clinical site agreement, if the data is collected in compliance with a protocol, informed consent, and proper record keeping. It is recognized that the diseases and conditions being treated in the United States are not significantly different than in other parts of the world.

The agency is strict about record keeping, record retention, and reporting requirements. All study information should be reported in the case report form or patient medical records, that is, written down or entered in the computer system.

Essential elements of trial designs that are considered by the agency to be adequate and well-controlled investigations include [2c]

1. a clear statement of the objective of the study and a summary of the methods of analysis of the trial results,
2. the design must permit a valid comparison with a control group (regulations define the five types of control groups that may be considered acceptable) to permit a quantitative assessment of the effect of the drug, and
3. the protocol for the study should precisely define the design, including the duration of the study, whether treatments are parallel or sequential, and issues related to sample size.

Clinical trials are generally organized into three phases: phase 1, which includes toxicological studies; phase 2, which includes dosage range studies; and phase 3, which includes pivotal clinical studies that allow the agency to approve the application.

5.12.1. Phase 1
Phase 1 studies are usually conducted in healthy volunteers (20–80 patients), to gain basic safety, metabolism, and pharmacological information in humans and to determine the preferred route of administration. This phase is sometimes referred to as "clinical pharmacology." Dosing is initiated at lower levels than what is expected to have a therapeutic effect in order to avoid unexpected adverse events, and it is gradually titrated upward until side effects begin to emerge.

5.12.2. Phase 2
After the safety of the drug has been established, the focus shifts to determining the drug's efficacy. Phase 2 trials involve a small group of subjects (few dozen to approximately 300). This phase aims to obtain preliminary data on whether the drug works in people who have a certain disease or condition. For controlled trials, patients receiving the drug are compared with similar patients receiving a different treatment—usually a placebo or a different drug. The subjects are randomized and a double-blind approach is used (i.e., neither the subject nor the investigator are aware of who is receiving the drug or a placebo). The safety of the drug continues to be evaluated as well as any short-term side effects. Based upon data gathered from clinical studies through phase 2 studies, a safe and effective dose of the drug is determined.

If the sponsor can show that the drug is clinically safe, either through data from drug use or a previous clinical study, phase 1 and possibly phase 2 trials may be bypassed in some cases.

Traditionally, the FDA offers an "end-of-phase 2" meeting to sponsors of a drug that is determined to provide a therapeutic benefit. The meeting allows FDA and the sponsor to review and agree upon an overall plan for conducting phase 3 trials and design specific studies. At the end of phase 2, the FDA and sponsors try to come to an agreement on how the large-scale studies in phase 3 should be conducted.

The Agency requires that sponsors inform FDA of any new patient safety-related information obtained during phase 2 or phase 3 studies in the form of IND amendments. The sponsor may also take advantage of IND amendments to update CMC information.

5.12.3. Phase 3
In phase 3 investigations, a drug is tested under conditions closely resembling those under which drug will be approved for marketing. Here, the investigational drug is administered to a significantly large patient population usually ranging from several hundreds to approximately 3000 subjects. These studies gather more information about the drug's safety and effectiveness, are performed in populations representative of the general population, study different dosages of the drug, and evaluate the drug in combination with other drugs. The goal of the studies is to gather information needed to evaluate the risk-benefit relationship of the drug and to provide an adequate basis for labeling. Phase 3 trials are often referred to as *pivotal trials* and provide critical information for FDA's evaluation of the drug. The key study criteria for pivotal study design are (a) controlled, (b) blinded, (c) randomized, and (d) adequate size. The study should be discontinued if a dangerous adverse effect is found or the drug lacks efficacy.

5.12.4. Phase 4
Because premarket review cannot identify all potential problems with a drug, the FDA continues to track approved drugs for adverse events through a postmarketing surveillance program. Under accelerated approval rules, if studies do not confirm the initial results, the FDA can withdraw approval of the drug.

Postmarketing study commitments, also called phase 4 commitments, are studies required of or agreed to by a sponsor that are conducted after the FDA has approved a product for marketing. The FDA uses postmarketing study commitments to gather additional information about a product's safety, efficacy, or optimal use. Phase 4 trials can involve head-to-head comparisons with other proven drugs to demonstrate either superior or equal efficacy, to determine the comparative side effect profile, or to determine long-term impact (such as functional status, quality of life, or probability of relapse). Off-label drug use also motivates sponsors to seek an additional indication for the drug (for example, serotonin reuptake inhibitors approved for depression have subsequently been approved for obsessive compulsive disorder, panic disorder, and posttraumatic disorder; aspirin used for mild to moderate pain management has also been discovered to be useful in management and prevention of heart attacks and strokes).

5.13. NDA

This is the formal step taken by a drug sponsor to request that the FDA consider approving a new drug for marketing in the United States. The NDA has changed measurably over time due to numerous amendments to the FD&C act. The sponsor is encouraged to check FDA guidance for current content and format requirements for an NDA. The act requires a new drug application to include the following information: (1) full reports of investigations that have been made to show whether or not such drug is safe for use and whether such drug is effective in use, (2) a full list of the articles used as components of such drug, (3) a full statement of the composition of such drug, (4) a full description of the methods used in, and the facilities and controls used for, the manufacture, processing and packing of such drug, (5) make available samples of such drug and of the articles used as components thereof, and (6) specimens of the labeling proposed to be used for such drug. Final printed labeling for the drug can be submitted later during the review process.

When an NDA is submitted, the FDA has 60 days to decide whether to file the NDA so that it can be reviewed. The FDA can refuse to file an application that is incomplete. In accordance with PDUFA, CDER expects to review and act on at least 90% of NDAs for standard drugs not later than 10 months after the applications are received. The review goal is 6 months for priority drugs (See "PDUFA"). The official review time is the length of time it takes to review a new drug application and issue an action letter that is an official statement informing a drug sponsor of the agency's decision.

5.14. Reviewing Application

A clear FDA priority in the review of the NDA is that the agency must be convinced that the drug is both safe and effective. FDA remains uncompromising in its responsibility for protecting the US public from unsafe and ineffective medicine. As this chapter is written, it is expected that the NDA/BLA/ANDA content, format, and requirement may undergo significant change; however, the basic principles behind obtaining approvals in a timely manner will remain the same. The purpose of the review is to evaluate whether the safety and toxicological, clinical, CMC, and labeling information provided by the sponsor is adequate to establish the safety, efficacy, and potency of the drug (Fig. 3).

The FDA review process is quite transparent. Once a new drug application is filed, an FDA review team consisting of medical doctors, chemists, statisticians, microbiologists, pharmacologists, and other experts evaluates whether the studies the sponsor performed demonstrate that the drug is safe and effective for its proposed use. No drug is absolutely safe; all drugs have side effects. "Safe" in this sense means that the benefits of the drug appear to outweigh the risks.

FDA reviewers analyze study results and look for possible issues with the drug. The reviewers also determine whether they agree with the sponsor's results and conclusions or whether they need any additional information to make a decision. Each reviewer prepares a written evaluation containing conclusions and recommendations about the application. These evaluations are then considered by team leaders, division directors, and office directors, depending on the type of application. Once the review is finalized, the review division director issues a "complete response" letter.

Sometimes, the FDA calls on advisory committees made up of outside experts, who help the agency make decisions on drug applications. Whether an advisory committee is needed depends on many things such as whether there are significant questions regarding the drug, if it is the first in its class, or the first for a given indication. Generally, the FDA follows the advice of advisory committees, but not always. The advisory committee's role is to advise the agency.

It is important to note that a sponsor must continue to submit any new safety information about a drug during the review process through "safety update reports." The labeling

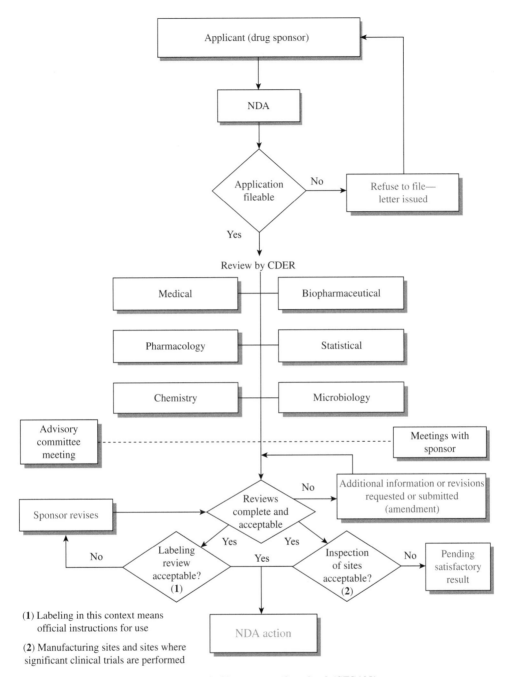

Figure 3. NDA review process (http://inside.fda.gov:9003/downloads/CFSAN/OfficeofFoodDefenseCommunicationdEmergencyResponse/CAERS/UCM030387.pdf).

is usually the final major consideration during the drug approval process. According to the Code of Federal Regulations (21 CFR 201), prescription drug labeling must be "informative and accurate ... contain summary of the essential scientific information needed for the safe and effective use of the drug..." and "... be based, whenever possible, on data derived from human experience." The labeling "negotiation process" through which a final approved labeling is agreed upon, can take several weeks to several months. The package insert is generally not given to the patients but is an educational tool for physician and pharmacists. A product is considered misbranded if anything in the labeling is "false or misleading in any particular way."

5.15. Bumps in the Road

If the FDA decides that the benefits of a drug outweigh the risks, the drug will receive approval and can be marketed in the United States. But, if there are problems with an NDA or if more information is necessary to make that determination, the FDA may decide that a drug is "approvable" or "not approvable."

A designation of approvable means that the drug can probably be approved provided that some issues are resolved first. A designation of "not approvable" describes deficiencies significant enough that it is not clear whether approval can be obtained in the future, at least not without substantial additional data.

The FDA outlines the justification for its decision in an action letter to the drug sponsor. CDER provides the sponsor a chance to meet with agency officials to discuss the deficiencies. At that point, the sponsor can choose to ask for a hearing, correct any deficiencies and submit requested information, or they can withdraw the application.

5.16. Quick and Complete Response

During or after completion of the NDA review, the FDA may have questions or comments regarding the submission. The questions and comments may be verbal or written. It is important for an applicant to submit a complete and accurate response in a timely manner.

5.17. Problem and Challenges

Most delays in drug approval are due to unacceptable product performance (for example, instability of the drug product, unacceptable level of impurities, and poor bioavailability) and poor submission (gaps or incomplete data). The Agency is advocating science-based regulation and review practices.

5.18. Sponsor Rights During NDA Review

Sponsor rights during the NDA review process fall into the following four categories:

1. right to timely review,
2. right to confidentiality,
3. right to request a meeting or conference with FDA, and
4. right to protest if the sponsor believes its rights are violated.

5.19. FDA Meetings

FDA allows several kinds of meetings with sponsors in order to assist and gather information: typical meetings include pre-IND, end-of-phase 2, and pre-NDA. FDA requests that sponsors submit a brief document containing relevant background information and questions to the agency at least 6 weeks prior to the requested meeting (see FDA guidance on formal meeting with sponsors).

5.20. The Clinical Site

The clinical site is a site where clinical studies are conducted. The responsibility of a clinical site is to enroll subjects, ensure their safety, and conduct the study according to the approved protocol. A principle investigator is present at each clinical site and is accountable for all practices.

5.21. The Clinical Research Subjects

Clinical research subjects are the subjects who participate in the clinical investigation. The subjects should be recruited ethically, stringently evaluated for appropriateness, and closely monitored to ensure well-being. The well-being of the subjects is of paramount importance in clinical research. For each study protocol, an established set of inclusion and exclusion criteria are systematically applied to evaluate individuals for participation.

5.22. Institutional Review Board

The IRB is a panel of scientists and nonscientists from medical and research institutions that possess the necessary competence to oversee specific clinical research. IRBs approve clinical trial protocols which describe the subjects who may participate in the clinical trial, review the content of informed consent documents, evaluate the schedule of tests and procedures, the medications and dosages to be studied, the length of the study, the study's objectives, and other details. IRBs ensure that the study is acceptable, that participants have given consent and are fully informed of their risks, and that researchers take appropriate steps to protect patients from harm. The definition and role of an IRB is provided in 21 CFR 56.102. The IRB's primary mandate is to protect the human subject.

5.23. Event Reporting

Companies must submit adverse event(s) to the FDA and report changes to the drug's formulation, manufacturing, or labeling. Health professionals and consumers are also encouraged to report adverse events to the Agency. If an approved product is deemed unsafe or if the company is in violation of regulations, the product may be withdrawn from the market or its distribution restricted, it may require "black box" labeling or the company may be required to perform Phase IV studies.

5.24. Inspections

To protect the rights and welfare of subjects in clinical trials and to verify the quality and integrity of data submitted, FDA inspects the study sponsor, contract research organizations (CROs), clinical investigators, and IRBs. Following the conclusion of an inspection, the FDA field investigator may issue an FDA-483 form that lists inspectional observations and violations. The FDA inspector will also write a report that is submitted to the Division of Scientific Investigation (DSI) for review and send one of the three regulatory action letters (no action indicated, voluntary action indicated, and official action letter). DSI also reviews the records of institutional review boards to ensure they are fulfilling their role in patient protection.

5.25. The Quality of Clinical Data

The FDA relies on data that sponsors submit to decide whether a drug should be approved. FDA has standards for conducting these tests that are set of regulations known as "good clinical practices" (GCP) or "current GCP." The principle intent of GCPs are to (1) dictate procedures that will help assure the quality and integrity of data obtained from clinical testing, (2) protect the rights and safety of clinical subjects, and (3) define responsibilities of the key figures involved in clinical trials and patient informed consent requirements.

5.26. Current Good Laboratory Practice (cGLP)

To ensure the quality of the data, FDA requires that all nonclinical laboratory studies designed to provide data for an IND or NDA to be conducted under cGLP. cGLP regulations first became a requirement on June 20 1979 due to the FDA's finding of serious problems. cGLP guidelines specify requirements for (1) organizational and personnel, (2) physical structure and facility, (3) standard operating procedures and protocols, (4) safeguard of specimens and records, and (5) methods to communicate test results. FDA conducts announced and unannounced inspections. The cGLP regulations apply to all facilities that conduct animal testing and analyses whose results will be used to support drug applications.

5.27. Current Good Manufacturing Practices (cGMP)

The cGMP and NDA approval are two pillars of federal law providing support of manufacturing quality for pharmaceutical drug products sold in the United States. The cGMP requirement focuses on manufacturing quality for the purpose of assuring that the drug product is safe with defined identity, quality, strength, and purity characteristics. Under section 501(a)(2)(B) of the Act, manufacturing, processing, packaging, and holding of a drug should be in conformance with current good manufacturing practice. The cGMP requirement applies to manufacturing facilities, methods, and controls used to perform the activities above. Products not manufactured in accordance with cGMP regulations may be adulterated or misbranded and subject to seizure (21CFR314.170).

The cGMP requirement assures that legally acceptable product attributes result from the manufacturing operation. The cGMP requirement also assures support for reduction of information to be provided in an application to be submitted to the FDA. Prior approval and periodic inspections of pharmaceutical manufacturing facilities for conformance to cGMPs is a specific requirement under FD&C section 510(h). Although thousands of legally marketed drugs in the United States are not the subjects of a new drug application (such as OTC products or pre-NDA drugs), they are not exempted from the cGMP requirement, thus, the cGMP requirement provides protection.

5.28. Defining New Drug

Pharmaceuticals that are subject to some aspect of FDA's new drug approval process include the following characteristics:

1. Drugs containing novel compounds as the active ingredient NCE.
2. Substances marketed in the other countries and naturally occurring compounds. A drug containing an active moiety that has never been used in the United States is also considered a new molecular entity (NME).
3. Previously approved drugs now proposed for new uses or indications.
4. Previously approved drugs that are proposed in a different dosage form or route of administration. This also includes previously approved prescription drugs proposed for OTC use.

NCEs receive significant attention from the Agency. Drugs previously approved by FDA will still be considered new and will require the submission of an NDA or supplemental NDA.

6. OTC DRUG APPROVAL [3]

OTC drug products are those drugs that are available to consumers without a prescription to conveniently and effectively self-treat a number of minor ailments. There are more than 80 therapeutic categories of OTC drugs, ranging from acne drug products to weight control. Nonprescription medications account for approximately 60% of all medications used in the United States and may be used to treat or cure approximately 400 ailments. As with prescription drugs, CDER oversees OTC drugs to ensure that they are properly labeled and that their benefits outweigh their risks.

Prior to 1938, technically all drugs could be marketed without a prescription. After 1938, FDA decided on case-by-case basis if the drug was to be considered prescription only or OTC. The key consideration was adequacy of the label. In 1951, the Durham–Humphrey amendments to the FD&C Act established the distinction between prescription drugs and OTC drugs. It stated that a drug requires a prescription if (a) it is habit forming, (b) it is not safe to administer without the supervision of a licensed practitioner, and (c) is approved for prescription only. Thus, drugs are nonprescription unless they require supervision of a licensed practitioner for safe use.

For an OTC drug, all the information that will allow a consumer to select and use the OTC product safely and effectively must appear on the outer package. The labeling of an OTC product must be clear, understandable, truthful, and contain adequate directions for proper use as well as warnings against unsafe use, side effects, and adverse reactions. FDA published a final regulation (21 CFR 201.66) establishing standardized content and format for the labeling of OTC drug products (http://www.fda.gov/cder/otc/label/default.htm). OTC drugs generally have the following characteristics:

1. Their benefits outweigh their risks.
2. The potential for misuse and abuse is low.
3. The consumer can use them for self-diagnosed conditions.
4. They can be adequately labeled.
5. Health practitioners are not needed for the safe and effective use of the product, such as special monitoring requirements or narrow therapeutic index drugs.

6.1. Route to FDA Approval of OTCs

Two regulatory mechanisms exist for the legal marketing of OTC drug products:

1. OTC drug monographs (21 CFR Part 330).
2. NDA (21 CFR Part 314).

6.2. The OTC Drug Monograph

The Kefauver–Harris Amendments (1962) required all drug products (existing and new) to be effective. This required FDA to retroactively evaluate thousands of products on market. In 1968, a Drug Efficacy Study Implementation (DESI) was established to study the effectiveness of approximately 400 products approved though the NDA process between 1938 and 1962 and were approved only on the basis of safety. Products failing to meet the efficacy requirement were ordered to be removed from the market.

In 1972, FDA initiated a scientific review of OTC product ingredients that were in use at that time to ensure that appropriate safety, effectiveness, and labeling standards were met. The review process consists of three phases: an advisory panel review, creation of a tentative monograph, and publication of a final monograph. This is a long and drawn out process that involves full notice and content rule making and to date has not been fully completed.

The DESI review classified the active ingredients into three categories: (a) Category I, generally regarded as safe and effective, (b) Category II, unsafe and/or ineffective, and (c) Category III, no sufficient data to permit classification. The initial OTC drug review (Fig. 4) determined that approximately 33% of available OTC ingredients were safe and effective for their intended uses. Approximately 40 primary product ingredients were also reclassified from prescription to OTC. The FDA required Class II and Class III ingredients to be taken off the market until sufficient proof dictated otherwise. The OTC review resulted in the publication of a series of ingredient-by-category monographs that explain the requirements to legally make and sell OTC human drug products without obtaining NDA approval. If a manufacturer follows the OTC monograph and manufacturer the product in full compliance with cGMP regulations, no further approval or clearance is required to make and market an OTC product. The manufacturing facility and product must be registered with the FDA. There is a provision for amending the existing monograph or creation of new monograph. OTC monographs provide a relatively straightforward path to legally bring OTC products to market. A company can also petition the FDA to switch a specific ingredient from prescription to OTC status via the monograph system.

Three additional avenues for marketing OTC drugs include (1) new NDA route: if new clinical research provides information that allows a drug to be approved for an OTC indication or at OTC dosage levels, a new NDA may be filed by the manufacturer, (2) reclassification of an existing drug: the FDA can reclassify a drug if it is determined that prescription status is not required for its safe use, and (3) supplement to an original NDA: if favorable postmarketing safety experience for a product provides evidence that the drug product may be safely used without supervision by a physician, a supplement to the original NDA may be filed.

6.3. NDA Route to OTC

The 505(b)(2) NDA may be used as a pathway to switch a product from prescription status to OTC for a product that differs from the OTC monograph. A drug product previously available only by prescription (Rx) can be marketed OTC under an approved "Rx-to-OTC switch" NDA. FDA approves an efficacy supplement to the NDA for an OTC drug product before that product can be marketed OTC.

A drug manufacturer can also submit data in an NDA demonstrating that the drug product is safe and effective for use by consumers without the assistance of a healthcare professional. An original NDA under section (505)(b)(1) or 505(b)(2) needs to be submitted if the sponsor plans to market either a new product OTC whose active substance, indication, or dosage form has never previously been marketed OTC. Thirty percent of new OTC drugs marketed today were products that had been changed from prescription to OTC status.

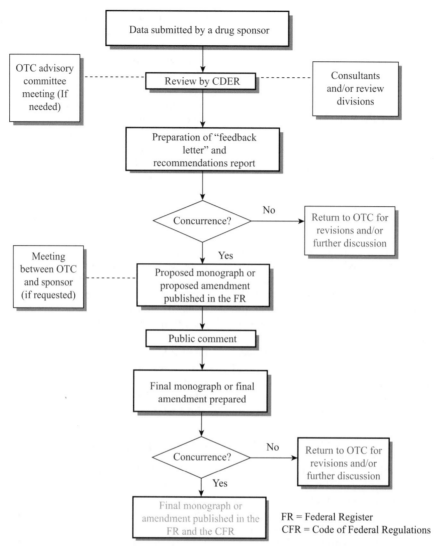

Figure 4. OTC drug monograph review process (http://inside.fda.gov:9003/downloads/CFSAN/ OfficeofFoodDefenseCommunicationdEmergencyResponse/CAERS/UCM030387.pdf).

7. GENERIC DRUG REVIEW AND APPROVAL [4] (http://www.fda.gov/cder/ogd/)

Generic drugs play an important role in the management and economy of the US health care. The term "generic drug product" means an "interchangeable multisource pharmaceutical product," that can be substituted for an innovator drug product based on the belief that the two products are pharmaceutically equivalent as well as bioequivalent.

7.1. Pharmaceutically Equivalent (PE)

To be considered pharmaceutically equivalent, two drug products must contain identical amounts of the same active ingredients in the same dosage form, be formulated to meet the same compendial or other applicable standards of quality and purity, and generally be labeled for the same indications. Pharmaceutically equivalent products may differ in the excipients they contain (i.e., fillers, flavors, and preservatives), their shape, scoring, packaging, and, in certain circumstances, labeling.

7.2. Bioequivalence (BE)

BE is the absence of a significant difference in the rate and extent to which the active ingredient in pharmaceutical equivalents becomes available at the site of drug action in the body when administered at the same molar dose under similar experimental conditions in an appropriately designed study.

7.3. Generic Drug Approval Prior to Hatch–Waxman (1938–1962)

Under the FD&C Act, the definition of drug applies to the drug product rather than to the active ingredient. Thus, a drug cannot be marketed without preclearance from the agency. To provide the mechanism for approval of a generic product based on a DESI evaluation, FDA established a procedure for submission of ANDA. Approval of an ANDA was based on the knowledge that the evidence of safety and effectiveness data necessary for approval had been provided and reviewed by the brand name company, and accepted during the DESI review process. Thus, the manufacturer of a duplicate of the drug not already holding an NDA was required to submit an ANDA to obtain approval to market such product. The information provided in the ANDA had to demonstrate that the generic drug product met appropriate standards of identity, strength, quality, and purity equivalent to the brand name drug whose safety and effectiveness has been established during the course of DESI review.

There was no statutory provision for an abbreviated approval process for generic version of brand name drug products introduced after 1962. Thus, FDA required submission of a full NDA in order to approve a generic version of a post-1962 brand name drug. Additionally, use of a reference drug to perform the necessary tests for generic approval constituted patent infringement if testing was performed prior to the expiration of the patent.

In 1970, FDA recognized that duplication of preclinical and clinical research was not an efficient use of scarce research resources and established the "paper NDA" whereby a company could obtain approval of a generic version of a brand name drug solely based on published scientific literature that provided safety and efficacy information in conjunction with bioequivalence data. Although this was an alternative route, its applicability was limited due to the lack of well-controlled studies to establish safety and efficacy in published literature. Only few (approximately 11) drug products were approved via this route. Thus, no clear mechanism for approval of a generic version of brand name drug products introduced after 1962 existed.

7.4. Hatch–Waxman Amendment (Post-1984)

The regulatory scheme for approval of generic drugs changed after the enactment of the Drug Price Competition and Patent Term Restoration Act of 1984, commonly known as Waxman–Hatch or Hatch–Waxman Act. The act was intended to balance the interests of consumers, pioneer pharmaceutical companies, and the generic drug industry. The Hatch–Waxman Act created an abbreviated approval pathway for generic drugs and patent term extension for brand pharmaceuticals. The patent term extension provision compensates brand name companies for time lost during FDA review of their NDA.

The Hatch–Waxman Act has been credited with creating the modern generic drug industry. The act statutorily established an abbreviated review process for post-1962 generic drugs approvals. Since enactment of the statute, generic drug market share has increased from 10% to approximately 60% of prescriptions filled [4b].

Major assumptions underlying the Hatch–Waxman Act are (1) the generic version of a pioneer drug will be the same as the innovator drug and (2) bioequivalence is an effective surrogate for safety and efficacy. Thus, the generic drug manufacturer need only demonstrate that their generic formulation has the same active ingredient, route of administration, dosage form, strength, and pharmacokinetic properties ("bioequivalence") as the corresponding brand name drug.

Enactment of the generic drug approval provision of the Hatch–Waxman Act ended the need for approvals of duplicate drugs through the FDA's paper NDA process by permitting approval under 505(j) of duplicates of approved drugs (listed drugs) on the basis of chemistry and bioequivalence data, without the need for evidence from literature of effectiveness and safety. Section 505(b)(2) of the act permits approval of applications other than those for duplicate products and permits reliance for such approvals on literature or on an agency finding of safety and/or effectiveness for an approved drug product.

The Hatch–Waxman Act also allowed generic companies to conduct research and development activities prior to expiration of the brand name drug patents without being liable for patent infringement. However, submission of an ANDA prior to patent expiration is considered an act of patent infringement. The Hatch–Waxman Act also requires an ANDA applicant to make a patent certification with respect to any patent for the innovator drug that is listed in the agency's publication, *Approved Drug Products with Therapeutic Equivalence Evaluations*, also known as the "orange book." The generic drug applicant must certify that the reference listed drug is either not patented (also known as a paragraph I certification), the patent has expired (a paragraph II certification), the applicant will not seek approval prior to patent expiration (a paragraph III certification), or the patent is invalid or will not be infringed (a paragraph IV certification). A paragraph IV certification imposes certain obligations on the applicant, patent owner, and FDA. An ANDA applicant must provide notice of a paragraph IV certification to the patent owner(s) and reference listed drug owner that includes a detailed explanation of why the patent is invalid or not infringed. Instead of a paragraph IV certification, an ANDA applicant may provide a method of use statement (under section 505(j)(2)(A)(viii) of the act) for a patent protected use of the drug for which they are not seeking approval.

FDA may immediately approve ANDAs submitted with only paragraph I or paragraph II patent certifications when all other requirements for approval have been satisfied. ANDAs filed with paragraph III patent certifications may be approved on effective date of the expiration of the patent(s) (and any pediatric exclusivity attaching to the patent) and successful conclusion of the review (Fig. 5). The paragraph IV certification imposes certain restrictions on ANDA approval (Fig. 6).

7.5. Delay in ANDA Approvals and Exclusivities

The Hatch–Waxman Act also created several exclusivities (Table 5) that further restrict ANDA approval and even submission. Both patents and exclusivities offer periods of market protections, but they should not be confused. A patent is granted by the United States Patent and Trademark Office and exclusivities are awarded by FDA. Exclusivities preclude FDA from approving generic version of reference listed drugs during the period of exclusivity.

In general, an ANDA may not be submitted within 5 years after approval of NDA for NCE. A 3-year exclusivity period may be granted for new clinical investigations and 6 months of exclusivity may be granted for pediatric studies. Pediatric exclusivity applies to both previously approved and new NDAs.

7.6. 180-Day Exclusivity and 30-Month Stay Provisions

The first generic applicant to file (and maintain) a paragraph IV certification is awarded a 180-day exclusivity period by the FDA. If a brand company brings a patent infringement suit against the ANDA applicant within 45 days of receiving notice of the paragraph IV certification, the FDA is prohibited from approving the ANDA for 30 months or until such time that the patent is found to be invalid.

7.7. Citizens Petition

A citizen petition may be submitted under 21 CFR 10.30 to request the agency to take a particular action. The citizen petition provides ordinary citizens a voice in the FDA approval process.

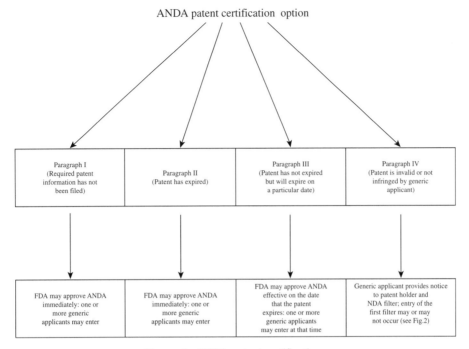

Figure 5. ANDA patent certifications.

7.8. Patent Term Extensions

The Hatch–Waxman Act established patent term extensions for innovator drugs to compensate for the patent term that was lost during the FDA approval process. The patent term extension was one of many changes aimed at benefiting consumers by increasing the supply of generic drugs while preserving pioneer drug companies' incentive to invest in research and development. Patent term extension or restoration may not exceed 5 years and may be shorter depending on the delay in regulatory approval of the NDA. Further, the patent term may not be extended more than 14 years after the date of approval of the NDA.

7.9. ANDA Review and Approval Process

The ANDA review and approval process is summarized in Fig. 7.

An applicant submitting an ANDA under Section 505(j) of the act (except Suitability Petitions submitted under 505(j)(2)(c) of the act) must demonstrate both PE and BE between the generic product and the listed innovator reference drug product. With acceptance of this documentation by FDA, along with other information, the generic product is deemed bioequivalent, therapeutically equivalent, and interchangeable with the listed reference drug product. The ANDA applicant's goal is to receive approval with an "A" rating in the orange book, which means that the FDA considers the generic drug to be bioequivalent to the brand name drug.

In order to obtain approval, the sponsor must file an ANDA for review by the Office of the Generic Drugs (OGD). Even though the application is abbreviated, it is not a short document. The ANDA must contain certain information in order to achieve an acceptable therapeutic equivalence rating in the orange book. The reader is advised to review the FDA Web site and guidance prior to preparing an ANDA.

An ANDA contains several sections comparable to an NDA (Table 6), including a BE section, but does not include preclinical, clinical safety and efficacy, and bioavailability (BA) studies (21CFR 314.94). The ANDA review process is also similar to the NDA review process. Firms located outside the United

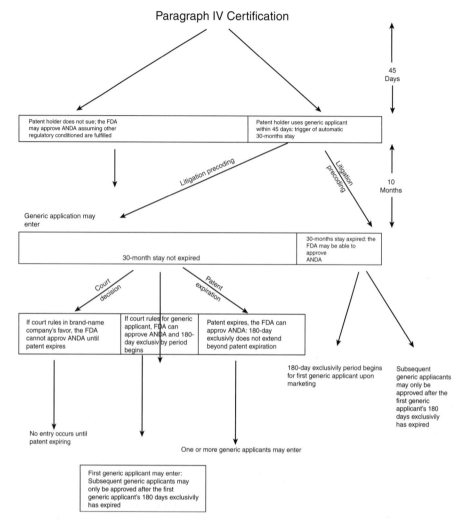

Figure 6. Paragraph IV certifications.

Table 5. Exclusivities Provided by Hatch–Waxman

Types of Statutory Exclusivities	Exclusivity Period
Patent term	20 years
New chemical entity	5 years
New clinical study exclusivity	3 years
Orphan drug exclusivity	7 years
Pediatric Exclusivity	6 months
First ANDA to file a paragraph IV certification	180 days

States are required to appoint an agent with an office in the United States. The ANDA application typically contains the following information: cover letter, 356h form, establishment information, patent and exclusivity certification, basis of submission, labeling, CMC, bioequivalence study or biowaiver request, microbiology, debarment certification, environmental assessment, and submit a field copy (copy to the FDA local office). The CMC section is a relatively small section of an NDA (~15–20%) but is often a significant portion of an ANDA (80–90%). The CMC section is important in postapproval management of the

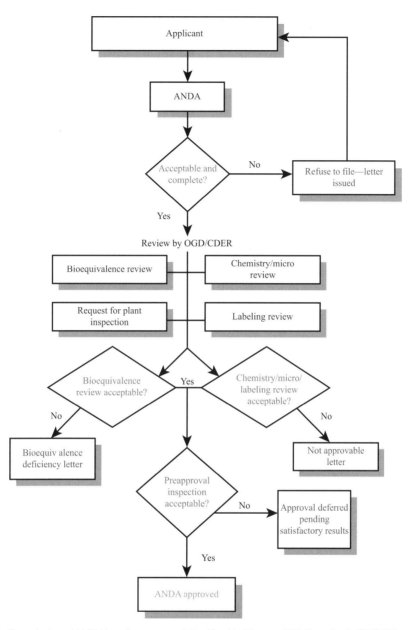

Figure 7. Generic drug (ANDA) review process http://inside.fda.gov:9003/downloads/CFSAN/OfficeofFoodDefenseCommunicationdEmergencyResponse/CAERS/UCM030387.pdf.

product. It is important to note that a partial or incomplete ANDA is not accepted for filing.

The goal of the CMC review of ANDA is to ensure that the generic product is appropriately designed (a pharmaceutical equivalent to the reference listed drug (RLD)) and that sponsors have methods and controls in place for the manufacture, processing, and packaging of a drug that are adequate for assuring and preserving the identity, strength, quality, and purity of the proposed drug product [4g–j]. To promote the technological innovation for enhanced public health promotion and protection, FDA is actively advocating a concept of

Table 6. Submission Requirements: NDA Versus ANDA

Brand Name Drug NDA Requirements	Generic Drug ANDA Requirements
Chemistry	Chemistry
Manufacturing	Manufacturing
Controls	Controls
Labeling	Labeling
Testing	Testing
Animal studies	
Clinical studies	Bioequivalence
Bioavailability	

quality by design (QbD), that is, "quality cannot be tested into products, but should be built in by design." The Pharmaceutical QbD refers to designing and developing formulations and manufacturing processes to ensure predefined product quality. Since 2007, OGD has been requesting ANDA applicants to submit their application in common technical document (CTD) format [4g]. The CTD is divided into five following modules (Fig. 8):

1. Administrative and prescribing information (M1).
2. Overview and summary of modules 3–5 (M2).
3. Quality (pharmaceutical documentation) (M3).
4. Safety (toxicology studies) (M4).
5. Efficacy (clinical studies) (M5).

The CTD quality module (M3) incorporates section on drug substance, drug product, and pharmaceutical development. The pharmaceutical development section provides an opportunity for the sponsor to demonstrate their

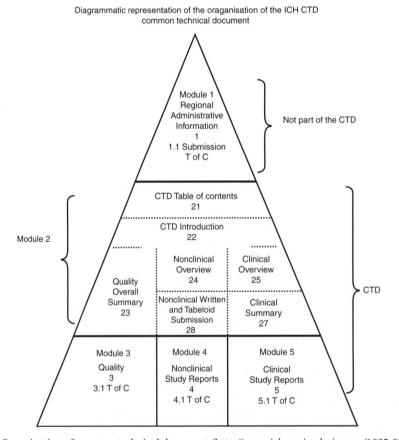

Figure 8. Organization of common technical document (http://www.ich.org/cache/compo/1325-272-1.html).

understanding of the critical quality attributes (e.g., raw material, chemistry, pharmaceutical formulation, in-process material) and process parameters (e.g., manufacturing process) to ensure product quality (e.g., product performance). Understanding of the critical quality attributes and critical process parameters are concepts of QbD and risk management [4i]. To assist sponsor implement QbD goals, the OGD has developed a question-based review (QbR) for its CMC evaluation of ANDAs [4j,4k] The QbR incorporates scientific and regulatory review questions that focus on critical pharmaceutical attributes essential for ensuring generic drug product quality. The questions for QbR are available on the FDA Web site http://www.fda.gov/cder/ogd/.

Bioequivalence data evaluation is a critical component of the ANDA review process [4d]. Bioequivalence for most oral tablet or capsule dosage forms is often demonstrated *in vivo* by comparing the rate and extent of absorption of the generic product with that of the innovator product. This is performed by measuring the concentration of the active ingredient (or active moiety and, when appropriate, active metabolites) in blood, plasma, serum, or other biological fluid over time for both the innovator and the generic product.

In vivo bioequivalence studies for approval of extended or controlled release (ER or CR) formulations should be designed to demonstrate that the product meets the ER claim, that is, that the product does not release the active ingredient too rapidly (dose dump) [4d]. Highly variable drugs (such as propafenone immediate release, verapamil, and nadolol) present unique challenges for bioequivalence testing, due to high intrasubject variability of these drugs. When it is not feasible to assess plasma concentration of the drug due to lack of analytical methods or the drug is not systemically available (for example laxatives, resin, etc.), the assessment may be based on pharmacodynamic parameters that measure the drug effect (21 CFR 320 1-63).

The FDA may waive the requirement for *in vivo* bioequivalence studies if data demonstrate that the generic and innovator formulations are identical and bioavailability is self-evident. For example, biowaivers may be granted for injectables, ophthalmic solutions, and oral solutions since they are already in solutions and dissolution concerns are not relevant or, in the case of solid oral dosage forms (excluding enteric-coated or controlled-release products) that have not been identified as having known or potential bioequivalence problems.

FDA's guidance for industry: Individual Product Bioequivalence, available on the Web, provides bioequivalence study design recommendations for specific drug products to support an ANDA. The FDA's dissolution methods database provides recommended dissolution test methods for drug products that do not have a United States Pharmacopeia dissolution test method (see http://www.accessdata.fda/scripts/cder/dissolution/).

Since 2007, OGD is requesting ANDA applicants to submit their application in CTD format. The quality overall summary (QOS), as required by CTD, may be submitted in the form of answers to questions. The bioeqivalence study data set, labeling, and QOS–QbR must be submitted electronically even if the remainder of the ANDA is submitted in paper format.

7.10. Labeling

An ANDA must contain labeling for the proposed generic product that is the same as the approved labeling for the RLD. The labeling review process ensures that the proposed generic drug labeling (package insert, container, package label and patient information) is identical to that of the reference listed drug except for differences due to changes in the manufacturer, distributor, exclusivity issues, or other characteristics inherent to the generic drug product (such as tablet size, shape or color). Furthermore, the labeling review serves to identify and resolve issues that may contribute to medication errors such as similar sounding or appearing drug names and the legibility or prominence of the drug name or strength.

7.11. Preapproval Inspection

A satisfactory recommendation from the Office of Compliance based upon an acceptable preapproval inspection is required prior to approval.

7.12. Approval

After all components of the application are found to be acceptable an approval or tentative approval letter is issued to the applicant. The letter details the conditions of the approval and allows the applicant to market the generic drug product. If the approval occurs prior to the expiration of any patents or exclusivities accorded to the reference listed drug product, a tentative approval letter is issued to the applicant which details the circumstances associated with the tentative approval of the generic drug product and delays final approval until all patent/exclusivity issues have expired. A tentative approval does not allow the applicant to market the generic drug product.

After all approval requirements (including patent and exclusivity issues) are satisfied, an ANDA will be approved. At the time of this writing, the median approval time is approximately 18.4 months from submission date. If the application is not approvable, the agency provides the sponsor with an opportunity to amend the application, to withdraw the application, or request a hearing.

7.13. Withdrawal of an Approved ANDA

FDA may withdraw or suspend approval of an ANDA if approval of the listed drug on which the ANDA is based has been withdrawn or suspended for safety or efficacy reasons. FDA may also withdraw or suspend approval of ANDA if it determines that the applicant violated the Generic Drug Enforcement Act of 1992 (FD&C A Act sections 335c(a)(1) and (2)) often called as "debarment act".

7.14. 505(b)(2)

A 505(b)(2) application is necessary for generic drugs that differ significantly from the reference listed drug (such as different salt form).

8. ORANGE BOOK [5] (http://www.fda.gov/cder/ob/docs/preface/ecpreface.htm)

The publication, *Approved Drug Products with Therapeutic Equivalence Evaluations,* is commonly known as the "orange book" based on the color of its cover when it was available in print format. The orange book identifies drug products approved on the basis of safety and effectiveness by the FDA under the FD&C Act and the PHSA. The orange book listing also includes approved OTC drug products. Drugs that were approved only on the basis of a safety review or pre-1938 drugs are not included in this publication. The orange book identifies the reference listed drug product that has been selected by the Agency as the reference standard for *in vivo* bioequivalence studies. A single reference listed drug is designated as the standard to which all generic versions must be shown to be bioequivalent to avoid possible significant variations among generic drugs and their brand name counterpart.

In addition, the orange book contains therapeutic equivalence evaluations for approved multisource (generic) prescription drug products. These evaluations are presented in the form of code letters that indicate the basis for the evaluation. These evaluations have been prepared to serve as a public information and advice to state health agencies, prescribers, and pharmacists to promote public education in the area of drug product selection and to foster containment of health care costs. The FDA draws a clear distinction between therapeutic equivalency evaluations and generic substitution, as FDA does not mandate which drug products may be prescribed, dispensed, or substituted for one another. The orange book publication is not an official legally binding regulation.

The orange book utilizes two letter codes to indicate equivalency. If the first letter of the code is an "A," then the product with this code is considered bioequivalent. If the first letter of the code is a "B," then the product with this code is not considered bioequivalent to each other. The second letter of the code represents the dosage form. The term "therapeutic equivalents" applies only to products that are pharmaceutical equivalents and bioequivalent to the innovator product and not to different therapeutic agents used to treat the same condition. Different salts and esters of the same therapeutic moiety (e.g., tetracycline hydrochloride, 250 mg capsules versus tetracycline phosphate complex, 250 mg capsules) are regarded as pharmaceutical alternatives. Anhydrous and hydrated entities, as well as different polymorphs, are considered pharmaceutical equivalents and must meet the same standards and, where necessary, as in the case of ampicillin and ampicillin trihydrate, their equivalence is supported by appropriate bioavailability and bioequivalence studies.

An addendum to the orange book contains drug patent and exclusivity information for prescription products as well as a list of discontinued drug products. The publication may also include additional information that the agency deems appropriate to disseminate.

A drug listing in the orange book is terminated if the drug's approval is withdrawn or suspended for safety or efficacy reasons. Likewise, FDA will not approve an ANDA if the reference listed drug has been withdrawn for safety or efficacy reasons or if it poses imminent hazard to the public health.

9. DRUGS DERIVED THROUGH BIOTECHNOLOGY [6]

Federal regulation of biologics dates back to 1902 with the passage of the Biologic Control Act also known as the Virus–Toxin Law that authorized the Hygiene Laboratory (a precursor of NIH) to regulate vaccines, serums, toxins and antitoxins, and similar products to ensure their safety, purity, and potency. In 1934, the efficacy requirement was added as a condition of granting new biologic licenses. The Biologic Control Act was incorporated into the Public Health Service Act (PHSA) in 1944 without significant substantive changes. The manufacturing and testing requirements for biologics were strengthened as a result of the "cutter incident" (defective polio vaccine batches resulted in hundreds of cases of polio). In 1972, biologics regulatory authority was transferred to the FDA and FDA promulgated regulations of various activities of biologics under the Food and Drug Act. The PHSA has remained the primary source of FDA authority to regulate biologics.

A biologic is a compound consisting of or derived from all or part of a living organism and used for therapeutic or diagnostic purposes. Biologics include a wide range of products such as vaccines, blood and blood components, allergenics, somatic cells, gene therapy, tissues, and recombinant therapeutic proteins. Biologics can be composed of sugars, proteins, or nucleic acids or complex combinations of these substances, or may be living entities such as cells and tissues. Biologics are isolated from a variety of natural sources—human, animal, or microorganism and may be produced by biotechnology methods and other cutting-edge technologies. Gene-based and cellular biologics, for example, often are at the forefront of biomedical research, and may be used to treat a variety of medical conditions for which no other treatments are available.

The overlapping nature of most biologics both as biologics and as drugs or devices poses unique regulatory challenges (Table 7). Biological products are approved for marketing under provisions of the PHSA. All biological products also meet the definition of "drugs" under the FD&C Act, thus they are also subject to regulation under FD&C Act and are approved and regulated in all of the FDA's organizational centers. For example, CDER has oversight of monoclonal antibodies, cytokines, growth factors, enzymes, and interferon. CBER has oversight of proteins intended for therapeutic use that are extracted from animals or microorganisms, immunotherapies, vaccines, blood and blood derivatives, gene and cell therapy and tissue implantation.

In principle, the standards for product quality and the evidence of effectiveness and safety that are required for all new drugs approved by FDA also apply to new biological products intended to be marketed as drugs in the United States. However, in contrast to most drugs (that are chemically synthesized with a known structure), most biologics are complex mixtures that are not easily identified or characterized. Biological products, including those manufactured by biotechnology, tend to be heat sensitive and susceptible to microbial contamination. The characteristics of biologics that contribute to their specific areas of regulatory concern include the fact that they are created by extraction from living organisms, are less physicochemically defined than pharmaceutical drugs (less pure), have higher molecular weights, are generally sterile injections, are less stable with complex degradation, and are often immunogenic. A combination of the above concerns also limits the relevance of preclinical testing in animals to human experience. Thus, there is more emphasis on the individual product's chemistry, manufacturing and control information.

9.1. Approval of Biological Products

The PHSA requires individuals or companies who manufacture biologics for introduction into interstate commerce to hold a license for the product. These licenses are issued by CBER. Biological products intended for veterinary use are regulated under a separate law, the Virus, Serum, and Toxin Act, which is administered by the U.S. Department of Agriculture.

The licensing of biologic products under the PHSA is very similar to the new drug approval process for human drugs. Following initial laboratory and animal testing, a biological product is studied in clinical trials in humans under an investigational new drug application (IND), Phase I studies (safety and toxicity), Phase II studies (Dose ranging), and Phase III studies (indication, safety, and efficacy). The sponsor's process technology is also evaluated for product quality assurance. For example, recombinant DNA technology used to manufacture new drug or biological products may result in products that differ from similar products manufactured with conventional methods. In some instances, the sponsor's process technology may result in a product with a molecular structure that differs from the structure of the active molecule in nature. For example, growth hormones manufactured using recombinant microorganisms have an extra amino acid, methionine, at the amino terminal. An important consideration is what is adequate quality control. For example, a mutation in the coding sequence of a cloned gene during fermentation could give rise to a subpopulation of molecules with anomalous primary structures and altered activity. Such differences in the process could affect the drug's activity or immunogenicity and consequently, the extent of testing required by the FDA for approval. This is a potential problem inherent in the production of polypeptides in any fermentation process. Due to the potential of formation of new structural features in the product resulting in product microheterogeneity or the introduction of a new contaminant (e.g., associated with cell substances), each of which may affect safety, efficacy, and stability of the product, a new marketing application will be required for most products manufactured using new biotechnology. Because of the potential differences in the product resulting from use of recombinant DNA technology, the resulting products may be "new" products. Each case is examined separately, and sponsors are urged to contact the FDA to establish the scope of information required. Reviewers of biologics license applications (BLAs) have a heightened concern for new safety information and previously unseen adverse events.

If the data generated by the studies demonstrate that the product is safe and effective for its intended use, the data are submitted to CBER as part of a BLA for review and approval for marketing. The fundamental standards for approval of a BLA (Table 7) are the same as for an NDA: the product is safe and effective for its labeled indication as demonstrated in clinical trials. A basic BLA includes all preclinical and clinical information concerning biologics as well as a full description of the manufacturing methods. The application is than reviewed by a multidisciplinary review team. To date, no BLAs for gene therapy have been submitted or approved.

The inspection of biologics manufacturing facilities and processes is critically important since the manufacturing process often defines biological products. A "Team Biologics" approach is used for inspection of licensed biologic manufacturers.

After a BLA is approved for a biological product, the product may also be subject to official lot release. If the product is subject to official release by CBER, the manufacturer is required to submit samples of each lot to CBER together with a release protocol that includes a summary of the manufacturing history of the lot and the results of all tests performed on the lot. CBER may also perform certain confirmatory tests on some products, such as viral vaccines, before releasing the lots for distribution by the manufacturer. In addition, CBER conducts laboratory research

Table 7. Comparison of NDA and BLA Regulatory Requirements [6b]

	BLA	NDA	Comments
Regulation	Approved under PHSA and FD&C. PHSA emphasizes the importance of manufacturing control for products that cannot be defined. PHSA provides authority to immediately suspend license, quickly cease distribution, or prepare or procure as the situation demands. The PHSA provides important flexibility in regulation of biotechnology products.	Approved under FD&C	Most biological products are complex mixtures that are not easily identified or characterized. They tend to be heat sensitive and susceptible to microbial contamination.
	PHSA section 610.2 specifies that FDA may require applicant to submit samples and CMC data for each lot for review (e.g., Botox).	N/A	This option does not apply to drugs.
	Licensed products may not be combined with other licensed or nonlicensable products.	Combination drug products are approved as NDA	Combination of biologics with other biologics or drug may have significant impacts.
	Potency	Strength	Due to the complexity and heterogeneity of the protein structure a potency assay (bioassay) is needed.
Preapproval inspections	Conducted by the review team.	Conducted by the field investigator	Due to complexity of the biologics, physiochemical characterization may be limited. The analytical methods can be specific and variable.
Content	Content of application is similar for both NDA and BLA. The BA and BE are specific to NDA.		
Postapproval	Significant amount of information is submitted to the Agency; annual reports are limited to the reporting of changes.	Most of the information is provided in the annual report. The requirement for supplemental application is described in 21 CFR 314.70-314.71	
Change in the ownership	License revocation and reissuance dependent on satisfactory compliance.	Notification to FDA	
Generics	N/A	505b(2) or 505(j)	Generics are not available for BLA.

related to regulatory standards for the safety, purity, potency, and effectiveness of biological products. FDA's requirements for manufacturing facility licensing and batch certification for "well-characterized" biologics are codified in the 1997 FDA Modernization Act.

9.2. Generic Biologics

FDA's ability to approve generic biologics is unsettled at present. From a statutory perspective, the PHSA does not contain any provisions that relate to generic biologics in the way that the Hatch–Waxman Amendments relate to small molecule drugs. In the case of biologic products, small CMC changes can create an entirely new product that would otherwise require approval as a full BLA. Therefore, comparability protocols have been traditionally used by BLA sponsors to gain approval of CMC changes without having to repeat clinical studies. These comparability protocols may be an attractive regulatory precedent for approval of a generic biologic. The absence of analytical standards (product characterization, identity, purity, and assessment of PK parameters) is also a major impediment to the approval of generic biologics. The most likely candidates for generic biologics will be those products regulated by CDER including fibrinogen, somatostatin, and insulin. The 505 (b)(2) NDA approval pathway has been considered another potential mechanism for approval of generic biologics without further legislation.

9.3. Advantages of Regulating Biologics Under the PHSA

The ability to regulate biologics through both the PHSA and the FD&C provision is important especially because of the difficult nature of characterization for some biologics. The PHSA emphasizes the importance of manufacturing control for products that cannot be defined. The potency, consistency, safety, efficacy, and stability of these products, especially vaccines, are dependent on adhering to the processes described in an application. The emphasis on manufacturing control and the ability to impose rigorous lot release requirements under PHSA enable FDA to fulfill the mandate of protecting the public health.

The PHSA also provides authority to immediately suspend licenses in situations where there exists a danger to public health. The PHSA provides important flexibility in the regulation of biotechnology products that facilitates the development and introduction of new medicines.

The FD&C Act provides the enforcements tools of injunction and seizure to complement the PHSA authority to immediately suspend licenses.

10. POSTAPPROVAL

The company's obligations do not end once an application is approved. A company must submit adverse events to the FDA postapproval and report changes in the drug formulation, manufacturing, and labeling. Health professionals and consumers are also encouraged to report adverse events. If a product is shown to be unsafe postapproval, or is linked to a high number of adverse events, the product may be withdrawn from the market, its distribution may be restricted, "black box" warnings or other labeling changes may be required, or FDA may require further studies (Phase IV studies). Postapproval, companies must adhere to the cGMP regulations for the manufacture, processing, and distribution of their product.

The FDA provides the following methods for reporting postapproval changes to applications (NDA, ANDA, and BLA): annual report (AR), changes being effected (CBE) supplement, CBE-30 day supplement, and prior approval supplement (PAS). The changes that have a greatest potential of affecting safety and effectiveness are reported as PAS and the ones with lowest risk are reported as AR.

11. UNITED STATES PHARMACOPOEIA AND NATIONAL FORMULARY (http://www.usp.org/)

Publication of the United States Pharmacopoeia (USP) in 1820 was a milestone in the history of drug regulation. It established a national compendium of drug standards to help respond to an increasingly adulterated (corrupt) drug supply. The first drug law, the

Importation Act of 1848, identifies the USP and other pharmacopoeias as the source of standards. The National Formulary, NF (1888), is a public standard for purity, quality, and strength for excipients, dietary supplements, vitamins and minerals, and compounding information in the United States. The USP and the NF evolved independently as compendia of medical agents and their format and general content were quite similar. In 1975, the USP acquired the NF, resulting in one compendium with two separate sections. The USP-NF is a private not for profit organization.

The USP-NF is a book of public pharmacopeial standards. It contains standards for medicines, dosage forms, drug substances, excipients, medical devices, and dietary supplements. Monographs for drug substances and preparations are featured in the USP. Monographs for dietary supplements and ingredients appear in a separate section of the USP. Excipient monographs are in the NF. A monograph includes the name of the ingredient or preparation; the definition; packaging, storage, and labeling requirements; and the specification. The specification consists of a series of tests, procedures for the tests, and acceptance criteria. The USP works closely with the FDA to establish standards for the quality, purity, strength, and consistency of these products that are critical to the public health. The sponsors of NDA/ANDA/BLA must provide data and information to demonstrate that their products, if compendial, are fully compliant with the compendial requirements.

Compendial standards along with technical information on testing are used as primary guidance in meeting many cGMP requirements, especially testing procedures. Currently USP-NF is recognized in federal statute as a drug standard-setting body and mandates that all marketed medicinal articles in US that are subject to USP-NF monographs must comply with all standards established in the monograph for the medicinal article. Noncompliance with the USP-NF monographs results in the article being considered adulterated (FD&C 1906). It should be noted that the USP requirements are expected quality attributes that must be met through the defined expiry period for the product.

In addition to monographs, the USP-NF contains general notices and requirements and General chapters that provide information that is applicable to all drug substances, inactive ingredients, drug products, and other official articles.

12. INTERNATIONAL CONFERENCE ON HARMONIZATION (http://www.ich.org/)

The International Conference on Harmonization of Technical Requirements for Registration of Pharmaceuticals for Human Use (ICH) brings together the regulatory authorities of Europe, Japan, and the United States and experts from the pharmaceutical industry in the three regions to discuss scientific and technical aspects of pharmaceutical product registration. The purpose of the ICH is to reduce or obviate the need to duplicate testing carried out during research and development of new medicines by recommending ways to achieve greater harmonization in the interpretation and application of technical guidelines and requirements for product registration. Harmonization would lead to more economical use of human, animal, and material resources, and the elimination of unnecessary delay in the global development and availability of new medicines while maintaining safeguards of quality, safety, and efficacy, and regulatory obligations to protect public health.

13. ANIMAL DRUG

Medications intended for use in animals are submitted to the Center for Veterinary Medicine in a New Animal Drug Application (NADA). These products are specifically evaluated for their use in food animals and their possible effect on food from animals treated with the drug.

14. FINAL THOUGHTS

Since its humble origin as a single chemist in the Department of Agriculture, the FDA has grown to more than 10,000 strong. It is responsible for protecting the public health by assuring the safety, efficacy, and security of human and veterinary drugs, biological products, medical devices, our nation's food supply, cosmetics, and products that emit radiation. Looking at the past 100 years of history, one can safely conclude that FDA has done a fairly good job of protecting the American people.

Since enactment, the FD&C Act of 1906 has gone through monumental changes through several amendments. The FDA is one of those rare institutions that relies on both science and law. The FD&C Act gives the FDA authority to regulate various industries and products and it is the science that provides the knowledge needed to develop and apply the regulations in the right way. One can safely argue that the history of the FDA reflects the progression of science. As science has progressed, it has allowed the FDA to move forward with new and improved ways of implementing food and drug laws.

The structure of the FDA and the drug review and approval process has significantly changed over the last century and will continue to evolve with the ever-expanding field of medicinal chemistry and the heterogeneity of treatment approaches, regulations, globalization of the marketplace and socioeconomic reasons, but the FDA's adamant adherence to safety, quality, and efficacy standards will remain constant.

ACKNOWLEDGMENTS

Encouragement and suggestions for preparation of the manuscript from following individuals is greatly appreciated: Mr. David Read and Drs. Paul Schwartz, Rajiv Agrawal, Pradit Akolkar, Chandrashekar Chaurasia, Harvey Greenberg, and Scott Furness. Special thanks to Ms. Susan Levine for critical reading.

REFERENCES

1. (a) www.fda.gov/~Ird/history. (b) http://www.fda.gov/oc/history/historyoffda/. (c) Berger MS. FDA enforcement manual, FDA group, Thompson Publishing group. Available at http://www.emedicinehealth.com/script/main/art.asp?articlekey=59287&page=1. (d) Borchers AT, Hagie F, Keen CL, Gershwin EM. The history and contemporary challenges of the US food and drug administration. Clin Ther 2007;29(1):1–16. (e) Mathieu MP, editor. New Drug Development: A regulatory Overview. Cambridge, MA: Parexel International Corporation; 1987. (f) Missinghoff GJ. Overview of the Hatch–Waxman Act and Its Impact on the Drug Development Process. Food Drug Law J 1999;54:187–194. (g) http://en.wikipedia.org/wiki/Food_and_Drug_Administration. (h) Hilts PJ. Protecting America's Health: The FDA Business, and One Hundred Years of Regulation. New York: Alferd A. Knopf (Division of Random House, Inc.); 2003. (i) Daemmrich A, Radin J, editors. Perspective on Risk and Regulation: The FDA at 100. Philadelphia, PA: Chemical Heritage Foundation; 2007. (j) Adulteration usually refers to mixing other matter of an inferior and sometimes harmful quality with food or drink intended to be sold, whereas misbranding usually refers to false or misleading labeling.

2. (a) www.fdareview.org/glossary.shtml. (b) Lipsky MS, Sharp LK. From idea to market: the drug approval process. JABFP 2001;14(5):362–367. (c) Russell K. FDA: evidentiary standards for drug development and approval. Neurotherapeutics 2004;1:307–316. (d) Rivas-Vazquez RA, Mendez CI. Overview of drug approval process. Prof Psycol Res Pr 2002;33(5):502–506. (e) Rzeszotarski WJ. The FDA and Regulatory Issues. Burger's Medicinal Chemistry, Drug Discovery and Development. 6th ed. (f) Berry IR, editor. Introduction to the Pharmaceutical Regulatory Process. Marcel Dekker, 2006; Also see Guarino RA, editor. New Drug Approval Process. 1–4 ed. Marcel Dekker. (g) Mathieu MP, editor. New Drug Development: A Regulatory Overview. Cambridge, MA: Parexel International Corporation; 1987. Murphy WJ, contributing editor. Washington DC: OMEC International, Inc.

3. (a) http://www.fda.gov/cder/handbook/. (b) http://www.fda.gov/cder/Offices/OTC/reg_mechanisms.htm. (c) Jacobs LR. Prescription to over-the-counter drug reclassification. Am Fam Physician 1998;57(9):2209–2214.

4. (a) http://www.fda.gov/cder/handbook/anda.htm. (b) Greb E, McCormick D, Van Arnum P. Pharmaceutical Technology's 2005 Manufacturing Rankings. Pharm Technol 2006;30(7):36–50. (c) Mossinghoff GJ. Overview of the Hatch–Waxman Act and its impact on the drug development. Food Drug Law J 1999;54:187–194. (d) Welage LS, Kirking DM, Ascione FJ, Gaither CA. Understanding scientific issue embedded in the generic

drug approval process. J Am Pharm Assoc 2001;41(6):856–867. (e) Chen ML, Shah V, Patnaik R, Adams W, Hussain A, Conner D, Mehta M, Malinowski H, Lazor J, Huang SM, Hare D, Lesko L, Sporn D, Roger W. Pharm Res 2001;18 (12):1645–1650. (f) Leon S, Isadore K, editors. Generic Drug Development: Solid Oral Dosage Forms. Marcel Dekker; 2005. (g) Organization of the common technical document for the registration of pharmaceuticals for human use: M4. Available at http://www.ich.org/cache/compo/ 1325-272-1.html. (h) Sheinin E, Williams R. Chemistry, manufacturing, and controls information in NDAs and ANDAs, supplements, annual reports, and other regulatory filings. Pharm Res 2002;19(3):217–226. (i) http://www.fda.gov/ cder/ogd/02-10_BCBS_gjb/sld005.htm. (j) Yu L. Pharmaceutical quality by design: product and process development, understanding, and control. Pharm Res 2008;25(4):781–791; ICH Pharmaceutical development Q8: http://www.ich.org/ LOB/media/MEDIA1707.pdf. (k) FDA OGD's white paper on question-based-review: white paper: http://www.fda.gov/cder/ogd/QbR_white_paper.htm. (l) Yu L, Raw A, Lionberger R, Rajagopalan R, Lee M, Holcombe F, Patel R, Fang F, Sayeed V, Schwartz P, Adamas R, Buehler G. US FDA question-based review of generic drugs: a new pharmaceutical quality assessment system. J Generic Med 2007;4(4):239–248.

5. (a) Vivian JC. Generic-substitution laws. US Pharm 2008;33(8):30–34. (b) Mahn TG. Patent and the orange book: a new wrinkle. Pharmaceutical Law and Industry, 2006;4(10):1–2; ISSN: 1542-9547.

6. (a) http://www.fda.gov/cber/faq.html#4. (b) Ho RJY, Gibaldi M. Comparative drug development of proteins and small molecules. Biotechnology and Biopharmaceuticals. New Jersey: John Wiley & Sons; 2003, 11–20; ISBN 0-471-20690-3. (c) Banait N. Follow-on biological product. Fenwick & West LLP 2005:1–4. Available at www. fenwick.com.

INTELLECTUAL PROPERTY IN DRUG DISCOVERY AND BIOTECHNOLOGY

Timothy P. Maloney
Richard A. Kaba
James P. Krueger
Rudy Kratz
Calista J. Mitchell
Fitch, Even, Tabin & Flannery, Chicago, IL

1. OVERVIEW

Intellectual property is the branch of law that protects and, indeed, encourages the creation of certain products of the human mind or intellect. This chapter is intended to provide a basic understanding and appreciation of intellectual property law, especially as it relates to patents, trademarks, and trade secrets, in the United States and, to a lesser extent, in the rest of the world.[1] Issues and concerns particularly related to the drug discovery and development process and the general biotechnology industry are emphasized.

By making effective use of the legal protection afforded by the intellectual property laws in the United States and elsewhere, the drug developer can protect its investment, enhance the value of the technology being developed, and earn a profit sufficient to allow and encourage further research into improving existing drugs and therapies as well as developing new drugs and therapies. By better understanding these intellectual property laws, the drug developer, together with experienced intellectual property counsel, can develop an effective intellectual property strategy. In this way, new and emerging technologies, as well as new drug discoveries, can be identified, managed, and protected as an integral part of an organization's research and development activities to create a strong intellectual property portfolio.

The rewards flowing from the development of a strong intellectual property portfolio can be significant. A patent allows the patent holder to exclude others from making, using, offering for sale, or selling the patented invention during the term of the patent. A carefully crafted intellectual property portfolio (including pending patent applications, issued patents, trademarks, and trade secrets) can also serve many other purposes. It can be used defensively to prevent others from patenting the invention. It can present legitimate barriers to competitors attempting to enter a new field. It can allow time for recouping investments and establishing market position and identity. It can be used to generate revenues through licensing arrangements or outright sales of the patents, trademarks and associated goodwill, or trade secrets. It can be useful in obtaining outside financing or entering into shared research arrangements, joint ventures, or cross-licensing arrangements. In many instances, a startup biotechnology company's only marketable asset may be its intellectual property. A carefully crafted and maintained patent portfolio can be an especially beneficial asset when seeking outside funding or negotiating an agreement with a large, well-established, and well-funded partner.

The application of intellectual property law to the field of drug discovery and biotechnology presents unique and challenging issues for the individual researcher, the research organization or company, and the intellectual property counsel. These issues arise mainly because of the fast-developing nature of the drug discovery and biotechnology fields, the enormous investment in time and money required in the current regulatory climate to develop a new drug or treatment process and

---

[1] For following up on specific points, more detail is available in several excellent treatises devoted to intellectual property. See, for example, Donald S. Chisum, *Chisum on Patents* (2008); Iver P. Cooper, *Biotechnology and the Law* (2008); Jerome Rosenstock, *The Law of Chemical and Pharmaceutical Invention: Patent and Nonpatent Protection* (2d ed., 2008); Martin J. Adelman, *Patent Law Perspectives* (2d ed. 2008); Kenneth J. Burchfiel, *Biotechnology and the Federal Circuit* (1995); John G. Mills et. al., *Patent Law Fundamentals* (2008); Stephen P. Ladas, *Trademarks and Related Rights: National And International Protection* (1975); R. Carl Moy, *Moy's Walker on Patents* (4th ed., 2008); J. Thomas McCarthy, *Trademarks And Unfair Competition* (4th ed., 2008); Melvin F. Jager, *Trade Secrets Law* (2008); Roger M. Milgrim, *Milgrim on Trade Secrets* (2008).

bring it to the marketplace, and the opportunities derived from the "biotechnology revolution" to achieve rapid breakthroughs in the health care area with the potential for substantial economic rewards. How the industry meets these challenges, and how the legal system evolves and adapts to this rapidly changing field, will significantly affect the development of the burgeoning biotechnology–pharmaceutical industry and the health care system in general. How well individual companies or research organizations protect their intellectual property will determine, to a significant degree, who will survive and prosper.

The development of a drug or treatment process from its conception until its introduction in the marketplace generally requires 6–10 years, sometimes even more. This delay is generally due to the time required for research and development, pilot plant studies, scale-up studies, animal studies, clinical studies, obtaining the necessary regulatory approvals (e.g., from the U.S. Food and Drug Administration (FDA)), marketing studies, and so on. A successful drug development program can cost hundreds of millions of dollars. The ability to protect that human and economic investment has become an increasingly important factor in the drug discovery and development process. A business organization, whether a startup company or an established pharmaceutical giant, often cannot justify the necessary investment if their intellectual property cannot be reasonably protected. Without such protection, a so-called free rider could offer the same or very similar drug or treatment process at a significantly lower cost. Without the ability to protect and recover one's investment and earn a profit, drug discovery and development, for all practical purposes, could only be carried out or sponsored by governments or large nonprofit organizations. This would severely limit the number of persons generating new ideas, decrease the number of new drugs entering the marketplace, and increase the time required for the development of new drugs or treatment processes.

Intellectual property law—that body of law that includes patents, trademarks, trade secrets, and copyrights—provides the framework and mechanism by which investment in intellectual property can be protected. The drug discovery process, especially as it has developed in response to federal regulation and the recent biotechnology revolution, faces new and difficult challenges and issues within the field of intellectual property law. The biotechnology–pharmaceutical industry must recognize and understand these challenges and issues in order to take advantage of the protection now offered and to be prepared to adapt to modifications that may be made in intellectual property law in the future. It is also important that the biotechnology–pharmaceutical industry participate in the ongoing debates and dialogs between society, academic institutes, industry, and government in order to participate in the continued development of suitable government polices, laws, and regulations related to the development and protection of intellectual property.

Patents are generally considered the strongest form of protection available for intellectual property and, therefore, should be the cornerstone of the intellectual property protection strategy. An effective patent strategy or program should identify new and emerging technologies and inventions. A significant part of this program is educating researchers and other employees about the ever-increasing importance of protecting intellectual property and providing mechanisms and incentives to encourage them to bring forward their ideas and innovations for appropriate evaluation. The patent strategy should provide a mechanism to evaluate the inventions and determine whether to file a patent application on a given invention and, if so, when and where to file throughout the world. It must also determine whether and when to update pending patent applications when new information and data become available. This will generally require a careful case-by-case evaluation of each invention, including the likelihood of patentability and success in the ongoing research and FDA approval processes. Unfortunately, such decisions must almost always be made without complete data or information and long before concrete assessments and estimates can be made concerning the ultimate technological and economic success of the invention.

Drug discovery technology has become increasingly complex and multidisciplinary. It is increasingly difficult for meaningful research to be carried out by individuals or even small research teams. Rather, large multidisciplinary teams bringing wide-ranging expertise to bear on a given problem are generally needed to stay ahead of the competition. The existence and requirements of such teams may have a significant effect on the patentability of drugs and treatment processes.

In addition to being new or novel, a product or process to be patentable must not have been obvious at the time the invention was made to "a person having ordinary skill in the art to which said subject matter pertains."[2] Just who is a person of ordinary skill in the drug discovery area? Clearly, a person of ordinary skill in the art of drug discovery is at least a highly skilled individual, probably with an advanced degree. Does that person have a Master's, Ph.D., or medical degree in the field to which the invention is most closely related? Or does that person have advanced-level knowledge in more than one field associated with the invention? If so, in how many fields? Is the person of ordinary skill a single individual or a mythical person having the combined knowledge and skill of a multidisciplinary team where each member possesses ordinary skill in a specific art? These questions remain for the U.S. Patent and Trademark Office (PTO), and ultimately Congress and the courts, to decide and/or modify in order to accommodate and appropriately encourage this rapidly developing scientific area.

Drug discovery has accelerated over the past several decades, and will likely continue to do so, due to the continuous and phenomenally rapid advance of the underlying technology as well as the potential benefits that should follow from the biotechnology revolution. The amount of information and data in the literature is enormous and is growing at an increasing rate as evidenced by the successful sequencing of the human genome and other dramatic developments that were unthinkable only a short time ago. What is nonobvious today to one of "ordinary skill in the art" may well be obvious tomorrow, next week, or next month. This rapidly expanding body of technical information dramatically increases the pressure to seek patent protection as early as possible—oftentimes before the invention is fully developed and its ramifications and significance are fully known.

Another unique aspect of the drug discovery industry is that the majority of research is directed toward a relatively limited number of well-known target diseases or disorders (e.g., AIDS, cancer, Alzheimer's disease) and enabling technologies (e.g., receptors) that are useful in drug discovery. Due to the importance of these diseases, disorders, and enabling technologies, and the potentially huge economic rewards, many research groups and organizations have turned their resources toward these relatively few targets.[3] While one hopes that this intense competition will lead to near-term breakthroughs in new drugs, methods of treatment, and cures, the intense competition makes it more difficult to protect inventions made along the way. Also, because of the large number of groups working and filing patent applications in the same or closely related biotechnology research areas, the number of potential invention priority contests in the PTO (i.e., patent interferences; see Section 3.7) is likely to be significantly higher than in other technologies. The increased possibility of interferences in the area of drug discovery and treatment processes contributes to the pressure to file as quickly as possible.

Throughout most of the world (with the notable exception of the United States), patents are granted to the first to file rather than the first to invent. In such countries, the failure to file quickly can result in loss of valuable patent rights. And, if others independently make the same invention and obtain a patent, the inventor who files late may be prevented from using the very technology upon which

---

[2]35 U.S.C.A. § 103 (West 2001 & Supp. 2008); see also Standard Oil Co. v. Am. Cyanamid Co., 774 F.2d 448, 454 (Fed. Cir. 1985).

[3]Congress has attempted to alleviate this problem somewhat with the Orphan Drug Act which provides, with some limitations, an exclusive 7-year right to market a drug for treatment of a disease affecting less than 200,000 individuals or for which there is "no reasonable expectation" that the developmental costs of the drug can be recovered through sales in the United States. 21 U.S.C. §§ 360aa–360ee (2007).

vast sums and significant human resources were spent.

The changing pathway for drug discovery also influences the way in which inventors or their assignees interact with the patent system.[4] Historically, drug discovery and development generally was carried out by large pharmaceutical companies. More and more, the basic discovery and initial stages of drug development are being carried out by university research teams and startup or relatively small biotechnology companies. These groups generally do not have the internal economic resources to seek worldwide patent protection or to carry a new drug or therapy through the clinical stages. Outside funding, strategic alliances, or licensing arrangements are usually necessary as the research and development progresses. In seeking funding or other business arrangements, researchers are generally required to disclose at least basic business and/or technical information. Extreme care should be taken to prevent public disclosure of inventions before the appropriate patent application is filed. The United States has a 1-year grace period in which a patent application can be filed after the first public disclosure, public use, or offer for sale of the subject matter of the invention. In most other countries, however, there is an "absolute novelty" requirement—public disclosure of the subject matter of the invention anywhere in the world prior to filing the patent application will likely preclude foreign patent protection (see Section 2.3). It is critical, therefore, that secrecy be maintained until the initial patent application is filed covering the invention. Once the initial patent application is filed (usually in the country where the invention is made or developed), corresponding applications can be filed within 1 year in most other countries under prevailing international agreements (i.e., the Paris Convention; see Section 5.1) claiming the benefit of the filing date of the initial application.

Secrecy may be very difficult to maintain if one is seeking outside funding. Most potential investors demand significant business and technical details before making the desired investment. To the extent possible, however, the amount of technical information provided should be strictly limited and its use and dissemination carefully controlled. Confidentiality agreements are helpful in maintaining secrecy and are highly recommended when seeking private funding or joint research arrangements. Public funding and offerings, which trigger the disclosure requirements of the securities act, the Securities Exchange Commission (SEC), and state blue sky laws, present even more difficult problems. Public offerings, when possible, should be delayed until patent applications are filed because public disclosure, even if accidental or in violation of a confidentiality agreement, can preclude patent protection in most countries of the world. Once again, there is great pressure for filing patent applications as quickly as possible.

The requirements set forth in the patent law and as dictated by procedures for obtaining research funding—both in the United States and the rest of the world—also strongly encourage filing a patent application covering a new drug or treatment process as soon as possible. In many cases, it may be desirable or even necessary to file the patent application before complete data are available. For example, a patent application covering a protein for which only a partial DNA coding sequence has been determined may be filed in the United States to establish an early priority date for the invention. This would permit such preliminary information to be disclosed with reduced risk of losing valuable patent rights. Once the sequence is complete, a continuation-in-part (CIP) application including the additional data may be filed in the United States (and, if appropriate, the original application abandoned). The new material added in the CIP application receives the priority date of the actual CIP filing date. In some cases, it may be desirable to file several CIP applications as new data and/or discoveries are made. Using such an approach, however, has risks. In the United States a patent application must be "enabling," that is, it must provide a "written description of the invention, and of the manner and process of making and using

---

[4]In the United States, patent applications must be filed in the name of the inventors. Inventors may assign their rights in the patent application and any patents that may issue therefrom to third parties (i.e., assignees). 37 C.F.R. §§ 1.41, 1.46 (2008).

it, in such full, clear, concise, and exact terms as to enable any person skilled in the art to which it pertains, or with which it is most nearly connected, to make and use the same."[5] If a patent application is filed too early (i.e., before sufficient data are available to allow an enabling disclosure), the application may be rejected or any resulting patent may be invalid. In the protein example, if it is determined that a complete DNA sequence is required for enablement, the first filed patent application would not be legally sufficient; the CIP application containing the full sequence, however, would be enabling.

One must take into account that the United States generally has a more stringent enablement requirement than many other countries. Thus, in the DNA sequencing example above, a partial sequence may be sufficient in other parts of the world. In such countries, a patent application having only a partial sequence and an earlier filing date may have priority over a second patent application filed by another party where the second application is based on a U.S. application, the filing of which was delayed because the full sequence was not yet complete.[6]

This chapter will present a discussion of provisional and utility patents, trademarks, and trade secrets and emphasizes their use in protecting intellectual property in the drug discovery and biotechnology areas. Other forms of intellectual property protection will be mentioned briefly. This chapter cannot, of course, provide the reader with sufficient detail to allow him or her to protect drug-related technology effectively and comprehensively. Moreover, intellectual property laws and regulations can change significantly and rapidly.[7] It is imperative to obtain competent legal counsel specializing in the area of intellectual property and technology transfer, preferably counsel with the appropriate drug research and biotechnology experience, as early in the research and development process as possible. This chapter should enable the reader to communicate and interact more effectively with counsel as they jointly fashion, within the ongoing research and development process, the appropriate legal protection for the particular technology.

2. PATENT PROTECTION AND STRATEGY

The U.S. Constitution provides that "Congress shall have [the] power ... [t]o promote the progress of science and useful arts by securing for limited times to authors and inventors the exclusive right to their respective writings and discoveries."[8] In exercising that power, Congress has established a system for granting utility patents,[9] design patents, and plant patents. Utility patents protect the structural and functional aspects of products or processes and are granted for a term ending twenty years from the earliest nonprovisional filing date of the utility application for which

---

[5]35 U.S.C.A. § 112 (West 2001).

[6]An export license is generally required to export technology developed in the United States. 35 U.S.C.A § 184 (West 2001); 37 C.F.R. § 5.11 (2008); *see also* International Traffic in Arms Regulations, 22 C.F.R. §§ 120–130 (2008); Export Administration Regulations, 15 C.F.R. §§ 730–744 (2008); Assistance to Foreign Atomic Energy Activities Regulations, 10 C.F.R. § 810 (2008). Thus, a patent application for an invention made in the United States generally cannot first be filed in another country with a lesser enabling requirement unless the appropriate foreign filing license is obtained.

[7]Indeed, as this goes to press, significant changes to patent laws and regulations are being considered. Where appropriate in this chapter, we mention and briefly discuss some of the major proposals under consideration and how they might impact intellectual property law. Biotechnology and related industries should not only should remain aware of such proposed changes but also participate in the debates and discussions concerning how intellectual property laws and regulations may be changed in the future to assist, encourage, and/or control future technological growth and development for the benefit of mankind.

[8]U.S. Const. art. I, § 8, cl. 8.

[9]Applicants can also file a provisional application that is described in more detail in Section 3.4. Such a provisional application cannot be converted into a utility patent application or mature into a utility patent. A utility patent application can be filed within 1 year of the provisional application and claim benefit of the provisional application filing date; such a utility patent application can, of course, mature into a utility patent.

benefit is claimed. Design patents protect the ornamental design or aspect of an article of manufacture and are granted for a term of fourteen years from issuance. Plant patents grant the right to exclude others from reproducing, selling, or using an asexually reproduced plant variety for a term of twenty years from the initial filing date. Certain sexually reproduced plant varieties can be protected under the Plant Variety Protection Act (as amended in 1994) for a term of twenty years from issuance; plant variety "certificates" under this program are issued by the U.S. Department of Agriculture. Although design and plant patents can, in some cases, be an important part of a company's patent portfolio, this chapter will concentrate on utility patents.

Utility patents (hereinafter "patents") are generally considered the strongest form of legal protection for intellectual property. They grant the patent holder a "legal monopoly" on the invention, effective on the actual issue date, and running for twenty years after the filing date of the earliest nonprovisional U.S. or international application relied upon.[10] During the term of the patent, the patent holder can prevent others from making, using, offering for sale, or selling the patented invention in the United States or importing the patented invention into the United States. In exchange for this limited right to exclude, the patentee must fully disclose the invention to the public; at the end of the patent term, the invention is dedicated to the public.[11]

Patent protection is limited geographically. For the most part, a U.S. patent does not provide legal protection from, or prevent, an act occurring outside the United States (and its territories and possessions) although that same act would fall within the scope of patent protection if carried out in the United States (the one exception is discussed in Section 4.1). The same is generally true for other countries. Thus, a comprehensive patent strategy should take into account the possibility of obtaining patent protection in all countries where an invention will be exploited (i.e., sold, manufactured, or used).

The U.S. patent system is designed to protect new and nonobvious products and processes. It protects the application of ideas and laws of nature; it does not protect the ideas or laws of nature themselves. Thus, Einstein's equation $E = mc^2$ would not have been patentable, even though Einstein or others might have obtained patent protection for a nuclear power plant, a nuclear rocket engine, or the myriad other products and processes derived from this basic principle. The idea or basic principle itself is available for all to use and develop.

In the United States, the 20-year term[12] of a patent is measured from its earliest nonprovisional filing date for which benefit is claimed.[13] Largely because of the time involved in the FDA approval process and associated clinical trials, a considerable portion of the patent term can elapse before a patented

---

[10]If the utility application is based on an international application (i.e., an application filed under the Patent Cooperation Treaty (PCT); See Section 5), the patent term is calculated from the filing date of the international application. The term "international application" does not include national or regional applications filed in other countries (i.e., applications filed in a national or regional patent office, such as the Japanese Patent Office or European Patent Office, respectively).

[11]Under the current U.S. law, almost all patent applications are published eighteen months after the filing date (generally well before any patent issues). Thus, the applicant generally only obtains an opportunity to seek such legal protection in exchange for disclosing the invention. The right to exclude others only begins after the patent, if any, actually issues. Potential infringers will generally be discouraged from using the invention disclosed in the pending application since any investment made would be at risk when and if a patent is obtained. The applicant can generally keep the application pending in the PTO for a considerable period of time using continuation applications in order to extend this period of uncertainty.

[12]The United States has adopted a 20-year patent term for all applications filed on or after June 8, 1995. Prior to this time, the patent term was 17 years as measured from the actual issue date. For patents issuing from applications filed before June 8, 1995, the patent term is the longer of 17 years from the issue date or 20 years from the U.S. or international filing date. The 20-year term is discussed in more detail in Section 2.4.

[13]Claiming priority from provisional applications or foreign applications does not begin the patent term.

drug can be sold in the marketplace. Thus, the period effectively available for the drug developer to recoup its investment and earn a profit can be considerably shorter than the 20-year patent term.

Since the early 1980s, Congress has taken a number of steps to significantly strengthen and improve the patent system. These steps include adding a reexamination process,[14] authorizing the formation of the U.S. Court of Appeals for the Federal Circuit to hear appeals from the PTO and in patent infringement cases, and providing for the extension and/or adjustment of the patent term for certain drug-related inventions and for certain delays in the PTO. Patent term may be extended and/or adjusted for qualifying patents which claim a product, method of using the product, or method of manufacturing the product in order to compensate for certain, but not all, delays in the FDA regulatory and approval process and/or in the PTO patenting process.[15] Patent term may also be extended for qualifying drugs, medical devices, food additives, and methods "primarily us[ing] recombinant DNA technology in the manufacture of the product."[16] The patent term may be extended using FDA-related delays only if the approval of the first commercial use of the patented product occurs during the original patent term and the FDA-related extension is applied for within 60 days of the approval. Due to the limited time in which to apply for the FDA-related extension, it is important that patent counsel be informed when approval of a drug or medical device is granted, so that the application for term extension can be filed within the required time period. The patent term extension process is jointly administered by the PTO and the FDA. The PTO determines whether a patent qualifies for the FDA-related extension, and the FDA determines the allowable extension term. As for adjustments based on PTO delays, the PTO determines the allowable adjustment of patent term (see Section 2.4).

2.1. Global Patent Strategy

Pharmaceutical and other high-technology industries are increasingly global in nature. More and more, competition in global markets requires effective global protection for intellectual property. As barriers to trade decrease and the legal mechanisms for protecting intellectual property are strengthened around the world, such protection will become even more important. It is an important part of any drug discovery organization's intellectual property strategy to determine how best to protect its intellectual property throughout its global market.

The simplest strategy would be to file patent applications for each invention in each and every country having a patent system. Such a strategy could, of course, be prohibitively expensive and, in most cases, not cost-effective. A cost-benefit analysis should be applied to each particular invention. It is essential to evaluate the market potential of the invention around the world and the ability to control that market based on patent protection and enforcement mechanisms available in key countries. For countries with interrelated markets, it may be possible to protect technology effectively in one country with a patent in another.[17]

Appropriate technical, business, marketing, and legal personnel should be involved in the decision-making process so that all relevant factors can be considered in determining whether to file a patent application on a given invention, and if so, when and where to file. The relevant factors will, of course, vary from case to case, as will their relative importance. For example, a startup company interested in developing a single drug or family of drugs may have a strong interest in obtaining as comprehensive patent coverage as possible. Unfortunately, such a startup company may not have the resources necessary to seek such comprehensive patent coverage. A pioneer in-

---

[14]The reexamination process was modified in 1999 thereby allowing a larger role to third parties in the process. See Section 3.8.
[15]35 U.S.C.A. § 156(a)(5)(A) (West 2001).
[16]35 U.S.C.A. § 156(a)(5)(B) (West 2001).

[17]For example, assume a product is protected by a U.S. patent but not by Canadian or Mexican patents. A competitor would be less inclined to make and offer the product in Canada or Mexico where it is not protected if they are not free to import the product into the United States.

vention (i.e., one which breaks new ground or provides an important technical breakthrough and will likely dominate a particular industry segment) generally warrants wider patent protection than an invention that provides only an incremental improvement in an existing technology. Patent protection around the world is further discussed in Section 5.

Improvements of previously patented inventions deserve special consideration in developing a patent strategy. Patents covering even relatively minor improvements can be important elements in expanding and extending protection of a basic technology. If the inventor or assignee of the improved invention is also the holder of the patent of the basic invention, there are generally two options. The improvement can be kept as either a trade secret or patented. Although a trade secret potentially has an unlimited lifetime (i.e., until secrecy is lost), the actual lifetime is likely to be much shorter, especially in the drug industry where detailed FDA disclosures are required. However, if the improvement is only an obvious variation of a basic invention, reliance upon trade secret law may be the only option.

Generally, patent protection of the improvement is the preferred option. A patent for an improved drug or process may allow for additional patent protection for commercially significant embodiments past the term of the basic patent on the original invention. Obtaining such improvement patents will also make it more difficult for competitors to penetrate or expand into a market. "Driving stakes in the ground" in the form of improvement patents all around the basic or core invention makes it more difficult for any potential competitor to carve out a niche in the market.

In most cases where the developer of the improvement does not hold the patent on the basic or core invention, keeping the improvement as a trade secret is not a realistic option unless the patent covering the basic invention is due to expire in a relatively short time. Therefore, seeking patent protection is generally the best option. Assume the basic drug X is protected by a patent held by Company A and that Competitor B develops and patents a significantly improved drug formulation Y containing drug X. Company A can continue to market drug X but cannot offer drug formulation Y. Competitor B cannot offer either drug X alone or in the form of formulation Y.[18] Competitor B may wish simply to license the improvement to Company A and collect revenues through a license. Or Competitor B can use its patent position on formulation Y as leverage in seeking access to the market. In many cases, Company A and Competitor B will agree to cross-license each other so that each can offer the improved formulation Y. Thus, for a competitor seeking to enter a market otherwise closed by another's patent, improvement patents can provide valuable leverage.

Several differences between the U.S. patent law and the patent law of almost every other country also significantly affect patent strategy, especially the determination of when to file a patent application in the United States. Some of the most important of these include the rules for determining priority and the requirement of "absolute novelty" essentially everywhere except the United States. These differences[19] are discussed below.

2.2. First to Invent Versus First to File

In our world of rapidly advancing technology, and particularly for very active research areas such as drug discovery and biotechnology, investigators at varying locations are often working in the same general area, often on the same specific research topic, and frequently discover essentially the same invention within a very short time of each other. Thus, the issue often arises as to which of two (or more) inventors is entitled to a patent on a contemporaneously discovered invention.

U.S. patent law establishing entitlement to a patent in the case of essentially simulta-

---

[18]It is important to understand that a patent does not give the patentee an affirmative right to practice the claimed invention. Other patents may prevent the patentee from practicing his or her own invention.

[19]The "best mode" requirement is another unique feature of the U.S. patent system. Under U.S. law, an inventor must disclose the best mode of making and using the invention in the patent application as of the filing date. This requirement is discussed further in Section 3.3.3.

neous invention is different from the law of substantially all other countries throughout the world. Nearly all countries award the patent to the first party who files a patent application (i.e., the first-to-file rule).[20] The United States, however, continues to follow the first-to-invent rule, although there has been, and continues to be, significant debate regarding the need for international harmonization of substantive patent law by changing U.S. patent law to the first-to-file system.[21] Under the current first-to-invent system, the first to invent is generally entitled to the patent even though he or she was not the first to file a patent application. Thus, it is possible that one party (i.e., the first to invent) who loses the race to the PTO may be entitled to patent protection in the United States while another party (i.e., the first to file) may be entitled to patent protection for the same invention in most other countries. This possibility increases the incentive for a party to file a patent application covering the invention as quickly as possible.

The PTO, on discovering that two or more parties have copending patent applications or a patent application and a recently issued patent claiming the same invention, may set up an interference proceeding to determine which party is the first inventor of the subject matter. Such a determination is not straightforward. In an attempt to make the procedure as predictable as possible, a great number of rules (both substantive and procedural) have been adopted by the PTO to govern the proceedings and the gathering of the evidence necessary to establish the facts surrounding the making of the invention by each party. These rules give the party who was the first to file (the senior party) substantive and procedural advantages that significantly increase the senior party's chances of prevailing in the interference proceeding.

Generally, in the United States, the party who is first to "reduce an invention to practice" is given priority and awarded the patent, unless another party who reduced the invention to practice at a later date can prove that he or she was the first to conceive the invention and worked diligently to reduce it to practice from a time prior to the other party's date of conception.[22] Reduction to practice may be an "actual reduction to practice" (physically making or carrying out the invention) or a "constructive reduction to practice" (filing a patent application). Therefore, in the United States, at least in theory, the filing date of a U.S. patent application may not control the outcome of the priority contest between parties who each actually reduced the invention to practice. As just noted, however, the party who files first has certain practical advantages in the interference proceedings.

Currently, there is considerable interest in the world community for the United States to harmonize its laws with the rest of the industrialized world and adopt the first-to-file system. Although efforts have been made in the United States to adopt the first-to-file rule, there has been considerable resistance to such a change. Should the United States adopt such a rule, it will likely insist that other major industrial countries enact changes in their laws to favor true international protection of patentable subject matter. The adoption of such a first-to-file rule in the United States may have only a relatively small practical effect on the drug discovery field, especially for large corporate entities and others involved in the global marketplace, since they already have significant incentives to file patent applications as quickly as possible. The adoption of the first-to-file rule in the United States would initially appear to have a significant effect on individual inventors or small organizations who may be interested almost

---

[20]Generally, most countries provide an exception to the first-to-file rule in cases of derivation (i.e., where the first to file, who is not actually an inventor, learns of the invention from the actual inventor who files second).

[21]There have been several attempts over the years to pass legislation that would change the U.S. patent system to a first-to-file system. The most recent attempt was the Patent Reform Act of 2009 (H.R. 1260 and S. 515). The Patent Reform Act of 2009 was introduced in both the House and Senate in March of 2009. The draft legislation is substantially similar to that of the Patent Reform Act of 2007 (with some exceptions) but it is too soon to speculate whether these bills will be passed and in what form at the time of this writing.

[22]35 U.S.C.A. § 102(g) (West 2001).

entirely in the U.S. market or who do not have adequate resources for quickly developing inventions or filing patent applications. However, such individuals and small organizations are generally at a significant disadvantage in any interference proceeding simply because of the cost involved. Such individual inventors and small organizations may not, therefore, be as deeply affected in a practical sense by a first-to-file rule as one might first imagine.

2.3. Absolute Novelty

In most countries, a public disclosure of an invention prior to filing a patent application precludes obtaining patent protection for the invention. This is in contrast to the United States, where an applicant has 1 year after publication, public use, or offer of sale of the subject matter of the invention in which to file a patent application. The effect of such a public disclosure is the same whether it is made by the inventor or by another.[23] Thus, if patent protection outside the United States is desired, a patent application must be filed in at least one Paris Convention country (see Section 5.1) before the public disclosure, followed by the filing of the corresponding applications in other countries within 1 year of the filing date. Public disclosure within the convention year does not adversely affect any later filed applications filed within the convention year.

Valuable patent rights can be lost because of early disclosure of the patentable technology. Such loss can be especially damaging to an organization involved in drug discovery because of the global market for drugs and drug-related technologies. Because of required public disclosure related to the FDA approval process, the patent rights associated with drug discovery are especially at risk through premature disclosure. Disclosure of technical information should be closely monitored and controlled as part of a comprehensive intellectual property program. All employees, including research, medical, technical, and business personnel, should be carefully educated in regard to the confidential nature of technical information and the consequences of premature disclosure.[24] An essential component of such a program is an evaluation procedure for all articles, abstracts, seminars, or presentations prior to actual submission and/or presentation. In addition to preventing premature disclosures, such an evaluation program can aid in educating personnel on the importance of protecting confidential information.

An intellectual property committee responsible for reviewing and approving all disclosures containing technical information, including FDA submissions, is highly recommended. In cases where disclosure is a potential problem, the committee should, if possible, delay publication until the appropriate patent applications are filed. To avoid delays in FDA submissions and scientific or technical publications, such submissions or publications should be reviewed by the committee as early as possible, to allow for sufficient time to prepare and file any necessary patent applications.[25] Due to the importance of FDA submissions and the amount of technical data involved, it may be highly desirable to have at least one individual responsible for FDA matters included on the committee to insure proper coordination between the functions.

The need for monitoring and controlling technical information does not end with the filing of the initial patent application. Disclosures relating to an invention claimed in an earlier patent application may also contain

[23] A notable exception to this general rule is found in Canada. Generally, an inventor has a 1-year grace period from the date of public disclosure in which to file a Canadian application if the public disclosure is made by the inventor or by a person who obtained the information from the inventor. There is no grace period (i.e., absolute novelty applies) if the disclosure is made by someone other than the inventor who did not obtain the information from the inventor. *Patents Throughout the World* 33-35 (Thomson/West eds. 2008).

[24] Since an "offer of sale" of the invention is considered a public disclosure in the United States, an ongoing educational program for sales personnel is recommended.

[25] The FDA submission will generally not be released to the public by the FDA upon receipt. Rather, at some later time, the FDA may make the information publicly available. However, one should not rely on the FDA delaying disclosure if valuable patent rights are at stake. See Section 7.5.

information concerning new inventions or improvements of the earlier claimed invention that may be the subject of later filed applications. In addition, control of the disclosure of inventions contained in a patent application may allow for filing patent applications in countries requiring absolute novelty after expiration of the convention year, should that become necessary or desirable. Failures to file such applications within the convention year may be intentional (i.e., too expensive or perceived lack of technical merit) or accidental. If funds become available, technical merit is established, or the accident is discovered, applications can be filed after the convention year if there has not been a public disclosure. Any application filed after the convention year, however, cannot rely on the earlier priority date; its effective filing date will be the actual filing date in the specific country. Loss of the priority date can be significant since additional prior art may be available against the application.

2.4. Patent Term

Over the past 15–20 years, the U.S. patent law has been significantly changed in order, at least in part, to conform it to various international trade negotiations and agreements.[26]

One major change in the U.S. law resulting from this so-called patent harmonization process involves moving from a fixed patent term beginning at the issue date and ending 17 years thereafter to a nonfixed patent term (the so-called 20-year term) beginning at the issue date of the patent and ending 20 years from the earliest nonprovisional U.S. or international filing date relied upon.[27]

Thus, all patents resulting from new applications, continuation applications, continuation-in-part applications, and divisional applications filed on or after June 8 1995, have a 20-year term. If an applicant claims the benefit of an earlier U.S. application filing date, the earliest application filing date relied upon will control and set the 20-year term running.[28] Further, an international application filed under the Patent Cooperation Treaty (PCT) and designating the United States has the effect of a national application regularly filed in the PTO, so that the filing date of such an international application would be used for the purpose of calculating the 20-year term. The filing dates for national or regional applications in other countries or provisional applications in the United States do not start the clock running on the 20-year term; the 20-year term is measured from the filing date of the regular U.S. application.[29] The old 17-year patent term and the new 20-year patent term are illustrated in Fig. 1.

As can be easily seen, if the period between the earliest filing date and the actual issue date is greater than 3 years, the 20-year term will generally provide a shorter term of patent protection since such protection only begins on the actual issue date. The effective shortening of the patent term will be especially significant in cases where one or more continuation type applications are required before a patent based on these applications actually issues. For example, assume an application is filed on January 1 2000, followed by a continuation application filed on January 1 2003 (at which

---

[26]The Uruguay Round Agreements Act of 1994, Pub. L. No. 103-465, 108 Stat. 4809 §§ 531-534 (codified as amended in scattered portions of 35 U.S.C.) (2000).

[27]35 U.S.C.A. § 154(a)(2) (West 2008). The end of the patent term under the 17- or 20-year term can be advanced by failure to pay maintenance fees or by a terminal disclaimer. The end of the patent term can be extended in some case by the patent term extension and patent term adjustment provisions.

[28]By not claiming benefit of the earlier filed application in a later filed application, the shortening of the patent term can be avoided. This strategy is not generally recommended since events occurring between filing dates of the earlier application and the later filed application could bar the granting of a valid patent. In many cases, the patentee may not become aware of such events until the time he or she attempts to enforce the patent. This strategy could be used, however, in the case of a continuation-in-part application where the newly claimed subject matter depends only on the newly added subject matter included in the continuation-in-part application and is patentably distinct from the subject matter disclosed in the earlier filed application.

[29]Patents "in force" on June 8, 1995 and patents resulting from applications filed before June 8, 1995 will have the longer of 17- or 20-year terms. The patent term will be "reset" for patents "in force" on June 8, 1995 which had a pendency period of less than 3 years. 35 U.S.C. §154(c)(1) (West 2001).

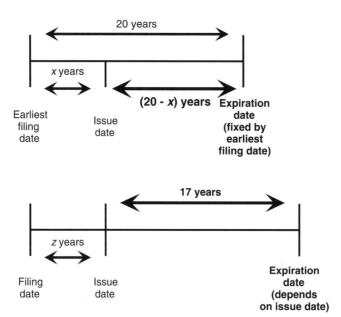

Figure 1. 17 vs. 20 year patent term.

point the original application is abandoned), followed by a second continuation application filed on January 1, 2006 (at which point the first continuation application is abandoned), and the second continuation application results in an issued patent on January 1, 2009, the actual life of patent term available to the patent owner would be from January 1, 2009 (the actual issue date) to January 1, 2020 (20 years from the filing date of the original application) or a period of only 11 years. If more continuation applications are necessary or if the time required to prosecute any of the applications in the series increased, the effective patent term could be reduced even further.

Recognizing that the 20-year patent term could result in a significant reduction of patent term, and thus patent protection, Congress has provided mechanisms which attempt to reduce, at least in part, this significant loss of patent term. These patent term extension and adjustment provisions should be especially important to the biotechnology and pharmaceutical industries since their inventions are routinely subject to extended delays in the prosecution in the PTO as well as delays caused by the FDA approval process. Patents issuing from applications filed between June 8, 1995 and May 28, 2000 may be eligible for patent term extension,[30] while patents issuing from applications filed on or after May 29, 2000 may be eligible for patent term adjustment.[31] The PTO-based extensions and adjustments are independent of premarketing regulatory review of the patented product after the patent issues.[32] The extensions provided for FDA regulatory review can be granted for up to a maximum of 5 years; there is no fixed time limit for extensions based on PTO delay.

For applications filed between June 8, 1995 and May 28, 2000, the patent term can be extended if issuance was delayed due to interference proceedings, secrecy orders, or appellate review by the Board of Patent Appeals and Interferences or by a federal court.[33] The term is extended for the sum of the periods of delay, assuming the delays are not overlapping, up to a maximum extension of 5 years.[34]

[30] 37 C.F.R. § 1.701 (2008).
[31] 37 C.F.R. § 1.702 (2008).
[32] 35 U.S.C.A. § 156 (West 2001 & Supp. 2008).
[33] 37 C.F.R. § 1.701(a)(1)–(3) (2008).
[34] 37 C.F.R. § 1.701(b) (2008).

For applications filed on or after May 29, 2000, the patent term can be adjusted based on three types or classes of delays during patent prosecution:the so-called "A delays" or "A periods" result from a failure of the PTO to comply with various statutory deadlines during prosecution;[35] (2) the so-called "B delays" result from a failure of the PTO to meet its "[g]uarantee of no more than 3-year application pendency";[36] and (3) the so-called "C delays" result from interferences, secrecy orders, and successful appeals[37] to the Board of Patent Appeals and Interferences or to the federal courts. The term adjustments are on a day-for-day basis; in cases where the delay periods overlap, double counting is not allowed (i.e., only 1 day is added for each calendar day).

The adjustment may not exceed the actual number of days that issuance was delayed, it may not adjust the term beyond an expiration date specified in a terminal disclaimer, and it is reduced by the time the applicant failed to use "reasonable efforts" to prosecute the application. The term adjustment is automatically determined by the PTO and reported to the applicant when the notice of allowance is issued.[38] The applicant may request reconsideration of the PTO's determination of the term adjustment by filing an application for patent term adjustment before payment of the issue fee.[39] The patent will then issue with the length of the adjustment based on either the initial determination or, if a request for reconsideration was made, the length determined in the reconsideration. If the applicant remains unsatisfied, the applicant must file a request for reconsideration of the term adjustment indicated in the issued patent within 2 months of the issue date[40] or appeal the determination in the U.S. District Court for the District of Columbia within 180 days of the issue date.[41]

As noted above, the term adjustment is determined in the case of overlapping periods to avoid double counting. The PTO has taken the position, with regard to overlapping periods, that "B-delays" are considered to cover the guaranteed 3-year pendency.[42] Thus, for example, if in addition to a "B-delay" there is also an "A-delay" during the initial 3-year period the patentee is only entitled to the longer of the "A-delay" or "B-delay" and not to the sum of the delays. In a recent case involving both an "A-delay" in the initial 3-year period as well as a "B-delay" due to an overall pendency of greater than 3 years, a

---

[35] 35 U.S.C.A. § 154(b)(1)(A) (West 2001). "A events" include failure of the PTO to provide a first office action or notice of allowance within fourteen months of the filing date, failure of the PTO to respond within four months of certain acts (e.g., applicant responding to an office action, applicant appeal), failure of the PTO to act on an application having allowable claims within four months of a decision by the Board of Patent Appeals and Interferences or by a federal court, and failure of the PTO to issue a patent within four months of the applicant meeting all requirements regarding the issuance of the patent.

[36] 35 U.S.C.A. § 154(b)(1)(B) (West 2001). "B delays" start 3 years after the filing date and end on actual issuance of the patent but exclude any time consumed by continued examination of the application (i.e., request for continued examination ("RCE") under 37 C.F.R. § 1.114), any time consumed by interferences, secrecy orders, appellate review, or most delays requested by the applicant.

[37] 35 U.S.C.A. § 154(b)(1)(C) (West 2001).

[38] In contrast, the premarket extension period is not automatically determined by either the PTO or the regulating agency (e.g., FDA). The patent holder must apply to the PTO for such a premarket extension (along with the specific time period requested) within a specific time period (60 days) after regulatory approval. 37 C.F.R. §§ 1.710–1.760 (2008).

[39] 37 C.F.R. § 1.705(b)–(c) (2008). The application for term adjustment may be based on an error in the PTO's calculation or by making a showing that, in spite of all due care, the applicant was unable to respond to a PTO communication regarding a rejection, objection, argument, or other request within three months of the mailing date of the PTO communication. 37 C.F.R. § 1.705(c) (2008).

[40] However, a request for reconsideration that raises issues that were previously raised, or that could have been raised, in a request for reconsideration filed before payment of the issue fee shall be dismissed as untimely. 37 C.F.R. § 1.705(d) (2008).

[41] 35 U.S.C.A. § 154(b)(4)(A) (West 2001 & Supp. 2008). A third party does not have the right to appeal the PTO's determination of the extension period. 35 U.S.C.A. § 154(b)(4)(B) (West 2001).

[42] 69 Fed. Reg. 34238 (2004).

District of Columbia court overruled the PTO's interpretation, concluding that periods of delay overlap only if they actually occur on the same calendar day.[43] Thus, the patentee was entitled to an adjustment equal to the sum of the two delays since there was no actual overlap. The court's decision was affirmed on appeal.[44]

According to the complex rules issued by the PTO[45] regarding term adjustment, the delay provisions have another significant limitation. When continuation applications[46] are used to obtain a patent, only the delays associated with a continuation application filed under 37 C.F.R. § 1.53(b) can be used to calculate the term adjustment. Assume that an original application was filed and that 4 years after the filing date, a continuation application was filed, which 3 years later, issued as a patent. Under the current rules, the patent would not be extended for the 1 year "B-delay" based on the original application but, instead, would only be adjusted for any PTO delays incurred during pendency of the continuation application (which in this example would be zero days). Moreover, there is often a significant lag (e.g., usually about 1–2 years) between filing a continuation application and substantive examination by the PTO.

If the applicant files a request for continued examination (RCE), the filing of the RCE cuts off the accrual of patent term adjustment for failure of the PTO to meet its guarantee of 3-year pendency. However, the applicant may still accrue patent term adjustments for "A-delays" and "C-delays." For example, if an application is pending for 2.5 years before the filing of an RCE and the patent ultimately issues 4 years after the initial filing of the application, the patent term will not be adjusted for the fourth year of pendency.[47] The applicant should consider such limitations on term adjustment when deciding whether an RCE or continuation application is more appropriate under the particular circumstances involved.

2.5. Publication of Patent Applications

Another major change in the U.S. law resulting from the patent harmonization process involves the publication of the U.S. patent applications filed on or after November 29, 2000.[48] Applications filed on or after this date, with some exceptions, are published 18 months[49] after the earliest filing date for which priority is sought. This change brought the U.S. system in line with most other countries with regard to publication of pending applications.

There are several—but limited—exceptions to the publication requirement. Applications which are not pending at the end of the 18-month period are not to be published.[50] Applications which are undergoing national security review or which have been subject to a secrecy order are not to be published.[51] Provisional and design applications are not to be published.[52]

In addition, an applicant can request nonpublication if the U.S. application is not to be filed in another country or under a multilateral international agreement that provides for publication. To qualify for this exemption, the applicant must certify at the time the application is filed that the invention disclosed in the

---

[43]*Wyeth v. Dudas*, No. 07-1492, 580 F. Supp. 2d 138 (D.D.C. 2008).

[44]*Wyeth v. Kappos*, 93 us PQ. 2d (Fed. Cir. 2010).

[45]37 C.F.R. §§ 1.701–1.705 (2008). The rules for original applications filed on or after June 8, 1995, but before May 29, 2000, are different for original applications filed on or after May 29, 2000. *Compare* 37 C.F.R. § 1.701 and 37 C.F.R § 1.702.

[46]There are effectively two mechanisms to continue prosecution of claims: (1) request for continuing examination (RCE) under 37 C.F.R § 1.114 and (2) a traditional continuation application under 37 C.F.R. § 1.53(b). The mechanism selected to continue examination can have a significant impact on the term of any patent issuing therefrom. See Section 3.6 for full details.

[47]35 U.S.C.A. § 1.54(b)(b)(B)(i).

[48]American Inventors Protection Act of 1999, Pub. L. No. 106–113, 113 Stat. 1501 (codified as amended in scattered portions of 35 U.S.C. (2000)).

[49]Applicants can request publication earlier than the end of the 18-month period. 37 C.F.R. § 1.129 (2008).

[50]37 C.F.R. § 1.211(a)(1) (2008).

[51]37 C.F.R. § 1.211(a)(2) (2008).

[52]37 C.F.R. § 1.211(b) (2008).

application has not been and will not be the subject of an application filed in another country or under a multilateral international agreement that provides for publication.[53]

In cases where the U.S. application contains more extensive information or description than the corresponding applications filed in another country or under a multilateral international agreement that provides for publication, the applicant may file a redacted copy of the application removing the new information or description for publication.[54]

Of course, publication will destroy any trade secrets contained in the application. In practice, however, the publication rules should not significantly affect an applicant's ability to protect trade secrets contained in the application since publication effectively occurs, assuming the proper certification is made and that PTO does not improperly publish the application, only if it would have been published in another country or under a multilateral international agreement that provides for publication.

Publication should provide the public with knowledge of the invention at an earlier time, thereby advancing the state of the art. In addition, the publication will qualify as prior art as of its publication date even if no patent is ultimately granted. Thus, inventions for which patents never issue will become part of the state of the art and be available to the public.

2.6. PTO Patent Prosecution Strategy

The changes discussed above—especially the introduction of the 20-year patent term—have fundamentally changed the U.S. prosecution practice. The potential loss of patent term has a significant impact on patent prosecution strategy and dramatically changes the patent prosecution environment from a relatively stable environment to one of a much more dynamic nature due to the tension between enforceability of the patent and length of the patent term.

Ideally, efforts should be made to shorten the time between filing the initial application and issuance in order to obtain the maximum patent term. Thus, it is generally in the applicant's best interest to speed up the prosecution process so long as the validity of the claims is not compromised or the scope of the claims unduly limited. Complicating the issue even further, however, is the recent decline in the overall patent allowance rate in the PTO. As discussed in more detail in Section 3.6, the PTO patent allowance rate has declined from an average of about 65% during the period of 1975–2000 to below about 45% in 2008.[55] Thus, it has become increasingly more difficult to obtain allowable claims in the PTO. Unfortunately, the use of multiple continuation applications to obtain an initial patent—with increased prosecution costs and potentially greater loss of patent term—is often necessary to obtain allowable claims in this new environment.[56]

To achieve a valid patent with reasonable claims and a reasonable term, patent prosecu-

---

[53]37 C.F.R. § 1.213 (2008). In some cases, an applicant may later determine that applications outside the United States are desired. In that case, the application directed to the invention can be filed in another country or under a multilateral international agreement that provides for publication. The applicant must, however, notify the PTO within 45 days that such filings have been made; failure to notify can result in abandonment of the U.S. application.

[54]37 C.F.R. § 1.217 (2008).

[55]Erik Sherman, *Applying for a Patent? Why You Need More Luck Than Ever*, IP Law & Business, December 2008, at 16, 16.

[56]The PTO recently attempted to adopt new rules that, among other things, significantly limit the number of continuation-type applications that can be filed in a given case. These rules were scheduled to go into effect on November 1, 2007. 72 Fed. Reg. 161, 46716–46843 (August 21, 2007). Based on a lawsuit filed against the PTO under the Administrative Procedure Act, a federal court has ruled that the PTO exceeded the scope of its rulemaking authority and voided the rules. *Tafas v. Dudas*, 541 F. Supp. 2d 805 (E.D. Va. 2008). The PTO appealed (May 7, 2008) to the Federal Circuit. The Federal Circuit affirmed in part, vacated in part, and remanded the case for further consideration by the district court. *Tafas v. Doll*, 559 F.3d 1345 (Fed. Cir. 2009). In a press release dated October 8, 2009 the PTO essentially reversed their position with regard to these rules and announced that the rule had been rescinded. See www.uspto/news/09-21.jsp.

tion should be modified in order to speed up, to the extent possible and reasonable, overall patent prosecution, including, for example, the filing of divisional, continuation, and continuation-in-part applications, and responses to office actions, while maintaining the high quality of the resulting product. In the pharmaceutical field, applications typically include claims to a family of new compounds, methods for making them, and one or more therapeutic uses. These types of claims are often considered by the PTO to be separate inventions and are often subject to multiple restriction requirements—meaning that the applicant must filed separate "divisional" applications for each invention for which patent protection is sought. Under the 17-year patent term, it was possible to obtain successive patents to each group of claims, each of which had a 17-year term from the date of grant. Thus, under the old system, the overall period of protection could be effectively extended by filing the divisional application just before the parent application issued. This technique is no longer effective since the term of each divisional application will end at the same time as the parent case (i.e., 20 years after the earliest U.S. nonprovisional filing date). In most cases, consideration should be given to filing divisional applications as early as possible in response to a restriction requirement in order to avoid potential loss of patent term in the divisional applications. Thus, the simultaneous prosecution of multiple divisional applications can help preserve the maximum patent term for each divisional application.[57]

Under the old 17-year term system, applications that may have been an improvement or modification to an invention on which an application was already pending, were filed as continuation-in-part applications claiming the benefit of the filing date of the earlier filed application. Claims of such continuation-in-part application were entitled to the parent application filing date if the claimed subject matter is disclosed in the parent application in the manner provided by 35 U.S.C. § 112, first paragraph, and if other requirements of 35 U.S.C. § 120 were satisfied; such a continuation-in-part application enjoyed its own 17-year patent term beginning with the issue date. But now, the patent term of such a continuation-in-part application will end 20 years from the filing date of its parent application. Hence, where the invention to be claimed is based on new matter, which is patentably distinct from the parent application, it may be appropriate to file a new application rather than a continuation-in-part application. By doing so, the new application will have its 20-year term calculated from its filing date rather than the priority date of the earlier application.

If the applicant meets certain conditions, he or she can attempt to speed up the PTO process by filing a petition with the PTO to have the application declared "special."[58] If special status is granted, the PTO should provide expedited examination of the application. Prior to August 25, 2006, applications could be advanced out of turn for examination upon the filing of a grantable petition based upon a variety of grounds.[59] For a petition to

---

[57]Under the current system, an applicant may question whether to even file such divisional applications. If broad and adequate protection can be provided in the parent case, divisional applications may seemingly not be necessary. For example, if broad composition of matter claims can be obtained in the parent application, method of use claims presented in a divisional application may offer little additional protection and offer no additional patent term. Thus, little additional protection would seem to be provided since the broad composition of matter claims will cover all uses of the compound, including a new use. In the event, however, that the composition of matter claims are later found to be invalid or unenforceable, the method of use claims could prove to be valid and enforceable. Thus, there are strong arguments that such divisional applications should be filed even under the current system.

[58]37 C.F.R. § 1.102 (2008).
[59]Grounds previously recognized by the PTO included (1) prospective manufacture, (2) infringement, (3) applicant's health or age, (4) the invention's contribution to environmental quality, (5) the invention's contribution to the discovery or development of energy resources, or efficient utilization and conservation of energy resources, (6) inventions relating to recombinant DNA, (7) inventions involving superconductivity, (8) inventions relating to HIV/AIDS and cancer, (8) inventions for countering terrorism, and (9) applications relating to biotechnology filed by a small entity applicant.

make special filed on or after August 25, 2006, the petition must be based on one of only a few grounds or upon a substantial showing that the claims presented in the application are patentable.[60]

In order to maximize the patent term, efforts should be made to speed up prosecution to the extent possible. In some cases, it may be desirable to present narrow claims in the parent application and prosecute these narrow claims first. In theory, it should be possible to obtain allowance quicker for narrow claims as opposed to broad claims. Support for such broad claims should, however, be included in the specification so that they can be presented and prosecuted in a continuation application. If allowance of the narrow claims is obtained relatively quickly in the parent application, they should provide both the maximum patent term and significant, but narrower, protection while one continues the prosecution of broad claims in appropriately filed continuation applications.[61] Care should be taken, however, to craft patentability arguments so as not to unnecessarily complicate the prosecution of broad claims in a later filed continuation.[62]

Regardless of the scope of the claims initially presented, timely and effective communications with the examiner at the PTO, including responding fully to official actions as quickly as possible, will likely help to speed up prosecution. Thus, one might respond to a first office action by significantly narrowing the claims by amendments in view of the prior art cited by the examiner in order to quickly obtain allowance. Thus, applications generally should be placed in the best possible position for allowance as quickly as possible in light of the first office action and the known prior art in order to minimize the prosecution period. Speeding up the process in this manner requires close cooperation between the attorney prosecuting the application and the client. In appropriate cases, the attorney should provide draft responses to the client as quickly as reasonably possible; likewise, the client should provide any input regarding the draft response as quickly as reasonably possible.

Another method of speeding up the prosecution process—and one that is highly recommended—involves the use of the interview process. In many cases, examiners can more quickly appreciate the invention, and its patentability, when the invention is presented orally by either a telephonic interview or a

---

[60]The petition may be based on the grounds of (1) applicant's age or health, or (2) the Patent Prosecution Highway ("PPH") program. The PPH is a cooperative arrangement between the PTO and certain patent offices around the world. An applicant who receives a ruling of patentability of at least one claim by a patent office participating in the PPH may petition that the examination of the corresponding application in the United States be expedited. As of this writing, the PPH is currently limited to the patent offices of Japan (JPO), Canada (CIPO), United Kingdom (UK IPO), Korea (KIPO), Australia (IPAU), Europe (EPO), Denmark (DKPTO), and Singapore (IPOS). For petitions not based on one of these two grounds, a grantable petition to make special under the accelerated examination program must be filed with the initial filing of the application to be expedited (i.e., the petition cannot be filed after the application has been filed). A grantable petition requires a significant showing by the applicant of the application's patentability, including, among other requirements, a comprehensive prior art search and an examination support document that details how each of the claims are patentable over the prior art. Compliance with the requirements of the accelerated examination program can substantially delay the filing of the application. Moreover, the detailed information required may raise significant "estoppel" issues in later infringement actions. See Section 4.4.

[61]Of course, the narrow claims should cover the contemplated commercial embodiment. Even if allowance of broader claims ultimately cannot be obtained, continued prosecution of the broader claims in pending continuation applications will indicate to a potential competitor that broader claims may be forthcoming in the future.

[62]The Court of Appeals for the Federal Circuit ruled that when a claim limitation is narrowed during prosecution, application of the doctrine of equivalents to that claim element is completely barred. *Festo Corp. v. Shoketsu Kinzoku Kogyo Kabushiki Co., Ltd.*, 234 F.3d 558 (Fed. Cir. 1999), *vacated*, 535 U.S. 722 (2002). Although the U.S. Supreme Court has vacated this decision, it could still have a significant impact on claim drafting; more specifically, claims could initially be drafted significantly more narrowly so as to avoid the need to amend claims during prosecution. See Section 4.4 for more details regarding *Festo's* potential impact.

personal interview. Such interviews are especially helpful after the first office action (assuming the claims are rejected) since the case for patentability can be presented in light of the art and arguments presented by the examiner in the office action. In many instances, the attendance of the inventor and/or presentation of demonstrations can be very helpful. Such interviews help more clearly define the issues of concern to the examiner and allow the next response (assuming agreement is not reached) to be more focused. Often the examiner may indicate amendments during the interview that would be considered favorably. In some cases, agreement can be reached. Even if agreement cannot be reached, the attorney generally has a much better idea of the examiner's concerns and the examiner generally has a much better understanding of the invention. Further prosecution is likely to be much more focused and directed. Indeed, narrowing the issues under consideration using interviews generally can avoid addressing many issues in writing and can, therefore, limit the potential effect of prosecution history estoppel during any subsequent litigation.

In appropriate cases, after final rejection or a second rejection on the merits, the applicant should consider going directly to the appeal process rather than filing continuation applications to continue prosecution before the PTO. Since the time taken for an appeal, if that appeal is ultimately successful, can be used to extend the 20-year patent term, it is likely that the appeal process will become even more important. In fact, in suitable cases, initiating the appeal process as soon as possible should be considered. Thus, in such cases, it is generally recommended that the claims be placed in the best possible form for appeal when responding to the first office action. Once again, a personal interview after the first office action can be very helpful in placing the claims in the best position for allowance or, if necessary, for appeal. If not satisfied with the Board's decision, the applicant may appeal that decision either to the Court of Appeals for the Federal Circuit based on the record before the PTO or to the U.S. District Court for the District of Columbia for a *de novo* review. If the examiner's position is overturned, the appropriate court can order the PTO to issue the patent. However, appeal to either the Federal Circuit or the D.C. district court destroys the secrecy of the application as well as that of the record of the proceedings within the PTO and, thus, destroys any trade secrets that may have been contained therein (assuming they have not already been lost through publication of the U.S. or corresponding foreign applications).

As suggested above, the U.S. patent system seems to be in a significant and, at least in our opinion, unsettling state of flux at the present time. Depending on various appeals currently before the courts in which the PTO is a party and/or potential changes in the patent allowance rate within the PTO, the patent landscape in the United States could change radically in the next few years. Whether such potential changes will improve or reduce the importance of patents remains to be seen. Clearly, prosecution strategy will need to be modified in light of such changes.

3. REQUIREMENTS FOR PATENTS

To obtain a patent on an invention in the United States, the inventor(s) must, as the initial step, file a patent application describing the invention, including the best mode, in such terms as to teach one of ordinary skill in the art how to make and use the invention and claiming the subject matter that the inventor(s) regards as the invention. The subject matter of the claimed invention must be within the statutory classes of patentable inventions. In addition, the claimed invention must have utility and be both new and nonobvious. The requirements for patentability (especially as to what constitutes patentable subject matter) can vary considerably throughout the world.

This section addresses the requirements for patentability in the United States and, to a much lesser extent, variations encountered in a few representative countries. The actual patenting procedure in the PTO will also be discussed.

3.1. Patentable Subject Matter in the United States

In the United States, patentable subject matter includes "any new and useful process, machine, manufacture, or composition of matter, or any new and useful improvement thereof."[63] An invention must be claimed so as to fit within one of the four statutory classes of inventions: process, machine, manufacture, and composition of matter. A process is essentially the means to achieve a desired end—e.g., a method to synthesize a drug or a method of using a drug to treat a specific condition. The other patentable classes are basically the end products themselves. These four classifications are broad and generally encompass the vast majority of technological advances. Examples of generally nonpatentable subject matter include laws of nature or abstract ideas, products of nature, algorithms,[64] and printed materials.

The subject matter of most inventions clearly falls within one of the four statutory classes. But for cases where the issue of patentable subject matter is raised, the line between what is patentable and what is nonpatentable often cannot be clearly drawn and must be evaluated on a case-by-case basis. Consider "products of nature": compounds occurring in nature normally cannot be patented. So a naturally occurring drug collected from, for example, a particular plant species is not generally patentable. However, if through human intervention the naturally occurring drug is produced in a purer or more concentrated form not naturally occurring, the new form of the drug may be patentable. Even if the actual drug is not patentable, a new and nonobvious use for that drug may be patentable. Or the combination of the drug with other active ingredients may be patentable. Likewise, a new method of preparing, concentrating, or purifying the drug (even if the drug is exactly the same as the naturally occurring drug) may also be patentable.

Patentable pharmaceutical inventions generally and broadly include drugs, diagnostics, intermediates, drug formulations, dosage forms, methods of treatments of animals and/or humans, kits containing the drug or diagnostic, methods of preparing the drug or diagnostic, and so on. Microorganisms, plasmids, cell lines, DNA, animals, and other biological materials have been found patentable if they are the result of human intervention or manipulation. Initially, the PTO resisted granting patents on living organisms on the basis that the claimed invention was a product of nature. The U.S. Supreme Court in the landmark case of *Diamond v. Chakrabarty*,[65] however, held that new life forms (i.e., bacteria altered genetically to digest crude oil) can be patentable subject matter. In 1987, the PTO formally issued a notice indicating that it "considers nonnaturally occurring, nonhuman, multi-cellular living organisms, including animals, to be patentable subject matter."[66] It is now clearly established that nonnaturally occurring biological materials, including microorganisms, plants, and animals, can be protected by patents in the United States.[67]

---

[63]35 U.S.C.A. § 101 (West 2001).

[64]A physical process using an algorithm can, however, be patented. See, for example, *In re Bilski*, 545 F.3d 943 (Fed. Cir. 2008) (*en banc*); *AT&T Corp. v. Excel Commc's, Inc.*, 172 F.3d 1352 (Fed. Cir. 1999); *State St. Bank & Trust Co. v. Signature Fin. Group, Inc.*, 149 F.3d 1368 (Fed. Cir. 1998).

[65]*Diamond v. Chakrabarty*, 447 U.S. 303 (1980).

[66]Donald J. Quigg, *Notice: Animals-Patentability*, 1077 Off. Gaz. Pat. & Trademark Off. 24 (April 21 1987). The PTO Notice reaffirmed that an "article of manufacture or composition of matter will not be considered patentable unless given a new form, quality, properties or combination not present in the original article existing in nature." The PTO Notice also added that a "claim directed to or including within its scope a human being will not be considered to be patentable subject matter."

[67]See, for example, *Ex parte Hibberd*, 227 U.S.P.Q. 443 (B.P.A.I. 1985) (corn plant with increased level of tryptophan was patentable); U.S. Patent No. 4,736,866 (filed June 22 1984) (first animal patent; transgenic mouse with cancer causing gene); U.S. Patent No. 5,183,949 (filed March 15 1989) (rabbit infected with HIV-1 virus). See also *U.S. Pat. & Trademark Off., Manual of Patent Examining Procedure*, §§ 2105–2106 (8th ed, 2001, latest revision 2008) available at http://uspto.gov (PTO guidelines regarding determination of patentable subject matter).

In the United States, it has been presumed that patents on human beings are precluded by the U.S. Constitution.[68] However, the definition of "human being" has become increasingly unclear with the advancement of cloning technologies. For example, a laboratory animal into which a single human gene has been transferred would not be regarded as human by most definitions of the term in common usage, and should not thereby be excluded from patent protection. Conversely, a human being whose somatic cells contain a single nonhuman gene, or multiple nonhuman genes, introduced for therapeutic purposes would still be considered human and thereby not patentable.[69] The question of how many characteristics may be transplanted before an animal is considered a "human being" may become an important consideration that will impact patentability.[70]

Sequences of human genes and their regulatory regions, either whole or partial, can be patented. Such patentable subject matter includes portions of genes such as expressed sequence tags (ESTs) and single nucleotide polymorphisms (SNPs).[71] Such partial sequences are potentially important in the detection and screening of their underlying genes. The PTO has indicated that such genes or portions thereof may be patented so long as the other statutory requirements are satisfied; thus, the claimed subject matter must have a "specific, substantial, and credible" utility.[72]

The Federal Circuit recently upheld a rejection of a claim directed to five ESTs by the PTO based on the utility requirement.[73] The claimed ESTs were fragments of genes of unknown structure and function. The alleged utility (i.e., detection and screening of underlying genes) was not deemed to be "specific" since all ESTs could be used in that general manner. Nor was the alleged utility "substantial" since the ESTs were essentially research tools that might be used to isolate the underlying genes for further studies.[74] Thus, at least one specific and substantial utility (preferably more than one and in as much detail as possible) should be explicitly identified in the application. Whether the structure and/or function of the underlying gene would be sufficient to satisfy the utility requirement remains to be seen.

The Federal Circuit has held that claims directed to business methods may constitute patentable subject matter.[75] More recently, the Federal Circuit clarified that a business method, like any other type of process, is suitable for patent protection if (1) it is tied to a particular machine or apparatus or (2) it transforms a particular article into a different state or thing.[76]

In addition to fitting into one of the statutory classes, a patentable invention must also possess utility (i.e., it must provide some useful function to society). The invention must be operable and accomplish some function, which is not clearly illegal.[77] A chemical compound for which the only known use is to make an-

[68]The Thirteenth Amendment to the Constitution provides that "Neither slavery nor involuntary servitude, except as punishment for crime whereof the party shall have been duly convicted, shall exist within the United States." U.S. CONST.amend.XIII.

[69]A claim directed to a method of treatment comprising transferring somatic cells containing such genes might be patentable.

[70]See also James P. Of mice and 'manimal': the Patent and Trademark Office's latest stance against patent protection for human-based inventions. J Intell Prop L 1999;7:99.

[71]66 Fed. Reg. 1092, 1095 (January 5 2001) (revised patent examination guidelines under 35 U.S.C. § 101). Although the patentability of genes is discussed in detail, the utility guidelines apply to all subject matter.

[72]Id.

[73]In re Fisher, 421 F.3d 1365 (Fed. Cir. 2005).
[74]Id. at 1371.
[75]State St. Bank & Trust Co. v. Signature Fin. Group, Inc., 149 F.3d 1368, 1373–1377 (Fed. Cir. 1998).
[76]In re Bilski, 545 F.3d 943 (Fed. Cir. 2008) (en banc).
[77]An invention that can be used for both "useful" and illegal or fraudulent purposes may be patented. Only an invention that could only be used for an illegal or fraudulent purpose would be rendered unpatentable because of the utility requirement. For example, a process for making tobacco only appear to be of higher quality or grade and having no other function or utility (i.e., the sole utility being to deceive the public) was not patentable because of lack of utility. Rickard v. Du Bon, 103 F. 868 (2d Cir. 1900).

other compound, which does not have a known use, is not patentable; if, however, the final product is useful, then the starting material is also useful and should have sufficient utility for patentability.[78] The patent specification must disclose at least one nontrivial utility for the invention. Thus, at least one practical utility for a new composition of matter (e.g., a new drug) must be disclosed. Where possible, however, it is generally recommended that several utilities are disclosed to reduce the risk that the PTO or later court will find the invention lacking utility.

Composition of matter claims covering the drug will generally protect all uses of the drug, even the ones not disclosed in and the ones not discovered until after the patent issues. The utility of the drug does not have to be patentable in its own right. The utility has to be neither developed nor discovered by the inventor of the composition of matter; the inventor merely has to disclose, and sometimes provide proof of, the utility.

Generally, if the utility disclosed in the patent specification is easily understood and is consistent with known scientific laws, the PTO will not question it. If, however, the disclosed utility clearly conflicts with general scientific principles or is incredible on its face (e.g., a perpetual motion machine), the PTO will presume the invention lacks utility and will require strong evidence supporting operability.[79] Thus, for example, it is likely that an invention asserting, as the only utility, a complete cure for leukemia, other cancer, or AIDS will require an especially strong showing of effectiveness.[80]

For inventions involving pharmaceuticals and methods of treatment of humans, the PTO often requires a relatively high level of proof for the disclosed utility. A utility likely to be deemed incorrect or unbelievable by one skilled in the relevant art in view of the contemporary knowledge of that art will require adequate proof. The proof required can generally be based on clinical data, *in vivo* data, *in vitro* data, or combinations thereof, where such data would convince one skilled in the art.[81] The data may be included in the patent specification as filed or provided (or, if appropriate, supplemented) in a later submitted declaration or affidavit.

The data necessary to support utility will, of course, vary as the relevant art advances. For example, proving the effectiveness of a new drug, which is the first of a new class of drugs for a particular disease, will require more convincing evidence than a later, but still new and nonobvious, drug of the same class. The data required to show utility for human use[82] are generally of the amount and type that those skilled in the art would consider acceptable for extrapolation to *in vivo* human effectiveness. Clinical data, while not required, will generally be preferred and, where available, should normally be provided. For the PTO to require clinical testing in all cases of pharmaceutical inventions would have a decidedly detrimental effect on the industry and its research efforts and, most likely, would be inconsistent with the overall goals of the patent system.

3.2. Patentable Subject Matter Outside the United States

Patentable subject matter in many countries (especially for health-related and biotechnology inventions) is often significantly restricted as compared to the United States. Some countries do not allow plants or animals to be patented. The United States, Japan, and the European Patent Office currently allow genetically engineered animals to be protected by patent.[83] Many countries do not allow phar-

---

[78]See *Brenner v. Manson*, 383 U.S. 519, 528–536 (1966).
[79]See, for example, *In re Langer*, 503 F.2d 1380, 1391–1392 (C.C.P.A. 1974); *Newman v. Quigg*, 877 F.2d 1575, 1581 (Fed. Cir. 1989).
[80]See, for example, *In re Jolles*, 628 F.2d 1322, 1326–1327 (C.C.P.A. 1980); *Ex parte Balzarini*, 21 U.S.P.Q. 2d 1892, 1897 (B.P.A.I. 1991); *Ex parte Kranz*, 19 U.S.P.Q. 2d 1216, 1218–1219 (B.P.A.I. 1991).

[81]See generally *U.S. Pat. & Trademark Off., Manual of Patent Examining Procedure*, § 2107.3 (8th ed. 2001, latest revision 2008) available at http://uspto.gov.
[82]Where appropriate, the invention might be claimed to cover uses for mammals (including humans).
[83]Patents Throughout the World (Thomson/West eds. 2008).

maceuticals or methods of medical treatment to be patented. Each country will, of course, have its own specific limitations and exceptions for patentable subject matter.[84] It is not possible in the present chapter to provide an even limited discussion of such patentable subject matter. Moreover, any details provided could very well be out of date in a relatively short time. However, a few examples for selected countries are helpful to illustrate the variations in patentable subject matter:

(1) *Australia.* Medicines that are mixtures of known ingredients are generally not patentable. Human beings and biological processes for their generation are not patentable.[85]

(2) *Canada.* Unicellular life forms are patentable; higher life forms and methods of medical treatment are generally not patentable.[86]

(3) *France.* Surgical or therapeutic treatment of humans and animals or diagnostic methods for humans and animals are generally not patentable (however, the first use of a known composition or substance for carrying out these methods may be patentable); plants and animals and biological processes for producing them (except microbiological processes and products) are generally not patentable.[87]

(4) *Germany.* Plants, animals, and biological processes for producing plants or animals (except microbiological processes and products) are not patentable; surgical, therapeutic, and diagnostic methods applied to humans or animals are generally not patentable although the products used in these processes may be patentable.[88]

(5) *Italy.* Plants and animals as well as the methods of producing them (except microbiological processes) are generally not patentable; surgical, therapeutic, and diagnostic methods applied to humans and animals are generally not patentable, although compositions for carrying out these methods may be; pharmaceuticals are patentable, but the compounding of medicine in a pharmacy is not.[89]

(6) *Sweden.* Methods for treating humans and animals are generally not patentable, although products used in these methods may be patentable; plants and animals and processes (except microbiological processes and products) for producing them are generally not patentable.[90]

(7) *United Kingdom.* Animals and plants and nonmicrobiological processes for producing them are generally not patentable; surgical, therapeutic, and diagnostic methods for humans and animals are generally not patentable, although compositions for use in such methods may be patentable.[91]

(8) *Japan.* Generally similar to patentable subject matter in the United States.[92]

Applicants interested in seeking worldwide protection for inventions, especially for the drug discovery and biotechnology industries, must take these differences in patentable subject matter into account. Even in jurisdictions where specific classes of inventions cannot be patented, it is often possible to claim the invention in a manner so as to be within patentable subject matter. For example, many countries that do not allow drugs to be patented may allow claims directed to a method of making the drug. Likewise, many countries that do not allow patent claims directed at methods for medical treatment of humans

[84]For a description of patent law and procedures in countries throughout the world, see generally *id.*
[85]Australian Governemnt, IPAustralia, Patent manual of practice & procedures §§ 2.7.1, 2.9.7 (2006); Australian Governemnt, IPAustralia, Australian Patents for biological inventions, available at http://www.ipaustralia.gov.au.
[86]Canadian Intellectual Property Office, Manual of Patent Office Practice, §§ 12.04.01, 12.04.02 (March 2007), available at http://www.cipo.ic.gc.ca.
[87]Patents Throughout the World, above. § 60:8.
[88]*Id.* § 64:8.
[89]*Id.* § 84:7.
[90]*Id.* § 159:8.
[91]*Id.* § 176:7.
[92]*Id.* § 86:7.

may allow claims directed to devices or compositions for carrying out the treatment processes. Generally, locally accredited patent counsel in the relevant country is retained to aid in the prosecution of the application because of the need for knowledge and understanding of substantive and procedural patent law in the relevant country. One of the tasks of such local counsel is to adapt the legal definition of the invention to local law and practice, especially in regard to patentable subject matter.

3.3. Patent Specification

The specification is that part of a patent application in which the inventor describes and discloses his or her invention in detail. In the United States, the specification "shall contain a written description of the invention, and of the manner and process of making and using it, in such full, clear, concise, and exact terms as to enable any person skilled in the art to which it pertains, or with which it is most nearly connected, to make and use the same, and shall set forth the best mode contemplated by the inventor of carrying out his [or her] invention."[93] Furthermore, the specification "shall conclude with one or more claims particularly pointing out and distinctly claiming the subject matter the applicant regards as his [or her] invention."[94] Thus, the specification must meet four general requirements: (1) provide a written description; (2) provide sufficient detail to teach persons of ordinary skill how to make and use the invention; (3) reveal the best mode of making and using the invention known by the inventor at the time the application is filed; and (4) provide at least one claim covering the applicant's invention. Each of these requirements will be discussed in turn below. Almost without exception, the requirements for a legally sufficient specification in other countries are less stringent than in the United States. Thus, a patent specification that is legally sufficient in the United States is usually sufficient for filing in other countries with only relatively minor modification to conform it to specific regulations and practices of the relevant country.[95]

3.3.1. Written Description The claimed invention must be described in the specification. In other words, the claims must be "supported" by the specification as originally filed. Thus, the claims cannot encompass or contain more or different elements, steps, or compositions than are described in the specification. The claims cannot be amended or new claims added during prosecution of the patent application in the PTO unless the portion added is found within or supported by the specification as originally filed. The test for whether amended or new claims are supported is generally "whether one skilled in the art, familiar with the practice of the art at the time of the filing date, could reasonably have found the 'later' claimed invention in the specification as [originally] filed."[96]

Although new matter cannot be introduced into the specification or claims by amendment during prosecution,[97] not all additions to the specification constitute new matter. For example, an applicant can generally supplement the specification by relying on well-known principles, prior art, and "inherency."[98] New matter, however, can be added by filing a CIP application claiming a priority date of the earlier filed application for the subject matter disclosed in the original application. The effective filing date for the newly added subject matter (i.e., the new matter) is the actual filing date of the CIP application. Nonetheless, it is generally recommended that the specification as originally filed be as complete as possible.

Care should be taken in preparing the patent specification to consider fully the ramifications of the invention and the possibility of extending or expanding the scope of the invention. Thus, if a specific new drug is developed, consideration should be given to structural and functional analogs. For example, in

---

[93]35 U.S.C.A. § 112, ¶ 1 (West 2001).
[94]Id. ¶ 2.
[95]See, for example, Samson Helfgott, A 'Global' Patent Application, 74 J. Pat. & Trademark Off. Soc'y 26 (1992).
[96]Texas Instruments v. Int'l Trade Comm'n, 871 F.2d 1054, 1062 (Fed. Cir. 1989).
[97]35 U.S.C.A. § 132 (a) (West 2001).
[98]See, for example, In re Wands, 858 F.2d 731 (Fed. Cir. 1988); Kennecott Corp. v. Kyocera Int'l, Inc., 835 F.2d 1419, 1422 (Fed. Cir. 1987).

many cases, the effectiveness of a drug will not be significantly affected by replacing a methyl group with an alkyl group containing, for example, two to six carbon atoms. Such variations, if allowed by the technology and the prior art, can considerably expand the scope of the claimed invention and the scope of protection afforded by the patent.[99]

3.3.2. Enablement In the United States, the specification, as filed, must teach one of ordinary skill in the art how to make and use the invention without undue experimentation. Enablement is essentially what the inventor gives to the public in exchange for the exclusive right afforded by the patent. The public must be able to understand the invention based on the specification in order to build upon the invention and develop new technology. Furthermore, the public is entitled to a complete description of the invention and the manner of making and using it so that the public can practice the invention after the patent expires. A specification that fails to teach one of ordinary skill in the art how to make and use the invention is not legally sufficient.

The enablement requirement does not mandate that each and every detail of the invention be included or that the specification be in the form of a detailed "cookbook" with every step specified to the last detail.[100] Rather, the skilled artisan must be able to practice the invention without "undue experimentation." Generally, "a considerable amount of experimentation is permissible, if it is merely routine, or if the specification ... provides a reasonable amount of guidance with respect to the direction in which the experimentation should proceed."[101]

Although the acceptable amount of experimentation will vary from case to case, the factors normally considered by the PTO and the courts in determining the level of permissible experimentation include (1) quantity of experimentation required, (2) amount of direction or guidance provided by the specification, (3) presence or absence of working examples, (4) nature of the invention, (5) state of the prior art, (6) relative skill of workers in the art, (7) predictability or unpredictability of the art, and (8) breadth of the claims.[102] The level of acceptable experimentation will vary as the state of the art advances and the level of skill in the art increases.

Working examples are one factor in determining whether the specification is enabling. They are not required if the specification otherwise teaches one skilled in the art how to practice the invention without undue experimentation. However, in appropriate cases and if drafted with sufficient detail, such examples provide a relatively easy and straightforward way in which to make the specification enabling. So-called "paper examples" (i.e., examples describing work that have not actually been carried out[103]) can be used to satisfy the enablement requirement if the level of predictability for the art and invention is sufficiently high.

Pharmaceutical patents, especially those involving microorganisms or other biological material, can raise significant enablement issues. These issues arise whether the microorganism or other biological material is simply used in an invention sought to be patented or is itself the subject matter sought to be patented. If the microorganism is known and readily available or can be prepared using a procedure described in the specification, the enablement requirement is fulfilled. Otherwise the applicant may need to take additional steps to comply with the enablement requirement. In such case, this requirement can normally be satisfied by the deposit of a micro-

---

[99]Such variations may fall within the scope of a claim under the doctrine of equivalents. See Section 4.4. It is, however, easier to prove literal infringement. Moreover, an original claim of properly expanded scope may have its own scope expanded, based on the doctrine of equivalents.

[100]*In re Gay*, 309 F.2d 769, 774 (C.C.P.A. 1962) ("Not every last detail is to be described, else patent specifications would turn into production specifications, which they were never meant to be.").

[101]*Ex parte Jackson*, 217 U.S.P.Q. 804, 807 (B.P.A.I. 1982).

[102]*Ex parte Forman*, 230 U.S.P.Q. 546, 547 (B.P.A.I. 1986); *In re Wands*, 858 F.2d at 737.

[103]Past tense—which might imply that the work was actually carried out—should not be used for paper examples. It should be clear from the text of the example that the work described was not carried out.

organism or other biological material in an International Depositary Authority established under the Budapest Treaty on the International Recognition of the Deposit of Microorganisms for the Purposes of Patent Procedure or a depository recognized as being suitable by the PTO.[104] Generally such a deposit can be made at any time before filing the application or during the pendency of the application.[105] "The depository must afford a term of at least 30 years[106] and the deposit must be readily accessible by the public once a patent is granted."[107] The depositor will generally have a continuing responsibility to replace the deposit during the enforceable life of the patent should the deposit become nonviable or otherwise unavailable from the depository.[108]

The Budapest Treaty enables an applicant to make a deposit of a microorganism or other biological material in a single International Depositary Authority and thereby satisfy the enabling requirements of the signatory nations to the treaty (and, generally, nonsignatory countries as well).[109] Where a deposit is required and patent protection outside the United States will be sought, a deposit under the Budapest Treaty will generally be recommended in order to minimize the number of deposits required; in such cases, the deposit generally should be made before the priority date of the U.S. application since some signatory nations may require earlier deposit dates.[110] Deposits made under the Budapest Treaty are generally available to certain "certified parties" 18 months after the priority date (i.e., after the patent application is published).[111] For purposes of U.S. patent law, deposits can be made outside of the Budapest Treaty so long as the conditions required by the PTO are ensured. Under current U.S. law, deposits are generally not required to be released to the public until the patent issues.[112] Should the applicant later determine that foreign protection will be sought, a non-Budapest Treaty deposit can generally be converted to a Budapest Treaty deposit. The major depository in the United States is the American Type Culture Collection (ATCC) in Manassas, Virginia. The ATCC is an International Depositary Authority under the Budapest Treaty; the ATCC also accepts non-Budapest Treaty deposits.

Deposits can also provide a convenient mechanism whereby a patent owner can monitor individuals and organizations obtaining samples of the deposit during the life of the patent. Most depositories will provide the submitter with notice of sample requests (in some cases, a relatively small fee may be required). Such information can be helpful in enforcing patent rights. Especially in cases where the deposited material is patented, the patent holder may wish to consider a simple "letter license"

---

[104]37 C.F.R § 1.803(a) (2008).

[105]37 C.F.R. § 1.804 (2008). If the original deposit is made after the filing date, a statement must be submitted from a person in a position to corroborate the fact that the deposited material is the same as that identified in the application as filed. *Id.* The PTO has proposed amendments to the rules so as to require the deposit to be made before the filing date or during the pendency of the application but before the publication date of the patent application. 73 Fed. Reg. 9258 (2008).

[106]37 C.F.R. § 1.806 (2008).

[107]37 C.F.R. § 1.808 (2008). The PTO has proposed amendments to the rules so as to require all restrictions on the availability of the deposit to be irrevocably removed on the earlier of the publication date of the patent application or the issue date of the patent. 73 Fed. Reg. 9258 (2008).

[108]37 C.F.R. § 1.805 (2008).

[109]Budapest Treaty on the International Recognition of the Deposit of Microorganisms for the Purposes of Patent Procedure (April 28, 1977, amended September 26, 1980) [hereinafter "Budapest Treaty"], available at http://www.wipo.int/treaties/en/. As of January 27 2009, 72 nations (including, for example, the United States, Japan, Australia, and most European countries) are contracting states of the Budapest Treaty.

[110]See, for example, European Patent Convention, Implementing Regulations, Rule 31(1)(a) (adopted December 7 2006) (requiring biological deposit to be made "not later than the filing date of the application").

[111]The "certified parties" entitled to obtain a deposit under the Budapest Treaty are determined by the laws and regulations of the country or countries in which a patent application is filed. See Budapest Treaty, Regulations, Rule 11 (adopted April 28 1977, amended on January 20 1981 and October 1 2002), available at http://www.wipo.int/treaties/int.

[112]37 C.F.R. § 1.805 (2008).

allowing the requester to use the deposited material only in a certain manner (e.g., research purposes only) and requiring the requester to report any commercial uses of, or derived from, the deposited material.

3.3.3. Best Mode

The specification must "set forth the best mode contemplated by the inventor of carrying out his [or her] invention."[113] The inventor cannot hide or conceal the best physical mode of making or using the invention. The purpose of this requirement "is to restrain inventors from applying for patents while at the same time concealing from the public preferred embodiments of their inventions."[114] Questions of failure to meet the best mode requirements are rarely raised during prosecution of a patent in the PTO. Best mode issues are more often raised during interference proceedings or during litigation to enforce the patent.

The inventor must disclose the best mode known to him or her at the time the application was filed. If the inventor was aware of a better mode at the time of filing the application but did not disclose it, the entire patent is invalid and no claims are enforceable. However, a better mode of carrying out the invention discovered after the filing date need not be disclosed. Inventors should be carefully advised about the importance of the best mode requirement and of the necessity of informing patent counsel of any improvements made or considered before the actual filing date. Of particular concern is the time period between preparation of the application and its filing date; any potential improvements made during this period must be carefully evaluated to determine whether they must be included in the specification to satisfy the best mode requirement. Any improvements made after the application is filed—especially those made shortly after the filing date—should be carefully documented in case it ever becomes necessary to prove they were made after the filing date.

Generally, the best mode requirement focuses on the inventor's state of mind at the time the application was filed. It is the best mode contemplated by the inventor, not anyone else, for carrying out the invention. It is generally immaterial whether the failure to include the best mode in the specification was intentional or accidental. The subjective test is whether the inventor knew of a better mode at the time of the filing date and, if so, whether it is adequately disclosed in the specification.[115] In some cases, deposit of biological materials may also be required to satisfy the best mode requirement. It is possible to include the best mode in the specification but to describe the best mode so poorly as to effectively conceal it. Because the potential penalty for failure to meet the best mode requirement is invalidity of the entire patent, considerable care should be taken to identify the best mode contemplated by the inventor and to describe it carefully and fully.

3.3.4. Claims

The specification must "conclude with one or more claims particularly pointing out and distinctly claiming the subject matter which the applicant regards as his [or her] invention."[116] Claims are the numbered paragraphs (each only one sentence long) at the end of the patent specification. The claims define the metes and bounds of the exclusive right granted by the patent. Each claim defines a separate right to exclude others from making, using, offering for sale, or selling embodiments within the scope of the specific claim.

The claims are critically important. They require careful drafting, preferably by highly qualified and experienced patent counsel, to cover the invention as broadly and comprehensively as appropriate. Normally, the claims should be drafted as broadly as the prior art and the specification allow. Allowable claim scope will, of course, depend in large part on the state of the art. Inventions in a crowded art (i.e., technology with a relatively large amount of closely related prior art) can generally only be claimed relatively narrowly. Inventions for which there is relatively little prior art can be claimed more broadly.

---

[113] 35 U.S.C.A. § 112 ¶ 1 (West 2001).
[114] *DeGeorge v. Bernier*, 768 F.2d 1318, 1324 (Fed. Cir. 1985), (citing *In re Gay*, 309 F.2d at 772).
[115] *Spectra-Physics, Inc. v. Coherent, Inc.*, 827 F.2d 1524, 1536 (Fed. Cir. 1987).
[116] 35 U.S.C.A. § 112 ¶ 2 (West 2001).

Generally, one attempts to draft one or more independent claims that describe the invention as broadly as the prior art will allow and then narrower, dependent claims specific to preferred embodiments, with dependent claims of intermediate scope in between. A dependent claim refers to an earlier claim and incorporates by reference all the limitations of the earlier claim and adds additional limitations. Thus, the scope of the claimed invention generally varies from the broader independent claim or claims to the narrower, more restricted dependent claims. Should a court later find, for example, that the broader claims are obvious over the prior art, the narrower claims, with added limitations or elements, may still be valid and enforceable.

Normally, it is also preferred that the invention be claimed so as to fit into as many of the statutory classifications as possible. For example, for a newly discovered drug, one might attempt to claim a method of making the drug, the drug itself, formulations containing the drug, and a method of using the drug to treat one or more conditions. For a new diagnostic, one might attempt to claim a method of making the diagnostic, the diagnostic itself, a method of using the diagnostic, and a kit containing the diagnostic. Other aspects of claims are discussed below.

3.4. Provisional Applications

Since mid-1995, inventors have been able to file a provisional application in the PTO. Provisional applications may offer significant advantages to at least some patent applicants. However, if not used carefully, the provisional application may, in some instances, lead to the loss of significant patent rights.

The provisional application provides a domestic priority document and thereby "levels the playing field" for domestic patent applicants relative to their non-U.S. counterparts. Applicants who have filed previously in countries outside the United States have been able to, and still can under the current statute, use their earlier filed foreign patent application to secure a priority date for the United States application (if the U.S. application is filed within 1 year of the first-filed foreign application).[117] A domestic patent applicant applying for a U.S. patent could not, prior to the advent of the provisional application, obtain a priority date earlier than the actual filing date of his or her first-filed U.S. application. Any applicant (domestic or foreign) may file a provisional application in the United States.[118] If the regular application (i.e., nonprovisional application) is filed within 12 months of the provisional application and claims benefit of the provisional application, the filing date for the provisional application can be used as the priority date for the regular application.[119] A provisional application can provide an earlier priority date for a domestic applicant just as an earlier filed foreign application can for a foreign applicant.

The filing of the provisional application does not trigger the 20-year patent term. Rather, only the later filed regular application starts the 20-year period.[120] Thus, by filing a provisional application, followed within 12 months by a regular application claiming priority from the provisional application, an applicant can effectively obtain a patent term of 21 years (as measured from the date of the provisional application).[121]

The priority date of the provisional application provides protection for the regular application against prior art dated after the provisional filing date and/or statutory bar events occurring after the provisional filing date.[122] Moreover, the provisional filing date can also be used in interference proceedings to obtain procedurally advantageous senior status and/or to prove an earlier invention date.

The provisional application can be especially useful in cases where an invention is undergoing active development. This use of provisional applications is illustrated in Fig. 2.

The top portion of Fig. 2 demonstrates a case where the invention is fully developed at the time the first provisional application

---

[117]35 U.S.C.A. § 119 (a) (West Supp. 2008).
[118]35 U.S.C.A. § 111(b) (West 2001).
[119]35 U.S.C.A. § 119(e)(1) (West 2001).
[120]35 U.S.C.A. § 154(a)(3) (West 2001).
[121]Although the ultimate patent cannot be enforced until the actual issue date, the provisional application allows the applicant to use the "patent pending" label on marketed products or promotional materials.
[122]35 U.S.C.A. §§ 102, 119(e)(1) (West 2001 & Supp. 2008).

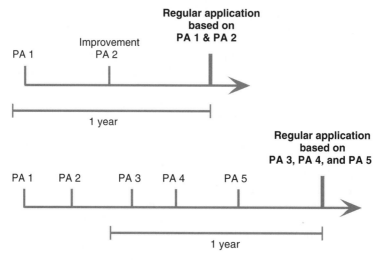

Figure 2. Multiple provisional applications.

1 (PA 1) is filed but a significant improvement is developed within the 1-year period. The significant improvement is then included in PA 2. When the regular application is filed within 1 year of PA 1, the benefit of both PA 1 and PA 2 are claimed. In this manner, both the basic invention (PA 1) and the improvement (PA 2) have the earliest possible priority dates.[123]

The bottom portion of Fig. 2 demonstrates a case where the invention is developed over a period greater than 1 year. At each point of a significant advancement, a provisional application is filed (i.e., PA 1 through PA 5). Close to the end of the 1-year period, as measured from PA 1, one should consider whether the regular application should be filed at that time. Generally, if at the 1-year anniversary it is determined that the earliest provisional application filed was fully enabled (and, thus, provides the benefit of an earlier filing date), it is recommended (and probably required if foreign applications are to be filed) that a regular application be filed and that regular application claim benefit of all earlier provisional applications. In that case, the regular application would be filed shortly after PA 5 (or perhaps in place of it) in the bottom portion of Fig. 2 and the benefit of all earlier provisional applications should be claimed.

In some cases, however, one may not be ready to file the regular application at the 1-year anniversary of PA 1. For example, the invention may still be undergoing development and not be at a stage where a decision can be made as to whether to maintain the invention as a trade secret. Or it may be determined that the earliest filed provisional applications were not enabled[124] and will not, therefore, provide the benefit of an earlier

---

[123]Of course, one could file the regular application containing the improvement at either the time of PA 2 in Fig. 2 or at the 1-year date. If filed at the time of PA 2, the 20-year term would begin at that date and, thus, the effective patent term would be reduced. If filed at the end of the 1-year period (and assuming PA 2 was not separately filed), the improvement would effectively lose its early priority date and may, therefore, be exposed to additional prior art during examination.

[124]As detailed in Section 3.4.1, a provisional application must meet the requirements of 35 U.S.C. § 112 by providing a written description that "enables" one of the ordinary skills in the art to make and use the invention.

priority date.[125] Or the development may take an unexpected turn, or the actual invention may have changed so dramatically, that the earliest provisional applications are simply not relevant to the invention disclosed and claimed in the regular application. The regular application can be filed at any time during the development process and claim the benefit of all provisional applications filed within the preceding year. In the example illustrated, the regular application would claim benefit of PA 3 through PA 5. The earlier provisional applications (PA 1 and PA 2) would have become abandoned at their respective 1-year anniversaries and would remain secret.[126] If the invention were fully developed earlier, one or more of these earlier provisional applications could have been claimed, if it fell within the 1-year period. In this manner, the earliest priority date can be obtained for each significant development in the inventive process. Of course, if the development is complete within 1 year of the first provisional application, then all provisional applications should be claimed when the regular application is filed.

3.4.1. Statutory Requirements The statutory requirements for a provisional application are somewhat less stringent than for a regular application. A provisional application, however, must fully comply with the requirements of the first paragraph of 35 U.S.C. § 112 in order to provide a priority date.[127] That is, the provisional application must include a written description of the invention in sufficient detail to allow or enable one of ordinary skill in the art to make and use the invention and it must include the "best mode" known to the inventors as of the provisional application filing date. Drawings, if required by the nature of the invention, must also be included. In addition, the provisional application must include a cover sheet identifying it as a provisional application, a list of the "inventors" of the disclosed subject matter, and payment of the required filing fee. Provisional application filing fees are relatively modest.[128]

Provisional applications do not require patent claims[129] or a declaration signed by the named "inventors."[130] Each named inventor must, however, have made a "contribution" to the subject matter disclosed in the provisional application.[131] A provisional application cannot claim the benefit of any earlier application (foreign or domestic).[132] In some cases, a regular application may be converted to a provi-

---

[125]Such a case could arise, for example, during gene sequencing. The initial sequencing data could provide the basis for an early provisional application with subsequent sequencing data being included in later provisional applications. Although it is hoped that the early sequence data would enable the gene and thus provide an early priority date, it may be determined later (before the 1-year anniversary) that the early data was not sufficient for enablement purposes. Thus, one could postpone the filing of the regular application until after the 1-year anniversary of the first provisional application since it would not provide any added benefit.

[126]This same strategy can also be used where there is some question as to whether the earlier provisional applications are fully enabled. Thus, if there is some concern whether PA 1 is enabled, claiming benefit of a series of provisional applications (the later ones which are assumed to be fully enabled) allows the earliest priority date to be obtained for the regular application.

[127]35 U.S.C.A. § 111(b)(1) (West 2001).
[128]Currently, the provisional filing fee is $220 for a large entity and $110 for a small entity; there is an additional charge of $270 for a large entity and $135 for a small entity for each 50 pages in excess of 100. Current fee structure for the PTO is available at http://www.uspto.gov/main/howtofees.htm.
[129]35 U.S.C.A. § 111(b)(2) (West 2001).
[130]35 U.S.C.A. § 111(b)(8) (West 2001).
[131]37 C.F.R. § 1.45(c) (2008).
[132]35 U.S.C.A. § 111(b)(7) (West 2001).

sional application.[133] The provisional application is not examined in the PTO and is maintained in secrecy within the PTO. One year after its filing date, the provisional application is automatically abandoned. A regular application must be filed within 1 year of the provisional application in order to claim benefit of the provisional application's filing date.[134] In other words, the applicant has 1 year in which to decide whether to go forward with a regular application if the benefit of the provisional application priority date is desired. Should the regular application not be filed, the provisional application will not be published. Thus, in appropriate cases and assuming corresponding foreign applications have not been filed, the applicant might maintain the invention disclosed in the abandoned provisional application as a trade secret. The regular application could also be filed after the abandonment (assuming no statutory bars) although the benefit of the provisional application filing date could not be claimed.

The provisional application can effectively give an applicant 2 years after a statutory bar (e.g., public use or sale of the invention) in which to decide whether to file a regular application. A statutory bar can be avoided if the provisional application is filed within 1 year of the statutory bar event.[135] If the regular application claiming benefit of the provisional application is filed within 1 year of the provisional application, it is entitled to benefit of the provisional application's filing date. Thus, the regular application could be filed up to 2 years after the otherwise disqualifying event.

3.4.2. Best Mode The provisional application must disclose the best mode known to the inventors as of the filing date.[136] If the regular application has a "substantive content" different from the provisional application, the best mode must be updated as of the time of the filing of the regular application.[137] Thus, in such cases, any improvements (if they form part of the best mode) developed between the filing dates of the provisional and the regular applications must be disclosed in the regular application.

In some cases, the use of a provisional application may allow an applicant to seek patent protection on a basic invention and to keep a later-developed improvement as a trade secret. If the regular application has the same "substantive content" as the provisional application, the PTO has acknowledged that updating the best mode may not be required.[138] Thus, by filing a "complete" provisional application (with a full set of claims),[139] an applicant appears to have the option of keeping later-developed improvements as trade secrets or including (and protecting) them in the later-filed regular application. A regular application that has the same "substantive content" as the provisional ap-

---

[133] 35 U.S.C.A. § 111(b)(6) (West 2001). A regular application can be converted to a provisional application within the first year; the effective date of the provisional application would be the original filing date of the regular application. If a regular application is filed (before the end of the original 1-year period), the applicant could take advantage of the delay in calculating the beginning of the 20-year term. Some have suggested that this procedure could be used to obtain a preview of the examiner's views of the invention by filing a regular application, obtaining the first office action, and then converting the regular application to a provisional application, and refiling the regular application (perhaps modified to reflect the results of the first office action). Prosecution on the second-filed regular application could then begin anew. This method relies upon receiving the first office action within 1 year of the regular application filing date. In most cases, no office action will be received within the first year, thereby rendering this method ineffective. Moreover, if such a practice became widespread, the PTO (or examiners on their own) might choose to delay most first office actions until at least 1 year after the filing date.
[134] 35 U.S.C.A. § 119(e)(1) (West 2001).
[135] 35 U.S.C.A. § 102 (West 2001 & Supp. 2008).
[136] 35 U.S.C.A. § 111(b)(1)(A) (West 2001).
[137] 60 Fed. Reg. 20195, 20209 (1995).
[138] *U.S. Pat. & Trademark Off., Questions and Answers Regarding the GATT Uruguay Round and NAFTA Changes to U.S. Patent Law and Practice*, (February 23 1995), available at http://www.uspto.gov/web/offices/com/doc/uruguay/QA.html.
[139] A "complete" provisional application is intended to mean an application that could have been filed, without changes, as a regular application.

plication would be analogous to a regular continuation application in which the best mode need not be updated.[140] The Court of Appeals for the Federal Circuit has not specifically addressed this issue. Because of this uncertainty, we still recommend in most cases that the regular application be updated to include any later-developed best mode and to claim the improvements.

3.4.3. Concerns The requirements for a provisional application appear deceptively simple. Indeed, the PTO initially urged that the provisional application would provide a quick and inexpensive method for inventors to obtain a priority date since claims were not required and, supposedly, the written description need not be as detailed as a regular application.[141] Almost immediately, however, it was recognized that the use of the provisional application might not be as easy as first suggested.[142] Nonetheless, it is also evident that the new provisional application offers significant benefits to the patent applicant. It establishes an early priority date, effectively tacks an extra year onto the 20-year patent term, has a relatively low filing fee, provides an easy way to avoid filing at least some incomplete regular applications (e.g., one without a signed declaration[143]), and, in at least some cases, may allow an applicant to keep later-developed improvements as trade secrets.

Failure to fully understand and appreciate the requirements and limitations of the provisional application can, at least in some cases, result in loss of valuable patent rights. As noted above, the provisional application disclosure must adequately enable the invention under 35 U.S.C. § 112. Only material adequately disclosed in the provisional application should be entitled to the priority date of the provisional application. The initial provisional application should, therefore, be as complete as possible. Material added in the regular application to "fill out" or complete the disclosure of the provisional application may not be able to rely on the date of the provisional application. Claims in the regular application supported only by the disclosure added in the regular application may, therefore, be unpatentable over prior art dated between the priority date and the filing date of the regular application or because of statutory bars the provisional application was thought to overcome. Thus, filing a "barebones" provisional application is generally not recommended.

A "barebones" provisional application may be useful when it is necessary to prepare and file a provisional application as quickly as possible—for example, to avoid an imminent statutory bar; even then, however, as complete a provisional application as time permits would still be preferred. Such a "barebones" provisional application might also be used to advantage in relatively simple mechanical inventions where "a picture may be worth a thousand words." To be effective in establishing a priority date, however, the "barebones" provisional application must still enable one of ordinary skill in the art to make or use the invention and must contain the best mode known to the inventors as of its filing date.

Enablement type issues may not arise during prosecution of the patent application; thus, an applicant may obtain a patent based on the provisional application's priority date. But during enforcement of the patent, a potential infringer will almost certainly argue that the relevant claims are not supported by the provisional application disclosure and are, therefore, not entitled to its filing date. An applicant with an otherwise strong and valuable patent could find that his or her patent is rendered valueless because of an attempt to save money by filing the "barebones" provisional application. Even if the potential in-

---

[140]*Transco Prods. Inc. v. Performance Contracting, Inc.*, 38 F.3d. 551, 557–559 (Fed. Cir. 1994).

[141]59 Fed. Reg. 63951, 63952 (1994); 60 Fed. Reg. 20195, 20196 (1995).

[142]See, for example, 60 Fed. Reg. 20195, 20201–20202 (1995).

[143]A regular application can be filed without signatures of the inventors. The PTO then issues a Notice to File Missing Parts requiring a fee and a signed declaration to be filed within a certain time period (normally 1–2 months with extensions of time possible). In the meantime, however, the 20-year period has already begun. A provisional application can be filed without signatures. If desired, the regular application could be filed as soon as the inventor's signature on the declaration is obtained or could be delayed until close to the 1-year period.

fringer is not successful, this type of defense may cause the patent holder to expend considerable time and expense during litigation to counter these arguments; from the patent holder's perspective, these resources may be better spent on other issues in the litigation.

Another potential problem is related to the claims, or lack thereof, in provisional applications. Claims are not required in the provisional application. Claim drafting is critical to patent protection and can be an expensive part of the preparation of the patent application. An applicant may wish to avoid claims in the provisional application in order to save money (or at least postpone such expenditures), especially if he or she can personally draft the rest of the application. But such an approach can be, we believe, hazardous. First, most patent professionals drafting an application begin with the claims (or at least the broad independent claims). The claims help in more fully understanding the invention and guide the drafting of the written description of the invention. Once drafted, the claims provide a framework for the specification and help to ensure that all claimed features of the invention are fully and adequately described in the application. A provisional application drafted without carefully considering the claims runs the significant risk of not providing adequate support to the later-drafted claims, which the applicant ultimately wishes to include in the later-filed regular application.

Lack of claims in the provisional application may also cause difficulty in naming the inventors. The "inventors" named in a provisional application are those who "contributed" to the disclosure. At this time, it is not clear who these "contributors" may actually be. It is possible that some so-called "contributors" may not be inventors of later-drafted claims. Indeed, individuals not even associated with the invention or the assignee (e.g., competitors who have published articles or issued patents that are incorporated into the provisional application disclosure by reference) could be considered as "contributors" and, thus according to the statute, required to be named as "inventors."[144] Although these individuals (as well as other "contributors" and noninventors of the claimed invention) should not be named as inventors in the regular application, their inclusion on the provisional application may cause significant problems. For example, an individual from outside the organization named on the provisional application would have no incentive to assign the provisional application to the organization, thereby making sale or transfer of the technology more difficult. An employee named as an inventor in a provisional application but then not listed as an inventor on the regular application can also present problems. If that employee has left the company after the filing of the provisional application (perhaps under strained circumstances), he or she (or a later potential infringer) may argue that the individual was removed only because of their departure and, thus, the removal was improper. Even if there are no legal challenges, employee morale could suffer when individuals are first listed and then removed as inventors.

3.4.4. Recommendations There appears to be at least one approach for utilizing provisional applications that avoids, or at least minimizes, these potential problems. Applicants using provisional applications should, in our opinion, file provisional applications having as complete a disclosure as possible, even to the extent of including claims. Ideally, this provisional application should be so complete that it can be filed without any changes as the regular application within the 1-year period. If the regular application is a virtual copy of the provisional application, a potential infringer cannot reasonably argue that the claims were not supported in the provisional application and, therefore, not entitled to the priority date (at least to any greater extent than the regular application). Moreover, including claims in the provisional application may simplify the naming of inventors. It seems reasonable in such a case that the named "contributors" of the provisional application could be selected

---

[144] 60 Fed. Reg. 20195, 20222 (1995). Of course, carried to the extreme, a "contributor" could be anyone who helped develop the background area of the invention. It seems very unlikely that Congress intended all such "contributors" to be named on a provisional application. The issue becomes where to draw the line in naming inventors.

based only on the claimed invention (rather than the much broader material disclosed). Thus, by including claims, the inventors of the provisional application could be selected by the same standards as inventors for the regular application. The Federal Circuit must, of course, ultimately decide whether such an approach is acceptable.

It is important to appreciate that the filing of a provisional application also starts the convention year under the Paris Convention for foreign applications. Thus, an applicant who files a regular application claiming benefit of a provisional application must also file the appropriate foreign applications within that same 1-year period. An applicant using a provisional application as a priority document must be alert so that his or her foreign applications are filed before the end of the convention year. Failure to file the appropriate foreign applications within the 1-year period will cause the loss of the priority date for any foreign applications and, perhaps, the loss of foreign patent rights. Docketing systems should be modified to incorporate provisional applications and their effects on foreign filing and other dates.

3.4.5. Applicants from Countries Other Than the United States Although provisional applications are designed to provide a priority document for domestic applicants, foreign applicants may use provisional applications to their advantage. A U.S. patent claiming benefit of an earlier filed foreign application has a prior art effect against other applicants only as of its U.S. filing date, not as of its foreign priority date. In other words, the foreign priority date provided a "shield" but not a "sword" against prior art for the foreign applicant.[145] It is possible (but until the Federal Circuit has spoken, not certain) that the provisional application can be used by foreign applicants to create an earlier patent-defeating date (i.e., the filing date of the provisional application). A foreign applicant could file a provisional application in the United States as quickly as possible (based on the export regulations of the relevant country) after filing his or her foreign application. Although the provisional application cannot claim the benefit of the foreign application, the regular application (filed within the convention year) could claim benefit of both the foreign application and the provisional application. The effective date of the regular application against prior art (and statutory bars) would be the foreign priority date; the effective date as an offensive "sword" against later filed U.S. applications (i.e., as prior art) would likely be the filing date of the provisional application. Thus, for a relatively low cost, a foreign applicant may be able to significantly increase the effect of his or her application as prior art against other patent applicants in the United States.

3.5. Invention Must Be New and Nonobvious

In the United States, an invention to be patentable must be both new[146] and nonobvious[147] over the prior art. Generally, the prior art is the body of existing technological information (i.e., the state of the art) against which the invention is evaluated. Prior art may include other patents from anywhere in the world, printed publications or printed patent applications from anywhere in the world, U.S. patent applications that eventually issue as patents, public use or offer for sale in the United States of the subject matter embodying the invention, and, depending on the circumstances, unpublished and unpatented research activities of others in the United States. These types of prior art are defined in § 102 of the Patent Statute.

Issued U.S. patents are effective as prior art as of their filing dates.[148] Patents from other countries, as well as publications from anywhere in the world, are effective as prior art in the United States as of their actual publication dates. Admissions by the inventor as to the content or status of the prior art, even if later shown to be incorrect, are also part of the prior art for that invention. Care should be taken, therefore, to insure that admissions against interest are not made in the patent

---

[145]*In re Hilmer,* 424 F.2d 1108 (C.C.P.A. 1970); *In re Hilmer,* 359 F.2d 859, 149 U.S.P.Q. 480 (C.C.P.A. 1966).

[146]35 U.S.C.A. § 102 (West 2001 & Supp. 2008).
[147]35 U.S.C.A. § 103 (West 2001 & Supp. 2008).
[148]If a U.S. patent is never issued by the PTO, the U.S. patent publication would still become prior art as of its actual publication date.

application or during prosecution of the application before the PTO. Unless it is absolutely clear that a given document qualifies legally as prior art against the invention, it should not be referred to as "prior art" or otherwise admitted to be "prior art."

3.5.1. Novelty under 35 U.S.C. § 102

The determination of novelty of an invention is generally straightforward: the issue is whether the invention is old or new. If a single prior art reference shows identically every element of the claimed invention (i.e., anticipation), the invention is not novel and, therefore, not patentable. The anticipating reference must, however, be enabling. In other words, it must teach one of ordinary skill in the art how to make and practice the invention. Thus, the inclusion of only the name or chemical structure of a compound in a reference, without providing a method of making the compound, will generally not be considered to have anticipated a later patent claiming that compound.[149] On the other hand, an anticipating prior art reference is not required to provide a use or utility for a compound.[150] Thus, a reference that teaches how to make a specific compound but does not disclose a use will still prevent a later inventor, who discovers a use, from claiming the compound itself.[151] The later inventor may be able to claim the use or a process taking advantage of a newly discovered property of the compound, if such use or process is new and nonobvious. In some cases, especially where new uses have been discovered, the effective amount required or the mode of administration of, for example, a drug may provide sufficient novelty for patentability.

An invention may also not be patentable under § 102 if certain events (so-called statutory bars) occur more than 1 year before the patent application is filed in the PTO. The statutory bars are designed to encourage the inventor to file his or her patent application in a timely manner. For example, a written description of the invention, a public use, or offer of sale of the subject matter of the invention in the United States more than 1 year before the filing date of the application bars the invention from being patented. An inventor has, therefore, a 1-year grace period after such a public disclosure in which to file a patent application in the United States.

Except for the United States, most other countries require absolute novelty. A written description, public use, or offer of sale that actually disclose the invention would, in most cases, prevent the inventor from obtaining patent protection outside of the United States. If worldwide patent protection is important, as is likely in almost all drug-related inventions, one should file a patent application covering the invention before public disclosure. Potential public disclosures should be carefully screened and, where appropriate, controlled to insure that valuable patent rights are not lost.

3.5.2. Obviousness Under 35 U.S.C. § 103

Even if the invention is novel under § 102, the invention may not be patentable under § 103 if "the differences between the subject matter sought to be patented and the prior art are such that the subject matter as a whole would have been obvious at the time the invention was made to a person having ordinary skill in the art to which the subject matter pertains."[152] The determination of obviousness under § 103 is considerably more difficult than the determination of novelty under § 102. Obviousness is determined using a three-part inquiry (the so-called *Graham v. John Deere Co.* test): (1) a determination of the scope and content of the prior art, (2) a determination of the differences between the prior art and the claimed invention, and (3) a determination of the level of ordinary skill in the art.[153] Additionally, so-called secondary considerations or "objective indices of nonobviousness" can often be used in determining whether an invention is obvious.[154] Secondary considerations

---

[149]If the method of making the compound is clearly within the skill in the art, such a reference could place the compound within the public domain. In such a case, the reference could anticipate a claim to the compound.
[150]*In re Schoenwald*, 964 F.2d 1122 (Fed. Cir. 1992).
[151]*In re Spada*, 911 F.2d 705 (Fed. Cir. 1990).
[152]35 U.S.C.A. § 103(a) (West 2001).
[153]*Graham v. John Deere Co.*, 383 U.S. 1, 17 (1966).
[154]*Id.* at 17–18.

that can be used to support nonobviousness and patentability include, for example, commercial success, a long-standing problem that resisted solution until the invention, failure of others to solve the problem, unexpected results, copying of the invention by others in the art, and initial skepticism in the art concerning the success or value of the invention.[155] Such secondary considerations provide evidence about how others in the art viewed the advance provided by the claimed invention and are often used in arguments presented to the PTO by the applicant in support of patentability. To support patentability, there must be a direct relationship between the secondary factor and the merits of the claimed invention. Therefore, to be useful in demonstrating nonobviousness, commercial success should be predominately the result of the benefits and merits of the invention rather than, for example, advertising or marketing considerations.[156] Independent and essentially simultaneous development of the invention by others in the art to solve the same or very similar problem is one secondary consideration that supports the conclusion that the invention may be obvious.

Until recently, the law concerning the obviousness determination was fairly well established. In 1986, the Court of Appeals for the Federal Circuit offered the following guidelines with regard to this three-part *Graham v. John Deere Co.* determination:

> The following tenets of patent law ... must be adhered to when applying § 103: (1) the claimed invention must be considered as a whole ... [because] though the difference between [the] claimed invention and prior art may seem slight, it may also have been the key to advancement of the art...; (2) the references must be considered as a whole and suggest the desirability and thus the obviousness of making the combination; (3) the references must be viewed without the benefit of hindsight vision afforded by the claimed invention; [and] (4) "ought to be tried" is not the standard with which obviousness is determined ...[157]

Based largely on this rationale and to provide more uniform decision making, the Federal Circuit developed and used a "teaching, suggestion, or motivation" (TSM) test that required the references to provide some suggestion or motivation to combine the prior art teachings in order to show obviousness under § 103.

In 2007, however, a unanimous Supreme Court in *KSR International Co. v. Teleflex Inc.*, held that the Federal Circuit applied the TSM test in a narrow, rigid manner that is inconsistent with § 103.[158] The Supreme Court found that the Federal Circuit had committed four specific errors in its analysis:

(1) "courts and patent examiners should look only to the problem the patentee was trying to solve";

(2) "assuming that a person of ordinary skill in the art attempting to solve a problem will be led only to those prior art elements designed to solve the same problem";

(3) "concluding that a patent claim cannot be proved obvious merely by showing that the combination of elements was obvious to try"; and

(4) "drew the wrong conclusion from the risk of courts and patent examiners falling prey to hindsight bias."

The Supreme Court added that "rigid preventative rules that deny recourse to common sense are neither necessary under nor consistent with [the Supreme] Court's case law."[159]

Prior to *KSR*, the relevant and applicable prior art for determining obviousness was normally found in the general technical field (or fields) to which the claimed invention was directed as well as analogous fields to which one of ordinary skill in that field or fields would reasonably turn or use to solve the problem. The Supreme Court concluded that "under the correct analysis, any need or problem known in the field and addressed by the

---

[155]*Id.* at 18.
[156]*Pentec, Inc. v. Graphic Controls Corp.*, 776 F.2d 309, 315–316 (Fed. Cir. 1985).
[157]*Hodosh v. Block Drug Co.*, 786 F.2d 1136, 1143 n.5 (Fed. Cir. 1986) (citations omitted).

[158]550 U.S. 398 (2007).
[159]*Id.* at 1742–1743.

patent can provide a reason for combining the elements in the manner claimed." This apparently opens the door for consideration of a potentially much wider range of prior art in determining obviousness.

The Supreme Court also noted that "it is common sense that familiar items may have obvious uses beyond their primary purposes, and a person of ordinary skill often will be able to fit the teachings of multiple patents together like pieces of a puzzle."[160] With regard to the use of the "obvious to try" standard, the Supreme Court noted that "[w]hen there is a design need or market pressure to solve a problem and there are a finite number of identified, predictable solutions, a person of ordinary skill in the art has good reason to pursue the known options within his or her technical grasp. If this leads to the anticipated success, it is likely the product not of innovation but of ordinary skill and common sense."[161]

KSR appears to forbid only the "rigid application" of preventive rules such as the TSM test in a manner that they "deny recourse to common sense." Indeed, the Court specifically notes that the TSM test captures a helpful insight:

> A patent composed of several elements is not proved obvious merely by demonstrating that each element was, independently, known in the prior art. Although common sense directs caution as to a patent application claiming as innovation the combination of two known devices according to their established functions, it can be important to identify a reason that would have prompted a person of ordinary skill in the art to combine the elements as the new invention does.[162]

Thus, it appears that although the motivation to combine such elements is still required for a finding of obviousness, it is not required to be found in the references themselves; such motivation can be found using "common sense."

The factors used to evaluate the level of ordinary skill, both before and after KSR, generally include (1) educational level of the inventor, (2) nature of problems generally encountered in the art, (3) solutions provided by the prior art for these problems, (4) rate of innovation in the art, (5) level of sophistication of the art, and (6) educational level of active workers in the art.[163] These factors, when applied to the drug discovery and development field, suggest a very high level of ordinary skill in this art.

Against this factual background of the three-pronged test as redefined or modified in KSR (and secondary considerations (if any)), the obviousness or nonobviousness of the invention is determined. The decision maker must step back in time and determine whether a person of ordinary skill would have found the invention as a whole obvious at the time the invention was made. But now the decision maker may use a broader range of prior art than prior to KSR, may use "common sense" to provide the motivation for combining the elements, and may use an "obvious to try" standard in assessing obviousness under § 103. Although not addressed by the Court in KSR, it still appears that a claim should not be rejected using "an improper amount" of hindsight to reconstruct the invention from the prior art only using applicant's own patent specification.[164] Just how much hindsight can be used by the examiner remains to be seen. Although much remains open regarding §103 determination, it appears clear that KSR allows an examiner wider latitude in rejecting claims.

In the PTO, obviousness rejections can be based on a single prior art reference or on the combination of two or more prior art references. Once a case of *prima facie* obviousness is made out by the PTO, the burden shifts to the applicant to rebut obviousness by showing that the invention had unexpected or surprising results. Unexpected results for a drug or

---

[160]Id. at 1742. "Common sense" is not defined; nor is the person supposedly having such "common sense" identified.
[161]Id. The Court did not define how many of "a finite number of identified, predictable solutions" that might be the basis of an "obvious to try" rejection.
[162]Id. at 1741.

[163]Envtl. Designs, Ltd. v. Union Oil Co., 713 F.2d 693, 696–697 (Fed. Cir. 1983).
[164]See, for example, Graham v. John Deere Co., 383 U.S. at 36 (courts should "resist the temptation to read into the prior art the teachings of the invention in issue"); Uniroyal, Inc. v. Rudkin-Wiley Corp., 837 F.2d 1044, 1051 (Fed. Cir. 1988); W.L. Gore & Assoc. v. Garlock, Inc., 721 F.2d 1540, 1553 (Fed. Cir. 1983).

drug formulation can include properties superior to those of prior art compounds including, for example, greater pharmaceutical activity than would have been predicted from the prior art, greater effectiveness, reduced toxicity or side effects, effectiveness at lower dosage, greater site-specific activity, and so on, as well as properties that the prior art compounds do not have. The unexpected results can be included in the patent application itself or presented by affidavit or declaration during prosecution of the patent application. If such unexpected results cannot be shown in cases of *prima facie* obviousness, the invention is not patentable. On the other hand, if a case of *prima facie* obviousness is not made out, it is not necessary to offer rebuttal evidence in the form of unexpected or surprising results for patentability. After *KSR*, the need for applicants to show unexpected or surprising results will likely become even more important.

Composition of Matter Claims. Prior art compounds that are structurally similar to the claimed compound or drug may render the claimed compound obvious and therefore unpatentable. But "[a]n assumed similarity based on a comparison of formulae must give way to evidence that the assumption is erroneous."[165] The Court of Appeals for the Federal Circuit reaffirmed this standard for *prima facie* obviousness as applied to composition of matter claims:

> This court . . . reaffirms that structural similarity between claimed and prior art subject matter, proved by combining references or otherwise, where the prior art gives reasons or motivation to make the claimed compositions, creates a *prima facie* case of obviousness, and that the burden (and opportunity) then falls on an applicant to rebut that *prima facie* case. Such rebuttal or argument can consist of a comparison of test data showing that the claimed compositions possess unexpectedly superior properties or properties that the prior art does not have.[166]

The court also affirmed that *prima facie* obviousness does not require, given structural similarity, the same or similar utility between the claimed composition and the prior art composition.[167]

Structural similarity to support a case of *prima facie* obviousness is often found in cases of homologous series of compounds, isomers, steroisomers, esters and corresponding free acids, and so on. Bioisosterism, which may be of particular interest to the medicinal and pharmaceutical researcher, can also give rise to *prima facie* obviousness for both compounds and methods of using such compounds.[168] Bioisosterism recognizes that substitution of an atom or group of atoms for another atom or group of atoms having similar size, shape, and electron density generally provides compounds having similar biological activity. For example, the Court of Appeals for the Federal Circuit found that the use of amitriptyline for treating depression was obvious in view of the closely related antidepressant imipramine.[169] Imipramine differs structurally from amitriptyline by replacement of the unsaturated carbon in the center ring of amitriptyline with a nitrogen atom. Another prior art reference showed that chloropromazine (a phenothiazine derivative) and chloroprothixene (a 9-aminoalkylene-thioxanthene derivative) had similar biological properties. Chloropromazine and chloroprothixene differ in the same manner as imipramine and amitriptyline (i.e., an unsaturated carbon versus nitrogen in the central ring structure) and are also closely related (but not as closely as imipramine and amitriptyline). In fact, this prior art reference "concluded that, when the nitrogen atom located in the central ring of the phenothiazine compound is interchanged with an unsatu-

---

[165]*In re Papesch*, 315 F.2d 381, 391 (C.C.P.A. 1963).
[166]*In re Dillon*, 919 F.2d 688, 692–693 (Fed. Cir. 1990) (*en banc*) (citations omitted). In light of *KSR*, however, the motivation to combine such references is not required to come only from the references themselves.

[167]*Id.* at 693.
[168]See, for example, *In re Merck & Co.*, 800 F.2d 1091, 1096–1098 (Fed. Cir. 1986) ("bioisosterism was commonly used by medicinal chemists prior to 1959 in an effort to design and predict drug activity"); *Imperial Chem. Indus. v. Danbury Pharm.*, 745 F. Supp. 998 (D. Del. 1990) (attempt to invalidate patent under 35 U.S.C. § 103 using bioisosterism theory).
[169]See generally *In re Merck & Co.*, 800 F.2d 1091 (Fed. Cir. 1986).

rated carbon atom as in the corresponding 9-amionalkylene-thioxanthene compound, the pharmacological properties of the thioxanthene derivative resemble very strongly the properties of the corresponding thiazines."[170]

As indicated above, the applicant has the opportunity to rebut the case of *prima facie* obviousness by presenting evidence showing unexpected or surprising results. In cases where it is known that the compound to be claimed is structurally similar to known compounds, it is generally preferred that the data supporting patentability be included in the application as filed. Even if the *prima facie* case cannot be overcome, it may be possible in some cases to obtain at least some patent protection by claiming a method of making the compound or a method of using the compound.

Process Claims. There are generally two types of process or method claims: (1) claims directed to a method of synthesizing or otherwise transforming a chemical or biological material and (2) claims directed to a method of using particular compounds to achieve a desired result. The same basic obviousness standard is generally applied to process claims as is applied to other claims, including composition of matter claims. The Court of Appeals for the Federal Circuit held that the use of novel starting materials and/or the production of novel products (even if the products are patentable in their own right) in a known process for making a compound may make the process new but does not necessarily make it nonobvious.[171] The PTO interpreted this case as effectively holding that an old process could never become nonobvious through the use of novel starting materials and/or the creation of novel products. The Court of Appeals for the Federal Circuit later suggested that this case merely "refused to adopt an unvarying rule that the fact that nonobvious starting materials and nonobvious products are involved *ipso facto* makes the process obvious" and added:

> The materials used in a claimed process, as well as the results obtained therefrom, must be considered along with the specific nature of the process, and the fact that new or old, obvious or nonobvious, materials are used or result from the process are only factors to be considered, rather than conclusive indicators of the obviousness or nonobviousness of a claimed process.[172]

The patentability of process claims, whether method of making or method of using, should be determined in the same way as any other claim, that is, by applying § 103 to determine whether the invention as a whole is obvious or nonobvious.

3.6. Procedure for Obtaining Patents in the PTO

Obtaining a patent in the United States involves an *ex parte* procedure solely between the inventor or the inventor's assignee and the PTO.[173] The proceedings between the applicant and the PTO are conducted in writing. The proceedings themselves and the written record of the proceedings, often called the file or prosecution history, are carried out and maintained in secret until the date the application publishes or the date the patent issues, whichever comes first. Pending utility patent applications are now published 18 months after their priority date unless, at the time of filing, the applicant requests that the application not be published and certifies that it has not and will not be filed in any country outside of the United States that provides for publication.[174] If and when the application is published or the patent actually issues, the secrecy of the proceeding ends and the entire written record is open to the public. Should a patent not issue (and if the applicant has made a request not to publish the application and

---

[170]*Id.* at 1095.
[171]763 F.2d 1406, 1410 (Fed. Cir. 1985).
[172]*In re Dillon*, 919 F.2d at 695 (dictum).
[173]The inventor or assignee is normally represented by a patent agent or attorney in proceedings before the PTO. Such agents or attorneys are admitted to practice before the PTO upon passing a bar examination given by the PTO. The inventor may, however, represent him or herself.
[174]American Inventors Protection Act of 1999, Pub. L. No. 106–113, 113 Stat. 1501 (codified as amended in scattered sections of 35 U.S.C.) (2000); 37 C.F.R. § 1.211(a) (2008).

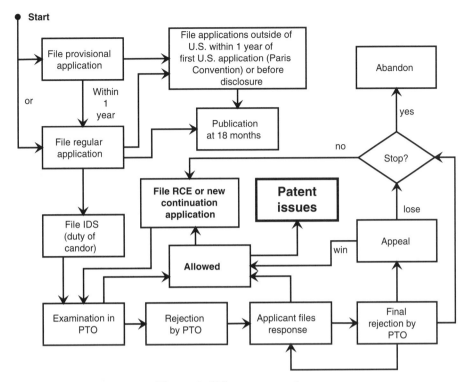

Figure 3. U.S. patent procedure.

met the other requirements), the record will be maintained in secret.[175] Thus, assuming corresponding patent applications were not filed in other countries and the patent application has not published in the United States, an abandoned U.S. patent application will remain secret and may be maintained as a trade secret. If a foreign patent application has been filed, however, retention of any trade secrets contained therein is possible only until the time the U.S. or foreign application is published (i.e., 18 months after the priority date).

For inventions developed in the United States, the patenting process generally begins with filing a patent application meeting the statutory requirements in the PTO. A

flow chart generally illustrating the U.S. patenting procedure is shown in Fig. 3. Once an application is filed in the United States, the applicant or assignee has 12 months in which to file patent applications throughout the rest of the world. Normally, the PTO will not have undertaken the initial examination of the application before the expiration of the 12-month period. Thus, one normally does not have the benefit of the PTO's initial patentability review before decisions must be made concerning filings elsewhere in the world.

Because the proceedings in the PTO are *ex parte* in nature, the applicant has a "duty of candor" to be forthcoming in dealing with the PTO throughout the prosecution of the patent application. This duty of candor includes providing the PTO with all information that may be material to the examination of applicant's patent application (e.g., potential prior art) of which the applicant is aware. The duty of candor does not require the applicant to carry

---

[175]The written record of an abandoned application is open for public inspection if a later U.S. patent formally relies upon, or is formally related to, the abandoned application.

out a literature or prior art search. However, if the applicant or the applicant's representative is aware of relevant prior art, it must be disclosed. Normally, such relevant prior art is provided to the PTO in an Information Disclosure Statement (IDS). The duty to disclose relevant prior art continues throughout the prosecution of the application. Therefore, if the applicant later becomes aware of relevant prior art, including prior art found by patent offices outside the United States during prosecution of the corresponding applications filed in other countries, that information must be provided to the PTO. Failure to comply with this duty of candor can result in any patent issuing from the application being held unenforceable.

Upon receipt of a patent application, the PTO assigns the application a serial number and filing date. Shortly thereafter and if disclosure of the content of the application would not be detrimental to national security, the PTO issues a foreign filing license that allows the application to be filed in other countries.[176] Thereafter, patent applications based upon the U.S. patent application can be filed elsewhere in the world; to claim benefit of the U.S. filing date, such foreign applications must be filed within 1 year of the U.S. filing date.

In some instances, the first response the applicant has from the PTO is a restriction requirement. The filing fee entitles the applicant to examination of one invention (e.g., a drug itself, a method of making the drug, and a method of using the drug may be considered separate inventions). If the application claims more than one invention, the PTO can issue a restriction requirement, thereby separating the claims into groups corresponding to the various inventions and requiring the applicant to choose (i.e., elect) one of the groups of claims for examination.[177] The other, nonelected inventions can be refiled as divisional applications while relying upon the priority date of the original application. Divisional applications must be filed before the original application or any intervening continuing applications are abandoned or issue as patents. Such divisional applications will expire 20 years from the filing date of the earliest filed U.S. nonprovisional or international application for which benefit has been claimed.[178]

A patent examiner at the PTO examines the patent application to determine if it meets the statutory requirements. This includes a search by the examiner of the prior art. Using the prior art uncovered in the search and information submitted by the applicant, the examiner evaluates the claims and determines whether the claimed invention is novel and nonobvious. Although rare, especially in drug discovery and biotechnology areas, the examiner may find the claims as submitted to be patentable. The patent would, on payment of the appropriate issue fees, then proceed to issue.

More likely, the examiner will initially find at least some of the claims to be anticipated or obvious over the prior art and will inform the applicant of this finding in an office action, along with the rationale for the rejection or rejections. The applicant has a limited time (normally 3 months[179]) in which to respond to the office action. The applicant can choose, especially if the examiner's arguments appear persuasive, to abandon the application and, if the application has not and will not be published, retain the invention as a trade secret. The applicant can attempt to amend some or all of the claims to overcome the rejections and/or present arguments demonstrating why

---

[176] 35 U.S.C.A. § 184 (West 2001); 37 C.F.R. §§ 5.11–5.25 (2001). Unless the PTO notifies the applicant otherwise, the applicant can file foreign applications 6-months after the U.S. filing date without a foreign filing license.

[177] Applicants can also attempt to convince the examiner that the restriction requirement is improper. Success in such an attempt is generally considered to be unlikely.

[178] Previously, divisional applications potentially allowed the applicant to obtain several patents in series on the same basic concept and thereby extend patent protection beyond the 17-year term of the first patent to issue. This is no longer possible under the 20-year term.

[179] Up to three additional months (total of 6 months) can be obtained by payment of extension fees.

the examiner has improperly rejected the claims. The applicant can also present, if appropriate, additional information or data, including information relating to secondary considerations, demonstrating the nonobvious nature of the invention. Such information or data is normally presented in an affidavit or declaration.

The examiner, on reconsideration of the claims in light of the applicant's response (and any supporting information or data), can repeat the rejections, submit new rejections based on other prior art or reasoning, or withdraw the rejections wholly or in part. If the examiner withdraws all the rejections and allows the claims, the patent will issue upon payment of the issue fee. If the examiner allows some claims but continues the rejection of other claims, the applicant has several options. The applicant could contest the rejections of the claims once again in the pending application. Or the rejected claims could be canceled and the application passed to issue with the allowed claims. After a second or final rejection, any rejected claims can be appealed to the Board of Patent Appeals and Interferences in the PTO. It is often preferred in cases having both allowed and rejected claims to allow a patent to issue with the allowed claims and file a continuation application with the rejected claims before the parent application actually issues as a patent. By effectively separating the allowed claims from the rejected claims, the risk that the Board of Patent Appeals and Interferences will undermine the decision on patentability of the already allowed claims on appeal of the rejected claims is significantly reduced.

If the applicant is not successful in overcoming the examiner's rejections and the examiner makes the rejections final, several options remain. Again the applicant may simply abandon the application and, if the application has not and will not be published, retain the invention as a trade secret. Or the applicant can either file a request for continued examination (RCE) under 37 C.F.R. § 1.114 or refile the application as a new divisional, continuation, or continuation-in-part application under 37 C.F.R. § 1.53(b) and continue prosecution in the PTO.[180]

An RCE maintains the same serial number and priority date as the application pending at the time of the request. Importantly, an RCE is not a new application, and prosecution of the application continues with the same examiner just as if the last office action had not been made final. In other words, the examiner will look at the application and any arguments made by the applicant in a relative short time (generally within a few months). The traditional continuation, divisional, or continuation-in-part application is treated as a new application and is assigned a new serial number and filing date (it does, however, maintain the priority of the parent application). Such a new application simply takes its place in a long line of new applications. Thus, the examiner (who may or may not be the original examiner) may not even see the new application for several years. And such long delays count against the patent term. To maximize the patent term, the RCE process should be used whenever possible.

The applicant may appeal the examiner's rejection to the Board of Patent Appeals and Interferences within PTO. If not satisfied with the Board's decision, the applicant may appeal that decision either to the Court of Appeals for the Federal Circuit based on the record before the PTO or to a federal district court for a *de novo* review. If the examiner's position is overturned, the Federal Circuit or the district court can order the PTO to issue the patent. Appeal to either the Federal Circuit or a federal district court destroys the

---

[180]The PTO recently attempted to adopt new rules that, among other things, significantly limit the number of continuation-type applications that can be filed in a given case. These rules were scheduled to go into effect on November 1 2007. 72 Fed. Reg. 161, 46716–46843 (August 21 2007). Based on a lawsuit filed against the PTO under the Administrative Procedure Act, a federal court has ruled that the PTO exceeded the scope of its rulemaking authority and voided the rules. *Tafas v. Dudas*, 541 F. Supp. 2d 805 (E.D. Va. 2008). The PTO appealed (May 7 2008) to the Federal Circuit. The Federal Circuit affirmed in part, vacated in part, and remanded the case for further consideration by the district court. *Tafas v. Doll*, 559 F.3d 1345 (Fed. Cir. 2009). In a press release dated October 8, 2009, the PTO announced that these rules would not be adopted. See www.uspto.gov/news/09-21.jsp.

secrecy of the application as well as that of the record of the proceedings within the PTO and, thus, destroys any trade secrets that may have been contained therein.[181]

The PTO, in the first instance and using the appropriate standards, rules, and statutes as described above, determines whether or not a patent will ultimately issue. How those standards, rules, and statutes are applied by the PTO will, of course, significantly affect the outcome. Since 2002–2003, the overall allowance rate for patent applications in the PTO has significantly dropped. After being relatively constant at an average of about 65% from 1975 to 2000, the allowance rate has dropped to a 2008 value of below 45%.[182]

The PTO argues that this drop in the allowance rate is largely due a decrease in the quality of patent applications being submitted.[183] Indeed, the PTO points out that its "patent allowance compliance rate" was 96.3% in 2008.[184] But not everyone agrees. Critics tend to point out that this patent allowance compliance rate is based on an internal review of only allowed applications by the PTO. Patent applications in which no claims have been allowed are not included in this review.[185] Patent examiners likely realize that if they reject claims, their work product will have a much smaller chance of being reviewed. And since the PTO appears to equate lower allowance rates with higher quality, the incentive to reject claims by examiners would appear to be increasing.

Regardless of the reason or reasons for the decline, it seems clear that it is becoming more and more difficult, and taking longer, to obtain patent protection. Moreover, it remains to be seen whether the patent allowance rate will remain relatively constant at about the current 45% rate or increase or even decrease further. Nonetheless, applicants should be aware of the decrease in the allowance rate in making decisions with regard to both filing and prosecution of patent applications. Generally, more aggressive responses to PTO rejections and delaying prosecution where possible (recognizing that patent term will be lost) are recommended in the immediate future in the hope that the PTO will quickly moderate its current position. As of early 2010, it appears that the PTO is, in fact moderating its position.

3.7. Interference

With the first-to-invent system in the United States, it is sometimes necessary to determine which of two or more inventors (or groups of inventors) first invented the subject matter that is claimed in common by the parties.[186] Interferences are the proceedings within the PTO for making such determinations. These proceedings, which are overseen by senior examiners within the PTO, are ultimately decided by the Board of Patent Appeals and Interferences in the PTO. The party who first conceives an invention and first reduces it to practice will normally be awarded priority and

---

[181]Of course, any trade secrets will be lost when the application is published by the PTO or any other patent office.

[182]Erik Sherman, *Applying for a Patent? Why You Need More Luck Than Ever*, IP Law & Business, December 2008, at 16, 16. Based on data supplied by the PTO, the yearly patent allowance rates have steadily dropped since 2000: about 70% (2000), about 70% (2001), about 66% (2002), about 67% (2003), about 62% (2004), about 58% (2005), about 53% (2006), about 51% (2007), and about 43% (2008). Id.

[183]Id. We are not aware of any other major patent office reporting a similar decline in quality of patent applications being submitted.

[184]U.S. Pat. & Trademark Off., *Performance and Accountability Report Fiscal Year 2008*, available at http://www.uspto.gov/web/offices/com/annual/2008/2008annualreport.pdf.

[185]If the PTO does evaluate applications wherein all claims are ultimately rejected (and which applications are ultimately abandoned by applicants), they do not appear to have published such data.

[186]There remains pressure for the United States to adopt a first-to-file system as adopted by most other nations. See, for example, Patent Reform Act of 2009 (H.R. 1260, S.515). In such a first-to-file system, such interference proceedings would not be necessary.

will be awarded the patent.[187] This is not the case, however, if another party, who reduced the invention to practice at a later date, can prove that he or she was the first to conceive the invention and proceeded diligently to reduce it to practice from a time before the other party's date of conception. The diligence of the first to reduce the invention to practice is normally immaterial in the priority contest.

Patent interferences are possible between two or more copending applications or between a pending application and a recently issued patent.[188] In the case of copending applications, an applicant will often be notified by the examiner that his or her application appears to be in allowable condition but that prosecution is being suspended for consideration of a potential declaration of an interference. In cases involving a pending application and an issued patent, the examiner may cite the issued patent as a reference and suggest that the applicant may wish to "copy" claims from the patent to provoke an interference, or the applicant may become aware of the issued patent and attempt to amend his or her application to contain claims from the issued patent to provoke an interference. In the case of a pending application and an issued patent, claims from the issued patent must be copied by the applicant within 1 year of the issue date of the patent in order to provoke an interference.[189] If the PTO determines that both parties have allowable claims directed to the same subject matter, a formal Declaration of Interference is issued.

An interference is a complex, multi-stage, *inter partes* procedure designed to determine which party has priority with respect to the patentable subject matter that is disclosed and claimed by two or more inventors. The interference normally proceeds through the following stages: (1) declaration of interference by the PTO, (2) motion period, (3) filing of preliminary statements, (4) discovery, (5) testimony period, (6) final hearing, (7) decision by Board of Patent Appeals and Interferences, and (8) appeal and court review. Each of these stages is governed by complex procedural and substantive rules with many potential pitfalls for the unwary. Failure to follow these rules carefully can result in an adverse ruling or, ultimately, judgment against the party violating them.

The party with the earlier priority date is designated the senior party and is presumed to be the first inventor. The other party is designated the junior party and has the burden of proving an earlier date of invention, generally by a preponderance of the evidence. If, however, the junior party's application was filed after the issuance of the senior party's patent, an earlier invention date must be proven beyond a reasonable doubt. Throughout the interference proceeding, the senior party retains significant procedural and substantive advantages.

Once an interference has been declared by the PTO, the parties are given an opportunity to redefine the interfering subject matter (i.e., the counts of the interference that are similar in form to patent claims) and, if possible, to assert an earlier effective filing date based on a related U.S. patent application or a corresponding patent application in another country. In redefining the interfering subject matter, each party will generally seek to amend the counts in a manner more favorable to itself (i.e., consistent with and supported by his or her evidence concerning conception and reduction to practice). The parties, if appropriate, may also raise issues that are not directly

---

[187]Such a party could lose priority if that party "abandoned, suppressed, or concealed" the invention. 35 U.S.C.A. § 102(g) (West 2001). For example, if the first party to conceive and reduce the invention to practice decided to keep the invention a trade secret and was only spurred to file an application upon learning of another's invention of the same subject matter, the first party's priority could be extinguished by its failure to file its patent application in a timely manner.

[188]On occasion the PTO may inadvertently issue two patents that claim the same subject matter. In such cases, the PTO does not have jurisdiction to institute an interference proceeding. If one of the patentees brings an appropriate civil action in a federal district court, the court could determine which patentee is entitled to a patent claiming the subject matter. 37 C.F.R. §§ 41.202, 41.207 (2008).

[189]35 U.S.C.A. § 135 (b)(1) (West 2001). Where a pending application has published, applicant must copy any claims in the published application within one year of the publication date. 35 U.S.C.A. § 135(b)(2) (West 2001).

related to the dates of conception or reduction to practice. For example, one party may allege that the other party derived the interfering subject matter from someone else or that the other party's application has deficiencies that render the interfering subject matter unpatentable to that party. Either party can also argue that the subject matter in question is simply not patentable to anyone (i.e., effectively that neither party is entitled to a patent).[190] Generally, however, the most significant issues relate to establishing the respective dates of conception and reduction to practice and, where appropriate, diligence in reducing the invention to practice.

During the interference, parties can rely on inventive activities occurring in the United States and, with certain limitations, a North American Free Trade Agreement Implementation Act (NAFTA) country or a World Trade Organization (WTO) country to prove a date of invention prior to the actual filing date.[191] The parties cannot rely on inventive activities within non-NAFTA or non-WTO countries to establish a date of invention; in such cases, the parties will be limited to their foreign priority date as the date of invention. Thus, in an interference proceeding, activities within the United States, independent of the date, can be used to establish an earlier date of invention. Additionally, a party can establish a date of invention based on inventive activities in a NAFTA member country on or after December 8, 1993 or in a WTO member country on or after January 1, 1996.[192]

Thus, for applications filed after January 1, 1996, applicants from other countries (at least those carrying out inventive activities in NAFTA and WTO countries) are, at least in theory, on equal footing with U.S. applicants in proving earlier dates of inventions in an interference. In many cases, however, applicants and researchers from other countries will need to modify their methods of research record keeping to take advantage of this provision regarding foreign activities. Applicants from other countries must also understand, and adapt their operations to, the types of proofs and evidence necessary to prove dates of conception, diligence, and actual reduction to practice under the U.S. law, as discussed below.

In adopting legislation regarding inventive activities outside the United States, Congress recognized that many countries do not provide similar opportunities for discovery during judicial proceedings as in the United States. Thus, an applicant from another country could potentially gain a significant advantage over his or her U.S. counterpart in an interference proceeding if the applicant from another country could use U.S. discovery procedures to gain information regarding the U.S. applicant's position while at the same time resisting such discovery based on his or her country's limited or otherwise inadequate discovery proceedings.

To address this potential problem, Congress has provided that "appropriate inferences" should be drawn when evidence regarding a date of invention cannot be obtained from a party (and perhaps even a third party) in a NAFTA or WTO country to the extent "such information could be made available in the United States."[193] "Appropriate inferences" are not defined and thus, depending on the circumstances, may range from mere "slaps on the wrist" to loss of rights in the proceeding. It is likely that the type of discovery that "could be made available" will depend on the type of proceeding involved. Thus, the effect of this requirement could, for example, be very different in an interference proceeding before the PTO (where the scope of discovery is

---

[190] A party who cannot prove an earlier filing date may prefer that no one obtain a patent so he or she can practice the invention without restriction. See, for example, *Perkin v. Kwon*, 886 F.2d 325 (Fed. Cir. 1989). Such a party (i.e., one who knows that priority cannot be proven) could also seek a favorable license (especially while the preliminary statements remain sealed) and thereby, at least to some degree, benefit from the protection offered by the other party's patent (if granted by the PTO). The other party may be willing to make concessions in the terms of such an agreement since it would effectively terminate the interference and eliminate the risk of losing the priority contest.

[191] See 35 U.S.C.A. § 104 (West 2001).

[192] *U.S. Pat. & Trademark Off., Manual of Patent Examining Procedure*, § 2138.02 (8th ed. 2001, latest revision 2008) available at http://uspto.gov.

[193] 35 U.S.C.A. § 104(a)(3) (West 2001).

limited) and a later appeal before the federal courts (where the scope of discovery is much broader). Due to these uncertainties, the practical effects of this provision are unclear.

Once the PTO has declared the interference, each party must file a preliminary statement by a date set by the PTO. Facts alleged in the preliminary statement must be proved later in the proceedings. The preliminary statement must provide the dates of invention each party will rely on. A party intending to rely on an invention date earlier than his or her filing date must allege the earlier invention date and identify facts supporting the earlier date in the preliminary statement. The parties are held strictly in their proofs to any dates alleged in the preliminary statement. Thus, a party able to introduce evidence showing an invention date earlier than alleged in the preliminary statement will still be held to the alleged date. The preliminary statement, therefore, should be carefully prepared and only allege dates and facts that the party can prove by clear and convincing evidence. A party relying on dates of conception and actual reduction to practice for the invention date must generally prove such dates by corroborated evidence. The testimony of an inventor or coinventor must be corroborated as it relates to priority of invention.

The preliminary statements are placed under seal and provided to the opposing party only at a later time set by the PTO. In the initial stages of the proceeding, neither party is aware of the alleged invention date of the opposing party or if the opposing party has even alleged an invention date earlier than its filing date. After filing the preliminary statements, the parties generally undertake discovery and testimony to develop their cases. Once all evidence periods have expired, the preliminary statements are opened and each party presents the formal record of evidence on which it wishes to rely. After the parties have submitted formal briefs enumerating their legal arguments and, where appropriate, rebuttal arguments, and after any oral arguments, the interference is decided by a board of three senior PTO examiners. The losing party may appeal the decision to the U.S. Court of Appeals for the Federal Circuit based on the interference record or bring a *de novo* civil action in the U.S. District Court for the District of Columbia.

Conception generally is considered to occur when the inventor forms a definite perception of the complete invention sufficient to allow one of ordinary skill in the art to understand the concept and reduce it to practice without further inventive steps. Reduction to practice can be either constructive (i.e., by filing a patent application meeting the statutory requirements) or actual (i.e., by making and testing the invention to demonstrate that it yields the desired result). Therefore, to achieve an actual reduction to practice for a pharmaceutical, there must be both an available process for making the chemical compound (i.e., the drug) and testing to establish its utility. To be useful in an interference proceeding, such evidence generally must be corroborated by one or more persons who are not inventors. Circumstantial evidence can also be used to help establish an actual reduction to practice. Often the best corroborative evidence is provided by laboratory notebooks used to record ideas and experiments associated with the invention. Preferably, laboratory notebooks should be signed and dated daily by the investigators actually doing the work and diligently witnessed in writing by at least one noninventor coworker who has read and understood the record. The witness should be able to understand the significance of the experiments being witnessed. Generally, the witness should not be actively involved in the work, so as to reduce the risk that the witness will later be found to be a coinventor. If a witness is later determined to be a coinventor, he or she cannot corroborate the work. For particularly important inventions, two witnesses should be considered so that even if one witness is effectively disqualified (e.g., one witness is later found to be a coinventor), there still remains a corroborating witness. Without independent corroborating evidence of the events surrounding the invention, including conception and the reduction to practice and, possibly, diligence, a junior party will most likely lose the priority contest even if he or she was the first to invent. Should a party need to prove diligence, a corroborated written record showing almost continuous activity from a time before the other party's date of conception up to the time of that party's own reduction to practice is generally required. If a party fails to work on the project for several successive weeks without adequate explanation, such inactivity can be sufficient to destroy the case for diligence. The activity of the inventor in developing another invention does not normally constitute an adequate excuse for such a period of inactivity.

A comprehensive intellectual property program should, therefore, give special attention both to implementing acceptable record-keeping procedures (including procedures for witnessing notebooks and other records of invention in a timely manner) and to ensuring that these record-keeping procedures are followed. Once again, the intellectual property committee is the logical focus for this task.

Interferences are long, costly, and complex procedures that are laden with procedural pitfalls for both the senior party and, especially, the junior party. The senior party is heavily favored to be awarded the right to a patent on the contested invention.[194] Based on the advantage of the senior party in such interference proceedings, patent applications should be filed in the United States as quickly as possible to increase the probability of achieving senior status in any interference that may be declared.

3.8. Correction of Patents

Issued patents are often found to contain errors of varying degrees. Minor errors (e.g., clerical or typographical errors and erroneous inclusion or exclusion of an inventor), if made without deceptive intent, can usually be corrected through a Certificate of Correction issued by the PTO at the request of the inventor or assignee.[195] Generally, a mistake is not minor if its correction would materially affect the scope or meaning of the patent claims.

Patents that are wholly or partly inoperative or invalid may, in some cases, be corrected by reissuing the patent. Reissue patents are generally sought because (1) the claims in the original patent are either too narrow or too broad, (2) the specification contains inaccuracies or errors, (3) priority from a patent application filed in another country was not claimed or was claimed improperly, or (4) reference to a prior copending application was not included or was improperly made.

In order to obtain a reissue patent, the patent owner must establish that (1) the patent is considered "wholly or partly inoperative or invalid" because of a "defective specification or drawing" or because the inventor claimed "more or less than he [or she] had a right to claim," (2) the defect arose "through error without any deceptive intention," (3) new matter is not introduced into the specification, and (4) the claims in the reissue application meet the legal requirements for patentability.[196] An application for a reissue patent that contains claims enlarging the scope of the original patent must be filed within 2 years of the date of grant of the original patent.[197] The term of a reissue patent is the unexpired portion of the original patent; a reissue patent cannot extend the duration of the original patent. An infringer may have a personal defense of intervening rights to continue an otherwise infringing activity if the activity or preparation for the activity took place before the grant of the reissue patent.[198] The reissue process is often used by patent owners to "clean up" a patent prior to embarking on litigation to enforce that patent.

---

[194]In two-party interference proceedings between fiscal years 1992 and 1994, the senior and junior parties prevailed in about 62% and 28% of the cases, respectively; the rest were split decisions. For proceedings decided by the Board, the senior and junior parties prevailed in about 53% and 32% of the cases, respectively; again the rest were split decisions. Of the 673 two-party interferences during this period, about 21% proceeded to a decision by the Board. Ian A. Calvert & Michael Sofocleous, *Interference Statistics for Fiscal Years 1992 to 1994*, 77 J. Pat. & Trademark Off. Soc'y 417 (1995).

[195]35 U.S.C.A. §§ 254–256 (West 2001).
[196]35 U.S.C.A. § 251 (West 2001); 37 C.F.R. §§ 1.171–1.179 (2001).
[197]*Id.*
[198]35 U.S.C.A. § 252 (West 2001).

A patent owner or any third party[199] can seek a review of an issued patent by the PTO on the basis of a substantially new question of patentability. The new question of patentability may be based on prior art considered during the original examination or newly discovered prior art. If the PTO determines that a substantially new question of patentability has been raised, the PTO will issue an order for the reexamination. Such reexamination procedures, although technically not mechanisms for correcting a patent, allow the PTO to determine the correctness of the original patent grant in light of such substantially new questions of patentability. There are two types of reexamination: (1) *ex parte* proceedings[200] that can be requested by the patent owner or a third party and (2) *inter partes* proceedings[201] that can be requested by a third party. A third party requestor for either type of reexamination must identify the real party in interest; that identification will be provided to the patent owner.

Using *ex parte* reexamination, the third party requestor is limited to responding to a patent owner's initial statement concerning patentability (if filed).[202] After responding to the initial statement, the third party requestor has no further input in the proceeding. The patent owner's initial statement can include arguments with regard to the substantially new question of patentability as well as new claims and/or amendments to the patent specification and existing claims.[203] The patent owner could delay presenting such new claims or amendments until later in the proceedings (after the time period for the third party requestor's reply to the initial statement has passed); in such a case, the third party requestor will have no opportunity to respond to such new claims and/or amendments. Only the patent owner has the right to appeal an adverse decision.[204]

Under *inter partes* proceedings, the third party requestor has the opportunity to respond to arguments and amendments made by the patent owner throughout the reexamination. The third party requestor in the *inter partes* proceeding is limited to issues raised by the patent owner or PTO and cannot raise new issues.[205] The third party requestor is, however, barred from later asserting invalidity in a civil action (e.g., an infringement suit based on the reexamined patent) on any grounds that were raised, or could have been raised, during the reexamination proceeding.[206] Either the patent owner or the third party requestor can appeal an adverse decision.[207] Third parties should carefully consider the potential risks (i.e., possible estoppel effects) and benefits (i.e., involvement throughout the proceedings) before requesting such an *inter partes* reexamination.

The original patent is not presumed valid during either type of reexamination proceeding. The PTO can reaffirm the original grant, substitute a new grant by allowing new or amended claims, or withdraw the original grant.[208] Nor is the legal presumption of va-

---

[199]Any person at any time may submit patents or printed publications he or she believes have a bearing on the patentability of any patent claim; if desired, a written explanation of the pertinency and manner of applying such prior art to the patent claim can also be included. The cited prior art and explanation (if submitted) are placed in the PTO file for the particular patent. The submission can be made anonymously if desired. 35 U.S.C.A. § 301 (West 2001).

[200]35 U.S.C.A. §§ 301–307 (West 2001 & Supp. 2008). The PTO Director can initiate an *ex parte* reexamination proceeding based on prior art of which he or she is aware or prior art identified by third parties. 35 U.S.C.A. § 303(a) (West Supp. 2008).

[201]35 U.S.C.A. §§ 311–318 (West 2001 & Supp. 2008). *Inter partes* proceedings are available only for patents that issue from an original application filed in the United States on or after November 29, 1999. 37 C.F.R. § 1.913 (2008).

[202]35 U.S.C.A. § 304 (West 2001).

[203]*Id.*

[204]35 U.S.C.A. § 306 (West 2001).

[205]35 U.S.C.A. § 314(b)(2) (West Supp. 2008).

[206]35 U.S.C.A. § 315(c) (West Supp. 2008).

[207]35 U.S.C.A. §§ 315(a) (b) (West 2001 & Supp. 2008). The patent owner or the third party may be a party to an appeal by the other party. The third party, if participating in an appeal taken by the patent owner, would be estopped from asserting invalidity for any claim finally determined to be valid based on any grounds the third party raised, or could have raised, during the inter partes proceeding in any later civil action. 35 U.S.C.A. § 315(c) (West 2008).

[208]35 U.S.C.A. §§ 307, 316 (West 2001).

lidity of the patent strengthened by the successful completion of the reexamination proceeding. However, the practical effect of the reexamination proceeding may considerably strengthen the patent in the eyes of a judge or jury in later litigation since a patent that has twice been found patentable by the PTO (i.e., once during the original prosecution in the PTO and then again during the reexamination proceedings) may be viewed favorably. And, of course, the estoppel effects possible in an *inter partes* reexamination proceeding could have a significant impact on a third party requestor's ability to defend against a later infringement action.[209]

On the other hand, the reexamination proceeding can provide an accused or a potential infringer the possibility of invalidating a patent outside of the court system at a considerably reduced cost. In making a request for reexamination, the third party requestor should carefully consider the differences between the two types of reexamination proceedings in terms of his or her involvement in the process and any potential estoppel effects of an adverse decision in later civil actions as well as the overall likelihood of success. Generally, it is recommended that third parties do not request reexamination unless the prior art is believed, to a high level of certainty, to render the patent invalid; moreover, the selection of an *inter partes* proceeding over an *ex parte* proceeding would normally require an even higher level of certainty.[210]

4. ENFORCEMENT OF PATENTS

Fundamentally, a patent provides the right to exclude others from practicing the invention covered by (i.e., "claimed in") the patent. Enforcement of patents is limited both temporally and geographically. A patent can be enforced only against acts occurring during its term.[211] Patent term is discussed in detail in Section 2.4.

4.1. Geographic Scope

Generally, a patent can be enforced only within the country in which it was granted. There is one exception to the general principle that a U.S. patent does not provide legal protection against acts outside its geographical borders. Under the U.S. law, the unauthorized importation, use, sale or offer for sale of a product in the United States is an act of infringement if the product was made outside the United States by a process patented in the United States.[212]

Thus, a U.S. patent covering a new and novel process for producing known products or drugs cannot be circumvented by manufacturing it outside the United States and then importing the product or drug into the United States. The U.S. patent holder cannot, of course, use its U.S. patent to prevent the use of that process overseas or the importation of the resulting products into other countries or the use of the same products made by a different process.

4.2. The Trial and Appellate Courts

Generally speaking, the U.S. civil justice system is comprised of the federal courts and the courts of each of the fifty individual sovereign states. The U.S. district courts are the trial courts of the federal court system. There are 94 federal judicial districts, including at least one in each state, the District of Columbia and Puerto Rico.[213]

In the United States, patent infringement is a federal cause of action. To enforce a patent, the patent holder normally brings suit against

---

[209] 35 U.S.C.A. § 315(c) (West Supp. 2008).
[210] One exception to this general rule is where a third party is precluded from entering a market by a blocking patent which, in the opinion of the third party, is invalid based on prior art not considered by the PTO. By requesting a reexamination, the third party would hope to invalidate the patent and clear the path for its entry into the market. If unsuccessful, the third part may wish to move on to other business opportunities.

[211] A suit alleging infringement during the term of a patent can generally be brought after the patent expires. Damages cannot, however, be recovered for infringing acts occurring more than 6 years prior to filing the complaint. 35 U.S.C.A. § 286 (West 2001).
[212] 35 U.S.C.A. § 271(g) (West 2001).
[213] The Virgin Islands, Guam and the Northern Mariana Islands, which are U.S. territories, also have district courts that hear patent and other federal cases.

the alleged infringer in a federal district court. Patent litigation may also be initiated by a party seeking to have the patent declared invalid, unenforceable, or not infringed if there is an actual dispute between the parties under the patent. Actions involving U.S. patents are conducted like all other civil actions in the federal courts, and are governed by the Federal Rules of Civil Procedure and the Federal Rules of Evidence.

Most federal judges have no formal training in the sciences or patent law. Because patent cases, which represent only a small percentage of federal cases, are sparsely distributed among the district court judges, the typical judge has very little experience handling the numerous complex issues involved.

The United States is unique in using juries to decide commercial civil trials, including patent disputes. The right to a trial by jury is enshrined in the Seventh Amendment to the U.S. Constitution and has been exercised readily by patent litigants. The incidence of jury demands in patent cases has grown from about 3% in the late 1960s to more than 60% in recent years.[214] The jury's role is to listen to the evidence presented at trial to determine the facts, and to determine disputed issues by applying the patent law to those facts. When the evidence presents underlying disputed fact issues, the jury plays a crucial role in determining patent validity, infringement, and damages.

There are strong opinions both for and against the increased resort to juries in patent cases. Opponents of patent jury trials argue that people of only average education, and infrequent technical education, cannot rationally and fairly resolve complex technology disputes involving highly specialized legal doctrines. Others have seemingly unfettered trust in the ability of juries to reach predictably rational and reliable outcomes when properly aided by the court and trial counsel. Those who support the jury system argue that juries are just as able as judges, most of whom also lack technical training and have little experience with patent law, to decide patent cases.

Two other forums for resolving patent infringement suits also deserve mention. The U.S. Court of Federal Claims has exclusive jurisdiction over patent infringement claims made against the federal government.[215] The U.S. International Trade Commission ("ITC") has authority to preclude importation of products covered by a U.S. patent or made abroad by a process covered by a U.S. patent.[216] Significant procedural and substantive differences exist between these forums and the district courts. For example, there is no right to trial by jury in the Court of Federal Claims, nor is injunctive relief available against the government or suppliers to the government. ITC investigations are conducted by ITC staff, whose determinations must consider, in addition to patent issues, other factors, such as the public welfare and competitive economic conditions.[217] Such additional facts generally do not align with the interests of at least one of the parties.

Patent owners have increasingly turned to the ITC in patent disputes in appropriate cases. The ITC offers a relatively fast proceeding and can issue preclusion orders blocking importation and sale of infringing products. The ITC remedy is of heightened importance in view of a recent U.S. Supreme Court case making it more difficult for patent owners to obtain injunctive relief in federal court infringement actions.[218] Under the *eBay* standard, a federal court may enjoin further sales only if the patent owner establishes that (1) it has suffered irreparable injury; (2) damages remedies are inadequate to compensate for that injury; (3) considering the balance of hardships between the plaintiff and the defendant, an injunction is warranted; and (4) the injunction would not disserve the public interest. The ITC, however, is not required to use the same four-factor test as the courts, and thus may be a better forum for a patent owner

---

[214] Paul R. Michel & Michelle Rhyu, *Improving Patent Jury Trials*, 6 Fed. Cir. B. J. 89 (1996); Kimberly A. Moore, *Jury Demands: Who's Asking?*, 17 Berkeley Tech. L. J. 847, 851 (2002).

[215] 28 U.S.C.A. § 1498 (West 1994).
[216] 19 U.S.C.A. § 1337 (West 1999 & Supp. 2008).
[217] 19 U.S.C.A. § 1337(d) (West 1999).
[218] *eBay Inc. v. MercExchange, L.L.C.*, 547 U.S. 388 (2006).

attempting to stop further importation and sale of infringing products.

The 94 federal judicial districts are organized into 12 regional circuits, each of that has its own court of appeals. With the Federal Courts Improvement Act ("FCIA") of 1982,[219] Congress created the Court of Appeals for the Federal Circuit and vested it with exclusive jurisdiction over patent cases. One of its objectives was to promote uniformity in the area of patent law and diminish the uncertainty created by inconsistent application of the patent law among the regional circuits.[220] The Federal Circuit also hears appeals of patent matters from the PTO, the Court of Federal Claims, and the ITC, as well as appeals of a variety of nonpatent matters. Matters involving issues of patent law make up a substantial portion of the court's docket, and the court is considered to have specialized expertise in the patent area.

A party may petition the Supreme Court to accept an appeal from an adverse decision of the Federal Circuit. The Supreme Court has discretionary control over its docket. It generally declines to hear the vast majority of cases it is asked to decide and will only rarely take a patent case. Thus, the Federal Circuit is effectively the court of last resort for most patent litigants.

4.3. The Parties

A patent owner or an assignee of all rights in a patent may bring suit. Such an assignee must possess the required patent rights at the time that the suit is filed.[221]

Where a patent is co-owned by multiple persons, all must jointly bring suit.[222] Corporations must therefore make certain that each of the joint inventors execute an assignment document to avoid subsequent difficulties in bringing an infringement action. A transfer of less than the entire patent, an undivided share of the patent, or all rights in the patent in a limited geographical region of the United States is ordinarily deemed a license. Such a license generally does not convey a right to sue.

A licensee may sue in its own name for patent infringement only if he exclusively owns "all substantial rights" under the patent, such that the license operates as a "virtual" assignment. In certain circumstances, a licensee with less than all substantial rights may have standing to sue as a coplaintiff.[223] The Courts look to the substance of the agreement to decipher the intent of the parties. To qualify as an exclusive license, an agreement must clearly manifest the patentee's promise not to grant anyone else a license in the area of exclusivity. License provisions expressly granting the licensee a right to sublicense without the licensor's approval, to exclude others from using the technology, to file suit for infringement and to defend the licensor's interest in such a suit may confer standing to a licensee as a coplaintiff.

Despite rather clear guidance from the courts regarding principles of standing in patent enforcement actions, disputes in this area are quite frequent. The varied contexts of reported cases regarding standing to sue emphasize the importance of assignments, licenses and other commercial documents that clearly delineate the parties' rights and obligations with respect to litigation against infringers.

Section 271 of the Patent Code provides a remedy against those who make, use, sell, or offer for sale the patented invention in the United States or import the patent invention into the United States.[224] These are each acts of direct infringement. Those who actively induce infringement by knowingly and actively aiding another's direct infringement can also be held liable.[225] A remedy may also be had for contributory infringement against

---

[219]Pub. L. No. 97–164, 96 Stat. 25 (codified as amended in scattered sections of U.S.C.).
[220]See H.R. Rep. 97-312 at 23 (1981) ("[T]he central purpose [of the FCIA] is to reduce the widespread lack of uniformity and uncertainty of legal doctrine that exists in the administration of patent law.").
[221]Mas-Hamilton Group v. LaGard, Inc., 156 F.3d 1206, 1210 (Fed. Cir. 1998).
[222]Willingham v. Lawton, 555 F.2d 1340, 1343 (6th Cir. 1977).
[223]Enzo APA & Son, Inc. v. Geapag A. G., 134 F.3d 1090, 1093 (Fed. Cir. 1998); Textile Prods., Inc. v. Mead Corp., 134 F.3d 1481, 1484 (Fed. Cir. 1998).
[224]35 U.S.C.A. § 271 (West 2001 & Supp. 2008).
[225]35 U.S.C.A. § 271(b) (West 2001).

one who sells a material part of the invention that he knew was particularly adapted to infringe the patent and that material part is not a staple item of commerce suitable for noninfringing use.[226] In order to prove inducement or contributory infringement, a patentee must prove at least one act of direct infringement.[227] The establishment of both active inducement and contributory infringement requires that the accused infringer knew, or should have known, of acts that constitute direct infringement. As opposed to the case of direct infringement, a truly "innocent infringer" cannot be liable either for actively inducing infringement or for contributory infringement. It is not necessary that the direct infringer be named as a party, although doing so often simplifies the task of obtaining evidence sufficient to prove direct infringement.

A patentee has a cause of action against those who supply components of a patented product in the United States intending that the components be assembled into an infringing product abroad.[228] A patentee may also sue one who imports, offers to sell or uses in the United States a product made by the patented method in a foreign country.[229] Those who purchase a patented product from the patent owner or an authorized licensee receive an automatic implied license to use and resell the product.[230]

For a composition of matter claim covering a drug, the unauthorized manufacturer who prepares the drug using any process, the drugstore that sells it, and the patient who takes it may each be liable as direct infringers; the doctor who prescribes it may be liable by actively inducing infringement by the patient. As a practical matter, however, a patent owner would typically pursue the infringement action against the drug manufacturer, who generally has the greatest commercial interest in the dispute, control over the most relevant evidence of infringement, and the strongest financial resources to satisfy a settlement or damage award. For a claim covering a process for making a specific drug, both the manufacturer who makes the drug in the United States using the patented process and the manufacturer who makes the drug outside the United States using the patented process and sells the drug in the United States would be liable as infringers. Manufacturing the drug by a different process would not, of course, infringe the process claim. For a claim covering a method of treatment using a specific drug, the ultimate user might be liable as a direct infringer; the manufacturer, druggist, and doctor might be liable if they actively induce infringement by the patient. Thus, from the patent holder's viewpoint, it is desirable to have as many different types of claims in the patent to protect against infringement.

There is a special patent infringement remedy for a patent owner when another party files an Abbreviated New Drug Application ("ANDA") or a "paper NDA" for a patented drug. Both ANDAs and paper NDAs permit companies to seek approval to market a generic drug before the patent coverage on the product has expired without the costly studies required for a so-called pioneer drug.

Pioneer drug applicants are required to provide the FDA with the identity and expiration date of any patent that claims the drug for which FDA approval is requested, or a method of using such drug.[231] ANDAs and paper NDAs must certify that (1) no such patents have been identified by the original pioneer drug applicant, (2) the identified patents have expired, (3) the date that patents will expire, or (4) that the patents are invalid or will not be infringed by the making, use or sale of the generic drug.[232] This certification determines the date on which FDA approval of the ANDA or paper NDA may be made effective and when marketing may begin. If no patent has been identified, or if an identified patent has expired, approval can be immediate. If unexpired patents are found, approval is delayed until the patents expire. If the applicant asserts invalidity or noninfringement, the generic applicant must provide the original

---

[226] 35 U.S.C.A. § 271(c) (West 2001).
[227] *Water Tech. Corp. v. Calco, Ltd.*, 850 F.2d 660, 668 (Fed. Cir. 1988).
[228] 35 U.S.C.A. § 271(f) (West 2001).
[229] 35 U.S.C.A. § 271(g) (West 2001).
[230] *Bandag, Inc. v. Al Bolser's Tire Stores, Inc.*, 750 F.2d 903, 921–926 (Fed. Cir. 1984).

[231] 21 U.S.C. § 355(b)(1) (2007).
[232] 21 U.S.C. §§ 355(b)(2)(A), 355 (j)(2)(A)(vii) (2007).

pioneer drug applicant a detailed statement supporting the assertion of noninfringement or invalidity. The original pioneer drug applicant may then bring suit for infringement within 45 days, causing delay of the approval until the earlier of 30 months or a court ruling that the patent is not infringed or is invalid. Monetary damages are not available until actual commercial manufacture and sale of the generic drug.

A party sufficiently threatened by another's patent may file an action in a district court seeking to have the patent declared invalid or not infringed. The court's jurisdiction over such actions derives from the Federal Declaratory Judgments Act of 1934.[233] A declaratory judgment action must be based on an "actual controversy" between the parties. Until recently, the party filing suit had to show (1) that it has or is prepared to produce the accused product or use the accused method and (2) that the patentee's conduct created an objectively reasonable apprehension that the accused infringer would eventually be sued if it continued its activity.[234] However, the Supreme Court in *MedImmune v. Genentech* repudiated the former test in favor of a more generalized standard.[235] Declaratory judgment jurisdiction now exists if there is a substantial controversy between parties with adverse legal interests of sufficient immediacy and reality under all circumstances. [236]

In applying this "all the circumstances" test, the Federal Circuit has declined to adopt a bright line test or define the outer boundaries of declaratory judgment jurisdiction.[237] However, it has indicated that declaratory judgment jurisdiction generally will not exist merely on the basis that a party learns of the existence of a patent owned by another or even perceives such a patent to pose a risk of infringement, without some affirmative act by the patentee. Cases in which the Federal Circuit has found an actual controversy under the *MedImmune* standard have involved situations in which the patent owner took overt actions directly and expressly implicating the accused infringer from which the existence of an actual, imminent controversy regarding a specific patent was apparent.

4.4. Determining Infringement

When speaking of infringement, it is the claims of the patent that are infringed. The claims set the legal bounds of the technical area in which the patent holder can prevent others from making, using, and selling or offering to sell or importing the patented invention. In other words, the claims define the metes and bounds of the exclusive right granted by the patent. Each claim defines a separate right; some claims may be infringed while others are not. Claims also define the bounds outside of which others can operate without infringing the patent.

The process for determining patent infringement has two main steps. First, the claims of the patent are interpreted to determine their scope and meaning. Next, the properly interpreted claims are compared to the accused product or process.[238] Patent claim interpretation is performed by the judge based on the words of the claims themselves interpreted in light of the written description and drawings, and the record of proceedings in the PTO. Words in a claim are interpreted to have their ordinary meaning in the applicable field unless they were given a special meaning in the written specification or during prosecution.[239] Especially when a term is used in an out of the ordinary way, consideration should be given to expressly defining the term in the patent specification.

The terms *comprising*, *consisting essentially of*, and *consisting of* are used in patent claims to link the preamble and the limitations of the claimed invention; these "terms of art" have special meanings in patent claims and can dramatically affect the scope or coverage of a patent claim. For example, a claim

---

[233] 28 U.S.C.A. § 2201 (West 2006).
[234] *Arrowhead Indus. Water, Inc. v. Ecolochem, Inc.*, 846 F.2d 731, 736 (Fed. Cir. 1988).
[235] 549 U.S. 118 (2007).
[236] *Id.* at 771.
[237] *Teva Pharm. USA, Inc. v. Novartis Pharm. Corp.*, 482 F.3d 1330, 1336 (Fed. Cir. 2007).

[238] *Markman v. Westview Instruments, Inc.*, 52 F.3d 967, 973 (Fed. Cir. 1995) (*en banc*), *aff'd*, 517 U.S. 370 (1996).
[239] *Id.* at 979–980.

directed to a "composition of matter *comprising* compound X" would be infringed by a formulation containing compound X regardless of the presence of other components; an infringing composition must contain the listed element and can contain any other elements. A claim directed to a "composition of matter *consisting essentially of* compound X and compound Y" would be infringed by a formulation containing both compounds X and Y and other components that do not materially affect the basic and novel characteristic of the invention. A claim directed to a "composition of matter *consisting of* compound X and compound Y" would be infringed by a formulation containing only compounds X and Y (and normal impurities, and so on); an infringing composition in this case must have only the listed elements and no others.

In the United States, patent infringement can occur either by literal infringement or under the doctrine of equivalents. Literal infringement of a patent requires that each and every limitation or element of at least one claim be found in the composition, device, or process as claimed. If a single limitation is not present, there is no literal infringement. However, infringement may still be found under the doctrine of equivalents where the differences between the claimed invention and the accused composition, device, or process are "insubstantial." Evidence that the accused composition, device, or process performs "substantially the same function, in substantially the same way, to achieve the same result" as the composition, device, or process defined in the patent claim may also support a finding of equivalency, but this test is generally more suitable for analyzing mechanical devices than chemical compositions or processes.

The doctrine of equivalents provides a limited ability to effectively expand the scope of a patent claim beyond the literal language of the claim. Otherwise, a party who copies a patent's inventive concept, but makes some trivial or obvious change, would avoid legal infringement. The doctrine of equivalents is a flexible concept. Pioneer patents will normally be entitled to a broader range of equivalents than patents claiming a relatively small advance over the prior art. As a technology develops and matures (i.e., becoming a "crowded art"), claims generally will be entitled to a smaller range of equivalents.

Balanced against this need to protect against invention pirating is the public's interest in knowing with reasonable certainty what a given patent claim covers. The doctrine of equivalents creates uncertainty as to the scope of patent rights and may thereby discourage legitimate efforts to design around existing patents. Important technical advancements often are the result of such design around activities. A proper balance between these competing considerations is crucial to a properly functioning patent system.

The courts have struggled to formulate and apply the doctrine in a manner that achieves this difficult goal. In 1995 the Federal Circuit addressed basic issues concerning the doctrine of equivalents and held that (1) the determination of equivalents is a fact issue to be resolved by the jury, (2) the test focuses on the substantiality of differences between the claimed and the accused products or processes, although other factors such as evidence of copying or designing around may also be relevant, and (3) equivalency is determined objectively from the perspective of a hypothetical person of ordinary skill in the relevant art.[240] Two years later, the Supreme Court confirmed equivalents infringement as a viable doctrine, but expressed concern that the doctrine had become "unbounded by the patent claims."[241] The Supreme Court required a "special vigilance against allowing the concept of equivalence to eliminate completely any [claim] elements without really clarifying the proper test."[242]

There are important limitations on the doctrine of equivalents intended to prevent patent rights from being extended too broadly. First, the doctrine is applied on an element-by-element basis. Thus, for each claim limitation there must be a corresponding element in the accused product or process that is identical or equivalent (i.e., only insubstantially different) to that limitation for there to be infringement

---

[240] *Hilton Davis Chem. Co. v. Warner-Jenkinson Co.,* 62 F.3d 1512, 1518–1523 (Fed. Cir. 1995) (*en banc*).
[241] *Warner-Jenkinson Co. v. Hilton Davis Chem. Co.,* 520 U.S. 17, 28–29 (1997).
[242] *Id.* at 40.

by equivalents. Stated differently, if a theory of equivalence would effectively read a claim limitation out of the claim altogether, then there is no infringement.

Further, a composition, device, or process cannot infringe a patent under the doctrine of equivalents where the broader interpretation of the patent claims necessary to cover the accused device or process would also cover "prior art" (i.e., the state of the technology before the filing of the patent application).[243]

A third limitation on the doctrine of equivalents is "prosecution history estoppel" that prevents a patent owner from obtaining coverage under the doctrine of equivalents of subject matter that was effectively relinquished during prosecution of the patent application. The purpose of prosecution history estoppel is to protect the notice function of the claims. Courts traditionally adopted a "flexible bar" approach under which the subject matter surrendered was determined by what a reasonable competitor would likely conclude from reviewing the prosecution file. However, in its 1999 *Festo* decision (*Festo I*), the Federal Circuit ruled that when a claim limitation is narrowed during prosecution, application of the doctrine of equivalents to that claim element is completely barred.[244] This complete bar approach attempted to eliminate the public's need to speculate as to the subject matter surrendered by a claim amendment. The *Festo I* decision also confirmed that prosecution history estoppel applies to all narrowing amendments made for any reason related to patentability, even those made for purposes of clarification and not in response to a prior art rejection.[245] Arguments made during prosecution, without a claim amendment, can also create an estoppel if they demonstrate a surrender of subject matter.[246]

Although the *Festo I* decision would lend certainty to the determination of patent scope, there was concern that it represents an overly rigid approach that will, in effect, severely restrict patent rights. The Supreme Court overturned the complete bar approach of *Festo I* and returned the case to the Federal Circuit for further consideration.[247] The claim scope actually surrendered by amendment will remain an important consideration in applying the doctrine of equivalents and prosecution history estoppel. However, according to the Supreme Court, the patent owner should bear the burden of overcoming presumptions that (1) amendments that narrow a patent claim, for any reason, create prosecution history estoppel and (2) all coverage between the originally proposed claim and the issued claim was surrendered by the patent owner.

In 2003, the Federal Circuit issued its *Festo II* opinion, clarifying its interpretation of the Supreme Court's ruling and providing further guidance for analyzing the doctrine of equivalents when prosecution history estoppel is at issue.[248] The court confirmed that any narrowing amendment made for a reason related to patentability may invoke prosecution history estoppel, even amendments not made to distinguish prior art. If the prosecution record does not reveal the reason for an amendment, then it is treated as an amendment that was made for a substantial reason related to patentability, which creates a second presumption that all equivalents for the amended claim limitation were surrendered. The patent holder may rebut the presumption by showing that the particular equivalent in question was not surrendered. The criteria for determining the scope of surrender may include whether a particular equivalent was foreseeable, whether the amendment was directly relevant to the alleged equivalent, and whether failure to expressly claim the equivalent was due to shortcomings of language. Other criteria may come to light as the district courts gain experience applying the reformulated doctrine.

4.5. Defenses to Infringement

Noninfringement of the patent claims is almost always raised as a defense in patent

---

[243] *Wilson Sporting Goods Co. v. David Geoffrey & Assoc.*, 904 F.2d 677, 684 (Fed. Cir. 1990).
[244] *Festo Corp. v. Shoketsu Kinzoku Kogyo Kabushiki Co.*, 234 F.3d 558, 563 (Fed. Cir. 1999); vacated, 535 U.S. 722 (2002).
[245] *Id.* at 566.
[246] *Id.* at 568.
[247] *Festo Corp. v. Shoketsu Kinzoku Kogyo Kabushiki Co.*, 535 U.S. 722 (2002).
[248] *Festo Corp. v. Shoketsu Kinzoku Kogyo Kabushiki Co.*, 344 F.3d 1359 (Fed. Cir. 2003).

infringement suits. The alleged infringer may assert that the accused product or process does not infringe the patent claims either literally or under the doctrine of equivalents. The burden is on the patentee to show that the claims cover the alleged infringing composition, device or process. The alleged infringer may also attempt to show that the patentee is estopped from expanding the claims under the doctrine of equivalents sufficiently to cover the alleged infringing composition, device or process either because of admissions or arguments presented during the prosecution of the patent before the PTO or because the broader reading of the claims would encompass and cover prior art.

The invalidity of patent claims can also be raised as a defense. Claims in an issued patent are presumed valid. Thus, the alleged infringer must prove a given claim is invalid by a clear and convincing standard of proof. Validity generally involves three components: novelty, utility, and nonobviousness. If the court finds that a claim lacks any one of the three attributes, that claim is invalid and cannot be infringed by the alleged infringer or anyone else. In other words, a final holding by a court that a patent claim is invalid prevents the patentee from asserting the specific claim against anyone. Other claims in the patent may, however, remain valid and enforceable.

The validity of a patent claim is often attacked by a challenger presenting prior art that, in the challenger's opinion, anticipates the patent claim or renders it obvious. If the prior art offered by the challenger is not materially different from that considered by the PTO during examination of the patent, the challenger will generally have a difficult time showing that the patent is not valid; in effect, the challenger must convince the judge or jury that the PTO failed to do its job properly in granting the patent. If the challenger can present prior art that was not before the PTO examiner, and that is more relevant to the invention than the prior art before the examiner, the challenger will have an easier time proving invalidity. In important cases, the challenger may be willing to go to great lengths to find additional prior art, including conducting extensive literature searches, hiring technical experts, interviewing other researchers in the field, and even posting reward for helpful information.

A challenger may also attempt to invalidate a patent by showing sales, offers of sale, or public use by the patentee or a third party of the invention more than 1 year prior to the effective filing date of the patent. Evidence of such activity rarely comes to the attention of the PTO during prosecution of the patent. Such evidence presented in an infringement suit can represent a serious challenge to patent validity. In other cases, a challenger may also attempt to invalidate a patent by showing that the specification is not enabling or that it fails to disclose the best mode of carrying out the invention that was known by the inventor (s) at the time the application was filed. These defenses are used less often and are considered less likely to persuade a jury at trial.

The challenger may also assert that the patentee acted improperly before the PTO in obtaining the patent. This inequitable conduct or fraud defense is raised in many patent infringement suits. The process of obtaining a patent is an *ex parte* proceeding between the applicant and the PTO. Because representations made by the applicant are generally accepted at face value, the courts have established a relatively high standard of candor and conduct for the applicant in the dealings with the PTO. A challenger asserting inequitable conduct will attempt to prove that the applicant was not entirely forthcoming during the *ex parte* proceedings before the PTO. Inequitable conduct might include, for example, an applicant failing to bring relevant and material prior art known to the applicant to the attention of the PTO, attempting to "bury" an especially relevant prior art document in a large collection of seemingly less relevant prior art, falsifying data, or otherwise misleading the PTO. The challenger must show by clear and convincing evidence that the alleged mischaracterization or other inequitable conduct was "material" and that the applicant acted with the required intent to mislead or to deceive the PTO.

The materiality and intent requirements of inequitable conduct are interrelated. For example, a high level of materiality can make it easier to show that the conduct was intentional, and a specific showing of intent can reduce the level of materiality required. Most often, information is considered material if there is a reasonable likelihood that a reasonable examiner would have considered it important in deciding whether to allow the claims. The necessary "intent," which can be shown by circumstantial evidence, is usually characterized as an intent to deceive the PTO. Simple negligence or an erroneous judgment made in good faith will generally not support a finding of the necessary intent.[249] Gross negligence may support a finding of the required intent. A specific showing of wrongful intent clearly provides the necessary intent.[250] In contrast to invalidity based on the other criteria discussed, a finding of unenforceability based on inequitable conduct affects all claims in the patent, not just the claims specifically related to the inequitable conduct. It may also render related continuing patents unenforceable.

There is no general "experimental use" exemption or defense against a charge of infringement.[251] There is, however, a so-called clinical trial exemption that is of special interest to the drug discovery and development industry. This clinical trial exemption generally provides that it is not infringement to make, use, or sell a patented invention "solely for uses reasonably related to the development and submission of information under a Federal law which regulates the manufacture, use, or sale of drugs or veterinary biological products."[252] Thus, a third party may use certain patented drug-related devices, compositions, or processes during the life of the patent in order to develop data and information reasonably necessary for use in, for example, the FDA approval process. This exemption is generally intended to allow a manufacturer to obtain FDA approval during the life of the patent in order to be able to offer a generic drug as quickly as possible after the patent on the drug expires. If the manufacturer is forced to delay clinical trials and similar data-generating activity until the patent term expires, the FDA approval could be delayed until well after the end of the patent term, thereby effectively extending the patent term and depriving the public of the alternative product. This provision, and the patent term extension provision for drug-related inventions, are generally designed to insure that the patentee obtains the full measure of protection normally offered by a patent, but no more.

The Supreme Court in *Merck v. Integra* broadly construed this safe harbor provision as extending to "all uses of patented inventions that are reasonably related to the development and submission of any information under the FDCA."[253] In doing so, it rejected the Federal Circuit's more narrow reading suggesting that the exception was limited to research relevant to filing an ANDA for approval of a generic drug and that animal testing was not included in the exemption.[254] The Supreme Court interpreted the statute as applying to any testing "reasonably related" to the FDA approval process, even if the test data is not ultimately submitted to the

---

[249]However, such intent may be sufficient for a finding of willful infringement if the materiality of the particular prior art is especially high. See, for example, *N.V. Akzo v. E.I. Du Pont de Nemours,* 810 F.2d 1148, 1153 (Fed. Cir. 1987).

[250]*See Kingsdown Med. Consultants Ltd. v. Hollister Inc.*, 863 F.2d 867, 876 (Fed. Cir. 1988); *Atlas Powder Co. v. E.I. Du Pont de Nemours,* 750 F.2d 1569, 1578 (Fed. Cir. 1984).

[251]A very limited "experimental use" exception allows otherwise infringing acts if the acts are "for amusement, to satisfy idle curiosity or for strictly philosophical inquiry." *Roche Prods., Inc. v. Bolar Pharm. Co.*, 733 F.2d 858, 862 (Fed. Cir. 1984). Commercial activities are very unlikely to fall within this exception.

[252]35 U.S.C.A. § 271(e) (West 2001 & Supp. 2008). This exemption specifically excludes from its coverage "a new animal drug or veterinary biological product ... which is primarily manufactured using recombinant DNA, recombinant RNA, hybridoma technology, or other processes involving site specific genetic manipulation techniques."

[253]*Merck KGaA v. Integra Lifesciences I, Ltd.*, 545 U.S. 193, 202 (2005).

[254]*Id.* at 206.

FDA.[255] It seems clear that both preclinical and clinical testing conducted after identification of a "lead" compound is exempt from infringement. It remains unclear, however, whether the exemption extends back to earlier stages of drug investigation.

4.6. Remedies for Infringement

If a patent claim is found infringed and not invalid, the court "shall award ... damages adequate to compensate for the infringement but in no event less than a reasonable royalty ... together with interest and costs..."[256] The purpose of such compensatory damages is to return to the patentee the value of the loss associated with the unauthorized use of the patented item or process. Where the patentee or exclusive licensee is exploiting the patent and lost or diverted sales can be demonstrated which, but for the infringement, would have gone to the patentee or exclusive licensee, lost profits (rather than a reasonable royalty) might be the appropriate estimate of damages.

To prove lost profits, the patentee must show (1) demand for the patented product; (2) ability to meet the demand; (3) absence of noninfringing substitutes; and (4) the amount of profit that would have been earned.[257] However, this is not an exclusive test, and the patentee may find other ways to prove with reasonable probability that it would have made all or some of the infringing sales.[258]

Lost profits may be proven based on diverted sales, eroded prices, or increased costs. In a two-supplier market, it is presumed that the patentee would have made the infringer's sales or charged higher prices without the infringing competition.[259] In markets with more than two competitors, market shares and growth rate projections may be considered.[260] Price erosion considerations can become significant to the damages calculation if the patentee can show it could have charged higher prices.

Lost profits on unpatented products that directly compete with the infringing product are recoverable if the lost sales were reasonably foreseeable. The patentee in *Rite-Hite Corp. v. Kelley*[261] made a product covered by its patent and a higher priced unit that was not covered. The allegedly infringing product was intended to compete with the unpatented unit. The Federal Circuit held that the patentee was entitled to recover profits it lost on diverted sales of the unpatented unit as compensation for the infringement.

A patentee may also recover damages based on unpatented components sold with a patented product, either on a lost profit or on a royalty theory. *Rite-Hite* further clarified that doing so requires proof that the unpatented components function together with the patent product in some manner to produce a desired result. Unpatented components sold with the patented product only for convenience or business advantage may not be included in the damage calculation.[262] However, the infringer who proceeds with the infringing activity, reasoning that the worst case outcome is having to pay a royalty on its sales, may be unpleasantly surprised when ordered to pay damages based on lost profits of the patentee's unpatented products and accessories.

If lost profits cannot be shown, the patentee is entitled to a reasonable royalty, which is calculated as the amount a reasonable person would have been willing to pay as a royalty in a hypothetical negotiation at the time infringement began. Some of the factors that may be

---

[255] The Court provided the following guidance regarding the applicability of the exemption: "At least where a drug-maker has a reasonable basis for believing that a patented compound may work, through a particular biological process, to produce a particular physiological effect, and uses the compound in research that, if successful, would be appropriate to include in a submission to the FDA, that use is 'reasonably related' to the 'development and submission of information under ... Federal law' ..." *Id.* at 207.

[256] 35 U.S.C.A. § 284 (West 2001).

[257] *Gyromat Corp. v. Champion Spark Plug Co.*, 735 F.2d 549, 549 (Fed. Cir. 1984).

[258] *Carella v. Starlight Archery & ProLine Co.*, 804 F.2d 135, 141 (Fed. Cir. 1986).

[259] *Lam, Inc. v. Johns-Manville Corp.*, 718 F.2d 1056, 1065 (Fed. Cir. 1983).

[260] *State Indus., Inc. v. Mor-Flo Indus., Inc.*, 883 F.2d 1573 (Fed. Cir. 1989).

[261] *Rite-Hite Corp. v. Kelley*, 56 F.3d 1538 (Fed. Cir. 1995).

[262] *Id.* at 1550–1551.

considered in determining the reasonable royalty include the infringer's expected profit, whether the license would have been exclusive, actual established royalties under the patent or in the particular field, and the parties' bargaining positions at the time the infringement began.[263]

Infringement can range from unknowing or accidental to deliberate or reckless disregard of the patentee's legal rights. For "willful infringement," damages can be increased up to three times the actual damages[264] and may, in exceptional cases, include an award of attorney fees.[265] The risk of up to triple damages in the case of willful infringement, and attorney fees in exceptional cases, are designed to deter infringement and encourage would-be infringers to evaluate their actions carefully in light of the claims of relevant patents.

Until recently, willful infringement decisions have focused on the notion that a potential infringer with notice of the patent has an affirmative duty of care to determine whether any of its actions would constitute infringement. This inquiry focused on the potential infringer's state of mind. This duty could generally be discharged by obtaining competent legal advice concerning the potential infringement.[266] In its decision in *In re Seagate Technology*, the Federal Circuit abandoned the affirmative duty of care standard and held that willful infringement requires a showing of "objective recklessness" on the part of the infringer.[267] The court further noted that under the new standard, there is no affirmative obligation to obtain opinion of counsel. To prove that the infringer's conduct has been objectively reckless, the patentee must show by clear and convincing evidence that the infringer acted despite an objectively high likelihood that its actions infringed a valid patent. In making this threshold determination, the infringer's subjective state of mind is irrelevant. The patentee must also show that the objectively high risk either was known or should have been known to the infringer.[268]

The *Seagate* decision represents a substantial change to the law of willful infringement; its full impact will not be known until the courts have had a chance to apply this new test. Defendants will probably continue to introduce written opinions of counsel in many cases as evidence of nonrecklessness, but other sources of evidence, such as testimony of legal and technical experts, may also play a significant role in the determination of willful infringement.

In addition to money awards, the court may grant an injunction to "prevent the violation of any right secured by patent, on such terms as the court deems reasonable."[269] Preliminary injunctions, by which the alleged infringer is required or forbidden to do certain specified acts before a full trial on the merits of the case, serve to preserve the status quo in the market during the suit. These are rarely granted. In order to obtain a preliminary injunction, the patentee must show (1) a reasonable likelihood of success on the merits, (2) irreparable harm to the patentee if the injunction is not granted, (3) hardships favoring the patentee, and (4) that the public interest will be favored by granting the injunction.[270] Such a showing is often difficult to make at the early stages of the proceeding.

Permanent injunctions forbidding further infringement during the remaining term of the patent have traditionally been routinely granted after a trial court has found the patent to be valid and infringed. However, as mentioned in Section 4.2, the recent Supreme Court decision in *eBay v. MercExchange* substantially limited the availability of perma-

---

[263] See *Fromson v. Western Litho Plate & Supply Co.*, 853 F.2d 1568, 1576–1578 (Fed. Cir. 1988); *Georgia Pacific Corp. v. United States Plywood Corp.*, 318 F. Supp. 1116, 1120 (S.D.N.Y. 1970).

[264] 35 U.S.C.A. § 284 (West 2001).

[265] 35 U.S.C.A. § 285 (West 2001). An award of attorney fees might be appropriate for an especially willful and egregious infringer or against a patentee who litigated in bad faith or committed fraud during the prosecution of the patent in the PTO. See, for example, *Mach. Corp. of Am. v. Gullfiber AB*, 774 F.2d 467, 472–473 (Fed. Cir. 1985).

[266] See *Underwater Devices, Inc. v. Morrison Knudsen*, 717 F.2d 1380, 1389–1390 (Fed. Cir. 1983).

[267] *In re Seagate*, 497 F.3d 1360, 1371 (Fed. Cir. 2007).

[268] *Id.*

[269] 35 U.S.C.A. § 283 (West 2001).

[270] *Atlas Powder Co. v. Ireco Chem.*, 773 F.2d 1230, 1231 (Fed. Cir. 1985).

nent injunctive relief against adjudicated infringers. In contrast to presumed appropriateness of such relief before the *eBay* decision, now a court may only enjoin further infringement if it is shown that (1) the plaintiff has suffered irreparable injury, (2) damages remedies are inadequate to compensate for that injury, (3) considering the balance of hardships between the plaintiff and the defendant, an injunction is warranted, and (4) the injunction would not disserve the public interest.[271] Requests for permanent injunctions have met with only limited success following the *Ebay* decision. Some decisions have resulted in the setting of a reasonable royalty on future sales as an alternative to injunctive relief.[272]

4.7. Commencement of Proceedings

A suit to enforce or attack a patent may be filed against a given defendant in any district court in which requirements of personal jurisdiction and venue are met. The rules governing venue in patent cases seek to prevent the defendant from having to litigate in an inconvenient forum. Generally, the proper venue for a patent infringement defendant is any district where the defendant resides or has committed acts of infringement and has an established place of business.[273]

The requirement that the court have power to exercise personal jurisdiction over the defendant depends on whether the defendant has established some minimum contact with the state in which the district court is located and, if so, whether the assertion of jurisdiction over the defendant is fair and reasonable in light of all of the circumstances.[274] All of the defendant's contacts with that state, such as offices, employees, bank accounts, product sales, advertising, telephone listings, and so on, may be considered. Even a non-U.S. company with no U.S. operations may be subject to jurisdiction if it places an infringing product into distribution channels that it reasonably knows will lead to product sales in the forum state.[275]

The complaint that commences the action must identify the parties, state the basis of the court's jurisdiction, and contain a "short and plain statement showing that the pleader is entitled to relief."[276] In a patent infringement complaint, the statement will normally allege that the plaintiff owns or otherwise has rights to enforce the patent, that the defendant is infringing the patent by making, using, selling, offering for sale, or importing products embodying the patented invention, and that the infringement has damaged plaintiff. The plaintiff may request preliminary and/or final injunctive relief and an accounting for damages. The complaint should also include a demand for jury trial if a jury is desired.

When suspected patent infringement activities occur across an entire industry, the patentee may proceed simultaneously in one or multiple suits against all competitors, or in some logical sequence. Often the first suit is brought against one competitor, in the hope that a ruling of infringement and validity will persuade others to accept a license. Although every defendant is entitled to its own day in court,[277] the Federal Circuit's holding in *Markman* that claim interpretation is an issue of law to be decided by the court, and not the jury, left it unclear whether a subsequent defendant may relitigate claim interpretation issues decided in a previous action to which it was not a party.[278] A previous claim construc-

---

[271]*eBay Inc. v. Merc Exchange, L.L.C.*, 547 U.S. 388, 391 (2008).
[272]See, for example, *Innogenetics N.V. v. Abbott Labs.*, 512 F.3d 1363 (Fed. Cir. 2008) (vacating permanent injunction and remanding to district court for determination of terms of a compulsory license).
[273]*See* 28 U.S.C.A. §§ 1391(c), 1400(b) (West 2006).
[274]See, for example, *Red Wing Shoe Co. v. Hockerson-Halberstadt, Inc.*, 148 F.3d 1355 (Fed. Cir. 1998).
[275]*See Beverly Hills Fan Co. v. Royal Sovereign Corp.*, 21 F.3d 1558 (Fed. Cir. 1994) (holding district court in Virginia had personal jurisdiction over Chinese corporation who manufactured accused product in Taiwan and sold to New Jersey importer because it placed the products in a "stream of commerce" knowing the likely destination of some of the products was Virginia).
[276]Fed. R. Civ. P. 8(a).
[277]*Blonder-Tongue Labs Inc. v. Univ. Ill. Found.*, 402 U.S. 313 (1971).
[278]*Markman v. Westview Instruments, Inc.*, 52 F.3d 967, 976–979 (Fed. Cir. 1995) (*en banc*), *aff'd*, 517 U.S. 370 (1996).

tion by a district court is normally binding on the patentee if the determination of claim scope was essential to a final judgment on the question of validity or infringement.[279] A prior interpretation by a district court is not binding on a later defendant who did not have a full and fair opportunity to litigate the issue in the first action.[280] It is less clear whether an accused infringer is bound by a prior Federal Circuit decision construing a patent's claims in a suit against a different infringer.

The complaint will sometimes name multiple competitors as defendants in the same action. The patentee's interest in doing so is to avoid the added expense and potential for inconsistent outcomes associated with piecemeal litigation. The competitor-defendants' interests in avoiding common discovery and trial proceedings involving commercially sensitive technical and financial information is also compelling, particularly the risk that the jury may confuse the features of the different accused products or methods or attribute the actions of one defendant to other defendants. There is also a public interest in not directing the resources of multiple federal judges toward the same patent.

The defendant must respond to the patent complaint by either filing an answer or bringing motions challenging the complaint on one or more procedural grounds. An answer must admit or deny each of the plaintiff's allegations, and raise any "affirmative defenses."[281] Affirmative defenses are matters outside the plaintiff's main case that are in the nature of avoidance, such as invalidity of any patent claim in suit, unenforceability of the patent, laches, and implied license. An accused infringer will normally bring a counterclaim seeking to have the patent declared invalid or not infringed. In a declaratory judgment action, the defendant patentee is usually seeking a counterclaim for patent infringement.

The most common motions filed in lieu of an answer in patent cases are those challenging the court's jurisdiction over the defendant, requesting transfer to a more convenient district court, or alleging that an indispensable party with an interest in the dispute has not been joined. Such motions rarely resolve cases, but may result in tactical advantages such as adding a party with greater access to important evidence or avoiding a court where fast disposal rates provide less time for an accused infringer to investigate the prior art or to design around the patent.

4.8. Discovery

Extensive discovery is a hallmark of U.S. patent litigation. The rules relating to discovery encourage full disclosure by the parties to avoid potentially unfair surprises at trial. The scope of permissible discovery is extremely broad, and allows a party to seek from its opponent or a nonparty any information that is not protected by a recognized privilege and either is relevant to an issue in the case or appears "reasonably calculated" to lead to admissible relevant evidence.

The patentee will seek evidence regarding the accused product or process, the profitability of the infringing activities, the accused infringer's awareness of and intent with regard to the patent, and the accused infringer's defenses. Discovery into the latter subject may encompass noninfringement defenses, prior art defenses, and equitable defenses such as undue delay. Persons requested to provide such information may include the management, research, engineering, and accounting personnel of the accused company, retained experts, customers and suppliers, former employees, and even competitors.

The accused infringer will generally seek evidence regarding the development of the invention and prosecution of the patent application, details of the patentee's commercial product or process, prior art to the asserted patent, the patentee's awareness of uncited prior art, and the profitability of the patented product or process. Discovery requests for such information may be directed to the inventors, the patentee's own engineering, marketing and financial personnel, the patent attorney involved in prosecution, and other companies with prior related technology. Obtaining discovery from a nonparty requires

---

[279] *In re Freeman*, 30 F.3d 1459, 1465–1466 (Fed. Cir. 1994).
[280] *Id.* at 1465.
[281] Fed. R. Civ. P. 8(c).

counsel to issue a subpoena requesting specific information or a deposition, which the district court may quash or modify if the subpoena request is overbroad or otherwise unreasonable.[282]

Two important limitations on discovery of relevant information in patent actions are the attorney–client privilege and the work product immunity doctrine. The former protects confidential communications between lawyer and client for the purpose of securing legal advice. The latter protects documents prepared by or for counsel in connection with litigation, and counsel's mental impressions.

The five discovery procedures commonly employed in patent actions are requests for production, interrogatories, requests to admit, depositions, and inspections. Production requests are used to obtain tangible evidence such as documents, drawings, photographs, product samples, and digitally stored data. A party may also request to inspect and test the accused or patented product or process. Interrogatories are written questions posed to the opposing party that must be answered in writing under oath, subject to any valid objections. The written answers may be used as evidence at trial. Requests to admit seek written admissions that certain facts are correct. Although used somewhat sparingly in patent cases, requests to admit can be effective tools for resolving issues, such as the genuineness of important documents or dates of important events, before trial.

The deposition process permits counsel to obtain sworn testimony from witnesses before trial. Depositions enable counsel to learn more about the underlying facts and events, explore an opponent's case, identify additional evidence and witnesses, and obtain admissions regarding important facts. Deposition testimony may be admissible at trial if the witness is unavailable to testify in person. Depositions are the primary mechanism of obtaining testimony from nonparty witnesses who reside outside the geographic limits of the court's trial subpoena power. A deposition transcript may also be used to contradict the testimony of a deponent who later testifies at trial. The deposition of a party may be used at trial for any purpose.[283] Depositions are also important opportunities for each party to assess the strengths and weakness of their case and the credibility and demeanor of key witnesses.

The discovery rules also require the parties to identify each expert witness they will call at trial and provide a report summarizing the expert's credentials and opinions.[284]

4.9. Summary Judgment

A court may dispose of an action without conducting a trial where the evidence presents no genuinely disputed issues of fact. A party may seek "summary judgment" at any time, although such requests are most commonly made after the completion of discovery. Courts are frequently requested to resolve patent cases in the context of summary judgment motions. If complex factual issues exist, and expert testimony is required to explain the technology to the court, such motions should be denied.[285] However, when the facts are not disputed, and an issue such as infringement depends only on a question of law, summary judgment becomes an effective way to resolve cases more quickly and efficiently.

Rulings commonly made on summary judgment motions include invalidity due to anticipation (i.e., lack of novelty), no literal infringement because of the absence of a claim element, or infringement when all elements are present. Where the court finds no literal infringement, the appropriateness of granting summary judgment of no infringement under the doctrine of equivalents depends on whether one of the legal limits on the doctrine applies, such as prosecution history estoppel or the prior art. Issues such as obviousness and unenforceability due to inequitable conduct often involve fact issues precluding the use of summary judgment.

---

[282]Fed. R. Civ. P. 45.
[283]Fed. R. Civ. P. 32. A corporation may be deposed through one or more designated representatives who speak on the corporation's behalf. Fed. R. Civ. P. 30(b)(6).

[284]Fed. R. Civ. P. 26(a)(2)(A)–(B).
[285]See *P. M. Palumbo v. Don-Joy Co.*, 762 F.2d 969, 975 (Fed. Cir. 1985).

4.10. Patent Claim Construction Proceedings

As mentioned above, the Federal Circuit's *Markman* decision resolved inconsistencies in prior precedent and held that the court, and not the jury, has the power and obligation to construe the meaning and scope of patent claims as a matter of law. Claim construction is common in patent cases, and the court's rulings on these issues often have a significant impact on validity and infringement issues.

The court is to construe disputed claim terms to have the meaning that would be understood by one of ordinary skill at the time of the invention. The starting point is the claim language, which is interpreted in a manner that is consistent with the invention described in the patent specification.[286] The court is required to walk a fine line between construing the claims consistent with the disclosed invention, without improperly limiting them to that disclosure. The court also considers the prosecution history record, and looks for statements defining claim terms or unmistakably surrendering claim scope.[287] A court has more limited discretion to consider evidence external to the patent and its prosecution history record. It may do so to educate itself about the invention and the relevant technology, but may not use such extrinsic evidence to arrive at a claim construction that contradicts the intrinsic evidence.[288]

The court will frequently conduct a separate evidentiary hearing, known as a "*Markman* hearing," prior to trial for this purpose. Such hearings may be particularly helpful to the judge when expert testimony is required to understand the relevant technology. The scope of evidence to be presented and the timing of the hearing are within the judge's discretion. Construing the claims early in the litigation process enables the parties' expert witnesses to predicate their opinions on a correct interpretation of the claims and allows the parties to better focus their efforts on pretrial discovery and dispositive motions. An early ruling regarding claim construction may also facilitate settlement. When the case involves particularly complex technology, the court's preference may be to defer issues of claim construction until the parties have completed sufficient discovery to present a full picture of the relevant technology and prior art.

4.11. Trial

The trial of the patent action involves the presentation of evidence, submission of the case to the trier of fact, followed by the entry of judgment. The purpose of the trial is to decide facts based on the testimony of witnesses and examination of evidence. In a jury trial, factual issues are determined by the jury and legal issues are decided by the judge.[289] If a jury is not requested, the judge also performs the fact-finding role. The judge presides over the trial to assure that the process is fair to the litigants. In carrying out this obligation, the judge will resolve the parties' disputes regarding the conduct of the trial, rule on objections to the admissibility of evidence or improper questions asked of witnesses, instruct the jury as to the findings it must make and the governing law, and control the conduct of the attorneys and witnesses. Prior to trial, each party will educate the judge with a trial brief outlining their case and what they intend to prove at trial.

Jury selection is also conducted prior to a trial. The judge and the attorneys question potential jurors to determine if they are able to be fair and impartial; this process is called *voir dire*. The judge may excuse an individual from service in a particular case for various rea-

---

[286] *Markman v. Westview Instruments, Inc.*, 52 F.3d 967, 976 (Fed. Cir. 1995) (*en banc*), *aff'd*, 517 U.S. 370 (1996); *Phillips v. AWH Corp.*, 415 F.3d 1303, 1314–1317 (Fed. Cir. 2005) (*en banc*).
[287] *Phillips*, 415 F.3d at 1317.
[288] *Vitronics Corp. v. Conceptronic, Inc.*, 90 F.3d 1576 (Fed. Cir. 1996).

[289] Examples of fact issues include disputes regarding the structure and operation of the accused device, the teachings of the prior art and level of skill in the art, and the amount of damages sustained on account of the infringement. Examples of legal issues include claim interpretation, and whether the prior art prevents infringement under the doctrine of equivalents.

sons. Each party may also use a limited number of "peremptory challenges" to excuse a potential juror for any reason except race or gender. The process of questioning and excusing jurors continues until eight to twelve persons are accepted as jurors for the trial. To the extent possible, the parties are also attempting during the jury selection process to learn the attitudes jurors have about patents, the parties involved in the dispute, and other factors that may affect their view of the dispute.

Another aspect of the court's role relates to expert testimony. Technical, economic, and financial experts often play a key role in patent cases. Often, the party with the more credible and convincing experts wins. A trial judge has broad discretion to consider whether a witness is qualified as an expert, and whether the expert's knowledge will assist the jury. This requires a preliminary assessment of whether the expert's reasoning and methodology is scientifically valid.[290] Technical or financial expert analysis that does not satisfy accepted standards in the expert's particular field may also be subject to exclusion.

Each party may present live witness testimony, deposition testimony, documents and demonstrative evidence such as charts, photographs, videos, and physical objects. Given the potential for jury confusion, trial counsel must carefully plan the sequence and manner of presenting the evidence that will aid the court and jury in understanding the complex technical and legal issues involved and lead the decision maker to the desired result. Generally, counsel will attempt to present a simple, straightforward case focusing on a few key themes and a logical "story" that the judge and jury can understand.

The patent owner has the burden of proving infringement by a preponderance of evidence, which requires showing that the defendant has more likely than not infringed. This may be done through the testimony of lay and expert witnesses, depending on the circumstances of a particular case. The patent owner must also prove damages, and will normally do so through the testimony of economic and financial experts, whose opinions will often be based on facts established through the testimony of lay witnesses.

After cross-examining the patentee's witnesses, the accused infringer will attempt to prove its defenses. The most common defenses are noninfringement and invalidity for obviousness or lack of novelty. Any challenge to patent validity must be proven by clear and convincing evidence. Since a patent claim is presumed to be valid, this standard is a greater burden than the patent owner's preponderance of evidence standard. This is because a patent is presumed to be valid.

Testimony of the defendant's own witnesses may help demonstrate differences between the invention and the accused device, or show that the invention represents only a small advance over the prior art. Proving the content of the prior art may require introducing documents and testimony from nonparty witnesses, often through deposition testimony. The defendant will also invariably introduce testimony from at least one technical expert in its attempt to demonstrate noninfringement and/or invalidity. The defendant will also offer its own fact and expert witnesses to counter the patentee's damages evidence.

The patentee will also have an opportunity to rebut the defenses offered, such as by challenging the disclosures of the prior art, showing the tremendous success of the invention, or that the differences in the accused composition, device or process are insubstantial and should not avoid infringement.

After the presentation of evidence by both parties, the trial counsel delivers closing arguments to the fact finder after which a decision is reached. Certain posttrial motions are available to the losing party, the most common being a motion to set a jury's verdict aside as unreasonable in view of the evidence.[291] Either party may appeal to the Court of Appeals for the Federal Circuit.

4.12. Appeal

Appeals from final decisions in patent infringement cases are to the Court of Appeals for

---

[290] *Daubert v. Merrell Dow Pharm., Inc.*, 509 U.S. 579 (1993).

[291] Fed. R. Civ. P. 50(a).

the Federal Circuit. The Federal Circuit must apply the standard of review that is appropriate to the issues presented on appeal. The standard of review both determines the power of the Federal Circuit to address claims for error and defines the degree of deference it will accord the findings of fact and conclusions of law made in the district court. The standards of review, ranging from broadest to narrowest, include *de novo* review, substantial evidence, clear error standard, and abuse of discretion.

The *de novo* standard of review provides the Federal Circuit the widest latitude in reviewing the lower court decision. Under this standard, the court exercises its own independent judgment on the evidence of record. Examples of issues that the Federal Circuit reviews *de novo* are claim construction, prosecution history estoppel and determinations of enablement, obviousness, claim indefiniteness.[292]

The Federal Circuit applies a more deferential "substantial evidence" standard of review to the jury's factual findings. Typically the "substantial evidence" standard is satisfied unless the jury decision requires something a reasonable mind would not accept. Some of the issues that the Federal Circuit reviews under the substantial evidence standard are literal infringement, equivalency under the doctrine of equivalents, and the factual underpinnings of a finding of inequitable conduct.[293]

Under the "abuse of discretion" standard, the Federal Circuit does not reverse the decision below unless it is grossly unsound, unreasonable, or is a clear error of judgment.

Examples of issues reviewed under this standard are decisions by a judge to grant enhanced damages and a judge's decision on whether to hold a patent unenforceable due to inequitable conduct based on the jury's fact findings.[294]

4.13. Alternative Dispute Resolution

Patent litigation is an expensive undertaking, and both parties face risks that are difficult to assess. Absent unique circumstances, litigants in the United States pay their own attorney fees and costs. In many other countries, the prevailing party may recover some or all of its fees and costs. Legal and expert witness fees, travel expenses, document expenses, and court fees can mount quickly.

Numerous other factors necessarily enter into the strategic planning of the patent owner and accused infringer. In the United States, strong remedies are available to the victorious patent owner. The accused infringer not only faces the risk of being enjoined from continuing the commercial activities in question but also may be required to pay compensatory damages, as well as enhanced damages and attorney fees. The financial consequences of an adverse decision of infringement are thus potentially catastrophic. A legal victory for the accused infringer can be a hollow one if it spends, in the worst case, so much defending its right to continue its commercial activities that it is not profitable to do so.

On the other hand, because patent invalidity is an absolute defense to an infringement charge, the patent owner risks investing substantial litigation expense and effort only to emerge from trial with a valueless, invalid patent and no means of preventing competitors from practicing the once protected technology.

Patent legal proceedings, which can last many years, can seriously disrupt business planning and strategy as well as divert human and economic resources away from the drug discovery and development process. The extremely high cost of litigation generally and especially in complex patent cases, coupled with unpredictability of outcome, the considerable time periods involved, and the likelihood of appeals resulting in even further de-

---

[292]*Markman,* 52 F.3d at 979 (claim construction); *Paragon Podiatry Lab, Inc. v. KLM Labs, Inc.*, 984 F.2d 1182, 1186 (Fed. Cir. 1993) (validity issues).
[293]*Miles Labs, Inc. v. Shandon Inc.*, 997 F.2d 870 (Fed. Cir. 1993) (literal infringement); *Warner-Jenkinson Co. v. Hilton Davis Chem. Co.*, 520 U.S. 17 (1997) (equivalency under the doctrine of equivalents); *Gerber Garment Tech., Inc. v. Lectra Sys., Inc.*, 916 F.2d 683, 686 (Fed. Cir. 1990) (factual underpinnings of inequitable conduct).
[294]*Cybor Corp. v. FAS Tech. Inc.*, 138 F. 3d 1448 (Fed. Cir. 1998) (enhanced damages); *Gerber,* 916 F.2d at 683 (judge's determination regarding inequitable conduct).

lays, may increase the desire of both parties to reach an acceptable settlement before or even during trial. A settlement reached on reasonable business terms can, in many cases, provide a more favorable and satisfactory outcome for both parties.

In some cases, alternative dispute resolution processes (e.g., negotiation, mediation, arbitration, neutral expert fact finding, and so on) can offer benefits over traditional litigation for resolving patent-related disputes, including infringement. Such benefits include, for example, faster resolution of the disputed issues, the ability to tailor the process to the needs of the parties, the ability to select fact finders or decision makers with the educational and technical backgrounds suitable for the technology, generally lower cost, increased predictability of outcome, and a finite and definite resolution of the dispute. Alternative dispute resolution processes are likely to be used with increasing frequency in patent-related cases.

5. WORLDWIDE PATENT PROTECTION

As the marketplace for the products from drug discovery and development has become global in nature, worldwide patent protection has become increasingly important. Seeking patent protection in many countries throughout the world can be very expensive. The cost of obtaining patent protection should be weighed against the benefits derived from patent protection on a country-by-country basis. The countries and regions most often chosen for patenting purposes include the United States, Canada, the European Community, Australia, and Japan. For some specific products or processes and marketing considerations, other countries may be as important, if not more important, than those just listed. Pharmaceutical companies, for example, may wish to file patent applications in most or all countries where they (or their subsidiaries, affiliates, or licensees) are likely to produce and/or market a new drug.

In most cases, it is simply too expensive to attempt to file patent applications in a majority of the countries of the world, much less in every country. The evaluation of where to file should be undertaken on a case-by-case basis taking into account the technology itself and the marketplace. In many cases, it may be possible to obtain adequate patent protection without seeking patent coverage in an excessively large number of countries. For example, if there are interrelated markets, a patent in one country often can offer practical and effective protection (but not legal, enforceable protection) against infringing acts in another country or countries in the market. Thus, for example, a competitor may be discouraged from offering a drug in Canada if that drug cannot be offered in the United States because of the existence of a blocking U.S. patent. In effect, the United States, Canada, and Mexico could form a single North American market.[295] Taking such market considerations into account throughout the world, it may be possible, depending on the specific technology, to obtain practical worldwide patent protection through patents in only a reasonable number of countries.

5.1. International Agreements

Having determined where to file, one faces the task of filing patent applications in the appropriate countries. International agreements have made this process much easier. Although it is possible to file separate patent applications in each of the countries selected, this procedure is rarely used. Rather, the procedures of various international intellectual property treaties can be used to simplify this administrative task considerably. The principal international treaty governing patents is the Paris Convention for the Protection of Industrial Property,[296] which has approxi-

---

[295]The North American Free Trade Agreement (NAFTA) between the United States, Canada, and Mexico has continued the trend toward such a single North American market. See generally Roger M. Milgrim, 2 *Milgrim on Trade Secrets* § 9.07 (2008); Paul, Hastings, Janofsky & Walker, North American Free Trade Agreement: Summary and Analysis (1993).

[296]Paris Convention of March 20, 1883, as amended at Stockholm on September 28, 1979, reprinted in Patents Throughout the World, App. 2(e) (Thomson/West eds. 2008) (earlier revisions are also included); also available from the World Intellectual Property Organization, at http://www.wipo.int.

mately 173 signatory member nations as of mid-2009. A patent application filed in any member nation creates a priority date for applications filed within 12 months (i.e., the convention year) in other convention nations. Thus, an applicant can file a patent application in the United States and then file separate patent applications in other member countries within the ensuing 12 months. Such applications have an effective filing date that is the same as the filing date of the U.S. application.[297] For non-Paris Convention countries (e.g., Taiwan), patent applications must be filed directly in the national patent offices; unless a particular country's laws provide otherwise, such patent applications must rely on their actual filing date in the particular country.[298] The filing procedure utilizing the Paris Convention still requires an application to be filed in every country in which protection is sought and is, therefore, still unwieldy if a large number of countries are selected.

The European Patent Convention (EPC)[299] allows for the filing of a single patent application designating selected member European countries that, following prosecution before and issuance by the European Patent Office (EPO), can become effective as national patents in the designated countries. For applications filed on or after April 1, 2009, an applicant pays a flat designation fee, and unless otherwise indicated, all available EPC contracting states will be designated. Ordinarily, an applicant will want to designate all states because the applicant cannot later seek to obtain a national patent in a country that was not designated. Further, designation of all states does not require the applicant to pursue a national patent in each of these states. Instead, upon issuance of a patent by the EPO, the applicant may seek validation in some, all, or none of the designated states.

Currently, 35 European nations (listed in Table 1) are members of the EPC. In addition, three other countries, although not members, can be reached through EPC procedures.[300] A patent issued by the European Patent Office is not a true European patent; rather it is a grant of separate national patents in the member countries designated, and validated, by the applicant, each of which is enforceable under the laws of the country in which it was granted. Although applications can be filed in individual countries, the use of the single EPC application has been widely accepted as a convenient and less expensive mechanism to obtain coverage when seeking patent protection in four or more of the European member countries. If protection is desired in only a few member countries (i.e., three or less), national applications in the individual countries may be a less expensive alternative. The decision of

Table 1. Members of the European Patent Convention (as of Mid-2009)

Austria	Lithuania
Belgium	Luxembourg
Bulgaria	Macedonia
Croatia	Malta
Czech Republic	Monaco
Cyprus	Netherlands
Denmark	Norway
Estonia	Poland
Finland	Portugal
France	Romania
Germany	Slovakia
Greece	Slovenia
Hungary	Spain
Iceland	Sweden
Ireland	Switzerland
Italy	Turkey
Liechtenstein	United Kingdom
Latvia	

---

[297] Applications can be filed in Paris Convention countries after the 12-month period if there has been no public disclosure. Generally, the priority date of the earlier filed application cannot be claimed. Under certain limited circumstances, however, an applicant may be able to seek a restoration of the right of priority if the application is filed within fourteen months of the priority date.

[298] Pursuant to Taiwanese law, however, Taiwan recognizes priority claims from Paris Convention member states.

[299] European Patent Convention 2000, effective December 13, 2007, available at http://www.epo.org.

[300] These countries include Albania, Bosnia and Herzegovina, and Serbia.

Table 2. Contracting States of the PCT (as of Mid-2009)

Africa	Algeria, Angola, Benin, Botswana, Burkina Faso, Cameroon, Central African Republic, Chad, Comoros, Congo, Cote d'Ivoire, Egypt, Equatorial Guinea, Gabon, Gambia, Guinea, Guinea-Bissau, Kenya, Lesotho, Liberia, Libyan Arab Jamahiriya, Madagascar, Malawi, Mali, Mauritania, Morocco, Mozambique, Namibia, Niger, Nigeria, Oman, Sao Tome and Principe, Senegal, Seychelles, Sierra Leone, South Africa, Sudan, Swaziland, Togo, Tunisia, Uganda, United Republic of Tanzania, Zambia, and Zimbabwe.
Americas	Antigua and Barbuda, Barbados, Brazil, Canada, Chile, Colombia, Costa Rica, Cuba, Dominica, Dominican Republic, Ecuador, El Salvador, Grenada, Guatemala, Honduras, Mexico, Nicaragua, Peru, Saint Kitts and Nevis, Saint Lucia, Saint Vincent and the Grenadines, Trinidad and Tobago, and the United States of America
Asia/ Pacific	Armenia, Azerbaijan, Australia, Bahrain, China, Democratic People's Republic of Korea, Georgia, India, Indonesia, Israel, Japan, Kazakhstan, Kyrgyzstan, Lao People's Democratic Republic, Malaysia, Mongolia, New Zealand, Papua New Guinea, Philippines, Republic of Korea, Singapore, Sri Lanka, Syrian Arab Republic, Tajikistan, Turkmenistan, United Arab Emirates, Uzbekistan, and Vietnam.
Europe	Albania, Austria, Belarus, Belgium, Belize, Bosnia and Herzegovina, Bulgaria, Croatia, Cyprus, Czech Republic, Denmark, Estonia, Finland, France, Georgia, Germany, Greece, Hungary, Iceland, Ireland, Italy, Latvia, Liechtenstein, Lithuania, Luxembourg, Malta, Monaco, Montenegro, Netherlands, Norway, Poland, Portugal, Republic of Moldova, Romania, Russian Federation, San Marino, Serbia, Slovakia, Slovenia, Spain, Sweden, Switzerland, The former Yugoslav Republic of Macedonia, Turkey, Ukraine, and United Kingdom.

whether to use a single EPC application or whether to file directly in individual countries depends on many factors and should be decided on a case-by-case basis.

Since its adoption in 1978, the Patent Cooperation Treaty[301] has become an increasingly important and useful mechanism to obtain patent protection throughout the world. Currently the PCT has 141 member states as of mid-2009, including the United States, Canada, Japan, Australia, and most European countries. The member states are listed in Table 2. The PCT allows the filing of a single international application that has the same effect as if separate applications were filed in each designated country. The PCT does not create an international patent and does not modify the substantive requirements for patentability in the member countries. It simply reduces the effort and resources necessary to file the patent application initially in multiple countries at the same time, thus proving an effective mechanism for filing an international application (especially as the convention year draws to a close and it becomes necessary to file applications very quickly).

Typical practice for an applicant based in the United States might involve filing an application in the U.S. PTO, and then filing applications in the desired countries throughout the world within the convention year. Alternatively, the procedures offered by the EPC and/or PCT could be used for filing such applications.[302] Filing a PCT application directly in the local national patent office (assuming it is a PCT-receiving office) designating most, if not all, of the PCT countries is increasingly becoming the preferred practice both in the United States and the rest of the world. Due to the increasing importance of the

---

[301]Patent Cooperation Treaty of June 19, 1970 (as amended to October 3, 2001), reprinted in Patents Throughout the World, App. 3(a) (Thomson/West eds. 2008); also available from the World Intellectual Property Organization, at http://www.wipo.int.

[302]An applicant may file a PCT application and enter the regional stage in the EPO within 31 months from the priority date. In other words, using the PCT does not prevent an applicant from later taking advantage of the EPC.

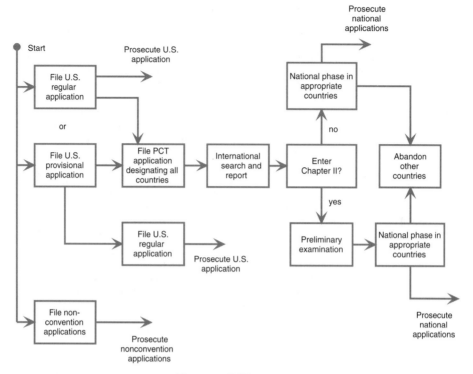

Figure 4. PCT practice.

role of the PCT in obtaining patent protection, it is important to have at least a basic understanding of PCT procedures in order to capitalize on the advantages it offers.

5.2. PCT Patent Practice

A single PCT application can be filed in a PCT-receiving office (the U.S. PTO is one such receiving office) designating all or only certain member states. For example, a U.S. applicant could file a PCT patent application in the English language in the U.S. PTO and designate the appropriate PCT member states[303] in which protection is desired. A flow chart generally illustrating typical PCT practice is shown in Fig. 4. Once the PCT application is filed, an international search is performed by the International Search Authority—through a national patent office or intergovernmental alliance (e.g., the U.S. PTO or the European Patent Office). The application is published 18 months after the effective filing date (i.e., the priority date). Therefore, 18 months after the priority date any trade secrets or other technical information contained in the application are disclosed to the public.

The PCT also provides for an optional international preliminary examination related to the patentability of the claimed invention ("Chapter II proceedings"). Although the results of the preliminary examination are not binding on individual member states, the results of preliminary examination as well as the international search report can offer significant insight and assistance for an applicant in determining the likelihood of ultimately obtaining patent protection and, thus, whether to proceed with the application in the individual designated states.

---

[303] For European countries, an applicant can designate an EPC application (which in turn designates the desired member states) or separate national applications in the individual European countries.

The applicant must elect whether or not to go forward in some or all of the designated countries (i.e., enter the national stage in the elected countries) within thirty or 31 months after the priority date, depending on the specific country.[304] For example, the deadline is 30 months for the United States and 31 months for the European Patent Office.

By delaying the final selection of the elected countries up to 30–35 months after the priority date, the applicant can postpone significant fees associated with entering the national stages, including the costs associated with preparing any required foreign translations. This delay may, at least in some cases, allow the applicant to have a better understanding of the patentability, marketability, and/or commercial potential of the invention. If the viability or importance of the invention has decreased, the number of countries where the application will be pursued can be appropriately reduced (even to zero) for a significant cost savings. The PCT process, therefore, allows additional time to consider the appropriate countries in which to pursue foreign patent protection. It is important to remember when considering the PCT process that, although the number of PCT countries in which patent protection is sought can be reduced, the PCT countries in which the national stage can be entered is limited to those listed or selected in the original PCT filing. For this reason, applicants routinely select all PCT countries at filing.

After the PCT application has been searched and published, it is transferred to the patent offices of the individual countries designated by the applicant for entry into the national stage. In the national stage, the individual countries proceed to grant or reject the application in accordance with their specific domestic laws.

5.3. Other Aspects of Patent Laws in Other Countries

The U.S. patent system differs in a number of ways from many other national patent systems. Some of these differences are discussed above, including patentable subject matter, priority of invention, and absolute novelty. Applicants evaluating whether and where to seek patent protection throughout the world should also be aware of working requirements and compulsory licensing requirements that are included in various forms in the patent laws of many countries. Applicants should also be aware of, and perhaps use in appropriate cases, opposition proceedings in certain countries whereby anyone, including competitors, can oppose the granting of a patent and, thereby, become involved in the patenting process.

A typical working provision provides that a patentee can lose patent rights if (1) the patentee does not use the invention or discovery within the relevant country within a fixed time period (usually 1–4 years) after the grant of the patent or (2) the patentee ceases the use of the invention or discovery for a fixed time period (usually 1–3 years consecutively) unless the patentee can justify the cause of the inaction. Thus in many countries the patent owner must either use the invention or run the risk of losing certain rights otherwise granted by the patent. A sufficient use must be determined on a case-by-case basis in light of each specific country's working provision. Use or manufacture on a commercial scale is sufficient in literally all cases to satisfy such working provisions. Where commercial use is not possible, production on a more limited scale with offers of sale of the product may be acceptable. Such uses should be carefully documented in the event the working of the invention is contested and must be proven. Where it is not possible to use the invention on even a limited scale, it may be possible to satisfy the working provision by offering licenses to parties within the country who

---

[304]As of mid-2009, three countries (Luxembourg, United Republic of Tanzania, and Uganda) and certain territories require entry into the national stage within 20 or 21 months from the priority date. An applicant, however, can extend this deadline by entering Chapter II proceedings. An applicant who enters Chapter II proceedings will have until thirty or 31 months within which to enter the national stage in these countries. Alternatively, an applicant can enter these countries following entry into the European Patent Office or the African Regional Intellectual Property Organization within 31 months.

would be reasonably interested in such licenses or by advertising the availability of such licenses in appropriate local or regional media. In most cases, the working of the invention by a licensee satisfies the working requirement. In some cases, and especially where countries have entered into agreements granting reciprocal rights, working in another country may satisfy the working requirement. Generally, the greater the demand for the patented invention within the country, the more difficult it will be to justify manufacture outside the country.

Additionally, in many countries, if the invention is not adequately worked within the country or if the public interest so requires, the law provides for and requires the patent holder to grant licenses under the patent upon application. Generally, such compulsory licenses are not available until 4 years after the filing of the application or 3 years after the grant of the patent, whichever is later. In many countries, compulsory licenses can be granted, regardless of any working requirement, when it is in the public interest. Inventions relating to food products, pharmaceuticals, and health-related products can fall within this public interest provision. In general, the royalty from such a compulsory license is agreed on by the parties. Should the parties not reach agreement as to an acceptable royalty, the compulsory licensing provisions generally provide that the royalty will be determined by the government. In general, the royalty set by the government to be paid by a domestic organization to a foreign patent holder is likely to be lower than the royalty determined in an arms-length negotiation between the licensor and the licensee.

Both the working requirements and the compulsory license provisions vary considerably from country to country. Thus, the laws of each relevant country must be reviewed to determine the potential effects of such provisions on specific inventions and patents. Such effects should also be taken into account in determining where to seek patent protection. Even in cases for which it is unlikely that the working requirement can be satisfied and that compulsory licenses might be required, it is still generally advisable to seek patent protection. It is possible that no one will actually seek such a compulsory license. Even if a compulsory license is granted, the patent owner may be able to object on the grounds that the delay in working was unavoidable for economic reasons and thereby frustrate or delay the grant of a compulsory license for an extended period of time. The granting of a compulsory license might, of course, provide a reasonable return in the form of royalty payments. Any royalty may be better than letting a competitor freely practice the invention. Finally, especially if the grant of the compulsory license is contested, the patent holder will generally have a number of years in which to develop the market for himself or herself before the compulsory license is finalized. A competitor coming into the market at that later time may find the market more difficult to penetrate.

For many years, prosecution of a U.S. patent application was conducted in secrecy, but this rule of secrecy has been changed.[305] Before this change, third parties were generally unaware that a patent application had even been filed, much less what, if any, claims would be granted until the patent actually issued. Now, nonprovisional patent applications in the United States are generally published 18 months after the priority date of the application. Publication may be avoided under certain circumstances, as discussed in Section 2.5. This change in the U.S. prosecution practice conforms to the practice of most other countries. In nearly all other countries, a patent application is published 18 months after the priority date.

The U.S. prosecution practice, however, still differs from that of many other countries following publication of an application or grant of a patent. In many other countries, once the patent application has been allowed by a national or regional patent office or the patent has been granted, the application or patent may be subject to an opposition proceeding. In general, such oppositions give third parties an opportunity to bring additional prior art or other factors affecting the grant of the patent to the attention of the pertinent national or regional patent office.

[305] See 35 U.S.C.A. § 122 (West Supp. 2008).

Oppositions generally must be filed within a fixed time (normally 1 to 12 months) after publication for opposition. Once filed, however, the opposition proceeding itself may take several years and include a lengthy appeal process from the national or regional patent office's decision. In many cases, the opposition procedure has made it even more expensive to obtain patents, especially important patents, in such countries as Japan and many European countries (e.g., Germany, the Netherlands, and the Scandinavian countries). In some countries, the patent is not granted or issued and is not enforceable until the opposition proceedings are complete and the patent office actually issues the patent. In Europe, this potential problem has been alleviated to some degree by the advent of the EPC. Although the EPC has a 9-month period in which an opposition can be brought, the opposition does not forestall the issuance of the individual national patents that, once issued, are immediately enforceable against an infringer who may be opposing the patent grant in the European Patent Office. There is some movement globally to prevent or at least reduce the ability of a competitor to stall interminably the issuance of key patents until late in the patent term (usually 20 years as measured from the priority date). Of course, the desirability of curtailing oppositions and their effect will depend on which side of the fence one is on. For a competitor, delaying or preventing the grant of such a patent may be important.

In the United States, the reexamination proceeding allows a limited form of opposition for issued patents. During a reexamination, the claims of an issued patent are examined in light of prior art that was not considered by the PTO in the original prosecution. In the past, the involvement of third parties in reexamination proceedings was strictly limited, but the U.S. practice has been changed to allow at least some participation of third parties in reexamination proceedings. The reexamination process is discussed in more detail in Section 3.8.

6. TRADEMARKS

Although laws regulating and protecting trademarks were mainly developed in the twentieth century, trademarks have been in use in one form or another (e.g., artisan's "potter's marks") for at least 4000 years.[306] The motivation for placing such a "potter's mark" on products probably arose from the artisan's pride in his work and the desire of the individual to take credit for what he had produced. For them, their marks were a means of identifying and distinguishing their products from similar works by other artisans. That concept still remains the primary function of a trademark and generally is the basis for trademark law and protection throughout the world.

Trademarks are the words, names, slogans, pictures, symbols, or designs that are used to identify the source or origin of goods. Similarly, service marks are the words, names, slogans, pictures, symbols, or designs that are used to identify and distinguish services. Trademarks and service marks are generally governed by the same legal principles. Throughout this section, references to *trademarks* or *marks* will generally include and apply to service marks as well.

Trademarks perform four basic functions: (1) identify one seller's goods or services and distinguish them from the goods or services of others, (2) indicate a common source of goods or services, (3) indicate a certain level of quality of the goods or services, and (4) assist in selling the goods or services (e.g., advertising). Trademarks do not provide protection for the underlying goods or services. That is the role of patents or trade secrets.

The ability to identify and distinguish the goods of one party from those of another is an essential prerequisite and function for a trademark. Contemporary trademark laws generally do not recognize property rights in a trademark *per se*. Rather, the "property" in which a trademark owner may claim a legitimate and protectable interest is the goodwill of the business symbolized by the trademark.

[306]Frank I. Schechter, The Historical Foundations of the Law Relating to Trademarks (1925). See also Sidney A. Diamond, *The Historical Development of Trademarks*, 65 Trademark Rep. 265 (1975).

It is the value of that goodwill that establishes the value and worth of the trademark. Thus, a competitor should be precluded from misappropriating another's goodwill by using the trademark that symbolizes that goodwill.

In the United States, the Lanham Act[307] as amended provides the framework for registration of trademarks in the PTO and for claims for infringement of federally registered trademarks as well as related claims for specific acts of unfair competition committed "in commerce." The Trademark Counterfeiting Act of 1984[308] provides remedies against parties who counterfeited federally registered trademarks, including *ex parte* seizures and criminal sanctions.

6.1. Trademarks as Marketing Tools

Trademarks are uniquely suited to facilitate the marketing of drugs. The value of a trademark is measured by the goodwill, which is generated from sales of products using that trademark. Goodwill, however, is dependent upon the consumer's favorable perception of the goods. If the customer is satisfied with the product, he or she is likely to form a favorable impression as to products bearing the same trademark, and may look for the same mark when purchasing similar products in the future (i.e., "brand loyalty"). Such brand loyalty is difficult to establish where products are essentially interchangeable and compete mainly through price. Generally, however, it is relatively easy to establish brand loyalty for drugs and other health-related products, especially when they are patented. During the patent term for a new drug, the consumer's ability to select an alternate medication may be limited or nonexistent. Consequently, patent protection can aid in the creation of brand loyalty that may continue even after the patent expires. Furthermore, successful medications act positively and directly upon the patient; they heal, relieve pain, lower blood pressure, ease breathing, or otherwise improve the health and comfort of the patient. Each successful use may result in a positive psychological reinforcement as to the importance and value of that particular medication. That value can translate directly into goodwill. If that goodwill can be successfully linked to a trademark in the mind of the consumer or prescribing doctor, the marketing potential for that trademark can be significantly increased.

Unlike a patent, the duration of a trademark is not limited to a fixed term. Hence, a trademark can be a valuable tool to help protect and maintain a market long after patent protection has expired. Individuals accustomed to buying a trademarked product during the life of a patent (when the patent owner or his licensee may be the only one offering the product) often will continue to seek that same product after the expiration of the patent. Such product and brand loyalty, along with its associated goodwill (if carefully developed and nurtured), can represent a significant obstacle for market entry and/or penetration by others even after the patent expires. Efforts to establish and promote the trademark undertaken during the exclusive period offered by the patent may pay dividends well into the future. Thus, in addition to seeking patent protection, drug companies should, at a very early stage, develop a marketing strategy to maximize and link the potential goodwill of a new drug with a particular trademark. Although generic drugs offered after the drug patent expires may cost less, many patients will request, and many doctors will prescribe, the trusted product they used in the past and can identify by its associated trademark.

6.2. Selection of Trademarks

Before bringing a new drug to the marketplace, careful consideration should be given to the selection of an appropriate trademark. Normally, a good trademark will not only identify and distinguish the drug in the marketplace but also secure a market advantage for the manufacturer and/or seller. Some words, however, can be more successfully developed as trademarks than others because of their distinctiveness. There is a "spectrum of distinctiveness" for potential trademarks. At one end is the arbitrary or fanciful word, in the middle are suggestive or descriptive words relating to the product or its characteristics,

---

[307]15 U.S.C.A. §§ 1051–1127 (West 1997, 1998 & Supp. 2008).
[308]18 U.S.C.A. § 2320 (West Supp. 2008).

and at the other end is the generic name of the product itself. An arbitrary or fanciful word is ideally suited for use as a trademark. A generic name cannot be used as a trademark for the product that it describes. Such a generic name, however, might function as a trademark for other, unrelated products (e.g., *mustang, jaguar,* and *cougar* for automobiles). For a new drug, however, it may be preferable to coin a new term or word as opposed to appropriating an existing word.

A suggestive term (i.e., one that suggests but does not directly describe something associated with the goods or services) may function as a trademark without further evidence of the public's recognition of its status as a trademark. A descriptive term (i.e., one that directly describes something associated with the goods or services) cannot function as a trademark until and unless it is established that, in addition to its descriptive meaning, it has also acquired a "secondary meaning" (i.e., a new meaning attached to the term relating to its use as a trademark, which identifies and distinguishes the goods of a particular manufacturer or merchant).[309]

For example, *mustang, jaguar,* and *cougar* are successful as trademarks for cars because such names have connotative meanings that *suggest* that the characteristics associated with these animals might also be found in the cars themselves (e.g., speed, agility, and power). However, if instead of merely suggesting the qualities or characteristics of a product, a word actually *describes* the qualities or characteristics, then it cannot be recognized as a trademark by the PTO until it has been shown to have acquired secondary meaning. Thus, *coupe, sedan, convertible,* and *fastback* are not likely candidates for acquiring secondary meaning and trademark status for automobiles, as they clearly describe types of automobiles. Such terms should be free to be used by all. Other descriptive terms, however, may become legitimate trademarks by acquiring secondary meaning.

Whether a particular term is suggestive or descriptive is a gray area in trademark law. In attempting to distinguish between suggestive and descriptive terms, the PTO has indicated that

> Suggestive marks are those that, when applied to the goods or services at issue, require imagination, thought or perception to reach a conclusion as to the nature of those goods or services. Thus, a suggestive term differs from a descriptive term, which immediately tells something about the goods or services. Suggestive marks, like fanciful and arbitrary marks, are registrable on the Principal Register without proof of secondary meaning... The great variation in facts from case to case prevents the formation of specific rules for specific fact situations. Each case must be decided on its own merits.[310]

The acquisition of secondary meaning for a descriptive mark usually requires a calculated effort by the mark's owner to establish a specific relationship or recognition between the mark and the particular goods or service. Thus, the mark's owner can attempt through marketing techniques to form an association in the relevant marketplace between the mark and the specific goods or services for which the mark is used. For example, Owens-Corning Fiberglass Corporation employed an advertising and marketing campaign (using the Pink Panther cartoon character) for its pink-colored insulation to establish that color as an identification of its insulation. This effort to strengthen the public's recognition and association of the desired mark with the product was so successful that Owens-Corning was able to establish secondary meaning and obtain a federal registration for the color pink for

---

[309] Secondary meaning is "acquired when the name and the business becomes synonymous in the public mind; and [it] submerges the primary meaning of the name ... in favor of its meaning as a word identifying that business." 2 J. Thomas McCarthy, McCarthy on Trademarks and Unfair Competition § 13.2 (4th ed. 2008), citing *Visser v. Macres,* 137 U.S.P.Q. 492 (Cal. Dist. Ct. App. 1963).

[310] *U.S. Pat. & Trademark Off., Trademark Manual of Examining Procedure* §§ 1209.01(a)–(b) (5th ed. 2007) (internal citations omitted), available at http://uspto.gov.

its insulation product.[311] Proof of secondary meaning, however, is not always easy to acquire. The Court of Appeals for the Federal Circuit held that, without more, several million dollars of advertising and many millions of dollars of sales under a particular term were insufficient to prove that the term had acquired secondary meaning in the marketplace.[312] The court stressed the need for consumer surveys to measure whether the sales and advertising have been successful in creating secondary meaning for a particular mark.

Descriptive and suggestive terms are generally not good candidates as trademarks for drugs. Because a new drug often represents a new discovery, it is useful to select or develop a new word or name to identify this product to the public. The selection process of a new trademark for a new drug (or any new product) must, however, take into account the fact that the Lanham Act prohibits certain items from ever being registered as trademarks. Registered trademarks cannot include, for example, "immoral, deceptive, or scandalous" matter, state or national flags or similar insignia, names of living individuals (except by written consent) or a deceased U.S. president during the life of his widow (except by written consent), "merely descriptive or deceptively misdescriptive" names, "primarily geographically descriptive or deceptively misdescriptive" names, and surnames.[313] The selection process should also take into account, especially if registrations in other countries are desired, the possible meaning of any potential mark in those other countries and languages. For example, an arbitrary and fanciful word in the English language may have a specific (perhaps negative, scandalous, or bizarre) meaning in another country, culture, or language.

A mark cannot be registered "which so resembles a mark registered in the U.S. Patent and Trademark Office, or a mark or trade name previously used in the United States by another and not abandoned, as to be likely, when used on or in connection with the goods of the applicant, to cause confusion, or to cause mistake, or to deceive."[314] Once a tentative choice of one or more potential trademarks is made, it is prudent, if not essential, to conduct a trademark search to protect the expected investment in the chosen mark. Normally, an initial computer search of the records in the PTO should be made to locate any similar marks that are registered or for which registration is pending before the PTO. This search of the PTO files is a relatively fast and inexpensive procedure by which possibly insurmountable problems (e.g., so-called knock-out marks that likely prevent registration) can be identified. If very similar marks are not found in the PTO search, a more comprehensive search, including databases containing state trademark registrations and common law marks, is generally undertaken to provide additional security that the mark in question is available for adoption. Preferably an outside search agency conducts the search using databases containing registered and unregistered names and marks in use throughout the United States. If the proposed trademark is intended for use on products or services that may be marketed in countries outside the United States, the search should be expanded to include trademark applications and registrations in such other countries. If no closely similar marks are found (or, in some cases, if such marks are used on very different types of goods), one may proceed to register the new mark in the PTO and, if appropriate, seek registrations in other countries as well.

6.3. Registration Process

In the United States, trademark rights arise from use of a mark, rather than from its registration in the PTO.[315] However, registra-

---

[311] *In re Owens-Corning Fiberglas Corp.*, 774 F.2d 1116, 1125–1126 (Fed. Cir. 1985).
[312] *Braun Inc. v. Dynamic Corp. of Am.*, 975 F.2d 815, 826–827 (Fed. Cir. 1992) (reversing jury's finding that Braun's hand held blender design had acquired secondary meaning because "[the] limited evidence as to advertising, sales and media attention, *standing alone*, is not sufficient to demonstrate that the consuming public identified the blender design with its maker.") (emphasis added).
[313] 15 U.S.C.A. §§ 1052(a)–(c), (e) (West 1997 & Supp. 2008).
[314] 15 U.S.C.A. § 1052(d) (West Supp. 2008).
[315] Registration of a mark is not necessary. A common law trademark can be created by using the mark in commerce. Generally, such a common law trademark affords protection to its owner only in the geographical area of actual use.

tion in the PTO provides significant advantages to a trademark owner. Furthermore, marks can be registered based either on actual use or upon a *bona fide* intent to use the mark with a particular product or a specific service.[316] Under the intent-to-use filing provisions, however, it is still necessary to establish actual commercial use of the mark before the actual registration will issue. One important benefit of the intent-to-use process is the recognition of the date of filing of an intent-to-use application as the priority date, on which trademark rights begin, rather than the date the mark is first used commercially on the product or for the service. The intent-to-use provisions can be especially useful in marketing a new drug because they provide a method for obtaining protection for a trademark before the drug actually enters the marketplace.

A trademark application must include both the mark for which registration is sought and a description of the goods and/or services upon which it is to be used. Once an application for registration of a mark has been filed in the PTO, the examination process is essentially the same whether the application is based on actual use or intent to use. A trademark examiner in the PTO reviews the application to determine whether it meets the statutory requirements. The examiner may issue a refusal to register the mark based on a failure to meet one or more of the statutory requirements. The applicant is given an opportunity to present arguments or to amend the application to overcome these objections. If the examiner finds the application to be in order, the mark will be published for opposition in the PTO's *Official Gazette*. Anyone who believes he or she may be injured by the registration of the mark may file a Notice of Opposition within 30 days of the date of publication unless an extension of time is granted.

If an opposition is not filed, an application based on actual use will mature into an issued registration or an intent-to-use application will be allowed. For an allowed intent-to-use application, the applicant has 6 months within which to establish actual use of the mark. If actual use of the mark is not established within this 6-month period, the applicant may request, upon payment of required fees, extensions of time (in 6-month increments up to a total of five extensions). If actual use of the mark in commerce has not been made by the end of this period (including any requested extensions), the application will be deemed abandoned. Thereafter, a new application for registration of the same mark may be filed, but the earlier priority date will be lost.

The initial term of a registered mark is 10 years. Between the fifth and the sixth years of the initial term, the registrant must file an affidavit or declaration stating that the mark is in use with the goods specified in the registration. Failure to timely file this affidavit or declaration will result in the cancellation of the registration. If the registrant files an affidavit or declaration confirming, in addition to the mark being in use, that it has been in continuous use for the preceding 5 years, the registration becomes "incontestable."[317]

As noted, trademark rights in the United States arise from use of the mark. Although not required, registration of a mark in the PTO provides the mark's owner many significant procedural advantages. First, a certificate of registration of a mark on the principal register constitutes "*prima facie* evidence of the validity of the registration ... and the registrant's exclusive right to use the registered mark in commerce on or in connection with the goods and services specified in the certificate."[318] The registrant's rights are not limited geographically to the locations in the United States where the mark has actually been used. The registrant's "exclusive" rights extend throughout the United States (and territories and possessions). Moreover, if the registration has become "incontestable" based on consecutive 5 years of continuous use after registration, then the registrant's "exclusive rights" are incontestable except upon certain limited grounds specified in the Lanham Act.[319] Such incontestable registration constitutes "conclusive evidence" of the registrant's "exclusive rights" in the registered mark.[320]

---

[316] The Trademark Law Revision Act of 1988 added a new "intent to use" provision, 15 U.S.C.A. § 1051(b) (West Supp. 2008).

[317] 15 U.S.C.A. §§ 1065 & 1115 (West 1997, 1998 & Supp. 2008).
[318] 15 U.S.C.A. § 1057(b) (West 1997).
[319] 15 U.S.C.A. § 1065 (West 1997 & Supp. 2008).
[320] 15 U.S.C.A. § 1115(b) (West Supp. 2008).

The registration is also "constructive notice of the registrant's claim of ownership" throughout the United States. Constructive notice means that, after a mark has been registered, others have legal notice of the registrant's trademark rights even if they are not aware of the registration.[321] Constructive notice prevents later users of a mark from relying on the common law defense of "good faith" adoption of the mark. To take advantage of this constructive notice provision, the registrant must use the mark with the proper registration notation.[322]

These procedural advantages are of considerable importance should the registrant attempt to enforce the trademark rights in court. The federal courts, rather than state courts, have original jurisdiction over causes of action arising from federally registered marks.[323] Furthermore, a registrant may claim an earlier priority date based on an intent to use application. The ability to claim an earlier priority date may be critical in a contest between rival claimants of the same or similar marks. A single day of priority over a junior party may be sufficient to sustain a senior party's rights and require the junior party to discontinue further use of his or her mark.

Registration also provides a relatively simple mechanism to prevent the importation of goods bearing a counterfeit or infringing copy of the mark into the United States. Upon the filing of a certified copy of a trademark registration with the U.S. Secretary of the Treasury, customs officers will undertake to exclude entry into the United States of products bearing an infringing copy of the Registrant's mark.[324] The presumed ability to determine infringement (i.e., that the alleged infringing mark is confusingly similar to the registered mark) by direct visual inspection allows for the simplified procedure.

Finally, registration of a trademark entitles the registrant to both injunctive and monetary relief for trademark infringement. Under the Lanham Act, monetary relief can include (1) the profits that an infringer has earned by marketing products under the infringing trademark and (2) any damages that have been sustained by the registrant from the infringement.[325] Moreover, "[i]n assessing profits [of the infringer] the plaintiff [registrant] shall be required to prove defendant's sales only; defendant must prove all elements of costs or deduction claimed."[326] Similarly, the court may increase the award of damages to the registrant to three times the amount of damages proven at trial, and the court may award attorney fees to the prevailing party.[327] The court may also order "that all labels, signs, prints, packages, wrappers, receptacles, and advertisements in the possession of the defendant bearing the registered mark ... shall be delivered up and destroyed."[328]

In an action for trademark infringement, the issue is whether a mark used by the defendant with particular goods or services is likely to cause confusion, mistake, or deception of the public. Generally the best evidence of likelihood of confusion is that which shows substantial actual confusion. However, actual confusion is not necessary to prove infringement of a federally registered trademark: "It has been well said that the most successful form of copying is to employ enough points of similarity to confuse the public with enough points of difference to confuse the courts."[329] The registration of a trademark in the PTO considerably strengthens trademark rights and makes them considerably easier to enforce.

6.4. Oppositions and Cancellations

Once the mark is published in the *Official Gazette*, anyone believing that he or she will be injured by the issuance of the registration may file a Notice of Opposition within 30 days following publication. Additionally, anyone

---

[321] 15 U.S.C.A. § 1072 (West 1997).
[322] *See* 15 U.S.C.A. § 1111 (West 1997).
[323] 15 U.S.C.A. § 1121 (West Supp. 2008).
[324] 15 U.S.C.A. § 1124 (West Supp. 2008).
[325] 15 U.S.C.A. § 1117 (West 1998 & Supp. 2008). The registrant is also entitled to recover the "costs" of the litigation against an infringer.
[326] *Id.*
[327] *Id.*
[328] 15 U.S.C.A. § 1118 (West Supp. 2008).
[329] *Boston Athletic Ass'n v. Sullivan*, 867 F.2d 22, 30 (1st Cir. 1989), quoting *Baker v. Master Printers Union*, 34 F. Supp. 808, 811 (D.N.J. 1940).

who believes he or she may be injured by maintenance of a registration on the Principal Register may file a Petition for Cancellation of that registration. Once either a Notice of Opposition or a Petition for Cancellation is filed, the application is transferred to the Trademark Trial and Appeal Board for a determination of the merits of the opposition or cancellation request. Opposition and cancellation proceedings are, in many ways, procedurally similar to a court trial. The only issues before the Board, however, are whether an applicant should be allowed to register the mark or whether a registrant can maintain the registration for the mark. Either party can appeal the Board's decision to the Court of Appeals for the Federal Circuit on the record established before the Board or to the U.S. District Court for the District of Columbia for a *de novo* review of the Board's decision.

6.5. Preserving Trademark Rights through Proper Use

Trademark rights can be diminished and even destroyed through improper use. A trademark is to be used as an adjective modifying the generic name of a product. The trademark owner should never use the trademark as the name of the product itself. The trademark owner should also attempt to prevent others from doing so.

Once established, the trademark must be protected and its usage carefully monitored and controlled. A trademark (even if arbitrary or fanciful when first coined) can be lost if it becomes a descriptive or generic name for the product itself and thus fails to identify the particular product offered by the trademark owner. The original *aspirin* trademark lost its status as a trademark in the United States when it lost its distinctiveness in identifying the particular Bayer Company product. The general public came to regard this term as identifying the type of drug rather than distinguishing the product of Bayer. Consequently, the term *aspirin* was held to have fallen into the public domain in the United States and not the exclusive property of Bayer.[330]

In addition to proper use through its own advertising and marketing, the trademark owner should police the use of the mark by others. For example, the trademark owner may wish to subscribe to a service that will search various media databases on a routine basis for improper uses of the mark. Or the trademark owner can carry out the search on their own. When improper uses are found, a letter or other notification can be sent explaining why the use is improper and requesting proper usage. Depending on the degree of misuse and the importance of the mark, other mechanisms, including, for example, advertising and marketing campaigns, might be used. Documentation of such policing should be maintained in the event it becomes necessary during litigation or cancellation proceedings to prove that the trademark owner acted in a proper and prudent manner to protect the trademark.

When used, the mark should be set apart, preferably in bold type, from the other words around it. If the mark has not been registered in the PTO, it should be followed whenever possible by the designation "TM"; if it has been registered, it should be followed by the proper registration notice:

> A registrant of a mark registered in the Patent and Trademark Office, may give notice that his mark is registered by displaying with the mark the words "Registered in U.S. Patent and Trademark Office" or "Reg. U.S. Pat. & Tm. Off." or the letter R enclosed within a circle, thus ®; and in any suit for infringement under this Act by such a registrant failing to give such notice of registration, no profits and no damages shall be recovered under the provisions of this Act unless the defendant had actual notice of the registration.[331]

6.6. Worldwide Trademark Rights

It is impossible here to discuss in any detail the plethora of laws of other countries regarding the establishment and protection of trademark rights, but a few comments may be in order. As noted above, when selecting a trademark it is prudent to verify whether the desired mark has some meaning in another lan-

---

[330] *Bayer Co. v. United Drug Co.,* 272 F. 505, 509–510 (S.D.N.Y. 1921).

[331] 15 U.S.C.A. § 1111 (West 1997).

guage or sounds similar to a word in another language. In a few instances, trademark owners have discovered to their chagrin that an English or made-up word used as a trademark is so similar to a word in another language having a negative or otherwise inappropriate connotation that it is impossible to use the trademark in some countries. It is far better to discover such problems before significant resources are expended to develop the goodwill associated with that mark.

In the global marketplace, trademark protection normally will be required in various countries throughout the world. However, like patent protection, trademark protection is limited geographically. Determining the appropriate countries in which to seek trademark protection should be an integral part of the intellectual property strategy. Due to the generally lower cost of obtaining trademark protection, it may be advantageous to obtain such protection in a larger number of countries than where one might seek patent protection. As a general rule of thumb, trademark protection should be perfected, at a minimum, in every foreign country in which patent protection is sought. Similarly, if licensing in other countries is contemplated, registration of the mark in those countries is advised. Licensees are likely to be interested in both the patent and the trademark rights. It may also be appropriate to obtain trademark registrations in countries having potentially significant markets even though patent protection will not be sought and/or licenses will not be granted. It is prudent to ensure that a well-known drug can be exported into such countries under its trademark, and that someone else does not acquire the rights to market a drug or other health-related product under that trademark.

When and to what extent trademark rights should be acquired in countries throughout the world must necessarily be determined on a case-by-case basis. In countries outside the United States, trademark rights generally arise from registration rather than from use of the mark. It is therefore prudent to file applications for registration of a selected mark as soon as practicable in those countries where substantial use of the trademark is anticipated. Therefore, when a drug is ready to be announced to the public under a particular trademark, even if it is not anticipated that sales under the mark will begin for quite some time, the strategy for establishing trademark rights in other countries should generally already be in place. Unfortunately, it is not uncommon for unrelated parties, upon learning that a new product will be marketed under a particular mark, to attempt to register the mark in at least some countries in advance of the originator of the trademark. The unrelated party could, for example, hold the mark for ransom or transfer the rights to the mark to a local company for marketing similar products in that country under the mark. Although redress may, in some instances, be achieved through the courts in appropriate countries, this can be expensive and may involve many years of litigation. Thus, in considering the development of trademark rights for new drugs in foreign countries, an "ounce of prevention" may be worth considerably more than several "pounds of cure."

One mechanism for pursuing trademark registrations in a number of foreign countries is through the Madrid Protocol.[332] The Madrid Protocol is an international registration agreement having 78 members, including the United States and the European Community, in which a single international registration creates a bundle of national registrations. Under the Protocol, a trademark application may be filed in a "home" member country and serve as the basis for an international registration. In turn, this international application is used as the basis for seeking registrations in other member countries. The Madrid Protocol also provides a way to convert the international registration into local registrations in the event that the home registration is challenged and invalidated.

6.7. Other Rights under the Lanham Act

The Lanham Act in § 43(a) also creates a federal cause of action for "false designation of origin and false description of goods."[333] Courts have construed § 43(a) as regulating

---

[332]The text of the Madrid Protocol is available from the World Intellectual Property Organization at http://www.wipo.int.
[333]15 U.S.C.A. § 1125(a) (West 1998 & Supp. 2008).

any act or representation that might cause the purchasing public mistakenly to believe that a product originating from one manufacturer or merchant originated from some other manufacturer or merchant; § 43(a) now forms the basis of the federal law of unfair competition.

Under § 43(a), some characteristics of a product can be protected from imitation by others. Thus, the color, shape, and size of pills have been held to be protectable. The court allowed a drug company, even after the patent on the drug expired, to market the particular drug exclusively in capsules of a particular color, shape, and size.[334] Other drug companies could, of course, market generic forms of the drug after the patent had expired, but not in the format in which the public had come to know and recognize the product. Thus, the exclusive right to market a drug in capsules of the same color, shape, and size that the public has come to associate with "the authentic product" can be a protectable property interest and a valuable marketing tool. Section 43(a) is not restricted to registered trademarks. Rather it embraces the broad panoply of "trade dress" for a product (i.e., the total visual combination of elements in which a product or service is packaged and offered to the public). The development of valuable rights that may be protected under this section depends primarily on how the product is marketed and presented to the public.

The protection offered by § 43(a) was expanded further by the Trademark Revision Act of 1988 to cover false representations made in regard to another's goods; it provides that any person who

> in commercial advertising or promotion, misrepresents the nature, characteristics, qualities or geographical origins of his or her or *another person's* goods or services or commercial activities shall be liable in a civil action by any person who believes that he or she is or is likely to be damaged by the act.[335]

---

[334]*Ciba-Geigy Corp. v. Bolar Pharm. Co.*, 547 F. Supp. 1095 (D.N.J. 1982), *aff'd*, 719 F.2d 56 (3d Cir. 1983); *but see Shire US Inc. v. Barr Labs., Inc.*, 329 F.3d 348, 358–359 (3d Cir. 2003) (affirming district court's finding that Shire had failed to show that its product configuration was nonfunctional).
[335]15 U.S.C.A. § 1125(a)(1)(B) (West 1998).

Thus false or misleading statements about one's own goods or another's goods can give rise to a cause of action.

7. TRADE SECRETS

Information that has value because it is not generally known (e.g., business and technical, patentable, and nonpatentable) can be protected as a trade secret against discovery by improper means or through breach of confidence. To be protected as a trade secret, the information must in fact be "secret" (i.e., not generally known by the industry). The duration of a trade secret is limited by the length of time the information is kept secret. Public disclosure of the information by the owner or anyone else results in the loss of the trade secret status of the information.

The practical role of trade secrets in the drug development and discovery industry may, of course, be significantly limited by the public disclosure of information and data required by the FDA approval process, publication of pending patent applications, and the large number of researchers working in this area who may independently make similar discoveries. Data and other information that the FDA publicly discloses or otherwise makes available to the public loses its status as a trade secret. Thus, where trade secret protection is not available due to the inability to maintain secrecy, patent protection may be the only viable form of protection available.

Trade secret protection can nevertheless be a valuable component of an overall intellectual property strategy for drug research and development activities. Trade secret protection can be especially important in protecting technology at its earliest stages of development (e.g., before publication of pending patent applications and/or release of data and information by the FDA). Of course, trade secrets can be used to protect technical or business information that is not disclosed to the FDA (or, if disclosed, not released to the public by the FDA) or to the public through, for example, published patent applications or other publications. In some circumstances, trade secrets can provide a viable alternative and/or adjunct to obtaining patent protection.

Thus, where the patentability of an invention is in doubt, trade secret protection can be used. Like patents and trademarks, trade secrets and general technical know-how can be sold outright or licensed.

Relying too heavily on trade secret protection, however, can have significant limitations and risks. The cornerstone of trade secrets is secrecy. Once secrecy is lost (regardless of how it is lost), the protection afforded by trade secret law is lost and competitors are free to use the technology. To reiterate, the duration of a trade secret is the length of time the information is kept secret. Public disclosure of the information by the trade secret owner or others (even by one who improperly obtained and/or disclosed the information) effectively terminates the protection offered by the trade secret. Furthermore, even if secrecy is maintained, the technology can be used by others who obtain the information by independent discovery or acceptable reverse engineering. Thus, it may be proper, for example, for a competitor to analyze a new drug and, based on the information obtained, seek FDA approval to market the drug (assuming there are no blocking patents and the drug sample was obtained properly). Therefore, the scope of protection available and risks associated with trade secrets must be carefully considered and evaluated, and procedures should be defined and implemented to protect and maintain the required secrecy before significant reliance is placed on trade secret protection as an alternative to patent protection. This is especially true for the drug discovery and development industry, where detailed technical disclosures to the FDA are generally required. Ideally, patent protection and trade secret protection should be carefully coordinated to provide maximum protection.

7.1. Trade Secret Definition

Trade secret protection is generally governed by state law; thus, its definition can vary from state to state. One common definition provides that a trade secret consists "of any formula, pattern, device, or compilation of information which is used in one's business, and which gives him an opportunity to obtain an advantage over competitors who do not know or use it."[336] Another definition, which a significant number of states[337] have adopted in some version, is provided by the Uniform Trade Secrets Act: "[I]nformation, including a formula, pattern, compilation, program, device, method, technique, or process that (i) derives independent economic value, actual or potential, from not being generally known to, and not being readily ascertainable by proper means by, other persons who can obtain economic value from its disclosure or use and (ii) is the subject of efforts that are reasonable under the circumstances to maintain its secrecy."[338] Therefore, trade secrets generally can include formulae for chemical compounds and drugs; processes for manufacturing, treating, and preserving materials; patterns and designs for a machine or other device; computer software; business strategies and plans; customer lists; and similar business and technical information having economic value.

7.2. Requirements for Trade Secret Protection

For a protectable trade secret to exist, generally it must meet four interrelated criteria: (1) it must fall within the type of information protectable as a trade secret), (2) it must not generally be known in the trade; (3) it must be of commercial value to the holder, and (4) it must be treated and maintained as a secret. Although the trade secret must be secret, novelty in the patent sense is not required. Thus, an obvious improvement in a drug manufacturing process, which could not be protected by a patent, could be retained as a trade secret so long as it is not known to others in the industry and the other above criteria are met. The third requirement, commercial value, is generally met if knowledge or use of the trade secret by the holder provides some competitive advantage. The fourth criterion essen-

---

[336]Restatement (First) of Torts, § 757 cmt. b (1939).

[337]As of 2004, approximately 44 states and the District of Columbia had adopted the Uniform Trade Secrets Act or some version thereof. See 1 Roger M. Milgrim, *Milgrim on Trade Secrets* § 1.01[3], n.36 (2008).

[338]Uniform Trade Secrets Act, § 1(4) (1985).

tially requires that the trade secret holder take reasonable measures to keep and maintain the information as a secret. The efforts to maintain secrecy will vary with the information and the financial resources of the organization. At a minimum, such reasonable efforts should include limiting access to the information to key employees who have a need to know, having employees sign confidentiality agreements, and alerting employees about the status of the sensitive information that is considered to be a trade secret. Practices such as appropriate labeling of documents as *confidential*, storing such documents in a secure manner, and marking process areas that are off-limits to unauthorized visitors are not only prudent to minimize the risk of inadvertent disclosure but also essential to establishing the information as a trade secret if legal action is necessary to enforce trade secret rights. Disclosure to outsiders should be limited to that necessary for business reasons and should be carefully controlled; such disclosure generally should be through confidentiality agreements.

The courts have applied a number of specific factors in determining the existence of a trade secret under these general principles. Some of the more frequently used factors are (1) the extent to which the information is known outside of one's business, (2) the extent to which it is known by employees and others involved in the business, (3) the extent of measures taken to guard the secrecy of the information, (4) the value of the information to the trade secret holder and, potentially, to competitors, (5) the amount of effort or money expended in developing the information, and (6) the ease or difficulty with which the information could be properly acquired or duplicated by others.[339]

In setting up a program to protect trade secrets, these factors should be carefully considered in order to maximize the probability of a court later finding that a protectable trade secret does in fact exist. For example, documents containing trade secrets should be labeled *confidential* or with a similar notation. Overuse of a *confidential* stamp, however, should be avoided. If all documents are routinely labeled *confidential* without regard to the trade secret content, a court might later determine that employees were not properly informed of the trade secrets or that trade secrets were treated no differently than other information. Moreover, if all documents are marked *confidential*, employees may not treat the trade secrets with the appropriate care, thereby increasing the risk that actual secrecy will be lost. In some instances, several classifications of information with varying degrees of control might be appropriate.

It is generally desirable to have a comprehensive and documented program for protection of trade secrets. This program can be invaluable in maintaining a competitive advantage in the marketplace, as well as providing a means to demonstrate to a court that protectable trade secrets existed and were treated in the appropriate manner. Such a program should have a mechanism for identifying trade secrets and then protecting and maintaining them. This trade secret program can form an integral component of an overall security program to maintain patentable inventions as secrets, at least until the appropriate patent applications are filed. Such a program may be implemented through an intellectual property committee responsible for general intellectual property matters.

7.3. Enforcement of Trade Secrets

Trade secrets generally protect only against wrongful disclosure or discovery of information by competitors or others. Thus, one might have a cause of action against an employee who leaks information to a competitor, or against a competitor who discovers a trade secret through improper industrial espionage. In addition, in appropriate cases, one might bring suit to prevent the improper disclosure of a trade secret.[340] For example, a key employee, who resigns to join a competitor, might be enjoined from disclosing trade secrets of his or her former employer to the new employer.[341]

Not all means of discovering a trade secret are actionable. For example, it is acceptable to learn of the trade secret by independent discovery, by proper reverse engineering, or by evaluation of products or data available

---

[339]Restatement (First) of Torts, § 757 cmt. b (1939).

[340]Uniform Trade Secrets Act, § 2 (1985).

[341]See, for example, *PepsiCo, Inc. v. Redmond*, 54 F.3d 1262 (7th Cir. 1995).

publicly. Thus, for example, a trade secret holder would not have a cause of action against a competitor who independently discovers the trade secret. In addition, one who properly obtains the trade secret without any obligation to maintain the trade secret in confidence is free to use it and, if desired, disclose it to the public, thereby destroying the original trade secret status. Indeed, one who independently and properly discovers an invention held as trade secret by another may be able to obtain a patent covering the invention. In such a case, the potential rights of the patentee and the trade secret holder relative to each other appear to depend upon the particular facts of each case.[342]

Remedies for misappropriation of a trade secret can include actual and punitive damages as well as injunctive relief. An injunction may only prevent the wrongdoer from using the trade secret information for a fixed length of time. Some courts will limit the length of the injunction to the estimated time it would take a hypothetical competitor to discover the trade secret by reverse engineering (a so-called "lead-time injunction"). Only the wrongdoer may be prevented from using the trade secret. Other competitors, as well as the general public, are generally free to use the trade secret to the extent that it has been publicly disclosed.

Although trade secrets potentially offer protection for an unlimited duration (i.e., so long as secrecy is maintained), in practice the time of protection is often relatively brief. Due to the intense competition, employee mobility, and FDA disclosure requirements, the lifetime of an average trade secret in the drug discovery and development industry may be even shorter. Even within such a short lifetime, however, trade secrets remain a useful adjunct to patent protection. For example, trade secret protection can be used to protect an invention prior to filing a patent application and while the application is pending before the PTO up until the time of publication. Trade secrets may also be used to protect later improvements in patented processes or materials that do not, in themselves, warrant filing separate patent applications.

7.4. Relationship of Trade Secrets and Patents

Trade secret and patent protection have coexisted in the United States for more than 200 years. The U.S. Supreme Court in 1974 made it clear that federal patent law does not preempt state trade secret law.[343] Nonetheless, the disclosure requirement of patent law and the secrecy requirement of trade secret law are often in conflict. The Patent Statute requires that patent applications include a specification that teaches one of ordinary skill in the art how to make and use the invention and discloses the best mode of carrying out the invention known to the inventor as of the application filing date. Any trade secrets disclosed in the patent specification lose their status as trade secrets once the patent application is published or issued as a patent. Failure to disclose in a patent application a trade secret that is necessary to satisfy these enablement or best mode requirements could prevent the application from being allowed, or may lead to an issued patent being challenged as invalid if ever enforced against competitors in a court action.

Although the issue is easily stated, it is considerably more difficult in practice to determine which trade secrets relating to an invention must be disclosed. Clearly, an applicant should not attempt to obtain patent protection for an invention while seeking to keep the commercial embodiment (the best mode) as a trade secret. Yet as noted, the best mode requirement relates to the applicant's knowledge at the time the application is filed. Improvements made before the filing date should normally be included in the original application. Improvements made after the filing date, even if they constitute a better method of practicing the invention, can be retained as trade secrets. However, such improvements

---

[342] For example, it is possible that the trade secret holder's activities may be § 103 prior art to the patent applicant. Roger M. Milgrim, 2 *Milgrim on Trade Secrets* § 8.02[3] (2002). See also Frank E. Robbins, *The Rights of the First Inventor—Trade Secret User as Against Those of the Second Inventor-Patentee (Part I)*, 61 J. Pat. Off. Soc'y 574 (1979); Karl F. Jorda, *The Rights of the First Inventor—Trade Secret User as Against Those of the Second Inventor-Patentee (Part II)*, 61 J. Pat. Off. Soc'y 593 (1979).

[343] *Kewanee Oil Co. v. Bicron Corp.*, 416 U.S. 470, 491 (1974).

may have to be disclosed in any subsequent continuation-in-part application adding new matter to the original specification.[344] This is because the CIP application, although related to the original parent application, must independently satisfy the enablement and best mode requirements as of its filing date.

Generally, it is not necessary to disclose trade secrets that are related to the invention but are not required for its operation and are not related to the best mode of operation known to the applicant as of the filing date of the patent application. In a close case, it may be preferable to err on the side of disclosing more than the required minimum to reduce the risk of a court later declaring the patent invalid as not providing an enabling specification or disclosing the best mode. After all, as discussed above, trade secrets may have only a limited lifetime given the disclosure requirements to the FDA or other federal agencies.

7.5. Freedom of Information Act

So-called "sunshine laws," including state and federal Freedom of Information Acts (FOIAs) and state right-to-know laws, can significantly impact the ability of pharmaceutical and drug discovery companies to retain the secrecy required for viable trade secrets. Such laws serve to increase the openness of governmental processes and decision making. Yet, release of information by the government under FOIA can destroy valuable trade secrets rights. Some of the information submitted to government agencies will be routinely released to the public as part of the functioning of the agencies. Other information may be released in response to requests by members of the public.

The federal FOIA mandates disclosure of official information of the administrative agencies of the federal executive branch (including, for example, FDA and EPA) unless the information falls within one of the nine statutory exemptions.[345] FOIA provides that "each agency, upon any request for records that (i) reasonably describes such records and (ii) is made in accordance with published rules stating the time, place, fees (if any), and procedures to be followed, shall make the records promptly available to any person."[346] Thus, any member of the public, including, for example, domestic or foreign competitors, can request, and often obtain, records through the FOIA. Such requests can include drug-related submissions to FDA and identifications of new chemical compounds submitted to EPA.

Under FOIA, the burden of proof for withholding information is on the government agency having possession of the information. Potentially a government agency can rely on two FOIA exemptions to withhold trade secret-type information. Exemption 3 allows an agency to withhold information exempted from disclosure by another statute, "provided that such statute (A) requires that the matters be withheld from the public in such a manner as to leave no discretion on the issue, or (B) establishes particular criteria for withholding or refers to particular types of matters to be withheld."[347] This provision, taken together with the Trade Secrets Act,[348] may provide a basis for exempting trade secrets from disclosure under FOIA. Exemption 4 provides that the disclosure requirements of FOIA do not apply to "trade secrets and commercial or financial information obtained from a person and privileged or confidential."[349] On their face, these exemptions appear to provide considerable protection against public disclosure for trade secrets disclosed to government agencies such as FDA and EPA. The courts, however, especially the U.S. Court of Appeals for the District of Columbia, have tended to read the exemptions narrowly. Furthermore, the Supreme Court has held that FOIA exemptions are permissive rather than mandatory:

> FOIA by itself protects the submitters' interest in confidentiality only to the extent that this interest is endorsed by the agency collecting the information. Enlarged access to gov-

---

[344]See, for example, *Transco Prods. Inc. v. Performance Contracting Inc.*, 38 F.3d 551 (Fed. Cir. 1994) (patentee not required to update best mode in a continuation application in which no new matter was added).

[345]5 U.S.C.A. § 552 (West 2007 & Supp. 2008). Virtually every state has adopted a corresponding state version of FOIA.
[346]5 U.S.C.A. § 552(a)(3)(A) (West 2007).
[347]5 U.S.C.A. § 552(b)(3) (West 2007).
[348]18 U.S.C.A. § 1905 (West. 2001).
[349]5 U.S.C.A. § 552(b)(4) (West 2007).

ernmental information undoubtedly cuts against the privacy concerns of nongovernmental entities, and as a matter of policy some balancing and accommodation may well be desirable. We simply hold here that Congress did not design the FOIA Exemptions to be mandatory bars to disclosure.[350]

Therefore, an agency retains the discretion to disclose information that falls within the exemption. Agencies may tend to grant more liberal disclosure simply to avoid lawsuits by requesters seeking to compel disclosure.[351]

Trade secret owners should consider taking precautionary steps when making disclosures of trade secrets to a government entity, including obtaining a formal agreement with the agency that prohibits disclosure of the trade secret, an agreement that the information submitted will not be considered "agency records" (and therefore not subject to FOIA), or an agreement that agency will give notice before disclosure of information is made.[352] These agreements provide additional enforceable mechanisms for preventing disclosure of information.

The details and nuances of the law governing FOIA disclosure are beyond the scope of this chapter. But any submitter of information, especially data and other information involving drug discovery and development relating to the public health, should be aware that the recipient government agencies may at some time release the information to the public at large or to individuals or organizations that submit specific requests. This also tends to elevate the importance of patent protection over trade secret protection in many contexts. The intellectual property strategy devised for the drug discovery and development organization should take these factors into account.

7.6. Trade Secret Protection Outside the United States

We do not here discuss in any detail the protection afforded to trade secrets in other countries. A few general comments, however, are provided.

Protection for trade secrets varies dramatically from country to country, ranging from essentially none or very little to protections exceeding those provided under U.S. law. Therefore, legal counsel in the relevant country should be consulted in the event that trade secret or related issues arise.

Even in countries that do not afford significant protection through laws directly covering trade secrets, protection may be possible through contracts or agreements to protect confidential information disclosed to other parties. Preferably and wherever possible, any required disclosure of the trade secrets or other confidential information to third parties (e.g., to employees, potential business partners, vendors, and so on)—whether in the United States or elsewhere—should be made under appropriate confidentiality or secrecy agreements. Improper disclosure of such trade secrets or information likely would be a breach of contract. In countries not recognizing trade secrets or offering little trade secret protection, redress for improper disclosure of a trade secret could be sought under contract law. In countries offering significant protection for trade secrets, redress could be sought under both contract law and trade secret law.

Trade secret holders interested in using their trade secrets throughout the world must protect their secrecy of the relevant information in each and every country in which it is used. The loss of secrecy anywhere in the world can affect, and often destroy, the trade secret around the world. Perhaps even higher safeguards could be maintained in countries that do not offer adequate trade secret protection because such countries might provide ideal havens for individuals or organizations seeking to discover trade secrets for their own use or for sale to others. Even where redress may be obtained in local courts for improper use or disclosure of a trade secret, such litigation can be expensive, and it is unlikely that a damage award could be obtained that would reasonably compensate the trade secret holder for loss of his or her trade secret throughout the world.

---

[350]*Chrysler Corp. v. Brown*, 441 U.S. 281, 293 (1979).
[351]A trade secret owner may seek, through a so-called "reverse FOIA" suit, to enjoin an agency from releasing information containing trade secrets. See generally Roger M. Milgrim, 3 *Milgrim on Trade Secrets* § 12.03[3][a] (2008).
[352]*Id.* at § 12.03[3].

8. OTHER FORMS OF PROTECTION

Other forms of protection for intellectual property available in the United States include copyrights, statutory invention registrations, and design patents. These methods of protecting intellectual property generally have only limited applicability to drug discovery and related technology. Copyrights, for example, protect a work of authorship; they do not protect inventions such as new drugs, diagnostic assays, or methods of treatment. Thus, although the copyright owner may prevent others from making copies of printed materials, such as, for example, advertising, a published article, manual, pamphlet, or a computer program, he or she cannot prevent others from using the ideas or data contained therein.[353] A copyright is created once the work of authorship is produced in any tangible form. Registering the copyright in the U.S. Copyright Office of the Library of Congress is optional. However, the copyright must be registered prior to enforcing the copyright in litigation.[354]

Limited protection can also be provided by the statutory invention registration (SIR) administered by the PTO.[355] This procedure provides for a patent-like publication that officially and affirmatively places the invention in the public domain for defensive purposes. The SIR is essentially a defensive publication that can be used when the inventor does not wish to obtain patent rights, yet wishes to be free of any later patents by others claiming the same invention. The SIR applicant must file a waiver of the applicant's right to receive a patent on the invention claimed in the SIR. The SIR is treated in the same manner as a patent for defensive purposes (i.e., both as prior art and as establishing a constructive reduction to practice in interference proceedings). Unlike a patent, the SIR cannot be used to prevent others from making, using, offering to sell, selling, or importing the disclosed invention or inventions.

The publication of the invention in a trade or scientific publication or other media has essentially the same defensive effect as a SIR, but the SIR undergoes a brief examination by the PTO to ensure that the SIR meets certain disclosure requirements before qualifying for publication.[356] However, the advantage of the SIR over a publication in a trade or scientific publication or other media is that the effective date of a publication is generally the actual date of publication, whereas the effective date of the SIR is the filing date.[357] The SIR procedure is sometimes used to disclose work done at federal research agencies for which the agency does not wish to seek patent protection. Anyone, however, can use the SIR procedure. The SIR program should be considered as an alternative to publication in a technical journal when one has determined not to seek patent protection for a specific invention but wishes to prevent others from obtaining patents covering that invention.[358] However, the limited use of the SIR[359] indicates that most applicants who intend to dedicate their inventions to the public simply file a nonprovisional patent application, which automatically publishes 18 months after the earliest priority date, and let that publication serve as prior art.[360]

---

[353]For instance, a computer program useful in DNA sequencing can be protected by copyright even though the ideas contained in the computer program and the actual DNA sequences determined using the program cannot be protected by copyright.
[354]See, for example, 17 U.S.C.A. §§ 411–412 (West 2005).
[355]35 U.S.C.A. § 157 (West 2001).
[356]The SIR must meet the requirements of 35 U.S.C. § 112. 37 C.F.R. § 1.294 (2008).
[357]Essentially the same effect can be obtained with an issued patent by simply not enforcing the patent or filing a disclaimer whereby the patent is dedicated to the public. 35 U.S.C.A. § 253 (West 2001); 37 C.F.R. § 1.321(a) (2001). Such dedication to the public does not affect the defensive aspects of the patent.
[358]The time lag between submission and the actual publication date in a scientific journal, especially in peer-reviewed journals, can be significant. Such delays may allow for another inventor to file a patent application covering the invention before the actual publication and ultimately obtain patent protection.
[359]Very few Statutory Invention Registrations have published in recent years: 34 in 2003; 27 in 2004; 14 in 2005; 41 in 2006; 27 in 2007; and 21 in 2008. *U.S. Pat. & Trademark Off., Performance and Accountability Report Fiscal Year 2008*, available at http://www.uspto.gov/web/offices/com/annual/2008/oai_05_wlt_11.html.
[360]Of course, an applicant could determine to pursue patent protection as long as the application remains pending before the PTO.

Design patents can be used to protect the ornamental design of an article of manufacture or portion thereof.[361] A design patent grants the holder the exclusive right to exclude others from making, using, and selling designs closely resembling the patented design. To be eligible for protection in this manner, the design must be novel, nonobvious, and ornamental. For example, a design patent could cover an ornamental packaging design for a drug or diagnostic kit or an ornamental design for a pill or capsule. Generally, design patents are governed by the same rules for validity as utility patents. However, they cover very different aspects of a given product: a utility patent relates to the functional aspects whereas a design patent relates only to the ornamental aspects of an article. Thus, it is possible to have a utility patent and a design patent covering the same product.

The term of a design patent is 14 years from the date of grant. Design patents do not have claims like utility patents. Rather, a design patent contains one or more drawings that define the scope of protection. The drawings are compared to the appearance of an alleged infringing design while also considering the prior art as a frame of reference to highlight the distinctions between the claimed design and the alleged infringing design.[362] Infringement is found if an ordinary observer, familiar with the prior art designs, would be deceived into believing the alleged infringing product is the same as the patented design.[363] For a drug manufacturer, design patents might be used to help establish brand loyalty and recognition in a manner similar to trademarks. Thus, where possible and appropriate (e.g., for a unique pill or capsule design), trademark protection may be preferred because its duration is generally limited only by the requirement for continued use.

9. CONCLUSION

Careful use of the intellectual property system, especially the patent system, in the United States and elsewhere in the world can enable those in the drug discovery and biotechnology industry to protect the fruits of their labor. By making effective use of the legal protections afforded by the intellectual property laws, a drug developer can protect its investment, enhance the value of its technology, and earn a profit sufficient to allow further research into improving existing drugs and therapies, as well as developing new drugs and therapies. Indeed, by providing such protection, the intellectual property system seeks to encourage the development of new and useful technology and products. For any industry, attention to the protection of intellectual property at the earliest stages of its development is of the utmost importance. This especially applies to the drug discovery and biotechnology industry because of the rapidly developing nature of the technology and the FDA submission requirements.

The formation of an intellectual property strategy that promotes the scientific and business objectives of the enterprise is extremely important. A centralized intellectual property function, charged with developing and overseeing the strategy on a continuing basis is highly recommended. Through the use of intellectual property committees or other similar mechanisms, sound strategies that are based on scientific and business expertise within the enterprise can be developed and implemented. Proactive organizations will work closely with qualified legal counsel, preferably patent counsel skilled in the relevant technology, to fashion internal mechanisms for protecting the technology, for seeking the appropriate legal protection, and for developing appropriate enforcement policies. In this manner, one can obtain the appropriate legal protection that the particular technology demands and deserves.

---

[361] 35 U.S.C.A. §§ 171–173 (West 2001).
[362] *Egyptian Goddess, Inc. v. Swisa, Inc.*, 543 F.3d 665, 677 (Fed. Cir. 2008) (*en banc*).
[363] *Id.* at 681.

POLYMORPHIC CRYSTAL FORMS AND COCRYSTALS IN DRUG DELIVERY (CRYSTAL ENGINEERING)

Ning Shan[1]
Michael J. Zaworotko[2]
[1] Thar Pharmaceuticals, Inc., Tampa, FL
[2] Department of Chemistry, University of South Florida, Tampa, FL

Active pharmaceutical ingredients, APIs, are most conveniently developed and delivered orally as solid dosage forms that contain a defined crystalline form of an API. This means that the pharmacokinetic profile of a dosage form is at the very least linked to the physicochemical properties of the crystal form that is selected for development. Furthermore, that crystal forms of new chemical entities are novel, lack obviousness, and have utility makes them patentable. Therefore, selection of a specific crystal form for a given API is a profoundly important step in drug development from clinical, legal, and regulatory perspectives. In this context, scientific developments that afford greater understanding of and diversity in the number of crystalline forms available for a given API, which have traditionally been limited to salts, polymorphs, and hydrates/solvates [1], are obviously of relevance to the pharmaceutical industry. The science of *crystal engineering* [2] focuses upon self-assembly of existing molecules or ions and it has evolved in such a manner that a wide range of new crystal forms can be generated without the need to invoke covalent-bond breakage or formation. This contribution will address the impact of crystal engineering upon our fundamental understanding of crystal form diversity and how physical properties of crystals can be customized via the emerging class of crystal forms that have been termed *pharmaceutical cocrystals* [3].

1. INTRODUCTION

The importance of crystallization and crystal forms to pharmaceutical science is the result of multiple practical considerations. In terms of processing, crystallizations tend to afford highly pure products, they are typically reproducible and scalable, and they are generally stable when compared to amorphous solids or solutions. They are therefore preferred by developers and regulatory bodies. Furthermore, although crystallization has been widely studied scientifically since at least the early nineteenth century, this does not mean that crystallization is predictable [4] or even controllable [5]. New crystal forms are therefore likely to be patentable in their own right since they meet the primary criteria for patentability: novelty, lack of obviousness and utility. Finally, it has been known for over 100 years that rate of dissolution of a solid is at least partly determined by thermodynamic solubility of a compound [6] and it is well recognized that solubility can significantly influence the bioavailability and pharmacokinetics of an API. Given that the majority of APIs currently under development fall into Biopharmaceutical Classification Scheme [7] (BCS) classification II (low solubility, high permeability), the importance of API crystal form screening and selection is, if anything, increasing in scope and importance. In short, the existence of multiple crystal forms of an API affords both challenges and opportunities to the pharmaceutical industry. In this context, the emergence of the concept of crystal engineering is timely and relevant.

Crystal engineering [2] was coined by R. Pepinsky [2c] in 1955 and brought to practice by G.M.J. Schmidt in the context of topochemical reactions [2d]. Crystal engineering has more recently matured into a paradigm for the understanding of existing crystalline solids and the design of new compounds with customized composition and physical properties. Indeed, crystal engineered materials have been studied in the context of host–guest compounds, nonlinear optical materials, organic conductors, and coordination polymers [8–11]. However, given that APIs are perhaps the most valuable crystalline substances known and their very nature (i.e., the presence of hydrogen-bonding functionality at their periphery) makes them predisposed toward crystal engineering, it is perhaps unsurprising that crystal engineering concepts are increasingly being applied to pharmaceutical science by both industrial and academic researchers [12–23]. It can be asserted that crystal engineering is finally realizing Desiraju's vision that crystal engineering is "the understanding of intermolecular interactions in the context of crystal packing and utilization of such understanding in the design of new solids with desired physical and chemical properties" [2e].

The range of crystal forms that are typically exhibited by APIs represents a microcosm of organic compounds although it would be fair to assert that APIs are more promiscuous than "typical" organic compounds because they contain multiple hydrogen-bonding sites and/or torsional flexibility. It is hydrogen bonding sites or, more specifically, the detailed understanding of the supramolecular chemistry of these hydrogen-bonding sites that is the key to understanding the structure–property relationships in crystal forms. The existence of multiple crystal forms for an API is therefore to be expected and they are typically categorized as follows: polymorphs, salts, solvates, hydrates, and cocrystals (Fig. 1).

- *Polymorphs*: Polymorphism, the existence of more than one crystal form for a compound, has been described as "the nemesis of crystal design" by one of the pioneers of crystal engineering, G.R. Desiraju. Indeed, there are probably many researchers in the pharmaceutical industry who would regard polymorphism as the nemesis of crystal form selection since the unpredictability of polymorphism complicates all aspects of crystallization from laboratory scale discovery through to industrial scale processing.

- *Salts*: Salts have long been an integral part of crystal form selection because they offer diversity of composition and can therefore exhibit a wide range of physicochemical properties. However, salts, especially chloride salts, tend to be prone to exist as hydrates, there are a

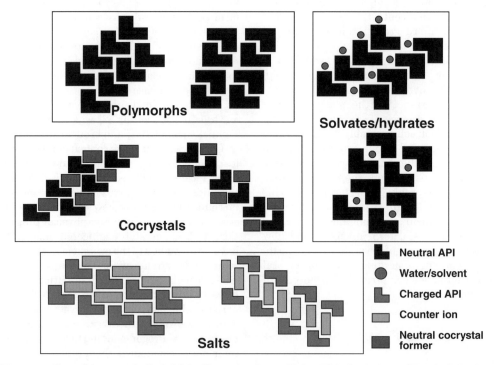

Figure 1. Crystal forms typically exhibited by molecular organics. (This figure is available in full color at http://mrw.interscience.wiley.com/emrw/9780471266945/home.)

INTRODUCTION

limited number of pharmaceutically acceptable counterions, and not all APIs are acidic or basic enough to form salts [24].

- *Hydrates and Solvates*: Solvates are crystalline compounds in which solute and solvent molecules coexist, normally but not always through interaction of noncovalent bonds such as hydrogen bonds. Likewise, hydrates are compounds that contain water bound within the crystal lattice. One might think that hydrates are typically prepared using water as a solvent but the ubiquitous presence of water means that they are most typically isolated through the presence of adventitious water molecules. Indeed, they represent more than 10% of the >500,000 crystalline organic compounds that have been archived in the Cambridge Structural Database, CSD. However, just as polymorphs are unpredictable, so are solvates and hydrates. Furthermore, solvates and hydrates are less likely to be selected as dosage forms because they tend to be prone to desolvation or dehydration in dry conditions.

- *Cocrystals*: Cocrystals represent a class of compounds that could reasonably be described as long known but little studied [25]. Indeed, to our knowledge the term cocrystal was not coined until 1967 [26] and it was not popularized in the context of small molecules until M.C. Etter used the term extensively in the 1980s [2a]. Furthermore, even today the term cocrystal is poorly defined and represents ambiguity or even controversy [27]. We define a cocrystal as following: a multiple component crystalline solid formed in a stoichiometric ratio between two compounds that are crystalline solids under ambient conditions. At least one of these compounds is molecular (the cocrystal former) and forms supramolecular synthons(s) with the remaining component(s) [3a–3e]. If one uses this definition then the first cocrystals were reported in the 1800s [25] and they have had various terms applied to them: addition compounds, organic molecular compounds, complexes, and heteromolecular crystals [28–35]. Cocrystals are also distinct from solvates, salts and inclusion compounds if one employs this definition. Nevertheless, the term pharmaceutical cocrystal, that is, a cocrystal between an API and a molecular cocrystal former, was not widely used until recent years. Pharmaceutical cocrystals were reported as far back in the 1930s [36], yet only in recent years has their diversity in terms of crystal form and physical properties been fully recognized in the context APIs.

Salt screening and selection is covered in a different chapter and solvates and hydrates tend to exhibit lower stability than polymorphs or pharmaceutical cocrystals. This chapter will therefore focus upon polymorphs and cocrystals with emphasis upon how they can be subjected to rationalization through crystal engineering. The key to crystal engineering in the context of APIs lies with understanding the hydrogen-bonding groups present in the API. Two approaches have been developed to analyze existing crystal structures with the view to utilize the structural knowledge thereby gained to rationalize and even control the composition or even structure of new crystal forms. These related and compatible approaches, *graph sets* and *supramolecular synthons*, were developed by Etter [2a] and Desiraju [2i], respectively. In both instances, there is reliance upon utilizing the Cambridge Structural Database [37], to gather statistical information about crystal packing and intermolecular interactions. We shall focus herein upon supramolecular synthons, which are defined as "a structural unit within the supermolecule that can be formed and/or assembled by known or conceivable intermolecular interactions." Supramolecular synthons focus upon functional groups rather than molecules and exist in two distinct categories: *supramolecular homosynthons* that are composed of identical complementary functional groups, for example, carboxylic acid dimers [38], amide dimers [39] (Fig. 2a and b); *supramolecular heterosynthons* composed of different but complementary functional groups such as acid-amide [40] and acid-aromatic

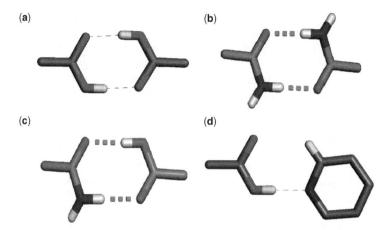

Figure 2. Prototypal supramolecular homosynthons (a) and (b) and supramolecular heterosynthons (c) and (d). (This figure is available in full color at http://mrw.interscience.wiley.com/emrw/9780471266945/home.)

nitrogen [41] (Fig. 2c and d). The aforementioned supramolecular synthons are particularly salient because carboxylic acids are present in 25 of the top 100 most prescribed drugs in the United States. Furthermore, they are frequently encountered in pharmaceutical excipients, salt formers and cocrystal formers.

2. CRYSTAL FORM TYPES

2.1. Polymorphs

The first observation of polymorphism can be attributed to Wöhler and von Liebig, who in 1832 reported that upon cooling a boiling solution of benzamide, needle-shaped crystals would initially form [42]. However, upon standing the needle-shaped crystals would slowly be replaced by rhombic crystals. This observation is a manifestation of Ostwald's step rule, that is, that the crystal form first obtained upon crystallization of a substance from a solution or a melt will be a metastable polymorph, a long recognized [43] and qualitative generalization about crystallization. However, despite a long history, it would be fair to say that, until recently, polymorphism has been more of a scientific curiosity than an urgent challenge of commercial relevance. Pharmaceutical science has been largely responsible for a change in this situation since most orally delivered APIs receive regulatory approval for a single crystal form or polymorph and novel crystal forms are patentable. Awareness of the matter heightened following a now classic patent litigation between Glaxo and Novopharm in which Glaxo defended its patent for the form II polymorph of ranitidine hydrochloride, the API in Zantac®. The Glaxo patent on form I of ranitidine hydrochloride (US patent 4,128,658) expired on December 5 1995, but the form II patent (US patent 4,521,431) did not expire until 2002. Although Novopharm ultimately prevailed, Glaxo retained exclusivity beyond the patent expiration of form I for several years on what was at the time the top-selling drug in the world. In addition to legal, regulatory and commercial considerations, polymorphism in drug substances can also have direct clinical implications since dissolution rates are sometimes impacted by polymorphism. However, although polymorphism might be long recognized and topically relevant to pharmaceutical science [44], this does not mean that polymorphs are predictable or that their discovery is routine despite McCrone's statement on the subject in 1965 [45]: "Every compound has different polymorphic forms and the number of forms known for a given compound is proportional to the time and energy spent in research on that compound." This provocative statement has often been debated and many solid-state scientists would be inclined to support such an assertion. However, McCrone's statement cannot realistically be proved through experiment and even today the number of publicly disclosed cases of polymorphism in organic compounds remains quite low based upon CSD statistics: only 8525 out of 195,222 organic compounds archived are polymorphic and since there must be at least two entries for each compound this represents ≤2.2% of organic compounds; only 667 out of 11,501 compounds with biological or pharmacological activity are polymorphic (selected using keywords "activity," "agent," "biological," "drug," "pharmaceutical," "pharmacological"), representing just ≤2.9% of this subset. One must also bear in mind that the CSD is unlikely to be representative in the context of polymorphs since entries are biased by the compounds that have been of interest to crystallographers at particular points in time. The focus until recently has been upon molecular structure rather than crystal structure and many polymorphs probably remain unpublished.

Although polymorphism might remain largely unpredictable it can be rationalized and categorized through understanding the molecular and supramolecular structure of the compound in question, allowing us to define at least two classes of polymorphism [45]: *conformational polymorphism* is the consequence of more than one conformer in the solid state (i.e., the shape of the molecule is different); *packing polymorphism* occurs when rigid molecules exhibit more than one packing arrangement. Packing polymorphs might be caused by different supramolecular synthons (i.e., the intermolecular connectivity is different) or they might retain their supramolecular synthons but exhibit different crystal packing. Such a situation might be termed *supramolecular synthon polymorphism*. Conformational polymorphism is exemplified by what is thus far the most promiscuous molecule in terms of the number of structurally characterized polymorphs, 5-methyl-2-[(2-nitrophenyl)amino]-3-thiophenecarbonitrile, a pharmaceutical intermediate that has been called ROY because its eight crystallographically characterized polymorphs are red, orange, or yellow in color [46,47]: ROY is illustrated in Fig. 3, which highlights the portion of ROY that is responsible for its conformational flexibility. Six room temperature polymorphs of ROY were reported by Yu et al. in 2000 [46] and two additional polymorphs, Y04 and YT04, were reported in 2005 [47]. Y04 was prepared from a melt at room temperature, and YT04 was obtained via solid-state transformation of Y04. Y04 and YT04 exemplify polymorphs that would likely be missed by solvent-based screening, highlighting the experimental

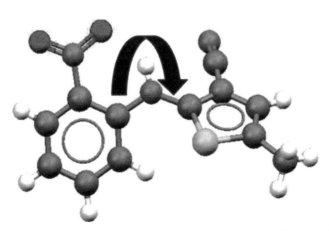

Figure 3. The molecular structure of the ON polymorph of 5-methyl-2-[(2-nitrophenyl)amino]-3-thiophenecarbonitrile, ROY (CSD refcode = QAXMEH) indicating the region of torsional flexibility. (This figure is available in full color at http://mrw.interscience.wiley.com/emrw/9780471266945/home.)

challenge of polymorph discovery. Packing polymorphism is exhibited by numerous APIs and exemplified herein by Piracetam and Aspirin. Piracetam, 2-oxo-pyrrolidineacetamide, is a nootropic drug that improves cognitive ability and it exhibits five structurally characterized polymorphs [48]. Although one of these polymorphs, the high-pressure form IV, is a conformational polymorph, forms I and II are examples of packing polymorphism caused by different supramolecular heterosynthons. Form I exists as a cyclic tetramer whereas form II forms infinite tapes in which amide–amide dimers are hydrogen bonded to adjacent dimers through amide-carboxamide N—H ... O hydrogen bonds. Aspirin had long been considered to represent an example of a compound that does not exhibit polymorphism. However, in 2005, metastable form II of aspirin was discovered during an attempted cocrystallization reaction [49] Forms I [50] and II are illustrated in Fig. 4, which reveal that both crystal forms of aspirin contain dimers that are sustained by the carboxylic acid supramolecular homosynthon. However, C—H ... O interactions between adjacent dimers are different and in turn cause different crystal packing. Subsequent work has suggested that forms I and II might coexist within the same crystal (Fig. 5) [51].

In conclusion, polymorphs can generally be rationalized through supramolecular concepts such as crystal engineering but this does not mean that they can yet be predicted from first principles. However, although one should not confuse crystal engineering with crystal structure prediction, crystal structure prediction using computer modeling has advanced considerably within the past decade [52].

2.2. Cocrystals

2.2.1. What is a Cocrystal? That there is not yet a recognized definition of the term "cocrystal" has engendered debate on the subject [27]. We have been using the following operating definition: a multiple component crystalline solid formed in a stoichiometric ratio between two compounds that are crystalline solids under ambient conditions. At least one of these compounds is molecular (the cocrystal former) and forms supramolecular synthons(s) with the remaining component (s) [3a–3e]. That all components are solids under ambient conditions has important practical considerations since synthesis of cocrystals can be achieved via solid-state methods (e.g., mechanochemistry) and chemists can execute a certain degree of control over the composition of a cocrystal since they can invoke molecular recognition, especially hydrogen bonding, during the selection of cocrystal formers. These features distinguish cocrystals from solvates and despite restrictions they still represent a broad range of compounds since most molecular compounds exist as solids under ambient conditions [53].

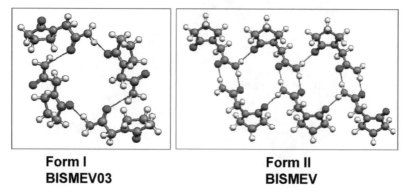

Form I
BISMEV03

Form II
BISMEV

Figure 4. Forms I and II of piracetam exhibit packing polymorphism because they exhibit different supramolecular synthons. Form I exists as a cyclic tetramer whereas form II forms infinite tapes. (This figure is available in full color at http://mrw.interscience.wiley.com/emrw/9780471266945/home.)

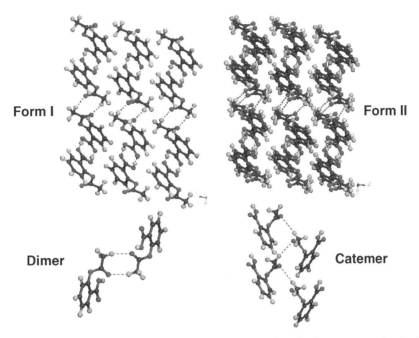

Figure 5. The two polymorphs of aspirin are both based upon carboxylic dimer supramolecular homosynthons. However, they differ in the manner in which adjacent dimers interact. In form I C—H...O dimers are formed whereas in form II the structure is sustained by C—H...O catemers. (This figure is available in full color at http://mrw.interscience.wiley.com/emrw/9780471266945/home.)

2.2.2. Why are Cocrystals of Interest to the Pharmaceutical Industry?

Pharmaceutical cocrystals, that is, cocrystals in which the target molecule or ion is an active pharmaceutical ingredient, API, and the cocrystal former is a pharmaceutically acceptable molecule or ion, are emerging rapidly because of a number of factors including the following:

- *Design*: Our scientific understanding of the noncovalent forces that sustain molecular organic crystals has advanced to the extent that control over the stoichiometry and composition of cocrystals can be asserted. This is not ordinarily the case for polymorphs and solvates for which high-throughput screening, which to a certain extent practices serendipity, tends to be relied upon rather than design, or for salts, which require an ionizable functional group.

- *Discovery*: That mechanochemistry can be utilized to synthesize cocrystals has been known since the first cocrystals were discovered in the 1840s by dry grinding [25a], but it has only recently been realized and accepted that "solvent-drop" or "liquid assisted" grinding are preferred methodologies [54]. Indeed, it is fair to assert that cocrystals are most readily accessible through solvent-free or solvent-reduced methods although other techniques such as slurrying [55] and solution [56] are complementary.

- *Diversity*: It has become apparent that pharmaceutical cocrystals always exhibit different physicochemical properties compared to the pure crystal form(s) of APIs, that a given API might form cocrystals with dozens of cocrystal formers and that some of these cocrystals might exhibit enhanced solubility or stability to hydration. Therefore, pharmaceutical cocrystals represent an opportunity to diversify the number of crystal forms of a given API and in turn fine-tune or even customize its physicochemical properties without the need for chemical (covalent) modification.

- *Development*: Whereas pharmaceutical cocrystals can be designed using crystal engineering strategies this does not mean that details of their crystal structures or physical properties can be predicted before they have been measured. Therefore, one might assume that it will be possible for pharmaceutical cocrystals of existing APIs to be patented as new crystal forms and, if they exhibit clinical advantages, developed as new drugs. This has implications for drug development because it abbreviates some aspects of drug development timelines and mitigates costs and risks related to discovery and toxicology of new APIs.

- *Delivery*: As mentioned earlier, being able to fine-tune solubility can be a critical factor that influences the clinical performance of an API if its bioavailability is affected by rate of dissolution. This is generally considered to be important for BCS Class II APIs [57], perhaps the most common classification for the current generation of APIs.

The August 2008 release of the CSD contains structural information on 456,628 organic, metal–organic, and organometallic crystal structures, but there is not a great deal of structural information on cocrystals. There are only two cocrystal entries prior to 1960 and even today there are only ca. 2083 (0.46% of the CSD) hydrogen-bonded cocrystals versus 50,019 hydrates (10.95% of the CSD). Therefore, it would be fair to summarize cocrystals as being a long known but little studied class of compounds. Nevertheless, the realization that there will be multiple cocrystal formers for a given API makes pharmaceutical cocrystals somewhat diverse in terms of their composition. The scope of available cocrystal formers is not yet set but even if it is limited to "generally regarded as safe" (GRAS) compounds or compounds that have already been approved by the federally mandated Food and Drug Administration (FDA) for use in formulation such as a "salt formers," there could be 100 or more possible pharmaceutically acceptable cocrystal formers for an API.

In terms of the pharmaceutical industry, perhaps the earliest example of a pharmaceutical cocrystal was reported in a 1934 French patent that disclosed cocrystals of barbiturates with 4-oxy-5-nitropyridine, 2-ethoxy-5-acetaminopyridine, N-methyl-α-pyridone, and α-aminopyridine [36]. In 1995, Eli Lilly and Co. patented complexes of cephalosporins and carbacephalosporins, a class of β-lactam antibiotics, with parabens and related compounds [57]. In terms of the scientific literature, there were few reports of pharmaceutical cocrystals until the past decade. However, Caira demonstrated that "old" drugs such as sulfonamides can form cocrystals [58] and also emphasized their potential in drug development.

2.2.3. Design of Cocrystals A crystal engineering experiment typically involves CSD surveys followed by experimental work to prepare and characterize new compounds that are sustained by *supramolecular synthons*. Supramolecular synthons facilitate understanding of the supramolecular chemistry of the functional groups present in a given molecule and are prerequisites for designing a cocrystal since they facilitate selection of an appropriate cocrystal former(s). However, when multiple functional groups are present in a molecule, the CSD rarely contains enough information to address the hierarchy of the possible supramolecular synthons. Fortunately, the hierarchy of the supramolecular synthons that can occur for common functional groups such as carboxylic acids, amides, and alcohols with emphasis upon *supramolecular heterosynthons* is becoming better defined [12d,e]. Furthermore, it is becoming evident that such interactions are key to implementing a design strategy for cocrystals in which a target molecule forms cocrystals with cocrystal formers that are carefully selected for their ability to form supramolecular heterosynthons with the target molecule.

The design aspect of cocrystals is illustrated if one focuses upon carboxylic acids, perhaps the most important and widely studied functional group in the context of pharmaceutical cocrystals since carboxylic acids represent ca. 25% of marketed drugs and carboxylic acids are commonly used as salt formers or excipients. The CSD enables statistical surveys of intermolecular contacts as well as intramolecular connectivity and it is therefore a powerful tool for addressing supramolecular chemistry in the solid state. A survey of the CSD revealed that there were 8154 organic carboxylic acids in the CSD as of August 2008. However, an analysis of intermolecular contacts in this subset revealed that only 1926 of these carboxylic acids exhibit the carboxylic acid dimer supramolecular homosynthon (Fig. 6) and that only 143 exhibit the carboxylic acid catemer motif. So what about the remaining 75% of carboxylic acids that have been crystallographically characterized? As revealed by Fig. 7, there is a tendency for carboxylic acids to form supramolecular heterosynthons with, for example, chloride anions and aromatic nitrogen moieties. Furthermore, the statistics seem to strongly favor these supramolecular heterosynthons over the corresponding supramolecular homosynthons. For example, there are 277 crystal structures that contain both a carboxylic acid and a chloride anion and 180 of them exhibit the carboxylic acid chloride supramolecular heterosynthon. In only one of this subset of 277 crystal structures does the carboxylic dimer exist. The statistics are similar for the carboxylic acid—pyridyl supramolecular heterosynthon. There are 606 crystal structures that contain both a carboxylic acid and a pyridyl moiety and 415 of them exhibit the carboxylic acid—pyridyl supramolecular heterosynthon. In only 25 of this subset of 606 crystal structures does the carboxylic dimer exist. In short, although these data are raw and une-

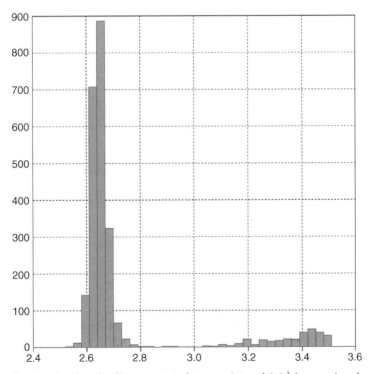

Figure 6. Distribution of carboxylic dimer contacts between 2.4 and 3.6 Å in organic only carboxylic acid crystal structures in the CSD. The distribution reveals 1926 H-bonded contacts between 2.55 and 2.80 Å. (This figure is available in full color at http://mrw.interscience.wiley.com/emrw/9780471266945/home.)

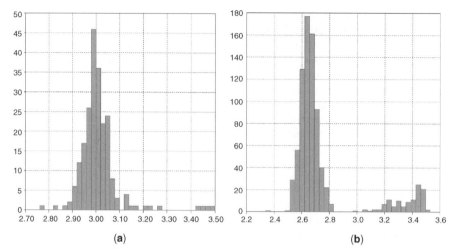

Figure 7. (a) Distribution of carboxylic acid—chloride anion contacts between 2.4 and 3.6 Å in organic only carboxylic acid crystal structures that also contain chloride anions. There are 180 short contacts between 2.7 and 3.3 Å. (b) Distribution of carboxylic acid—aromatic nitrogen contacts between 2.4 and 3.5 Å in organic only carboxylic acid crystal structures that also contain aromatic nitrogen moieties. (This figure is available in full color at http://mrw.interscience.wiley.com/emrw/9780471266945/home.)

dited, it strongly suggests that if the relevant functional groups are in different molecules then a cocrystal involving supramolecular heterosynthons is likely to occur over the corresponding single component structures that would be sustained by supramolecular homosynthons. This principle is exemplified by several of the case studies presented herein.

2.2.4. Polymorphs, Solvates, and Hydrates of Cocrystals There remains a dearth of systematic structure and property information on cocrystals. However, at this point there is no reason to believe that pharmaceutical cocrystals will be more or less promiscuous than single component APIs when it comes to crystal form diversity. For example, both conformational and packing polymorphs have been observed in cocrystals. Figure 8 reveals that rotation around the central C—C bond in 4,4'-biphenol can afford conformational polymorphism in the 2:1 cocrystal of 4-cyanopyridine and 4,4'-biphenol [59]. Figure 9 reveals how a model cocrystal based upon the pyridine-carboxylic acid supramolecular synthons [60], a supramolecular synthons that is particularly relevant to APIs, exhibits packing polymorphism. In this case, packing polymorphism manifests itself through networks and interpenetration in the polymorphs of the 3:2 cocrystal of 4,4'-bipyridylethane and trimesic acid [61].

3. CASE STUDIES THAT DEMONSTRATE HOW CRYSTAL FORMS CAN IMPACT PHYSICOCHEMICAL PROPERTIES AND/OR BIOAVAILABILITY

3.1. Case Studies of Polymorphs

The impact of polymorphism on solubility was addressed by Pudipeddi and Serajuddin, who collated data on 81 polymorphic pairs [62]. The majority of these polymorphs (63/81) were observed to exhibit a solubility ratio of ≤2 and only one pair of polymorphs exhibits a solubility ratio of >10. This outlier, premafloxacin (Fig. 10), is a broad-spectrum antibiotic initially developed for veterinary use by Pharmacia and Upjohn, Inc. and it is chemically known as $[S$-$(R^*,S^*)]$-1-cyclopropyl-6-fluoro-1,4-dihydro-8-methoxy-7-{3-[1-methylamino) ethyl]-1-pyrrolidinyl}-4-oxo-3-quinolinecarbo-

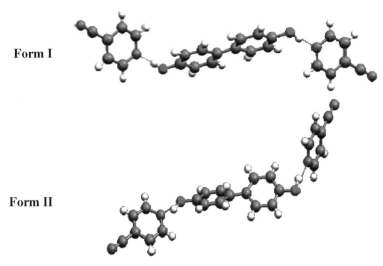

Figure 8. The conformational polymorphs exhibited by the 2:1 cocrystal of 4-cyanopyridine and 4,4'-biphenol. (This figure is available in full color at http://mrw.interscience.wiley.com/emrw/9780471266945/home.)

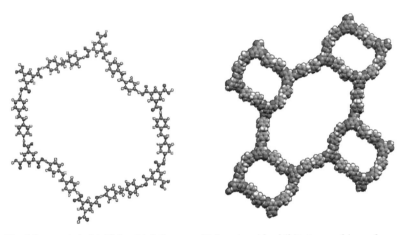

Figure 9. The 3:2 cocrystal of 4,4' bipyridylethane and trimesic acid exhibits two packing polymorphs: a (6,3) honeycomb network that with 3-fold parallel interpenetration and a (10,3)-*a* 3D network with 18-fold interpenetration. (This figure is available in full color at http://mrw.interscience.wiley.com/emrw/9780471266945/home.)

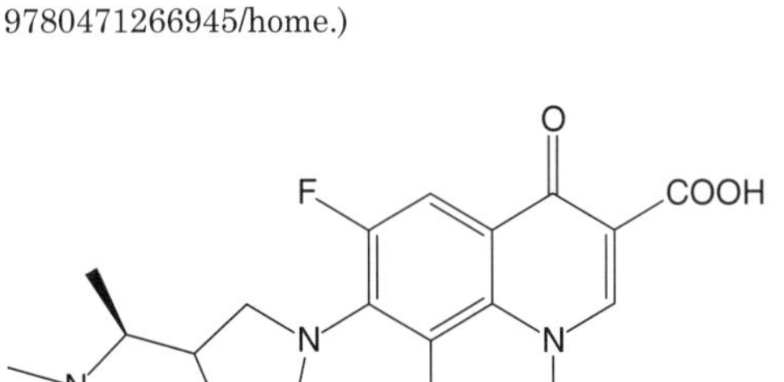

Figure 10. Molecular structure of premafloxacin.

xylic acid. This fluoroquinolone derivative has activity against a wide range of veterinary pathogens with equivalent activity to similar antibiotics such as ciprofloxacin against Gram-negative bacteria but enhanced MICs (minimum inhibitory concentration) against Gram-positive bacteria [63]. This API exhibits polymorphism and five crystal forms have been reported although they are not yet archived in the CSD [64]. Schinzer et al. analyzed interconversion of these crystal forms through PXRD, thermomicroscopy, DSC, isothermal calorimetry, and dynamic moisture sorption gravimetry [64]. It was concluded that three anhydrous forms, a hydrate and a methanol solvate exist. Form I was made by desolvating the methanol solvate and form II was a metastable form that occurred through a melt-recrystallization process at 140–150 °C. When heated further, another phase transition occurred at 165–180 °C and resulted in form III. The hydrate was formed by exposing form I to 80% relative humidity (RH). Form III can be grown directly from ethylacetate but not from methanol, which results in the methanol solvate. Form I was observed to convert to form III at temperatures as low as 40 °C in the presence of moisture but at a RH of 51%, the rate of conversion was almost two orders of magnitude lower than at 75% RH. Form III is the most stable form and it sorbed less water than form I at all humidity conditions tested. Solubility in ethyl acetate was determined to be 3.23 mg/mL for form I and 0.14 mg/mL for form III (i.e., solubility ratio between the most soluble form and least soluble form is 23.1), easily the largest solubility ratio difference of the 81 polymorphic pairs analyzed [62].

3.2. Case Studies of Pharmaceutical Cocrystals

Perhaps, the earliest examples of pharmaceutical cocrystals were described in a series of studies conducted in the 1950s by Higuchi and his coworkers [65,66], who studied complex formation between macromolecules and certain pharmaceuticals; for example, complexes of polyvinylpyrrolidone (PVP) with sulfathiazole, procaine hydrochloride, sodium salicylate, benzylphenicillin, chloramphenicol, mandelic acid, caffeine, theophylline, and cortisone were isolated [65,66]. However, these compounds would not be classified as pharmaceutical cocrystals according to the criteria applied herein. Perhaps, the first application of crystal engineering to the generation of pharmaceutical cocrystals was described in a series of papers by Zerkowski et al. [67], who reported the use of substituted barbituric acids, including barbital, and melamine derivatives to generate supramolecular "linear tape," "crinkled tape," and "rosette" motifs sustained by robust supramolecular synthons with three-point hydrogen bonding [67]. In spite of their success in cocrystal formation, the focus of these studies was not so much the physical properties of the resulting cocrystals but rather the supramolecular functionality of barbitals and their complementarity with melamine. Nevertheless, these studies highlighted the potential diversity of forms that can exist for a particular API as more than 60 cocrystals were structurally characterized in this series of studies. Undoubtedly, such a diversity of forms could offer an exciting opportunity to produce, patent and market novel API crystalline forms with improved physical properties of clinical relevance. Herein we have selected a series of case studies (Table 1) that illustrate how pharmaceutical cocrystals can significantly alter the physicochemical properties of APIs.

3.2.1. Pharmaceutical Cocrystals of Carbamazepine (Tegretol®)

Carbamazepine (CBZ) has been used as an important anti-epileptic drug for over three decades. The oral administration of CBZ encounters multiple challenges, including low water solubility with high dosage for therapeutic effect (i.e., >100 mg/day), dissolution-limited bioavailability, and autoinduction for metabolism. In contrast to its simple molecular structure, CBZ exhibits complex behavior in the context of its crystal forms [15b,68]. A CSD analysis on CBZ reveals that it has four fully characterized polymorphs [69], a dihydrate [70], 14 solvates (i.e., acetone, furfural, dimethyl sulfoxide, trifluoroethanol, dimethylformamide, N-methylpyrrolidone, nitromethane, acetic acid, formic acid, butyric acid, formamide, trifluoracetic acid, tetrahydrofuran, N,N'-dimethyl acetamide) [12b,71], two ammonium salts [72], and a solid solution with dihydrocarbamazepine [73]. In addition, Hilfiker et al. [74] have identified three new polymorphic forms and a dioxane solvate using high-throughput screening. It is noted that, in the crystal structures of all CBZ polymorphs and solvates, the self-complementary nature of the amide group manifests itself in a predictable manner.

Table 1. A Summary of Pharmaceutical Cocrystals in the Patent and Scientific Literature by February 2009

	API	Cocrystal Former	Solubility Study	Stability Study	Pharmacokinetic Study	References
1	(2R-trans)-6-Chloro-5-[[4-[[(4-fluorophenyl)methyl]-2,5-dimethyl-1-piperazinyl]carbonyl]-N,N,1-trimethyl-alpha-oxo-1H-indole-3-acetamide	HCl form: urea, arginine Free base form: salicylic acid, 4-aminosalicylic acid, benzoic acid				Chiarella RA. WO 2008013823 A2.
2	5,5′-Diphenylhydantoin	1-(4-Bromophenyl)-4-dimethylamino-2,3-dimethyl-3-pyrazolin-5-one				Uno T, Shimizu N. *Acta Crystallogr* 1980; B36:2794.
3	5-Methoxysulfadiazine	Acetylsalicylic acid				Caira MRJ. *Chem Crystallogr* 1994;24:695.
4	AMG 517	Benzoic acid, *trans*-cinnamic acid, 2,5-dihydroxybenzoic acid, glutaric acid, glycolic acid, *trans*-2-hexanoic acid, 2-hydroxycaproic acid, L-lactic acid, sorbic acid, L-tartaric acid	Dissolution		Pharmacokinetic study	[80a] [80b]
5	Aspirin	4,4′-Bipyridine				Walsh RBD, Bradner MW, Fleishman S, Morales LA, Moulton B, Rodriquez-Hornedo N, Zaworotko MJ. *Chem Commun* 2003;186.
6	Barbiturates	5,5′-Diethylbarbituric acid:*N*,*N*′-bis(4-t-butyl-phenyl)melamine 5,5′-Dimethylbarbituric acid:*N*,*N*′-bis(4-*t*-butyl-phenyl)melamine				[67]
7	Barbiturates	5,5′-Diethylbarbituric acid:acridine 5,5′-Diethylbarbituric acid:1,10-phenanthroline				Vishweshwar P, Thaimattam R, Jaskolski M, Desiraju GR. *Chem Commun* 2002;1830.

(*continued*)

Table 1 (*continued*)

	API	Cocrystal Former	Solubility Study	Stability Study	Pharmacokinetic Study	References
8	Barbiturates	5,5-Diethylbarbituric acid:2,4-diamino-5-(3,4,5-trimethoxybenzyl)pyrimidine				Shimizu N, Nishigaki S, Nakai Y, Osaki Y. *Acta Crystallogr* 1982;B38:2309.
9	Barbiturates	5,5-Diethylbarbituric acid: N,N'-diphenylmelamine Barbituric acid:melamine				Zerkowski JA, Seto CT, Wierda DA, Whitesides GM. *J Am Chem Soc* 1990;112:9025.
10	Barbiturates	Amobarbital:salicylamide				Hsu IN, Craven BM. *Acta Crystallogr* 1974;B30:843.
11	Barbiturates	Barbital N'-(p-cyanophenyl)-N-(p-iodophenyl)melamine 5,5-Dimethylbarbituric acid:N,N'-diphenylmelamine Barbital 2-amino-4-(m-bromophenylamino)-6-chloro-1,3,5-triazine 5,5-Dibromobarbituric acid: melamine Barbituric acid:melamine				Zerkowski JA, McDonald JC, Whitesides GM. *Chem Mater* 1994;6:1250.
12	Barbiturates	Barbital:2-aminopyrine				Kiryu S. *J Pharm Sci* 1971;60:699.
13	Barbiturates	Barbital:9-ethyladenine				Voet DJ. *Am Chem Soc* 1972;94:8213.
14	Barbiturates	Phenobarbital:(4,4'-bipyridine-N,N'-dioxide)$_{0.5}$ Barbituric acid:4,4'-bipyridine-N,N' dioxide:H$_2$O				Reddy LS, Babu NJ, Nangia A. *Chem Commun* 2006;1369.
15	Barbiturates	Phenobarbital:8-bromo-9-ethyladenine				Kim SH, Rich A. *Proc Natl Acad Sci USA* 1968;60:402.

#	Compound	Coformers	Property	Reference
16	Beta-lactam compounds	Cephalexin:methyl paraben:H$_2$O Cephradine:Methyl Paraben:H$_2$O Cefaclor:methyl paraben:3H$_2$O Loracarbef:methyl 3-hydroxybenzoate:5H$_2$O Loracarbef:methyl paraben Loracarbef:ethyl paraben Loracarbef:propyl paraben Loracarbef:butyl paraben		Amos JG, Indelicato JM, Pasini CE, Reutzel SM. EP0637587
17	Bicalutamide (Casodex)	4,4′-Bipyridine, *trans*-1,2-bis(4-pyridyl)ethylene		Bis JA, Vishweshwar P, Weyna D, Zaworotko M. *J Mol Pharm* 2007;4:401.
18	Caffeine	Formic acid, acetic acid, trifluoroacetic acid, adipic acid		Trask AV, van de Streek J, Motherwell WDS, Jones W. *Cryst Growth Des* 2005;5(6):2233.
19	Caffeine	Glutaric acid, saccharin, salicylic acid		[54b]
20	Caffeine	Methyl gallate	Mechanical stability	Lu E, Rodriguez-Hornedo N, Suryanarayanan R. *CrystEngComm* 2008;10(6):665.
21	Caffeine	Oxalic acid, malonic acid, maleic acid, glutaric acid	Stability	Sun CC, Hou H. *Cryst Growth Des* 2008;8(5):1575.
22	Carbamazepine	4,4′-bipyridine, 4-aminobenzoic acid, 2,6-pyridinedicarboxylic acid		[16d]
23	Carbamazepine	Aspirin		[76]
24	Carbamazepine	Benzoquinone, terephthalaldehyde, saccharin, nicotinamide, butyric acid, trimesic acid, 5-nitroisophthalic acid, adamantane-1,3,5,7-tetracarboxylic acid		[49] [12b]

(*continued*)

Table 1 (*continued*)

	API	Cocrystal Former	Solubility Study	Stability Study	Pharmacokinetic Study	References
25	Carbamazepine	Glutaric acid, salicylic acid, urea.				Lu E, Rodriguez-Hornedo N, Suryanarayanan R. *CrystEngComm* 2008;10(6):665.
26	Carbamazepine	Glycolamide, lactamide				Ras E. WO 2008108639 A1.
27	Carbamazepine	Nicotinamide and saccharin (cocrystal form II)				[75]
28	Carbamazepine	*p*-Phthalaldehyde				[12c]
29	Carbamazepine	Quinoxaline *N,N'*-dioxide, pyrazine *N,N'*-dioxide				[13b]
30	Carbamazepine	Saccharin	Dissolution		Pharmacokinetic study	[68]
31	Carbamazepine	Succinic acid, benzoic acid, ketoglutaric acid, maleic acid, glutaric acid, malonic acid, oxalic acid, adipic acid, (+)-camphoric acid, 4-hydroxybenzoic acid, salicylic acid, 1-hydroxy-2-naphthoic acid, DL-tartaric acid, L-tartaric acid, glycolic acid, fumaric acid, DL-malic 2 acid, L-malic acid (total of 27 forms)	Dissolution	Stability in water		[55]
32	Celecoxib	Nicotinamide, 18-crown-6	Dissolution	Stability		[12c] Remenar JF, Peterson ML, Stephens PW, Zhang Z, Zimenkov Y, Hickey MB. *Mol Pharm* 2007;4(3):386.
33	C-glycoside (1*S*)-1,5-anhydro-1-[3-(1-benzothien-2-ylmethyl)-4-fluorophenyl]-D-glucitol	L-proline		Stability	Pharmacokinetic study	[83a]

34	Fenbufen	Nicotinamide		Berry DJ, Seaton CC, Clegg W, Harrington RW, Coles SJ, Horton PN, Hursthouse MB, Storey R, Jones W, Friscic T, Blagden N. *Cryst Growth Des* 2008;8(5):1697.
35	Fluconazole maleate	Maleic acid hydrate		McMahon J, Peterson M, Zaworotko MJ, Shattock T, Bourghol Hickey M. WO 2006007448 A2.
36	Fluoxetine HCl	Fumaric acid, succinic acid, benzoic acid	Dissolution	[14c]
37	Flurbiprofen	4,4′-Bipyridine, 1,2-bis(4-pyridyl)ethylene		Walsh RBD, Bradner MW, Fleishman S, Morales LA, Moulton B, Rodriquez-Hornedo N, Zaworotko MJ. *Chem Commun* 2003;186.
38	Flurbiprofen	Nicotinamide		Berry DJ, Seaton CC, Clegg W, Harrington RW, Coles SJ, Horton PN, Hursthouse MB, Storey R, Jones W, Friscic T, Blagden N. *Cryst Growth Des* 2008;8(5):1697.
39	Gabapentin	3-hydroxybenzoic acid	Solubility Stability	Reddy LS, Bethune SJ, Kampf JW, Rodriguez-Hornedo N. *Cryst Growth Des* 2009;9(1):378.
40	Gemfibrozil	Hydroxy derivatives of t-butylamine		Cheung EY, David SE, Harris KDM, Conway BR, Timmins P. *J Solid State Chem* 2007;180:1068.
41	Gossypol	C1–8 carboxylic acid, C1–8 sulfonic acid (example only shows acetic acid solvate)		Wang S, Chen S. WO 2005094804 A1.
42	Ibuprofen	4,4′-bipyridine		Walsh RBD, Bradner MW, Fleishman S, Morales LA, Moulton B, Rodriquez-Hornedo N, Zaworotko MJ. *Chem Commun* 2003;186.
43	Ibuprofen	Nicotinamide		Berry DJ, Seaton CC, Clegg W, Harrington RW, Coles SJ, Horton PN, Hursthouse MB, Storey R, Jones W, Friscic T, Blagden N. *Cryst Growth Des* 2008;8(5):1697.
44	Imipramine HCl	(+)-Camphoric acid, fumaric acid, 1-hydroxy-2-naphthoic acid	Dissolution	Childs SL. WO 2007067727 A2.
45	Indomethacin	Saccharin	Dissolution Stability	Basavoju S, Bostroem D, Velaga SP. *Pharm Res* 2008;25(3):530.

(*continued*)

Table 1 (continued)

	API	Cocrystal Former	Solubility Study	Stability Study	Pharmacokinetic Study	References
46	Isovaleramide	Citric acid, gentisic acid, glutaric acid, maleic acid, mandelic acid				Oliveira M, Peterson M. WO 2006116655 A2.
47	Itraconazole	Fumaric acid, L-tartaric acid, succinic acid, D-tartaric acid, L-malic acid, DL-tartaric acid	Dissolution		Pharmacokinetic study	[15a]
48	Lamivudine	Zidovudine, 3,5-dinitrosalicylic acid, 4-quinolinone				Bhatt PM, Azim Y, Thakur TS, Desiraju GR. *Cryst Growth Des* 2009;9:951–957.
49	Merck compound I N-{(1S)-1-[2-(1-{[(3S,4R)-1-tert-butyl-4-(2,4-difluorophenyl)pyrrolidin-3-yl]carbonyl}piperidin-4-yl)-5-chlorophenyl]ethyl}acetamide	HCl cocrystal of HCl salt (rigorously not a cocrystal)	Solubility			Peresypkin A., Variankaval N, Ferlita R, Wenslow R, Smitrovich J, Thompson K, Murry J, Crocker L, Mathre D, Wang J, Harmon P, Ellison M, Song S, Makarov A, Helmy R. *J Pharm Sci* 2008;97(9):3721.
50	Merck monophosphate salt	Phosphoric acid	Solubility	Stability	Pharmacokinetic study	[84]
51	Mestanolone	Salicylic acid				Takata N, Shiraki K, Takano R, Hayashi Y, Terada K. *Cryst Growth Des* 2008;8(8):3032.
52	Metronidazole	Gentisic acid, gallic acid	Dissolution			Childs SL. WO 2007067727 A2.
53	Modafinil	Malonic acid, succinic acid, fumaric acid	Dissolution		Pharmacokinetic study	[12c]
54	Naproxen	trans-1,2-bis(4-Pyridyl)ethylene, 1,2,4,5-tetracyanobenzene				Peterson M, Bourghol Hickey M, Oliveira M, Almarsson O, Remenar J. US Patent 2,005,267,209 A1.
						Weyna DR, Zaworotko MJ. Abstracts, 59th Southeast Regional Meeting of the American Chemical Society, 2007, GEN-015.
						Koshima H, Ding K, Chisaka Y, Matsuura T. *Tetrahedron* 1995;6:101.
55	Norflaxacin	Isonicotinamide (salts formed with malonic acid, and maleic acid)	Solubility			Basavoju S, Bostroem D, Velaga SP. *Cryst Growth Des* 2006;6:2699.

56	Norfloxacin	Norfloxacin saccharinate-saccharin dihydrate cocrystal	Velaga SP, Basavoju S, Bostroem D. *J Mol Struct* 2008;889(1–3):150.
57	Olanzapine	Nicotinamide	[12c]
58	Ornidazole	3,5-Dinitrobenzoic acid	Anderson KM, Probert MR, Whiteley CN, Rowland AM, Goeta AE, Steed JW. *Cryst Growth Des* 2009; 9:1082.
59	Paracetamol	Morpholine	Oswald IDH, Motherwell WDS, Parsons S, Pulham CR, *Acta Crystallogr* 2002;E58(11):o1290.
60	Phosphodiesterase-IV Inhibitor	L-Tartaric acid	Variankaval N, Wenslow R, Murry J, Hartman R, Helmy R, Kwong E, Clas S-D, Dalton C, Santos I. *Cryst Growth Des* 2006;6:690.
61	Piracetam	Gentisic acid, 4-hydroxybenzoic acid	[12a]
62	Piroxicam	L-Tartaric acid, citric acid, fumaric acid, adipic acid, succinic acid, L-malic acid, glutaric acid, DL-malic acid, oxalic acid, (+)-camphoric acid, ketoglutaric acid, benzoic acid, 4-hydroxybenzoic acid, malonic acid, salicylic acid, glycolic acid, 1-hydroxy-2-naphthoic acid, gentisic acid, DL-tartaric acid, maleic acid, caprylic acid, hippuric acid, L-pyroglutamic acid (50 cocrystals)	[14a]
63	Piroxicam	Saccharin	Solubility Bhatt PM, Ravindra NV, Banerjee R, Desiraju, GR. *Chem Commun* 2005;1073.
64	Pyrazinamide	2,5-Dihydroxybenzoic acid	[76]
65	Quercetin	Isonicotinamide	Solubility Zaworotko MJ, Clarke H, Kapildev A, Kavuru P, Shytle Roland D, Pujari T, Marshall L Ong TT. WO 2008153945 A2.

(*continued*)

Table 1 (*continued*)

	API	Cocrystal Former	Solubility Study	Stability Study	Pharmacokinetic Study	References
66	Salicylic acid	Nicotinamide				Berry DJ, Seaton CC, Clegg W, Harrington RW, Coles SJ, Horton PN, Hursthouse MB, Storey R, Jones W, Friscic T, Blagden N. *Cryst Growth Des* 2008;8(5):1697.
67	Sildenafil	Aspirin	Solubility			Zegarac M, Mestrovic E, Dumbovic A, Devcic M, Tudja P. WO 2007080362 A1.
68	Sodium-channel blocker [4-(4-chloro-2-flourophenoxy)phenyl]pyrimidine-4-carboxamide	2-Glutaric acid	Dissolution		Pharmacokinetic study	[79]
69	Stanolone	L-Tartaric acid				Takata N, Shiraki K, Takano R, Hayashi Y, Terada K. *Cryst Growth Des* 2008;8(8):3032.
70	Sulfadimidine	2-Aminobenzoic acids, 4-aminobenzoic acid				Caira MR. *J Crystallogr Spectrosc Res* 1991;21:641.
71	Sulfadimidine	Aspirin, 4-aminosalicylic acid				Caira MR. *J Crystallogr Spectrosc Res* 1992;22:193.
72	Sulfadimidine	*p*-Chlorobenzoic acid				Lucaciu R, Ionescu C, Wildervanck A. Caira MR. *Analytical Sciences: X-Ray Structure Analysis Online* 2008;24(5):87.
73	Sulfadimidine	Trimethoprim, 2-aminobenzoic acid, 4-aminobenzoic acid, aspirin, 4-aminosalicylic acid, 2-aminobenzoic acid				58
74	Sulfamethazine	Indole-2-carboxylic acid, 2,4-dinitrobenzoic acid				Lynch DE, Sandhu P, Parsons S. *Aust J Chem* 2000;53:383.
75	Sulfamethazine	Nicotinamide, saccharin, salicylic acid				Lu E, Rodriguez-Hornedo N, Suryanarayanan R. *CrystEngComm* 2008;10(6):665.
76	Sulfamethoxypyridazine	Trimethoprim				58
77	Sulfathiazole	Sulfanilamide				Shefter E, Sackman P. *J Pharm Sci* 1971;60:282.

78	Sulfathiazole	Trimethoprim		Giordano F, Bettinetti GP, La Manna A, Ferloni P. *Farmaco Ed Sci* 1977;32:889.
79	Temozolomide	4,4'-Bipyridine-*N*,*N*'-dioxide Carbamazepine, 3-hydroxypyridine-*N*-oxide		Babu NJ, Reddy LS, Aitipamula S, Nangia A. *Chem An Asian J* 2008;3(7):1122.
80	Tenofovir disoproxil	Hemi-fumaric acid	Pharmacokinetic study	Dova E, Mazurek JM, Anker J. WO 2008143500 A1.
81	Theophylline	5-Chlorosalicylic acid		Shefter E. *J Pharm Sci.*1969;58:710.
82	Theophylline	Glutaric acid, nicotinamide, saccharin, salicylic acid and urea		Lu E, Rodriguez-Hornedo N, Suryanarayanan R. *CrystEngComm* 2008;10(6):665.
83	Theophylline	L-Tartaric acid, DL-tartaric acid		Friscic T, Fabian L, Burley JC, Jones W, Motherwell MDS. *Chem Commun* 2006;5009.
84	Theophylline	Malonic acid, glutaric acid, maleic acid, oxalic acid	Stability	[16c]
85	Theophylline	Phenobarbital		Nakao S, Fujii S, Sakaki T, Tomita K. *Acta Crystallogr* 1977;B33:1373.
86	Theophylline	*p*-Nitrophenol		Aoki K, Ichikawa T, Koinuma Y, Iitaka Y. *Acta Crystallogr* 1978;B34:2333.
87	Theophylline	Sulfathiazole		Shefter E, Sackman P. *J Pharm Sci* 1971;60:282.
88	Trimethoprim	5,5'-Diethylbarbituric acid		Shimizu N, Nishigaki S, Nakai Y, Osaki K. *Acta Crystallogr* 1982;B38:2309.
89	VX-950	Salicylic acid, 4-aminosalicylic acid, oxalic acid 4-Hydroxybenzoic acid, 4-amino salicylic acid, phenylalanine, threonine, tartaric acid, adipic acid, succinic acetate, proline, me 4-hydroxybenzoate, anthranilic acid, and d-biotin	Solubility Suspension stability	Connelly PR. WO 2007098270 A2. Connelly PR, Kadiyala I, Stavropolus K, Zhang Y, Johnston S, Bhisetti GR, Jurkauskas V, Rose P. WO 2008106151 A2.
90	Zidovudine	2,4,6-Triaminopyrimidine		Bhatt PM, Azim Y, Thakur TS, Desiraju GR. *Cryst Growth Des* 2009;9:951–957.

Therefore, CBZ has been used as an ideal candidate to demonstrate how APIs can be converted to pharmaceutical cocrystals and how these cocrystals could offer optimized physicochemical properties over existing forms of an API [12b,68], Two strategies have been adopted for cocrystal formation of CBZ. One crystal engineering strategy is to employ the peripheral H-bonding capabilities that are not engaged in the pure form of CBZ. A second strategy for cocrystallization of CBZ involves breakage of CBZ amide–amide dimer and formation of a supramolecular heterosynthon between CBZ and a cocrystal former [12b]. Both strategies have proven to be successful and have afforded a number of CBZ cocrystals that exhibit improved physicochemical properties. Crystal structures of 16 CBZ cocrystals including cocrystal hydrate/solvates and cocrystal polymorphs [12b,13b, 49,75,76] have been determined and deposited in the CSD. As further crystal form studies of CBZ continue, Childs et al. [55] have demonstrated the preparation of 27 unique solid phases of CBZ utilizing 18 carboxylic acids as cocrystal formers together with four different screening methods.

CBZ perhaps has more reported cocrystals than any other API and some of these cocrystals have also been studied in terms of their dissolution and bioavailability. For example, the CBZ:saccharin cocrystal shows significantly improved physical stability, that is, between two polymorphic cocrystal forms [68,75] that have been identified, the stable form I [68] can be reliably prepared and have equivalent chemical stability to the anhydrous polymorph. In addition, the CBZ:saccharin cocrystal form I possesses favorable dissolution properties and suspension stability. One dissolution study shows that, within the initial 10 min, the API concentration in water solution generated by slurry of the CBZ:saccharin cocrystal form I is twice as much by slurry of the pure CBZ. In the further study of pharmacokinetics using dog models, the CBZ:saccharin cocrystal form I prototype exhibits comparable oral absorption profile with the marketed immediate release formulation [68]. In summary, the CBZ:saccharin cocrystal form I appears to be superior to existing crystal forms of CBZ in many respects.

3.2.2. Pharmaceutical Cocrystals of Fluoxetine Hydrochloride (Prozac®)

The availability and marketability of a variety of APIs as chloride salts is long known and recently an approach to utilize such chloride salts, specifically fluoxetine hydrochloride (fluoxetine HCl), to generate cocrystals of an amine hydrochloride salt via a chloride mediated carboxylic acid supramolecular synthon has been reported [14c]. That chloride is perhaps the most preferred anion for salts APIs makes generating cocrystals of fluoxetine HCl prototypal for many other APIs. Fluoxetine HCl is the active pharmaceutical ingredient found in the common antidepressant drug Prozac. It is a solid under ambient conditions, only one crystalline phase is known, and is available in the salt form. Childs et al. have demonstrated the preparation of cocrystals of fluoxetine HCl using pharmaceutically acceptable carboxylic acids that form hydrogen bonds with the chloride ions. In addition, the resulting cocrystals of fluoxetine HCl, while still retaining the hydrochloride salt of the API, exhibit dramatically different physical properties compared to the original API [14c]. Fluoxetine HCl cocrystals are the first cocrystal examples of an HCl salt.

Fluoxetine HCl was cocrystallized with benzoic acid (1 : 1), succinic acid (2 : 1), and fumaric acid (2 : 1) from solution evaporation. For all three cocrystals, the carboxylic acid was found to hydrogen bond to the chloride ion that in turn interacted with the protonated amine, thus generating, in all three cases, an amine hydrochloride salt hydrogen bonding to an additional neutral molecule [14c]. Powder dissolution experiments were carried out in water for these three novel cocrystals resulting in a spread of dissolution profiles (Fig. 11). The fluoxetine HCl:benzoic acid cocrystal was found to have a decrease in aqueous solubility by ca. 50% and the fluoxetine HCl:fumaric acid cocrystal had only a slight increase in aqueous solubility. However, the fluoxetine HCl:succinic acid cocrystal exhibited an approximately 2-fold increase in aqueous solubility after only 5 min. The complex formed between succinic acid and fluoxetine

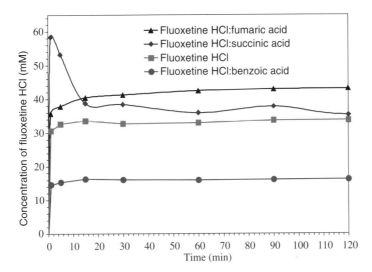

Figure 11. Dissolution profiles for novel forms of fluoxetine HCl. (This figure is available in full color at http://mrw.interscience.wiley.com/emrw/9780471266945/home.)

HCl falls apart in solution to generate its pure components after about 1 h. The intriguing factor in this study is that by simply hydrogen bonding a hydrochloride salt of an API with similar cocrystal formers one can generate such distinctively different dissolution profiles [14c].

3.2.3. Pharmaceutical Cocrystals of Itraconazole (Sporanox®)
Itraconazole is a triazole antifungal agent that is extremely water insoluble, that is, aqueous solubility of itraconazole is estimated to be ca. 1 ng/mL at neutral pH and ca. 4 μg/mL at pH 1 [77]. It is administered both orally and intravenously for patients with fungal infections [15a]. To achieve the required oral bioavailability, the oral formulation of itraconazole is the amorphous form coated on the surfaces of sucrose beads, and marketed as the Sporanox capsule. In addition, coadministration of acidified beverages with Sporanox capsules is required to achieve the maximal absorption of the API, even though such a coadministration could cause diarrhea [15a,78]. Interestingly, no crystalline salt of itraconazole has been reported in the patent literature, despite that salt formation using itraconazole and an acidic salt former seems to be a logical approach to improve the absorption properties of the API. To improve the absorption of the API and maintain the form crystallinity/stability, the pharmaceutical cocrystal approach has been evaluated in the formulation of itraconazole. As successfully demonstrated in the previous examples, crystalline phases of itraconazole can be engineered by introduction of additional molecules to match hydrogen-bond donors and acceptors [15a,78]. A number of stable pharmaceutical cocrystals of itraconazole and 1,4-dicarboxylic acids were synthesized and characterized [15a]. The cocrystals each contain two API molecules and one acid cocrystal former hydrogen bonded together to form a trimeric assembly. The aqueous dissolution of itraconazole cocrystals was studied to assess their potential impact on bioavailability of the API. The dissolution of itraconazole cocrystals was observed to behave more similarly to Sporanox form than to the crystalline form of the pure API. In particular, it was noted that the itraconazole:L-malic acid cocrystal exhibits a similar dissolution profile to that of the marketed formulation [15a]. In a further pharmacokinetic study of itraconazole cocrystals, it was revealed that cocrystal for-

mulation of the API gives similar oral bioavailability to the Sporanox form in the animal trial using a dog model [78]. In short, this study demonstrates the use of pharmaceutical cocrystals for the improvement of solubility and bioavailability without compromising crystallinity and stability.

3.2.4. Pharmaceutical Cocrystals of 2-[4-(4-Chloro-2-Fluorophenoxy)Phenyl]Pyrimidine-4-Carboxamide, a Sodium-Channel Blocker

2-[4-(4-Chloro-2-fluorophenoxy)phenyl]pyrimidine-4-carboxamide (CFPPC) is an active pharmaceutical compound that belongs to the pharmacologic class of sodium-channel blockers; CFPPC was developed as a potential drug candidate useful for treating or preventing surgical, chronic, and neuropathic pain [79]. The pharmacokinetic study in dogs shows that the oral bioavailability of CFPPC is very low due to its extremely low aqueous solubility (i.e., <0.1 μg/mL). Based on the calculated octanol/water partition coefficient ($c \log P$ 2.9), it is suspected that CFPPC is a compound of BCS Class II. To identify a solid form with better bioavailability, both pharmaceutical salts and amorphous materials of CFPPC have been investigated, yet the attempt was proven to be unsuccessful. As an alternative choice of development, the potential of forming CFPPC pharmaceutical cocrystals with higher dissolution rate has been examined [79].

A cocrystal screening was carried out employing both melt crystallization and supersaturated solution crystallization. A total of 26 carboxylic acids has been used in the screening while cocrystals of CFPPC and glutaric acid with 1:1 molecular ratio was successfully obtained and characterized. The CFPPC:glutaric acid cocrystal that can be scaled up in gram quantities, is nonhygroscopic and chemically and physically stable to thermal stress. An additional dissolution study revealed that the intrinsic dissolution rate of CFPPC in the cocrystal form showed an 18-fold increase compared to that of the original API in water at 37 °C. Single-dose pharmacokinetic evaluations for the CFPPC:glutaric acid cocrystal has also been performed. At the 5 mg/kg dose, the use of cocrystal significantly improved in vivo exposure in dogs, as the cocrystal achieved a mean plasma AUC (i.e., area under curve) of 1234 ng·h/mL from an original value of 374 ng·h/mL for the free base. In addition, the use of CFPPC:glutaric acid cocrystal also exhibits a significant increase of the AUC value using a dosage of 50 mg/kg CFPPC equivalent. Clearly, this case study exhibits how in vivo exposure of the original API could be significantly increased by the pharmaceutical cocrystal approach [79].

3.2.5. Pharmaceutical Cocrystal of AMG 517

AMG 517 is a transient receptor potential vanilloid 1 antagonist that was developed by Amgen, Inc. for the treatment of acute and chronic pain [80]. It is observed that AMG 517 has several isolated crystal forms, that is, two polymorphs and a number of crystalline solvates including a monohydrate that is stable for at least 3 years at ambient conditions. The free base of AMG 517 is practically insoluble in water and in physiological pH buffer solutions. Naturally, development of pharmaceutical salts for AMG 517 has been attempted whilst the resulting forms were found unstable in aqueous solutions, that is, they either converted to the monohydrate or decomposed at lower pH conditions. As a result, AMG 517 was formulated as a suspension in 10% (w/v) Pluronic F108 in the OraPlus® at lower doses and satisfying in vivo exposure in animal studies have been observed. However, absorption at higher doses was limited by the low solubility of AMG 517. Interestingly, a further investigation of the AMG 517 suspension revealed the unexpected in situ formation of AMG 517:sorbic acid cocrystals. Physical characterization including a solubility study was carried out for the cocrystal of AMG 517 and sorbic acid. The solubility study in FaSSIF (fasted state simulated intestinal fluid, pH 6.8) [81] showed that the AMG 517:sorbic acid cocrystal achieved an API concentration almost 10 times that of AMG 517 free base at 1.1 h. After prolonged slurry, it was observed that the cocrystal converted back to the free base monohydrate form. The pharmacokinetic study using Sprague-Dawley rats was also carried out. At 500 mg dose, the peak plasma concentration (C_{max}) of AMG 517 achieved by oral administration of the cocrystal was approximately 7.7 times that of the free base. Meanwhile, the *in vivo* exposure of the cocrystal formulation, as indicated by $AUC_{0\text{-inf}}$, increased almost 10 times compared to that of the free base formulation. In reality, the 30 mg/kg dose AMG 517:sorbic acid cocrystal formulation has a comparable exposure to a 500 mg/kg dose free base formulation. AMG 517 has also been found to be capable of cocrystallizing with 10 additional carboxylic acids, that is, benzoic acid, *trans*-cinnamic acid, 2,5-dihydroxybenzoic acid, fumaric acid, glutaric acid, glycolic acid, *trans*-2-hexanoic acid, 2-hydroxycaproic acid, L-lactic acid, and L-tartaric acid. The physicochemical properties such as particle size, solubility, stability, hygroscopicity, thermal behavior, and structural characteristics of these cocrystals were studied in details. Good correlation between the melting point of cocrystal formers and AMG 517 cocrystals has been observed; while no direct correlation was found between melting point and solubility of the AMG 517 cocrystals [80].

3.2.6. Pharmaceutical Cocrystals of Sildenafil (Viagra®)
Sildenafil is a drug used in the treatment of pulmonary arterial hypertension, congestive heart failure, atherosclerosis, conditions of reduced blood vessel patency and peripheral vascular disease, as well as male erectile dysfunction and female sexual disorders [82]. Sildenafil selectively inhibits cyclic guanosine monophosphate (cGMP) specific phosphodiesterase type 5 that is responsible for degradation of cGMP in the corpus cavernosum, leading to smooth muscle relaxation in the corpus cavernosum, and resulting in increased inflow of blood and an erection. Sildenafil citrate, with moderate water solubility, has been commercially developed and marketed by Pfizer, Inc. and is available under the trademark Viagra [82]. It has been observed that sildenafil in a pharmaceutical cocrystal form could provide an improved solubility of the API under acidic conditions. In addition, such an improvement of solubility of sildenafil could be particularly advantageous for its orally administrable formulation. Sildenafil has been successfully cocrystallized with acetylsalicylic acid (1 : 1 molar ratio) by slurry or under reflux conditions [82]. The crystal structure of the cocrystal of sildenafil and acetylsalicylic acid has been determined by single-crystal X-ray diffraction [82] and in addition the composition of matter was confirmed by powder X-ray diffraction and infrared spectrometry. Moreover, the differential scanning calorimetry and thermogravimetric analyses indicate that the melting point of the cocrystal is approximately 143 °C [82]. An intrinsic dissolution study in simulated gastric fluid (pH 1.2) shows that the sildenafil:acetylsalicylic acid cocrystal exhibits an intrinsic dissolution rate (IDR) of ca. 11.75 mg/min·cm^2. Within just 10 min, the IDR of sildenafil:acetylsalicylic acid cocrystal exhibits approximately twice that of sildenafil citrate under the same conditions [82].

3.2.7. Pharmaceutical Cocrystal of a C-Glycoside Derivative
Recently a *C*-glycoside derivative, (1*S*)-1,5-anhydro-1-[3-(1-benzothien-2-ylmethyl)-4-fluorophenyl]-D-glucitol (ABYFG), has been developed as an active pharmaceutical compound to inhibit Na$^+$-glucose cotransporter for the treatment and prevention of diabetes, such as insulin-dependent diabetes (type 1 diabetes) and non-insulin-dependent diabetes (type 2 diabetes), insulin resistance diseases, and obesity [83]. The crystal of ABYFG forms a clathrate hydrate that reversibly transform from an anhydrous compound to a nonstoichiometric hydrate depending on hygrothermal condition. Because of its physical instability, ABYFG is difficult to retain a constant quality as a drug substance used for preparing pharmaceuticals. To avoid the formation of clathrate hydrate, investigation of novel crystal forms of ABYFG has been attempted using various solvents or solvent mixtures. It was observed that, while some solvents still produced the clathrate, others led to the formation of solvates that contain hazardous solvents in the crystal lattices. Pharmaceutical salt formation was also considered. Given the fact that ABYFG is present as a nonionic compound in an ordinary pH range, however, preparation of a pharmaceutically acceptable salt of ABYFG is impossible. As a result, pharmaceutical cocrystal approach has been used to explore for novel crystal forms of ABYFG with consistent quality and superior storage stability.

Thirty-five amino acids were used in the cocrystal form screening with ABYFG. Consequently, cocrystals of ABYFG and L-proline form at a 1:1 molar ratio from water/alcoholic solutions. It is noted that the ABYFG:L-proline cocrystal is one of the first pharmaceutical cocrystals comprising a sugar derivative and a zwitterions [82]. In the storage stability test, the ABYFG:L-proline cocrystal showed no form transformation in the condition of no less than 7 days at 25 °C at 63.5–84% RH. The cocrystal was also physically stable for at least 2 months at 40 °C (75% RH, open vial), 60 °C (uncontrolled humidity) or 80 °C (uncontrolled humidity). In addition, the cocrystal, with only of 0.7% or less moisture contents, showed no moisture absorption between 5% and 95% RH range. Moreover, the *in vivo* pharmacological study using nonfasted mice as test animals showed that oral administration of the ABYFG:L-proline cocrystal with a 1 mg/kg dose enabled a strong antihypoglycemic action. In summary, this case study demonstrates the use of a zwitterion as a coformer in the pharmaceutical cocrystallization such that consistent quality and superior storage stability can be achieved in a crystalline form of the original API [83].

3.2.8. Pharmaceutical Cocrystal of a Monophosphate Salt I

Compound **I** was in the drug development pipeline of Merck, Inc. As the development of a crystalline form of compound **I** was not successful, an amorphous bis-HCl salt was initially selected for early development. Such an amorphous form, however, was proven unsuitable for further development as an oral dosage form due to its hygroscopicity and chemical instability [84]. After 1 week storage at 40 °C and 80 °C (both at ambient RH), the amorphous HCl salt exhibits 7 and 40% degradation, respectively. Extensive efforts have been taken to identify a crystalline form for compound **I**. As a result of the high-throughput screening, the only crystalline form produced was compound **I** with two phosphoric acids. A more careful analysis of the crystal structure revealed that, in such a crystal structure, half of phosphoric acids are ionized while the other half remain neutral; clearly this molecular complex is a cocrystal of compound **I** monophosphate salt and phosphoric acid (Fig. 12) [84].

The physicochemical properties of this cocrystal were characterized. It was observed that the cocrystal exhibited a high melting point of ca. 235 °C, plate-like morphology and good powder flow properties. No degradation has been detected for the cocrystal within 8 weeks of storage at 40 °C/75% RH and 60 °C. In addition, the cocrystal was found highly soluble in water and showed an excellent *in vivo* performance. The cocrystal structure was proven to be stable as no cocrystal polymorph was obtained from high-throughput screening. Naturally, this cocrystal was selected as the optimal solid form for further development. Cocrystal of compound **I** monophosphate salt and phosphoric acid is the first example of pharmaceutical cocrystals formed between an API phosphate and a phosphoric acid. Such an example sheds light on the use of

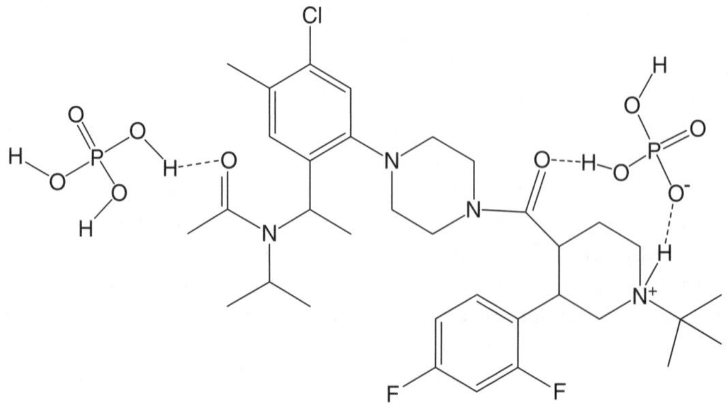

Figure 12. Phosphoric acid cocrystal of compound **I** phosphate salt.

an inorganic acid as a coformer in the pharmaceutical cocrystal approach for exploring suitable solid dosage forms in pharmaceutical development [84].

3.2.9. Cocrystal of Melamine and Cyanuric Acid

In early 2007, the FDA received complaints from owners of more than 4000 pets regarding the deaths of animals after taking food that was later recalled; it was reported that majority of those deadly incidents were caused by acute renal failure [85]. At first, melamine that was observed in the tainted products was the suspected contaminant, since this particular chemical could be intentionally added to raise the apparent protein content of the food. However, melamine is considered relatively nontoxic, that is, the acute toxicity of melamine in rats has reported oral lethal doses 50 (LD_{50}) of 3100 mg/kg (male) and 3900 mg/kg (female) [85]. Also, the quantity of melamine observed in those incidents was not at levels that would normally kill. In the course of the pet food recall investigation, cyanuric acid, another relatively nontoxic compound, was also identified in the pet food as a cocontaminant. Although melamine and cyanuric acid are relatively safe individually, no data could be found in the literature that has determined the potential toxicity of melamine and cyanuric acid in combination [85]. From the crystal engineering viewpoint, melamine and cyanuric acid (1:1 molar ratio) form extensive two-dimensional network in the solid state based on robust three-point molecular recognition, and it was observed that the resulting melamine:cyanuric acid cocrystal is highly insoluble in water [85,86]. As reported by a recent investigation, the combination of melamine and cyanuric acid can result in the intratubular precipitation of melamine:cyanuric acid cocrystals in the kidney, even though the mechanism associated with renal damage are not fully understood to date [85]. A study conducted at the Bergh Memorial Animal Hospital in New York revealed that cocrystals blocked the tubes leading from the kidneys to the bladder in one cat [85] and a toxicology assessment of melamine and cyanuric acid indicated that a single oral exposure of cats to the melamine:cyanuric acid cocrystal at a concentration of 32 mg/kg body weight can result in acute renal failure. It seems clear that the formation of a low solubility cocrystal of melamine and cyanuric acid is responsible for these incidents. Perhaps this case study of melamine:cyanuric acid cocrystal is the first example showing how cocrystals can significantly alter the relevant physical properties in a negative manner.

4. CONCLUSION

The science of crystal structure prediction continues to evolve [52] and the legal and regulatory aspects of API crystal forms are also moving targets. Nevertheless, the relevance of crystal forms to oral delivery, intellectual property and regulatory control is unlikely to diminish when one considers the impact of pharmaceutical cocrystals upon crystal form diversity and the resulting opportunity to customize the physicochemical properties of APIs. In this context, the "state of the art" concerning pharmaceutical cocrystals can be summarized as follows:

- Cocrystals were discovered at least as early as 1844 but they are underrepresented in the CSD (ca. 0.5% of structures). In short, they might be long known but they are little studied.
- In principle, the range of cocrystal formers for an API can include excipients, salt formers, food products, and nutraceuticals, that is, pharmaceutical cocrystals will ultimately offer more crystal form diversity than polymorphs, solvates, hydrates, and salts combined.
- Unlike polymorphs, solvates, hydrates, and salts, pharmaceutical cocrystals are amenable to a level of design from first principles, that is, by exploiting the supramolecular heterosynthon strategy.
- Pharmaceutical cocrystals can profoundly change the physicochemical properties of an API by using noncovalent bonds only, that is, without making derivatives of the API.
- Although there are limited data on solubility and bioavailability, it is becoming apparent that pharmaceutical cocrystals

can afford unique pharmacokinetic profiles because of the complex mechanisms of dissolution.

- Pharmaceutical cocrystals can be prepared via multiple methods (e.g., supercritical fluids, solution, mechanochemistry, melt, slurry) and their discovery is not as amenable to high-throughput screening as, for example, polymorphs and solvates.
- There remain a number of legal and regulatory uncertainties because there are few if any precedents.

The overall situation is that pharmaceutical cocrystals represent a vehicle to fine-tune the physicochemical properties of APIs, especially in terms of solubility and stability. It should therefore be unsurprising that they are being studied extensively by pharmaceutical companies in preclinical research and their more commonplace usage in drug products seems to be imminent. From a crystal engineering perspective it is now feasible to view pharmaceutical cocrystals as a mechanism to address control and/or customization of properties to a particular need, that is, we are now able to "engineer crystals." The almost 50-year old dream of physicist and Nobel Laureate Richard Feynman is therefore being realized: "I can hardly doubt that when we have some control of the arrangement of things on a small scale we will get an enormously greater range of possible properties that substances can have, and of different things that we can do" (Richard P. Feynman lectures, December 29, 1959).

REFERENCES

1. (a) Byrn SR, Pfeiffer RR, Stowell JG. Solid State Chemistry of Drugs. 2nd ed. West Lafayette, IN: SSCI, Inc.; 1999. (b) Haleblian JK. J. Pharm. Sci. 1975;64:1269–1288.
2. (a) Etter MC. J Phys Chem 1991;95:4601–4610. (b) Zerkowski JA, MacDonald JC, Seto CT, Wierda DA, Whitesides GM. J Am Chem Soc 1994;116:2382–2391. (c) Pepinsky R. Phys Rev 1955;100:971. (d) Schmidt GMJ. Pure Appl Chem 1971;27:647–678. (e) Desiraju GR. Crystal Engineering: The Design of Organic Solids. Amsterdam: Elsevier; 1989. (f) Moulton B, Zaworotko MJ. Chem Rev 2001;101: 1629–1658. (g) Braga D. Chem Commun 2003;22:2751–2754. (h) Hosseini MW. Coord Chem Rev 2003;240:157–166. (i) Desiraju GR. Angew Chem Int Ed Engl 1995;34:2311–2327.
3. (a) Almarsson Ö, Zaworotko MJ. Chem Commun 2004; 1889–1896. (b) Vishweshwar P, McMahon JA, Bis JA, Zaworotko MJ. J Pharm Sci 2006;95:499–516. (c) Peterson ML, Hickey MB, Zaworotko MJ, Almarsson O. J Pharm Pharm Sci 2006;9:317–326. (d) Zaworotko M. Am Pharm Outsourc 2004;5:16–23. (e) Shan N, Zaworotko MJ. Drug Discov Today 2008;13:440–446. (f) Blagden N, de Matas M, Gavan PT, Yoark P, Crystal engineering of active pharmaceutical ingredients to improve solubility and dissolution rates Adv Drug Del Rev 2007;59:617–630. (g) Nangia A. Cryst Growth Des 2008;8:1079–1081.
4. Maddox JR. Nature 335:1988; 201 Ball P. Nature 1996;381:648–649.
5. Dunitz JD, Bernstein J. Acc Chem Res 1995;28:193–200.
6. Noyes AA, Whitney WR. J Amer Chem Soc 1897;23:698
7. (a) Amidon GL, Lennernas H, Shah VP, Crison JR. Pharm Res 1995;12:413–420. (b) Lobenberg R, Amidon GL. Eur J Pharm Biopharm 2000;50:3–12.
8. Host–guest complexes: (a) Atwood JL, Davies JED, MacNicol DD. Inclusion Compounds. London: Academic Press; 1984. (b) Weber E. Molecular Inclusion and Molecular Recognition-Clathrates I and II: Topics in Current Chemistry. Berlin: Springer. Vols140 and 149. 1987, 1988. (c) Atwood JL, Barbour J, Jerga A. Angew Chem Int Ed Engl 2004;43:2948–2950. (d) Thallapally PK, Lloyd GO, Wirsig TB, Bredenkamp MW, Atwood JL, Barbour J. Chem Commun 2005; 5272–5274. (e) Atwood JL, Barbour J, Jerga A. Science 2002;296:2367–2370. (f) MacGillivray LR, Atwood JL. J Am Chem Soc 1997;119:6931–6932. (g) Benito JM, Gomez-Garcia M, Ortiz Mellet C, Baussanne I, Defaye J, Garcia Fernandez JM. J Am Chem Soc 2004;126:10355–10363. (h) Iyengar S, Biewer MC. Cryst Growth Des 2005;5:2043–2045. (i) Dalgarno SJ, Atwood JL, Raston CL. Cryst Growth Des 2006;6:174–180.
9. NLO materials: (a) Zelichenok A, Burtman V, Zenou N, Yitzchaik S, Di Bella S, Meshulam G, Kotler Z. J Phys Chem B 1999;103:8702–8705. (b) Ishow E, Bellaiche C, Bouteiller L, Nakatani K, Delaire JA. J Am Chem Soc 2003;125:

15744–15745. (c) Wang Y-T, Fan H-H, Wang H-Z, Chen X-M. Inorg Chem 2005;44: 4148–4150. (d) Lemaitre N, Attias A-J, Ledoux I, Zyss J. Chem Mater 2001;13:1420–1427. (e) Nemoto N, Miyata F, Nagase Y, Abe J, Hasegawa M, Shirai Y. Chem Mater 1997;9:304–311. (f) Pal T, Kar T, Bocelli G, Rigi L. Cryst Growth Des 2003;3:13–16. (g) Gao F, Zhu G, Chen Y, Li Y, Qiu S. J Phys Chem B 2004;108:3426–3430. (h) Huang KS, Britton D, Etter MC, Byrn SR. J Mater Chem 1997;7:713–720.

10. Coordination polymers: (a) Moulton B, Zaworotko MJ. Curr Opi Solid State Mater Sci 2002;6:117–123. (b) McManus GJ, Perry IV JJ, Perry M, Wagner BD, Zaworotko MJ. J Am Chem Soc 2007;129:9094–9101. (c) Biradha K, Sarkar M, Rajput L. Chem Commun 2006;40:4169–4179. (d) Blake AJ, Champness NR, Hubberstey P, Li WS, Withersby MA, Schroder M. Coord Chem Rev 1999;183: 117–138. (e) Halder GJ, Kepert CJ, Moubaraki B, Murray KS, Cashion JD. Science 2002;298: 1762–1765. (f) Chui SSY, Lo SMF, Charmant JPH, Orpen AG, Williams ID. Science 1999;283:1148. (g) Batten SR, Robson R. Angew Chem Int Ed Engl 1998;37:1460–1494.

11. Organic Conductors: (a) Huang J, Kertesz M. J Am Chem Soc 2003;125:13334–13335. (b) Bandyopadhyay P, Janout V, Zhang L-H, Sawko JA, Regen SL. J Am Chem Soc 2000;122: 12888–12889. (c) Goutenoire F, Isnard O, Retoux R, Lacorr P. Chem Mater 2000;12: 2575–2580. (d) Batail P. Chem Rev 2004;104: 4887–4890. (e) Laukhina E, Tkacheva V, Chekhlov A, Yagubskii E, Wojciechowski R, Ulanski J, Vidal-Gancedo J, Veciana J, Laukhin V, Rovira C. Chem Mater 2004;16:2471–2479. (f) Anthony JE, Brooks JS, Eaton DL, Parkin SR. J Am Chem Soc 2001;123:9482–9483.

12. (a) Vishweshwar P, McMahon JA, Peterson ML, Hickey MB, Shattock TR, Zaworotko MJ. Chem Commun 2005;36:4601–4603. (b) Fleischman SG, Kuduva SS, McMahon JA, Moulton B, Walsh RB, Rodriguez-Hornedo N, Zaworotko MJ. Cryst Growth Des 2003;3:909–919. (c) Almarsson Ö, Bourghol Hickey M, Peterson M, Zaworotko MJ, Moulton B, Rodriguez-Hornedo N. 2004, PCT Int. Appl., 489 pp. WO 2004078161 A1. (d) Weyna DR, Shattock T, Vishweshwar P, Zaworotko MJ. Cryst Growth Des 2009;9:1106–1123.

13. (a) Reddy LS, Babu NJ, Nangia A. Chem Commun 2006;13:1369–1371. (b) Babu NJ, Reddy LS, Nangia A. Mol Pharm 2007;4(3): 417–434.

14. (a) Childs SL, Hardcastle KI. Cryst Growth Des 2007;7(7): 1291–1304. (b) Childs SL, Hardcastle KI. CrystEngComm 2007;9:363–366. (c) Childs SL, Chyall LJ, Dunlap JT, Smolenskaya VN, Stahly BC, Stahly GP. J Am Chem Soc 2004;126:13335–13342.

15. (a) Remenar JF, Morissette SL, Peterson ML, Moulton B, MacPhee JM, Guzmán HR, Almarsson Ö. J Am Chem Soc 2003;125(28): 8456–8457. (b) Morissette SL, Almarsson Ö, Peterson ML, Remenar JF, Read MJ, Lemmo AV, Ellis S, Cima MJ, Gardner CR. Adv Drug Del Rev 2004;56:275–300.

16. (a) Caira MR. Mol Pharm 2007;4(3): 310–316. (b) Jones W, Motherwell WDS, Trask AV. MRS Bull 2006;31:875–879. (c) Trask AV, Motherwell WDS, Jones W. Inter J Pharma 2006;320: 114–123. (d) Trask AV, Motherwell WDS, Jones W. Cryst Growth Des 2005;5(3): 1013–1021. (e) Trask AV. Mol Pharm 2007;4:301–309.

17. (a) Seefeldt K, Miller J, Alvarez N, Rodriguez-Hornedo N. J Pharma Sci 2007;96:1147–1158. (b) Rodriguez-Hornedo N, Nehm SJ, Seefeldt KF, Torres Y, Falkiewicz CJ. Mol Pharm 2006;3:362–367.

18. (a) Dabros M, Emery PR, Thalladi VR. Angew Chem Int Ed Engl 2007;46:4132–4135. (b) Stahly GP. Cryst Growth Des 2007;7: 1007–1026. (c) Rizzo P, Daniel C, Girolamo Del Mauro A, Guerra G. Chem Mater 2007;19: 3864–3866. (d) Reddy LS, Bhatt PM, Banerjee R, Nangia A, Kruger GJ. Chem Asian J 2007;2:505–513.

19. (a) Jayasankar A, Good DJ, Rodriguez-Hornedo N. Mol Pharm 2007;4:360–372. (b) Linden A, Gunduz MG, Simsek R, Safak C. Acta Crystallogr C 2006;62:o227–o230. (c) Seefeldt K, Miller J, Alvarez N, Rodriguez-Hornedo N. J Pharm Sci 2007;96:1147–1158.

20. (a) Aakeroy CB, Salmon DJ, Smith MM, Desper J. Cryst Growth Des 2006;6:1033–1042. (b) Cowan JA, Howard JAK, Mason SA, McIntyre GJ, Lo SMF, Mak T, Chui SSY, Cai J, Cha JA, Williams ID. Acta Crystallogr C: 2006;62: 157–161. (c) Bucar D-K, MacGillivray LR. J Am Chem Soc 2007;129:32–33.

21. (a) Zhu S, Jiang H, Zhao J, Li Z. Cryst Growth Des 2005;5:1675–1677. (b) Bhogala BR, Basavoju S, Nangia A. Cryst Growth Des 2005;5: 1683–1686. (c) Bhogala BR, Basavoju S, Nangia A. CrystEngComm 2005;7:551–562. (d) Aakeroy CB, Desper J, Leonard B, Urbina JF. Cryst Growth Des 2005;5:865–873.

22. (a) Zhang XL, Chen XM. Cryst Growth Des 2005;5:617–622. (b) Hammond RB, Ma C, Roberts KJ, Ghi PY, Harris RK. J Phys Chem B 2003;107:11820–11826. (c) Lynch DE, Singh M,

Parsons S. Cryst Eng 2000;3:71–79. (d) Tukada H, Mazaki Y. Chem Lett 1997; 441–442.

23. (a) Bis JA, McLaughlin OL, Vishweshwar P, Zaworotko MJ. Cryst Growth Des 2006;6: 2648–2650. (b) Bis JA, Zaworotko MJ. Cryst Growth Des 2005;5:1169–1179. (c) Bailey WRD, Bradner MW, Fleischman S, Morales LA, Moulton B, Rodriguez-Hornedo N, Zaworotko MJ. Chem Commun 2003;9:186–187. (d) Sharma CVK, Zaworotko MJ. Chem Commun 1996; 2655–2656.

24. (a) Delori A, Suresh E, Pedireddi VR. Chem Euro J 2008;14:6967–6977. (b) Li ZJ, Abramov Y, Bordner J, Leonard J, Medek A, Trask AV. J Am Chem Soc 2006;128:8199–8210.

25. (a) Wöhler F. Annalen 1844;51:153. (b) Kitaigorodskii AI. Mixed Crystals. New York; Springer-Verlag; 1984. (c) Huang CM, Leiserowitz L, Schmidt GMJ. J Chem Soc Perkins Trans 1973;2(5): 503–508. (d) Leiserowitz L, Nader F. Acta Crystallogr 1977;B33: 2719–2733.

26. Schmidt J, Snipes W. Int J Radat Biol Relat Stud Phys Chem Med 1967;13:101–109.

27. (a) Desiraju GR. CrystEngComm 2003;5: 466–467. (b) Dunitz JD. CrystEngComm 2003; 4:506 (c) Bond AD. CrystEngComm 2007;9: 833–834. (d) Aakeröy CB, Fasulo MF, Desper J. Mol Pharm 2007;4:317–322. (e) Stahly GP. Cryst Growth Des 2007;7:1007–1026. (f) Zukerman-Schpector J, Tiekink ERT. Zeit Fur Kristallogr 2008;223:233–234. (g) Parkin A, Gilmore CJ, Wilson CC. Zeit Fur Kristallogr 2008;223: 430

28. Ling AR, Baker JK J Chem Soc Trans 1893;63:1314–1327.

29. Anderson JS. Nature 1937;140:583–584.

30. Hoogsteen K. Acta Crystallogr 1963;16: 907–916.

31. Wenner W. J Org Chem 1949;1:22–26.

32. Buck JS, Ide WS. J Am Chem Soc 1931;53:2784–2787.

33. Vanniekerk JN, Saunder DH. Acta Crystallogr 1948;1:44

34. Hall B, Devlin JP. J Phys Chem 1967;71: 465–466.

35. Pekker S, Kovats E, Oszlanyi G, Benyei G, Klupp G, Bortel G, Jalsovszky I, Jakab E, Borondics F, Kamaras K, Bokor M, Kriza G, Tompa K, Faigel G. Nat Mater 2005;4:764–767.

36. von Heyden F, et al. French Patent 769586. 1934.

37. Allen FH, Kennard O. Chem Des Automat News 1993;8:31–37.

38. Carboxylic acid dimer examples: (a) Duchamp DJ, Marsh RE. Acta Crystallogr 1969;B25:5–19. (b) Cate ATT, Kooijman H, Spek AL, Sijbesma RP, Meijer EW. J Am Chem Soc 2004;126: 3801–3808. (c) Bruno G, Randaccio L. Acta Crystallogr 1980;B36:1711–1712. (d) Bailey M, Brown CJ. Acta Crystallogr 1967;22: 387–391. (e) Ermer O. J Am Chem Soc 1988;110:3747–3754.

39. Amide dimer examples: (a) Leiserowitz L, Hagler AT. Proc R Soc London A 1983;388:133–175. (b) Weinstein S, Leiserowitz L, Gil-Av E. J Am Chem Soc 1980;102:2768–2772. (c) Leiserowitz L, Tuval M. Acta Crystallogr 1978;B34: 1230–1247.

40. Acid-amide supramolecular heterosynthon examples: (a) Aakeroy CB, Beatty AM, Helfrich BA, Nieuwenhuyzen M. Cryst Growth Des 2003;3:159–165. (b) Aakeroy CB, Beatty AM, Helfrich BA. Angew Chem Int Ed Engl 2001;40:3240–3242. (c) Vishweshwar P, Nangia A, Lynch VM. Cryst Growth Des 2003; 3:783–790. (d) Reddy LS, Nangia A, Lynch VM. Cryst Growth Des 2004;4:89–94. (e) Videnovaadrabinska V, Etter MC. J Chem Crystallogr 1995;25:823–829.

41. Acid-aromatic nitrogen supramolecular heterosynthon examples: (a) Arora KK, Pedireddi VR. J Org Chem 2003;68:9177–9185. (b) Steiner T. Acta Crystallogr 2001;B57:103–106. (c) Vishweshwar P, Nangia A, Lynch VM. J Org Chem 2002;67:556–565. (d) Etter MC, Adsmond DA. J Chem Soc Chem Commun 1990; 589–591. (e) Aakeroy CB, Beatty AM, Helfrich BA. J Am Chem Soc 2002;124:14425–14432. (f) Bhogala BR, Vishweshwar P, Nangia A. Cryst Growth Des 2002;2:325–328.

42. Wöhler F. Liebig J. Ann Pharm 1832;3:249–282.

43. Ostwald W. Zeit fur Phys Chem 1897;22: 289–330.

44. (a) Brittain HG. Polymorphism in Pharmaceutical Solids. Informa HealthCare; 1999. (b) Bernstein J. Polymorphism in Molecular Crystals. Oxford University Press; 2003. (c) Nangia A. Acc Chem Res 2008;41:595–604.

45. McCrone, WC. Polymorphism. In: Fox D, Labes MM, Weissberger A, editors. Physics and Chemistry of the Organic Solid State. Vol. II. New York: Interscience; 1965. p. 725–767.

46. Yu L, Stephenson GA, Mitchell CA, Bunnell CA, Snorek SV, Bowyer JJ, Borchardt TB, Stowell JG, Byrn SR. J Am Chem Soc 2000;122: 585–591.

47. Chen S, Guzei IA, Yu L. J Am Chem Soc 2005;127:9881–9885.

48. (a) Admiraal G, Eikelenboom JC, Vos A. Acta Crystallogr 1982;B38:2600–2605. (b) Galdecki Z, Glowka ML. Pol J Chem 1983;57:1307–1312. (c) Louer D, Louer M, Dzyabchenko VA, Agafonov V, Ceolin R. Acta Crystallogr 1995; B51:182–187. (d) Fabbiani FPA, Allan DR, Parsons S, Pulham CR. CrystEngComm 2005;7:179–186. (e) Fabbiani FPA, Allan DR, David WIF, Davidson AJ, Lennie AR, Parsons S, Pulham CR, Warren JE. Cryst Growth Des 2007;7:1115–1124.

49. Vishweshwar P, McMahon JA, Oliveira M, Peterson ML, Zaworotko MJ. J Am Chem Soc 2005;127:16802–16803.

50. (a) Wheatley PJ. J Chem Soc 1964; 6036–6048. (b) Kim Y, Machida K, Taga T, Osaki K. Chem Pharm Bull 1985;33:2641–2647. (c) Wilson CC. New J Chem 2002;26:1733–1739.

51. Bond AD, Boese R, Desiraju GR. Angew Chem Int Ed Engl 2007;46:618–622.

52. (a) Price SL. Int Rev Phys Chem 2008;27:541–568. (b) Day GM, Motherwell WDS, Ammon HL, Boerrigter SXM, Della Valle RG, Venuti E, Dzyabchenko A, Dunitz JD, Schweizer B, van Eijck BP, Erk P, Facelli JC, Bazterra VE, Ferraro MB, Hofmann DWM, Leusen FJJ, Liang C, Pantelides CC, Karamertzanis PG, Price SL, Lewis TC, Nowell H, Torrisi A, Scheraga HA, Arnautova YA, Schmidt MU, Verwer P. Acta Crystallogr B 2005;61:511–527.

53. Ulrich J. Cryst Growth Des 2004;4:879–890.

54. (a) Shan N, Toda F, Jones W. Chem Commun 2002; 2372–2373. (b) Trask AV, Motherwell WDS, Jones W. Chem Commun 2004; 890–891. (c) Trask AV, Jones W. Top Curr Chem 2005;254:41–70. (d) Karki S, Friscic T, Jones W, Motherwell WDS. Mol Pharma 2007;4:347–354.

55. Childs SL, Rodríguez-Hornedo N, Reddy LS, Jayasankar A, Maheshwari C, McCausland L, Shipplett R, Stahly BC. CrystEngComm 2008;10(7): 856–864.

56. Rodriguez-Hornedo N, Nehm SJ, Seefeldt KF, Torres Y, Falkiewicz CJ. Mol Pharma 2006;3:362–367.

57. Amos JG, Indelicato JM, Pasini CE, Reutzel SM. *Bicyclic beta-lactam/paraben complexes.* US5412094 (A1), JP7048383 (A), FI943081 (A), BR9402561 (A), and EP0637587 (B1), (1995).

58. Caira MR. Mol Pharma 2007;4(3): 310–316.

59. Bis JA, Vishweshwar P, Weyna D, Zaworotko MJ. Mol Pharm 2007;7:401–416.

60. Shattock TR, Arora KK, Vishweshwar P, Zaworotko MJ. Cryst Growth Des 2008;8:4533–4545.

61. Shattock TR, Vishweshwar P, Wang Z, Zaworotko MJ. Cryst Growth Des 2005;5:2046–2049.

62. Pudipeddi M, Serajuddin AT. J Pharm Sci 2005;94:929–939.

63. Watts JL, Salmon SA, Sanchez MS, Yancey RJ Jr. Antimicrobial Agents and Chemother 1997;41:1190–1192.

64. Schinzer WC, Bergren MS, Aldrich DS, Chao RS, Dunn MJ, Jeganathan A, Madden LM. J Pharm Sci 1997;86:1426–1431.

65. Higuchi T, Roy K. J Am Pharm Assoc 1954;43: 393–97.

66. Higuchi T, Roy K. J Am Pharm Assoc 1954;43: 398–401.

67. Zerkowski JA, Seto CT, Whitesides GM. J Am Chem Soc 1992;114:5473–5475.

68. Hickey MB, Peterson ML, Scoppettuolo LA, Morrisette SL, Vetter A, Guzman H, Remenar JF, Zhang Z, Tawa MD, Haley S, Zaworotko MJ, Almarsson O. Eur J Pharma Biopharma 2007;67(1): 112–119.

69. CBZ four polymorphs: (a) Reboul JP, Cristau B, Soyfer JC, Astier JP. Acta Crystallogr 1981; B37:1844–1848. (b) Lowes MMJ, Cairo MR, Lotter AP, van der Watt JG. J Pharm Sci 1987;76:744–752. (c) Grzesiak AL, Lang M, Kim K, Matzger AJ. J Pharm Sci 2003;92: 2260–2271. (d) Lang M, Kampf JW, Matzger AJ. J Pharm Sci 2002;91:1186–1190.

70. CBZ dihydrate: Gelbrich T, Hursthouse MB. CrystEngComm 2006;8:448–460.

71. CBZ solvates: (a) Chang CH, Yang DSC, Yoo CS, Wang BL, Pletcher J, Sax M. Acta Crystallogr 1981;A37:c71. (b) Johnston A, Florence AJ, Kennedy AR. Acta Crystallogr 2005;E61: o1777–o1779 (c) Johnston A, Florence AJ, Kennedy AR. Acta Crystallogr 2005;E61: o1509–o1511 (d) Lohani S, Zhang Y, Chyall LJ, Mougin-Andres P, Muller FX, Grant DJW. Acta Crystallogr 2005;E61:o1310–o1312 (e) Fernandes P, Bardin J, Johnston A, Florence AJ, Leech CK, David WIF, Shankland K. Acta Crystallogr 2007;E63:o4269. (f) Fabbiani FPA, Byrne LT, McKinnon JJ, Spackman MA. CrystEngComm 2007;9:728–731. (g) Johnston A, Johnston BF, Kennedy AR, Florence AJ. CrystEngComm 2008;10:23–25.

72. Reck G, Thiel W. Pharmazie 1991;46:509–512.

73. Florence AJ, Leech CK, Shankland N, Shankland K, Johnston A. CrystEngComm 2006;8: 746–747.

74. Hilfiker R, Berghausen J, Blatter F, Burkhard A, De Paul SM, Freiermuth B, Geoffroy A, Hofmeier U, Marcolli C, Siebenhaar B, Szelagiewics

M, Vit A, von Raumer M. J Ther Anal Calorim 2003;73:429–440.
75. Porter WW III, Elie SC, Matzger AJ. Cryst Growth Des 2008;8(1): 14–16.
76. McMahon JA, Bis JA, Vishweshwar P, Shattock TR, McLaughlin OL, Zaworotko MJ. Zeit Fur Kristallogr 2005;220(4): 340–350.
77. (a) Peeters J, Neeskens P, Tollenaere JP, Remoortere PV, Brewster ME. J Pharm Sci 2002;91:1414–1422. (b) Francois M, Snoeckx E, Putteman P, Wouters F, Proost ED, Delaet U, Peeters J, Brewster ME. AAPS PharmaSci 2003;5:1–5.
78. Remenar JF, MccPhee M, Peterson ML, Morissette S, Almarsson Ö.US Patent 20,050, 070,551 2008.
79. McNamara DP, Childs SL, Giordano J, Iarriccio A, Cassidy J, Shet MS, Mannion R, O'Donnell E, Park A. Pharm Res 2006;23:1888–1897.
80. (a) Bak A, Gore A, Yanez E, Stanton M, Tufekcic S, Syed R, Akrami A, Rose M, Surapaneni S, Bostick T, King A, Neervannan S, Ostovic D, Koparkar A. J Pharm Sci 2008;97:3942–3956. (b) Stanton MK, Bak A. Cryst Growth Des 2008;8(10): 3856–3862.
81. Dressman JB, Amidon GL, Reppas C. Shah VP. Pharm Res 1998;15:11–12.
82. Zegarac M, Mestrovic E, Dumbovic A, Devcic M, Tudja P.WO 2007/080362 A1.
83. (a) Imamura M, Nakanishi K, Shiraki R, Onda K, Sasuga D, Yuda M.WO 2007114475 A1. (b) Gougoutas JZ.US Patent 6,774,112 B2.
84. Chen AM, Ellison ME, Peresypkin A, Wenslow RM, Variankaval N, Savarin CG, Natishan TK, Mathre DJ, Dormer PG, Euler DH, Ball RG, Ye Z, Wang Y, Santos I. Chem Commun 2007;4:419–421.
85. Puschner B, Poppenga RH, Lowenstine LJ, Filigenzi MS, Pesavento PA. J Vet Diagn Invest 2007;19:616–624.
86. Ranganathan A, Pedireddi VR, Rao CNR. J Am Chem Soc 1999;121:1752–1753.

PRODRUGS: STRATEGIC DEPLOYMENT, METABOLIC CONSIDERATIONS, AND CHEMICAL DESIGN PRINCIPLES

Paul W. Erhardt
Rahul Khupse
Jeffrey G. Sarver
Jill A. Trendel
Center for Drug Design and Development, Department of Medicinal and Biological Chemistry, The University of Toledo College of Pharmacy, Toledo, OH

1. INTRODUCTION

This chapter is meant to complement the information provided in the related chapter entitled *Retrometabolism-Based Drug Design and Targeting* that has been written by Nicholas Bodor and Peter Buchwald. In addition to covering soft drugs, the latter provides a thorough discussion about using chemical delivery systems (CDS) to achieve selective drug action at targeted sites. Thus, even though the use of CDS represents an important aspect of prodrug design that continues to grow in both its practical applications and its technical sophistication, this topic will be addressed only briefly herein to avoid repetition.

1.1. Definition

The term "pro-drug" was first introduced by Albert [1,2] about 50 years ago when he used it to describe compounds that require metabolic biotransformation in order to exhibit their pharmacological effects, such conversions being either an inherent property of the parent compound ("accidental" pro-drug), or a property intentionally incorporated into an otherwise active drug by specific design. Emphasizing the latter, the term "drug latentiation" was also introduced at nearly the same time by Harper [3,4] wherein this type of design strategy was defined as "chemical modification of a biologically active compound to form a new compound which upon *in vivo* enzymatic attack will liberate the parent compound." Kupchan et al. [5] subsequently expanded the strategy to further include drug modifications that relied upon nonenzymatic processes to regenerate the parent compound within an *in vivo* setting, such as spontaneous hydrolysis of the modification as a function of pH. Because "latentiation" implies a time-lag for the regeneration of the active substance and this type of event is not always required to be part of a prodrug's most desirable pharmacokinetic (PK) profile, such terminology has not been sustained. Eventually, the term "prodrug" (without the former hyphen mark as shown above) became universally accepted.

Despite congruence about what term to finally adopt, its intended meaning in a given usage has continued to vary, often being taken to include a much wider context of drug-related profiles and design strategies that, in some cases, are not even appropriate. Thus, we prefer to convey a strict definition as provided below. To further appreciate the rich history of this field, readers are encouraged to purview a few of the excellent review articles, monographs, or texts that have been dedicated specifically toward prodrugs across this period in time by several distinguished experts: Higuchi and Stella (1975) [6] from which portions of our historical section were derived; Roche (1977) [7]; Bungaard (1985) [8]; Lin and Lu (1997) [9]; and Testa and Mayer (2003) [10].

We will define prodrugs as being compounds that require one or more structural changes to occur after their administration in order to elicit their desired pharmacological effects, such changes not including simple shifts in the equilibria of ionizable groups or the interchange of various salt forms that may be applicable to the compound. Thus, within the domain of their advantageous administration protocols, prodrugs are inactive with regard to their desired effects, although not necessarily so for any unwanted side-effects, until they undergo bioactivation by either a chemically driven conversion, an enzymatically driven biotransformation, or any multiple combination of these processes. According to this strict definition, efficaciously active agents whose metabolites are similarly active or contribute to the desired pharmacological profile by another mechanism will not be considered to be prodrugs. While such compounds have sometimes been referred to as "limited prodrugs," the present authors would rather retain their classification as "drugs" no matter how short-lived their observable activity might be *in vivo*. That such drugs also happen to have active metabolites does not then alter their initial classification according to our strict definition. Finally, sometimes confused in the literature because they can also be used as a strategy to target selected biological compartments, "soft drugs" should not be regarded as prodrugs. Indeed, soft drugs are efficaciously active agents that have been intentionally modified so as to program a specified route and rate for metabolic degradation that leads to efficaciously silent and nontoxic metabolites. Typically, these metabolites, in turn, are readily excreted. The relationships between a drug, prodrug and soft drug are displayed schematically in Fig. 1.

1.2. Principal Uses and Some Early Examples

In order to succeed in the marketplace, a drug must exhibit an overall profile that favorably encompasses many additional factors beyond its optimized interaction with the biological surface responsible for therapeutic efficacy. These additional features derive from interactions with a diverse array of other pharmacological media and biological surfaces that begin with a drug's formulation and subsequently include its absorption, distribution, metabolism, excretion and toxicity, constituting a drug's so-called "ADMET" profile. Obtaining optimized ADMET features for a "drug wannabe" [11] often represents a significant hurdle during the course of drug discovery and development programs. This situation is reflected by the high attrition rate historically observed during initial clinical studies when ADMET parameters were not rigorously studied during the early preclinical development stages of such investigations [12].

Taking an orally administered central nervous system (CNS) agent as an example, Fig. 2 delineates the various pharmacological media and biological surfaces that the compound will need to traverse if it is to ultimately display efficacy at its desired site of action within the brain. Although this circuitous path was likened to a "random walk" by Hansch [13], he also recognized that a compound's profile *in vivo* is governed by its gross physicochemical properties, as well as by its specific structural detail [14]. Deriving equations to address the impact of these properties upon a drug's activity, Hansch pioneered the still evolving area known today as quantitative structure–activity relationship (QSAR) studies. During drug discovery and development, attempts are now made to optimize a compound's profile in each of the ADMET behaviors so that its therapeutic benefit can be maximized. Historically, such optimizations have striven to retain the desired efficacy-related actions within the modified "active analogs." Only when this could not be accomplished were various prodrug strategies contemplated. As can be seen from the CNS drug example, several "AD" and early "M" hurdles occur prior to the interaction of the agent with its efficacious site. These "front-side" hurdles represent ideal points for potential use of a prodrug approach since the so-modified compounds still have time to be converted back to their active drug forms before, or upon arrival at, their desired destination. Alternatively, the "MET" hurdles continue to occur throughout the "back-side" course of a drug's desired action. Although they would appear to pose a more difficult scenario for use of a prodrug approach, these "back-side" problems can also be addressed by prodrug strategies that involve modification of early "DM" behavior. As additionally shown in Fig. 1, controlling the back-side of a drug's action can also be attempted by adopting a soft drug strategy rather than a prodrug strategy (see the complementary chapter by Bodor and Buchwald for a thorough discussion about soft drugs). A list of the specific issues that may be able to be addressed by a prodrug strategy relative to oral and parenteral drugs is provided in Table 1. Similar issues pertaining to other routes of administration are likewise applicable. Each of the major issues (indicated by *italic* style print in Table 1) is briefly discussed below and accompanied by descriptions of a few classic drug examples.

From both a technical and a commercial viewpoint, considerable success has been achieved by using prodrug approaches to solve

INTRODUCTION

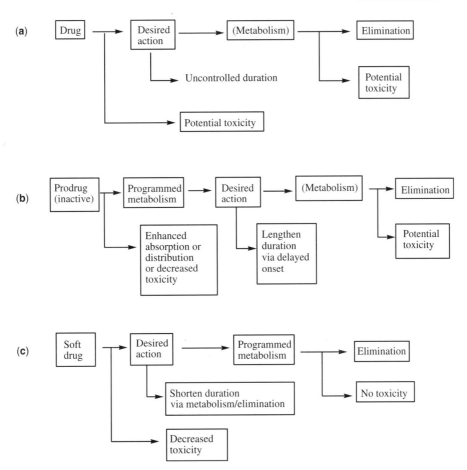

Figure 1. Contrasting dispositions of a drug, prodrug and soft drug. Panel (a) depicts a generalized version of a standard drug's pattern of observed activities. Panel (b) depicts how a prodrug approach can be used to modify the entry-side portion of a given drug's overall profile of actions. Panel (c) depicts how a soft drug approach can be used to modify the elimination-side portion of a given drug's overall profile of action. Both prodrugs and soft drugs can be used to decrease toxicity. Note that while "Programmed Metabolism" events are typically devised for both prodrug and soft drug strategies, this key event can also utilize chemically driven (nonenzymatic) conversions that have been programmed into the molecule to take advantage of specific physiological or pathophysiological chemical environments that are encountered after administration.

formulation and *administration issues* for parenteral drugs that have low aqueous solubility. When an ionizable group is present in a drug, different salt forms are immediately possible. As a first approach, a systematic exploration of this practical type of manipulation may be able to readily address limitations across the entire spectrum from low aqueous solubility to high hygroscopicity for a raw drug material whose manufacturing process happens to be so hampered [15]. Recall, however, that by our strict definition, different salt forms of the same parent structure do not constitute prodrugs. All salts, as well as the parent drug itself, become the same set of equilibrium species when dissolved in, for example, the acidic nature of the stomach fluid or the pH 7.4 environment maintained by the body's systemic buffering system. Alternatively, it is when this first approach does not solve the problem, or when it is not available because the compound does not contain an ionizable group, that a prodrug strategy becomes extremely valuable. Fosphenytoin

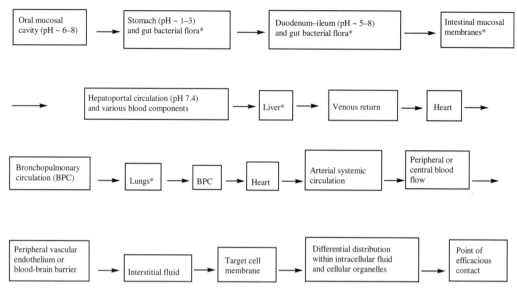

Figure 2. "Random walk" taken by an oral drug on route to its point of efficacious interaction [13]. In this example, the latter is a human target cell within the brain that becomes the target for a CNS drug. This continuum of interactions between a drug and various biological surfaces is typically divided into categories associated with ADMET and efficacy. Biological milieu marked with an asterisk (*) represent compartments having particularly high drug metabolizing capabilities in general. Blood is notably high in only esterase capability.

Table 1. Issues that may be able to be Addressed by Using a Prodrug Strategy

Formulation
  Physical form of raw drug is not conducive to manufacturing processes
  Inadequate aqueous solubility for liquid dosage forms
  Inadequate shelf-life stability for solid or liquid dosage forms
Administration
  Oral form has unpleasant taste or is an irritant*
  Injectable or intravenous form causes local irritation*
Absorption (Primarily Oral Dosage Form)
  Requires improved stability in stomach
  Requires higher dissolution rate in stomach or intestine
  Requires higher lipophilicity to enhance passive transport
  Need to avoid or overcome P-glycoprotein transport back into GI-tract
  Requires improved stability toward "first-pass effects" (see discussion)
Distribution
  Need to decrease plasma protein binding or deposition in lipophilic compartments
  Need to better traverse blood–brain barrier (CNS drug)
  Need to enhance local penetration at desired target site overall or relative to other tissues so as to decrease side-effect toxicity (CDS approaches)*
Metabolism and Excretion
  Need to avoid or localize (chemotherapeutic agent) metabolic activation to toxic species*
  Need to prolong half-life for therapeutic effect
Toxicity
  See entries above marked with asterisk (*)
  Need to temporarily mask a reactive (inherently toxic) functional group

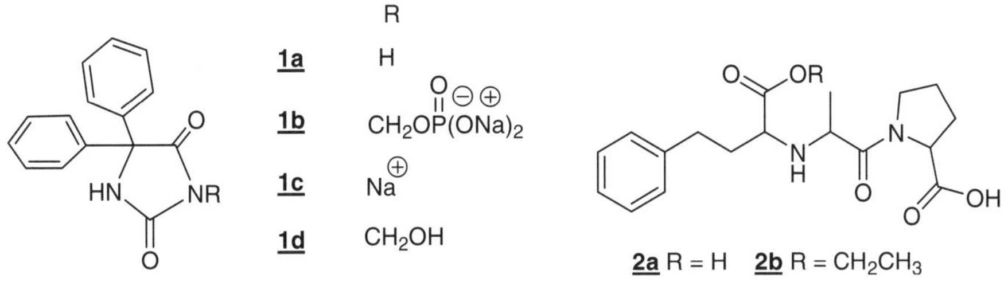

Figure 3. Examples of using prodrugs to either increase or decrease hydrophilicity. Fosphenytoin sodium, **1b**, has increased water solubility for parenteral administration while enalapril, **2b**, has increased lipophilicity for oral administration. See text for details.

sodium, **1b** in Fig. 3, represents a useful example of a highly water soluble, phosphate prodrug of the anticonvulsant phenytoin (**1a**) that otherwise suffers from erratic oral and parenteral bioavailability [16]. Although the simple, nonprodrug sodium salt form, **1c**, derived from the weakly acidic NH proton that is flanked by two carbonyl groups in **1a**, also exhibits good water solubility, formulation of this correspondingly strong conjugate base results in a pH of ca. 12 and this, in turn, can cause local irritation at the site of injection during parenteral administration. Alternatively, **1b** exhibits excellent water solubility in formulations having acceptable pH ranges that are nonirritating. While stable at neutral pH values, **1b** is rapidly hydrolyzed *in vivo* by serum phosphatases to the unstable "aminol" intermediate **1d** that immediately collapses to the active drug **1a**. The half-life for this two-step, metabolic and chemical, bioreversible prodrug process takes only about 10–15 min in humans.

A frequent use of prodrugs is to increase a parent drug's lipophilicity in order to enhance *absorption* across the gastrointestinal tract (GI-tract) after oral administration. A classical example involves the "block-buster" prodrug enalapril, **2b** in Fig. 3, which is the ethyl ester of the angiotensin converting enzyme (ACE) inhibitor parent drug enalaprilat, **2a** [17]. A discussion that encompasses these types of compounds is further provided in Section 4. As we shall see from both a strategic design perspective, and our survey of recently approved drugs/prodrugs provided in Section 4, the use of an ester to mask either a carboxylic acid functionality, as was done in **2b**, or a hydroxyl group, as can be accomplished most simply by using a "reversed ester," is a very fruitful prodrug strategy [18].

A long-standing [19] but still interesting case of "directing *distribution*" by deploying a prodrug strategy subsequently coupled with additional metabolism-related interventions, is afforded by levodopa (L-DOPA), **3b**, which serves to deliver dopamine (DA, **3a**) to the brain as a treatment for Parkinsonism [20]. The underlying principles for this example are highlighted in Fig. 4. After systemic administration, DA cannot cross the blood–brain barrier (BBB) whereas L-DOPA is able to take advantage of an amino acid transporter (AAT) that is present for the uptake of L-tyrosine. Because L-tyrosine can be converted by an aryl hydroxylase to L-DOPA that, in turn, is converted to DA by dopa decarboxylase, the arrival of L-DOPA in the brain feeds directly into the normal biosynthetic pathway for the production of DA within the CNS. Thus, this prodrug strategy relies upon dopa decarboxylase rather than hydrolytic enzymes for its biotransformation to the active agent. Although desirable CNS effects were obtained, the initial dosage protocol still required ca. 3–6 g to be administered each day. This was largely due to premature decarboxylation of L-DOPA while still in the peripheral circulation by dopa decarboxylase present in the liver, lungs, heart, and kidneys. Thus, a dopa decarboxylase inhibitor, namely carbidopa (**4**) that does not cross the BBB, was subsequently devised for coadministration with L-DOPA. Marketed today as a single product having a fixed ratio of 1 part **4** to 10 parts **3b**, a ca. 80% reduction in dose accompanied by fewer side effects has been achieved. Finally,

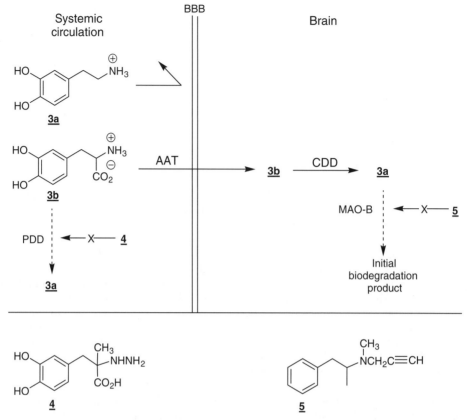

Figure 4. Prodrug approach toward delivering dopamine into the CNS. While dopamine, **3a**, cannot itself cross the blood–brain barrier, levodopa, **3b**, is able to take advantage of a specific amino acid transporter (AAT). Once inside the CNS, **3b** undergoes biochemical conversion to **3a** via a central dopa decarboxylase (CDD). To prevent premature decarboxylation of **3b** within the systemic circulation, the peripheral dopa decarboxylase (PDD) inhibitor carbidopa, **4**, is administered as a combination product. To better sustain the brain levels of **3a** derived from **3b**, the centrally acting monoamine oxidase B inhibitor selegiline, **5**, is sometimes additionally administered as an adjuvant product.

because monoamine oxidase (MAO) readily degrades DA within the CNS, administration of the centrally active monoamine oxidase B (MAO-B) inhibitor selegiline, **5**, is often additionally deployed as an adjunct therapy to better sustain the brain DA levels derived from L-DOPA. Further reductions in dose, longer durations and less accompanying side effects are again typically observed.

The use of prodrugs to attenuate overly rapid elimination of the active agent via *metabolism* and/or *excretion* is not as common as the previous categories because these hurdles are generally overcome by an active analogs campaign or by the development of controlled-release formulations. Masking the functional groups responsible for this behavior via a prodrug approach, however, represents an additionally useful strategy toward addressing such issues. For example, the bronchodilator terbutaline, **6a** in Fig. 5, typically requires multiple oral dosing throughout the course of a day because its phenolic hydroxyl groups are subject to rapid metabolic reactions to form primarily the sulfate, and to a lesser extent the glucuronide, conjugates [21]. Temporarily masking these metabolically susceptible groups by converting them to

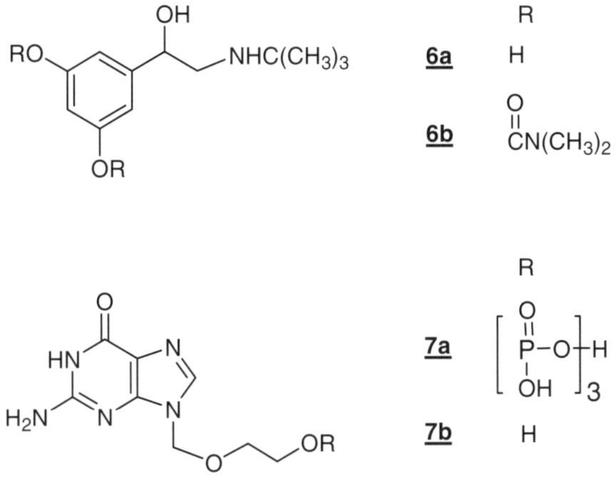

Figure 5. Examples of prodrugs designed for metabolic removal or addition of chemical substituents. Bambuterol, **6b**, relies upon hydrolases to slowly unmask a requisite pharmacophoric functional group that otherwise causes the parent drug, **6a**, to be readily eliminated during absorption from the GI-tract. Acyclovir, **7b**, relies upon viral enzymes to selectively add a requisite pharmacophoric functional group to yield the active form, **7a**, only within the biological compartment where the activity is wanted.

carbamates as shown within the prodrug bambuterol, **6b**, affords once-daily dosing. The carbamate functionalities are only slowly hydrolyzed by nonspecific cholinesterases [22]. In this particular case, it should also be noted that the presence of **6b** can lead to drug–drug interactions with other agents that rely on plasma cholinesterase for their clearance, such as suxamethonium [23].

As indicated in Table 1, addressing *toxicity* issues can sometimes be accomplished by prodrug approaches that encompass alterations of one or more of the aforementioned ADME parameters. Of particular interest are Ehrlich's early notions about "magic bullets" [24]. While this concept has never lost its appeal for medicinal chemists, it still constitutes a very elusive therapeutic goal. Today, advances in molecular biology have led to a rapidly expanding knowledge about signaling pathways and their uniquely aberrant behavior in a given pathology. This knowledge can help to address toxicity issues by serving in either a direct or an indirect manner. When a distinct target along a signaling pathway is revealed, it can be exploited directly toward achieving a highly selective therapeutic intervention. Similarly, when overexpression of a specific signaling factor is revealed, it can be exploited indirectly as an "address" to which an attached cargo molecule can be guided for selective delivery of its therapeutic "message" [25] even though the cargo, in itself, may not otherwise exhibit any inherent selectivity. In the end, both strategies can ultimately result in a reduction of side effects.

Prodrug strategies in themselves, or in combination with agents targeting such sites either for direct intervention or for indirectly as useful "address" systems, clearly represent an exciting opportunity for today's practicing medicinal chemists. Many compounds and a wealth of ongoing research activities can be cited as interesting examples for these prodrug combination strategies, including those derived from the CDS approaches discussed in the complementary chapter. Here, we will recite just one example where selectivity for chemotherapy of viral-infected cells has been achieved. Acyclovir, **7b** in Fig. 5, belongs to the long standing, general class of moderately polar, "antimetabolite" prodrugs [26]. Subsequent to their "random walk" and eventual cellular uptake, members from this special class of prodrugs require intracellular phosphorylation by kinases to produce the active nucleotide analog forms. Because of their extremely high polarity, the active forms cannot, themselves, cross cell membranes. Once generated inside the cell, the activated drugs inhibit DNA synthesis immediately or become erroneously incorporated during DNA synthesis only to inhibit it shortly thereafter. Previously, nonselective antiviral agents have relied upon both native and viral-induced kinases to accomplish the bioactivating phosphorylations. Advances in molecular biology coupled with advancing protein X-ray diffraction techniques, however, have allowed characterization of the viral enzymes to the extent that they can be selectively targeted for bioconversion of these types of prodrugs. Specifically, **7b** was designed to be phosphorylated only by kinases coded from the invading virus and, in turn, its resulting activated form **7a** can demonstrate selective chemotherapeutic action toward viral-infected cells compared to healthy cells [27]. Desirable toxicity is obtained selectively compared to unwanted side-effect toxicity. Hence, **7b** represents a "magic bullet" that was derived by a rather elegant prodrug design strategy involving an interdisciplinary mix of "old" and "new" technologies.

1.3. Role During Drug Discovery

1.3.1. Classical Paradigm The typical approach taken during the long-standing ("classical") drug discovery paradigm, shown in Fig. 6 [28], generally deployed prodrug strategies only after a reiterative active analog synthesis campaign had not been able to circumvent a particular ADMET hurdle. Experts in the prodrug and drug metabolism fields have expressed different views on this practice. These range from a strong endorsement for using prodrug strategies only "as a last resort" relative to solving complex oral bioavailability issues [18], all the way to advocating the early adoption of prodrugs into a "two-step" process of optimizing efficacy at the target site independent of any PK concerns, followed by immediate deployment of prodrug strategies to enhance ADMET-driven PK properties [29]. An interesting aspect of the latter strategy is that it perfectly preserves what will be the bioregenerated pharmacophore as present in the already efficaciously-optimized active compound. These authors suggest that the two-step path should be run in parallel with the normal course of active analog development such that, if needed, the prodrug strategies would have already been initiated and partly evaluated in an ongoing fashion. They refer to the combined process as an "ad hoc prodrug approach." Some of these same experts, as well as many other authors, have advocated the use of prodrug strategies in a more flexible manner that would lie somewhere between these two extremes [28,30–33]. This "flexible" view is nicely summarized by the modified (in parentheses) but otherwise apt quote from Ettmayer et al. [33]: *"medicinal chemists (should) be creative* (relative to the adoption of prodrug strategies) *when they face situations where the SARs for the drug target are incompatible with PK objectives or* (whenever) *a prodrug approach for rationally designed tissue/cell targeting* (can be readily undertaken) *early in a drug discovery program."*

1.3.2. Today's Drug Discovery Paradigm Today's drug discovery paradigm is shown in Fig. 7 [11,12,28]. Clearly, the marriage of combinatorial chemistry and compound library technologies to high throughput screening (HTS) has added a shot of vigor to the drug discovery process by providing a new means of identifying efficacious "hit" and "lead" compounds in a very rapid manner. Likewise, that screening for "drug-like" properties [34] via HTS ADMET assays is also now being done at earlier stages of the overall drug discovery process, most certainly represents an attempt to take a huge step forward [11,12,28,30–32,34–36]. However, when these pieces of the new drug discovery paradigm are taken together in wrote practice without continuing to emphasize the elements of "knowledge generation" about the problems at hand [12,28,36] or the "creativity" that might be expended toward alternate approaches that could also be envisioned to overcome them [33], the new paradigm can detract from both the role of rationally-devised active analog development and the role that prodrug approaches might play during drug discovery. This is because there is a strong tendency to keep "falling-back" to HTS and compound library (albeit then further directed in structural space) surveys in an attempt to identify another drug-like hit compound that this time

INTRODUCTION

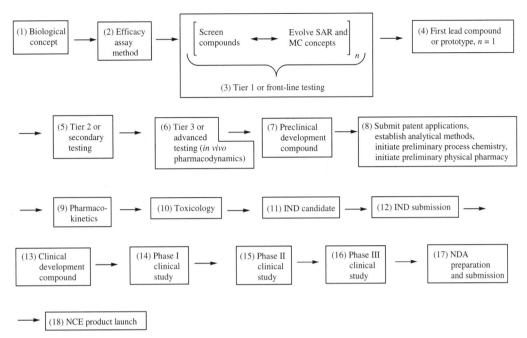

Figure 6. Classical drug discovery and development paradigm [28]. This model portrays interactions with US regulatory agencies and uses terms related to those interactions for Steps 11–17. All other terms typify generic phrases that have been commonly adopted by the international pharmaceutical community. While some of these activities can be conducted in parallel or in an overlapping manner, the stepwise, sequential nature of this paradigm's overall process is striking. Furthermore, whenever a progressing compound fails to meet criteria set at an advanced step, the process returns to Step 3 for another reiteration. Numerous reiterations eventually identify a compound that is able to traverse the entire process. A successful passage through the entire process to produce just one product compound has been estimated to require about 10–15 years at a total cost of about $1 billion. While the largest share of these time and cost requirements occur during the later steps, the identification of a promising preclinical development compound in Step 7 can be estimated to take about 4 years from the time of initiating a therapeutic concept. Step 1 is typically associated with some form of physiological or pharmacological notion that intends to amplify or attenuate a specific biological mechanism so as to return some pathophysiology to an overall state of homeostasis. Step 2 typically involves one or two biochemical level assay(s) for the interaction of compounds intending to amplify or attenuate the concept-related mechanism. Steps 3 and 4 have historically reflected key contributions from medicinal chemistry and typically use all sources of available information to provide for compound efficacy hits, for example, everything from natural product surveys to rational approaches based upon X-ray diffractions of the biological target. Step 5 generally involves larger *in vitro* models (e.g., tissue level rather than biochemical level) for efficacy-related selectivity. Step 6 generally involves *in vivo* testing and utilizes a pharmacodynamic (observable pharmacologic effect) approach toward assessing compound availability and duration of action. Step 7 typically derives from a formal review conducted by an interdisciplinary team upon examination of all data obtained to that point. Step 8 specifies parallel activities that are typically initiated at this juncture by distinct disciplines within a given organization. Step 9 begins more refined pharmacokinetic (PK) evaluations by utilizing validated GLP compliant analytical methods for the drug itself to address *in vivo* availability and duration of action. Step 10 represents short-term (e.g., 2-week) dose-ranging studies to initially identify toxic markers within one or more small animal populations. Expanded toxicology studies typically progress while overlapping with Steps 11–14. Steps 11–13 represent formalized reviews undertaken by both the sponsoring organization and the U.S. FDA. Step 14 is typically a dose-ranging study conducted in healthy humans. Steps 15 and 16 reflect efficacy testing in sick patients, possible drug-interactions, and so on. Step 17 again reflects formalized reviews undertaken by both the sponsoring organization and the FDA. The FDA's *fast-track* review of this information is said to have been reduced to an average of about 18 months. It is estimated that it costs at least $200 K each day that a compound spends in clinical development. Step 18 represents the delivery of a new chemical entity, NCE, to the marketplace wherein it can finally be stated that a new drug has been truly discovered [11,12].

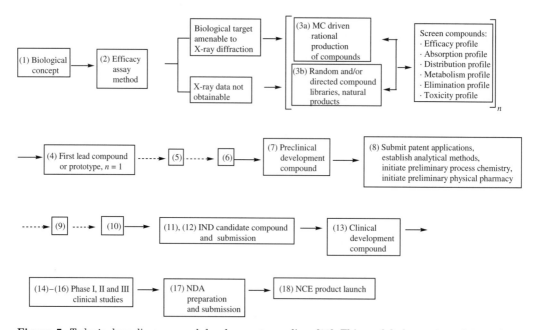

Figure 7. Today's drug discovery and development paradigm [28]. This model also portrays interactions with US regulatory agencies and uses terms related to those interactions for Steps 11–17. All of the other terms typify generic phrases that are commonly used by the international pharmaceutical community. The battery of simultaneous profiling included in Step 3 represents a striking contrast to the classical drug discovery paradigm (Fig. 6). Furthermore, all of these screens are typically of a high-throughput nature such that huge numbers of compounds can be automatically tested in extremely short time periods. As the predictive value of the resulting profiles improve, selected compounds will have a higher and higher likelihood to successfully proceed through Steps 5, 6, 9, and 10 (same as in Fig. 6) to the point that these assays may become more of a confirmatory nature or may even be completely omitted (hence their dotted lines). The efficiency of successfully traversing the various clinical testing Steps 14–16 will also be improved, but their complete removal from the overall process is highly unlikely. After the initial investments to upgrade Step 3 and enough time has passed to allow for the generation of knowledge from Step 3's raw data, the overall timeframe and cost for a single NCE to traverse the new paradigm should be considerably reduced from the estimates provided in the accompanying Fig. 6. Step 1 is likely to be associated with some type of genomic and/or proteomic derived notion that intends to amplify or attenuate a specific biological mechanism such as an aberrant signal transduction pathway, so as to return some pathophysiology to an overall state of homeostasis. Step 2 is typically a high-throughput assay derived from using molecular biology and bioengineering techniques. Because Step 3a exploits actual X-ray "pictures" of what type of structural arrangements are needed to interact with the biological targets, this approach toward identifying new compound hits will continue to operate with high efficiency. However, because of this same efficiency, the biological targets that lend themselves to such experimental depiction (by affording crystals suitable for X-ray diffraction) could become depleted much faster than the range of more difficult biological targets. Step 3b represents various combinatorial chemistry-derived libraries, natural product collections, and elicited natural product libraries. Steps 4–18 are similar to the descriptions in Fig. 6 [11,12].

does not collide with one or more of the ADMET hurdles. An unfortunate but all-too-often accurate generalization can be postulated that larger organizations, independent of whether corporate, academic, or public (e.g., "NIH Roadmap" initiative) are the most prone to this "falling-back" to HTS tendency. This is because they typically have very large compound libraries and highly robust HTS capabilities, making it all the more likely they will immediately return to these resources as soon as problems are encountered. The unfortunate side effect is that this prevents the opportunity to discover new approaches that arise when engaging in creative medicinal chemistry investigations.

Faced with this present trend, where might medicinal chemists want to advocate for the consideration of prodrug strategies to be most effectively deployed during today's drug discovery paradigm? Certainly their upfront consideration and actual use in an interdisciplinary manner during the very early stages of any program directed toward enhancing selectivity seems clearly advantageous. Alternatively, the timing for their deployment to address various PK and toxicity issues remains debatable. Reemphasizing the importance of promoting an environment conducive to creative medicinal chemistry, perhaps the best time-point for initially contemplating the use of prodrugs to potentially solve later ADMET problems, would be during Steps 3a and 3b in Fig. 7. This is the time when directed compound libraries are typically generated as a follow-up to an initial hit compound. Ideally at this point, structural space can then be explored with an eye toward not only generating the most efficacious drug-like molecule possible, but also toward defining the regions in space that are the most amenable for future modifications, including the potential introduction of chemical handles for subsequent incorporation of prodrug strategies. While the systematic delineation of "efficacy-positive" structural space would thus remain a hallmark endeavor for medicinal chemists at this juncture, initial exploration of "neutral" and "negative" space relative to potential prodrug manipulation would also be undertaken to further contribute to the overall SAR map of the hit compound [11,12,28,36]. Carefully note, however, that full deployment of the prodrug strategy would then be strategically held until additional data about one or more of the active analog's *in vivo* behavior was obtained during Steps 5 and 6. The reason for this key, strategic pause is further discussed in the next section.

2. GENERAL DESIGN PRINCIPLES

2.1. Drug Targeting

As discussed above, advances in molecular biology coupled with an interdisciplinary approach toward exploiting the resulting knowledge to directly (drug-derived) or indirectly (prodrug-delivered) achieve selective action at targeted sites, constitutes an exciting arena for today's medicinal chemists to participate in the pursuit of "magic bullets." The complementary chapter by Bodor and Buchwald provides a thorough discussion of both general and specific details pertaining to the use of "retrometabolic strategies" that can be undertaken during the design of chemical delivery system, or CDS, approaches toward drug targeting. Principles associated with only some of the more recent prodrug trends that rely upon advancing molecular biology will be additionally mentioned here.

The use of mutated endogenous enzymes and even the use of exogenous enzymes introduced for a specific purpose, represent the most recent trends to be receiving attention. An example of the former case is the specific modification of human carboxylesterases (CESs) for improved prodrug activation [37]. Examples of the second case include [33] (i) virus-directed enzyme prodrug therapy ("VDEPT") wherein delivery of the gene for an exogenous enzyme is first accomplished by deploying a viral vector; (ii) chemical-directed gene delivery agents ("GDEPT") wherein gene delivery utilizes a nonviral chemical vector; and (iii) antibody-directed enzyme prodrug therapy ("ADEPT") wherein the exogenous enzyme or enzyme-delivery system is additionally coupled to a monoclonal antibody ("mAb") that can selectively recognize an "address" being overexpressed by the target cell to gain specific localization or uptake by target cell. In all three examples, prodrugs that have been designed to depend exclusively upon the delivered enzymes for their selective bioactivation are then subsequently administered.

Clearly, the most important principle to emphasize within the arena of drug targeting is to gain a thorough interdisciplinary knowledge about the process that is to be exploited (or created) within the target area. The methods recited above demonstrate how our growing knowledge in the area of molecular biology continues to contribute to the field of small molecule prodrug design, as well as toward more selective, directly active drug design.

2.2. Addressing ADMET Issues

Just as for targeted prodrugs, understanding as many of the biological and chemical details of the problem as possible represents the most important principle for prodrug applications directed toward ADMET issues. For example, if a selected "lead compound" that has been optimized for efficacy and perceived drug-like profile based upon the *in silico* and *in vitro* models deployed in Step 3 (Fig. 7), proceeds to exhibit low oral bioavailability in Step 6, then it still remains to be experimentally determined as to what are the exact reasons for this confrontation with the "A" in the desired ADMET profile. Only then can the most appropriate prodrug strategies be devised to properly reconcile the situation. Assuming inherent chemical stability is not an issue, one needs to first determine if it is the agent's low solubility and/or slow dissolution rate in the gut that is at fault versus some problem with diffusion across a barrier. If it is the latter, is the oral bioavailability problem due to low passive transport because of too much polarity, or perhaps so much lipophilicity that once into the gut lining the agent does not want to leave? Alternatively, is the agent a welcome substrate for the P-glycoprotein transporter that resides in the gut membrane just waiting to pump compounds, particularly small molecule xenobiotics, back into the intestine rather than allowing them to pass into the hepatoportal circulation (Fig. 2)? Finally, is it instead some metabolic degradation event that occurs either within the gut membrane itself, the immediate hepatoportal circulation, or the liver through which all oral xenobiotics must first pass (hence the so-called "first-pass effect") in order to subsequently reach the heart and finally arrive in the general systemic circulation? While the liver has seemingly evolved to represent the body's metabolic "clearing house" for small molecule xenobiotics that happen to become ingested as part of a human's normal diet, both the gut membrane, rich in certain classes of metabolizing enzymes, and the blood, rich in certain esterases, should all be considered as constituting the barrier to drug absorption known as the first-pass effect. Similar sets of sequential questions can be posed with regard to a shortcoming observed at any of the other DMET-related hurdles.

It can be noted that some of the questions within the aforementioned example may have been answered quite well and thus already ruled out, by the preliminary screening activities (Step 3) that utilized high-throughput types of *in vitro* models. However, many of these screens, and certainly most of the *in silico* methods, are not foolproof in terms of largely eliminating the high attrition rate for a "drug wannabe" [11], particularly as the latter ultimately proceeds into the clinic for the first time [11,12,28,36]. Prudence dictates that follow-up studies in the case of a failing lead compound are usually very much still in order relative to at least many of these questions. Once the ADMET problem, or a given targeting strategy for that matter, has been defined to the fullest extent possible, it then becomes much more feasible to embark on a compound-efficient prodrug design strategy. Hence the key "strategic pause" recommended earlier regarding when to actually begin to deploy potential prodrug strategies within the context of today's drug discovery paradigm. Coupling the follow-up knowledge gained about the detailed biological aspects of the problem with the previously gained knowledge about the chemical feasibility for potentially devising various prodrug molecular constructs within the structural space that was mapped out during the initial hit-follow-up campaign, the medicinal chemist now becomes ideally poised to entertain the remaining prodrug-related general principles discussed below. It can also be pointed out at this point that upon devising these new molecular constructs, the latter are likely to have the added benefit of also creating new compositions of matter in an intellectual property (IP) or overall patent landscape theme.

2.3. Fundamental Metabolic Considerations

A fundamental set of metabolism-related principles that must be adhered to during prodrug design is illustrated in Fig. 8 [31]. First and foremost, if the rate for conversion of the prodrug to the active agent, arrow (i), is not faster than the rate of the latter's further metabolic degradation or clearance by

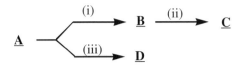

Figure 8. Key relationships between metabolism and prodrugs [31]. Prodrug **A** and metabolites **C** and **D** represent molecules that are not active at the desired site. **B** represents the active parent drug. Conversion (i) represents a metabolic pathway or chemically driven bioconversion that accomplishes the desired activation while (ii) and (iii) represent other metabolic conversions and/or elimination pathways. The goal in any prodrug design strategy is to have the rate of (i) > (ii) + (iii). *Note*, however, that while the rate for (i) must be greater than the rate for (ii), it need not necessarily be greater than the rate for (iii).

excretion, arrow (ii), then the parent drug and its desired therapeutic action will never be observed. In opposition to what a medicinal chemist generally targets during the design of inhibitors for a given enzymatic pathway, the programmed bioactivation process for a prodrug strategy cannot be the rate-limiting step in any sequence of metabolic events by which the active agent may become degraded. By their nature, targeting strategies are able to address this requirement by selectively generating the active agent only where it is supposed to act, and further, by doing so in a manner where the parent drug's action precedes any inherent chemical instability or other local metabolism possibilities present within the target area. Alternatively, prodrug strategies aimed at addressing ADMET issues do not have this inherent benefit, and so they must deal with this absolute requirement in some other manner.

Simultaneously, conversion of prodrug to a metabolite other than the parent compound, or diversion by some other process of xenobiotic elimination (both encompassed in arrow (iii)), should be slower than the prodrug's conversion to the desired active agent. The higher the competition down these nonproductive routes, the lower the efficiency of the prodrug system from a mass balance point of view, as well as from a therapeutic point of view. This second, fundamental principle of metabolic considerations presents a challenge to both targeting and ADMET-issue-resolving prodrug design strategies. In fact, it can be even more of an issue for the targeting agents because their bioconversion is supposed to be "latentiated" (to draw from an old term) so as to occur only at the target site. Again just the opposite from the design of enzyme inhibitors, in this case the programmed bioactivation process contemplated during the design of a prodrug strategy is better-off not being the rate-limiting step among the composite of other metabolic events that can precede formation of the active agent. Taking these principles together, both types of prodrug usage must devise an overall system wherein the rate for step (i) is always greater than that for step (ii) and, ideally, also is larger than the sum of all of the competing pathways via (ii) and (iii).

By way of example, recall the earlier case shown in Fig. 4 where peripheral dopa decarboxylase (PDD) was able to effectively compete with central dopa decarboxylase (CDD), resulting in premature metabolism of L-DOPA preventing delivery across the BBB. Although this event constitutes conversion to the active compound, it disrupts the desired "distribution" sequence strategically devised for the prodrug, analogous to the case wherein arrow (iii) is meant to represent any alternative metabolic or elimination pathway that leads to a diminished production of active compound via arrow (i). In this case, (i) is present within the CNS locale where the desired bioconversion needs to occur. Also from this example, it can be seen that subsequent degradation by central MAO-B can significantly detract from the beneficial effects of liberated dopamine, reflecting a negative impact of arrow (ii) on such strategies even when it is not faster than (i). In this case, both of these shortcomings were partially addressed by the simultaneous administration of co- and adjuvant-agents that served to block each of these competing metabolic pathways, respectively. Another "classical problem in prodrug chemistry" has been the "interesting scientific conundrum" initially presented by attempts to mask the carboxylic acid moiety present in aspirin, **8a**, so as to lesson the GI-tract toxicity often encountered upon chronic administration [38]. Active itself, and thus not a prodrug by our definition, **8a** also produces salicylic acid, an important active metabolite. As shown in Fig. 9, this transformation normally occurs slowly by butyrylcholinesterase (BuChE) (half-life ca. 2 h) even though the latter abounds in human plasma. Because aspirin's reduced affinity for BuChE is due to the presence of its carboxylate anion (at physiological pH), when the carboxylic acid functionality is masked as simple ester derivatives, **8b**, which remain neutral in charge, these attempted prodrugs become "super substrates" for removal of the acetyl group by BuChE (half-lives typically in the range of minutes). Thus, competition by pathway (iii) in Fig. 8 overwhelms the desired pathway (i) (corresponding to the dotted-line in Fig. 9) such that little, if any, of **8a** becomes liberated. Notable successes for this second example were finally achieved by Bundgaard et al. using N,N-disubstituted glycolamide esters such as **8c** [39] and amidomethyl esters such as **8d** [40]. These prodrugs displayed half-lives that became more competitive for the desired conversion to **8a**, such as ca. 10 min for enzymatic bioconversion in human plasma, and ca. 10 s for spontaneous chemical hydrolysis in aqueous solution at pH 7.4 and 37°C, respectively. More recent successes have also been noted by others attempting to incorporate nitric oxide releasing moieties so as to simultaneously deliver two different types of therapeutic agents [38,41].

Given these fundamental challenges associated with prodrug design, it should be clear that the choice to be pursued for programming pathway (i) in an optimal manner becomes "all-important." While this step is often dictated very specifically by the distinct compo-

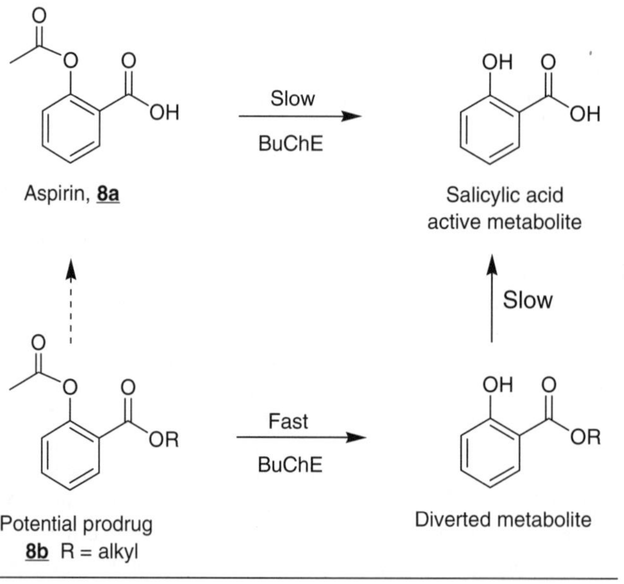

Figure 9. Prodrug "conundrum" presented by aspirin, **8a** [38]. Because the acetyl group becomes metabolically labile when the carboxylic acid moiety in **8a** is masked as a simple ester (**8b**), it has remained an interesting challenge to successfully deploy these otherwise straightforward strategies toward reducing the GI-tract irritation often observed for **8a** with its chronic administration. More elaborate approaches such as in **8c** and **8d**, however, have been able to effectively compete for the desired hydrolytic bioconversion process depicted as the dotted line. See text for details.

nents associated with a given prodrug targeting strategy, the choices to be considered during an ADME-rescue strategy remain wide-open across all of the metabolic pathways present within humans, including even the possibility of tailoring a new pathway by genetic modification (e.g., using some of the recent "XDEPT" technologies that were alluded to earlier). In addition to needing pathway (i) to dominate the metabolic competition from nonproductive pathways, it is also desirable that pathway (i) is not subject to individual variation (genetic polymorphism) or to the potential for becoming involved in serious drug–drug interactions.

Arguably, one of the most functionally-prominent and diversity-tolerant broad category of human xenobiotic metabolizing enzymes are the hydrolases. Further, standing tall from a prodrug design target within this broad class would be the subcategory of carboxylesterases. While there are some notable exceptions such as a simple catechol moiety, when nonsterically hindered carboxylic esters are compared across a wide variety of numerous other types of chemical functional groups that are often present in xenobiotics, a far greater proportion of the esters are likely to be unable to survive the first-pass effect, let alone reach more distal destinations within the body for any prolonged periods of time. In general, the carboxylesterases are metabolically aggressive toward a variety of ester arrangements while being subject to attenuation by "tailored" manipulation of their steric environments. They are ubiquitously distributed in plentiful amounts, which generally reduces their potential to become saturated and for their substrates to become involved in drug–drug interactions, even though an example of the latter was intentionally noted for the case of bambuterol, **6b**. Finally, they do not appear to be subject to high genetic polymorphisms in a manner that might compromise their eventual exploitation within the clinic. Thus, carboxylesterases represent an excellent starting point for pathway (i) during the early stages of considering a prodrug design strategy. Even the "highly evolved" cytochrome P450 (CYP-450) enzymes, which are so prominent in the liver and to lesser extents in the linings of the intestine and lung, cannot be relied upon with the same degree of certainty for a successful prodrug strategy. History reveals that the carboxylic ester pathways have, indeed, been repeatedly pursued and often successfully exploited for such purposes [10], as do more recent surveys [33], including the one conveyed herein (see Section 4) relative to new chemical entities (NCEs) entering the marketplace over the past 5 years. Beginning with the next section and continuing throughout the rest of this chapter, the focus will be on this particularly interesting subclass of hydrolytic enzymes.

3. CARBOXYLESTERASES

3.1. Classification

The esterase family of enzymes includes the serine hydrolases such as cholinesterase (EC 3.1.1.7 and EC 3.1.1.8), arylesterase (EC 3.1.1.2), and carboxylesterase (EC 3.1.1.1). The EC numbers are assigned by the International Union of Biochemistry and Molecular Biology, which categorizes classes of enzymes and subclasses according to the reaction catalyzed. While all of these hydrolases have carboxylesterase function, this section will focus on the carboxylesterase that are the most prominent in the bioactivation of prodrugs. The latter are encompassed by the family of genes denoted as CES1–5 within the EC 3.1.1.1 categorization. The CES family of serine hydrolases consists of glycoproteins that hydrolyze ester, thioester, carbamate, and amide bonds, all of which have been incorporated into prodrug design.

CESs can be classified into five major groups (CES1–CES5) according to sequence homology to human CES1A1 (hCE-1). Most of the known CESs belong to the CES1 or CES2 group of enzymes. Members of the CES1 group have recently been further divided into eight subfamilies (CES1A–CES1H). CES1A contains the major forms of human CES1, while the other subfamilies include the major CES1 forms from dog, rabbit, rat, mouse, pig, and cat. Human CES2 (hCE-2, hiCE) belongs to the subfamily CES2A and is designated as CES2A1. The family of CES3A contains two CESs, human hCE-3 and mouse esterase 31. CES4 is divided into three subfamilies

CES4A–CES4C and contains mammalian CAUXIN gene that is a form of CES secreted in the urine. CES5A and CES5B have the least homology to CES1 and contain three members of the arylacetamide deacetylase (AADAC) gene for human, rat, and mouse [42,43].

3.2. Function and Distribution

Carboxylesterases function as a protective and clearing mechanism for xenobiotic substances, detoxifying or metabolically activating drugs, environmental toxicants, and carcinogens. These enzymes are usually found in the liver, but various isoforms of CESs are also found in other tissues. For example, human CES1 is highly expressed in the liver, but is also detected in the colon, heart, kidney, lung, spleen, brain, stomach, and testes. Human CES2, contrarily, is highly expressed in the small intestine and colon, and at much lower levels than CES1 in the liver [42–46]. Human CES3 has been detected at very low levels in the liver and gastrointestinal tract [47]. Although esterase activity has been reported in blood, Li et al. [48] found that human CESs are not responsible for hydrolyzing susceptible drugs in human blood. However, CESs found in the blood of mouse, rat, rabbit, horse, and cat do hydrolyze susceptible drugs.

Since its isolation from human samples over 60 years ago, human CES1 has been listed in the literature under several names including serine esterase, carboxylic ester hydrolase, liver CE, alveolar esterase, brain CE, egasyn, HU1, and hCE-1 [46,49–52]. Mammalian CES1 from dog, guinea pig, rat, rabbit, mouse, horse, pig, cat, and several other species have been compared to human CES1, with significant differences noted between species. For example, rabbit CES1 hydrolyzes irinotecan (Table 2) 100–1000 times more efficiently than the human form of CES1 [53]. Comparison of the hydrolyses of p-nitrophenyl acetate, butanilicaine, and isocarboxazid (Table 2) between human liver microsome CES and mouse, hamster, guinea pig, rabbit, monkey, pig, cow, and dog liver microsome CES indicates that human CES1 has less activity on all of these substrates (4–30-fold) depending on the species compared [54,55]. These differences indicate that care must be taken when using different species to predict the effectiveness of prodrug hydrolysis by human CES1.

Differences in hydrolysis of substrates between human CES1 and CES2 have also been observed. Although there is 60% sequence homology between CES1 and CES2, their substrate specificities are slightly different. CES1 prefers compounds with large acyl and small alcohol groups because it has a large opening in the pocket to allow binding of the larger acyl group. On the other hand, CES2 prefers binding molecules with large alcohol groups and small acyl groups. For example CES2 hydrolyzes irinotecan and heroin with a higher efficiency than CES1 [47,57,58]. CES1 also exhibits stereoselectivity, examples being its preferential hydrolysis of the L-isomer of methylphenidate [60] and the R-form of both flurbiprofen ethyleneglycol and propionyl propranolol [51] (Table 2). It is important to note that single nucleotide polymorphisms (SNPs) in both CES1 and CES2 may alter the activity of the enzymes in specific individuals [63–65], although as suggested earlier in this chapter, the impact of these differences may be less significant clinically than that for the polymorphisms displayed by several of the cytochrome P450 family members.

3.3. Molecular Biology

With the use of recombinant DNA technology and protein expression, human CES1 and CES2, along with other mammalian CESs, have been studied extensively. Human CES1 and CES2 are about 60 kDa in size with pIs of 5.8 [66] and 4.9 [57], respectively, although CES1 has been described as having a range of pIs from 5.5 to 5.9 [67] and two liver isoforms have been reported as having pIs of 5.3 and 4.5 corresponding to human CES1 and human CES2 [52]. CES2 is typically expressed as a monomer, whereas CES1 has been shown by both Western blot and crystal structure to be a monomer, dimer, trimer, and hexamer [68–70]. There is 60% amino acid homology between the two CES isoforms with several conserved amino acids.

As with other esterases, the catalytic triad contains Ser203, Glu335, and His448, all of which are essential for esterase activity.

Table 2. Selected Substrates of CES1 and CES2

Substrates	Structure (dashed Line Indicates Bond Cleaved by CES)	K_{cat}/K_M (mM^{-1} min^{-1})	
		CES1	CES2
Cocaine (methyl ester) [56]		0.5	0
Cocaine (benzoyl ester) [56,57]		0.5	18.4
Meperidine [56]		0.35	0
4-Methylumbelliferone acetate [57]		2,000	60,000
Irinotecan [58]		0.73	47
Capecitabine [59]		14.7	12.9

Table 2. (*Continued*)

Substrates	Structure (dashed Line Indicates Bond Cleaved by CES)	K_{cat}/K_M (mM^{-1} min^{-1})	
		CES1	CES2
Heroin [57]		69	314
Delapril [52]		63	0
6-Acetylmorphine [57]		0.024	22
Temocapril [52]		370	75
Methylphenidate (L-form D-form) [60]		7.7, 2.1	0, 0
		(µmol/min/mg protein)	
Procaine [52]		0	0.009

Table 2. (*Continued*)

Substrates	Structure (dashed Line Indicates Bond Cleaved by CES)	K_{cat}/K_M (mM^{-1}min^{-1})	
		CES1	CES2
Propionyl propranolol (*R*-form *S*-form) [61]		23.4, 0.22	29.8, 44.7
p-Nitrophenyl acetate [61]		5020	2330
Butanilicaine [62]		0.13	ND
		(nmol/min/mg protein)	
Isocarboxazid [62]		1.05	ND
Flurbiprofen propyleneglycol ester (*R*-form *S*-form) [61]		42, 9	0.2, 0.5

Catalytic rate or activity is included to show specificity for CES1 or CES2 when available. K_{cat} is the number of times each enzyme site converts substrate to product per unit time. K_M is the Michaelis–Menten constant, the substrate concentration needed to achieve half-maximal enzyme velocity. Specific activity defined as μmol/min/mg protein or nmol/min/mg protein. * indicates location of chiral center responsible for differences in enantiomeric conversion rates. ND indicates value not determined.

Mutations to any of these residues decrease activity [46,49,51,71]. As depicted in Fig. 10, this catalytic triad is located at the bottom of a deep cleft that has a large flexible pocket on one side and a small rigid pocket on the other side. The crystal structure of CES1 indicates that the large flexible pocket allows for various substrates to bind, while the small rigid pocket is more selective [42,72]. The crystal structure for CES2 has not yet been solved. The conserved Gly-Gly124-125 residues generate an "oxyanion hole" that stabilizes the

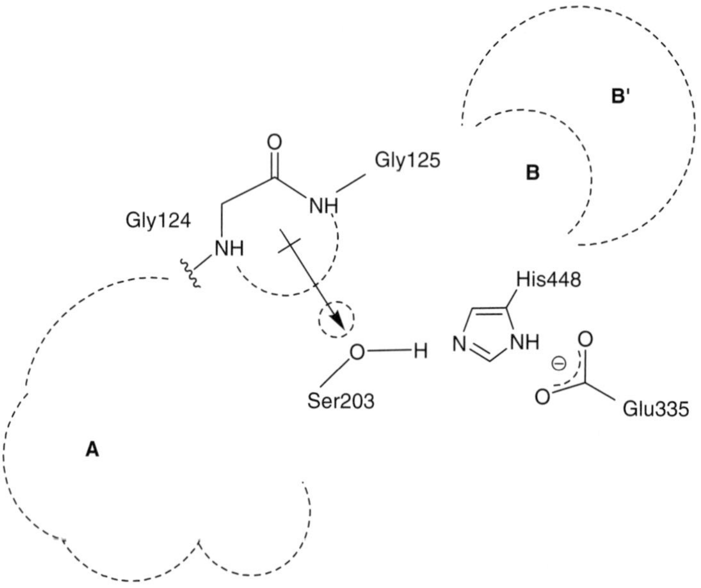

Figure 10. Aerial view of the CES substrate cleft and active site. This cartoon derives from data primarily obtained for CES1 [45]. The "catalytic triad" having Ser203, His448 and Glu335 is thought to be located within a deep cleft that is flanked by a large flexible pocket **A** and a comparatively smaller and more rigid pocket **B**. In our depiction, pocket **A** accommodates the "acyl-side" of an ester-containing substrate and its multiple binding sites can account for stereochemical dependent hydrolysis rates. Pocket **B** and its slightly larger version **B'** represent regions in CES1 and CES2, respectively, that in our depiction can accommodate the "alcohol-side" of an ester-containing substrate. Also shown are the amide hydrogens of Gly-Gly124-125 that work together to establish an "oxyanion hole" that helps to stabilize the hydrolytic transition state from which it takes its name. This partial positive region also creates a dipole (shown by arrow) with the electronegative oxygen atom of Ser203 for complementary recognition and orientation upon initial binding of the substrate's carbonyl moiety in a reversed dipole manner. Once so arranged, the Ser203 oxygen begins the hydrolysis reaction by waging a nucelophilic attack upon the substrate's carbonyl carbon. See Fig. 16 and the text for a detailed description of this overall mechanism.

tetrahedral arrangement of the transition state (TS) [55,73]. There are four Cys residues that are conserved, which are involved in disulfide bonds. Conserved glycosylation sites are located near the C-terminus. In order to be retained in the endoplasmic reticulum, both enzymes contain the N-terminal signal peptide sequence, NXT and have the C-terminal KDEL consensus sequence, HXEL, where X is I and T for CES1 and CES2, respectively.

3.4. Substrates and Inhibitors

Design of a prodrug for a particular CES, namely CES1 or CES2, should take into consideration the active sites of the two enzymes. While the crystal structure of CES2 has not yet been determined, some interesting insights can be gleaned by comparing the hydrolysis of some common substrates by these two enzyme forms, summarized in Table 2. First, the sterospecificity of CES1 yields some clues as to how well an enantiomer fits into the active site and affects the rate of hydrolysis. For example, the L- and D-forms of methylphenidate both fit into the active site of CES1, but the catalytic efficiency is much higher for the L-enantiomer. This is probably due to the ability of the two enantiomers to generate different conformations of the active site allowing faster exit of the product in the L-enantiomer conformation [60]. Comparison of irinotecan hydrolysis by CES1 and CES2 indicates that the catalytic efficiency is much higher for CES2 than CES1 (0.047 versus 0.00073 μM^{-1}

min^{-1}) [58]. Finally, examination of the catalytic efficiency of capecitabine, a prodrug of 5-fluorouracil(5-FU), between CES1 and CES2 shows that, although only slightly higher, CES1 had a higher K_{cat}/K_m of 14.7 mM^{-1} min^{-1} versus 12.9 mM^{-1} min^{-1} [59]. Table 2 lists the catalytic rate or activity of some of the substrates tested against CES1 and CES2 to assess the structure of the substrate and the ability of the enzymes to hydrolyze the ester bond.

Table 3 offers a summary of common CES inhibitors. In 1953, Gomori [74] used a battery of 27 different phenolic and naphtholic esters to detect human esterases in serum, urine, liver, and pancreas extracts. Taurocholate (TCH) was used as an inhibitor of esterase and activator of lipase. For liver extracts and serum samples, hydrolysis of all the esters occurred and was inhibited to some extent by TCH, indicating esterases as being responsible for hydrolysis. In the pancreas, however, hydrolysis occurred in the presence of TCH indicating pancreatic lipases involvement in hydrolysis. In the urine, only five of the substrates were hydrolyzed in the presence of TCH (recall that some carboxylesterases, such as CES4, are secreted in the urine) and an increase of hydrolysis was observed in six of the substrates in the presence of TCH. Since this time, several groups have generated alternate forms of substrates and inhibitors of CESs in order to build a pattern for the mechanism of hydrolysis [47,75–80]. In Table 3, some of the inhibitors that have been tested against CES1 and CES2 are listed along with the K_i or IC$_{50}$, whenever such data was available. The information provided in Tables 2 and 3 is intended to serve as a general database or background material for the series-specific SAR treatments that would need to be derived for each new scaffold upon entering into a prodrug design program.

3.5. Preclinical Assay Considerations

Interspecies differences make it difficult to predict how humans will react to a prodrug moiety hydrolyzed by CES. However, using human recombinant enzymes or human tissue extracts to test substrates can reduce the interspecies variability and indicate tissue specific hydrolysis. Comparisons of hydrolyses of substrates between CES1 and CES2 from mammalian species and from human liver and intestine may aid in the design of new drug entities by examining differences in amino acids and enzyme structure. Examination of CES1 from human liver homogenates and CES2 from human intestinal homogenates has indicated differences in optimal pH and isoelectric points that may affect substrate hydrolysis [83]. Danks et al. [84] introduced recombinant rabbit CES into human rhabdomyosarcoma cells that increased sensitivity to irinotecan eightfold. Side-by-side comparison of recombinant rabbit CES and recombinant human CES indicated that rabbit CES is more efficient at hydrolyzing irinotecan than human CES [53]. Xie et al. [85] examined recombinant human, rat, and mouse CESs and found that the rodent CESs had higher enzyme activity on p-nitrophenylbutyrate than the human CESs. In addition to structural differences, variability of CES expression in human tumor samples may contribute to the response, or lack thereof, of the tumor to chemotherapeutics [86].

Figures 11 and 12 provide minimal and more robust testing paradigms to be undertaken during the early stages of any prodrug assessment program. Specific esterase assays can take advantage of the substrates and inhibitors listed in Tables 2 and 3, respectively. In terms of *in vivo* metabolic and PK studies, as mentioned in Section 3.2, the species differences noted at the *in vitro* level also become significant when finally embarking upon advanced *in vivo* testing. While no species can be thought of as being optimal across the various carboxylesterases and certainly not for a broad range of structural classes or prodrug constructs, our own previous data suggests that mongrel dogs may be as close to such a designation as we may be able to find [31], short of specifically cloning a novel species with a broad complement of the hydrolases along with the most relevant CYP-450 human enzyme forms involved in xenobiotic metabolism. While the latter may represent a novel species, it by no means reflects a novel concept within this field, as such pursuits have been underway for several years.

Table 3. Selected Inhibitors of CES1 and CES2

Inhibitor	Structure	CES1 K_i (nM)	CES2 K_i (nM)
Taurocholate monosodium [74]		ND	ND
Benzil [79]		148	14.7
2,3-Dinitrophenylphosphonic acid [81]		ND	ND
Strawberry extract [82]	Multiple components	ND IC$_{50}$ (mM)	ND IC$_{50}$ (mM)
Loperamide [59,78]		0.44	0.000038
Dolasetron [59]		0.33	0.50
Docetaxel [59]		0.68	0.79
Atropine [59]		7.0	9.0

Inhibitory concentration included when available. K_i is the inhibition constant, the unbound inhibitor concentration that reduces intrinsic clearance by 50%. IC$_{50}$ is the half-maximal inhibitory concentration, which indicates the effectiveness of a compound in inhibiting biological or biochemical function (mM). ND indicates value not determined.

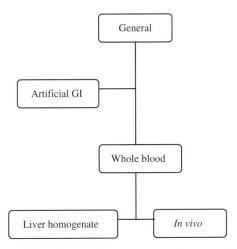

Figure 11. Minimal testing paradigm for a prodrug assessment program. See text for discussions about species differences and the use of standard substrates and inhibitors.

4. MARKETED PRODRUGS

4.1. Prevalence

Due to the large and ever-changing number of marketed drugs, the relative role of prodrugs is difficult to determine and not often reported in the literature. A review of pharmacologically active (i.e., excluding imaging or diagnostic agents) new molecular entities (NMEs) approved by the U.S. FDA between 2004 and 2008 shows that there are 5 prodrugs out of the 83 total NMEs entering the market during this time period (see Appendix for details). This gives a prodrug prevalence of 6% for these recently approved agents, which is very similar to the value of 7% estimated by Ettmayer et al. for all marketed drugs in Germany [33]. The fact that these figures are nearly identical for recent US drugs and all German drugs may offer some indication that the prevalence of

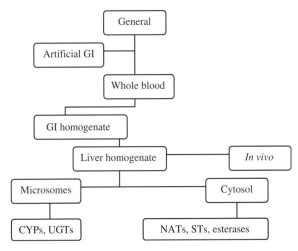

Figure 12. Robust testing paradigm for a prodrug assessment program. See text for discussions about species differences and the use of standard substrates and inhibitors. Enzyme abbreviations: CYPs: cytochrome P450 enzymes, UGTs: uridinediphospho-glucuronosyltransferases, NATs: *N*-acetyltransferases, STs: sulfotransferases.

prodrugs has not changed dramatically over the past several decades, and is reasonably consistent across multiple world markets.

4.2. Recent FDA Approvals

The five prodrugs approved by the U.S. FDA between 2004 and 2008 are summarized in Table 4. These agents demonstrate the wide range of uses and benefits of the prodrug approach. They include four different types of administration: oral, ocular, inhalation, and intravenous (i.v.). These new prodrugs also display many types of activity (antiinflammatory, antimuscarinic, stimulant, and sedative) as well as indications (ocular pain, asthma, ADHD, overactive bladder, and sedation). Finally, they are all examples of hydrolase bioactivation, a very common approach in prodrug design, with about half of all marketed prodrugs being activated via hydrolysis [33]. The focus of this chapter on esterases in general, and carboxylesterase in particular, is additionally born out by the fact that these enzymes likely play at least a partial, if not predominant, role in bioactivating four of the five new prodrugs: cyclesonide by CES1 in the lungs [46], fesoterodine by CES1 and/or CES2 in the intestines and liver following oral delivery [46], lisdexamfetamine by the amidase activity of liver/intestinal CES1 and/or CES2 [87], and nepafenac by the amidase activity of esterases/carboxylesterases in the eye [88–90].

4.3. Selected Examples

The role of ester prodrugs in the development of marketed drugs will be further reviewed by consideration of a few examples of important classes of drugs where a prodrug approach has provided valuable benefits. As in the other sections of this chapter, these examples will focus on cases that involve bioactivation by CES1 and/or CES2 enzymes.

As previously indicated in Section 1.2, diacidic angiotensin converting enzyme (ACE) inhibitors marketed as monoester prodrugs are one of the more successful classes of ester prodrugs. Numerous examples of these ACE inhibitor prodrugs are summarized in Table 5. The presence of two acidic groups is important for ACE inhibiting activity, but these groups also cause the active forms to have poor intestinal absorption after oral delivery. Masking one of the acidic groups with an ethyl ester has proven to be an effective means of improving the oral bioavailability of these molecules [18]. This is reflected by the bioavailability data in Table 5. These improvements are likely due to the combination of increased lipophilicity achieved by masking one of the acid groups, as well as the monoacid forms being substrates for intestinal uptake transporters [103–108]. It should be noted that even with the improvement in bioavailability provided by the prodrug approach, quite a few of the ACE inhibitors in Table 5 have less than 50% oral bioavailability, which has traditionally been considered the minimal benchmark for effective oral delivery. Further, the oral bioavailability of many of these ACE inhibitors can also be quite variable with food intake and other factors [109]. The fact that these agents remain successful as oral drugs in spite of these bioavailability issues is likely because they are generally characterized by a flat dose–response curve and are typically well tolerated with few serious adverse effects [109]. There is some evidence that these agents are preferentially activated by CES1 rather than CES2 enzymes [52,110], indicating that the primary site of bioactivation is likely to be in the liver, which has significantly higher levels of CES1 expression than the intestines.

Prodrugs of β-lactam antibiotics are another class of drugs for which ester prodrug designs have been successfully employed to improve oral bioavailability. Prodrugs of this class most commonly have a double ester design, as illustrated by three of the four ampicillin prodrugs summarized in Table 6. Other types of ester prodrugs have also been developed, such as the cyclic ester ampicillin prodrug lenampicillin. As shown in Table 6, this particular prodrug construct is a cyclic carbonate rather than a carboxylic ester. Furthermore, the carbonate in lenampicillin is in a "vinylogous" relationship with the "oxymethylene" unit in its structure. This makes lenampicillin electronically similar to the double ester "oxymethylene" unit present in the other three structures in Table 6. Thus, all of these prodrugs are bioactivated in an analogous manner through the cleavage of two

Table 4. Prodrugs Approved by the U.S. FDA During the 5-Year Period 2004–2008

Prodrug, Approval Date (Active Form), Activating Enzyme	Prodrug Structure (Dashed Lines Indicate Bonds Cleaved During Bioactivation)	Route (Classification), Indication	Benefits of Prodrug Form
Nepafenac, August 19, 2005 (Amfenac), ocular hydrolase		Ocular (NSAID), cataract surgery	Improved penetration of cornea and ocular tissues. Prolonged duration of activity [91,92]
Ciclesonide, October 20, 2006 (desisobutyryl-ciclesonide), lung esterase		Inhalation (corticosteroid antiinflammatory), asthma	Reduced systemic exposure. Reduced risk of adverse effects [93,94]
Lisdexamfetamine, February 23, 2007 (D-amphetamine), intestinal or hepatic hydrolase[a]		Oral (stimulant), ADHD	Prolonged active drug release. Reduced potential for abuse [95,96]
Fesoterodine, October 31, 2008 (5-hydroxymethyl tolterodine), plasma esterase		Oral (antimuscarinic), overactive bladder	Reduced interpatient pharmacokinetic variability [97–99]
Fospropofol, December 12, 2008 (propofol), endothelial alkaline phosphatase[b]		Intravenous (sedative), anesthesia	Water soluble, aqueous delivery. Reduced injection site pain. Prolonged duration of activity [100–102]

[a] Tissue where bioactivation occurs has not been conclusively determined.
[b] First step (①) of bioactivation by alkaline phosphatase, second step (②) occurs spontaneously.

Table 5. Diacidic Angiotensin Converting Enzyme Inhibitors Marketed as Ethyl Ester Prodrugs for Improved Oral Bioavailability

Prodrug (Active Form)	Prodrug Structure (Dashed Lines Indicate Bonds Cleaved During Bioactivation)	Active Form Bioavailability for Oral Prodrug (%)	Active Form Bioavailability for Oral Diacid (%)
Benazapril (benazaprilat) [111–113]		36–44	4[a]
Cilazapril (cilazaprilat) [114]		57	19
Delapril (delaprilat) [18,115]		55	Not reported Expected to be low
Enalapril (enalaprilat) [116,117]		40	3
Imidapril (imidaprilat) [118,119]		15	~1/7 of Imidapril[a]
Moexipril (moexiprilat) [18,109,120]		13–22	Not reported Expected to be low

Table 5. (Continued)

Prodrug (Active Form)	Prodrug Structure (Dashed Lines Indicate Bonds Cleaved During Bioactivation)	Active Form Bioavailability for Oral Prodrug (%)	Active Form Bioavailability for Oral Diacid (%)
Perindopril (perindoprilat) [121]		66	19
Quinapril (quinaprilat) [113,122]		46–60	Too low for clinical use except via i.v. delivery
Ramipril (ramiprilat) [18,123]		44	Not reported Expected to be low
Spirapril (spiraprilat) [18,124]		50	Not reported Expected to be low
Temocapril (temocaprilat) [18,125]		37	Not reported Expected to be low
Trandolapril (trandolaprilat) [18,126]		40–60	Not reported Expected to be low

Oral bioavailability of the active form when administered as the prodrug or as the diacidic active form are provided where available.
[a] Measured in animal studies, no human data available.

Table 6. Ampicillin Ester Prodrugs Designed for Improved Oral Bioavailability

Prodrug (Active Form)	Prodrug Structure (Dashed Lines Indicate Bonds Cleaved During Bioactivation)	Active Form Bioavailability for Oral Prodrug (%)	Active Form Bioavailability for Oral Active Form (%)
Bacampicillin (ampicillin) [127,130–132]		86–87	32–47
Lenampicillin (ampicillin) [131–134]		50–60	32–47
Pivampicillin (ampicillin) [127,132]		89–92	32–47
Sultamicillin[b] (ampicillin and sulbactam) [18,128–132]		75–80 ampicillin, 60–80 sulbactam	32–47 ampicillin, sulbactam very low

Oral bioavailability of the active form when administered as the prodrug or as the active form are provided where available. Each of these prodrugs becomes bioactivated through the cleavage of two bonds, with the first cleavage (①) being enzymatically driven, while the second cleavage (②) generally occurs spontaneously. [b] Prodrug bioactivation yields two different active forms.

bonds, with the first cleavage being enzymatically driven, and the second being a chemical cleavage that occurs spontaneously. The specific mechanism of the lenampicillin bioactivation process is illustrated in Fig. 13. Activation of ampicillin ester prodrugs is essentially complete following intestinal absorption and/or first-pass metabolism [127], indicating that CES1 and/or CES2 likely play a significant role in the bioactivation process. All of these ester prodrug designs increase oral bioavailability of ampicillin relative to direct oral ampicillin dosing, as indicated by the bioavailability data in Table 6. Bioactivation of sultamicillin also releases a second active antibiotic agent, sulbactam, which has very low oral bioavailability by itself [18], but is 60–80% bioavailable via this dual prodrug approach [128,129].

Table 7 offers two examples of anticancer prodrugs that utilize CES bioactivation to yield improvements in the mode of active

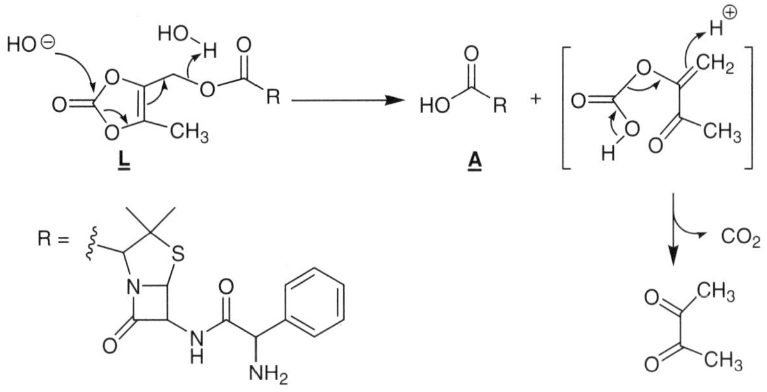

Figure 13. Enzymatic hydrolysis of a "vinylogous cyclic carbonate." The enzymatic reaction has been simplified to show just the entry of a hydroxyl group. The specific prodrug shown is lenampicillin, **L**, which also contains an "oxymethylene" as part of its vinylogous relationship to the parent ester. This prodrug, as well as the simpler "diester" types of "oxymethylene" prodrugs listed in Table 6, all undergo spontaneous collapse of the oxymethylene unit after the initial enzymatic hydrolysis has occurred to eventually liberate ampicillin shown here as **A**, after a concerted reaction depiction that simultaneously includes both steps. See Fig. 17 for a detailed, two-step depiction of these hydrolysis-spontaneous collapse reactions.

agent delivery. Irinotecan is a water-soluble carbamate prodrug of the camptothecin analog SN-38, which has very limited water solubility. Use of the prodrug allows intravenous delivery via an aqueous injection solution, avoiding injection site discomfort and other adverse effects associated with nonaqueous injection formulations. Irinotecan is bioactivated by both CES1 and CES2 [47]. Capecitabine is a reverse carbamate prodrug, which allows for oral delivery, and is specifically designed to avoid CES2 intestinal cleavage in order to minimize intestinal adverse effects caused by intermediate products of capecitabine activation. Following intestinal absorption, capecitabine undergoes the first of three steps of bioactivation via CES1 in the liver to form 5'-deoxy-5-fluorocytidine (5'-d-5-FCyd). This intermediate is converted to 5'-deoxy-5-fluorouridine (5'-d-5-FUrd) by cytidine deaminase, which for normal tissues is expressed primarily in the liver, but has also been shown to be expressed at elevated levels in many types of tumor tissues [45]. The final activation step is the conversion to the active form 5-fluorouracil by thymidine phosphorylase, which has been demonstrated to have significantly higher levels of expression in tumor cells than in corresponding normal cells for a large variety of tissues [45]. This series of bioactivation steps is shown in Fig. 14. The prodrug design of capecitabine thus allows for effective oral delivery, along with some level of tumor targeting via the final two steps of bioactivation.

4.4. Takeaway Messages

Recently approved prodrugs and other marketed prodrug examples discussed in this section clearly indicate the continuing value and versatility of the prodrug approach in developing successful drug products. As mentioned in the introduction to this chapter, prodrugs can enable a more convenient route of administration, enhance absorption and tissue penetration, prolong the duration of activity, reduce unwanted side effects, and in some cases allow targeting of specific tissues. This survey demonstrates that these benefits can be achieved for a wide array of pharmacological activities and parent drug chemical forms. The selected examples provided here also illustrate the need for proper prodrug design to assure bioactivation occurs in the desired tissue location at an appropriate rate. While the design and development of the prodrugs described in this section has in many cases been a costly and time consuming endeavor, these efforts have been invaluable in surmounting

Table 7. Anticancer Agents Marketed as Ester Prodrugs with Bioactivation Involving Carboxylesterases (CESs)

Prodrug (Active Form), Bioactivation	Prodrug Structure (Dashed Line Indicates Metabolic Product After Bioactivation by CES Enzymes)	Benefits of Prodrug Form
Irinotecan (SN-38), single-step activation by hepatic CES1 and CES2		Water-soluble prodrug allows aqueous intravenous formulation [47]
Capecitibine (5-fluorouracil), three-step activation, first step by hepatic CES1[a]		Allows oral drug delivery. Avoids intestinal side effects of partially activated intermediates. Final step of activation via enzyme localized within tumor offers cytotoxic targeting [45]

[a] The first step of bioactivation actually involves the breaking of two bonds: CES1 hydrolysis of bond (①), followed by spontaneous decarboxylation of bond (②).

ADMET hurdles that might have otherwise precluded getting new and improved drug products approved for marketing.

5. SPECIFIC DESIGN PRINCIPLES

5.1. Drug Metabolism's "Rule-of-One"

As appreciated long ago by Hansch, one of the key paths traversed during a given drug's "random walk" is that of xenobiotic metabolism which, in turn, is governed by the drug's overall physicochemical properties coupled to its distinct chemical structure [13,14]. By analogy to Lipinski's "rule-of-five" that pertains specifically to a drug's ability to permeate membranes by passive processes [135], an elementary and generally useful "rule-of-thumb" related to physicochemical properties can likewise be applied to drug metabolism. Because this particular "rule" operates with such a high degree of certainty compared to all other rules that might be devised for predicting metabolism of a given xenobiotic, it has previously been referred to as "drug metabolism's rule-of-one" [11,12]. The rule pertains to the impact of steric features upon drug metabolism rates. From the start, it has been cast in conjunction with the hydrolysis of esters because of the dominant role that the latter typically play during biotransformation when compared to other chemical functional groups that may be found on a given xenobiotic [28,30]. Within the context of esters, the portions of this "rule" that are the most relevant for the present discourse are as follows:

(1) If there is a simple ester present in the molecule, it will typically get hydrolyzed as the predominant metabolic pathway unless it is sterically hindered.

(2) All metabolic processes are exquisitely sensitive to steric features wherein the rate for any given event tends to be inversely proportional to the immediate steric environment at its biotransformation site.

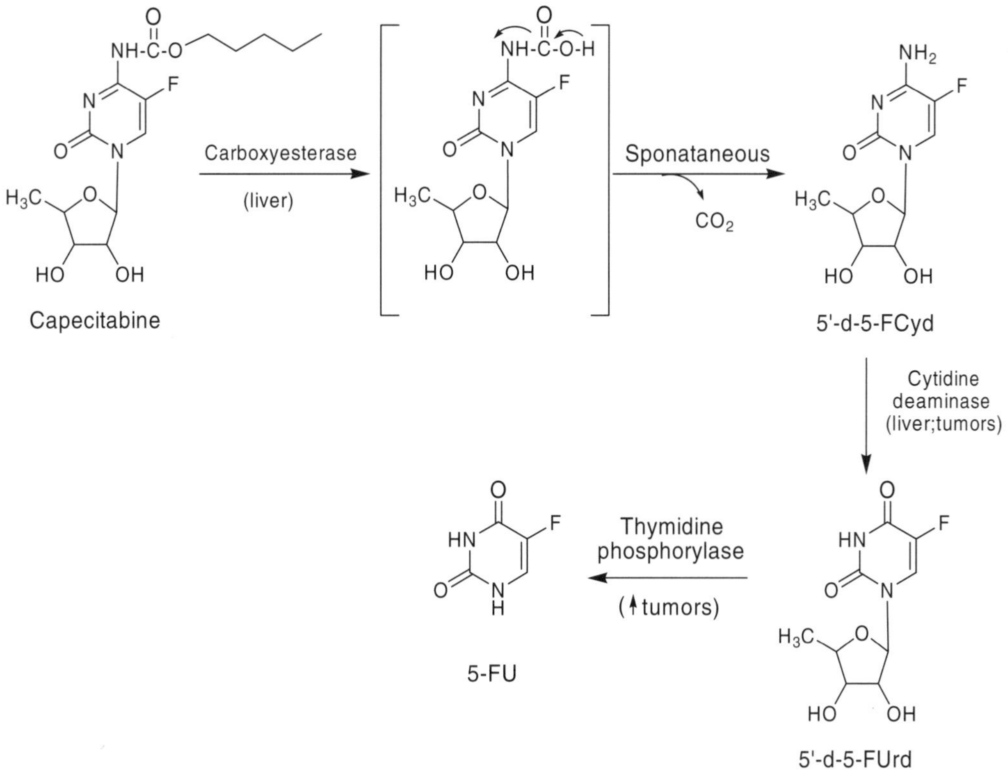

Figure 14. Bioactivation of capecitabine to 5-fluorouracil. This three-step process begins in the liver where CES1 hydrolyzes the "ester side" of the carbamate prodrug moiety. The resulting product is unstable and spontaneously collapses to 5′-deoxy-5-fluorocytidine (5′-d-5-FCyd). This intermediate is deaminated by cytidine deaminase in the liver and in some tumor cells to form 5′-deoxy-5-fluorouridine (5′-d-5-FUrd). The latter is then finally converted to the active agent 5-FU by thymidine phosphorylase, which is overexpressed in many types of tumor cells.

The simplicity of this rule should be readily apparent in that it derives from the fundamental physical law that "no two bodies can occupy the same space at the same time." It is the rule's subtleties that become important for a medicinal chemist embarking upon a prodrug campaign. Thus, after constructing an ester moiety within a parent "drug wannabe" based upon the previous delineation of an amenable structural space from the directed library that was assessed during the hit-follow-up chemistry, "metabolism's rule-of-one" suggests that it generally becomes quite feasible to fine-tune the hydrolysis rate according to systematic structural manipulation of the ester's immediate steric environment. While it should also be apparent that one can often slow down reactions, including a given metabolic process, by steric inhibition, perhaps not so obvious is that it remains possible to deploy the logic of this rule to increase the hydrolysis rate of an added ester moiety. This interesting situation is further exemplified in the next section within one of the "case studies." Continuing more specifically at this point, as indicated in Section 3, not only can the metabolic rate at a given enzyme be altered by a substrate's overall steric features, but the selectivity for one esterase versus another can also be influenced by where such steric factors are explicitly constructed within the substrate. For example, CES1 versus CES2 selectivity can be so manipulated in a preferential manner by focusing upon the "alcohol side" or the "carbonyl side" of the ester's acyloxy moiety, respectively.

Because other practical considerations during the design of prodrugs can be simultaneously influenced by the steric environment constructed around the moiety destined to drive path (i) in Fig. 8, it becomes worthwhile to additionally track these variables while systematically defining the SAR between metabolic bioactivation rate and steric features. These factors include sufficient chemical stability to enable process and eventual manufacturing scale synthesis, "user-friendly" purification and formulation of drug substance, as well as convenient storage and shelf-life of the final product [18]. Likewise, since steric impediments can be assembled with varying degrees of hydrophilicity that are very much under the control of the medicinal chemist, it becomes advantageous to simultaneously consider this physicochemical parameter during prodrug design so as to enhance, rather than detract from, the range of net polarity wanted for optimum aqueous solubility and/or membrane penetration properties.

In terms of polarity, as long as lipophilic or hydrophilic extremes are not positioned within the immediate vicinity of the enzyme's active site, the resulting impact upon metabolic rate should still track well to correlations directed primarily toward the systematic exploration of steric features. Alternatively, prodrugs containing such groups will need to be investigated as their own distinct series relative to the specific enzyme associated with their pathway (i) process of bioactivation. For example, incorporation of a carboxylic acid group (which can ionize at physiological pH) too close to a carboxylester prodrug moiety, represents a hydrophilic extreme that can dramatically attenuate the metabolic rates for substrates that are dependent upon carboxylesterases (recall stability of aspirin's phenolic-acetyl moiety when designing toward butyrylcholine esterase to serve as pathway (i)). Alternatively, incorporation of a phosphatyl moiety, even so close as to be part of the target ester linkage itself, is still subject to rapid hydrolysis when designing toward serum phosphatases to serve as pathway (i) (recall rapid bioactivation of fosphenytoin sodium). Finally, close incorporation of amides and simple amines capable of protonating at physiological pH, can sometimes lead to dramatic increases in metabolic rates relative to certain carboxylesterases. Examples for the latter include the simple aminoalkyl arrangement $RCO_2(CH_2)_2N(CH_2CH_3)_2$ that is normally present in procaine, an ultrashort acting local anesthetic agent that can be thought of as an "accidental" soft drug (to again borrow from some "old" terminology). Analogs of this arrangement have been explored as prodrug moieties for various carboxylic acid-containing drugs such as ibuprofen, **9a** in Fig. 15, wherein significantly accelerated metabolic rates were then observed [136]. Thus, to be on the "safest ground" in terms of establishing steric-related SAR for a targeted carboxylesterase within the structural context of a given parent scaffold responsible for the desired efficacy, it remains prudent to construct the "prodrug wannabe" ester appendages in a manner that does not place ionizable groups too close to the prodrug moiety. Each type of "prodrug wannabe" ester and each substituted version of a given ester type having an additional ioizable group can then be examined as their own series by adopting the same constraints and using the preceding results as a growing baseline for rigorously assessing the importance of steric factors in a systematic manner.

5.2. Potentially Useful Prodrug Moieties

Tables 8–10 list some of the moieties that can be contemplated by medicinal chemists to prepare prodrugs of parent compounds having a carboxylic acid group, a hydroxyl group, or an amino group, respectively [10,29]. Most frequently, at least one of the latter three types of groups will constitute an inherent feature of the parent compound's pharmacophore so that when masked as the prodrug the desired activity is lost, and then when subsequently unmasked by biotransformation the desired efficacious activity is regained. Alternatively, it is also possible to add these groups to an innocuous region of the compound relative to the pharmacophore for the explicit purpose of providing a synthetic handle for the subsequent construction of a prodrug moiety. While the latter deployments are more frequently encountered during prodrug targeting strategies, they can also be imagined for use in addressing one or more ADMET hurdles. By our definition, for these latter

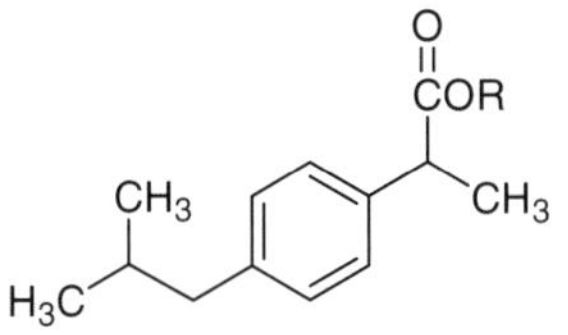

SPECIFIC DESIGN PRINCIPLES

	R	$t_{1/2}$ (min)
9a	H	—
9b	$CH_2CH_2N(CH_3)_2$	1.7
9c	$CH_2\overset{O}{\overset{\|}{C}}N(CH_3)_2$	1.5
9d	CH_2CH_3	5580

Figure 15. Alkylamino (**9b**) and alkylamido (**9c**) prodrugs of ibuprofen, **9a**. Metabolic half-lives ($t_{1/2}$) are shown in minutes for the active (*S*)-enantiomers. Note the dramatic enhancement in metabolic hydrolysis rates for **9b** and **9c** when compared to the simple ethyl ester prodrug **9d**.

instances, the appendage used as part of the prodrug moiety must deactivate the parent compound since masking an innocuously added functionality would not, in itself, be expected to lead to inactivation of the parent compound. Hence the value in defining not only the classical "positive" SAR associated with desired efficacy during a hit follow-up campaign so as to track enhancements of potency, but also to further assess a given

Table 8. Prodrug Moieties for Carboxylic Acid Groups

Category	Arrangement	R'
1. Simple ester	R–CO–O–R'	Alkyl, aryl, aralkyl
2. Functionalized ester	R–CO–O–R'	R' (above) substituted with any display of additional functional groups that do not combine to become an unacceptable extreme of hydrophilicity or lipophilicity for the path (i) enzyme
3. Amide	R–CO–N–R'R''	Typically R' is H and R'' is remainder of an amino acid; one or both R' and R'' are same as 1 or 2
4. Double ester	R–CO–O–CH$_2$–O–CO–R'	Same as 1 or 2
5. Amide ester	R–CO–NH–CH$_2$–O–CO–R'	Same as 1 or 2
6. Ester amide	R–CO–O–CH$_2$–NH–CO–R'	Same as 1 or 2

Table 9. Prodrug Moieties for Aliphatic or Phenolic Hydroxyl Groups

Category	Arrangement	R'
7. Simple/functionalized ester	R–O–CO–R'	Same as 1 or 2 in Table 8, or remainder of an amino acid
8. Carbamate/carbonate	R–O–CO–X–R' X = O, NH	Same as 1 or 2 in Table 8
9. Phosphate ester	R–O–PO(OR')$_2$	Typically at least one of R' is H, and the other is same as 1 or 2 in Table 8
10. Ester/amide methylether	R–O–CH$_2$–X–CO–R' X = O, NH	Same as 1 or 2 in Table 8
11. Carbonate/carbamate methylether	R–O–CH$_2$–O–CO–X–R'	Same as 1 or 2 in Table 8
12. Reversed carbamate	R–O–CH$_2$–NH–CO–O–R' X = O, NH	Same as 1 or 2 in Table 8

Table 10. Prodrug Moieties for Aliphatic or Aromatic Amino Groups[a]

Category	Arrangement	R'
13. Simple/functionalized amide	N–CO–R'	Same as 1 or 2 in Table 8, or remainder of an amino acid
14. Carbamate	N–CO–O–R'	Same as 1 or 2 in Table 8
15. Ester/amide methylamine	N–CH$_2$–X–CO–R'	Same as 1 or 2 in Table 8, or remainder of an amino acid
	X = O, NH	
16. Carbonate/carbamate methylamine	N–CH$_2$–X–CO–Y–R'	Same as 1 or 2 in Table 8
	X = O, Y = O, NH or X = NH, Y = O	
17. Phosphoryl methylamine/methylcarbamate	N–X–CH$_2$–O–PO–(OR')$_2$	Typically at least one of R' is H, and the other is the same as 1 or 2 in Table 8
	X = absent or CO–O	
18. Ester/amide methylcarbamate	N–CO–O–CH$_2$–X–CO–R'	Same as 1 or 2 in Table 8, or remainder of an amino acid
	X = O, NH	
19. Mannich base	N–CH$_2$–N(R')$_2$	Same as 1 or 2 in Table 8
20. Imine amine	N=CH–N(R')$_2$	Same as 1 or 2 in Table 8

[a] Although generally applicable to both primary and secondary amines, prodrug moieties are most frequently deployed for primary amines, although 19 would likely be restricted to secondary amines due to inherent stability issues.

scaffold's structural space in terms of "neutral" SAR and, importantly, even "negative" SAR regions [12,28,36]. As mentioned earlier, however, such broad knowledge-generating SAR investigations are typically not undertaken during today's drug discovery paradigm as the HTS numbers are almost always directed toward quickly achieving optimal potency among perceived drug-like compound library members.

5.3. Bioconversion Mechanisms

As mentioned in Section 3, CESs belong to the family of enzymes that rely upon a serine amino acid residue located within their active sites to participate in their hydrolysis reactions. Also contributing to this enzymatic process are a histidine residue and an ionized glutamic acid that together form the so-called "catalytic triad." Initially thought to involve a proton relay or charge-transfer of the serine hydroxyl proton to the glutamate anion via the histidine, which in a concerted mechanism would remain neutral, it is now thought that the glutamate remains ionized while stabilizing an imidazolium that instead forms and is retained as a transient cationic species during the reaction [137]. As shown in Fig. 16, after substrate recognition and binding (a), the first step (b) of the overall reaction leads to an acylated enzyme (e) after collapse (d) of the transition state (c), wherein the carbonyl carbon has temporarily adopted a tetrahedral arrangement. As also mentioned earlier, initial binding (a) and subsequent stabilization of the "oxyanion" formed (c) additionally involve the backbone amide hydrogens associated with a sequential pair of glycine residues located in an antiparallel dipole relationship to the substrate's carbonyl moiety (depicted in Fig. 10 as a standard dipole arrow). As a second step of the overall process, a stoichiometric water molecule attacks (f) the acyl group so as to regenerate the active form of the enzyme (i) via collapse (h) of a similarly stabilized tetrahedral transition state species (g).

The majority of the possible prodrug moieties shown in Tables 8–10 are subject to cleavage by the CESs via the mechanism detailed above, with notable exception made for the phosphate types that would instead be

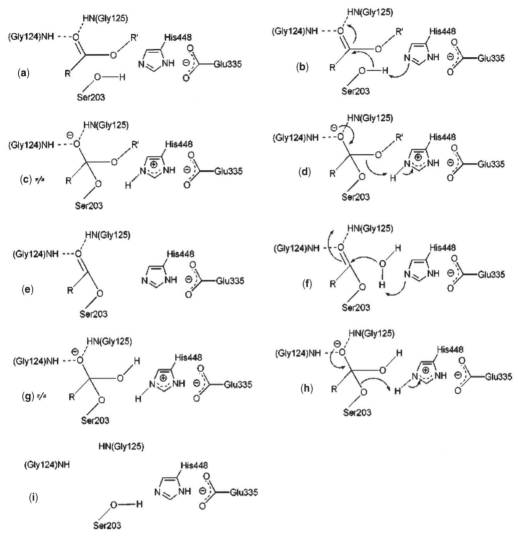

Figure 16. Detailed depiction of the CES hydrolytic mechanism. Panel (a) shows proper orientation of an ester substrate upon binding across the CES catalytic triad (also see Fig. 10). Panel (b) shows the "push-pull" type of attack upon the ester carbonyl carbon as waged by the Ser203 oxygen atom so as to pass over transition state (c) via the latter's collapse according to (d) and eventually become a temporarily "acylated enzyme" (e). Having accomplished its job, the CES is rejuvenated when an imidazole-activated water molecule attacks the carbonyl of the acylated enzyme (f) and passes over a similar transition state (g) by collapsing down path (h) so as to yield (i). See text for additional details.

subject to hydrolysis by the phosphatases. Similarly, the dotted lines in Tables 2 and 4–7 convey where the cleavages are likely to occur via this same type of enzymatic process so as to ultimately regenerate the active drug compounds. In this same regard, however, another key reaction mechanism that additionally needs to be appreciated is that of the spontaneous collapse of the di-heteroatom-methylene units (X–CH$_2$–Y) that are present as part of so many of the prodrug options. This system becomes chemically labile after enzymatic hydrolysis of an adjoining ester, amide, carbamoyl, or carbonate functionality as whatever the case may be for a given prodrug construct. In addition to being listed as design options within Tables 8–10, this frequently deployed system can be found in example **1b**, the last entry in Table 4, and all four of the entries within Table 6. Alternatively, this mechanism is likely not operable in the cases of **8c** and **8d** that can be considered to be "functionalized ester" arrangements having enhanced ester cleavage rates similar to that previously discussed for examples **9b** and **9c**. The mechanism for the collapse of these systems is exemplified in Fig. 17 for the specific case of a diester type of linkage. In general, among the various options that might be available for initial hydrolysis, any nonsterically hindered ester should be the preferred substrate over amides, and so on ("rule-of-one"). Chemical collapse of the resulting product will eventually produce formaldehyde as shown below. Most of the latter becomes quickly oxidized into the body's normal formate cycle and a small amount may become reduced to methanol. Either way, the initial formation of formaldehyde usually does not pose undue toxicity issues.

5.4. Structure–Metabolism Relationships

As cautioned by Testa and Mayer [10], attempts to establish broadly applicable structure–metabolism relationships (SMR) *remains a perilous exercise*. Alternatively, as we have suggested above, the specific delineation of the relevant physicochemical properties within a close series of prodrug analogs is quite feasible. Nevertheless, there are a few properties that do have some general applicability, and these may be useful as a "rule-of-thumb" to assist in the initial selection and prioritization of what prodrug constructs should be pursued and what early analogs might then be made within the context of one's given parent molecular system. For example, the important impact of *steric properties* has already been elaborated earlier. Thus, if a metabolic reaction needs to be attenuated, direct deployment of steric impediments can be a "ready fix." Alternatively, if the parent compound's amino, alcohol, or carboxylic acid functionalities that are to be masked happen to already be buried within a sterically hindered environment, R, that could compromise subsequent removal of their prodrug forms, then perhaps a di-heteroatom-methylene-containing prodrug construct should be pursued from the onset such that the desired enzymatic hydrolysis reaction can be directed to a location at least several atoms removed from the parent structure's steric impediment [11,12,28,30,35,36,138–141]. Another interesting consequence of incorporating an X–CH_2–Y system within a prodrug construct relates to how the substrate may then position itself across the catalytic triad of a given CES. For example, if the parent drug's "handle" is an alcohol, then the only difference becomes an extension of the proposed ester linkage

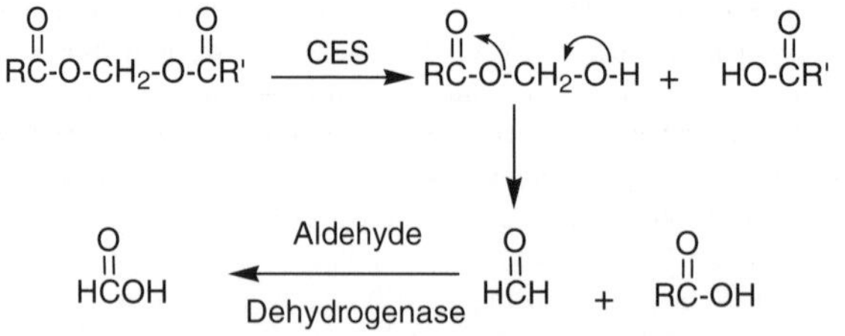

$$RC(=O)\text{-O-}CH_2\text{-O-}CR'(=O) \xrightarrow{CES} RC(=O)\text{-O-}CH_2\text{-O-H} + HO\text{-}CR'(=O)$$

$$\downarrow$$

$$HCOH \xleftarrow{\text{Aldehyde Dehydrogenase}} HCH(=O) + RC(=O)\text{-OH}$$

Figure 17. Spontaneous chemical collapse of the di-heteroatom-methylene unit X–CH_2–Y. Exemplified in this case is the common methylene diester system where X = Y = O. Also note that the formation of formaldehyde as a side-product is quickly further oxidized by aldehyde dehydrogenase to formic acid so that formaldehyde toxicity in not observed. A small amount of the formaldehyde may also become reduced to methanol. Levels of these three side-product species in metabolic equilibrium do not surpass the normal range for the endogenous "methanol–formate" cycle.

away from R. Alternatively, if the synthetic handle is an amine, then incorporation of this system also allows formation of an ester to be used in pathway (i) as opposed to an amide, that is, $RNHCH_2OCOR'$ as opposed to $RNHCOR'$. Finally and perhaps most interesting, if the synthetic handle is a carboxylic acid, then the entire construct will need to "flip" 180° in space so as to expose the resulting enzymatic target ester $-OCOR'$ in the proper three-dimensional orientation on "top" of the catalytic triad (note orientation specified by Fig. 16) that would be analogous to the case for a simple ester prodrug, that is, $R'CO_2CH_2OCOR$ becoming the orientation analogous to the simple ester RCO_2R'. Such a "flip" would be mandated even for the likely result that the functionalized "double ester" derivative would be expected to preferentially engage a different CES altogether. This represents another example of where the physicochemical properties established for one type of prodrug series may be difficult to track to another type of prodrug utilized for the same parent compound. However, even in this complicated case, if the same CES remains involved as the prominent enzymatic partner, one might still be able to approach such a relationship by comparing changes made in R to those made in R' between the two series rather than making R to R and R' to R' types of comparisons. These interesting ester metabolism relationships, reminiscent of the long-standing "left-hand" versus "right-hand" analogy, are further depicted as the third pair of possibilities in Fig. 18.

Equally informative for SMR is the preceding discussion about the hydrolytic mechanism and the exquisite arrangement achieved within the active site to catalyze the reaction by lowering the net energy of the transition state. Instead of the ionic nature of the latter presenting itself as a high-energy species, it forms a "salt" relationship with an imidazolium side-chain group after being initially attenuated in character by its interactions with the two sequential glycine NH groups located along the amino acid chain backbone. Evolved slowly over time to best accomplish these hydrolyses, a second "rule-of-thumb" thus pertains to fundamental *electronic properties*. This rule suggests that it is probably not worth one's time to try to help "Mother Nature" in the hydrolytic process by devising "better leaving groups" [142–145] or, alternatively, to try to slow "her" down by adopting a poorer leaving group whenever either strategy is driven only by electronic considerations (rather than being driven by simple steric considerations). Interestingly, one can further note in this same regard that this situation is quite the opposite of cytochrome P450-mediated xenobiotic metabolism events, where the latter are typically highly subject to even very subtle alterations in the electronic properties of their substrates[11,12,30,31].

Finally, as discussed above, hydrolysis by esterases generally increases with a substrate's overall *lipophilic properties*, and additionally, the presence of a full-blown anionic species such as a carboxylate group near the vicinity of the substrate's ester can profoundly decrease hydrolysis by CESs [39]. However, for the last "rule-of-thumb" it can be taken a bit further that while a full-blown anionic species is clearly intolerable, the net balance of lipophilicity near the ester cleavage site should also not become too extreme in either direction.

In concluding this section, a word of caution of a different nature must be added. Although the suggestion to pursue specific SMR under well-defined experimental conditions stands, the latter usually means such data is generated in well-controlled *in vitro* studies. This, in itself, can present later problems during drug development. For example, if a human CES has been modeled during the *in vitro* testing scheme as part of Step 3 in Fig. 7, once the prodrug is tested in an *in vivo* model (Step 6), even when the latter has been cloned to express the human form for that CES, the prodrug will now become exposed to the entire compliment of esterases present in that living system. It will then be this composite of hydrolytic capabilities that will attempt to attack the substrate [30,31]. As has been pointed out earlier, it is remarkable that in addition to the frequently noted differences encountered when studying SMR between different species, the differences when studying SMR between *in vitro* and *in vivo* tests within the same species can often be just as dramatic. When this occurs, it can become an even more

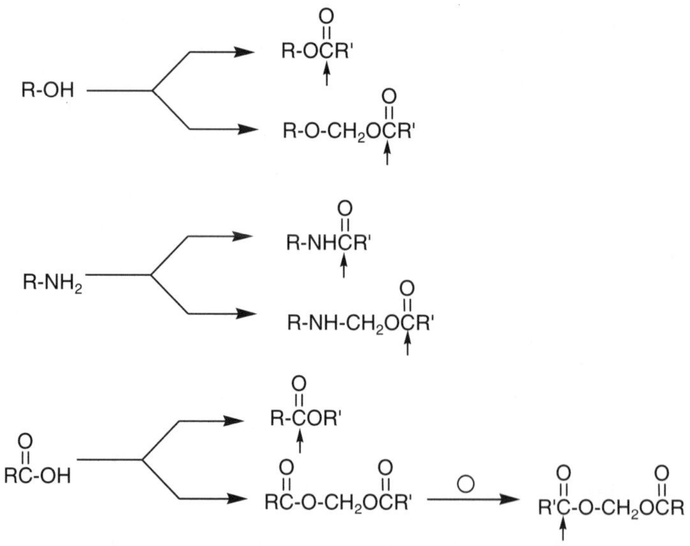

Figure 18. Use of an oxymethylene ester system to reposition the location of attack by CES. Note that the first two comparisons are done relative to formation of a simple acyl prodrug construct. For an alcohol- or amine-containing drug, this construct provides a simple ester or amide that would be attacked as depicted by the vertical arrows on the top pathway showing hydrolysis of the corresponding simple prodrugs for each of the parent drugs. As shown for the last entry, a carboxylic acid-containing drug would likewise become an ester in the simplest case wherein an alcohol is used for the prodrug construct. When an oxymethylene is also added to these simple constructs for the cases of an alcohol- and amine-containing parent, it can be noted that the point of attack by CES (lower pathway for each system) then becomes shifted further away from the parent drug, and that for the amine case one now has an ester rather than an amide. The latter can be expected to result in a dramatic increase in bioconversion rate. Most interesting is the case for the carboxylic acid-containing parent wherein such a maneuver now creates a prodrug construct that must turn 180° so as to properly align its second ester carbonyl into the cleft of the CES active site (assuming that the R'-acyloxy arrangement becomes the preferred substrate over the R-acyloxy arrangement). See text for a more detailed discussion of this strategy.

confounding factor upon trying to establish predictive correlations to select the optimal compound for deployment within the clinic [30,31]. Thus, during a thorough prodrug examination at the *in vitro* level, if the goal is to exploit that data in a practical manner for direct use in the clinic, then it behooves the investigators to also select representative structural examples from the systematic compound series and examine them in an *in vivo* model as an ongoing part of the overall investigation. While reasonable correlations exist between *in vivo* rodent and human PK profiles across a wide range of marketed drugs [146,147], a dog model probably represents the best choice for clarifying this situation among ester-containing compounds as suggested in Section 3. That such parallel studies should be a component for a prodrug program attempting to solve an ADMET issue was previously suggested when it was indicated that "as much about the problem at hand should be gleaned at the onset," as well as during the course of any such program. Alternatively, a dual *in vitro/in vivo* testing strategy is unlikely to be part of a sophisticated targeted delivery or CDS prodrug program wherein only the "optimized" leads from the complex strategy are typically taken forward into "tier two" (Step 6) types of testing. Many of these general points will be seen to be applicable within the next section wherein specific details are presented within the context of three case studies.

6. CASE STUDIES

This section contains three case studies. Paying tribute to what could be considered modern medicine's earliest example of a prodrug, the initial case study provides a brief account of the discovery of prontosil, which took place before the phrase "accidental prodrug" was introduced into the literature. This case study is entitled "More than Just a Gutsy Play" for two reasons, both of which are better revealed by simply allowing this truly remarkable story to quickly unfold. The next two studies are somewhat longer because they attempt to convey how several of the key principles delineated in the earlier sections of this chapter were actually implemented in the laboratory during the course of drug discovery. The second study takes place within a small company, and the third within a "big pharma" setting. Because one of this chapter's authors participated in these drug discovery programs, both of the latter are able to convey first-hand the prodrug-related considerations and, importantly, the creative thinking processes that occurred at various steps along the course of their overall investigations. To emphasize the practice of medicinal chemistry, all three case studies are conveyed in a "story narrative" form rather than as a strictly technical discourse.

The second case study covers the story behind the discovery of esmolol. Although esmolol is a soft drug rather than a prodrug, the chemical logic that allows a medicinal chemist to exploit the relevant enzymatic systems is essentially the same for both types of drug design strategies even though their end products become different. The esmolol case study is entitled "A Soft Drug Before Its Time" because similar to prontosil, discovery of esmolol occurred before its corresponding term was eventually popularized by Bodor. Because esmolol has come to represent a prototype for soft drug molecules, it is often cited as an example for new soft drug design strategies. At times, this situation has led to a certain degree of confusion by leaving readers with the erroneous impression that esmolol, too, was designed by that particular new strategy. Here, you are afforded the true story behind the discovery of esmolol as related first-hand by one of its original inventors.

The final case study covers the events behind the early deployment of a prodrug strategy to address issues associated with a development compound that was designated as "CK-2130." The prodrug was able to effectively hurdle the compound's lack of oral bioavailability. This case study is entitled "A Cardiotonic Drug Wannabe" because CK-2130 then proceeded to fall short during late-stage preclinical drug development due to another ADMET-related issue. Thus, while this discussion will relate some creative prodrug chemistry and its resulting "thrill of victory," it will additionally convey the "hard-knocks" that also accompany the overall business of drug discovery and, more often than not, result in the "agony of defeat." The take-home lessons in both regards are equally important to those that can be gleaned from the aforementioned prontosil and esmolol success stories.

6.1. Prontosil: More than Just a Gutsy Play[1]

In the mid-1930s, a German physician named Domagk was studying a series of bright-red azo-linked dyes and their effects upon streptococcal infections. Domagk noted that one of these dyes, namely prontosil (Fig. 19), was very active when administered orally to streptococcal-infected mice. However, he was not able to confirm this activity in an *in vitro* experiment when the streptococcal bacteria were systematically studied within the controlled environment of a culture dish. Unable to impress pharmaceutical companies into further exploring these "highly ambiguous" and "curious" findings, Domagk took a huge leap across all of today's drug development paradigm steps and, as a last resort for his own daughter who was succumbing to an incurable streptococcal infection, administered prontosil to her on his own. His daughter underwent a "miraculous recovery," from which further research ensued and unambiguously established this chemical substance does indeed have extremely valuable efficacious properties. Since the clinical observation ultimately became the driver for such experiments, this scenario can be considered to represent an

---
[1] This narrative draws closely from that provided for this same topic in Ref. [11].

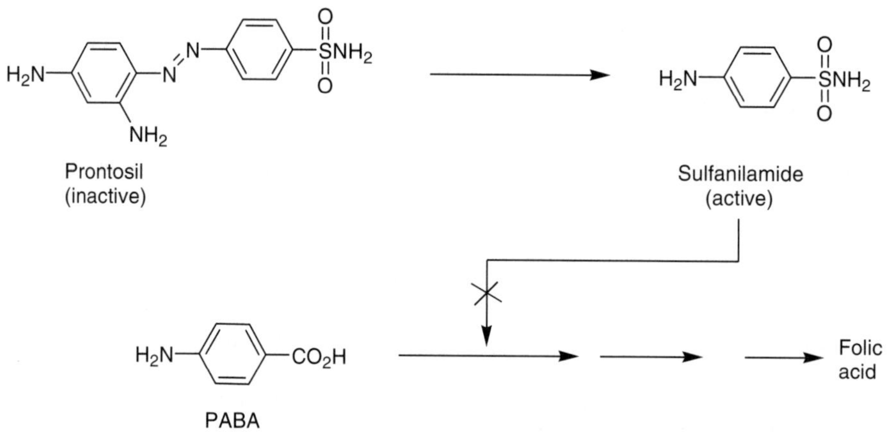

Figure 19. Bioactivation of prontosil to sulfanilamide. After oral administration of prontosil, this reductive metabolic event is accomplished by nonpathogenic microorganisms normally residing within the human GI-tract and often referred to as the "gut microflora." Also shown is how sulfanilamide, after absorption into the systemic circulation, then competes with *para*-aminobenzoic acid during biosynthesis of folic acid. While the latter is a required biosynthetic pathway within the invading infectious type of microorganisms, humans are able to absorb folic acid directly as part of our normal diets. Thus, there is an inherent selectivity afforded by the mechanism of action for the sulfonamide class of antibiotics.

example of what is now called "translational medicine" [11], although it could also be argued that one could instead cite Domagk's initial pharmacological observation in mice as an example of what is now called "translational research" [11]. Regardless of which term is chosen, everyone must agree that when Domagk then proceeded to personally skip all of today's key "translational-related" steps, he truly accomplished a rather remarkable feat in itself.

As shown in Fig. 19, prontosil is itself inactive. Upon oral administration prontosil undergoes metabolic reduction by the "gut" microflora to become sulfanilamide. Sulfanilamide is an active antimicrobial that is effective against streptococcal infections upon absorption into the body. Once this pathway of prodrug bioactivation was understood, the birth of the sulfonamide antibiotics occurred. While both Domagk and his daughter lived "happily-ever-after," sulfanilamide went on to become the prototypical sulfonamide pharmacophore that not only served as a pioneer drug, but also helped to usher in the field of modern antibiotic therapy.

Eventually it was shown that the structure of sulfanilamide resembles that of *para*-aminobenzoic acid (PABA) because its sulfonamide group has a weakly acidic proton that mimics the PABA carboxylic acid. PABA is a requisite biochemical building block for folic acid that microorganisms (MOs) are forced to biosynthesize. Unlike humans, microorganisms are not able to absorb this key vitamin intact. Thus, when sulfanilamide blocks the synthesis of folic acid, its toxicity is selectively directed toward the infecting microorganisms rather than toward the human host. This still constitutes an exquisite mechanism for achieving a "magic bullet's" selectivity. So, *"Thanks"* to Domagk for having the "guts" to take such a chance, although we, as well as the FDA, would never want anyone to try to do such a thing today.

6.2. Esmolol: A Soft Drug Before Its Time[2]

6.2.1. Introduction During the early-1960s, the clinical successes of propranolol (**10** in Fig. 20) were serving to establish it as an effective antihypertensive agent and the "pioneer drug" within the category of beta-adrenergic receptor blocking agents [148]. The field of cardiovascular preventative medicine,

---
[2]This narrative draws closely from that provided in Ref. [145]. An earlier and differently focused discourse can additionally be found in Ref. [143].

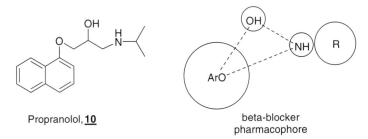

Propranolol, **10**

beta-blocker pharmacophore

Figure 20. Propranolol and the early beta-blocker pharmacophore. As the "pioneer" (first successful) clinical agent for this important class of adrenergic drugs, propranolol (**10**) also served as the structural prototype for establishing early versions of the "aryloxypropanolamine" pharmacophore needed to antagonize beta-1 and beta-2 adrenergic receptors (itself being a nonselective beta-adrenergic blocking agent). This pharmacophore is grossly depicted as a single plane (dashed lines) established by three spatial regions consisting of: (i) a lipophilic aryloxy-moiety (ArO); (ii) a secondary hydroxyl group (OH); and, (iii) a secondary amine bearing a lipophilic substituent (NH R). While this topographical map has gradually evolved into a much higher level of sophistication, these three key regions in space as conveyed by this early cartoon have basically remained the same over the years. (This figure is available in full color at http://mrw.interscience.wiley.com/emrw/9780471266945/home.)

just beginning to "bud" at about that time, would soon be joining these clinical successes rallied by this early "beta-blocker." As a result, numerous pharmaceutical companies were already deploying "analog-based drug discovery" (ABDD) [149] to pursue additional compounds having enhanced pharmacological profiles. For most cardiac indications, the preferred profile typically being sought during this highly competitive era of active beta-blocker analog campaigns was (i) selectivity for beta-1 receptors over alpha- and beta-2 receptors; (ii) low partial agonist or "intrinsic sympathomimetic activity" (ISA); (iii) low membrane depressant or stabilizing activity (MSA); and, (iv) a long duration of action conducive to once-a-day dosing.

Within the specific setting of acute myocardial infarction (MI), however, the seemingly beneficial use of beta-blockers to preserve ischemic cardiac tissue by lowering the energy and oxygen requirements was accompanied by considerable reservation. This was because unhealthy cardiac muscle often relies upon beta-1 adrenergic drive to sustain the inherent contractility, which during the early use of beta-1 blockers occasionally resulted in heart block and death [150]. Thus, in this indication, the clinical challenge became the need for an effective level of drug-induced beta-receptor antagonism to be finely balanced against the need to have a critical level of inherent, endogenous-driven beta-agonism. Further complicating this challenge was the fact that this combination was also in a dynamic state of flux associated with an emergency situation wherein the precise level of adrenergic tone across the heart was very difficult to ascertain. This clinical paradigm is illustrated in Fig. 21.

From this backdrop, in the mid-1970s a small company called Arnar-Stone Laboratories (ASL), whose niche business focus was directed toward emergency medications[3], set

---

[3]Arnar-Stone Laboratories was staffed by four practicing, drug discovery-related scientists. Originally part of American Hospital Supply Corporation (AHSC), the latter soon renamed ASL to American Critical Care (ACC). As esmolol began to undergo serious development, AHSC was purchased by Baxter-Travenol who, in turn, sold ACC to Dupont in the mid-1980s. Interestingly, "the purchase of ACC by Dupont was made for $425 million, $190 million of which was dependent upon the progression of esmolol into the clinic and marketplace (C&E News p. 13, January 12, 1987). As a result, accounts of esmolol's discovery by such a small company are often referred to as a "Cinderella Story." Eventually, Dupont merged with Merck and among these giants the original ASL niche strategy of critical care therapeutics would no longer be pursued. Esmolol, already doing well in the clinic, was eventually sold-off for others to market and, after a series of such transactions, ultimately wound up back in the hands of Baxter and its affiliates.

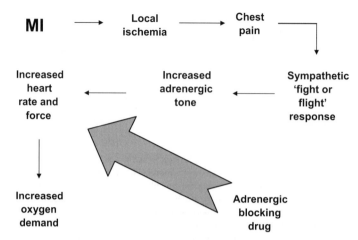

Figure 21. Pathophysiology of acute myocardial infarction and its potential treatment with a beta-adrenergic receptor blocking agent. Note that the body's natural "fight or flight" response can exacerbate the ischemic (low oxygen) damage during an MI because it increases the cardiac muscles' demand for oxygen. Alternatively, by attenuating heart rate and contractile force, a beta-blocker drug may be able to preserve cardiac tissue because oxygen demand will also be reduced. However, because the cardiac muscle is not healthy and its intrinsic beating properties may be compromised, it is important to maintain a certain level of beta-adrenergic tone (agonistic activity) in order to sustain heartbeats. Thus, in this critical setting, a delicate balance must be maintained between agonist drive and antagonist blockade of an adrenergic system that is in a very dynamic state of flux due to the ongoing pathophysiology and the body's misguided attempt to deal with it.

out to discover and develop a beta-blocker compound that could be used safely in the critical setting of MI. Not knowing that it would eventually come to be regarded as a "Cinderella company" [150], the remainder of this case study conveys the details behind the eventual success that ASL was able to achieve toward addressing this challenging clinical "conundrum."

6.2.2. Pharmacological Target It was clear that the desired balance between inherent adrenergic tone and its antagonism could only be achieved and continually adjusted if it were possible to titrate a beta-blocker's effects against some readily measurable parameter of cardiac function. This called for a short acting version of propranolol that could be administered by the intravenous route, while the measurement of heart rate (HR) as a noninvasive "biomarker" was regarded as being a useful way to continually monitor for the effects of such dosing. Not only could the level of blockade be readily adjusted for such a compound, if it were administered in too large an amount for a given patient, the short duration would allow for the antagonistic effects to be quickly removed by simply stopping the i.v. infusion [150]. From clinical experience using dopamine, a short acting, renal vasodilator, to effectively treat renal failure by i.v. infusion during late-stage shock, a half-life of ca. 10 min was envisioned to be ideal to precisely titrate the desired level of effect. The structure of dopamine has been shown previously as **3a** in Fig. 4 wherein its use was for a CNS indication. Since the term "soft drug" had not yet been popularized, ASL's new type of cardiovascular agent was dubbed an "ultrashort-acting" beta-blocker.

6.2.3. Chemical Target

Internal Esters At this point another pioneer drug served as an additional starting point for drug design, this time to pursue an

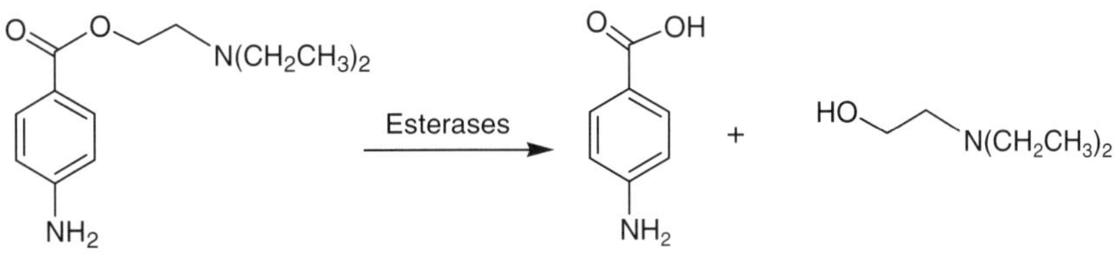

CASE STUDIES   261

Figure 22. Metabolic hydrolysis of procaine. Esterases sever the requisite pharmacophore needed for its anesthetic action by converting procaine into two inactive fragments. Since this process is accomplished very rapidly in the body by these ubiquitous enzymes, procaine exhibits an ultrashort duration of action after parenteral administration.

ultrashort-acting PK profile. As shown in Fig. 22, procaine is an ultrashort-acting anesthetic agent that is rapidly deactivated by esterase-mediated hydrolysis because its aryl and amino pharmacophoric elements become completely severed from their requisite contiguous arrangement. Combining the relevant "structural space" from each of these two pioneer drugs, namely 10 for its efficacy and procaine for its ultrashort duration, resulted in the "double ABDD" target compound 11 shown below in Fig. 23.

By analogy to procaine, it was anticipated that metabolic hydrolysis of 11's ester would fragment the requisite beta-blocker pharmacophore. Previously portrayed in Fig. 20, the key components of this pharmacophore are an aryloxy moiety, a secondary alcohol, and a secondary amino group, all displayed in a distinct 3D spatial arrangement usually specified by connection to a flexible three-carbon scaffold [148]. Disruption of this pharmacophore would rapidly deactivate such compounds. Because an ester linkage was to be incorporated within the requisite "aryloxypro-panolamine" pharmacophore, these types of target molecules were called *internal esters* [143,145]. Previous SAR indicated that an increase in the oxypropyl chain's length by one carbon did not dramatically alter the interaction with beta-receptors. Thus, this useful bit of "neutral SAR" data gained from the literature suggested that the addition of a carbonyl moiety to

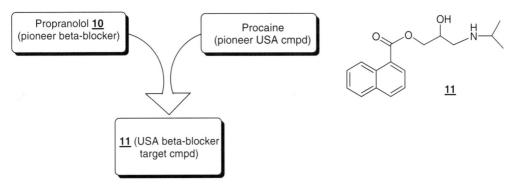

Figure 23. Initial design of beta-blockers having an internal ester. The unique starting point for the rational design of these compounds (cmpds) was able to draw from two different prototypical structures: propranolol (10 in Fig. 20) for its beta-blocking activity; and, procaine (shown in Fig. 22) for its ultrashort-acting ("USA") duration of action.

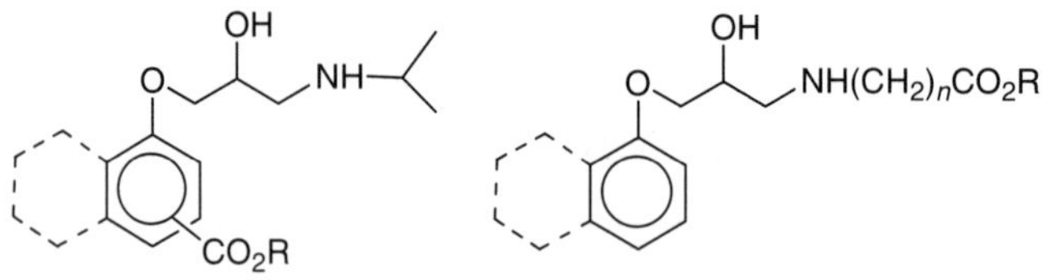

Figure 24. General external ester target molecules. For both the aryl- (**12**) and the N-external esters (**13**), the dashed ring is intended to also represent a naphthalene system that would then be most analogous in design to the prototypical beta-blocker propranolol (**10** in Fig. 20).

create the desired ester linkage should be an acceptable modification.

Unfortunately, these seemingly simple types of structures proved stubborn to synthesize. This was due, in part, to their propensity to undergo rapid side-product reactions that become problematic whenever the amine is not protonated. Although these side-product problems would eventually be solved [151], at this early point in ASL's history as a struggling small pharma company, an immediate solution to these difficulties was needed, including the possibility of pursuing different design constructs.

External Esters The following medicinal chemistry logic and hypothesis led to an alternative design strategy. By virtue of their lipophilic nature, it was assumed that the aryloxy and typical N-aliphatic substituent portions of the beta-blocker pharmacophore interacted with the beta-receptor in a hydrophobic manner, for example, within distinct pockets that preferred lipophilic groups. Thus, while the presence of a "neutral" ester moiety placed in either of these areas of the parent drug molecule might be tolerated, it was hypothesized that the acid resulting from ester hydrolysis, upon further ionization at physiological pH and then likely accompanied by solvation with a cluster of water molecules, would represent an "extreme" that would not be well tolerated. These new target molecules are shown in Fig. 24 as compounds **12** and **13**. Since their ester moieties had now been placed outside of the requisite beta-blocking pharmacophore, such compounds were called *aryl-* and *N-external esters*, respectively. Because later reviews have not always conveyed this key conceptual design step in a clear manner, it becomes worthwhile to reemphasize here that this critical aspect of the overall strategy pertaining to the discovery of esmolol was based upon what can be called a "lipophilic versus hydrophilic hypothesis." It had nothing to do with the metabolism observed for metoprolol (an ether-containing beta-blocker that has a long duration of action) or with any resemblance to "external" amide appendages as were present in practolol (which has a several hour half-life) [152,153]. Eventually proving to be a successful "umpolung" of a physicochemical property, this fundamental design concept has useful ramifications for prodrug/soft drug design constructs of today and for the future.

6.2.4. Structure–Activity Relationships

"Square Pegs and Round Holes" Although **13** provided certain synthetic challenges in terms of controlling the degree of N-alkylation [154], several compound representatives for both of these new series were able to be prepared quickly [142,155,156]. The resulting increase in the overall lipophilicity for these types of esters when placed on a naphthyl system analogous to **10**, however, proved to lower aqueous solubility to the point that biological assessment became problematic. The single phenyl-ring system was therefore deployed as an alternative scaffold because it provided an aryl pharmacophoric element having higher aqueous solubility. However, when a single-ring system was left unsubstituted, significant ISA became problematic. Given the ear-

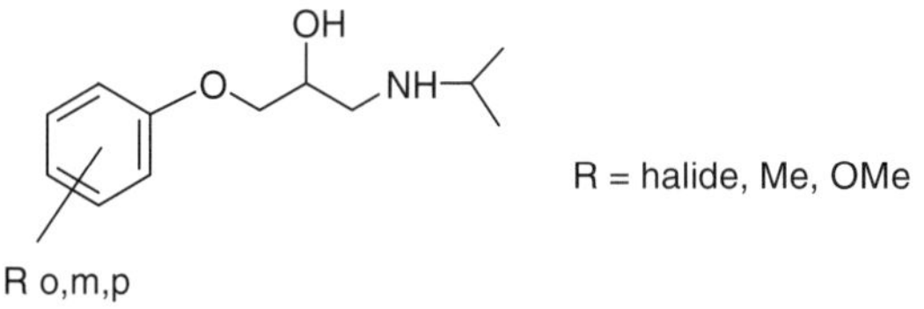

Figure 25. Identification of a surrogate scaffold for exploring the external ester concept. Approximately a dozen compounds were prepared and tested for beta-blocking potency, relative lipophilicity via partition coefficient, and intrinsic sympathomimetic activity effects. An "*ortho*-methylphenyl" system was selected because of its high beta-blocker activity, acceptably low relative lipophilicity and absence of ISA. This scaffold was then deployed with the N-external ester types of target series previously shown as **13** (Fig. 24) and ultimately demonstrated for the first time that external ester constructs could indeed be utilized to create ultrashort-acting beta-blockers.

lier synthetic difficulties encountered with the internal esters ("strike one"), at this point the poor behavior toward testing the external esters encountered first with the low-aqueous solubility napthyl series and immediately followed by the second, ISA-problematic series of N-external esters ("strikes two and three") almost caused ASL's entire ultrashort-acting beta-blocker program to be dropped. After nearly a year into this program with nothing to show, ASL management had begun to consider the effort to obtain ultrashort-acting drugs to be equivalent to trying to "put a square peg into a round hole." Rightly so, management wanted literature precedent indicating that the design and successful obtainment of a compound could be accomplished for any case of an ultrashort-acting compound. Still "ahead of its time" in this regard, however, the only examples to be found for this type of PK profile related to naturally short acting compounds ("accidental soft drugs" so to speak) rather than to any intentionally devised agents.

Surrogate Scaffolds for Testing Purposes and a "Glimmer of Hope" To address the ISA issues, a series of substituted-phenyl aryloxypropanolamine systems was quickly prepared and examined to find a more suitable scaffold to allow for true exploration of the uncharted structural space associated with the N-external constructs. This series is depicted in Fig. 25. After measuring beta-blocker potency, overall lipophilicity via partition coefficient, and level of ISA, an *ortho*-methyl-phenyl ring was chosen for series **13**. Note that the ring in **12** is inherently substituted via the ester placement and so it was not showing significant problems with ISA. Importantly, in both **12** and **13**, these types of esters were found to retain their potency as beta-blockers.

Even more important for the fundamental hypothesis and novel design strategy, however, their carboxylic acid (anticipated metabolite) versions were found to be essentially inactive. As mentioned for **13**, a substituted phenyl-ring was required to eliminate ISA for the N-external esters. In both **12** and **13**, the effects of aromatic substitution followed the pattern where *ortho*-substituents enhanced potency greater than meta-substituents that were more active than those placed in a *para* position. Furthermore, the fall-off in potency was much greater for beta-2 receptors than for beta-1 receptors. Thus, while the *para*-substituted series had lower potency overall, they were more selective for beta-1 in their actions. This efficacy-related SAR trend was in accord with that suggested by the existing literature [148]. In addition, another SMR was beginning to be conceptualized and would soon lead to a second important medicinal chemistry-derived hypothesis that can still be very useful for today's prodrug/soft drug design constructs. This concept stems from early observations associated with a comparison of the half-life data for some of the initial aryl external ester compounds, namely **14** and **15** as shown in Fig. 26. Although ASL's management now acknowledged that the 30 min half-life was "close," it was reiterated that the

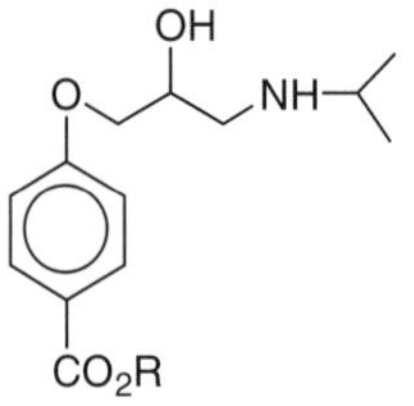

Figure 26. Early stage aryl-external esters. Half-lives measured in anesthetized dogs (five animals per reading) are shown in minutes. In addition to suggesting that this type of construct could be used to create ultrashort-acting beta-blockers, these particular compounds prompted a key hypothesis about the potential importance of steric effects in relation to metabolic hydrolysis rates. Eventually manipulated in a manner that proved to be successful, this hypothesis was later formalized as the key feature within what we have come to call "metabolism's rule-of-one."

	R	$t_{1/2}$ (min)
14	CH_3	30
15	CH_2CH_3	60

target value was 10 min, and were quick to point out that there was still "no cigar." The programmatic "chop-block" continued to loom ominously over the project.

A "Goldilocks" Compound Called Esmolol For **14** and **15**, the half-lives for beta-blocking activity in an *in vivo* dog model were found to be 30 and 60 min, respectively. Speculating that this difference reflected steric effects, it was hypothesized that hydrolysis might be even quicker for esters having less steric hindrance. But how can one construct a simple ester system smaller than the methanol partner that is already present as part of this arrangement in **14**? Since that is not possible from the alcohol side of the ester, one might instead focus attention upon the carbonyl side of the ester. And so came the notion that it might be advantageous to further distance the entire ester linkage from its close proximity to the neighboring (bulky) phenyl-ring. To address this hypothesis, compound series **16** was prepared wherein methylene "spacers" were sequentially added between the ester's carboxyl group and its attached phenyl-ring. These compounds and their corresponding half-life data are shown in Fig. 27 [142]. The *Goldilocks* nature[4] of **16**

where only $n = 2$ is "just right" for the desired duration, likely reflects the removal of the steric impediment caused by the aryl-ring and thus supports the hypothesis behind the design of this series. Likewise, the paradox observed for **16** when $n = 3$ could reflect conformational folding of the *para*-substituent such that its ester group tends to reside closer to the overall bulk displayed by the entire aryloxypropanolamine system. Alternatively, drawing from today's increased knowledge about carboxylesterases, **16** might be so long as to become sterically hindered by the "rear" wall of pocket **A** in Fig. 10. Either way, at this point in the esmolol story, the "Goldilocks compound" was clearly demonstrating much more than just a "glimmer of hope" for the overall program, and the image of the "chop-block" was beginning to fade away.

Ultimately, the desired 10 min half-life was reproduced in humans with $n = 2$ in compound **16**, which became esmolol. Further illustrating the importance of this exquisite inverse relationship uncovered between enzymatic hydrolysis rates and steric bulk is the fact that all of the attempts to increase metabolic hydrolysis rates by the construction of "good (electronic) leaving group" esters had only a negligible influence upon this process [142,155]. Presumably, once "Mother Nature" is freed from steric constraints and gains proper access to a potential substrate, "she" needs no further assistance lowering the transition state energy for effecting an enzymatic conversion with high precision and efficiency.

---

[4]Given the unique half-life for esmolol (**16** with $n = 2$; 10 min) relative to either the higher or lower homologs ($n = 3$ or $n = 1$ being too long), this result has been referred to as the "Goldilocks effect." The term "Goldilocks" is derived from an Anglo-Saxon fairytale, and in this context means "just right."

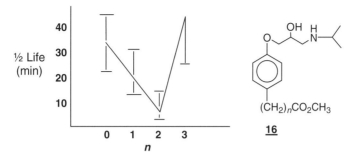

Figure 27. Relief of steric impediment caused by the phenyl ring system. The correlation between biological half-life in minutes and added methylene unit spacers (n) was assessed by measuring 80% recovery from 50% beta-blockade of isoproterenol-induced tachycardia (agonist effect) after i.v. administration to anesthetized dogs (five animals per reading). Error bars indicate ± standard deviation. Note that when $n = 0$ then **16** is the same as **14**, and when $n = 2$ then **16** becomes what later is to be called esmolol.

6.2.5. Pharmacology and Clinical Profile

The pharmacological profile exhibited by esmolol is depicted in Fig. 28, which also clearly shows its difference from propranolol [142]. Upon i.v. infusion, esmolol quickly produces a pseudosteady state of beta-blockade and when the infusion is stopped, the effects of esmolol rapidly disappear with a half-life of about 10 min. Alternatively, propranolol requires nearly 2 h to reach a pseudosteady state of beta-blockade, and then dissipates only very slowly with a half-life of nearly 4 h. In the clinic, esmolol has a distribution half-life of 2 min and an elimination half-life of 9 min.

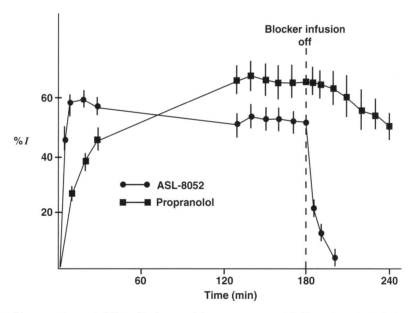

Figure 28. Pharmacodynamic PK profile for esmolol versus propranolol. Three-hour i.v. infusion of esmolol (designated above as "ASL-8052") versus propranolol in dogs (average of five animals for each drug) wherein percent inhibition (% I) reflects beta-blockade of a standardized isoproterenol-induced tachycardia and the x-axis reflects time after the start of infusion. Note that the infusion for each agent was stopped at 180 min, such that the subsequent fall-off in activity of each curve reflects the half-life of each compound. The ultrashort-acting beta-blockade exhibited by esmolol compared to propranolol is striking and its half-life achieves the pharmacological target value of 10 min. Error bars again indicate ± standard deviation.

According to the most recent compilation of reports used for the package insert, "esmolol hydrochloride is rapidly metabolized by hydrolysis of its ester linkage, chiefly by esterases in the cytosol of red blood cells."[5] Its volume of distribution is 3.4 L/kg and its total clearance is 285 ml/(kg/min), "which is greater than cardiac output; thus the metabolism of esmolol is not limited by the rate of blood flow to metabolizing tissues such as the liver, nor is it affected by hepatic or renal blood flow" (see footnote 5). As expected from such a "high rate of blood-based metabolism, less than 2% of the drug is excreted unchanged in the urine" (see footnote 5). Within 24 h after infusion, approximately 73–88% of the total dose can be accounted for in the urine as the acid metabolite. Esmolol becomes about 55% bound to plasma proteins, while the acid metabolite is only 10% bound. "The acid metabolite has been shown in animals to have about 1/1500th the activity of esmolol." In normal volunteers blood levels do not correspond to any significant level of beta-blockade. "The acid metabolite has an elimination half-life of about 3.7 h and is excreted in the urine with a clearance approximately equivalent to the glomerular filtration rate" (see footnote 5). The other metabolic by-product is methanol. Monitoring methanol in patients receiving esmolol for up to 6 h (300 μg/kg/min) and 24 h (150 μg/kg/min), amounts in the body are similar to endogenous levels and remain less than 2% of the level associated with methanol toxicity (see footnote 5).

6.2.6. "Esmolol Stat"[6] Esmolol is used to control adrenergic tone across the heart in emergency room (ER) situations during critical care medicine of very young, as well as adult patients (see footnote 6). It also finds use for alleviating the adrenergic-burst-prompted trauma that can be experienced during intubation prior to surgery. During the later stages of esmolol clinical development, this type of drug, that is, drugs having a deliberately appended functional group that is constructed to program a specified metabolic deactivation pathway and rate, were come to be known as *soft drugs* [28]. Thus, in the end, esmolol itself has become a *pioneer soft drug* for which the template portrayed below in **bold**-faced atoms within Fig. 29 can now serve as an effective starting point for *soft ABDD*.

6.2.7. Summary and Some Take-Home Lessons for Today The discovery of esmolol initially took advantage of a combination of ABDD from two different types of pioneer drugs, namely propranolol for efficacy and procaine for an ultrashort duration of action. During soft drug design and experimental investigations, a novel medicinal chemistry hypothesis pertaining to a "lipophilic umpolung" led to a significant departure from procaine's internal ester structure and prompted the external ester constructs. The latter, in turn, eventually led to esmolol as the "Goldilocks" arrangement upon subsequent exploration of another key hypothesis pertaining to the inverse relationship between enzymatic ester hydrolysis rates and steric factors (now part of "metabolism's rule-of-one"). In addition to demonstrating several of the chemical and metabolic principles that are useful for the design of soft drugs and prodrugs, some of the hurdles traversed during the discovery of esmolol provide useful lessons for today's medicinal chemists who often are confronted by problems associated with a lead compound's overall PK profile [28]. These final comments are listed below in a summary fashion under headings common to the jargon of today's drug discovery process [28]. Taken together, these comments convey the important role that creative medicinal chemistry can play toward establishing useful fundamental principals, as well as toward the practical improvement or specific tailoring of selected therapeutic entities.

Compound Libraries While the attributes of a *drug-like* profile are in no way to be minimized, the first and foremost properties for

---

[5]Quote and details taken directly from package insert as provided by Bedford Laboratories™ for "Esmolol Hydrochloride Injection ready-to-use 10 mL vials."

[6]Esmolol enjoys a certain level of name recognition among the lay-public in that its common use within emergency room settings has prompted the long-running television series entitled "ER" often to open its first scene with a physician shouting "Esmolol, STAT!", that is, "STAT" means immediately.

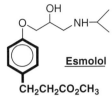

Figure 29. "Short acting" pharmacophore derived from esmolol. The substructure in **bold** represent a generally useful soft drug design construct. It can be deployed as a potential design element for ultrashort activity as long as the arrangement can be placed within a parent compound in an efficaciously innocuous manner (retention of activity), coupled with the likelihood that upon hydrolysis the parent molecule's pharmacophore is either severed or the conversion leads to an "external" carboxylic acid moiety that is not tolerated by the biological surface associated with efficacy. Similarly, this short acting pharmacophore can also be generally useful for the design of prodrug constructs whenever its arrangement can be introduced in a manner that inactivates the parent drug coupled with the knowledge that subsequent hydrolysis will occur to effect bioactivation back to the parent molecule. A simple example of the latter would be to mask a nonsterically hindered primary alcohol (required as part of a parent drug's efficacious pharmacophore) as its ester wherein the carboxylic acid to be deployed is a phenylpropionic acid, perhaps further substituted on the phenyl-ring to achieve some other beneficial physicochemical property for the overall system. In this prodrug example, the methyl alcohol portion of the esmolol ester would thus be replaced by the aliphatic primary alcohol of the parent compound. Once a baseline is established for this metabolic pharmacophore deployment within a new structural context as either a soft drug or a prodrug construct, then the additional use of steric features to fine-tune the observed hydrolysis rate can be further manipulated to allow for final hybridization of the new system to a desired PK profile.

a proposed test series during early drug discovery is certainly their *assay-likable* profile. That is, if a compound cannot be tested, you would not get a result, or even worse, it may be perceived that one has an inactive compound and erroneously derive a negative SAR data point. Indeed, for the case of esmolol, the entire research program was almost halted when its first two series of target compounds were either too insoluble or too plagued by ISA to be able to be effectively screened.

Likewise, while *preferred scaffolds* and *privileged structures* generally refer to the attributes of such molecules relative to their potential clinical performance as drug candidates, this concept need not be so restrictive. For example, it may be useful to think in terms of the various preclinical testing steps to exploit attributes to examine a specific hypothesis that will ultimately be reapplied to the actual lead series. This was done for esmolol when the N-external esters served as important testing surrogates to help confirm the merits of the overall "external esters hypothesis," and was then firmly established that it is not necessarily required to fragment a pharmacophore to eliminate activity.

Although the redundancy of homologs within a compound library may detract from *molecular diversity*, their inclusion should not be precluded altogether, and certainly their pursuit should not be regarded as a mundane operation once a lead has been selected. Additional comments in this regard are provided below in the SAR category.

Biological Testing While HTS can work its way through huge libraries of test compounds in short-order, it also pays to periodically check template representatives in an *in vivo* setting as early as possible during the screening process. In this case study, the discovery of esmolol actually relied upon a dog model to immediately ascertain overall metabolic stability. Although time consuming, this allowed the entire *in vivo* hydrolytic complement to be discerned, rather than quickly conducting a biochemical battery of discrete enzymatic assays. Had the latter been undertaken, SAR would have been very likely to have crossed back-and-forth from one test system to another for various molecular series, perhaps even to the extent of being noninterpretable relative to the key hypotheses being explored.

SAR The sole production of efficacy-associated *positive hit* data as typically derived from HTS methods can seriously detract from creative problem-solving medicinal chemistry. Note that verification of the key hypothesis leading to pursuit of the external esters was actually gleaned from the novel negative SAR data found for the proposed carboxylic acid metabolites, with the anticipated neutral SAR data associated with the esters also falling nicely into place.

While hydrophilic/hydrophobic and/or substituent electronic properties may be important for various selected interactions with a given biological surface, steric effects are by far the most reliable and universally applicable parameter when trying to monitor and exploit SMR. Recall that the esmolol ester family was essentially nonresponsive to the electronics of the "leaving group." Alternatively, it was exquisitely sensitive to localized steric parameters in terms of enzymatic hydrolysis rates. Indeed, the inverse relationship between the degree of steric bulk and the rate of drug metabolism within such localized vicinities is so much more generally reliable compared to any other attempted physicochemical correlation, that it does indeed deserve to be considered as "metabolism's rule-of-one," just as was advocated during several of the earlier discourses within this chapter.

Finally, the importance of pursuing methyl, ethyl, propyl, and so on as a medicinal chemist fine-tunes a lead is worth mentioning again in this category, that is, the *Goldilocks* nature of esmolol with its distinct 10 min half-life was found to reside only at $n=2$ within series **16**. With esmolol still saving lives today in the ER, what more can be said about the potential merits of carefully conducting such systematic molecular scrutiny in a deliberately methodical manner?

6.3. CK-2130: A Cardiotonic Drug "Wannabe"[7]

6.3.1. Introduction In the late-1970s, numerous pharmaceutical companies had rallied around the initial clinical observations for amrinone and its close structural analog milrinone [157,158]. Shown as **17** and **18** in Fig. 30, the selective action of these compounds as "cyclic adenosine monophosphate phosphodiesterase-III (cAMP PDE-III) inhibitors"[8] suggested that they could potentially be deployed as "cardiotonic" drugs to treat congestive heart failure (CHF). At that time, CHF was regarded to be "the most prevalent cause of death in hospitalized patients" [159]. Hence the excitement for embarking upon major expeditions in "search of the digitalis replacement" along the "cAMP cascade" [160]. In this cascade, cAMP flows across cardiac cells as a "second messenger" (a bit of "signal transduction" before its time) to increase the force of muscle contraction (inotropy). Because PDE-III biodegrades cAMP, inhibitors of this enzyme increase cAMP levels, which can then cause a positive inotropic effect that could potentially alleviate the deteriorating "pump performance" observed in CHF.

Perhaps in some way a tribute to the 200th anniversary of Withering's introduction of digitalis to treat "dropsy" (CHF in the 18th and 19th centuries) [161], by the mid-1980s several active analogs of amrinone were either already in early clinical trials or were involved in late-stage preclinical development studies according to the paradigm displayed in Fig. 6. A review at that time lists more than a dozen of such compounds, all of which contained an acylated heteroatom system (typically cyclic) linked to an aryl system that was either itself heteroaromatic or was further substituted

---

[7]This narrative draws partly from that provided for this same topic in Ref. [144].

[8]At lease five distinct families of PDE isozymes had been characterized by the start of the 1990s. Their classification varied in the literature and it was recommended to employ a nomenclature based upon primary protein and cDNA sequence information as well as substrate preference and regulatory properties. The previous designations were commonly referred to as Peak I, II, or III and correspond to elution times during chromatography where Peak I PDE is activated by ca-calmodulin and generally has a preference for hydrolysis of c-GMP, Peak II is sensitive to c-GMP and hydrolyzes both c-GMP and c-AMP, while Peak III has a low K_m for c-AMP as its substrate and is inhibited by c-GMP. The column classification that appears to most closely correspond to the recommended nomenclature PDE-III would be Peak III. The recommended nomenclature has been used herein.

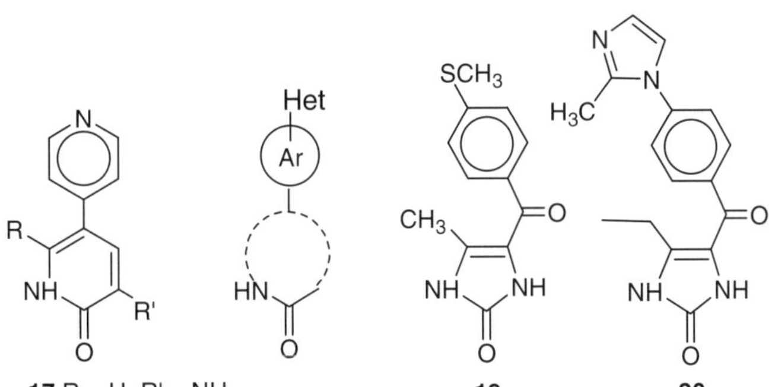

Figure 30. Some early "cardiotonic" phosphodiesterase inhibitors. Amrinone (**17**) represented a "pioneer drug" and prototype structure for this class of compounds which in the early 1980s had hoped to rely upon positive inotropic effects and after-load reduction (by peripheral vasodilation) to become cardiotonic agents that could replace digitalis in the treatment of CHF. Also shown is a very general pharmacophore for this compound class and two specific analogs, namely milrinone (**18**) and enoximone (**19**). Within the pharmacophore: the dotted semicircle is meant to imply that in most cases these southernmost heteroatoms were part of a ring system; Ar indicates an aromatic system and usually a heteroaromatic system; Het indicates that when Ar was not already a heterocycle, additional heteroatoms were typically present as substituent systems. Finally, "CK-2130" (**20**) represented an exciting lead compound that was able to display very selective positive inotropic effects without raising oxygen consumption (energy requirements) within cardiac muscle.

with heteroatoms [160]. This generalized pharmacophore is depicted in Fig. 30, along with one representative structure called enoximone **19** [162,163]. Also caught up in this expedition was Berlex Laboratories in Cedar Knolls (hence CK),[9] who were also exploring the imidazolone system present in **19**. Contrary to what was being reported by others, Berlex had found that a basic moiety placed on the aryl-ring was not only active but also endowed some additionally interesting properties [164,165]. The "lead compound" designated as "CK-2130" is also shown in Fig. 30 as number **20**. In terms of its profile among the many analogs that were prepared, only **20** was able to demonstrate its positive inotropic effect without concomitant peripheral blood pressure (BP) lowering effects and reflex tachycardia (i.e., "chronotropic" or increased heart rate). The latter was regarded by most as an unwanted, energy-wasting response.

[9]During the later stages of this program, Berlex Laboratories, located in Cedar Knolls as a US-based East Coast subsidiary of Schering A.G. (Germany), was merged with two small companies that had been purchased on the West Coast. Largely relocated to the West Coast, the merged organization is now called Berlex Biosciences.

Indeed, some authors had even begun to suggest that to induce any added activity, inotropic or chronotropic, across the heart muscle in an attempt to increase its performance within the setting of CHF, was equivalent to "whipping a sick horse" [166]. Thus, a bit of a "dark cloud" was already beginning to appear across the horizon of the cAMP cascade "hey day" and it was felt that any new target's pharmacological profile certainly needed to exclude an increase in heart rate.

Testifying to the unique selectivity of **20**, however, Berlex was able to demonstrate in an *in vivo* pig model that the positive inotropic effects for the compound were obtained without any significant chronotropic effects and without any increase in oxygen demand. This was a truly remarkable feature which none of the other competitors' agents were able to do [167,168]. Thus, Berlex was poised to embark upon preclinical compound development activities with what appeared to be a "one-of-a-kind" "energy sparing" selective positive inotropic agent.

6.3.2. Welcome to the "A" in ADME As a final prelude to moving into serious preclinical drug development, Berlex conducted a quick oral dose pharmacodynamic study in dogs.

The previous parenteral administration studies had been promising in that the several hour duration of subsequently observed activity suggested that there were no serious issues due to rapid metabolism and/or excretion of the active agent. Unfortunately, it did not fare nearly so well by the oral route. The exciting "one-of-a-kind" agent instead proved to be "plagued by erratically poor oral bioavailability (<20%)" (see footnote 7). Considerable SAR studies directed toward active analogs and receptor topographical mapping [164,165,167–173], plus preliminary attempts to devise specialized formulations, all indicated that a prodrug approach to this dilemma represented the most reasonable strategic course of action at this juncture.

6.3.3. What, After All, Is a Functional Group?

The commonly displayed "functional group" concept is an extremely valuable construct for physical chemists to track chemical behavior, synthetic chemists to track chemical reactions, and everyone to converse about such matters. There can, however, also be situations where a medicinal chemist's consideration of a drug's interaction with a biological surface can benefit by going beyond this type of a traditional conception. These two surfaces actually approach each other in terms of

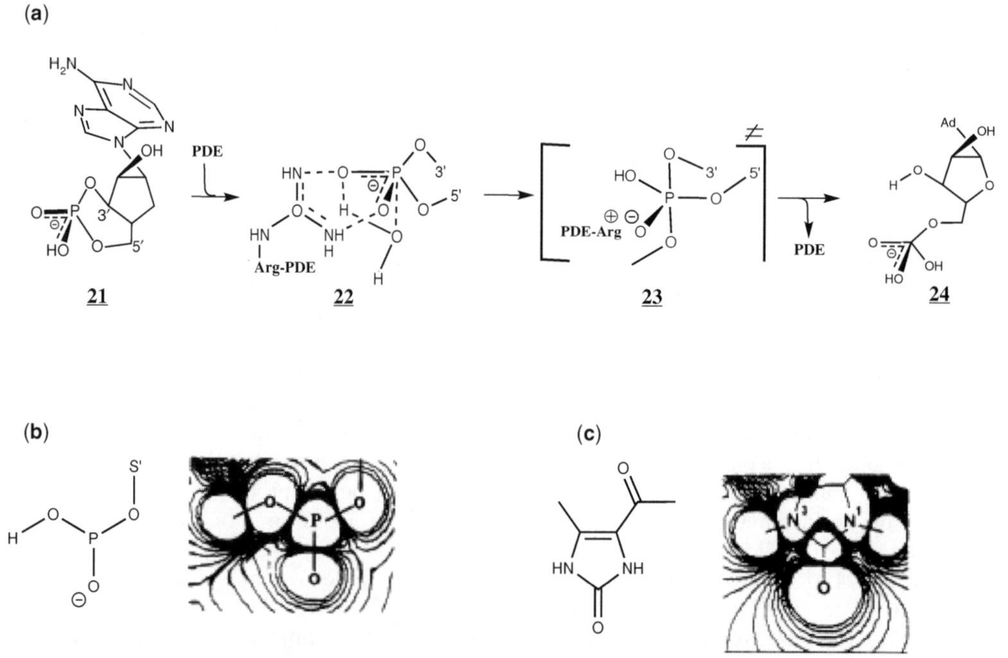

Figure 31. Electrostatic potential considerations. This figure depicts the conformational and electrostatic potential topographies of c-AMP phosphodiesterase III (PDE-III) active site ligands [171–173]. In Panel (a) compound **21** represents a c-AMP substrate with its adenine (Ad) and ribose moieties in an *anti*relationship. Interaction **22** depicts binding of the phosphate portion using an arginine residue and a water molecule that was initially associated with Mg^{2+} in a stoichiometric relationship. Complex **23** depicts S_N2 attack of phosphorous by H_2O with formation of a trigonal bipyramid transition state. Compound **24** represents 5′-AMP as its inverted product. The indicated electronic charges conserve the net charge overall and across the transition state (TS). Panel (b) represents an overhead view of the atoms in the single plane of the TS that forms the common base for the two pyramids of the TBP system. Also shown is the proposed electrostatic potential map for the same atoms. Panel (c) shows a portion of the PDE-III inhibitor ligand **20** (structure shown completely in Fig. 30) and the AM1 derived in-plane molecular electrostatic potential map of its imidazolone ring. Because of the very notable similarity between these two electrostatic potential maps, it has been proposed that these types of compounds, as well as several other heterocycles that have electron-rich heteroatoms in analogous locations, act as TS inhibitors of PDE-III [171–173].

their electrostatic potential maps, that is, more like an initial X-ray diffraction pattern before it has been data-reduced to the convenient "ball-and-stick" construct. For example, upon close inspection, the electrostatic surface of the imidazolone ring in **20** can be likened to the planar portion of a cyclic monophosphate group as the latter traverses a trigonal bipyramid (TBP) transition state during hydrolysis by cAMP PDE. Thus, as shown in Fig. 31, **20** can be viewed as a transition state analog type of enzyme inhibitor for PDE [171–173].

So how might the body's other biological media and surfaces, especially those associated with the process of drug absorption, be "looking at" this particular region of the overall molecule? Coupled with the aforementioned type of conceptualization about what the imidazolone "looks like," perhaps the "give-away" clue for this answer is the fact that experimentally the imidazolone's pK_a was approximately 10. Thus, it was very similar in acidity to a phenolic hydroxyl group. Might the body be "looking at" the electronic surface around one of the imidazolone NH groups as if it were also looking at a phenol? Taking the latter as a working hypothesis, it was therefore imagined constructing some simple acyl prodrugs to "mask" the NH. Initially, these simple constructs could have two potentially beneficial effects: (i) analogous to similarly masked phenolic hydroxyl groups (recall the earlier example of bambuterol, **6b**, in Fig. 5), the NH would no longer be subject to rapid metabolic conjugation reactions such as formation of the glucuronide while trying to traverse the "gut wall" and pass through the liver associated with "first pass" into the body; and (ii) the relative increase in lipophilicity for the acylated system compared to the bare NH should lead to enhanced penetration of the gut wall by simple passive diffusion. Equally important and by continued analogy to a nonsterically hindered phenolic ester, "metabolism's rule-of-one"

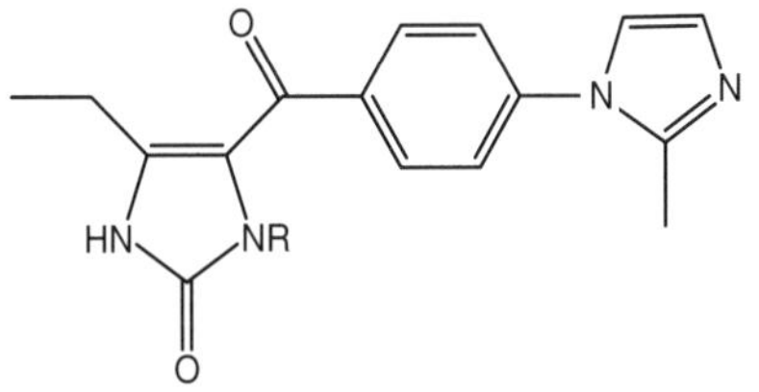

	R	Solubility pH 7.4 (µM)	Partition Coefficient	Conversion to CK-2130 ($t_{1/2}$)		
				pH 1.5 (h)	pH 7.4 (h)	Plasma (min)
20	H (CK-2130)	44	7	-	-	-
25	COCH$_3$	40	26	21	8	30
26	COCH$_2$CH$_3$	30	52	17	6	30
27	COC$_6$H$_5$	23	92	263	25	38
28	COCH(CH$_3$)$_2$	10	>200	132	8	47

Figure 32. Prodrugs of CK-2130. Structures, relevant chemical properties and dog plasma-associated metabolic hydrolysis rates for several acyl prodrugs of the lead compound are shown within the tabulated portion. Note that the units for the various time columns are not the same, namely chemical stability being in hours and metabolic stability being in minutes. The benzoyl prodrug **27** was selected as the lead for further development based upon its optimal blend of all of these properties. The half-life of **27** was later determined to be even faster in human plasma, where $t_{1/2}$ was found to be ca. 10 min.

would then dictate that the acyl should be readily removed by the ubiquitous esterases to quickly bioactivate the prodrug "back" to **20**.

After defining some interesting chemistry that ultimately places the acyl group on the more thermodynamically stable N-substituted product position, the series of prodrugs that were prepared to test this hypothesis are shown in Fig. 32.

6.3.4. An Immediate Success with a True Prodrug

The benzoyl-derivative **27** immediately stood out. It exhibited (i) adequate retention of aqueous solubility; (ii) a significant increase in lipophilicity as measured by partition coefficient; (iii) good stability across a range of pH conditions, in particular that of simulated gastric fluid; and, (iv) an acceptable rate of hydrolysis in dog plasma that should regenerate **20** and might even prolong the *in vivo* half-life. Further study showed that **27** appeared to retain an attenuated level of PDE-III inhibitory activity when examined at the biochemical level and so, by definition, it might not have been a true prodrug. However, the possibility could not be ruled out that some bioconversion might be happening even in that *in vitro* setting. Furthermore, it was not mandated that an inactive compound in actually needed to be obtained. An active analog that worked by itself and then led to an active metabolite would be acceptable from a therapeutic point of view even if it added to the burden of preclinical and clinical development activities by complicating the interface with regulatory guidelines. In the end, when appropriate doses were administered *in vivo* by the i.v. route, no activity was observed until bioactivation had occurred to regenerate **20**. Thus, **27** appeared to meet all the criteria of a "true prodrug" by the strictest of definitions for such.

The "good news," of course, was that prodrug **27** demonstrated excellent oral bioavailability that, just as expected, was followed by bioconversion to **20** and nicely demonstrable efficacious activity. Indeed, bioavailability was greater than 75% and very consistent, which represented a successful outcome when compared to the less than 20% and erratic behavior displayed by the parent compound **20**. These results are shown in Fig. 33. Finally, the half-life of **27** was only about 10 min in human plasma, suggesting that the ubiquitous esterases would become an important player in the bioactivation process.

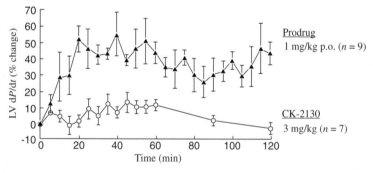

Figure 33. Pharmacodynamic comparison of CK-2130 to its lead prodrug (**27**). Note that while little cardiac effect was observed for CK-2130 (**20**) after oral administration to dogs, significant effects were observed for **27** at only one-third of the CK-2130 dose. Also note that there was little variation in **27** effects across the nine dogs that were used in this study. The quantitative difference that was observed between the oral bioavailabilities for these two compounds was >75% for **27** versus <20% for **20**. Remarkably, **27** was able to dramatically increase the force of cardiac muscle contraction (positive inotropic effect) as measured by the increase in left ventricular pressure over time (LV dP/dt), while having minimal effects on heart rate, blood pressure, and net oxygen consumption (OC). Specifically, for a 50% increase in LV dP/dt, there was less than a 5% increase in HR and a less than 10% decrease in BP with no detectable change in OC.

6.3.5. Welcome to the "T" in ADMET (Said "the Gatekeepers")

With the composite of *in vitro* and *in vivo* pharmacology and preliminary *in vivo* PK data all now looking very promising, in accord with Fig. 6, Berlex proceeded to conduct moderate level scale-up chemistry guided by validated GLP compliant analytical methods coupled to quality control (QC) measures. During this period in time and its regimented sequential approach to drug discovery, the importance of having this level of corporate enthusiasm behind a particular compound coupled with all QC measures being in-place, was essentially mandated before a compound would be moved into any type of serious *in vivo* toxicology studies. Investigators working in toxicology, in turn, stood as "gatekeepers" who defined their job as looking for reasons to deny a compound from continuing its progression toward the clinic, and who defined their success upon uncovering a reason to issue a denial. Today this situation has changed considerably in most organizations and we will return to discuss that evolution at the end of the present case study.

Unfortunately, at this point in the "story," the gatekeepers did succeed in uncovering what appeared to be a rather insidious toxicity. Upon 5-day oral dosing of **27** in dogs, markers of hepatotoxicity had climbed well above normal levels. Even more discouraging was the subsequent similar finding upon 3-day i.v. dosing of **20**. Thus, this toxicity was not associated with the prodrug but was instead inherent to the exciting "one-of-a-kind" parent compound. Thank goodness the medicinal chemists had been so quickly successful in their pursuit of a prodrug approach to solve the earlier oral bioavailability issue, or the latter would have only added further "insult" to the present "injury" of the overall program.

6.3.6. "All the King's Horses and All the King's Men"[10]

That the dog might be the only susceptible species was quickly negated when the insidious liver toxicity was next observed in rats, and then, perhaps as being the closest to humans, in monkeys. By deploying appropriate fragments of CK-2130 (**20**), "structure toxicity relationship" or "STR" studies [11,12,28] were undertaken and quickly revealed that the toxicity was associated with the imidazole ring system and not with the imidazolone ring system. Although certainly disappointing, this was not a total surprise. Some had already begun to think in terms of what we would today call "toxicophores" or "nondrug-like" features that should at least be "flagged" during drug discovery programs if not avoided altogether [11,12,28]. What came to mind in this setting was the antifungal agent ketoconazole shown as **29** in Fig. 34. As depicted, **29** resembles lanosterol that undergoes demethylation by a specific cytochrome P450 enzyme to become a requisite building block within fungal membranes. But **29** also contains an imidazole group and serves to poison the heme portion of the CYP-450 upon binding to this enzyme. Hence, the antifungal activity exhibited by **29**. In our minds it was clear that the imidazole as present in CK-2130 was also serving to "poison" the key CYP-450 enzymes that abound in the liver. While various substitutions on the imidazole, as well as surrogate (bioisosteric) systems, were able to eliminate the liver toxicity, none of such analogs was able to sustain the "one-of-a-kind" selective inotropic and energy sparing profile originally observed for **20**. Thus, in the end, CK-2130 "could not be put together again" within the context of a toxicity-free construct, and ultimately this program was dropped altogether.

6.3.7. So Did Anyone Ever Win this Expedition? The "Long and the Short of It"

"No." In addition to the growing reservations about potentially "whipping a sick" cardiac muscle in CHF when administering positive inotropic agents (which, by the way, do not appear to have ever again been evolved to the same level of selectivity that was so eloquently displayed by CK-2130), the arrival of the angiotensin converting enzyme inhibitors demonstrated, quite dramatically, just how therapeutically beneficial it really was to instead "unload" the heart muscle by reducing vascular resistance and thus improve both pump function and a patient's clinical condition. Interestingly, as

---

[10]The phrase "all the king's horses and all the king's men" is derived from an Anglo-Saxon fairytale and in this context means that even though all possible resources were put to the task at hand, the goal could not be accomplished.

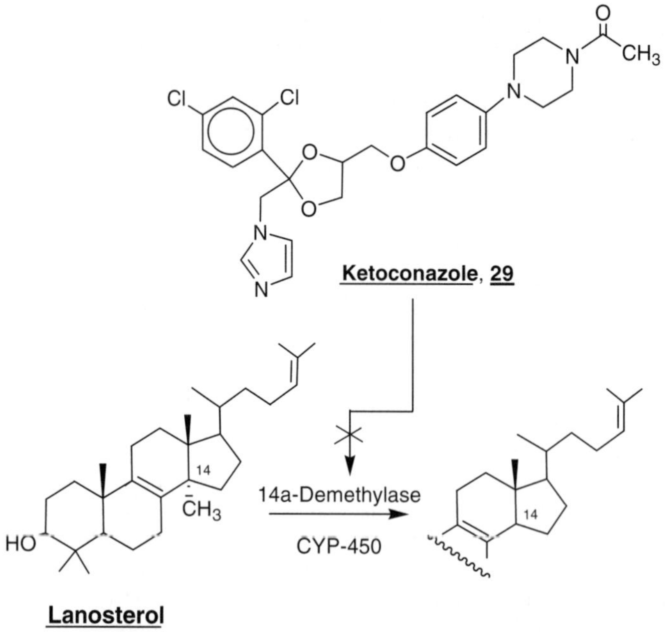

Ketoconazole, 29

Lanosterol

14a-Demethylase
CYP-450

Figure 34. Ketoconazole and its interaction with CYP-450. Ketoconazole, **29**, can inhibit 14-α-demethylase initially by virtue of its close structural analogy to the latter's normal substrate lanosterol (competitive inhibition), and then by virtue of its imidazole which can poison this enzyme's CYP-450 partner (noncompetitive inhibition). Because the biochemical demethylation of lanosterol within fungi is needed to produce a key building block, ketoconazole serves as an effective antifungal agent. From the human systemic circulation, **29** can also exhibit a certain degree of liver toxicity due to the abundant level of important CYP-450 enzymes that are present in this organ. The interaction between imidazoles and CYP-450 enzymes involves an association of the former with the latter's iron atom and is quite general. Thus, the presence of an imidazole moiety is now typically regarded to be problematic when pursuing "drug-like" scaffolds according to today's drug discovery paradigm.

shown earlier in this chapter, as a drug class the ACE inhibitors eventually came to rely heavily upon prodrug approaches to achieve adequate oral bioavailability.

"And yes." As a result of the broad exploration of other PDE inhibitors, the following should certainly be noted as a closely related success even if completely accidental [174]. The discovery of sildenafil (**30** in Fig. 35) began with a similar pursuit of a PDE V inhibitor. PDE V biodegrades cyclic-guanosine-monophosphate (cGMP) which, like cAMP, also serves as an intracellular messenger. In vascular tissue cGMP causes muscle relaxation. Thus, selective inhibition of PDE V would be expected to increase intracellular cGMP that, in turn, could potentially result in vasodilation that might be useful to treat hypertension or angina pectoris. Indeed, sildenafil was taken into clinical testing twice: first to be examined as an antihypertensive agent; and when that "fell short," second to be examined as a therapy for angina pectoris that also "fell short." Alternatively, another vascular bed was observed to be significantly dilated during these clinical studies and it translated into effects that were quite the opposite of falling "short." Hence the eventual development of **30** as a treatment for erectile dysfunction and the birth of a rather staggering multibillion-dollar market that is still going "strong" today.

6.3.8. Take-Home Lessons for Today Several points from this last case study are relevant to today's drug discovery paradigm and these merit emphasis. They are listed below.

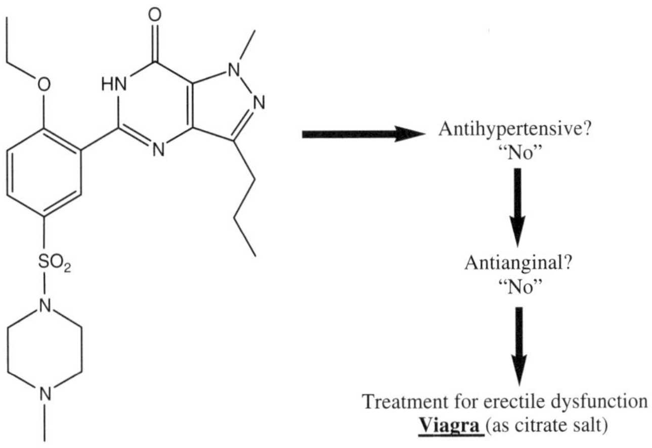

CASE STUDIES

Sildenafil (30)

→ Antihypertensive? "No"
→ Antianginal? "No"
→ Treatment for erectile dysfunction
Viagra (as citrate salt)

Figure 35. The discovery of viagra. Sildenafil, **30**, is a selective cyclic-guanosine-monophosphate phosphodiesterase V (PDE V) inhibitor that can thus raise intracellular levels of cGMP in various vascular beds so as to relax muscle and result in vasodilation. It was initially tested in the clinic as a potential antihypertensive agent (peripheral vasculature), and when this was not successful, it was tried clinically as an antianginal agent (cardiac vasculature), which also proved not to be successful. During these studies, however, it was noted that **30** does have significant effects upon another distinct vascular bed that finally became a useful therapeutic indication (see text for details).

- Drug metabolism can help or hurt. The deployment of a prodrug strategy was clearly a success for obtaining oral bioavailability. This prodrug also allowed for convenient multiday dosing types of toxicity testing to be done such that a key "go/no go" point could be reached and a decision made relative to the further progression of the program's lead compound series. In the end, it was a different type of metabolism issue that struck the final blow to this program.
- Medicinal chemists should always be ready to think in terms of "electrostatic potentials" as well as about classical functional groups and bioisosteres when contemplating the interactions between a xenobiotic and various biological surfaces. Such conceptualizations can be extremely useful for both the further pursuit of active analogs in a rational manner and the creative design of novel prodrug constructs.
- "Metabophores" need to be given their due, right along with "pharmacophores," particularly when they also have the potential to become "toxicophores" within the context of a programmatic structural theme.
- Ascertain indicators of anticipated ADMET profile as early as possible. Of course, but in addition to examining acute toxicity within HTS *in vitro* models, it is highly advisable to take a representative example of the scaffold system, and absolutely so if it is a potential lead compound that is still clinging to an "*in silico*-toxicity-flagged" functionality, into a multiday *in vivo* dosing paradigm so as to potentially uncover insidious toxicity.

Finally, a note about the evolved role of toxicology in today's drug discovery paradigm, and within the field overall for that matter: While still retaining the key "watch-dog" responsibility laid on their shoulders to detect potential toxicity issues before they become toxicity problems, in most pharmaceutical companies a toxicologist will now have joined the interdisciplinary drug discovery teams so as to help identify these types of concerns at the earliest stage possible and, furthermore, to then contribute proactively toward attempting to scale such hurdles when they do arise.

7. CONCLUDING REMARKS

The use of prodrugs within the field of medicinal chemistry, particularly in its applications toward drug discovery, has an extremely rich history that continues to be relevant today. Although strategies for when to deploy prodrug approaches can vary, an important point to emphasize is to ensure the parameters set in place for hit-identification and follow-up chemistry activities are conducive to *creative medicinal chemistry explorations*. Because the latter are beneficial for pursuing active analogs and prodrugs, an appropriately combined effort on both fronts can be highly complimentary and advantageous during the earliest stages of drug discovery, rather than waiting until a development compound has crashed into an ADMET hurdle. It is suggested here that after establishing an early baseline for prodrug chemistry, incorporation of a strategic pause in the synthesis of new analogs allows time for key metabolic considerations and chemical design principles to be optimized before progressing too far down the drug development pathway.

In terms of metabolic considerations, two points have been emphasized, namely the prominent role that esterases can play to convert an appropriately designed prodrug back to the active agent, and how substrate steric features can likely be manipulated to influence the rate of conversion. Both of these points have been captured within what has come to be called *metabolism's rule-of-one*. When the latter is cast against the fundamental requirements pertaining to competing metabolic pathways, its practical value within the context of prodrugs becomes eminently clear.

As reviewed herein, there is a wide repertoire of very interesting functional groups available for the chemical design of prodrugs, even when focusing upon just the utilization of esterase pathways. When considering such constructs, however, the most important point to be emphasized is to not consider their union with the parent compound only in terms of the display of functional group atoms, but rather as their *display of an electrostatic surface potential map* and how the latter, in turn, may be perceived by the surface of an enzyme's active site. Of course, the *impact of steric features upon metabolic rates* is again applicable here. Once this type of SMR is established for a given series, its manipulation can be done under exquisite control by the operating medicinal chemist.

It is hoped that the three case studies offered at the end of this chapter have been additionally useful toward demonstrating how prodrug strategies and principles can be adopted amidst the technical and administrative nuances that become entwined with the practice of drug discovery. To be true to the latter, both a successful story and an unsuccessful program have been provided for proper perspective.

APPENDIX A

The analysis employed to determine the relative prevalence of prodrugs among recently approved pharmacologically active products on the US market is described here. A complete list of all new molecular entities approved by the U.S. FDA between January 2004 and December 2008 was compiled from the drug approval reports available on the Drugs@FDA website hosted by the U.S. FDA Center for Drug Evaluation and Research (www.accessdata.fda.gov/scripts/cder/drugsatfda/). Products used solely for imaging or diagnostic purposes without significant biological or pharmacological activity in the body were excluded from the analysis. Each remaining product was then individually assessed as to whether it would be classified as a prodrug according to the definition provided earlier in this chapter. Table A1 lists the 83 NMEs included in this analysis, with the five classified as prodrugs indicated by *italic* style print.

Table A1. Pharmacologically active novel molecular entities (NMEs) approved by the U.S. FDA from 2004 to 2008

Drug Name	Chemical Name	Approval Date	Route of Delivery	Classification	Indication
Spiriva	Tiotropium	1/30/2004	Inhalation	Bronchodilator	COPD
Alimta	Permetrexed	2/4/2004	IV	Antineoplastic	Lung cancer
Sensipar	Cinacalcet	3/8/2004	Oral	Calcimimetic	Hypercalcemia
Ketek	Telithromycin	4/1/2004	Oral	Antibiotic	Pneumonia
Tindamax	Tindazole	5/17/2004	Oral	Antiprotozoal	Amebic infection
Vidaza	Azacitidine	5/19/2004	SC/IV	Antineoplastic	Leukemia
Xifaxan	Rifaximin	5/25/2004	Oral	Antibiotic	GI Infection
Sanctura	Trospium	5/28/2004	Oral	Antimuscarinic	Overactive bladder
Campral	Acamprosate	7/29/2004	Oral	Alcohol deterrent	Alcoholism
Cymbalta	Duloxetine	8/3/2004	Oral	Antidepressant	Depression/anxiety
Calcium pentetate	Calcium pentetate	8/11/2004	IV/Inhal	Chelator	Radiation exposure
Zinc pentetate	Zinc pentetate	8/11/2004	IV/Inhal	Chelator	Radiation exposure
Fosrenol	Lanthanum carbonate	10/26/2004	Oral	Phosphate binder	Hyperphosphatemia
Tarceva	Erlotinib	11/18/2004	Oral	Antineoplastic	Lung cancer
Vesicare	Solifenacin succinate	11/19/2004	Oral	Antimuscarinic	Overactive bladder
Lunesta	Eszopiclone	12/15/2004	Oral	Sedative	Insomnia
Macugen	Pegaptanib	12/17/2004	Ocular	Antiangiogenic	Macular degeneration
Enablex	Darifenacin	12/22/2004	Oral	Antimuscarinic	Overactive bladder
Clolar	Clofarabine	12/28/2004	IV	Antineoplastic	Leukemia
Prialt	Ziconotide	12/28/2004	Intrathecal	Analgesic	Severe pain
Ventavis	Iloprost	12/29/2004	Inhalation	Vasodilator	Pulmonary hypertension
Lyrica	Pregabalin	12/30/2004	Oral	Analgesic	Neuropathic pain
Mycamine	Micafungin	3/16/2005	IV	Antifungal	Candida infection
Symlin	Pramlintide	3/16/2005	SC	Antidiabetic	Type 1/2 diabetes
Baraclude	Entecavir	3/29/2005	Oral	Antiviral	Hepititis B
Byetta	Exenatide	4/28/2005	SC	Antidiabetic	Type 2 diabetes
Tygacil	Tigecycline	6/15/2005	IV	Antibiotic	Abdominal/skin infection
Levemir	Insulin detemir	6/16/2005	SC	Antidiabetic	Type 1/2 diabetes
Aptivus	Tipranavir	6/22/2005	Oral	Antiviral	HIV
Rozerem	Ramelteon	7/22/2005	Oral	Sedative	Insomnia
Nevanac	Nepafenac	8/19/2005	Ocular	Antiinflammatory	Cataract surgery
Arranon	Nelarabine	10/28/2005	IV	Antineoplastic	Leukemia
Exjade	Deferasirox	11/2/2005	Oral	Chelator	Iron overload
Nexavar	Sorafenib	12/20/2005	Oral	Antineoplastic	Renal cancer

277

Table A1. (*Continued*)

Drug Name	Chemical Name	Approval Date	Route of Delivery	Classification	Indication
Revlimid	Lenalidomide	12/27/2005	Oral	Antianemia	Myeloma
Vaprisol	Conivaptan	12/29/2005	IV	Diuretic	Hyponatremia
Sutent	Sunitinib	1/26/2006	Oral	Antineoplastic	GI/Renal cancer
Ranexa	Ranolazine	1/27/2006	Oral	Antianginal	Chronic angina
Amitiza	Lubiprostone	1/31/2006	Oral	Laxative	Constipation
Eraxis	Anidulafungin	2/17/2006	IV	Antifungal	Candida infection
Dacogen	Decitabine	5/2/2006	IV	Antineoplastic	Leukemia
Chantix	Varenicline	5/10/2006	Oral	Nicotinic agonist	Smoking cessation
Azilect	Rasagiline	5/16/2006	Oral	MAOI	Parkinson's disease
Prezista	Darunavir	6/23/2006	Oral	Antiviral	HIV
Sprycel	Dasatinib	6/28/2006	Oral	Antineoplastic	Leukemia
Noxafil	Posaconazole	9/15/2006	Oral	Antifungal	Immunocompromisation
Zolinza	Vorinostat	10/6/2006	Oral	Antineoplastic	Lymphoma
Januvia	Sitagliptin	10/16/2006	Oral	Antidiabetic	Type 2 diabetes
Omnaris	*Ciclesonide*	10/20/2006	Inhalation	Antiinflammatory	Asthma/allergies
Tyzeka	Telbivudine	10/25/2006	Oral	Antiviral	Hepititis B
Invega	Paliperidone	12/19/2006	Oral	Antipsychotic	Schizophrenia
Vyvanse	*Lisdexamfetamine*	2/23/2007	Oral	Stimulant	ADHD
Tekturna	Aliskiren	3/5/2007	Oral	Antihypertensive	Hypertension
Tykerb	Lapatinib	3/13/2007	Oral	Antineoplastic	Breast cancer
Altabax	Retapamulin	4/12/2007	Topical	Antibiotic	Impetigo
Neupro	Rotigotine	5/9/2007	Transdermal	Antiparkinson	Parkinson's disease
Torisel	Temsirolimus	5/30/2007	IV	Antineoplastic	Renal cancer
Letairis	Ambrisentan	6/15/2007	Oral	Vasodilator	Hypertension
Selzentry	Maraviroc	8/6/2007	Oral	Antiviral	HIV
Somatuline	Lanreotide	8/30/2007	IV	GF Downregulator	Acromegaly
Doribax	Doripenem	10/12/2007	IV	Antibiotic	Abdominal/UT infection
Isentress	Raltegravir	10/12/2007	Oral	Antiviral	HIV
Ixempra	Ixabepilone	10/16/2007	IV	Antineoplastic	Breast cancer
Tasigna	Nilotinib	10/29/2007	Oral	Antineoplastic	Leukemia
Bystolic	Nebivolol	12/17/2007	Oral	Beta-Blocker	Hypertension
Intelence	Etravirine	1/18/2008	Oral	Antiviral	HIV
Pristiq	Desvenlafaxine succinate	2/29/2008	Oral	Antidepressant	Depression
Treanda	Bendamustine	3/20/2008	IV	Antineoplastic	Leukemia

Lexiscan	Regadenoson	4/10/2008	IV	Vasodilator	Cardiovascular imaging
Relistor	Methylnaltrexone	4/24/2008	SC	Opiod antagonist	Opiod constipation
Entereg	Alvimopan	5/20/2008	Oral	Opiod antagonist	Intestinal surgery
Durezol	Difluprednate	6/23/2008	Ocular	Antiinflammatory	Ocular pain
Cleviprex	Clevidipine butyrate	8/1/2008	IV	Antihypertensive	Hypertension
Xenazine	Tetrabenazine	8/15/2008	Oral	Dopamine blocker	Huntington's disease
Rapaflo	Silodosin	10/8/2008	Oral	Alpha blocker	BPH
Vimpat	Lacosamide	10/28/2008	Oral/IV	Anticonvulsant	Partial seizures
Toviaz	*Fesoterodine*	10/31/2008	Oral	Antimuscarinic	Overactive bladder
Banzel	Rufinamide	11/14/2008	Oral	Anticonvulsant	Lennox–Gastaut syndrome
Promacta	Eltrombopag	11/20/2008	Oral	Thrombopoietic	Thrombocytopenia
Tapentadol	Tapentadol	11/20/2008	Oral	Analgesic	Severe pain
Lusedra	*Fospropofol*	12/12/2008	IV	Sedative	Anesthesia
Mozobil	Plerixafor	12/15/2008	SC	Stem cell mobilizer	Stem cell transplantation
Degarelix	Degarelix	12/24/2008	SC	Antiandrogen	Prostate cancer

[a] approved and marketed in prodrug form are shown in *italic*.

REFERENCES

1. Albert A. Chemical aspects of selective toxicity. Nature 1958;182:421–422.
2. Albert A. Selective Toxicity. New York, NY: John Wiley and Sons Inc.; 1964. p 57–63.
3. Harper NJ. Drug latentiation. J Med Pharmaceut Chem 1959;1:467–500.
4. Harper NJ. Drug latentiation. Prog Drug Res 1962;4:221–294.
5. Kupchan SM, Casy AF, Swintosky JV. Drug latentiation. Synthesis and preliminary evaluation of testosterone derivatives. J Pharmaceut Sci 1965;54:514–524.
6. Higuchi T, Stella V. Pro-drugs as Novel Drug Delivery Systems. Washington DC: American Chemical Society; 1975.
7. Roche B. Design of Biopharmaceutical Properties Through Prodrugs and Analogs. Washington DC: Academy of Pharmaceutical Sciences; 1977.
8. Bundgaard H. Design of Prodrugs. Amsterdam: Elsevier; 1985.
9. Lin JH, Lu AY. Role of pharmacokinetics and metabolism in drug discovery and development. Pharmacol Rev 1997;49:403–449.
10. Testa B, Mayer JM. Hydrolysis in Drug and Prodrug Metabolism. Zurich, Switzerland: Verlag Helvetica Chimica Acta; 2003.
11. Erhardt PW. Drug discovery. In: Bachmann K, Hacker M, Messer W,editors. Advanced Pharmacology. Oxford, UK: Elsevier; 2009.
12. Erhardt PW, Proudfoot JR. Drug discovery: historical perspective, current status and outlook. In: Taylor JB, Triggle DJ,series editors. Comprehensive Medicinal Chemistry II. Vol. 1.Oxford, UK: Elsevier; 2007. p 29–96.
13. Hansch C, Fujita TJ. P-σ-π analysis. A method for the correlation of biological activity and chemical structure. J Am Chem Soc 1964;86:1616–1626.
14. Hansch C, Leo A. Substituent Constants for Correlation Analysis in Chemistry and Biology. New York, NY: John Wiley & Sons; 1979.
15. Stahl PH, Wermuth CG. Handbook of Pharmaceutical Salts. Zurich, Switzerland: Verlag Helvetica Chimica Acta; 2002.
16. Gerber N, Mays DC, Donn KH, Laddu A, Guthrie RM, Turlapaty P, Quon CY, Rivenburg WK. Safety, tolerance and pharmacokinetics of intravenous doses of the phosphate ester of 3 -hyrdroxymethyl-5,5-diphenylhydantoin: a new prodrug of phenytoin. J Clin Pharmacol 1988;28:1023–1032.
17. Kubo SH, Cody RJ. Clinical pharmacokinetics of the angiotensin converting enzyme inhibitors: a review. Clin Pharmacokinet 1985;10:377–391.
18. Beaumont K, Webster R, Gardner I, Dack K. Design of ester prodrugs to enhance oral absorption of poorly permeable compounds: challenges to the discovery scientist. Curr Drug Metabol 2003;4:461–485.
19. Cotzias GC, Papavasiliou PS, Gellene RN. Modification of Parkinsonism—chronic treatment with L-dopa. New Engl J Med 1969;280 (7): 337–345.
20. Bradley DA. Prodrugs for improved CNS delivery. Adv Drug Deliv Rev 1996;19:171–202.
21. Parfitt K, Sweetman SC, Blake PS, Parsons AV, editors. Tertbutaline sulphate. In: Martindale: The Complete Drug Reference. London: Pharmaceutical Press; 1999. p 764–765.
22. Tunek A, Levin E, Svensson LA. Hydrolysis of 3H-bambuterol, a carbamate prodrug of tertbutaline, in blood from human and laboratory animals in vitro. Biochem Pharmacol 1988;37:3867–3876.
23. Parfitt K, Sweetman SC, Blake PS, Parsons AV, editors. Bambuterol hydrochloride. In: Martindale: The Complete Drug Reference. London: Pharmaceutical Press; 1999. p 749.
24. Erlich P. Principles of experimental chemotherapy. Angew Chem 1910;23:2–8.
25. Metzger TG, Paterlini NG, Portoghese PS, Ferguson DM. Application of the message-address concept to the docking of naltrexone and selective naltrexone-derived opioid antagonists into opioid receptor models. Neurochem Res 1996;21:1287–1294.
26. Parfitt K, Sweetman SC, Blake PS, Parsons AV, editors. Aciclovir. In: Martindale: The Complete Drug Reference. London: Pharmaceutical Press; 1999. p 602.
27. Richards DM, Carmine AA, Brogden RN, Heel RC, Speight TM, Avery GS. Aciclovir: a review of its pharmacodynamic properties and therapeutic efficacy. Drugs 1983;26:378–438.
28. Erhardt P. Medicinal chemistry in the new millennium: a glance into the future. Pure Appl Chem 2002;74:703–785.
29. Testa B, Caldwell J. Prodrugs revisited: the "ad hoc" approach as a complement to ligand design. Med Res Rev 1996;16(3): 233–241.
30. Erhardt P. Drug metabolism data: past and present status. Med Chem Res 1998;8(7,8): 400–421.
31. Erhardt P,editor. Drug Metabolism: Databases and High Throughput Testing During

Drug Design and Development. Geneva: IUPAC/Blackwell; 1999.

32. van de Waterbeemd H, Smith DA, Beaumont K, Walker DK. Property-based design-optimization of drug absorption and pharmacokinetics. J Med Chem 2001;44:1313–1333.

33. Ettmayer P, Amidon GL, Clement B, Testa B. Lessons learned from marketed and investigational prodrugs. J Med Chem 2004;47(10): 2393–2404.

34. Kerns EH, Di L. Drug-like Properties: Concepts, Structure Design, and Methods. New York, NY: Academic Press; 2008.

35. Erhardt P. Metabolism prediction. In: Darvas F, Dorman G, editors. High-Throughput ADMETox Estimation: In Vitro & In Silico Approaches. Westborough, MA: Eaton Publishing; 2002.

36. Erhardt P. Medicinal chemistry in the new millennium: a glance into the future. In: Chorghade MS, editor. Drug Discovery and Development Vol. 1 Drug Discovery. Hoboken, NJ: John Wiley & Sons; 2006. p 17–102.

37. Hatfield JM, Wierdl JM. Modifications of human carboxylesterase for improved prodrug activation. Exp Opin Drug Metabol Toxicol 2008;4(9): 1153–1165.

38. Moriarty LM, Lally MN, Carolan CG, Jones M, Clancy JM, Gilmer JF. Discovery of a "true" aspirin prodrug. J Med Chem 2008;51: 7991–7999.

39. Nielsen NM, Bundgaard H. Evaluation of glycolamide esters and various other esters of aspirin as true aspirin prodrugs. J Med Chem 1989;32:727–734.

40. Bundgaard H, Nielsen NM, Buur A. Aspirin prodrugs: synthesis and hydrolysis of 2-acetoxybenzoate esters of various N-(hydroxyalkyl) amides. Int J Pharmacol 1988;44:151–158.

41. Del Soldato P, Sorrentino R, Pinto A. NO-aspirins: a class of new anti-inflammatory and antithrombotic agents. Trends Pharmacol Sci 1999;20:319–323.

42. Satoh T, Hosokawa M. Structure, function and regulation of carboxylesterases. Chemico-Biol Interact 2006;162:195–211.

43. Hosokawa M. Structure and catalytic properties of carboxylesterase isozymes involved in metabolic activation of prodrugs. Molecules 2008;13:412–431.

44. Inoue M, Morikawa M, Tsuboi M, Yamada T, Sugiura M. Hyrdolysis of ester-type drugs by the purified esterase from human intestinal mucosa. Jpn J Pharmacol 1979;29:17–25.

45. Miwa M, Ura M, Nishida M, Sawada N, Ishikawa T, Mori K, Shimma N, Umeda I, Ishitsuka HE. Design of a novel oral fluoropyrimidine carbamate, capecitabine, which generates 5-fluorouacil selectively in tumours by enzymes concentrated in human liver and cancer tissue. J Cancer 1998;34(8): 1274–1281.

46. Satoh T, Taylor P, Bosron WF, Sanghani SP, Hosokawa M, LaDu BN. Current progress on esterases: from molecular structure to function. Drug Metabol Dispos 2002;30(5): 488–493.

47. Sanghani SP, Quinney SK, Fredenburg TB, Davis WI, Murry DJ, Bosron WF. Hydrolysis of irinotecan and its oxidative metabolites, 7-ethyl-10-[4-N-(5-aminopentanoic acid)-1-piperidino] carbonyloxycampthothecin and 7-ethyl-10-[4-(1-piperidino)-1-amino]-carbonyloxycampthothecin, by human carboxylesterases CES1A1, CES2, and a newly expressed carboxylesterase isoenzyme, CES3. Drug Metabol Dispos 2004;32(5): 505–511.

48. Li B, Sedlacek M, Manoharan I, Boopathy R, Duysen EG, Masson P, Lockridge O. Butyrylcholinesterase, paraoxonase, and albumin esterase, but not carboxylesterase, are present in human plasma. Biochem Pharmacol 2005;70: 1673–1684.

49. Augusteyn RC, de Jersey J, Webb EC, Zerner B. On the homology of the active-site peptides of liver carboxylesterases. Biochim Biophys Acta 1968;171:128–137.

50. Lee W, Wheatley W, Benedice WG, Huang C, Lee EYHP. Purification, biochemical characterization, and biological function of human esterase D. Proceed Natl Acad Sci USA 1986; 83:6790–6794.

51. Imai T. Human carboxylesterase isozymes: catalytic properties and rational drug design. Drug Metabol Pharmacokinet 2006;21(3): 173–185.

52. Takai S, Matsuda A, Usami Y, Adachi T, Sugiyama T, Katagiri Y, Tatematsu M, Hirano K. Hydrolytic profile for ester- and amide-linkage by carboxylesterase pI 5.3 and 4.5 from human liver. Biol Pharmaceut Bull 1997;20:869–873.

53. Danks MK, Morton CL, Krull EJ, Cheshire PJ, Richmond LB, Naeve CW, Pawlik CA, Houghton PJ, Potter PM. Comparison of activation of irinotecan by rabbit and human carboxylesterases for use in enzyme/prodrug therapy. Clin Cancer Res 1999;5:917–924.

54. Hosokawa M, Maki T, Satoh T. Characterization of molecular species of liver microsomal carboxylestrases of several animal species and humans. Arch Biochem Biophys 1990;277(2): 219–227.

55. Satoh T, Hosokawa M. The mammalian carboxylesterases: from molecules to functions. Annu Rev Pharmacol Toxicol 1998;38: 257–288.
56. Zhang J, Burnell JC, Dumaual N, Bosron WF. Binding and hydrolysis of meperidine by human liver carboxylesterase hCE-1. J Pharmacol Exp Therapeut 1999;290(1): 314–318.
57. Pindel EV, Kedishvili NY, Abraham TL, Brzezinski MR, Zhang J, Dean RA, Bosron WF. Purification and cloning of a broad substrate specificity human liver carboxylesterase that catalyzes the hydrolysis of cocaine and heroin. J Biol Chem 1997;272(23): 14769–14775.
58. Humerickhouse R, Lohrbach K, Li L, Bosron WF, Dolan ME. Characterization of Irinotecan hydrolysis by human liver carboxylesterase isoforms hCE-1 and hCE-2. Cancer Res 2000;60: 1189–1192.
59. Quinney SK, Sanghani SP, Davis WI, Hurley TD, Sun Z, Murry DJ, Bosron WF. Hydrolysis of capecitabine to 5'-deoxy-5-fluorocytidine by human carboxylesterases and inhibition by loperamide. J Pharmacol Exp Therapeut 2005;313(3): 1011–1016.
60. Sun Z, Murry DJ, Sanghani SP, Davis WI, Kedishvili NY, Zou Q, Hurley TD, Bosron WF. Methylphenidate is stereoselectively hydrolyzed by human carboxylesterase CES1A1. J Pharmacol Exp Therapeut 2004;310(2): 469–476.
61. Imai T, Taketani M, Shii M, Hosokawa M, Chiba K. Substrate specificity of carboxylesterase isozymes and their contribution to hydrolase activity in human liver and small intestine. Drug Metabol Dispos 2006;34(10): 1734–1741.
62. Satoh T, Hosokawa M. Molecular aspects of carboxylesterase isoforms in comparison with other esterases. Toxicol Lett 1995;82/83:439–445.
63. Kim S, Nakamura T, Saito Y, Sai K, Nakajima T, Saito H, Shirao K, Minami H, Ohtsu A, Yoshida T, Saijo N, Ozawa S, Sawada J. Twelve novel single nucleotide polymorphisms in the CES2 gene encoding human carboxylesterase 2 (hCE-2). Drug Metabol Pharmacokinet 2003;18(5): 327–332.
64. Wu MH, Chen P, Wu X, Liu W, Strom S, Das S, Cook EH Jr, Rosner GL, Dolan ME. Determination and analysis of single nucleotide polymorphisms and haplotype structure of the human carboxylesterase 2 gene. Pharmacogenetics 2004;14:595–605.
65. Marsh S, Xiao M, Yu J, Ahluwalia R, Minton M, Freimuth RR, Kwok P, McLeod HL. Pharmacogenomic assessment of carboxylesterases 1 and 2. Genomics 2004;84:661–668.
66. Brzezinski MR, Abraham TL, Stone CL, Dean RA, Bosron WF. Purification and characterization of a human liver cocaine carboxylesterase that catalyzes the production of benzoylecgonine and the formation of cocaethylene from alcohol and cocaine. Biochem Pharmacol 1994;18(9): 1747–1755.
67. Ketterman AJ, Bowles MR, Pond SM. Purification and characterization of two human liver carboxylesterases. Int J Biochem 1989;21(12): 1303–1312.
68. Junge W, Heymann E, Krisch K. Human Liver Carboxylesterase. Arch Biochem Biophys 1974;165:749–763.
69. Bencharit S, Morton CL, Hyatt JL, Kuhn P, Danks MK, Potter PM, Redinbo MR. Crystal structure of human carboxylesterase 1 complexed with the Alzheimer's drug tacrine: from binding promiscuity to selective inhibition. Chem Biol 2003;10:341–349.
70. Fleming CD, Bencharit S, Edwards CC, Hyatt JL, Tsurkan L, Bai F, Fraga C, Morton CL, Howard-Williams EL, Potter PM, Redinbo MR. Structural insights into drug processing by human carboxylesterase 1: tamoxifen, mevastatin, and inhibition by benzyl. J Mol Biol 2005;352:165–177.
71. Potter PM, Wadkins RM. Carboxylesterases—detoxifying enzymes and targets for drug therapy. Curr Med Chem 2006;13(9): 1045–1054.
72. Bencharit S, Morton CL, Xue Y, Potter PM, Redinbo MR. Structural basis of heroin and cocaine metabolism by a promiscuous human drug-processing enzyme. Nat Struct Biol 2003;10(5): 349–356.
73. Frey PA, Whitt SA, Tobin JB. A low-barrier hydrogen bond in the catalytic triad of serine proteases. Science 1994;264:1927–1930.
74. Gomori G. Human esterases. J Lab Clin Med 1953;42(3): 445–453.
75. Nishi K, Hyang H, Kamita SG, Kim I, Morisseau C, Hammock BD. Characterization of pyrethroid hydrolysis by the human liver carboxylesterases hCE-1 and hCE-2. Arch Biochem Biophys 2006;445(1): 115–123.
76. Hyatt JL, Wadkins RM, Tsurkan L, Hicks LD, Hatfield MJ, Edwards CC, Ross CR II, Cantalupo SA, Crundwell G, Danks MK, Guy RK, Potter PM. Planarity and constraint of the carbonyl groups in 1,2-diones are determinants for selective inhibition of human carboxylesterase 1. J Med Chem 2007;50: 5727–5734.

77. Yoon KJP, Hyatt JL, Morton CL, Lee RE, Potter PM, Danks MK. Characterization of inhibitors of specific carboxylesterases: development of carboxylesterase inhibitors for translation application. Mol Cancer Therapeut 2004;3(8): 903–909.

78. Brzezinski MR, Spink BJ, Dean RA, Berkman CE, Cashman JR, Bosron WF. Human liver carboxylesterase hCE-1: binding specificity for cocaine, heroin, and their metabolites and analogs. Drug Mctabol Dispos 1997;25(9): 1089–1096.

79. Wadkins RM, Hyatt JL, Wei X, Yoon KJP, Wierdl M, Edwards CC, Morton CL, Obenauer JC, Damodaran K, Beroza P, Danks MK, Potter PM. Identification and characterization of novel benzyl (diphenylethane-1,2-dione) analogues as inhibitors of mammalian carboxylesterases. J Med Chem 2005;48: 2906–2915.

80. Huang TL, Szekacs A, Uematsu T, Kuwano E, Parkinson A, Hammock BD. Hydrolysis of carbonates, thiocarbonates, carbamates, and carboxylic esters of α-napthol, β-napthol, and p-nitrophenol by human, rat, and mouse liver carboxylesterases. Pharmaceut Res 1993;10 (5): 639–648.

81. Tabata T, Katoh M, Tokudome S, Hosakawa M, Chiba K, Nakajima M, Yokoi T. Bioactivation of capecitabine in human liver: involvement of the cytosolic enzyme on 5'-deoxy-5-fluorocytidine formation. Drug Metabol Dispos 2004;32(7): 763–767.

82. Van Gelder J, Deferme S, Annaert P, Naesens L, de Clercq E, van den Mooter G, Kinget R, Augustijns P. Increased absorption of the antiviral ester prodrug tenofovir disoproxil in rat ileum by inhibiting its intestinal metabolism. Drug Metabol Dispos 2000;28(12): 1394–1396.

83. Inoue M, Morikawa M, Tsuboi M, Ito Y, Sugiura M. Comparative study of human intestinal and hepatic esterases as related to enzymatic properties and hydrolyzing activity for ester-type drugs. Jpn J Pharmacol 1980;30(4): 529–535.

84. Danks MK, Morton CL, Pawlik CA, Potter PM. Overexpression of a rabbit liver carboxylesterases sensitized human tumor cells to CPT-11. Cancer Res 1998;58(1): 20–22.

85. Xie M, Yang D, Liu L, Xue B, Yan B. Human and rodent carboxylesterases: immunorelatedness, overlapping substrate specificity, differential sensitivity to serine enzyme inhibitors, and tumor-related expression. Drug Metabol Dispos 2002;30(5): 541–547.

86. Xu G, Zhang W, Ma MK, McLeod HL. Human carboxylesterase 2 is commonly expressed in tumor tissue and is correlated with activation of irinotecan. Clin Cancer Res 2002;8: 2605–2611.

87. Rooseboom M, Commandeur JNM, Vermeulen NPE. Enzyme-catalyzed activation of anticancer prodrugs. Pharmacol Rev 2004;56:53–102.

88. Bora PS, Guruge BL, Bora NS. Molecular characterization of human eye and heart fatty acid ethyl ester synthase/carboxylesterase by site-directed mutagenesis. Biochem Biophys Res Commun 2003;312:1094–1098.

89. Lee VH, Chang SC, Oshiro CM, Smith RE. Ocular esterase composition in albino and pigmented rabbits: possible implications in ocular prodrug design and evaluation. Curr Eye Res 1985;4:1117–1125.

90. Cheng-Bennett A, Chan MF, Chen G, Gac T, Garst ME, Gluchowski C, Kaplan LJ, Protzman CE, roof MB, Sachs G, Wheeler LA, Williams LS, Woodward DF. Studies of a novel series of acyl ester prodrugs of prostaglandin F. Br J Ophthalmol 1994;78:560–567.

91. Ke TL, Graff G, Spellman JM, Yanni JM. Nepafenac, a unique nonsteroidal prodrug with potential utility in the treatment of trauma-induced ocular inflammation: II. In vitro bioactivation and permeation of external ocular barriers. Inflammation 2000;24:371–384.

92. Gamache DA, Graff G, Brady MT, Spellman JM, Yanni JM. Nepafenac, a unique nonsteroidal prodrug with potential utility in the treatment of trauma-induced ocular inflammation: I. Assessment of anti-inflammatory efficacy. Inflammation 2000;24:371–384.

93. Derendorf H. Pharmacokinetic and pharmacodynamic properties of inhaled ciclesonide. J Clin Pharmacol 2007;47:782–789.

94. Boero S, Sabatini F, Silvestri M, Petecchia L, Nachira A, Pezzolo A, Scarso L, Rossi GA. Modulation of human lung fibroblast functions by cyclesonide: evidence for its conversion into the active metabolite desisobutyryl-cicleonide. Immunol Lett 2007;112:39–46.

95. Blick SKA, Keating GM. Lisdexamfetamine. Pediatr Drugs 2007;9:129–135.

96. Suma K, Pennick M, Stark J. Metabolism, distribution and elimination of lisdexafetamine dimesylate. Clin Drug Invest 2008; 28:745–755.

97. Ney P, Pandita RK, Newgreen DT, Breidenbach A, Stohr T, Andersson KE. Pharmacological characterization of a novel investigational antimuscarinic drug, fesoterodine, in

vitro and in vivo. BJU Int 2008; 101: 1036–1042.
98. Michel MC, Hegde SS. Treatment of the overactive bladder syndrome with muscarinic receptor antagonists—a matter of metabolites. Naunyn-Schmiedeberg's Arch Pharmacol 2006;374:79–85.
99. Michel MC. Fesoterodine: a novel muscarinic receptor antagonist for the treatment of overactive bladder syndrome. Exp Opin Pharmacother 2008;9:1787–1796.
100. Fechner J, Ihmsen H, Hatterscheid D, Schiessl C, Vornov JJ, Burak E, Schwilden H, Schuttler J. Pharmacokinetics and clinical pharmacodynamics of the new propofol prodrug GPI 15715 in volunteers. Anesthesiology 2003;99: 303–313.
101. Fechner J, Ihmsen H, Hatterscheid D, Jeleazcov C, Schiessl C, Vornov JJ, Burak E, Schwilden H, Schuttler J. Comparative pharmacokinetics and pharmacodynamics of the new propofol prodrug GPI 15715 and propofol emulsion. Anesthesiology 2004;101:626–639.
102. Fechner J, Schwilden H, Schuttler J. Pharmacokinetics and pharmacodynamics of GPI 15715 of Rospropofol (Aquavan Injection)—a water soluble propofol prodrug. Handbook Exp Pharmacol 2008;182:253–266.
103. Friedman DI, Amidon GL. Intestinal absorption mechanism of dipeptide angiotensin converting enzyme inhibitors of the lysyl-proline type: lisinopril and SQ29852. J Pharmaceut Sci 1989;78:995–998.
104. Friedman DI, Amidon GL. Passive and carrier-mediated intestinal absorption components of two angiotensin converting enzyme (ACE) inhibitor prodrugs in rats: enalapril and fosinopril. Pharmaceut Res 1989;6:1043–1047.
105. Hu M, Zheng L, Chen J, Liu L, Zhu Y, Dantzig H, Stratford RE. Mechanisms of transport of quinapril in Caco-2 cell monolayers: comparison with cephalexin. Pharmaceut Res 1995;12:1120–1125.
106. Swaan PW, Stehouwer MC, Tukker JJ. Molecular mechanism for the relative binding affinity to the intestinal peptide carrier. Comparison of three ACE-inhibitors: enalapril, enalaprilat, and lisinopril. Biochim Biophys Acta 1995;1236:31–38.
107. Thwaites DT, Cavet M, Hirst BH, Simmons NL. Angiotensin-converting enzyme (ACE) inhibitor transport in human intestinal epithelial (Caco-2) cells. Br J Pharmacol 1995;11:981–986.
108. Zhu T, Chen XZ, Steel A, Hediger MA, Smith DE. Differential recognition of ACE inhibitors in *Xenopus laevis* oocytes expressing rat PEPT1 and PEPT2. Pharmaceut Res 2000; 17:526–532.
109. Song JC, White CM. Clinical pharmacokinetics and selective pharmacodynamics of new angiotensin converting enzyme inhibitors. Drug Dispos 2002;41:207–224.
110. Imai T, Imoto M, Skamoto H, Hashimoto M. Identification of esterases expressed in Caco-2 cells and effects of their hydrolyzing activity in predicting human intestinal absorption. Drug Metabol Dispos 2005;33:1185–1190.
111. Waldmeier F, Schmid K. Disposition of [^{14}C]-benazepril hydrochloride in rat, dog and baboon: absorption, distribution, kinetics, biotransformation and excretion. Arzneimittel-Forschung 1989;39:62–67.
112. Waldmeier F, Kaiser G, Ackermann R, Faigle IW, Wagner J, Barner A, Lasseter KC. The disposition of [^{14}C]-benazepril HCl in normal adult volunteers after single and repeated oral dose. Xenobiotica 1991;21:251–261.
113. Gengo FM, Brady E. The pharmacokinetics of benazepril relative to other ACE inhibitors. Clin Cardiol 1991;14:IV45–IV50.
114. Williams PE, Brown AN, Rajaguru S, Francis RJ, Walters GE, McEwen J, Durnin C. The pharmacokinetics and bioavailability of cilazapril in normal man. Br J Clin Pharmacol 1989;27:181S–188S.
115. Onoyama K, Nanishi F, Okuda S, Oh Y, Fujishima M, Tateno M, Omae T. Pharmacokinetics of a new angiotensin I converting enzyme inhibitor (delapril) in patients with deteriorated kidney function and in normal control subjects. Clin Pharmacol Therapeut 1988;43:242–249.
116. Ulm EH. Enalapril maleate (MK-421), a potent, nonsulfhydryl angiotensin-converting enzyme inhibitor: absorption, disposition, and metabolism in man. Drug Metabol Rev 1983;14:99–110.
117. Irvin JD, Till AE, Vlasses PH, Hichens M, Rotmensch HA, Harris KE, Merrill DD, Ferguson RK. Bioavailability of enalapril maleate. Clin Pharmacol Ther 1984;35:248.
118. Tagawa K, Mizobe M, Noda K. a saturable tissue-angiotensin I converting enzyme (ACE) binding model for the pharmacokinetic analysis of imidapril, a new ACE inhibitor, and its active metabolite in human. Biol Pharmaceut Bull 1995;18:64–69.
119. Yamada Y, Endo M, Kohno M, Otsuka M, Takaiti O. Metabolic fate of the new angiotensin-converting enzyme inhibitor imidapril in

animals. Arzneimittel-Forschung 1992;42: 457–465.
120. Stimpel M, Bonn R, Koch B, Dickstein K. Pharmacology and clinical use of the new ACE-inhibitor moexipril. Cardiovasc Drug Rev 1995;13:211–229.
121. Lees KR, Green ST, Reid JL. Influence of age on the pharmacokinetics and pharmacodynamics of perindopril. Clin Pharmacol Therapeut 1988;44:418–425.
122. Breslin E, Posvar E, Neub M, Trenk D, Jahnchen E. A pharmacodynamic and pharmacokinetic comparison of intravenous quinaprilat and oral quinapril. J Clin Pharmacol 1996;36:414–421.
123. Van Griensven JMT, Schoemaker RC, Cohen AF, Luus HG, Seibert-Grafe M, Rothig HJ. Pharmacokinetics, pharmacodynamics and bioavailability of the ACE inhibitor ramipril. Eur J Clin Pharmacol 1995;47:513–518.
124. Hayduk K, Kraul H. Efficacy and safety of spirapril in mild-to-moderate hypertension. J Cardiovasc Pharmacol 1999;34:S19–S23.
125. Puchler K, Sierakowski B, Roots I. Single dose and steady state pharmacokinetics of temocapril and temocaprilat in young and elderly hypertensive patients. Br J Clin Pharmacol 1998;46:363–367.
126. Wiseman LR, McTavish D. Trandolapril: a review of its pharmacodynamic and pharmacokinetic properties, and therapeutic use in essential hypertension. Drugs 1994;48:71–90.
127. Ehrnebo M, Nilsson SO, Boreus LO. Pharmacokinetics of ampicillin and its prodrugs bacampicillin and pivampicillin in man. J Pharmacokinet Biopharmaceut 1979;7:429–451.
128. Rogers HJ, Bradbrook ID, Morrison PJ, Spector RG, Cox DA, Lees LJ. Pharmacokinetics and bioavailability of sultamicillin estimated by high-performance liquid chromatography. J Antimicrob Chemother 1983;11: 435–445.
129. Friedel HA, Campoli-Richards DM, Goa KL. Sultamicillin: a review of its antibacterial activity, pharmacokinetic properties and therapeutic use. Drugs 1989;37:491–522.
130. Bergan T. Pharmacokinetic comparison of oral bacampacillin and parenteral ampicillin. Antimicrob Agents Chemother 1978;13:971–974.
131. Jusko WJ, Lewis GP. Comparison of ampicillin and hetacillin pharmacokinetics in man. J Pharmaceut Sci 1973;62:69–76.
132. Loo JCK, Foltz EL, Wallick H, Kwan KC. Pharmacokinetics of pivampicillin and ampicillin in man. Clin Pharmacol Therapeut 1974;16:35–43.
133. Saito A, Nakashima M. Pharmacokinetic study of lenampicillin (KBT-1585) in healthy volunteers. Antimicrob Agents Chemother 1986;29:948–950.
134. Sum AM, Sefton AM, Jepson AP, Williams JD. Comparative pharmacokinetic study between lenampiciilin, bacampicillin and amoxicillin. J Antimicrob Chemother 1989;23:861–868.
135. Lipinski CA, Lombardo F, Dominy BW, Feeney PJ. Experimental and computational approaches to estimate solubility and permeability in drug discovery and development settings. Adv Drug Deliv Rev 1997;23:3–25.
136. Mork N, Bundgaard H. Stereoselective enzymatic hydrolysis of various ester prodrugs of ibuprofen and flurbiprofen in human plasma. Pharmaceut Res 1992;9:492–496.
137. Dodson G, Wlodawer A. Catalytic triads and their relatives. Trends Biochem Sci 1998;23: 347–352.
138. Wilbury T. Comparison of commercially available metabolism databases during the design of prodrugs and codrugs. In: Erhardt P, editor. Drug Metabolism: Databases and High Throughput Testing During Drug Design and Development. Geneva: IUPAC/Blackwell; 1999.
139. Erhardt P. Statistics-based probabilities of metabolic possibilities. In: Erhardt P, editor. Drug Metabolism: Databases and High Throughput Testing During Drug Design and Development. Geneva: IUPAC/Blackwell; 1999.
140. Erhardt P. Metabolism prediction. In: Darvas F, Dorman G, editors. High-Throughput ADMETox Estimation: In Vitro & In Silico Approaches. Westborough, MA: Eaton Publishing; 2002.
141. Erhardt P. Using drug metabolism databases during drug design and development. In: Chorghade MS, editor. Drug Discovery and Development, Vol. 1 Drug Discovery. Hoboken, NJ: John Wiley & Sons; 2006. p 273–293.
142. Erhardt PW, Woo CM, Anderson WG, Gorczynski RJ. Ultra-short-acting b-adrenergic receptor blocking agents. 2. (Aryloxy) propanolamines containing esters on the aryl function. J Med Chem 1982;25:1408–1412.
143. Erhardt PW. Esmolol. In: Lednicer D, editor. Chronicles of Drug Discovery. Washington, DC: ACS Books; 1993. p 191.
144. Erhardt P. Case studies: a prodrug and a softdrug. In: Erhardt P, editor. Drug Metabolism: Databases and High Throughput Testing

During Drug Design and Development. Geneva: IUPAC/Blackwell; 1999.

145. Erhardt P. Case study: 'Esmolol stat'. In: Fischer J, Ganellin R, editors. Analog-Based Drug Discovery. Weinheim, Germany: Wiley-VCH; 2006. p 233–246.

146. Sarver J, White D, Erhardt P, Bachmann K. Estimating xenobiotic half-lives in humans from rat data: influence of log P. Environ Health Perspect 1997;105:1204–1209.

147. Ward K, Erhardt P, Bachmann K. Application of simple mathematical expressions to relate the half-lives of xenobiotics in rats to values in humans. J Pharmacol Toxicol Methods 2005;51:57–64.

148. Erhardt PW, Matos L. Selective beta-adrenergic receptor blocking agents. In: Fischer J, Ganellin R,editors. Analog-Based Drug Discovery. Weinheim, Germany: Wiley-VCH; 2006. p 193–232.

149. Fischer J, Ganellin CR. Analog-Based Drug Discovery. Weinheim, Germany: Wiley-VCH; 2006.

150. Zaroslinski J, Borgman RJ, O'Donnell JP, Anderson WG, Erhardt PW, Kam ST, Reynolds RD, Lee RJ, Gorczynski RJ. Ultra-short-acting beta-blockers: a proposal for the treatment of the critically ill patient. Life Sci 1982;31: 899–907.

151. Kam ST, Martier WL, Mai KX, Barcelon-Yang C, Borgman RJ, O'Donnell JP, Stampfli HS, Sum CY, Anderson WG, Gorczynski RJ, Lee RJ. [(Arylcarbonyl) oxy]propanolamines. 1. Novel beta-blockers with ultrashort duration of action. J Med Chem 1984;27:1007–1016.

152. Bodor N, Buchwald P. Soft drug design: general principles and recent applications. Med Res Rev 2000;20:58–101.

153. Gringauz A. Introduction to Medicinal Chemistry. How Drug Act and Why. New York: Wiley; 1997. p 435.

154. Erhardt PW. Benzylamine and dibenzylamine revisited. Syntheses of N-substituted aryloxypropanolamines exemplifying a general route to secondary aliphatic amines. Synth Commun 1983;13:103.

155. Erhardt PW, Woo CM, Gorczynski RJ, Anderson WG. Ultra-short-acting beta-adrenergic receptor blocking agents. 1. (Aryloxy) propanolamines containing esters in the nitrogen substituent. J Med Chem 1982;25:1402–1407.

156. Erhardt PW, Woo CM, Matier WL, Gorczynski RJ, Anderson WG. Ultra-short-acting beta-adrenergic receptor blocking agents. 3. Ethylenediamine derivatives of (aryloxy) propanolamines having esters on the aryl function. J Med Chem 1983;26:1109–1112.

157. Farah AE, Alousi AA. New cardiotonic agents: a search for digitalis substitute. Life Sci 1978;22:1139–1147.

158. Weishaar RE, Quade M, Boyd D, Schenden J, Marks S, Kaplan HR. The effect of several "new and novel" cardiotonic agents on key subcellular processes involved in the regulation of myocardial contractility: implications for mechanism of action. Drug Dev Res 1983;3: 517–534.

159. Braunwald K, Mock MB, Watson JT, editors. Congestive Heart Failure: Current Research and Clinical Applications. New York: Grune and Stratton; 1982.

160. Erhardt PW. In search of the digitalis replacement. J Med Chem 1987;30:231–237.

161. Withering W. An account of the foxglove and some of its medical uses: with practical remarks on dropsy and others diseases. In: Mosby CC, editor. Cardiac Classics. St. Louis (original publication in 1785); 1941. For review, see (1985) *Journal of Clinical Pharma- cology* 25.

162. Schnettler RA, Dage RC, Grisar JM. 4-Aroyl-1,3-dihydro-2H-imidazol-2-ones, a new class of cardiotonic agents. J Med Chem 1982;25: 1477–1481.

163. Shah PK, Amin DK, Hulse S, Shellock F, Swan HJ. Inotropic therapy for refractory congestive heart failure with oral fenoximone (MDL-17,043): poor long-term results despite early hemodynamic and clinical improvement. Circulation 1985;71:326–331.

164. Davey D, Erhardt PW, Lumma WC Jr, Wiggins J, Sullivan M, Pang D, Cantor E. Cardiotonic agents 1. Novel 8-aryl substituted imidazol[1,2-a] and [1,5-a]-pyridines, and imidazo[1,5-a] pyridinones as potential positive inotropic agents. J Med Chem 1987;30: 1337–1342.

165. Hagedorn AA III, Erhardt PW, Lumma WC Jr, Wohl RA, Cantor E, Chou YL, Ingebretsen WR, Lampe JW, Pang D, Pease CA, Wiggins J. Cardiotonic agents 2. (Imidazolyl) aroylimidazolones, highly potent and selective positive inotropic agents. J Med Chem 1987;30: 1342–1347.

166. Lopaschuk GD, Rebeyka IM, Allard MF. Metabolic modulation: a means to mend a broken heart. Circulation 2002;105:140–142.

167. Pang DC, Cantor E, Hagedorn A, Erhardt P, Wiggins J. Tissue specificity of cAMP-phosphodiesterase inhibitors: rolipram, amrinone,

milrinone, enoximone, piroximone and imazodan. Drug Dev Res 1988;14:141–149.

168. Erhardt PW, Hagedorn AA III, Davey D, Pease CA, Venepalli BR, Griffin CW, Gomez RP, Wiggins JR, Ingebretsen WR, Pang D, Cantor E. Cardiotonic agents 5. Fragments from the heterocycle-phenyl-imidazole pharmacophore. J Med Chem 1989;32:1173–1176.

169. Davey D, Erhardt P, Cantor E, Greenberg S, Ingebretsen WR, Wiggins J. Novel compounds possessing potent cAMP and cGMP phosphodiesterase inhibitory activity. Synthesis and cardiovascular effects of a series of imidazo[1,2-a]-quinoxalinones and imidazo[1,5-a]-quinoxalinones and their aza analogues. J Med Chem 1991;34:2671–2677.

170. Lampe JW, Chou YL, Hanna R, DiMeo SV, Erhardt PW, Hagedorn AA III, Ingebretsen W, Cantor E. (Imidazolylphenyl) pyrrol-2-one inhibitors of cardiac cAMP phosphodiesterase. J Med Chem 1993;36:1041–1047.

171. Erhardt PW, Hagedorn AA III, Sabio M. Cardiotonic agents 3. A topographical model of the cardiac c-AMP phosphodiesterase receptor. Mol Pharmacol 1988;33:1–13.

172. Erhardt PW. Second generation phosphodiesterase inhibitors: structure–activity relationships and receptor models. In: Beavo J, Housley M, editors. Cyclic Nucleotide Phosphodiesterases: Structure, Regulation and Drug Action. New York: John Wiley & Sons; 1990.

173. Erhardt PW, Chou YL. A topographical model for the cAMP phosphodiesterase III active site. Life Sci 1991;49:553–568.

174. Kling J. From hypertension to angina to viagra. Mod Drug Discovery 1998;1:31–38.

PROCESS DEVELOPMENT OF PROTEIN THERAPEUTICS

James N. Thomas
S. Sam Guhan
Dean K. Pettit
Amgen, Inc., Seattle, WA

1. BIOTECHNOLOGY: INTRODUCTION

The term "biotechnology" dates from 1919, when Karl Ereky, a Hungarian engineer, defined it as "any product produced from raw materials with the aid of living organisms" [1]. One modern definition proposed by the United Nations Convention on Biological Diversity defines biotechnology as "any technological application that uses biological systems, living organisms, or derivatives thereof, to make or modify products or processes for specific use" [2].

The discovery of the structure of DNA by Watson and Crick in 1953 resulted in an explosion of research in molecular biology and genetics. Molecular biologists devised methods to isolate, identify, and clone genes as well as to mutate, manipulate, and insert them into other species.

Today, biotechnology has far-reaching applications in health care, environmental sciences (bioremediation), agriculture, and synthesis of products such as biofuels and biodegradable plastics. From its relatively humble beginnings in the late seventies and early eighties, the biotech industry is a thriving business today. According to BIO (Biotechnology Industry Organization) as of Dec 31 2005, there were 1415 biotechnology companies in the United States of which 329 were publicly held [3]. Biotechnology companies, in the context of the pharmaceutical industry, are those companies that focus primarily on large-molecule protein therapeutics, often referred to as "biologics."

1.1. Process Development of Protein Therapeutics

Biologics are often large molecules (usually proteins) and much more complex compared to traditional small molecular weight pharmaceuticals. A typical biologic is hundreds of times larger than the compounds found in most pills or tablets. Moreover, most currently available biologics have to be directly introduced into the blood stream, either via subcutaneous injections or via intravenous infusions to achieve their intended affects.

In contrast to small-molecule therapeutics that can be made through organic synthesis, large-molecule proteins must be made using biological systems. Proteins, in addition to their primary structure (amino acid sequence), have secondary, tertiary, and quaternary structures. This three-dimensional structure is often critical for the activity of the protein and is an important difference between small molecular weight drugs and protein therapeutics. Figure 1 schematically illustrates the size difference between small-molecule and protein therapeutics, illustrating the significant difference in complexity between these drug classes.

The development of a manufacturing process for the production of a biological is very complex and is typically comprised of four parts:

(1) *Upstream Development.* Describes the activity of integrating a gene of interest (cDNA) into a host and generating a recombinant cell line or strain. It includes subsequent steps to isolate a single cell clone or microbial colony and to preserve such recombinant cells in the form of frozen cell banks. Upstream development also refers to the process through which the product is made, typically by fermentation or cell culture using live organisms (e.g., bacteria, mammalian cells, etc.).

(2) *Downstream Process.* In this step, the product of interest is isolated from the various contaminants and impurities present in the cell culture or fermentation broth and concentrated to the level required.

(3) *Formulation Development.* This describes the development of the optimum product containing solution, and its delivery device.

(4) *Analytical Development.* Development of analytical methods and technologies that are used throughout process development in order to provide critical

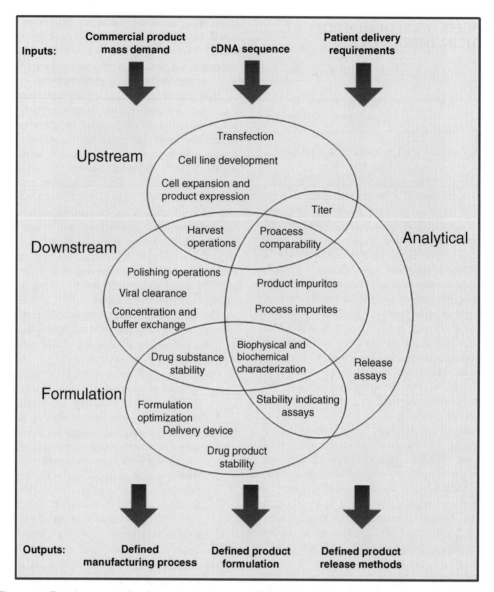

Figure 1. Protein process development requires coordinated integration of upstream, downstream, formulation, and analytical activities. Primary functional activities are indicated within each circle and areas of overlapping circles indicate shared activities. The major inputs required to initiate protein process development and the key deliverables or outputs from these activities are indicated at the top and bottom of the figure, respectively. (This figure is available in full color at http://mrw.interscience.wiley.com/emrw/9780471266945/home.)

information on the quantity and quality of the product.

The integration and overlap of the four basic components of protein process development are shown in Fig. 2.

A challenge to the organization and execution of protein biologics process development is the interdependence of each part of the process on the other. The upstream process has a large impact on the downstream process. It is not just the quantity or type of

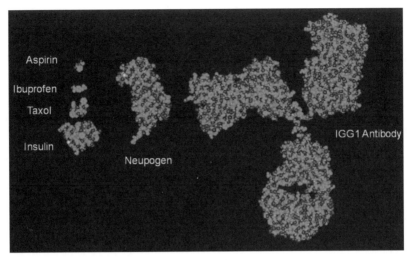

Figure 2. Size comparison of small-molecule and protein therapeutics, illustrating the significant difference in complexity between these drug classes. (See color insert.)

contaminants that need to be removed, but also the challenges created when the upstream process produces more product than the downstream system can handle. The downstream/purification process is highly dependent on the final formulation as well. The last step of the downstream process delivers the drug substance, preferably, in the final formulation. The use of appropriate analytical tools is important to monitor product quality during process development [4]. Any treatment during processing may subject the protein to stress (pH, concentration, salt, etc.), which would potentially impact chemical and/or structural characteristics.

The following sections discuss each of these individual parts of process development in detail.

2. UPSTREAM PROCESSING

Since the first gene cloning experiments were successfully performed by Boyer and Cohen at the University of San Francisco [5,6], various expression systems have been used to produce recombinant proteins of potential therapeutic value. The variety is staggering in complexity, ranging from simple microbes such as *Escherichia coli* to complex transgenic whole animal systems. While a complete review of all the available systems is beyond the scope of this chapter, a few examples will be cited to underline the numerous expression tools available to the modern scientist. The majority of marketed recombinant proteins, especially glycoproteins, are expressed using mammalian cell culture systems. With this in mind, the bulk of this section will be devoted to a more in-depth discussion of current mammalian cell culture technology.

2.1. A Plurality of Expression Systems

Many prokaryotic expression systems have been evaluated over the years for their ability to make recombinant proteins. Some of these include various strains of *Bacillus* and *Streptomyces* [7,8] and of course the workhorse system *E. coli* [9–11]. The desire has been to develop host strains with the ability to secrete properly folded protein, but yields employing this approach are still relatively low. Prokaryotes such as *E. coli* produce enormous quantities of recombinant protein, especially if the protein is protected from host proteolytic degradation within inclusion bodies. Protein in inclusion bodies is usually misfolded and insoluble, so one of the first downstream processing steps involves refolding the protein into a correctly folded and active form. The yields associated with refolding steps are usually low for complex proteins containing several disulfide bonds. *E. coli* is also missing cellular enzymes for adding carbohydrate to proteins. Many human proteins are glycoproteins and require the proper amount and type of carbohydrate for full in vivo activity. All prokaryotic and most eukaryotic systems are incapable of proper glylcosolation and are best used for expressing proteins that are not dependent on carbohydrate for potency or optimum pharmacology. Recent attempts have been made to address some of the drawbacks of using microbial systems such as *E. coli* by optimizing secretion signal sequences, coexpressing chaperones, foldases, selecting protease deficient mutants [8,12], optimizing expression vectors, codon usage [13], and engineering cell physiology [14].

A number of eukaryotic expression systems have been used to produce recombinant proteins during the last several years. The most popular of these have been yeast systems that include *Pichia pastoris, Saccharomyces cerevisiae, Kluyveromyces lactis,* and *Hansenula polymorpha* [8]. Yeast has rapid doubling times, which can grow to high cell densities in fermentors and can secrete complex proteins. *S. cerevisiae* (baker's yeast) has been a popular expression host for 25 years, but *P. pastoris* has recently become the yeast system of choice due primarily to its ability to secrete complex proteins [15–20]. Expression levels in yeast can be very high for some proteins (>10 g/L), but in general these systems have greater difficulty expressing complex glycoproteins than mammalian-based expression systems. While yeast adds carbohydrate to proteins, these organisms tend to attach more primitive high mannose species causing glycoproteins to be cleared more rapidly in vivo. A traditional method for getting around this concern is to remove N-linked glycosylation sites from the protein if this can be done without negatively impacting activity or immunogenicity. Still another, more sophisticated approach has been developed by GlycoFi, a wholly owned subsidiary of Merck. By knocking out 4 genes in *P. pastoris* responsible for glycosylation, and inserting 14 heterologous genes, GlycoFi has successfully expressed human glycoproteins with fully complex, terminally sialyated *N*-glycans [21]. This allows the expression of essentially fully "human" glycoproteins, with the added advantage of greater homogeneity of glycoforms.

Another eukaryotic expression system used for expression of a variety of glycoproteins is based on efficient transduction of insect cells with baculoviruses [17,22]. The host range of these viruses is restricted to invertebrates, primarily insect cells from moths and butterflies [7], making the system safe for use in manufacturing. Because insect cells are higher eukaryotes, cellular machinery is available for performing proper posttranslation modifications in most cases. Glycosylation patterns are still somewhat different than human, although some have addressed this issue by engineering insect cell lines to express enzymes necessary for proper mammalian carbohydrate structure [23]. Another challenge is the transient nature of the system; cells need to be transduced in every production phase. This requires preparation of both the cell inoculum and the baculovirus used for transduction [7].

Transgenic systems have been explored over the past several years for use as expression vehicles for recombinant proteins, especially those needed in very large amounts (metric tons) [24]. A wide variety of transgenic systems exist and include expression into the milk of transgenic mammals, the eggs of transgenic chickens, the cocoons of silkworms, and the seeds, leaves, or tubers of transgenic plants [24–29]. The core of the technology lies in the molecular approach for introduction and expression of the transgene in the host organism. Expression is frequently targeted to tissues or organs, such as the mammary glands of cattle or goats, designed for making enormous amounts of milk protein, and easily accessible for harvesting. Because of the longer development cycles and perceived regulatory hurdles of transgenic systems, some companies are developing molecules using transgenics in parallel to their primary development route, (i.e., mammalian cell culture).

While improved yields of conventional production technologies have made many transgenic systems less attractive than in the past, transgenic plants may still hold some long-term promise as production systems of protein therapeutics. Like mammalian transgenic systems, expressed proteins are targeted into plant organs, often concentrated, and readily accessible for harvest. Depending on the location of concentration in the plant, this offers some advantages for downstream processing [30]. If the transgenic protein is expressed in seed crops such as rice, wheat, maize, or peas; stability of the unprocessed seeds can be extremely favorable (>1 year), providing an excellent opportunity for optimizing the efficiency of downstream processing steps [31]. While plants can perform complex posttranslational processing of glycoproteins, they can also add immunogenic plant carbohydrate species at the same N-linked sites used by mammalian systems [30]. The same general strategies are used with plant systems as with organisms such as yeast or insect cells to control glycosylation patterns; either removal of the N-linked sites or expression of mammalian glycosyltransferases in the host plant to humanize plant N-glycans [30,32]. Time will tell whether these approaches are successful and how well these systems can compete with ever improving conventional expression systems.

In the 30 years since the beginning of the modern biotechnology industry, several mammalian expression systems have been used extensively to produce recombinant glycoproteins. Some of the major workhorse systems have been developed from Chinese hamster ovary (CHO), baby hamster kidney (BHK), mouse myeloma (NS0), human embryonic kidney (HEK-293), and human retina (PER.C6) cell lines [33–35]. It is important to note that these cell lines, while derived from normal tissues, are all transformed and can theoretically be passaged indefinitely in culture as opposed to normal diploid cells. The genomes of these cell lines are generally stable, but definitely aneuploid and more adaptable than normal diploid cells. Each cell line can be grown in suspension culture making scale-up in conventional bioreactor systems straightforward. Important qualities in a host cell line are several: the ability to grow in suspension culture to high cell densities with high viability; relatively fast doubling times (24 h or less); easy to transfect with foreign genes; ability to grow in chemically defined media; low protease production; not susceptible to transformation by viruses that are human pathogens; and ability to perform posttranslational modifications, such as glycosylation, similar to humans.

Of the cell line options, CHO cells have become the overwhelming choice of many biotechnology companies for producing complex glycoproteins, including recombinant antibodies. NS0 myeloma cells have been used extensively for the production of recombinant antibodies, but less frequently for the production of other complex glycoproteins due to a less desirable carbohydrate profile [36].

There are several CHO cell lines in use today in the industry, originally derived from CHO-K1 cells characterized by Kao and Puck in the 1960s [37]. Their laboratory created auxotrophic mutants for nutrients such as proline and glycine to study mammalian genetics [38–41], and performed the foundational work needed for the development of the *dhfr*- (dihydrofolate reductase-deficient) selection and amplification system used so widely today. Urlaub and Chasin [42] extended this work by deriving variants of CHO-K1 deficient in expression of *dhfr* by mutating CHO-K1 cells. The popular DXB11 cell line was derived from these mutational studies and while this cell line can revert to *dhfr*+, the frequency is very low. Further work by this laboratory [43] created other CHO-K1 mutant cell lines such as DG44, a double deletion mutant for *dhfr*, containing no functional copies of the hamster gene. CHO cells can be further adapted for enhanced expression or improved growth under a variety of culture conditions [44,45].

The importance of having a good host cell line was foundational for the beginning of modern pharmaceutical biotechnology, and particularly for the production of complex glycoproteins. CHO cells have remained a popular expression system because of the many positive qualities mentioned previously. Foreign genes can be efficiently transfected, selected and then amplified using *dhfr* as a selectable and amplifiable marker [46]. DXB11 or DG44 cells deficient in *dhfr* expression cannot be cultured in selective media lacking glycine, hypoxanthine, and thymidine [42]. Functionally, *dhfr* is the enzyme responsible for the formation of intracellular tetrahydrofolic acid, a cofactor for one-carbon transfers important in several cellular biosynthetic reactions. The role of *dhfr* is particularly important in the formation of precursors for DNA synthesis. Cotransfection of *dhfr* deficient CHO cells with both the foreign gene of interest and the gene coding for *dhfr*, followed by culturing these cells in selective media lacking glycine, hypoxanthine, and thymidine will select for cells expressing both *dhfr* and the foreign gene of interest. Only cells coexpressing *dhfr* and the foreign gene of interest will survive. After selection, amplifying the number of foreign gene copies can be accomplished by exposing the cells to methotrexate (MTX), an efficient binder and inhibitor of *dhfr*. Increasing the concentration of MTX, usually in a stepwise fashion, forces cells to express higher levels of the gene for survival. Since the *dhfr* gene and the foreign gene will generally integrate into the same region of the cell genome, amplifying *dhfr* will also result in coamplification of the foreign gene.

Other selectable and amplifiable markers have been used to increase the expression of foreign genes in CHO and other cell lines. For example, glutamine synthetase (GS) has been used to express foreign genes in cell lines such as CHO-K1 and NS0 myeloma [47,48]. This enzyme is essential for the conversion of glutamate and ammonia into glutamine, a nutrient for most transformed cells. While some cells can be adapted to grow in the absence of glutamine by increasing the endogenous expression of glutamine synthetase, other cells, such as myeloma, appear to have an absolute requirement for exogenous glutamine. The GS system is particularly useful in these cell lines as only those transfected with exogenous GS will survive in media lacking glutamine. Gene amplification can be accomplished by exposing the transfected cells to methionine sulphoximine (MSX), a specific inhibitor of GS. A similar selectable and amplifiable marker is asparagine synthetase (AS), based on the *E. coli* gene for this enzyme that catalyzes the conversion of aspartic acid to asparagine [49]. This system has been used in cell lines that express endogenous AS by using albizziin for initial drug selection. The exogenous bacterial AS is more resistant to inhibition by albizziin than the endogenous mammalian AS gene, so selection of transfected cells in albizziin containing medium favors the exogenous marker over the internal. Amplification is accomplished using a second AS inhibitor more effective against the *E. coli* enzyme. Other selectable and amplifiable markers have been used to enhance recombinant gene expression in animal cells, but most have not found widespread practical application in the industry.

2.2. Creating Recombinant Production Cell Lines

Now we will turn to a more specific discussion of the approaches used to create recombinant production cell lines using the components of the CHO system.

2.2.1. Expression Vectors Transcription of mRNA is generally not rate limiting for the expression of recombinant proteins in most current production cell lines [50]. This is due to advances in understanding gene regulation at the molecular level, and the design of powerful vectors used for driving expression.

cDNA, coding for the gene of interest, is packaged with other DNA sequences to facilitate efficient transcription and translation in the host cell. This DNA package is called the expression vector or plasmid and also contains sequences to facilitate replication in bacteria (generating sufficient plasmid for transfection). Our understanding of the regulation of gene expression in mammalian cells has advanced over the past 30 years, but there is still much we do not understand. Our knowledge has come from the study of how viruses replicate in host organisms. Some of the most critical elements for achieving high expression of foreign proteins in mammalian cells are viral promoter regions that influence transcription. They contain components such as the Goldberg-Hogness box (TATA sequence) approximately 25–35 bp upstream of the RNA initiation site, as well as an enhancer region that stimulates transcription rate. Some promoters have host specificity, working well in one cell line to drive expression but ineffective in others [51]. Promoters such as human cytomegalovirus immediate-early gene promoter and enhancer (huCMV P/E) are powerful drivers of transcription in multiple cell hosts. In CHO cells, the huCMV P/E has been used to drive expression of both the foreign gene of interest and the selectable marker from the same transcript [52,53]. This promoter can be transactivated by certain adenovirus proteins such as E1A, so one strategy for enhancing expression has been to include the gene coding for E1A in the expression vector [47]. Early adenovirus proteins function as transcription enhancers; so, when coexpressed with the gene of interest, they can serve as powerful stimulators of protein production from just a few integrated gene copies.

Expression vectors usually contain other elements to enhance either transcription or translation of mRNA in the host cell. The cap and poly(A) tail are regulatory determinants that establish the translational efficiency of mRNA [54]. The IRES sequence (internal ribosomal entry site) from encephalomyocarditis virus also enhances efficiency of translation [55]. Other elements, such as the EASE (expression augmenting sequence element) sequence, increase the speed of obtaining high expressing cell clones after transfection, and increase the stability of integration in the absence of selective pressure [53]. EASE is a CHO genomic sequence discovered by mapping the flanking sequence of a single integrated gene copy expressing relatively high levels of a recombinant protein. There are likely many such elements in the mammalian genome that serve as regulatory elements for gene expression [56–61]. Vectors can also be directed to a known active transcriptional locus in the cell genome by incorporating a targeting sequence to facilitate integration into the active site through homologous recombination [33,50,62–64].

2.2.2. Transfection and Gene Amplification

Efficient translocation of a plasmid through the cell membrane is essential for creating high-yielding cell lines. This process is called transfection, and a variety of methods have been employed to accomplish this over the years. The three most widely used methods for transfection of plasmid DNA are calcium phosphate precipitation, electroporation and the use of liposomes (lipofection) [65,66].

Calcium phosphate transfection involves the coprecipitation of calcium chloride, phosphate buffer and plasmid DNA to facilitate uptake of DNA via endocytosis [67,68]. This technique has been used extensively with a number of different cell types, and has been modified to improve its efficiency.

Electroporation is a simple method that uses an electrical field to create reversible pores in the plasma membrane of cells [69]. While this method is applicable to a variety of cell types, it must be carefully optimized for best results. Efficient transfection while maintaining high cell viability is a challenge for this technique.

Lipofection is based on the properties of cationic lipids (modification of lipids with quaternary amines) to package negatively charged DNA for transport through nonpolar membranes. Transport is achieved through either endocytosis or direct membrane fusion [70]. Optimization of the ratio of DNA to lipid is probably the most important aspect of this method.

Transfection can be accomplished by placing both genes on the same vector or by using multiple vectors and cotransfecting. Fortunately, even cotransfected genes tend to integrate into the same site on a chromosome, creating the opportunity of coamplification of the amplifiable marker gene with the foreign gene of interest. This happens because large regions of the chromosome near the amplifiable marker are replicated during drug selection.

Due to the complexity of gene regulation in the mammalian genome and the random nature of integration, the selection process prior to amplification must be stringent. Parachuting foreign DNA into a complex mammalian genome subjects foreign gene expression to a variety of positional influences. Various methods have been used to minimize these influences including the use of matrix attachment regions and anti-chromatin repressor elements to insulate against positional effects [58–60,63]. As described earlier in this chapter, the *dhfr* system functions as a highly stringent selection and amplification method due to the availability of *dhfr*-cell lines and the critical role this enzyme plays in the synthesis of cellular DNA. After initial selection in medium lacking glycine, hypoxanthine and thymidine, cells are usually exposed to increasing concentrations of MTX in a stepwise fashion [46]. Copies of the integrated foreign DNA can number in the thousands [46,65] after amplification. Integration is generally into one site or locus in the genome where, after exposure to selective pressure, gene amplification occurs causing elongation of the integration site chromosome [71]. By FISH (fluorescence in situ hybridizationanalys) is other "satellite sites" can be observed in the genome, probably arising through homologous recombination. Upon removal of selective pressure (MTX), these "satellite sites" usually disappear suggesting they are unstable.

One approach to ensuring high levels of stable expression from a small number of gene copies has been the use of targeted integration methods. Since some loci in the genome are more actively transcribed than others, the ability to target integration into a highly transcribed locus becomes attractive. Two primary systems have been successfully used in this approach, the Cre/loxP system and the Flp/FRT system [62–64,66]. The former system is from the bacteriophage PI and the latter is from yeast. Identification of the active transcriptional locus is accomplished by screening clones after transfecting with a vector containing the target site (lox or FRT), amplifiable marker and reporter gene. Selection of cells with high expression of the reporter gene provides evidence of transcriptional activity. Co-integration of the recombination target site facilitates future targeting of that site using recombinase enzymes (either Cre or Flp) that catalyze excision/integration of the gene of interest.

2.2.3. Cloning and Banking the Production Cell Line After amplification, cell pools expressing the desired quantity and quality of the therapeutic protein are cloned. This has been traditionally accomplished by limiting dilution of the transfected cell pools into 96-well plates. The objective is to dilute the pool to obtain a single cell per well, which can be confirmed by microscopically examining each well for colony growth. Typically clones are screened for growth and productivity, but the trend is to perform increasingly rigorous product quality analysis at this stage as well. Selecting the best clone may be the single most important activity in process development for ensuring high productivity and acceptable product quality. Due to the importance of this step, other methods have been developed to improve the selection of the best clone. Many of these are based on fluorescence-activated cell sorting (FACS) techniques as a way of visualizing and sorting the best expressing cell clones [72,73].

Since the cell line is a controlled reagent and a critical part of the manufacturing process, it must be banked and rigorously tested to confirm identity, purity, and suitability [74]. Banks are qualified by testing for the presence of adventitious agents including bacteria, fungi, mycoplasmas and viruses. Since recombinant cell lines are most often employed for the production of biotherapeutics, genetic stability is also an important characteristic and is determined at the molecular and cellular level.

A two tiered cell banking system is generally used for the manufacture of most biotherapeutic products [74]. This system is composed of a master cell bank (MCB) and a working cell bank (WCB). The MCB is rigorously characterized and serves as the core repository of the recombinant cell line. Many WCBs can be made from the MCB, ensuring plenty of well-characterized cell stock for the lifetime of drug manufacturing. Because the WCB is derived from the MCB after only a few passages, a much smaller subset of tests are required for qualification. An additional cell bank is usually established at the limit of *in vitro* cell age for determining genetic stability that spans the production window of the manufacturing process.

2.3. Upstream Process Formats

The primary objective of any manufacturing process should be to produce a high-quality product in the most cost effective way to meet market demands. As a way to achieve this, a wide variety of manufacturing formats have been used in the biotechnology industry. The choice of production format is influenced by a number of factors including the cells being cultured, the molecule being produced and the overall productivity required of the process. A complete review of all process formats used to culture mammalian cells is beyond the scope of this chapter, but a few of the most popular systems will be covered for illustration.

Every process format must meet the basic requirements for culturing mammalian cells including physical demands such as dissolved oxygen and temperature as well as nutritional and hormonal requirements. A cloned recombinant cell line will consistently produce a known amount of product/cell/unit time. To increase the concentration of secreted product in the cell culture medium therefore requires an increase in viable cells per volume of medium per unit of time. This can be achieved in a variety of ways and has therefore led to multiple choices in upstream process formats.

Some cells require attachment to a matrix or surface for growth, while other cells grow readily in suspension. This characteristic will have a significant influence on the format used for production.

2.3.1. Attachment Dependent Cultures Many cell types can be cultured in suspension attached to small particles called microcarriers [75]. This method was originally developed for normal diploid cells requiring attachment to a matrix for growth [76–78], but has also been used to culture transformed cell lines such as BHK, VERO, MDCK, and CHO cells. At one time, an enormous variety of microcarriers were available that varied in material from macroporous gelatin to ceramic or glass. One of the original and most popular materials is diethylaminoethanol (DEAE)—Sephadex A50 ion exchange resin, optimized for both charge density and the size to support efficient cell attachment and growth [79]. One advantage of microcarriers is the ability to culture attachment dependent cells in modified STRs (stirred tank bioreactors). Special impellors and sparging systems are required to minimize shear in bioreactors used for microcarrier culture. Another potential advantage of this method is the increased sedimentation velocity achieved when cells are attached to microcarriers. This allows easy separation of culture supernatant from cells, making this type of culture amenable to repeat batch or continuous processing.

Other formats for culturing cells attached to a surface include roller bottles, hollow fiber bioreactors and fluidized-bed systems to name a few. Roller bottle culture is perhaps the simplest system in its basic form; involving the attachment of cells to the surface of a rotating disposable plastic bottle [78]. This system can be scaled-up using an automated format, but due to the complexity and size of the automation, this system should only be used if needed to insure production of a product with a specific product quality profile or a product with very low production requirements.

Hollow fiber reactors have been used for many years to produce molecules such as antibodies from hybridoma cells. This format is particularly useful when only a few grams of product are needed, but can be especially challenging when hundreds of kilograms are required. Cells are cultured in the extracapillary space while medium is circulated through the intracapillary space or lumen to deliver oxygen and nutrients [80]. This system is relatively simple and will concentrate the product with the cells in the extracapillary space. Oxygen and nutrients will diffuse through the capillary to the cells while CO_2 and other metabolites will diffuse away from the cells into the perfused medium [81,82]. Scale-up can be a challenge due to the formation of significant diffusional gradients along the length of the hollow fiber reactor, particularly at the distal end.

Fluidized bed systems involve the immobilization of cells on some type of matrix and then "fluidizing" this matrix within the confines of a usually cylindrical reactor. The advantages of such system are improved mass transfer and mixing within the reactor containing as much as 50% volume as matrix and cells. One such system employed macroporous collagen microcarriers weighted to a specific gravity of about 1.6 with metal particles. Cells grew both inside the approximately 500 µm macroporous carriers and on their surface and at the proper flow rate remained fluidized within the reactor [83]. Fluidization was not a result of perfusion, but was due to flow through an external loop recirculating medium through the reactor. Fresh medium was added and spent medium was removed through a second pumping system. The recirculation loop also contained an oxygenator and inline oxygen probes to control the level of oxygen in the reactor and help exchange out excess CO_2. This reactor format can be scaled-up more easily than hollow fiber reactors, but although eloquent in design, is complex to operate and may not be suitable for a commercial manufacturing environment.

2.3.2. Suspension Cultures Perhaps the simplest and most widely used process format is batch culture of suspended cells. This is a process where cells are cultured in bioreactors of increasing size in order to inoculate the full scale production reactor, usually a STR (stirred tank reactor). Originally cell culture STRs were modified versions of bacterial fermentors, but over time the design has evolved to better meet the environmental needs of mammalian cells. This includes low shear impellors and baffling for improved mixing, special spargers for delivering dissolved oxygen and robust probes and ports for the measuring and sampling needs of mammalian cells. The integrity of the sterile envelope is critical for any mammalian cell culture system, so very sophisticated utilities for CIP (clean in place) and SIP (steam in place) are usually associated with cell culture STRs. For batch culture, cells are inoculated into the production STR and allowed to grow and produce product for several days before harvesting. Cells will grow until they reach either limiting concentrations of an essential nutrient or inhibiting concentrations of a metabolic byproduct. At this point, cells will begin to lose viability and the volumetric production rate of the culture will decrease. A modification of this approach is the fed-batch format. In this case cells are inoculated into the production STR and cultured in batch format; at periodic intervals additional nutrients are added that increase the total number of cells per volume and the length of the culture period. Other approaches (such as adjustment of temperature and pH) are also used to limit the production of metabolic byproducts in order to increase the total number of viable cell days in the production reactor.

There are many variations in the use of STRs for culturing mammalian cells, but the principle of optimizing the total number of viable cells per unit volume of medium per unit time (cells/volume/time), is the same. Some process formats constantly perfuse fresh medium into the production STR while retaining, through various retention methods, very high cell density cultures in the bioreactor [83].

Over the past several years, the most popular system for production of recombinant proteins has been a fed-batch process using CHO cells cultured in large scale STRs of several thousand liters. The productivity of these systems continues to improve, and has now surpassed the capacity of many older commercial downstream processing facilities without substantial facility modifications. While continuous processes are more eloquent in design and better mimic an *in vivo* environment for cultured mammalian cells, the relative simplicity and high productivity of current fed-batch systems will be difficult to supplant in commercial manufacturing operations.

2.4. Upstream Process Development

Cell culture process development of large molecules attempts to integrate the properties of the production cell line, the process format and the physical and chemical environment of the cell culture into a robust manufacturing process. Some fundamental goals of upstream process development, particularly as applied to batch and fed-batch formats are as follows:

- To create an optimum environment for culturing cells to high viable cell densities for an extended period of time in the production bioreactor.
- To optimize conditions for expressing high levels of a therapeutic protein with the desired critical quality attributes (CQAs).
- To facilitate efficient harvesting and downstream processing.

Specific productivity has increased several-fold over the past 25 years due to advances in molecular biology and our understanding of gene expression. The importance of creating a high expressing cell line with robust growth suitable for commercial manufacturing cannot be overemphasized. A survey of the essential elements for doing this have already been covered in this

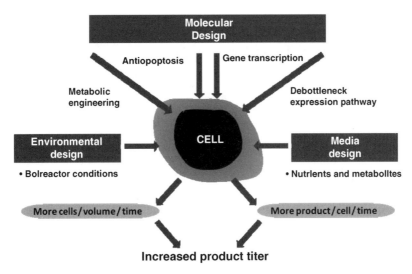

Figure 3. Greater productivity requires improved fundament scientific understanding leading to more cells/volume/time and/or more product/cell/time. (This figure is available in full color at http://mrw.interscience.wiley.com/emrw/9780471266945/home.)

chapter, so the next section will be devoted to a short discussion of considerations for optimizing culture conditions for increasing viable cells per unit volume per unit time. Figure 3 denotes the two dominant approaches for improving productivity.

2.4.1. The Physical Environment As already mentioned, the bioreactor must create the optimum physical environment for culturing mammalian cells. It should be recognized that although cell lines used to produce biotherapeutics today are transformed from a genotypic and phenotypic perspective, they still require a physical environment that is similar to that found *in vivo* in the animal of origin. In other words, the pH, temperature, dissolved oxygen, dissolved carbon dioxide, osmolality, and other physical parameters must be carefully optimized and controlled for the best performance of the culture. In some cases, there are advantages to controlling some of these parameters outside the normal range to improve viability, product quality or specific productivity; but this must be carefully defined during the course of process development.

2.4.2. Medium Design Complex cell culture media, the nutrient source for cells, are composed of over 50 different components including amino acids, vitamins, trace elements, salts, lipids, carbohydrates, growth factors, buffering components and shear protecting agents [84]. These factors provide the building blocks and stimulus for driving cell growth and the protein product being made. In addition, medium components are interlinked with the physical environment of the culture influencing pH, osmolality, the tolerance to bubble and mechanical shear. Requirements are similar, although not quite the same as those found in the host animal. For example, 13 amino acids are generally required in cell culture while only about 8 or so are required for the host animal [85]. The past 50 years have witnessed a progression of cell culture medium from an ill-defined mixture to one that is chemically defined (i.e., all chemicals are known). There are many commercial media available that can serve as starting points for optimization [84], but the ultimate medium used in a process will depend on the desired cell density and process format. In the case of fed-batch cultures, the goal is to provide an enriched base medium and feeds to support the highest possible cell density for the desired culture period. This is a complex optimization exercise composed of understanding the utilization rate for nutrients, their concentration change relative to other nutrients, the conversion rate into metabolic byproducts and the solubility limits for nutrients in both the base medium and the feed medium.

Amino acids that serve as the building blocks for proteins must be carefully optimized to maximize culture performance. Amino acid transporters in cell membranes transport multiple amino acids, providing a competitive situation whenever differential utilization rates cause significant imbalance over time [86]. Metabolic byproducts of amino acids may also be toxic for some cells, and in batch or fed-batch cultures the concentration of these toxic compounds will build over time. Due to the complexity of optimizing the nutritional and physical environment of cell culture processes, the industry is beginning to rely more heavily on DoE (design of experiments) and other statistical approaches and tools. While the work continues to be challenging, if done well, it provides an opportunity to push cell culture densities to much higher levels once nutrient utilization, metabolic production kinetics, and interactions are well understood.

2.4.3. Understanding Metabolism

Understanding metabolism is closely linked to medium optimization as metabolic byproducts are simply a conversion of available nutrients and oxygen into other chemical forms. Metabolic flux through various biochemical pathways within the cell can, to some degree, be regulated or influenced by a variety of controllable factors within the culture environment [86,87]. The complexity of optimizing batch or fed-batch cultures stems from the cell's expectation that the system contains a detoxifying organ, such as a liver, to convert or repackage metabolic byproducts into either a usable form or a detoxified form for elimination. Controlling cell culture inputs that are nutritional, physical and sometimes hormonal provides a mechanism for preventing metabolic byproducts from reaching toxic or inhibitory levels. The basis for successfully performing this type of work resides in understanding physiology and intermediary metabolism in the whole animal. It is only from this context, and from an understanding of the tweaks to normal physiology brought about by cell transformation, that one can productively approach this area.

2.5. Future Trends

While there are many potential areas for improving technology associated with production of biotherapeutics, one that promises significant benefit is continued improvement of the expression host. The complexity of biological systems, even at the cellular level, can be daunting. This poses a significant barrier to understanding and manipulating the host cell line for improved performance. Even so, there is growing interest and work in this area as academic and industrial laboratories use genomic and proteomic tools to improve host cell lines. The fundamental premise is that understanding the limitations or metabolic bottlenecks in cells will provide an opportunity for removal or improvement through cellular engineering [88].

In some cases experimental observations have led to engineering of host cells to improve cell viability by preventing or decreasing apoptosis [89–91]. The level of lactic acid in cultures can be controlled by transfecting and over expressing the GLUT5 transporter in CHO cells [92]. CHO cells are normally cultured with glucose as the source of carbohydrate, but when engineered with GLUT5 they can be cultured with fructose, and the slower rate of transport prevents the buildup of lactate in the culture. Culturing CHO cells at low temperatures (30–33°C) induces many "cold stress" genes and can induce recombinant protein expression. Over expressing a specific cold stress protein (CIRP) in a recombinant cell line and culturing at 37°C led to a 40% increase in recombinant protein production in suspension culture [93].

Another approach for this work is to compare gene expression under different cell performance conditions by genomic analysis, proteomic analysis or by using a combination of both approaches [94–97]. Analysis of this data is often complex as one finds multiple genes upregulated or downregulated in a given culture condition. Interpretation of expression patterns must rely on a fundamental understanding of molecular and cellular biology, cell physiology, and the biochemical pathways of metabolism. While this can be a challenging exercise, the power of understanding how cells behave at the molecular level under a variety of conditions will provide critical information for designing better cell hosts in the future.

3. DOWNSTREAM PROCESSING

3.1. Overview of Downstream Processing

The production of protein therapeutics with living cells, such as bacteria or mammalian cells, poses different challenges for purification than small molecule NCEs (new chemical entities) made using organic synthesis. Impurities such as solvents or isomers found in the manufacture of an NCE are "inherent" to the process. Impurities found during the production of biopharmaceuticals are more varied in composition and come from many sources, including the host itself. An added complication is that the protein product is susceptible to environmental conditions such as heat and pH, and can become unstable either physically (e.g., denaturation and aggregation) or chemically (e.g., deamidation).

The use of bioseparation for isolation of the target protein from a fermentation broth or cell culture supernatant [98] is called downstream processing. The material produced upstream is generally dilute and contains the protein in low concentrations (e.g., in a mammalian system, the product may only be present at a concentration of 1 g/L of cell culture fluid). The majority of cell culture fluid is an aqueous solution containing undesirable dissolved components such as soluble proteins from the host cell, imperfect forms of the product (e.g., aggregates), lipids, nucleotides, cell culture reagents, bacterial endotoxins, and other components. Downstream processing must successfully remove these impurities and contaminants, including a significant portion of the water, to render a concentrated final product at the desired purity. Typically the final product is formulated at about 10–100 g/L, corresponding to a 10–100-fold increase in concentration. Viral inactivation/removal steps are also required to ensure a low risk of viral infectivity for mammalian derived products [4]. Additional key attributes of a successful downstream process include scalability, regulatory compliance, and favorable economics. Most importantly, the process should be robust and be insensitive to small perturbations [99]. A variety of separation techniques are used for purification. Chief among these are centrifugation and filtration for primary (crude) separations, chromatography, and membrane filtration for polishing operations (fine separations). Together, chromatography and filtration membranes enable the downstream scientist to provide purified product at the target concentration and formulation conditions, and with the necessary product quality [100].

Some separation techniques play a key role in both analytical and purification process developments. In its analytical role, a separation technique is used to quantify and characterize the protein of interest. In purification, the chief aim is to isolate and purify the product. This difference in the goal often leads to very different operating conditions for the same separation technique. As an example, the chromatographic process used to characterize a protein employs a smaller column, typically 2–5 mm in diameter, with an emphasis on high resolution and high throughput. In preparative or process chromatography, the column sizes are much larger, ranging from 30 cm to 2 m in diameter, and the focus is on maximizing product recovery at a desired purity level while minimizing cost.

A good purification process must produce the final product at the desired purity and concentration, and be cost efficient. Studies indicate that downstream costs can be as high as 80% of the total production costs [101–103]. An increase in cell culture titer will shift this cost even further to downstream processing [4,104]. An efficient downstream process becomes crucial at this stage, underscoring the importance of continuous process improvements [105].

3.1.1. Typical Impurities and Contaminants

A more complete description of the classes of impurities that must be separated from the recombinant therapeutic protein is summarized in Table 1. Impurities can be classified as those that are process related versus those that are product related. Process related impurities typically require clearance to prescribed levels that are based on product safety. These include large cell debris and fragments that are removed during primary separation

Table 1. Common Impurities Encountered During Protein Process Development That may Require Removal and/or Analytical Assay

Process Related Impurities	Purification Modalities	Analytical Assay[a]
Cell debris and fragments	Depth filtration and centrifugation	Varied
DNA	Ion exchange (IEX) chromatography, flocculation	QPCR, others
Host cell proteins	IEX chromatography, affinity, hydroxyapatite (HA)	Immunoassay
Protein A	IEX chromatography	Immunoassay
Process reagents (e.g., antifoam, media components, etc.)	Varied	Varied
Virus	Low pH inactivation, nanometer filtration, chromatography	Varied, transmission electron microscopy (TEM)
Endotoxin	IEX chromatography	Limulus amebocyte lysate (LAL)
Salts, buffers	Ultrafiltration/diafiltration (UF/DF)	Liquid and gas chromatography
Potential product related impurities		
Aggregates and protein particles	IEX chromatography, hydrophobic interaction (HIC), HA, precipitation	Chromatography and other biophysical methods
Glycosylation variants	IEX chromatography	Chromatography
Misfolded forms of the protein	HIC, IEX chromatography	Varied
Structural isoforms	Refolding, chromatography (various)	Varied biophysical methods
Posttranslational modifications (e.g., deamidation, oxidation, clips etc.)	Chromatography (various)	Mass spectrometry and other biochemical methods

[a] See Section 5 for discussion of analytical methodologies.

processes; DNA derived from host cells; proteins that are encoded and coexpressed by the host cell; material leached from chromatography resins; reagents that are added as processing aids such as antifoam or media components; viruses that are potential contaminants in animal cell cultures, and can be inadvertently introduced via raw material sources [106]; endotoxins introduced either by *E. coli* derived fermentation systems or by inadvertent bacterial contamination during processing; and salts and buffers.

Product related impurities are those that occur as a result of physical or chemical modification of the primary species in the product. Whether these species are considered as impurities or simply product variants depends on their impact on product safety and efficacy. For example, a product isoform that impacts relative potency and pharmacokinetic profile would be considered an impurity whereas a chemically modified species that does not measurably impact product performance may not. Product impurities require removal, whereas product variants or isoforms require thorough characterization. Some potential product related impurities or variants include aggregates and protein particles that are undesirable; glycosylation variants such as alpha-galactose (a glycan linkage found in animal cells that is immunogenic in humans) and others that may impact product circulation half-life; misfolded forms of the protein that have the same primary sequence as the product of interest, but do not have the correct secondary and tertiary structure; structural isoforms that may occur naturally or as a result of improper disulfide bridging; and other posttranslational modifications including deamidation, oxidation, cyclization, and others. The topic of product characterization will be discussed in greater detail in later sections of this chapter.

3.1.2. Design of Downstream Processing A good purification process must take into account properties of the target protein, properties of impurities, and contaminants (both product related and nonproduct related), scalability, and process fit into a manufacturing facility. An immense number of possible combinations [107] exist, but heuristics can be used to provide a good starting point such as [98]

- perform the crude separations first,
- understand the properties of the molecule (e.g., the protein's pI will decide which pH will be best for separation (propensity to aggregate, etc.),
- lower the volume of the feed stream as early as possible,
- use high-resolution separation steps as late in the process as possible, and
- use orthogonal steps to maximize the separation of molecules with diverse physical and chemical characteristics (e.g., affinity, ion exchange, and hydrophobic interactions).

3.1.3. Key Classifications in Downstream Processing Downstream processing is generally classified as primary recovery (also known as harvest operations) and secondary recovery (also known as purification or polishing steps). The objective of primary recovery is to bring the feedstock into a state suitable for the application of more refined purification methods by removal of cells and cell debris [108]. The distinction between primary recovery and purification is at times unclear (e.g., when highly discriminating bioaffinity methods such as protein A chromatography are used to selectively extract the product from a crude feedstock). However, even in these applications primary recovery is required to prevent unrefined feedstock from fouling costly bioaffinity materials. Polishing steps take the product stream from the primary recovery operation and isolate the protein of interest using orthogonal techniques.

These operations determine the quality attributes of the final product (e.g., aggregate levels, product related impurities). The two key technologies used in this area are chromatography and filtration membrane processes.

An important concern of primary recovery is the removal of particulate materials that are incompatible with downstream operations. A most common method for removing particulates of a defined size is filtration through membranes of an absolute pore size. However, the soft and easily deformable solids from biological production can create virtually impermeable filter cakes leading to very low capacity on absolute filters. Open three-dimensional structure depth filters are well suited for clarification of biological feedstocks. However, they can become overwhelmed by the debris from high-density cultures typical of today's processes. Centrifugation and microfiltration have therefore been adopted in biotechnology to perform an initial removal of cellular solids [109]. In these operations, the fragile nature of cells poses a challenge as the mechanical shear generated tends to micronize a portion of the cells. In microfiltration, mechanical stress caused by repeated recirculation of the retentate can lead to rapid fouling of the microfiltration membranes. Most mammalian cell-based processes therefore employ centrifugation, where shear forces caused by rotational velocity of the unit will cause generation of fine particles, but within a range that allows subsequent removal by depth filtration.

The most common approaches used for purification and polishing are chromatography and membranes. Process chromatography, in particular, continues to be the workhorse and forms the basis of most downstream operations in biopharmaceutical production [110]. Chromatography (termed in 1906 by Mikhail Tswett) [100] literally means "color drawing" [111] since it was originally used for separation of natural pigments on filter paper. From this relatively humble beginning, chromatography has become the central technology for both protein characterization and downstream processing. The wide use of chromatography is due to the large variety of stationary phases or chromatographic supports available. For the most part, these can be broadly categorized as either affinity or nonaffinity approaches [112].

Affinity chromatography refers to the specific interaction of a ligand to a well-defined site on the protein [112]. This is usually an excellent method to effect separation since the interactions are specific and strong. However, there are trade-offs in terms of high cost of the ligand and, in some cases, availability of the specific ligand at commercial scale. The most common use of affinity chromatography in therapeutic protein production is the use of protein A for the purification of monoclonal antibodies and Fc fusion proteins.

Nonaffinity techniques include a host of chromatographic modes such as ion exchange, hydrophobic interaction, hydoxyapatite, metal chelate, and size exclusion chromatography that use a wide range of stationary phases. The three most common of these are detailed below.

Ion exchangers (IEX) are the most common adsorbents used in process chromatography in biotechnology. They exploit the net charges of the protein at a given pH. If the operating pH is below the pI of the protein, the protein has a net positive charge and *vice versa*. The most common ion exchangers are anion (positively charged support) and cation (negatively charged support). Within these categories are a variety of subcategories (e.g., weak cation exchangers, strong cation exchangers, etc.). In general, in a bind and elute mode, the protein is loaded under low salt (conductivity) conditions resulting in its binding to the stationary phase, and is then eluted using a higher salt concentration. The choice of salt, conductivity and pH of the operation are critical parameters when optimizing these separations. In a "flow through" mode, the protein of interest flows through the column while impurities are held behind. In general, the throughput (amount of protein purified per time per liter of resin) is higher for the flow through mode but resolution may not be as good as the bind and elute mode.

Hydrophobic interaction chromatography (HIC) exploits the interactions of hydrophobic patches on the surface of the molecule with hydrophobic adsorbents. In analytical chromatography and in small molecule separations, reverse phase chromatography is frequently used, where the interaction of the product with the stationary phase is so strong that it requires organic solvent to elute the product from the column. These are typically harsh conditions for protein isolation and may cause irreversible loss of activity. In protein purification, the use of HIC is performed at more gentle conditions. The protein is typically adsorbed on the stationary phase using high salt conditions and is eluted off the column in a decreasing salt gradient. Just as in IEX, HIC can be performed in either bind and elute mode or flow through mode.

Hydroxyapatite chromatography (HA) or ceramic hydroxyapatite (CHT) uses calcium phosphate as both the ligand and the base matrix. CHT is a mixed-mode support with functional groups consisting of calcium, phosphate and hydroxyl groups $(Ca_5(PO_4)_3OH)_2$. In a bind and elute mode, the protein is loaded at or near neutral pH in a low ionic strength phosphate buffer and eluted with a higher phosphate buffer or salt (depending on the separation mechanism). CHT is generally used to decrease the level of aggregate, DNA, host proteins and endotoxin.

The chromatography system may be operated in a number of ways. The most common are step gradients and linear gradients, where the material is bound to the column at conditions where it is held by the chromatographic stationary phase, and then eluted using an appropriate buffer (e.g., higher salt, change in pH, etc.), removing the protein of interest from the column. There are other modes such as displacement chromatography [113], but these are less commonly used in commercial protein production.

3.2. Protein Purification in *E. coli* Systems

Heterologous protein accumulation in *E. coli* systems often appears as inclusion bodies (IBs) due to the high expression levels and inability of bacterial cells to properly fold recombinant human proteins [114–116]. If protein renaturation is simple, IBs offer a source of highly concentrated product and can provide an excellent starting point for a very straightforward downstream process. On the other hand, problematic protein renaturation can make downstream process development difficult [117].

There are excellent examples of protein folding techniques and advances in this area [114,115,117–121]. De Bernardez Clark et al. [122] suggest the low yields across the refold step may be due to formation of inactive misfolded species, particularly aggregates, and the authors detail various techniques to "inhibit aggregation side reactions to ensure efficient *in vitro* protein folding." Cowgill et al. [123] present a practical introduction to protein refolding and discuss scale-up effects. They also discuss the emerging and promising technology of high pressure refolding. Another good review on this topic is the publication by Sahdev et al. [124], where the authors discuss soluble protein expression. It must be noted, however, that finding the optimum conditions for correct protein refolding is still relatively empirical and considered somewhat of "an art."

The quality of the IBs is critical in overall process yield and product quality. Wong et al. [125] have shown that maximizing the IB recovery during centrifugation does not guarantee high overall yield. The contaminants present in the IB "paste" can lead to a lower yield, depending on the nature of the contaminants.

Once the protein is refolded, it undergoes chromatographic purification to remove both product-related impurities (misfolds and others) and contaminants (e.g., host cell proteins). A typical flow chart describing purification of a protein from an *E. coli* host is detailed in Table 2.

Table 2 is only one example of a process path that can be used for purification of a protein from IBs. There are many examples where a different strategy has been employed to obtain the final product produced in an

Table 2. Typical Flow Chart Describing Purification of Protein from *E. Coli*

Purification Process	Description
Cell product intermediate (CPI)	Washed Inclusion Bodies
↓	
Solubilization	Dissolution of washed inclusion body cell product intermediate (CPI) to release the desired protein into solution. Denaturation of the desired protein to eliminate elements of secondary, tertiary, and quaternary structure. Reduction of protein disulfide bonds to ensure desired protein is fully denatured and present in monomeric form.
↓	
Refold	Renaturation of the desired protein to achieve the correct secondary, tertiary, and quaternary structure. Typically the most chemically complex operation in the purification process but necessary to eventually produce biologically active protein from inclusion bodies.
↓	
UF/DF	Concentration of the relatively dilute clarified refold solution. Reduction of low molecular weight solutes derived from CPI. Reduction of low molecular weight chemical components that would inhibit desired precipitation of impurities.
↓	
Precipitation	Species separation based, predominantly, on solubility. Condensation of impurities into relatively large precipitates that readily sediment during centrifugation or filtration enabling their removal.
↓	
Capture column (chromatography -1)	Volume reduction and gross removal of DNA, host protein, and product related impurities
↓	
Polishing columns (generally 2)	Removes DNA, host proteins, and other product related impurities to deliver product at desired purity.
↓	
Formulation UF/DF	Concentrate and buffer exchange the protein to formulation buffer.

E. coli system. As an example, Khalizadeh et al. [126] discuss an approach for purification of recombinant human interferon-γ expressed in E. coli that involves purification of the protein in a denatured state over two columns and a final column purification after the refold step.

3.3. Protein Purification in Mammalian Systems

Monoclonal antibodies are the most common class of therapeutic proteins expressed in mammalian systems. These molecules share common features such as framework of the Fab region and the Fc region. Due to structural similarity, many companies are adopting platform approaches to process development and manufacturing, applying a predefined sequence of unit operations and analytical methodologies to multiple molecules. For example, monoclonal antibodies bind with high specificity via the Fc region to protein A derived from *Staphylococcus aureus*. This property is exploited by employing protein A impregnated resins resulting in one the most powerful purification steps in monoclonal antibody processing. This step provides high specificity and high yield and forms the backbone of the downstream platform process approach.

While the isoelectric points (pI) of antibodies generally vary from slightly acidic through the basic region, polishing steps can generally be developed using ion exchange chromatography. Other polishing options that lend themselves to a platform approach include hydrophobic interaction chromatography and hydroxylapatite.

A platform approach to process development does not mean that the purification processes will be identical for all antibodies. The properties of the molecule are varied enough (different pI, hydrophobic nature, etc.), that the same purification process may not work across multiple molecules, and may not be the optimum process. The platform is used to define the overall scheme, provides ranges of operating conditions for each unit operation and limits the experimentation required to develop the final process.

3.3.1. Harvest Operations The most common technique used in the harvest step of mammalian cell culture processes is the use of centrifugation for cell removal [108]. Centrifuges exploit the density differences between the solid particles, the surrounding liquid medium, and the centrifugal force to achieve separation. Although many large-scale centrifuges are available, the most common are disk-stack separators, which have proven suitable for clean-room operation and have sufficient clarification performance to remove smaller sized cell fragments. An added benefit is the ability of disk-stack centrifuges to remove solids from the centrifuge bowl either continuously or intermittently, allowing uninterrupted clarification of large volumes of cell culture feed in relatively short time periods.

The performance of the centrifuge is determined by the interaction of multiple factors. The most common operational control parameters are rotational velocity of the bowl, residence time, and feed interval. In addition, the centrifuge design itself (e.g., pump type, back pressure control, mechanism of discharge, etc.) can affect the separation performance. Perhaps, the most important factor is how the feedstock fluid is accelerated as it enters the bowl. This is a complex hydrodynamic problem that is influenced by feed zone design, bowl speed and the feed flow rate. All of these factors affect the degree of cell disruption that can result in the generation of particles too small to settle in the centrifuge.

Centrifugation is a good primary separation step due to its rapid, crude separation ability, but is not the complete harvest solution. Typically, small particles that escape the centrifuge must be removed by a subsequent depth filtration step. In fact, the depth filter is where purification scientists often experience difficulty (e.g., high filter areas required), even if the cause lies upstream in the process.

It is difficult to develop good scale-up and scale-down models for centrifuges. Shear effects can cause large deviations from theoretical centrifuge capacity and the changes in such effects are little understood during scaling. Furthermore, the role of the cell culture on centrifuge performance is not well understood, but is likely significant. An ongoing area of research is in understanding scale-down performance of centrifuges [127,128].

One way to address harvest capacity is by changing the character of the debris (i.e., particle size). The use of flocculants has a long history in diverse industries such as wastewater, chemicals and food. This technology is now being applied to the cell culture harvest. Use of flocculants can result in a range of performance changes to the process; these vary from being a "patch" on an existing process to significant purification of impurities not previously affected during recovery (e.g., virus, HCP, and aggregates). Chitosan (a polycationic linear polymer) has been shown to improve clarification throughput of a large-scale cell culture harvest without negatively impacting mAb recovery or purity [129].

Tangential flow microfiltration (MF) is another technique used for clarification of mammalian cell culture [108]. An advantage over centrifugation is a particle-free harvest stream that requires minimal additional filtration prior to further processing [130]. Fine particle generation during MF is no worse than during centrifugation, but leads to failure of the operation by fouling the MF membranes. The situation is exacerbated in that the usual mode of overcoming membrane fouling, increasing the cross-flow velocity to sweep the membrane free of debris, only amplifies the problem by increasing the intensity of shear and hence particle generation.

A depth filter is a porous medium that can retain particles throughout its matrix [108]. While the primary mode of action of depth filtration is to trap particles, many modern filters have been shown to have an additional adsorptive capability, useful for host cell protein and DNA reduction. Therefore, depth filtration could potentially combine filtration for particulate removal with adsorptive binding for the removal of contaminants [109]. Some understanding exists in the use of adsorptive properties of these depth filters, but that knowledge remains largely empirical at present and further development is required.

3.3.2. Polishing Steps Table 3 describes a common process used for purification of monoclonal antibodies. The development, operation, and validation of chromatographic processes used in monoclonal antibody purification at industrial scale has been recently reviewed by Fahrner et al. [131]. In this paper, the focus is on a commonly used three-column purification (protein A, followed by cation exchange and flow through anion exchange).

Table 3. Typical Process Used for Purification of Monoclonal Antibodies

Purification Process	Description
Cell product intermediate	Harvested cell culture supernatant. The supernatant has undergone primary purification through centrifugation and depth filtration. Removal of cells and cell debris. Some impurity removal on the depth filter.
↓ Protein A	Affinity chromatography. Able to obtain high purity (>95%) and lower volumes (up to three to ten times concentration) across this step. Gross removal of DNA and host cell protein.
↓ Viral inactivation	Typically done at low pH; inactivates enveloped viruses such as XMuLV.
↓ Depth filter	For clarification of feed stream in preparation for downstream polishing chromatography steps.
↓ Polishing chromatography step 1 and step 2	An example is CEX chromatography (bind AND elute) and AEX chromatography flow through [131]. Reduction of DNA, host cell proteins, leached protein A, aggregates, and process reagents as well as viral clearance.
↓ Viral filtration	Typically a parvovirus filter (~20 nm). Removes viruses, both large (XMuLV) and small (MMV).
↓ Formulation UF/DF	Concentrate and buffer exchange the protein to formulation buffer.

Protein A affinity chromatography is a very powerful first step for the purification of monoclonal antibodies. The use of protein A as the ligand is common for this class of proteins due to its affinity for the Fc portion of antibodies, with an affinity constant K_D reported to be 70 nM [4,132]. Product purities of >95% (by SEC-HPLC) are possible after just one pass across this column.

Affinity of protein A for antibodies functions across a broad pH range, and conductivity is not a critical parameter; therefore, loading the column directly from the cell culture fluid is possible [4,133]. A benefit of the high-affinity interaction is that one can employ a broad range of washing conditions to ensure high purity of the eluted product. These wash steps can have the dual advantage of increasing product purity as well as extending the life time of these expensive resins [134]. In general, the product is eluted off the column using an acidic pH (typically 3–4). The column is then stripped (acidic solution around pH ~ 2) and regenerated.

A variety of protein A resins are available for commercial use. Hober et al. [135] detail, a summary of the most commonly used protein A affinity media in antibody production. Resins differ in their backbone chemistry, the two commonly used base matrixes are polymer (agarose based) and silica glass (controlled pore glass).

Due to the nature of the protein-based ligand, the protein A resin cannot be regenerated using harsh cleaning agents, potentially impacting the lifetime of the resin. Recent advances have provided more base-stable resins, which allows for cleaning using caustic solutions, such as 0.1–0.5 M NaOH [134,136,137].

Although the protein A step is extremely powerful, it is not without development challenges. The need to elute the column at low pH may induce aggregation in some mAbs. Shukla et al. [132] provide data from 14 different molecules that indicate problems of aggregation and precipitation occur frequently during protein A chromatography. It is also important to ensure that leached protein A from the column is captured and removed in the subsequent purification steps [138]:

Although the product from the protein A step has a purity of >95%, additional steps are required to achieve target purity. Additional chromatographic steps are generally required focused on reducing host cell protein impurities, aggregates, clipped species, DNA, leached protein A, and other contaminants.

These column steps are generally chosen from unit operations such as cation exchange (CEX) chromatography, anion exchange (AEX) chromatography, HIC, and less commonly, HA. A common process used in mAb production is the use of a cation exchange column followed by an anion exchanger [139]. Pete Gagnon [140] has detailed the various polishing methods available for mAb purification. Typically, one of these steps are run in flow through mode (where the product does not bind to the resin but flows through, while certain impurities are retained behind).

While similar, monoclonal antibody products are, of course, not identical. Specific chromatography resins and operating conditions are chosen based on intrinsic molecule properties and impurities to be separated. Differences in subclass and variable region sequences contribute to variations in molecular properties (e.g., charge, hydrophobicity, and other heterogeneity) that will impact purification process design [141]. A cation exchange step, for instance, can be used for depletion of HCP and leached protein A, but will not be as efficient as HIC for aggregate removal.

Viruses are potential contaminants of concern in animal cell cultures, and can be inadvertently introduced via sources such as serum-derived raw materials, contaminated proteins added to the nutrient broth, infected production cell lines, or accidental contamination during bioprocessing. The downstream process must ensure that these potential contaminants are effectively cleared by removal and inactivation [106]. Even in cases where no virus is introduced, manufacturers still need to demonstrate through validation that the process has significant capability to remove both enveloped and nonenveloped viruses.

The FDA Q5A guidance document [142] requires "demonstration of the capacity of the production process to inactivate or remove viruses" in order to assure safety of products produced by mammalian cell culture. The guidance document mentions that effective clearance may be obtained from a combination of inactivation and separation steps.

Chromatographic steps used in the downstream process each provide some degree of viral clearance. For example, greater than 4 logs of retroviral and parvoviral clearance have been achieved for monoclonal antibodies by use of AEX chromatography flow through steps [132,143]. In addition, the process also typically includes specific viral clearance by low pH viral inactivation and nanometer filtration. The depth filter step in the process also results in some viral clearance. Although this is currently not recognized by regulatory agencies as a robust orthogonal method for viral clearance, it does provide an additional safety margin [144].

Viral inactivation steps must ensure a permanent reduction of viral infectivity [4]. Many methods are possible to effect viral inactivation—chemical (low pH), heat, UV irradiation, etc. A comparison of different techniques used is detailed in other reviews on this topic [4,145]. Due to its operational simplicity, low pH (acid) inactivation is the most commonly used technique in antibody purification, the mechanism of inactivation is most effective against large enveloped viruses.

The most robust unit operation currently used to remove viruses is size-based nanometer filtration. It has been shown to be scalable and robust [146]. Typically, 50 and 20 nm pore sizes are used in the process—although the smaller pore size is being increasingly used due to its ability to remove both large viruses (e.g., x-MuLV) and small virus particles (e.g., MMV) [144].

The capability of each unit operation is validated by viral clearance studies using appropriate scale-down models. In general, process validation is performed in the commercial manufacturing facility. However, for safety, financial and technical reasons, viral clearance validation is performed in small-scale models ($\sim$1/1000 scale) taking care to ensure that relevant parameters are appropriately scaled to those used at full-scale production [147]. Model viruses that can be detected and quantified are used to characterize the capacity of the downstream process to clear adventitious viral agents. Xenotropic murine leukemia virus (x-MuLV) and murine minute virus (MMV) are common model viruses used to test the viral clearance capacity of each unit operation [144]. Zhou and Tressel [148] reviewed the operations for viral clearance in mAb downstream processing operations, including viral clearance strategies used in early stage (or Phase I processes) to commercial (biological license application) filings.

After purification, the product is delivered in the final formulation buffer by buffer exchange [132]. This is typically done by using UF/DF membranes [149]. Several parameters are important in the development of this step including the membrane type, the transmembrane pressure (TMP), cross-flow rate, final product concentration, etc. In the past, final product concentrations were lower (typically less than 30 mg/mL), but concentrations are increasing to 70–150 mg/mL. This has implications for solubility and viscosity of proteins and needs to be dealt with on a product-by-product basis. As an example, salts may be used to decrease the viscosity in some cases, but may not be an optimum solution in all instances.

3.4. Future Trends

Typically, downstream scientists develop the process at bench-scale by using either batch binding experiments or small chromatographic columns of $\sim$1–2 cm in diameter (or $\sim$5–15 mL column volume). This is, by necessity, a batch operation and limits the total number of data points that can be obtained during development. Advances in instrumentation technology as well as use of platform conditions allow for high-throughput purification development. Industry uses high-throughput 96-well plate formats for rapid purification development using a minimum of resources [150–153].

To make therapeutic protein at commercial scale the downstream process should be easily scalable. The process should also be capable of utilizing the facility in the most optimum way to maximize asset utilization and lower costs. While this is a strategic business objective, the solution will come from science and engineering. As an example, processes that utilize a minimal number of buffer tanks at lower volumes with faster cycle times will contribute to lower costs and smaller capital investment. The term "process intensification" is defined as the ability of the process to utilize the production facility in the most optimum manner. A specific example would be the development and use of a resin that supports higher capacity and can operate at higher flow rates.

Another example of process intensification would be the use of a two-column process versus the standard three-column process. The two-column process reduces raw material costs, supports a smaller plant footprint, and will utilize less pieces of capital equipment such as buffer tanks [154]. In addition, having fewer unit operations allows for fewer errors and improved success rates. Finally, the use of a two-column purification process may enable an easier semi-continuous (tandem process) approach. There are several examples of successful application of the two-column approach [141,155,156] "Connected processing" or tandem operation is one possible future direction. This is the connection of two consecutive unit operations without an in-process pool and/or tank between the two. This has been demonstrated in both microbial and mAb purifications [157].

In microbial systems, a key area of focus is the protein refold or renaturation step. This unit operation is not only a critical part of the purification process but also one of the least understood and modeled areas. Research on increasing refold concentrations and yields [158,159] is necessary to make microbial systems more productive and to decrease cost of goods manufactured (COGM).

Protein A has been well established as the affinity resin of choice in mAb purification. However, the technology development around this resin is not stagnant. Many companies are engaged in research to improve protein A resin as well as develop alternatives to this affinity adsorbent. There is on-going work to develop "protein A mimetics" that are synthetic ligands that mimic the interactions of the protein product with the natural ligand [112,160]. The use of more conventional resins, such as ion exchangers and mixed mode resins, has also been utilized as the first step in mAb purifications [154,160].

There have long been arguments that the cost of purchasing and operating protein A resin is high [161]. While costs may be more significant for early stage programs the selectivity and platform capability that protein A offers is essentially unmatched. Base stable resins and development of effective column regeneration processes commonly allow these columns to be used for >100 cycles in commercial applications, significantly reducing the per batch cost. For the time being, protein A remains the resin of choice as the capture step for most monoclonal antibody purifications.

Techniques such as simulated moving beds (SMBs) [162] and sequential multicolumn chromatography (SMCC) have the potential to utilize lower solvent (buffer) consumption and deliver higher productivities. There is research ongoing for the extension of this technology in bioprocessing, especially in the capture step of mAb purification. This technology evaluation is in its early days and needs to be evaluated both at lab-scale and during scale-up.

Disposable technologies have long been of interest in the industry as they can provide cost and operational efficiency without cleaning, lifetime and storage validations. The most impactful benefit is the use of disposables to increase the rate of equipment turnaround in highly scheduled plants. While single-use disposable systems are familiar in filtration, they are not common in packed-bed chromatography. A recent area of interest is membrane absorbers that may be used in a single-pass approach [163].

A popular alternative to ion exchange chromatography is membrane chromatography in flow through mode. The use of this technology has been demonstrated in mAb purifications. It is a rapid, cost-effective unit operation that is easy to scale up [104,164]. In contrast to bead chromatography, the binding capacity of these systems is independent of flow rate since membranes are not diffusion limited [130,165]. This technology is still under evaluation in terms of its overall benefits versus costs compared to the packed bed approach. An advantage with single-use membrane chromatography is reduced development and validation costs since there is no column packing or cleaning validation involved [166]. A more detailed evaluation of this technology is provided elsewhere [163,166,167].

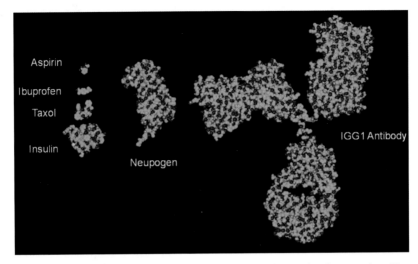

Figure 2 (Chapter 7). Size comparison of small-molecule and protein therapeutics, illustrating the significant difference in complexity between these drug classes.

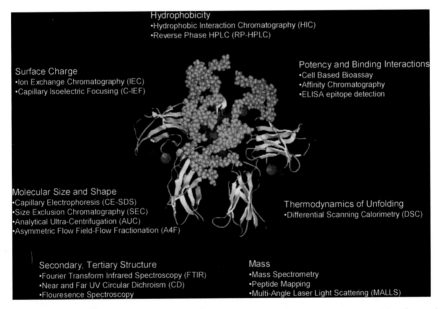

Figure 4 (Chapter 7). Depicted in this figure is a glycoprotein. The protein, defined by the amino acid sequence, folds to form secondary structures such as alpha-helices (red cylinders) and beta-strands (yellow ribbons) which are connected by ordered loops (green threads). The glycosylation attached to the protein is shown by blue and green spheres. Some common physical and chemical attributes of recombinant glycoproteins and an array of analytical tools that are utilized to characterize these proteins are listed.

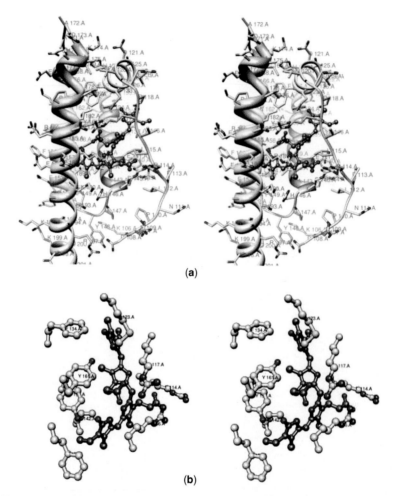

Figure 5 (Chapter 13). Stereo diagram illustrating the topology of subdomain IB and the placement of bilirubin within the subdomain IB-binding crevice. Bilirubin is bound in extended conformation stabilized by salt bridges from arginines 114, 117, and 184 to the proprionic carboxyls of bilirubin. The elimination of key salt bridge from Arg 114 by the substitution of Gly, accounts for the reduced affinity of Yanomama-2 to bilirubin and the fatal consequence in neonates as noted by Putnam and colleagues [14].

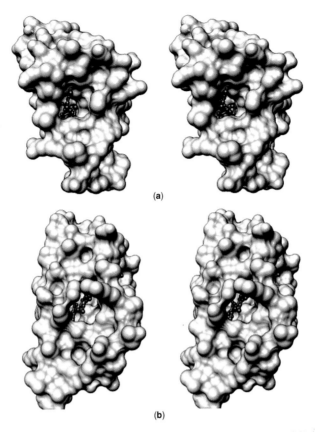

Figure 6 (Chapter 13). Stereo views of the surface illustrating the proximal (a) and distal (b) openings to the IB bilirubin complex. Gray: carbon; red: oxygen; blue: nitrogen.

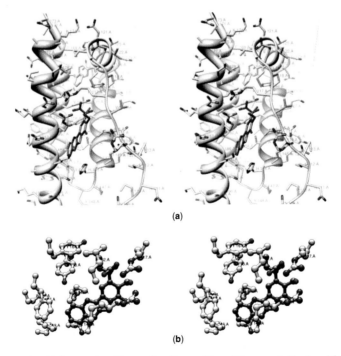

Figure 8 (Chapter 13). (a) Stereo view illustrating the position of the camptothecin within IB (distal view) and (b) details of its binding interaction with Arg 117 and Arg 186.

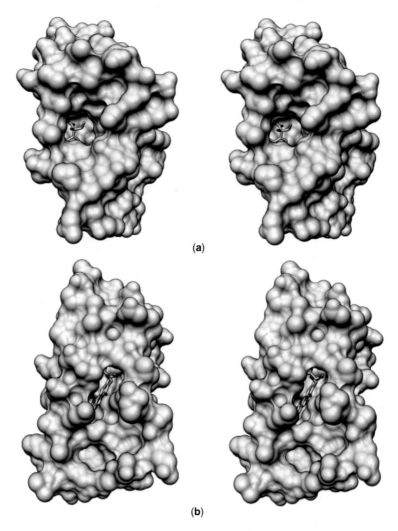

Figure 9 (Chapter 13). Stereo views of the surface illustrating the proximal (a) and distal (b) openings to the IB camptothecin complex. Gray: carbon; red: oxygen; blue: nitrogen; stereo gray: carbon; red: oxygen; blue: nitrogen.

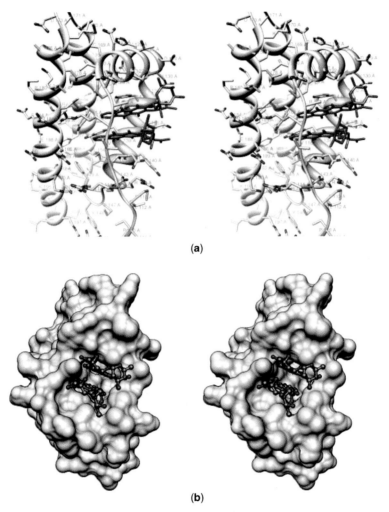

Figure 10 (Chapter 13). Stereo views of (a) the IB idarubicin dimer complex showing the binding site location at the proximal site; and the packing interactions with residues: Tyr 138, Tyr 161, Phe 134, Pro 118, and Leu 115. Note the close association of the pyranosides that suggests the potential for intermolecular hydrogen bonding interaction, possibly through bridging water molecules not identified by the present structure resolution. (b) Stereo view of the IB molecular surface illustrating the enlargement of the proximal site to accommodate the idarubicin dimer.

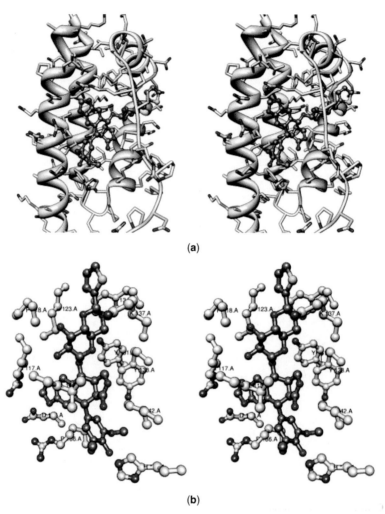

Figure 11 (Chapter 13). Stereo views of (a) the teniposide complex showing the binding site location within the IB-binding site and (b) a selection of key residues surrounding the teniposide molecule.

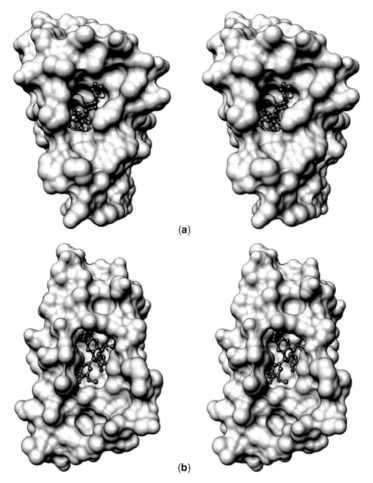

Figure 12 (Chapter 13). (a) Stereo view of the IB molecular surface with views of teniposide within the proximal site and (b) from the distal site. Note the unusually large access to the distal opening created by the rotation of Arg 114 side chain away from the ligand toward L1.

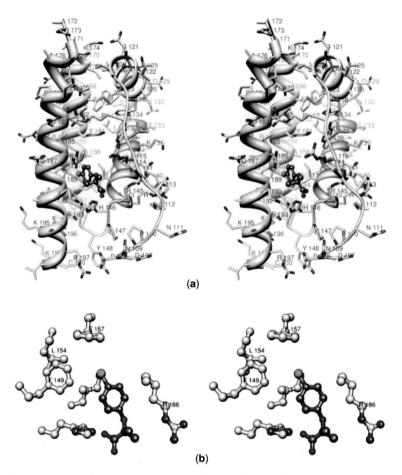

Figure 13 (Chapter 13). (a) Stereo views of the clofibric acid complex showing the binding site location within subdomain IB and (b) the detailed interactions of clofibric acid with His 146 and other key residues.

While chromatography has been the workhorse in preparative separations, it does suffer from certain limitations (e.g., resin availability, column packing, batch operations, etc.). There has been significant research in alternate bioseparation operations [164,168]. However, at least for recombinant antibodies, the current platform is capable of production of up to 10 tons/year [155]. A paradigm shift toward routine use of nonconventional approaches for monoclonal antibody purification in the near future is unlikely.

There is an increasing need to deliver high-concentration formulations. As an example, there are monoclonal antibody products being formulated at >150 g/L. This represents a greater than fivefold increase in concentration from just a few years ago. This trend toward high concentration causes significant challenges to downstream processing with respect to solubility, filter capability, particle formation, and yield optimization. Collaboration with formulation and manufacturing is needed to understand and solve these issues.

4. FORMULATION AND DELIVERY

4.1. Introduction

Protein drug formulation is a challenging and time consuming process. The formulation scientist is faced with the requirement of stabilizing a protein pharmaceutical in an acceptable liquid or lyophilized formulation for storage periods of up to 2 years. This long term stability must be achieved despite the stresses that are exerted on the protein during manufacturing (e.g., copurification with enzymes and other destabilizing compounds, and shear stresses exerted on the protein by vial filling operations), shipping and handling (e.g., agitation and temperature excursions), storage (e.g., temperature, light, excipient/buffer and pH exposure), and delivery (e.g., dilution and short term exposure within an IV bag). The formulated protein must also be compatible with a primary container over this storage period. These contact materials (typically glass vials or syringes) introduce concerns for protein adsorption and denaturation, and can also potentially release extractables and leachables such as rubber stopper monomers and initiators, ions from glass materials, and silicone coatings. The formulation scientist is also challenged by the fact that while candidate formulations may be decided based on data collected from accelerated storage conditions, such as incubation across a range of pH solutions or exposure to temperature extremes over a relatively short period of time, only real-time processing and storage can confirm the stability of a final protein formulation. These facts can lead to extended timeframes for formulation development relative to other process development activities. Review articles and book chapters have discussed general considerations for protein formulation development [169], and more recent literature has focused on the formulation of monoclonal antibodies [170–172].

4.2. Protein Degradation Pathways

The biological activity of proteins is defined by their primary, secondary, tertiary, and quaternary structure. Changes in any of these parameters may lead to diminished product potency, altered pharmacokinetic behavior, or potential increases in immunogenicity. During manufacture and long term storage, proteins can degrade by chemical and/or physical pathways. Early reviews in the literature describe a range of physical and chemical degradation mechanisms that need to be considered when formulating proteins [173–175]. The most likely pathways for degradation of a particular protein cannot be generalized or predicted in advance. Rather, degradation depends on the specific amino acid sequence and structural characteristics of the protein pharmaceutical that requires probing for potential degradation "hot spots".

Chemical degradation of proteins can be classified as either covalent or noncovalent. Covalent degradation pathways that have been described in the protein formulation literature include isomerization and deamidation, succinimide formation, oxidation, cyclization, photodegradation, cysteinylation, glycation, disulfide scrambling, hydrolysis, beta-elimination, carbamylation, and enzymatic degradation from proteases copurified in the drug substance. The primary noncovalent degradation pathways for proteins include denaturation,

Table 4. Some Common Protein Degradation Pathways and Potential Formulation Remedies

Degradation Pathway	Primary Cause	Potential Formulation Remedies
Deamidation	Solution pH; local peptide sequence	Lower formulation pH [178]; lyophilization [218].
Isomerization	Solution pH; local peptide sequence	Raise formulation pH [178]; lyophilization [218].
Succinimide formation	Solution pH; local peptide sequence	Neutral formulation pH [176,177]; lyophilization [218].
Oxidation	Exposure to oxygen, metal ions, solvents	Limit exposure to oxidants and catalysts during processing and long-term storage [181,182]; add antioxidants to formulation (e.g., methionine) [185].
N-Terminal cyclization	Solution pH	Control upstream processing conditions; raise formulation pH [186]; lyophilization [218].
Photodegradation	Exposure to light	Limit exposure to light; add antioxidants [187].
Glycation	Upstream processing conditions	Eliminate reducing sugars from upstream process and formulation [190–192].
Cysteinylation/adducts	Upstream processing conditions	Control upstream and downstream processing conditions (e.g., nutrients, and redox reagents) [188].
Aggregation	Surface hydrophobicity	Control processing conditions; addition of surfactants to formulation [197,206].
	High concentration	Addition of salt or other exdipients (histidine) to formulation [205]; Protein crystal formulation [247].
	Denaturation	Appropriate excipient choice [202,203].

aggregation, and precipitation. Several of the more common degradation pathways will be discussed below and summarized along with potential formulation remedies in Table 4.

Deamidation and isomerization result from the hydrolysis of side-chain amine groups from asparagine or glutamine amino acid residues. The chemical reaction proceeds through a succinimide intermediate to the deamidated or isomerized forms as dictated by pH conditions and local amino sequence. The stability of reactive succinimide intermediates have been described in the formulation literature [176,177]. Deamidation and isomerization have been characterized in great detail, in part because they can have significant impact on protein activity, and in part because they can be controlled by the selection of appropriate pH in the buffering solution (lower pH disfavors deamidation, whereas higher pH disfavors isomerization), or through the development of a lyophilized versus a liquid formulation (low levels of residual moisture disfavors deamidation) [178].

Early work from Patel and colleagues with model peptides showed the effect of amino acid sequence on the rates of deamidation [179]. More recent work describes the impact that local conformational flexibility, imparted by amino acid sequence, can have on the extent of deamidation [180]. The fact that primary sequence can play a large role in deamidation/isomerization may impact decisions on selecting candidate molecules to develop as protein pharmaceuticals.

Oxidation of therapeutic proteins is also known to occur during protein processing and under long term storage conditions. Protein oxidation can occur through several amino acids, including methionine, tryptophan, tyrosine, phenylalanine, and cysteine residues. There are several known catalysts for protein oxidation including heat, photochemical energy, and metal ions [181,182]. Researchers have found that the extent of exposure of labile amino acids can impact rates of oxidation [183] and that oxidation of amino acid residues can lead to conformational changes in proteins [184] that may limit bioactivity. Knowledge about the propensity of a protein to undergo oxidation may influence selection of a container closure system or lead to filling containers with oxygen depleted headspace (both approaches are intended to limit exposure to oxygen), or may call for the addition of an excipient such as methoinine that is intended to serve as an oxygen scavenger [185].

One common instability known for antibodies (since most recombinant antibodies contain glutamic acid or glutamine at their N-terminus) is the cyclization of N-terminal glutamate residues forming pyroglutamate (pGlu) [186]. This reaction may occur during cell culture production as a post-translational modification, or may occur during long term storage in formulation buffer. Pyroglutamate formation is favored at pH 4 and 8, but is less common at neutral pH. The impact of pGlu formation on the biological activity of proteins is likely limited due to the fact that this instability is localized on the N-terminus of the protein in the antibody framework.

Photodegradation of proteins can also occur as a result of light exposure during manufacturing or long term storage. Photoinstability of proteins may lead to degradation including photooxidation that can result in protein backbone cleavage and other byproducts. The major protein photodegradation pathways were recently reviewed by Kerwin and Remmele [187]. Knowledge of photoinstability of a protein product under development may lead to storage recommendations that limit exposure to light.

Two other covalent degradation pathways for proteins that can be induced by the manufacturing process include cysteinylation, and glycation. Cysteinylation occurs as a result of free cysteine binding to unpaired cysteine residues on proteins (other adducts may also be formed through unpaired cysteine) [188]. Structural changes induced by cysteinylation have been shown to impact the rate of protein aggregation and negatively impact bioactivity [189]. These results suggest that cell culture processes where free cysteine is added as a nutrient may need to be monitored and optimized when producing proteins with free cysteines to minimize the potential for protein cysteinylation. Glycation is a condensation reaction between the aldehyde groups of reducing sugars and the primary or the secondary amines of proteins, primarily lysines and the N-terminal amines of proteins. This may occur during cell culture processes that are fed with glucose, or during long term storage of proteins in solutions that contain reducing sugars such as sucrose [190–192].

Protein denaturation describes the process of unfolding of proteins, or perturbing them from their native conformation. Denaturation can occur as a result of stresses such as shear forces applied during a vial or syringe filling process, exposure to a liquid–air interface, or exposure to a polymer IV bag or glass vial surface [193]. Often protein denaturation is a reversible process, however, under some threshold conditions the association of multiple denatured molecules may ultimately lead to protein aggregation, a process known as nucleation. Whether denatured proteins nucleate to form aggregates or reverse back into properly folded monomers depends on solution conditions such as temperature, pH, protein concentration, salt types and concentrations, excipients, surfactants, and other factors [194].

Protein aggregation is arguably the most common and troubling degradation pathway for the process development scientist, encountered in almost all stages of protein drug development. Numerous review articles have been published describing fundamental understanding of the mechanisms by which proteins aggregate [195,196]. Protein aggregation, along with other physical and/or chemical instabilities of proteins, remains one of the major barriers to rapid development and commercialization of potential protein biopharmaceutical candidates. One recent report describes the importance of understanding where aggregation occurs during protein manufacturing and processing and how these processes can be controlled to reduce protein aggregation [197]. Reports from the formulation development literature have also detailed approaches to the prevention of aggregate formation during long term storage. These approaches include the control of solution pH, and the use of excipients, salts, and surfactants [196,198].

Larger aggregates of proteins may even form precipitates that may be visible to the eye and sediment upon centrifugation. Visible and subvisible precipitates and particles have become a topic of increasing interest. In a recent publication from Carpenter et al., the authors suggest that particles may pose immunogenicity concerns and that these particles should be measured and characterized by biopharmaceutical manufacturers [199]. The concern for monitoring particles in protein formulations on the basis of immunogenicity is supported by some reports in the literature [200,201]. Current particle counting instrumentation in the subvisible size range (0.1–10 µm) is limited and characterization of these particles will be challenged by the exceedingly small proportion of protein mass contained within these particles at typical biopharmaceutical concentrations.

4.3. Stability Considerations

Typically, the formulation scientist works to develop a stable protein formulation in stages. In the early stages of development, a period often called "preformulation," some basic properties of the protein are evaluated such as primary sequence (looking for chemical features that may lead to degradation "hot spots"), isoelectric point, glycosylation patterns, size distribution patterns, and potential structural features that may present challenges to protein stability. As a result of this early analytical evaluation, in later stages of development accelerated and real time stability testing is conducted to further determine the primary pathways of degradation. These accelerated studies may involve temperature or pH stressing, agitation, or other means to evaluate pathways of degradation. Knowledge of these accelerated degradation pathways allows for testing of candidate formulations, which have been designed to reduce degradation under real-time storage conditions. Finally, these candidate formulations must be tested for functionality in manufacturing processes such as syringe or vial filling, and in the patient delivery setting.

4.3.1. Liquid Formulations The formulation additives or excipients available for stabilization of proteins include buffers, bulking agents, salts, surfactants, and preservatives [202,203].

The formulation scientist will pay considerable attention to selecting appropriate pH levels and buffer types during formulation development. Optimization of pH and buffer strength has been shown to significantly impact structural and chemical stability of proteins. Investigators have also realized that while providing stabilizing effects, buffers can cause deleterious effects such as stinging upon injection [204], and other stability issues during freezing (e.g., sodium phosphate can crystallize out of the protein amorphous phase during freezing resulting in large shifts in pH). Salts can affect the physical stability of proteins by shielding charged species on the surface of the protein, thereby impacting protein–protein associations. Salts have also been shown to be useful to reduce solution viscosity [205] and ensure osmotic balance. Surfactants are often added to protein formulations in order to minimize protein denaturation and aggregation at interfaces [206]. Preservatives such as benzyl alcohol have also been added to protein formulations to allow for multidose parenteral administration, however, preserved formulations can be challenging to develop since preservatives may also destabilize proteins [207].

The stability of additives themselves must also be considered when selecting these additives for formulation development. For instance, surfactants such as polysorbate are known to degrade during storage [206,208]. Histidine, when used as a buffer or additive is known to oxidize and discolor on exposure to light. Additives such as sucrose (used as a stabilizer against freeze/thaw induced degradation) are known to undergo hydrolysis under acidic conditions.

Another challenge faced by the protein formulation scientist is the development of high-concentration formulations. This is especially the case for antibody therapeutics that may require drug product concentrations approaching or exceeding 100 mg/mL. At such high concentrations, unique formulation challenges emerge that result in several manufacturing, stability, analytical, and delivery challenges [209–211]. High-concentration challenges include solubility limitations, buffering capacity limitations, concentration dependent aggregation, and high solution viscosity that may complicate manufacturing and delivery by injection. Some solutions to these problems have been proposed including self-buffered formulations with antibodies [212], and histidine [213] or high salt containing formulations [214] to reduce solution viscosity.

4.3.2. Lyophilized Formulations Lyophilized ("freeze-dried") formulations have also been developed for protein biopharmaceuticals that may not be stable in liquid formulations for extended periods of time. Several recent reviews have summarized the challenges of developing lyophilized formulations for protein biopharmaceuticals [215–218]. The development of a lyophilized formulation can be complex and requires sophisticated processing equipment; however, once developed, these formulations can be particularly robust to long term storage stability stresses. The lyophilization process consists of two phases: freezing of a protein solution and drying under vacuum. The drying phase is further divided into primary and secondary drying. Primary drying removes the frozen water and secondary drying removes the nonfrozen "bound" water. The levels of residual moisture are monitored following the lyophilization process to assure appropriate dehydration. Excess residual moisture allows excessive molecular mobility and chemical degradation processes to occur, and inadequate residual moisture can disfavor native protein conformation [219]. Rates of freezing, drying and the level of residual moisture must be carefully controlled such that the protein retains its native conformation throughout the lyophilization process. Lyophilized formulations typically contain a buffer, to optimize conformational stability of the protein, a cryo-protectant designed to protect the protein during freezing (e.g., sucrose or trehalose), and a bulking agent designed to create a pharmaceutically acceptable cake (e.g., mannitol or glycine).

4.4. Interactions with Container Closure Materials

The selection of appropriate container closure systems for biopharmaceutical products is important, and depends to a large degree on patient delivery issues as well as product compatibility. Contact materials such as glass and plastics may delaminate or leach following long term storage of formulations. Significant differences in glass characteristics and performance are known among suppliers of glass vial and syringe materials. Detailed forensic investigations of vial surfaces from multiple suppliers have demonstrated delamination (flaking) and pitting in the presence of a parenteral solution on some of these surfaces [220].

Proteins may also adsorb or denature on contact with container closure materials, requiring the addition of surfactants to reduce surface interactions. For example, insulin has been shown to be destabilized by adsorption at hydrophobic interfaces [221]. Protein destabilization at hydrophobic surfaces is thought to result from nucleation, that is, formation of small intermediate aggregates that serve as precursors to larger precipitates. Specialty glass coatings have been developed to reduce surface hydrophobicity and limit product interactions [222]. Alternative materials have also been developed to minimize protein binding, (e.g., plastic CZ resin vial material may be a suitable candidate for packaging parenteral protein formulations since it offers significantly less protein binding compared with glass vials) [223].

Finally, silicone oil used to lubricate vial stoppers, syringe walls and plungers has long been known to interact with proteins resulting in denaturation [224]. Early work with insulin has highlighted the potential hazard of plastic insulin syringes coated with silicone oil [225]. More recent reports suggest that silicone oil may catalyze protein aggregation. Jones and coworkers suggested that silicone oil could induce conformational changes as measured by circular dichrosim and derivative UV spectroscopy that may lead to protein aggregation [226].

4.5. Alternate Modes of Delivery

Protein biopharmaceuticals are typically delivered via parenteral injection. In order to improve patient comfort and compliance, efforts are underway to deliver proteins through alternate routes, including oral, mucosal, transdermal, and inhalation. However, each of these routes of administration presents tremendous challenges. Oral delivery of proteins has been considered by many to be the "holy grail" of biopharmaceutical drug delivery. Drug delivery scientists have attempted to reduce the impact of digestive enzymes in the gastrointestinal tract through coadministration of enzyme inhibitors, absorption enhancers, encapsulation in liposomes, microemulsions, biodegradable particles, and the modification of proteins through the attachment of chemical moieties [227,228]. Mucosal delivery of proteins has achieved some modest success. In particular, low molecular weight proteins and peptides have been delivered through the nasal mucosa and across the oral cavity with the aid of bioadhesive polymers [229,230]. Transdermal delivery of proteins has been equally challenging. Peptides have been driven across the skin with the aid of skin perturbation devices; however, bioavailability for this route remains low [231]. Pulmonary delivery of proteins has also met with mixed success. Appropriately sized particles (1–5 μm) may be inhaled and deposited in the alveolar space, and it has been demonstrated that proteins such as insulin have measurable bioavailablity and efficacy through this route of administration [232].

For parenteral injection, many advances in drug delivery modalities have been considered and implemented including, degradable depots, PEGylation (polyethylene glycol conjugation), needle free injectors, and microneedle systems [233–237]. Depot systems fashioned from degradable polylactic coglycolic acid (PLGA)-based polymers have been introduced into the market, including gonadotropin-releasing hormone (GnRH) agonist (Lupron Depot®) and human growth hormone (Nutropin Depot™) [238,239]. Protein release kinetics from these systems can be controlled through the appropriate selection of polymers and fabrication techniques [240]. Protein PEGylation has been utilized to enhance the circulation half-life of peptides and proteins and reduce dosing frequency and improve patient compliance. Commercially successful PEGylated products include PEGylated G-CSF (Neulasta®) and PEGylated interferon alpha-2a (Pegasys®) [241]. Needle free injectors have been developed to improve patient compliance, and safety. These devices operate through pressurization of the drug container in order to force the aqueous formulation to exit an orifice at high velocity [242]. Finally, microneedles have been developed to pierce the skin with blunt or hollow needles fabricated with a range of sizes, shapes and materials. Microneedle delivery systems take advantage of the highly vascularized nature of skin for drug absorption while minimizing the pain and safety issues for patients associated with traditional needles [243].

4.6. Future Trends

The task of protein formulation development requires the evaluation of a large number of formulation conditions, sample testing and considerable analytical characterization to support the selection of a final formulation. Since this process also requires the collection of real-time stability data that can require 2 years after the manufacturing process has been developed, there is considerable pressure to develop formulations rapidly. Some efforts to increase protein formulation development throughput have been considered and reported in the literature [244,245]. A high-throughput formulation (HTF) platform is based on the use of microplates and robotics systems. Microplates allow for the incubation and analysis of multiple formulations simultaneously under stressing conditions and robotics systems automate the process of sample preparation and analysis. For a high-throughput system to be effective the formulation scientist will need to consider what should be the appropriate stress conditions of the protein that will accurately predict stability (agitation, temperature, pH, others) and what assays are stability indicating for the protein being tested under these conditions (size exclusion chromatography to measure aggregation, mass spectrometry to measure biochemical degradation, others). While HTF appears challenging given the current state of technology, the integration of high-throughput tools coupled with insightful testing strategies will undoubtedly yield robust formulations in a more rapid time frame in the near future.

Another area of increasing interest in protein formulation is in the production of protein crystals designed to be delivered as suspensions. In recent years, investigators have produced protein crystals that are suitable for long-term storage and delivery [246,247]. Protein crystals provide one solution to the problem of storage of high concentrations of biopharmaceuticals for prolonged periods of time. When delivered as a suspension, protein crystals assume lower viscosity than solubilized protein of the same concentration, alleviating high-concentration delivery issues. Also, investigators have recently begun to engineer protein crystals, controlling their dissolution characteristics such that their delivery rates may be controlled. Govardhan and coworkers demonstrated that crystals of human growth hormone coated in poly(arginine) could be injected once per week and achieve the same results in rat growth studies as daily injections of soluble growth hormone [248]. Despite the challenges in producing protein crystals at a reasonable commercial manufacturing scale and unknown immunogenicity issues, this technology may provide significant advances in protein formulation, storage, and delivery in the future.

5. ANALYTICAL

5.1. Introduction

The utilization of appropriate analytical methods is crucial in order to evaluate protein products and in-process intermediates that are produced throughout the process development cycle. Analytical methods are used to measure protein concentration, product quality attributes, process and product related impurities, product identity, and product potency. These product quality attributes are continuously monitored during the development of the upstream and downstream processes in order to assess the impact of individual processing steps on the physical and chemical characteristics of the protein. Over the course of the development process, analytical methods are refined and improved to provide relevant information regarding the consistency of the process and key attributes of the protein product. Finally, throughout the entire development process, the Analytical Scientist must consider that many of the methods developed to support process and formulation development will need to be utilized in a quality control environment in order to release the product.

In-process assays used in the development of upstream and downstream processes typically provide critical information to Process Development Scientists such as protein concentration, levels of aggregate and charge heterogeneity, levels of impurities such as in-process contaminants and host cell proteins, and bioactivity [249]. Early development stage assays are often suitable for development purposes but require further "qualification" to demonstrate they are fit for purpose and "validation" to ensure that they are useful as product release methods [250]. Product release specifications are set based on knowledge of key product attributes and key process attributes and an understanding of how controlling the process will produce a product of desired characteristics [251]. Common in-process and protein product release test methods include size exclusion chromatography (product purity) and ion exchange chromatography (product purity and identity), SDS-capillary electrophoresis or SDS-PAGE (product purity), immunoassay (identity and host cell protein impurities), DNA testing (impurities), and bio- or binding assay (potency). Table 1 summarizes the utility of many of these methods for the detection of process impurities.

Analytical methods are also used for product characterization and product or process comparability. Throughout the development process greater refinement of a product's analytical characteristics are determined. During early stages of development, features such as protein primary sequence, size characteristics, charge isoforms, biophysical structure, and chemical degradation products are determined. Later stages of development allow for greater refinement in product characterization as the analytical characteristics that impact product attributes such as bioactivity are better understood. When processes are modified, for example, when changing a commercial process in order to take advantage of potential titer, purity, or yield improvements, analytical characterization methods are essential in determining product comparability. Typically, appropriate characterization methods are developed in order to ensure that the new process generates material that is comparable to the original process. Analytical methods that are commonly used for product characterization and comparability testing include chromatographic analysis, peptide maps with mass spec analysis (to monitor post translational modifications, instabilities, impurities, product heterogeneities, disulfide bonding patterns, etc.), oligosaccharide maps (glycan heterogeneity), isoelectric focusing (charge heterogeneity), bioactivity, and biophysical methods (structural characterization). Figure 4 illustrates the application of product characterization methods for highly complex glycoproteins that are used to support process development.

Appropriate analytical methods and characterization techniques are also essential for the development of protein formulations and the evaluation of protein product stability. As upstream and downstream processing steps are developed analytical methods are assessed for "stability indication" (i.e., analytical methods that can quantitatively demonstrate changes in critical product parameters under normal and accelerated storage conditions). The development of stable protein formulations supports the overall development of release methods and characterization assays by providing key input on determining which analytical parameters are useful to monitor.

5.2. In-Process and Product Release Testing

5.2.1. Chromatography Chromatographic separations are the workhorse techniques of protein isolation and characterization. Chromatography is particularly well suited to analysis of proteins because of the ability to separate based on physical heterogenities that are common to biopharmaceutical products such as size, charge, and hydrophobicity and biochemical specificities such as ligand binding affinity. Chromatographic techniques can also be developed into rapid, high-resolution, high-throughput, and robust methods making them well suited in a quality control environment. Finally, chromatographic techniques

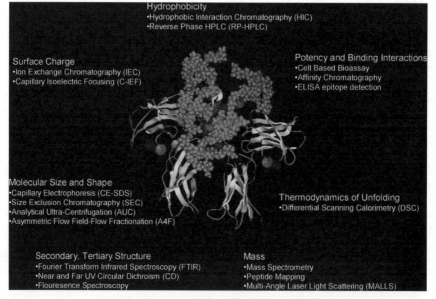

Figure 4. Depicted in this figure is a glycoprotein. The protein, defined by the amino acid sequence, folds to form secondary structures such as alpha-helices (red cylinders) and beta-strands (yellow ribbons) which are connected by ordered loops (green threads). The glycosylation attached to the protein is shown by blue and green spheres. Some common physical and chemical attributes of recombinant glycoproteins and an array of analytical tools that are utilized to characterize these proteins are listed. (See color insert.)

are desirable because of the ability to couple these separation methodologies with other high-resolution techniques such as mass spectrometry for in depth characterization.

Size Exclusion Chromatography Chromatographic separation based on size and shape, called size exclusion chromatography (SEC), is the most commonly utilized chromatographic technique for the characterization of aggregates in protein biopharmaceuticals. In SEC, separation is achieved in an aqueous mobile phase by molecular sieving (ideally), whereby larger species are excluded from penetration into porous resins and elute more rapidly relative to smaller species that do penetrate the pores of resins [252].

Frequently, protein products are produced with monomeric, dimeric, and potentially larger aggregated species that can be separated and quantified by SEC. Several examples from the literature demonstrate how SEC methods can be developed to provide a simple, direct measure of protein aggregates in a format that is suitable for evaluation of process development intermediates, and product stability [253,254]. The ability to quantitate aggregate levels in a protein product is thought to be critically important because of the potential for these species to induce immune reactions [255]. In order to further characterize sizes of protein species orthogonal techniques have been utilized to estimate levels of soluble aggregate including dynamic light scattering (DLS), analytical ultracentrifugation (AUC), and asymmetrical flow field flow fractionation (AF4) (these techniques are described later).

Ion Exchange Chromatography Analytical scale ion exchange chromatography (IEC) is used to separate and characterize proteins on the basis of surface charge [256]. Anion exchangers interact with negatively charged protein ions and cation exchangers interact with positively charged protein ions. In either case, the strength of the ionic interactions between the protein surface and the ion exchanger dictates protein retention times and leads to charge based separations. Typically, ion exchange assays are developed by testing strong or weak ion exchangers with aqueous mobile phases containing appropriate levels of salts or buffer pH conditions that selectively bind and elute or separate charged species of interest.

Some examples where IEC has proven to be useful to separate proteins on the basis of surface charge heterogeneity include deamidation, charge associated with glycosylation, phosphorylation, and truncated or oxidized forms [257–261]. Charge-based separations can be highly selective with even a single amino acid substitution, if the change causes significant modification of the charge distribution on the surface of the protein, resulting in complete separation by IEC [261,262]. Also, since IEC depends on surface charge interactions, changes in conformation that result in burying of charged regions of proteins may be detected by IEC [263]. Because the ion exchange profiles of proteins can be highly specific for protein sequence, this assay has been utilized by some biopharmaceutical manufacturers as an identity test. Characterization of the individual peaks that contribute to an ion exchange profile has also proven useful in understanding important product heterogeneities [261].

Reverse Phase HPLC Reverse phase chromatography operates on the principle of hydrophobic interactions, i.e. hydrophobic regions of proteins interact with the hydrophobic resins to a degree that can be impacted by resin side chain length, and aqueous phase conditions, resulting in separations. Reverse phase HPLC (RP-HPLC) is highly resolving and therefore has been utilized to characterize peptide fragments, and specific chemical modifications. RP-HPLC of proteins is carried out under denaturing conditions and therefore the entire unfolded protein sequence is capable of interaction with the stationary phase. Because of the high resolving power of RP-HPLC it is also commonly coupled to other detection systems such as mass spectrometers.

RP-HPLC has been shown to be a useful technique to separate proteins with heterogenities in carbohydrates, oxidation, and deamidation. Reverse phase is also very useful for separation of smaller peptides, such as peptide maps for characterization [264] and for implementation as a release assay [265,266]. Dillon and coworkers showed that reverse phase methods could distinguish between intact and fragmented monoclonal antibodies, and structural heterogeneity of IgG2 [267].

Quantitative methods suitable for quality control have also been developed from RP-HPLC methods [268,269].

Hydrophobic Interaction Chromatography HIC is complementary to RP-HPLC in the sense that HIC separates proteins on the basis of surface hydrophobicity whereas RP-HPLC separates proteins based on surface and core hydrophobic residues that are exposed during denaturation of the protein. The relative hydrophobicity of column materials used for HIC will influence the ability to separate hydrophobic species. Mobile phase buffer selection also plays an important role in the development of analytical HIC methods. In particular, salts impact the relative hydrophobicity of exposed protein surfaces.

Because of the selectivity of HIC for surface hydrophobicity of proteins this technique has been used to distinguish properly and improperly folded isoforms. Two examples are described by Scheich and Aizawa who developed HIC methods to assay for proper protein refolding from inclusion bodies produced in *E. coli.* [270,271].

Affinity Chromatography Affinity chromatography takes advantage of the binding of proteins to specific ligands such as occurs with the binding of protein A or protein G to the Fc domain of IgG. Affinity resins are commercially available with preconjugated ligands for many applications (e.g., MabSelect™ protein A resin from Amersham Biosciences). Affinity chromatography is carried out by passing analytes of interest such as heterogeneous populations of proteins over affinity columns. Analytes bound to the affinity resins are then eluted under high salt or low pH conditions. Affinity chromatography has been commonly practiced to quantitate levels of immunoglobulin, immunoglobulin fragments, or Fc fusion proteins bound to protein A [272]. This technique forms the basis for Fc protein quantitation from a complex mixture as occurs during production of monoclonal antibodies.

5.2.2. Electrophoretic Techniques Electrophoretic techniques such as polyacrylamide gel electrophoresis (PAGE) and capillary electrophoresis (CE) are commonly used to separate proteins on the basis of size or charge. In size-based applications sodium dodecyl sulfate (SDS) is added to the native or denatured protein solution prior to separation. The SDS, which carries negative charge, becomes associated with the protein and draws the protein through the gel under the force of the applied electrical gradient. For many years, SDS-PAGE has been the technique of choice for size-based protein separations. This is due to the relatively small mass of material that can be detected (1 μg following silver staining), the convenience of commercially available PAGE gels, and the relative simplicity and transferability of the technique.

In recent years, capillary electrophoresis (CE-SDS) has become more widely utilized in the analysis of biopharmaceuticals because of the advances in the robustness of instrumentation, the ability to quantitate peaks directly through absorbance measurements, and the relative speed and reproducibility of the technique [273–275]. Biotechnology manufacturers have successfully tested and released their products based on CE-SDS methodologies [276]. The superior resolving power of CE-SDS also lends itself to product characterization applications [277,278], and the ability to couple CE with mass spectrometers is increasingly finding application in glycoprotein analysis [279].

For charge-based separations electrophoretic methods can be utilized to focus proteins according to isoelectric point in a slab gel or capillary subjected to a voltage gradient. Focusing is achieved by preequilibrating the gel or capillary with polyampholytes that establish a pH gradient in response to the applied voltage between the anode and the cathode. As a technique isoelectric focusing is capable of extremely high resolution with proteins differing by a single charge being focused into distinct bands. Recent work from Han and coworkers demonstrated the utility of capillary isoelectric focusing (C-IEF) to monitor upstream and downstream processes, the analysis of in-process samples, and the characterization of soluble interleukin-1 receptor (IL-1R) [280].

5.2.3. Nucleic Acid Testing The ability to analyze residual levels of host cell DNA is important in order to minimize the potential of aberrant gene expression by insertion of

DNA into the genes of patients receiving recombinant proteins. Regulatory guidelines recommend that the final product contains no more than 100 pg cellular DNA per dose [281]. Traditional assays for the measurement of DNA include threshold, slot blot, and picogreen dye assays, however, these methods are relatively time consuming and error prone. More recently, the quantitative polymerase chain reaction (QPCR) has been developed for the detection of host cell DNA [282,283]. With this method host cell DNA is quantified using primers targeting repeat regions of the host cell genome. The sensitivity of QPCR is in the fg/ul range that is about a million-fold more sensitive than traditional DNA assays. Recent publications from Lovatt and Kubista and coworkers reviewed the QPCR method for quantitation of residual DNA in protein therapeutics, and biosafety applications such as the detection of the level of virus removal in process validation viral clearance studies [284,285].

5.2.4. Immunoassays Immunoassays are highly specific and highly sensitive biochemical techniques, with limits of quantitation in the range of 1–10 part per million (ppm). The immunoassay system typically consists of a "capture" antibody to bind the antigen of interest and a secondary antibody for the purpose of detection. Immunoassays are most typically carried out in a 96 or 384 well-plate format and are suitable for use in a quality control environment. In protein process development, immunoassays are often used as a test for product identity. Product identity immunoassays require the production of monoclonal, and/or polyclonal antibodies that specifically detect the product of interest without cross-reacting with other products in development. Immunoassays are also commonly used for the detection of process-related impurities. These impurities include host cell protein contaminants and often protein A, which may be leached from purification columns in processes developed for products that contain immunoglobulin Fc domains. Kits for the detection of protein A by immunoassay are commercially available from several companies (e. g., Repligen and Assay Designs, Inc.).

The development of immunoassays for measurement of host cell protein contaminants has gained considerable attention in the literature [286]. Host cell protein contamination consists of a complex mixture of proteins and peptides, and measuring all them with a single assay poses a challenge [287]. Given the fact that any single host cell protein antigen may be weakly immunogenic or even nonimmunogenic, an immunoassay may be insensitive to elevated levels of these particular antigens. Therefore, host cell protein immunoassays can only provide relative data, rather than absolute quantiation of host cell contaminants. Nonetheless, recent studies by Krawitz et al. [288] suggest that a generic assay, developed to detect host cell proteins from a single cell line, may be useful in detecting impurities when this same cell line is used to produce other products.

Other recent advances in immunoassay techniques include the Gyrolab from Gyros, AlphaLISA (Amplified Luminescent Proximity Homogeneous Assay) from Perkin Elmer, and Octet from ForteBio [289–291]. In the Gyrolab application the immunoassay process has been automated and simplified with the use of a compact disc that is preloaded with capture and secondary antibodies and detection reagents. Spinning the compact disc at predetermined speeds allows sequential mixing of analyte, capture antibodies, and secondary antibodies with incubation steps at appropriate intervals. In the bead-based AlphaLISA application, donor and acceptor beads conjugated with ligand and detection antibodies are utilized to amplify antigen–antibody interactions through chemiluminescence. The Octet application relies on the use of a 96-well-biosensor system that utilizes interferometry to detect analytes bound to the biosensor tip for rapid and label free protein quantitation.

5.2.5. Bioactivity Assays Potency assays are critically important for the assessment of the bioactivity of protein pharmaceuticals, both to ensure that product potency is maintained during processing and storage and for the purpose of product characterization. Typically potency assays fall into two categories: (1) bioassays, which utilize cells to directly assess the ability of protein products to "activate" or block target pathways, and (2) binding assays, which assess the ability of protein products to bind to receptors or ligands in a manner that is known to represent the mechanism of action of that particular protein product. For agonistic products, it is typical to use cell-based bioassays since both binding to a target receptor or ligand and signaling into the cell for the desired activity are required. The activity of antagonistic products may be assayed with binding assays if the desired mechanism of action can be demonstrated. A recent review article from Meager described the measurement of cytokine potency by bioassays [292]. Binding assays for product characterization have included ligand binding-based systems as well as surface plasmon resonance technology (Biacore™, from GE Healthcare) that is sensitive to changes in binding affinity between an analyte of interest and a surface immobilized ligand [293].

Potency assay scientists must develop release assays that are quantitative and able to be qualified for product release testing. In order to be useful for release testing bio- and binding assays should be sensitive to conformational and chemical characteristics of the protein product. Since the impact of degraded proteins on potency may not necessarily be demonstrated by other means (e.g., chromatography or immunoassay) it is essential to include a bioassay or binding assay in product release testing [294]. Also because of the inherent variability of bio- and binding assays, a thorough development process will be required to set useful specifications [295,296]. For certain products, bioactivity must include multifunctional assessment, for instance, a monoclonal antibody that relies on its ability to bind a specific target ligand as well as binding to Fc receptors for effector function [297].

Potency assays are also critical for the characterization of protein products. Cases have been shown where chemical modification of a single amino acid, such as deamidation of an Asp residue in the CDR of an antibody to human IgE, entirely inactivated these proteins [298] and other cases where degraded species such as aggregates became inactive or even hyperactive [299]. New technologies for bioassays include the use of "platform" systems such as the introduction of reporter gene elements to take advantage of common cell signaling pathways (including STAT and NF-κB transcription factors) for biopharmaceutical compounds with different mechanisms of action [300].

5.3. Mass Spectrometry and Biophysical Characterization

High-resolution and specialty analytical methods are powerful techniques that are often utilized during product characterization, determination of product comparability, or in support of product investigations. These techniques include methods for biochemical and biophysical analysis. Often, the results of these analyses provide insight into protein processing steps that cannot be measured by typical release methods or lower resolution techniques described in the sections above.

5.3.1. Mass Spectrometry Mass spectrometry (MS) is perhaps the single most powerful technique used in the biopharmaceutical industry for detailed biochemical characterization of proteins. MS instrumentation includes an ion source that transforms molecules from a sample into ionized fragments, a mass analyzer that separates the ions based on their mass-to-charge ratio (m/z), and a detector that measures the abundance of each of the ions present. MS is typically coupled with online chromatographic separations (LC-MS), which is particularly useful for analysis of separation products. Recent review articles provide exhaustive descriptions of common MS techniques and applications for the analysis of protein pharmaceuticals [301,302].

As a technique MS is well-suited to protein pharmaceutical characterization because of the wide range of problems that can be solved. So-called "top–down" analysis describes MS for intact proteins. Applications for top down analysis include whole mass analysis, assessment of overall product heterogeneity, identification of labile posttranslational modifications, and confirmation of N- and C-terminal sequences. Bottom–up analysis describes MS for protein digests, or peptide maps, which are produced by incubation of samples in enzymes such as trypsin, Lys-C, Asp-N, and Glu-C. Applications for bottom–up analysis include characterization of glycosylation patterns, determination of disulfide linkage patterns, and identification of posttranslational modifications such as Met and Trp oxidation, Asp isomerization, Asn deamidation, glutamate cyclization, glycation, fragmentation, and others.

As examples of upstream applications MS has provided useful information on the stability of master cell banks as measured by glycosylation profiles in product expressed from BHK 21A cells at 0 and 62 generations [303]. MS has been utilized to monitor the degradation of oligosaccharides during processing [304] and to demonstrate that during protein expression from CHO cells the culture conditions can have a significant impact on oligosaccharide profiles [305]. MS has also proven useful to map disulfide pair heterogeneity in IgG2 produced by CHO cell culture processes [263,306,307].

During the purification process MS has proven to be a useful aid in the identification of product impurities and instabilities from intermediate hold steps. Senderoff and colleagues utilized laser desorption MS to identify racemization of glucagon-like peptide-1 (rGLP-1) and to influence the purification scheme and the in-process storage and handling conditions [308]. More recently, Weinberger and colleagues described the use of protein biochip arrays carrying functional groups typical of those employed for chromatographic materials [309]. Application of proteins to these chips followed by typical test wash conditions resulted in a "retentate" bound to the chip that was then analyzed by surface-enhanced laser desorption-ioniziation (SELDI) MS.

During the development of protein formulations and storage of proteins over prolonged periods MS is frequently utilized to characterize degradation products. Among these, degradation products characterized by MS to support formulation development include covalent or disulfide linked aggregates, Met oxidation, Asn deamidation, and Asp isomerization following storage of formulated bulk solutions or products under moderate and accelerated conditions [310–312].

5.3.2. Biophysical Techniques Numerous biophysical techniques are utilized by the analytical scientist during protein process development and protein characterization in order to assess the structural characteristics of proteins. This panel of techniques includes the following methods to assess: (1) molecular size and size distribution of proteins such as light scattering, assymetric flow field flow fractionation (A4F), and AUC; (2) secondary protein structural assessment such as Fourier transform infrared spectroscopy (FTIR) and far-UV circular dichroism (CD); and (3) tertiary protein structure and conformational assessment such as near-UV CD, intrinsic and extrinsic fluorescence spectroscopy, and a calorimetric technique known as differential scanning calorimetry (DSC). As a biophysical "tool box," these techniques form a powerful set of analytical capabilities that complement chromatographic, biochemical, immunochemical, and other methodologies for product characterization and investigative purposes. Biophysical techniques have also been widely applied during protein formulation development where the ability to assess structural stability is considered to be a key factor in predicting and limiting aggregation behavior. Although several other structural/biophysical techniques such as nuclear magnetic resonance (NMR), and X-ray crystallography are highly resolving and are useful in some circumstances, these techniques have not traditionally been as widely utilized during protein process development.

Biophysical Assessment of Size and Size Distribution Static light scattering (SLS) measures the average light scattering intensity as a function of sample concentration. For a large size molecule, multiangle laser light scattering spectroscopy (MALLS) is commonly used. The combination of MALLS with a chromatographic method such as SEC has been used to obtain the accurate molecular masses of protein species such as monomers, dimers, etc., independent of chromatographic retention times [313]. Often, SEC leads to complicated profiles that result from the existence of dimers, multimers, and other species in solution that may have affinities for the size exclusion resin. In these examples, MALLS can provide a relatively rapid and orthogonal assessment of size [314]. For example, MALLS has been utilized in combination with SEC to estimate the size of polyethylene glycol conjugated (PEGylated) proteins and the size of protein–ligand complexes [315].

DLS is based on the principle of detecting fluctuations of scattered light due to the Brownian motion of molecules in solution, independent of the protein concentration in solution. Dynamic light scattering is used to measure the hydrodynamic radius, or size distribution of molecules in solution [316]. DLS measurements have greater utility for protein mixtures with a wide range of masses (chromatographic separations not required).

A4F is a separation technique that relies on the application of a perpendicular physical flow field to a solution mixture. Unlike chromatography, A4F does not rely on interactions with a stationary phase, rather, separation is achieved by differing diffusion coefficients of the various components in the field [317,318]. For A4F it is possible to inject samples containing precipitated material without any pretreatment (SEC requires filtration to avoid column clogging). A4F can cover a broader span of molecular weights than SEC, which is limited by the exclusion limit of the column and this technique is capable of separating aggregates and particles from 0.001 to 50 μm in size. Despite the advantages of A4F, widespread use of this technique has been limited by the difficulty in the use of the instrumentation and the complexity of data analysis [319].

AUC allows real-time UV absorption measurement of the sedimentation of a macromolecular sample as a result of an applied centrifugal field. Mathematical analysis of the sedimentation equilibrium (SE) or sedimentation velocity (SV) data through specialized software (SEDFIT) [320] can provide very high-resolution size information about the molecules in a given sample [321]. The kinds of information that can be obtained from an analytical ultracentrifuge include the gross shape, conformational changes, and size distributions of macromolecular samples. For macromolecules such as proteins that exist in chemical equilibrium with different noncovalent complexes, the number and subunit stoichiometry of the complexes and equilibrium constants can be studied. In addition, AUC allows sample testing to be conducted in the exact liquid formulation or reconstituted liquid formulation of the biopharmaceutical in the vial. AUC is a particularly useful technique for the measurement of protein aggregation as it provides an orthogonal measurement to other size-based techniques such as SEC, and A4F [319,322–324]. Light scattering, A4F and AUC can measure absolute levels of aggregates, excluding artifacts that can be induced by buffer and column interactions in the SEC methodology. Studies have demonstrated that SEC can underestimate soluble aggregate relative to measurements by AUC and A4F [325].

Biophysical Assessment of Secondary Structure FTIR has been utilized to assess protein secondary structure in a range of environments, including in solution and when adsorbed to solid surfaces. There is a wealth of information that can be used to derive structural information by analyzing the shape and position of bands in the amide I region of the FTIR spectrum, including α-helix, β-sheet, and β-turns. As a versatile biophysical characterization tool FTIR has been utilized to characterize inclusion bodies within bacterial cells [326], to measure the melting temperature (T_m) of proteins as an aid to formulation development [327,328], and as a method that allows direct analysis of the conformation of proteins at high concentration without dilution that allowed the construction of empirical phase diagrams [329].

CD spectroscopy is based on the differential absorption of left- and right-handed circularly polarized light. Far-UV CD can reveal important characteristics of protein secondary structure. In particular, CD spectra can be readily used to estimate the fraction of a molecule that is in the α-helix, β-sheet, β-turn, and other nonstructured conformations [330,331].

Biophysical Assessment of Tertiary Structure and Conformational Change Near-UV CD spectra can provide additional detail about the tertiary structure of proteins. The signals obtained in the near-UV spectra (from 250 nm to ~300 nm) are due to the absorption, dipole orientation and the nature of the surrounding environment of phenylalanine, tyrosine, cysteine, S-S-disulfide bond and tryptophan amino acids. Prolonged exposure of antibodies to low pH is unavoidable for protein A purification and viral inactivation. In order to minimize structural instability and aggregation, CD and other biophysical techniques have been used to assess the structural integrity of proteins during these low pH processing steps [332]. Numerous CD applications have been described in the literature with complementary biophysical techniques in order to characterize the structural integrity of antibody therapeutics and confirm the structural characteristics of Fc and Fab domains [333,334].

Intrinsic fluorescence spectroscopy describes the ability of tryptophan, tyrosine, and phenylalanine to absorb and fluoresce light in a manner that is characteristic of the local environment of these amino acids. Structural perturbations of proteins can be measured by shifts in fluorescence intensity and wavelength. Intrinsic fluorescence has also proven useful in analyzing binding interactions between ligands that shifts the fluorescence intensity of tryptophan, tyroisine, or phenylalanine at the interface between the binding partners [335].

Extrinsic fluorescence spectroscopy describes the addition of fluorescent tags to proteins that produce characteristic fluorescence emission profiles. Extrinsic fluorescence spectroscopy has been reported in the literature to characterize folding intermediates, measure surface hydrophobicity, and detect aggregation [336,337]. During protein formulation development extrinsic fluorescence spectroscopy has proven useful as a tool to assess the impact of excipients on the unfolding of proteins [338].

DSC relies on the ability to measure thermodynamic parameters, specifically, the amount of heat required to maintain the temperature of a given sample relative to a reference. DSC has the potential to measure thermodynamic parameters including the change in enthalpy (ΔH) and heat capacity (ΔCp), as well as the glass transition temperature (T_g) and the melting transition temperature (T_m). The main application of DSC in biopharmaceutical analysis is the measurement of the T_g of lyophilized cakes or the T_m of proteins in solution [339]. T_g, the temperature at which a glassy solid becomes soft upon heating, is often utilized to optimize formulation buffers and operating conditions for product lyophilization. A thorough understanding of the Tg is critical in order to reduce the likelihood of collapse of a lyophilized "cake" and to retain long-term product stability. Recent scientific reviews describe the factors involved in stabilizing folded proteins during the development of lyophilized formulations and lyophilization processes [340,341]. For proteins, considerable effort has been given to the measurement of T_m in order to describe protein unfolding and subsequent propensity to aggregate. Recent work with soluble IL-1 receptor and human IgG describes the relationship between T_m and the kinetics of aggregation [342,343]. In these studies the choice of formulation buffers used in the protein unfolding experiments were shown to impact T_m. In another study from Garber and Demarest, T_m values from the Fab domains of 17 human or humanized antibodies were found to vary from 57 to 82°C [344]. In this study, antibodies with lower T_m values were found to aggregate and to express poorly in cell culture. The measurement of T_m has also been utilized to evaluate the impact of antimicrobial preservatives on proteins in solution since antimicrobial preservatives may be physically destabilizing to proteins [345].

5.4. Other Support for Process Development

5.4.1. Comparability As mentioned, during stages of protein product development a process or formulation is modified in order to increase production titer, improve purification yield, make a change to the production scale or site of production, or to improve the storage or delivery characteristics of the product. However, even minor variations brought about by these process changes can lead to clinically relevant changes in efficacy and/or safety of the end product [346,347]. Analytical comparability is used to assure that changes in a process do not change the analytical attributes of that product. Appropriate analytical methods for comparability testing include critical product attributes, such as those described by biochemical and biophysical characterization tests, potency assays, and purity assays [348–350].

Process development scientists published results for the scale-up of a fed-batch bioreactor process going from 3 L to 2500 L scale (Epratuzumab) [351]. In this work, biochemical assays including N-linked oligosaccharide profiling, C-terminal lysine analysis, and tryptic peptide mapping showed that the scale-up process did not impact the analytical characteristics of this compound. Similar results from Meuwly and coworkers demonstrated analytical comparability following conversion of a CHO cell culture process from perfusion to fed-batch process [352]. Published results with Synagis® (a monoclonal antibody product) showed that there was a different pattern of glycosylation during the early stages of bioreactor culture, however, no other changes in microgeterogeneity were apparent for the other culture conditions studied [353]. Based on the results of these experiments critical processing attributes could be controlled to assure that product characteristics were consistent during process changes and production. Insulin has provided an excellent model system to illustrate many important considerations when dealing with comparability exercises for biotechnology products undergoing formulation changes [354]. The bioactivity of insulin is known to be impacted by structural and biochemical attributes and the influence of formulation changes must be demonstrated through suitable testing methodologies of the drug product.

Analytical testing has also been carried out in support of the comparability of biosimilar products [355]. While this information may be supportive of the analytical "quality" of a biosimilar product, since the manufacturing process, analytical methods, and formulation of the biosimilar are not the same as the innovators product, clinical studies will likely be required to assure suitable product performance characteristics such as safety/immunogenicity and efficacy [356].

5.4.2. Process Reagent Clearance and Excipient Testing
During protein process development, analytical methods may be required for determination of the clearance of process reagents with potential safety concern (PSC) [357]. Whether or not process reagents require demonstration of removal depends on an evaluation of their potential impact to product safety, an evaluation of capability of the process to remove the impurities, and an assessment of the overall impact of the impurities to the product quality. Generally, the ability of a process to clear process reagents is affected by the process step at which they are introduced, their solubility through subsequent processing steps, and their relative concentrations. Upstream and downstream process reagents that will likely not require in-process testing (depending on concentration) include those substances that are generally recognized as safe (GRAS) [358].

Upstream process reagents that may require monitoring include media components such as insulin and other growth factors, serum proteins, soy proteins, and buffers (e.g., HEPES and MES); expression inducers such as isopropyl-β-D-thiogalactopyranoside (IPTG), and methotrexate [359]; antibiotics; cell shear protectants such as pluronic F-68; and antifoam agents simethicone [360]. Downstream process reagents that may require testing include buffers, acids, and solvents such as CAPS, MOPS, acetone, and formic acid; protein refolding agents such as dithiothreitol (DTT), urea, and guanidine [361]; and column leachables such as protein A.

Excipients (preservatives, surfactants, salts, buffers, and tonic agents) may also require analytical testing to ensure these are present at appropriate levels to achieve their intended effect on product formulations [362]. Preservatives must be routinely identified and quantified at release. Excipient assays are as diverse as the compounds that are measured and may rely on HPLC, gas chromatography or other chemically specific methodologies [363].

5.5. Future Trends

Analytical technologies will continue to develop and evolve, leading to new techniques and methodologies that will be applied to protein process development. Three specific areas where improvements in analytical approaches will benefit protein process development include (1) greater focus on determining and measuring critical quality attributes, (2) development of higher throughput techniques, and (3) development of higher resolution techniques.

While analytical methods are currently available to measure a myriad of biochemical and biophysical properties of proteins it is most important to measure and control those attributes that are known to impact product safety (e.g., immunogenicity), efficacy (e.g., bioactivity and pharmacokinetics), and stability (e.g., aggregation). The notion of determining and measuring critical product quality attributes is consistent with the concept of quality by design (QbD) [364]. Given the current state of biopharmaceutical development it is not possible to assess the impact of all product heterogeneities on product safety, efficacy and stability; however, work needs to be conducted to do exactly this. For instance, if the presence of protein particles in drug product is indeed a critical product quality attribute because of the propensity of particles to be immunogenic, then quantitative methods beyond simple particle visualization need to be developed and utilized in quality control environments. Conversely, if we can determine that product quality attributes do not impact product safety, efficacy, or stability then these parameters may not require measurement or control during product manufacturing.

With the advances in high-throughput technologies that have been applied to upstream and downstream process development, the analytical scientist is faced with an increasing burden of analyzing large numbers of samples in a rapid manner. High-throughput systems have been developed in practically all areas of analytical testing including chromatography (ultraperformance liquid chromatography resins) [365], mass spectrometry (automated sampling from a microwell plate) [366], and capillary electrophoresis (a combinatorial approach to protein separation and peptide mapping) [367]. High-throughput robotic systems are gaining more widespread use as automated liquid sample handlers, especially for ELISA, and multiwell-plate-based applications. Londo and colleagues describe a robotic system used to accelerate process development through on-line chromatographic analysis of purification yields [368]. High-throughput systems are also being developed as a result of the miniaturization of analytical techniques, for instance, the use of disposable microfluidics systems that take advantage of computer chip fabrication technology to develop high-throughput devices that require small sample volumes, and short transit distances [369].

Complementing the high-throughput approaches are the advances in high-resolution analytics. Mass spectrometry in particular has continued to improve in its resolving power and mass accuracy. Fourier transform ion cyclotron resonance mass spectrometry is currently capable of greater than 500,000 resolution ($m/\Delta m$) with less than 1 ppm mass accuracy [370,371]. Through evolution of this field, what are considered today as lower resolution detectors are now more economical and commonly available. These "low resolution" mass spec decetors are more commonly used as an adjunct detector for chromatographic applications, enhancing the utility of many chromatography analyses. Finally, although time consuming, the analysis of protein crystal structure has the potential to provide useful structural information on engineered proteins [372], and product heterogeneities. Developments are underway to improve the ability to generate protein crystals that may be more widely utilized in the future to support protein process development.

6. CONCLUSION

The development of processes for the production of biological compounds is highly complex and requires functional expertise in many diverse fields including molecular biology, cellular biology, cell physiology, chemical engineering, chromatography, filtration science, virology, nucleotide chemistry, protein chemistry, physical chemistry, and immunochemistry. To successfully develop protein production processes, experts from each of these fields must work together in an integrated fashion with an awareness of the impact that individual processing steps or functions may have on the final product. For instance, upstream titer or product quality attributes may significantly impact the requirements for downstream purification processes; other critical product quality attributes delineated by analytical assessments may dictate the selection of upstream clones or cell lines; and downstream purification limitations may potentially dictate upstream processes or formulation presentations. Protein process development is ultimately an exercise in process optimization, trading variables such as processing speed for product titer and product quality.

This review summarizes the current state of the protein process development field. Much progress has been made toward the production of biologics over the last 30 years. Cell lines have been developed that can produce protein titers 100-fold higher than a few decades ago, advances in resins and filters have allowed purification processes to handle these titers with relatively high yields, protein formulations have improved with respect to stability and convenience of delivery, and many analytical testing developments have been made to enable higher resolution and higher throughput testing and characterization. However; despite these advances, the development of protein processes can still be empirical and unpredictable at times. Given the state of the art in protein process development and the current economic environment in the health care industry, the process development scientist will need to continue to develop technologies that will drive greater efficiencies. In the coming years, we can look forward to decreasing development costs, shorter development timelines, and quality improvements for the many patients who benefit from these products.

REFERENCES

1. Murphy A, Perrella J. Overview and brief history of biotechnology. 1993. Available at http://www.woodrow.org/teachers/bi/1993/intro.html.
2. The convention on biological diversity. Article 2. Use of terms. 1992. Available at http://www.cbd.int/convention/articles.shtml?=cbd-02.
3. Biotechnology Industry Orgnization—2007. 2007. Available at http://bio.org/speeches/pubs/er/BiotechGuide.pdf.
4. Jacobi A, Eckermann C, Ambrosius D. Developing an antibody purification process. In: Subramanian G, editor. Bioseparation and Bioprocessing. Weinheim: Wiley-VCH; 2001. p 431–457.
5. Cohen SN, Chang AC, Boyer HW, Helling RB. Construction of biologically functional bacterial plasmids in vitro. Proc Natl Acad Sci USA 1973;70(11):3240–3244.
6. Morrow JF, Cohen SN, Chang AC, Boyer HW, Goodman HM, Helling RB. Replication and transcription of eukaryotic DNA in Escherichia coli. Proc Natl Acad Sci USA 1974; 71(5):1743–1747.
7. Yin J, Li G, Ren X, Herrler G. Select what you need: a comparative evaluation of the advantages and limitations of frequently used expression systems for foreign genes. J Biotechnol 2007;127(3):335–347.
8. Schmidt FR. Recombinant expression systems in the pharmaceutical industry. Appl Microbiol Biotechnol 2004;65(4):363–372.
9. Balbas P, Bolivar F. Back to basics: pBR322 and protein expression systems in *E. coli*. Methods Mol Biol 2004;267:77–90.
10. Ernst S, Garro OA, Winkler S, Venkataraman G, Langer R, Cooney CL, Sasisekharan R. Process simulation for recombinant protein production: cost estimation and sensitivity analysis for heparinase I expressed in *Escherichia coli*. Biotechnol Bioeng 1997;53(6):575–582.
11. Graumann K, Premstaller A. Manufacturing of recombinant therapeutic proteins in microbial systems. Biotechnol J 2006;1:164–186.
12. Sorensen HP, Mortensen KK. Soluble expression of recombinant proteins in the cytoplasm of *Escherichia coli*. Microb Cell Fact 2005;4:1–8.
13. Jana S, Deb JK. Strategies for efficient production of heterologous proteins in *Escherichia coli*. Appl Microbiol Biotechnol 2005;67(3):289–298.
14. Chou CP. Engineering cell physiology to enhance recombinant protein production in *Escherichia coli*. Appl Microbiol Biotechnol 2007;76(3):521–532.
15. Li P, Anumanthan A, Gao XG, Ilangovan K, Suzara VV, Duzgunes N, Renugopalakrishnan V. Expression of recombinant proteins in *Pichia pastoris*. Appl Biochem Biotechnol 2007;142(2):105–124.
16. Balamurugan V, Reddy GR, Suryanarayana VVS. *Pichia pastoris*: a notable heterologous expression system for the production of foreign proteins—vaccines. Indian J Biotechnol 2007; 6:175–186.
17. Reyes-Ruiz JM, Barrera-Saldana HA. Proteins in a DNA world: expression systems for their study. Rev Invest Clin 2006;58(1):47–55.
18. Macauley-Patrick S, Fazenda ML, McNeil B, Harvey LM. Heterologous protein production using the *Pichia pastoris* expression system. Yeast 2005;22(4):249–270.

19. Jahic M, Veide A, Charoenrat T, Teeri T, Enfors SO. Process technology for production and recovery of heterologous proteins with Pichia pastoris. Biotechnol Prog 2006;22(6):1465–1473. Erratum: Biotechnol Prog 2007; 23(2):516.

20. Daly R, Hearn MT. Expression of heterologous proteins in *Pichia pastoris*: a useful experimental tool in protein engineering and production. J Mol Recog 2005;18(2):119–138.

21. Hamilton SR, Davidson RC, Sethuraman N, Nett JH, Jiang Y, Rios S, Bobrowicz P, Stadheim TA, Li H, Choi BK, Hopkins D, Wischnewski H, Roser J, Mitchell T, Strawbridge RR, Hoopes J, Wildt S, Gerngross TU. Humanization of yeast to produce complex terminally sialylated glycoproteins. Science 2006;313(5792):1441–1443.

22. Hu YC. Baculovirus as a highly efficient expression vector in insect and mammalian cells. Acta Pharmacol Sin 2005;26(4):405–416.

23. Donald J, Jarvis DL. Developing baculovirus-insect cell expression systems for humanized recombinant glycoprotein production. Virology 2003;310:1–7.

24. Farid SS. Established bioprocesses for producing antibodies as a basis for future planning. Adv Biochem Eng Biotechnol 2006;101:1–42.

25. Houdebine LM. Transgenic animal bioreactors. Transgenic Res 2000;9(4–5):305–320. Comment.

26. Kamihira M, Nishijima K, Iijima S. Transgenic birds for the production of recombinant proteins. Adv Biochem Eng Biotechnol 2004;91: 171–189.

27. Sang HM. Transgenics, chickens and therapeutic proteins. Vox Sang 2004;87(Suppl 2):164–166.

28. Lillico SG, McGrew MJ, Sherman A, Sang HM. Transgenic chickens as bioreactors for protein-based drugs. Drug Discov Today 2005;10 (3):191–196.

29. Fischer R, Stoger E, Schillberg S, Christou P, Twyman RM. Plant-based production of biopharmaceuticals. Curr Opin Plant Biol 2004;7 (2):152–158.

30. Lienard D, Sourrouille C, Gomord V, Faye L. Pharming and transgenic plants. Biotechnol Ann Rev 2007;13:115–147.

31. Schillberg S, Twyman RM, Fischer R. Opportunities for recombinant antigen and antibody expression in transgenic plants–technology assessment. Vaccine 2005;23(15):1764–1769.

32. Ko K, Koprowski H. Plant biopharming of monoclonal antibodies. Virus Res 2005;111 (1):93–100.

33. Wurm FM. Production of recombinant protein therapeutics in cultivated mammalian cells. Nat Biotechnol 2004;22(11):1393–1398.

34. Butler M. Animal cell cultures: recent achievements and perspectives in the production of biopharmaceuticals. Appl Microbiol Biotechnol 2005;68(3):283–291.

35. Morrow KJ Jr. Advances in antibody manufacturing using mammalian cells. Biotechnol Annu Rev 2007;13:95–113.

36. Werner RG, Noe W, Kopp K, Schluter M. Appropriate mammalian expression systems for biopharmaceuticals. Arzneimittelforschung 1998;48(8):870–880.

37. Kao FT, Puck TT. Genetics of somatic mammalian cells. VII. Induction and isolation of nutritional mutants in Chinese hamster cells. Proc Natl Acad Sci USA 1968;60(4):1275–1281.

38. Kao FT, Puck TT. Genetics of somatic mammalian cells. XIV. Genetic analysis in vitro of auxotrophic mutants. J Cell Physiol 1972; 80(1):41–50.

39. Puck TT, Kao FT. Genetics of somatic mammalian cells. V. Treatment with 5-bromodeoxyuridine and visible light for isolation of nutritionally deficient mutants. Proc Natl Acad Sci USA 1967;58(3):1227–1234.

40. Kao F, Chasin L, Puck TT. Genetics of somatic mammalian cells. X. Complementation analysis of glycine-requiring mutants. Proc Natl Acad Sci USA 1969;64(4):1284–1291.

41. Kao FT, Puck TT. Induction and isolation of auxotrophic mutants in mammalian cells. Methods Cell Biol 1974;8(0):23–39.

42. Urlaub G, Chasin LA. Isolation of Chinese hamster cell mutants deficient in dihydrofolate reductase activity. Proc Natl Acad Sci USA 1980;77(7):4216–4220.

43. Urlaub G, Mitchell PJ, Kas E, Chasin LA, Funanage VL, Myoda TT, Hamlin J. Effect of gamma rays at the dihydrofolate reductase locus: deletions and inversions. Somatic Cell Mol Genet 1986;12(6):555–566.

44. Sinacore MS, Charlebois TS, Harrison S, Brennan S, Richards T, Hamilton M, Scott S, Brodeur S, Oakes P, Leonard M, Switzer M, Anagnostopoulous A, Foster B, Harris A, Jankowski M, Bond M, Martin S, Adamson SR. CHO DUKX cell lineages preadapted to growth in serum-free suspension culture enable rapid development of cell culture processes for the manufacture of recombinant proteins. Biotechnol Bioeng 1996;52(4): 518–528.

45. Rasmussen B, Davis R, Thomas JN, Reddy P. Isolation, characterization and recombinant protein expression in Veggie-CHO: a serum-free CHO host cell line. Cytotechnology 1998;28:31–42.
46. Kaufman RJ, Sharp PA. Amplification and expression of sequences cotransfected with a modular dihydrofolate reductase complementary DNA gene. J Mol Biol 1982;159(4):601–621.
47. Cockett MI, Bebbington CR, Yarranton GT. The use of engineered E1A genes to transactivate the hCMV-MIE promoter in permanent CHO cell lines. Nucleic Acids Res 1991;19(2):319–325.
48. Bebbington CR, Renner G, Thomson S, King D, Abrams D, Yarranton GT. High-level expression of a recombinant antibody from myeloma cells using a glutamine synthetase gene as an amplifiable selectable marker. Biotechnology 1992;10(2):169–175.
49. Cartier M, Chang MW, Stanners CP. Use of the *Escherichia coli* gene for asparagine synthetase as a selective marker in a shuttle vector capable of dominant transfection and amplification in animal cells. Mol Cell Biol 1987;7(5):1623–1628.
50. Birch JR, Racher AJ. Antibody production. Adv Drug Del Rev 2006;58(5–6):671–685.
51. Weymouth LA, Barsoum J. Genetic engineering in mammalian cells. In: Thilly WG, editor. Mammalian Cell Technology. Stoneham: Butterworth Publishers; 1986. p 9–62.
52. Lucas BK, Giere LM, DeMarco RA, Shen A, Chisholm V, Crowley CW. High-level production of recombinant proteins in CHO cells using a dicistronic DHFR intron expression vector. Nucleic Acids Res 1996;24(9):1774–1779.
53. Aldrich TL, Thomas JN, Morris AE. Improved bicistronic mammalian expression vectors using expression augmenting sequence element (EASE). Cytotechnology 1998;28:9–17.
54. Gallie DR. The cap and poly(A) tail function synergistically to regulate mRNA translational efficiency. Genes Dev 1991;5(11):2108–2116.
55. Jang SK, Wimmer E. Cap-independent translation of encephalomyocarditis virus RNA: structural elements of the internal ribosomal entry site and involvement of a cellular 57-kD RNA-binding protein. Genes Dev 1990;4(9):1560–1572.
56. Running-Deer J, Allison DS. High-level expression of proteins in mammalian cells using transcription regulatory sequences from the Chinese hamster EF-1alpha gene. Biotechnol Prog 2004;20(3):880–889.
57. Allison DS. Hamster EF-1α transcriptional regulatory DNA. ICOS Corporation 1999; 847218 [US 5888809].
58. Benton T, Chen T, McEntee M, Fox B, King D, Crombie R, Thomas TC, Bebbington CR. The use of UCOE vectors in combination with a preadapted serum free, suspension cell line allows for rapid production of large quantities of protein. Cytotechnology 2002;38:43–46.
59. Kwaks TH, Barnett P, Hemrika W, Siersma T, Sewalt RG, Satijn DP, Brons JF, van Blokland R, Kwakman P, Kruckeberg AL, Kelder A, Otte AP. Identification of anti-repressor elements that confer high and stable protein production in mammalian cells. Nat Biotechnol 2003;21(5):553–558. Erratum: Nat Biotechnol 2003;21(7):822.
60. Kim JM, Kim JS, Park DH, Kang HS, Yoon J, Baek K, Yoon Y. Improved recombinant gene expression in CHO cells using matrix attachment regions. J Biotechnol 2004;107(2):95–105.
61. Beitel LK, McArthur JG, Stanners CP. Sequence requirements for the stimulation of gene amplification by a mammalian genomic element. Gene 1991;102(2):149–156.
62. Hollis G, Mark GE, Homologous recombination antibody expression system for murine cells. PCT/IB95/00014:1995; 23.
63. Bode J, Gotze S, Ernst E, Hüsemann Y, Baer A, Seibler J, Mielke C. Architecture and utilization of highly expressed genomic sites. In Bernardi G (Ed.). New Comprehensive Biochemistry. Amsterdam: Elsevier 2003; 38:551–572. Makrides S (Volume ed.). Gene Transfer and Expression in Mammalian Cells, Chap. 20.
64. Wilson TJ, Kola I. The LoxP/CRE system and genome modification. Methods Mol Biol 2001; 158:83–94.
65. Kingston RE, Kaufman RJ, Bebbington CR, Rolfe MR. Amplification using CHO cell expression vectors. In: Ausubel FM (Ed): Current Protocols in Molecular Biology. New York: John Wiley & Sons. 2002. Chap. 16: Unit 16.23.1–16.12.13.
66. Shen AY, Van de Goor J, Zheng L, Reyes A, Krummen LA. Recombinant DNA technology and cell line development. In: Ozturk SS, Hu W-S, editors. Cell Culture Technology for Pharmaceutical and Cell-Based Therapies. New York: CRC Press; 2006. p 15–40.
67. Graham FL, van der Eb AJ. A new technique for the assay of infectivity of human adenovirus 5 DNA. Virology 1973;52(2):456–467.

68. Wilson SP, Smith LA. Addition of glycerol during DNA exposure enhances calcium phosphate transfection. Anal Biochem 1997;246(1):148–150.
69. Canatella PJ, Karr JF, Petros JA, Prausnitz MR. Quantitative study of electroporation-mediated molecular uptake and cell viability. Biophys J 2001;80(2):755–764.
70. Felgner PL, Ringold GM. Cationic liposome-mediated transfection. Nature 1989;337(6205):387–388.
71. Wurm FM, Pallavicini MG. Effects of methotrexate on recombinant sequences in mammalian cells. In: Kellems RE,Editor. Gene Amplification in Mammalian Cells: A Comprehensive Guide. Boca Raton: CRC press; 1993. p 85–94.
72. Weaver JC, Bliss JG, Powell KT, Harrison GI, Williams GB. Rapid clonal growth measurements at the single-cell level: gel microdroplets and flow cytometry. Biotechnology 1991;9(9):873–877.
73. Gray F, Kenney JS, Dunne JF. Secretion capture and report web: use of affinity derivatized agarose microdroplets for the selection of hybridoma cells. J Immunol Methods 1995;182(2):155–163.
74. Schiff LJ. Review: production, characterization, and testing of banked mammalian cell substrates used to produce biological products. In Vitro Cell Dev Biol 2005;41(6):5–70.
75. Van Wezel AL. Growth of cell-strains and primary cells on micro-carriers in homogeneous culture. Nature 1967;216:64–65.
76. Hu W-S, Giard DJ, Wang DIC. Serial propagation of mammalian cells on microcarriers. Biotechnol Bioeng 1985;27:1466–1476.
77. Hu W-S, Wang DIC. Selection of microcarrier diameter for the cultivation of mammalian cells on microcarriers. Biotechnol Bioeng 1987;30:548–557.
78. Sambanis A, Hu W-S. Cell culture bioreactors. In: Rehm H, Reed G,Editors. Biotechnology. New York: VCH; 1993. p 106–126.
79. Himes VB, Hu W-S. Attachment and growth of mammalian cells on microcarriers with different ion exchange capacities. Biotechnol Bioeng 1987;29:1155–1163.
80. Warnock JN, Al-Rubeai M. Bioreactor systems for the production of biopharmaceuticals from animal cells. Biotechnol Appl Biochem 2006;45:1–12.
81. Piret JM, Cooney CL. Mammalian cell and protein distributions in ultrafiltration hollow fiber bioreactors. Biotechnol Bioeng 1990;36:902–910.
82. Piret JM, Cooney CL. Model of oxygen transport limitations in hollow fiber bioreactors. Biotechnol Bioengin 1991;37:80–92.
83. Fenge C, Lullau E, Cell culture bioreactors. In: Ozturk SS, Hu W-S, editors. Cell Culture Technology for Pharmaceutical and Cell-Based Therapies. Boca Raton: CRC Press; 2006. p 155–224.
84. Burgener A, Butler M. Medium development. In: Osturk SS, Hu W-S, editors. Cell Culture technology for Pharmaceutical and Cell-Based Therapies. Boca Raton: CRC Press; 2006. p 41–79.
85. Thomas JN. Nutrients, oxygen and pH In: Thilly WG,Editor. Mammalian Cell Technology. Stoneham: Butterworth Publishers; 1986. p 109–130.
86. Thomas JN. Mammalian cell physiology. Bioprocess Technol 1990;10:93–145.
87. Godia F, Cairo JJ. Cell metabolism. In: Osturk SS, Hu W-S,Editors. Cell Culture Technology for Pharmaceutical and Cell-Based Therapies. Boca Raton: CRC Press; 2006. p 81–112.
88. Seth G, Hossler P, Yee JC, Hu W-S. Engineering cells for cell culture bioprocessing: physiological fundamentals. Adv Biochem Eng Biotechnol 2006;101:119–164.
89. Majors BS, Betenbaugh MJ, Chiang GG. Links between metabolism and apoptosis in mammalian cells: applications for anti-apoptosis engineering. Metabol Eng 2007;9:317–326.
90. Chiang G, Sisk W. Bcl-xL mediates increased production of humanized monoclonal antibodies in Chinese hamster ovary cells. Biotechnol Bioeng 2005;91(7):779–792.
91. Arden N, Betenbaugh MJ. Life and death in mammalian cell culture: strategies for apoptosis inhibition. Trends Biotechnol 2004;22(4):174–180.
92. Wlaschin KF, Hu W-S. Engineering cell metabolism for high-density cell culture via manipulation of sugar transport. J Biotechnol 2007;131:168–176.
93. Tan H, Lee M, Yap M, Wang DIC. Overexpression of cold-inducible RNA-binding protein increases interferon-γ production in Chinese-hamster ovary cells. Biotechnol Appl Biochem 2008;49:247–257.
94. Korke R, De Leon Gatti M, Lau ALY, Lim JWE, Seow TK, Chung MCM, Hu W-S. Large scale gene expression profiling of metabolic shift of mammalian cells in culture. J Biotechnology. 2004;107:1–17.

95. Yee JC, de Leon Gatti M, Philp RJ, Yap M, Hu W-S. Genomic and proteomic exploration of CHO and hybridoma cells under sodium butyrate treatment. Biotechnol Bioeng 2007;99(5):1186–1204.
96. Griffin T, Seth G, Xie H, Bandhakavi S, Hu W-S. Advancing mammalian cell culture engineering using genome-scale technologies. Trends Biotechnol 2007;25(9):401–408.
97. Wlaschin KF, Seth G, Hu W-S. Toward genomic cell culture engineering. Cytotechnol 2006;50:121–140.
98. Leser EW, Asenjo JA. Rational design of purification processes for recombinant proteins. J Chromatogr A 1992;584(1):43–57.
99. Wolk B, Bezy P, Arnold R, Blank G. Characteristics of a good antibody manufacturing process. Bioprocess Int 2003;1(9):50–52, 54,56-58.
100. Curling J, Gottschalk U. Process chromatography: five decades of innovation. Biopharm Int 2007;1(1):70–94.
101. Farid SS. Process economics of industrial monoclonal antibody manufacture. J Chromatogr B 2007;848(1):8–18.
102. Ransohoff TC, Murphy MK, Levine HL. Automation of biopharmaceutical purification processes. BioPharm 1990;3:20–24.
103. Roque ACA, Silva CSO, Taipa MA. Affinity-based methodologies and ligands for antibody purification: advances and perspectives. J Chromatogr A 2007;1160(1–2):44–55.
104. Farid SS. Economic drivers and trade-offs in antibody purification processes. BioPharm Int 2008;21(Suppl 3):37–42.
105. Werner RG. Economic aspects of commercial manufacture of biopharmaceuticals. J Biotechnol 2004;113(1–3):171–182.
106. Isaacs IJ. The strategy employed for purification of protein-based biopharmaceutical products. Biotechnol 1996;6(3):88–92.
107. Graslund S, Nordlund P, Weigelt J, Hallberg BM, Bray J, Gileadi O, Knapp S, Oppermann U, Arrowsmith C, Hui R, Ming J, dhe-Paganon S, Park HW, Savchenko A, Yee A, Edwards A, Vincentelli R, Cambillau C, Kim R, Kim SH, Rao Z, Shi Y, Terwilliger TC, Kim CY, Hung LW, Waldo GS, Peleg Y, Albeck S, Unger T, Dym O, Prilusky J, Sussman JL, Stevens RC, Lesley SA, Wilson IA, Joachimiak A, Collart F, Dementieva I, Donnelly MI, Eschenfeldt WH, Kim Y, Stols L, Wu R, Zhou M, Burley SK, Emtage JS, Sauder JM, Thompson D, Bain K, Luz J, Gheyi T, Zhang F, Atwell S, Almo SC, Bonanno JB, Fiser A, Swaminathan S, Studier FW, Chance MR, Sali A, Acton TB, Xiao R, Zhao L, Ma LC, Hunt JF, Tong L, Cunningham K, Inouye M, Anderson S, Janjua H, Shastry R, Ho CK, Wang D, Wang H, Jiang M, Montelione GT, Stuart DI, Owens RJ, Daenke S, Schutz A, Heinemann U, Yokoyama S, Bussow K, Gunsalus KC. Protein production and purification. Nat Methods. 2008;5(2):135–146.
108. Shukla AA, Kandula JR. Harvest and recovery of monoclonal antibodies from large-scale mammalian cell culture. BioPharm Int 2008;21(5):34–36 38, 40, 42, 44–45.
109. Yigzaw Y, Piper R, Tran M, Shukla AA. Exploitation of the adsorptive properties of depth filters for host cell protein removal during monoclonal antibody purification. Biotechnol Prog 2006;22(1):288–296.
110. Ladisch MR, Velayudhan A. Scale-up techniques in bioseparation processes. Food Feed Chem J 1995;10:113–138.
111. Scopes RK. The protein purification laboratory. In: Cantor CR, editor. Protein Purification Principles and Practice. 3rd ed. New York: Springer-Verlag; 1994. p 1–21.
112. Shukla AA, Yigzaw Y. Modes of preparative chromatography. In: Shukla AA, Etzel MR, Gadam S,Editors. Process scale bioseparations for the biopharmaceutical industry. Boca Raton: CRC Press; 2007. p 179–275.
113. Shukla AA, Cramer SM. Bioseparations by displacement chromatography. Separ Sci Technol 2000;2:379–415.
114. Swartz JR. Advances in *Escherichia coli* production of therapeutic proteins. Curr Opin Biotechnol 2001;12(2):195–201.
115. Baneyx F. Recombinant protein expression in *Escherichia coli*. Curr Opin Biotechnol 1999;10(5):411–421.
116. Carrio MM, Villaverde A. Construction and deconstruction of bacterial inclusion bodies. J Biotechnol 2002;96(1):3–12.
117. Lilie H, Schwarz E, Rudolph R. Advances in refolding of proteins produced in *E. coli*. Curr Opin Biotechnol 1998;9(5):497–501.
118. Sadana A. Protein refolding and inactivation during bioseparation: bioprocessing implications. Biotechnol Bioeng 1995;48:481–489.
119. Arakawa T, Tsumoto K, Kita Y, Chang B, Ejima D. Biotechnology applications of amino acids in protein purification and formulations. Amino Acids 2007;33(4):587–605.
120. Singh SM, Panda AK. Solubilization and refolding of bacterial inclusion body proteins. J Biosci Bioeng 2005;99(4):303–310.

121. Geng X, Wang C. Protein folding liquid chromatography and its recent developments. J Chromatogr B 2007;849(1–2):69–80.
122. De Bernardez Clark E, Schwarz E, Rudolph R. Inhibition of aggregation side reactions during in vitro protein folding. Methods Enzymol 1999;309:217–236.
123. Cowgill C, Ozturk AG, St John R. Protein refolding and scale up. In: Shukla AA, Etzel MR, Gadam S,Editors. Process scale bioseparations for the biopharmaceutical industry. Boca Raton: CRC Press; 2007. p 123–178.
124. Sahdev S, Khattar SK, Saini KS. Production of active eukaryotic proteins through bacterial expression systems: a review of the existing biotechnology strategies. Mol Cell Biochem 2008;307(1–2):249–264.
125. Wong HH, O'Neill BK, Middelberg AP. Centrifugal recovery and dissolution of recombinant Gly-IGF-II inclusion-bodies: the impact of feed rate and re-centrifugation on protein yield. Bioseparation 1996;6(3):185–192.
126. Khalilzadeh R, Shojaosadati SA, Maghsoudi N, Mohammadian-Mosaabadi J, Mohammadi MR, Bahrami A, Maleksabet N, Nassiri-Khalilli MA, Ebrahimi M, Naderimanesh H. Process development for production of recombinant human interferon-gamma expressed in Escherichia coli. J Ind Microbiol Biotechnol 2004;31(2):63–69.
127. Hutchinson N, Bingham N, Murrell N, Farid S, Hoare M. Shear stress analysis of mammalian cell suspensions for prediction of industrial centrifugation and its verification. Biotechnol Bioengin 2006;95(3):483–491.
128. Maybury JP, Mannweiler K, Titchener-Hooker NJ, Hoare M, Dunnhill P. The performance of a scaled down industrial disc stack centrifuge with a reduced feed material requirement. Bioprocess Eng 1998;18:191–199.
129. Riske F, Schroeder J, Belliveau J, Kang X, Kutzko J, Menon MK. The use of chitosan as a flocculant in mammalian cell culture dramatically improves clarification throughput without adversely impacting monoclonal antibody recovery. J Biotechnol 2007;128(4):813–823.
130. van Reis R, Zydney A. Membrane separations in biotechnology. Curr Opin Biotechnol 2001;12(2):208–211.
131. Fahrner RL, Knudsen HL, Basey CD, Galan W, Feuerhelm D, Vanderlaan M, Blank GS. Industrial purification of pharmaceutical antibodies: development, operation, and validation of chromatography processes. Biotechnol Genet Eng Rev 2001;18:301–327.
132. Shukla AA, Hubbard B, Tressel T, Guhan S, Low D. Downstream processing of monoclonal antibodies: application of platform approaches. J Chromatogr B 2007;848(1):28–39.
133. Fahrner RL, Iyer HV, Blank GS. The optimal flow rate and column length for maximum production rate of protein A affinity chromatography. Bioprocess Eng 1999;21(4):287–292.
134. Gronberg A, Monie E, Murby M, Rodrigo G, Wallby E, Johansson HJ. A strategy for developing a monoclonal antibody purification platform. Bioprocess Int 2007;5(1):48–50, 52–55.
135. Hober S, Nord K, Linhult M, Protein A chromatography for antibody purification. J Chromatogr B 2007;848(1):40–47.
136. GE-Healthcare Instructions. 11-0026-01 AB affinity chromatography MabSelect SuRe™. 2005; 1-23.
137. Ghose S, Allen M, Hubbard B, Brooks C, Cramer SM. Antibody variable region interactions with protein A: implications for the development of generic purification processes. Biotechnol Bioeng 2005;92(6):665–673.
138. Carter-Franklin JN, Victa C, McDonald P, Fahrner R. Fragments of protein A eluted during protein A affinity chromatography. J Chromatography A 2007;1163(1–2):105–111.
139. O'Leary R, Feuerhelm D, Peers D, Xu Y, Blank GS. Determining the useful lifetime of chromatography resins. BioPharm 2001;14(9):10–18.
140. Gagnon P. Polishing methods for monoclonal IgC purification. In: Shukla AA, Etzel MR, Gadam S,Editors. Pocess scale bioseparations for the biopharmaceutical industry. Boca Raton: CRC Press; 2007. p 491–505.
141. Vedantham G, Vunnum S, Tressel T, Hubbard B. mAb purification processes: current trends and future directions. IBC's 3rd BioProcess International™ Asia Pacific Conference; 2008.
142. U.S. FDA. Guidance for industry Q5A viral safety evaluation of biotechnology products derived from cell lines of human animal origin. 1998. Available at http:///www.fda.gov/cber/guidelines.htm.
143. Curtis S, Lee K, Blank GS, Brorson K, Xu Y. Generic/matrix evaluation of SV40 clearance by anion exchange chromatography in flow-through mode. Biotechnol Bioeng 2003;84(2):179–186.
144. Zhou JX, Dehghani H. Viral clearance: innovative versus classical methods: theoretical and practical concerns. Am Pharm Rev 2006;9(5):74–82.

145. Walter JK, Nothelfer F, Werz W, Validation of viral safety for pharmaceutical proteins. In: Subramanian G, editor. Bioseparation and Bioprocessing. Weinheim: Wiley-VCH; 1998. p 465–496.
146. Wu Y, Ahmed A, Waghmare R, Genest P, Issacson S, Krishnan M, Kahn DW. Validation of adventitious virus removal by virus filtration: a novel procedure for monoclonal antibody processes. BioProcess Int 2008;6(5):54 56, 58–60.
147. Cartwright T. Protein purification applications. National Library 2001;2(2):49–71.
148. Zhou JX, Tressel T. Current practice and future strategies for viral clearance in mAb downstream production. Am Pharm Rev 2008;11(1):109–110, 112, 114–121.
149. van Reis R. New frontiers in bioseparation technology. Adv Filtr Sep Technol 13A:1999; 4–7.
150. Coffman JL, Kramarczyk JF, Kelley BD. High-throughput screening of chromatographic separations. I. Method development and column modeling. Biotechnol Bioeng 2008;100 (4):605–618.
151. Kramarczyk JF, Kelley BD, Coffman JL. High-throughput screening of chromatographic separations. II. Hydrophobic interaction. Biotechnol Bioeng 2008;100(4):707–720.
152. Wensel DL, Kelley BD, Coffman JL. High-throughput screening of chromatographic separations. III. Monoclonal antibodies on ceramic hydroxyapatite. Biotechnol Bioeng 2008;100 (5):839–854.
153. Kelley BD, Switzer M, Bastek P, Kramarczyk JF, Molnar K, Yu T, Coffman J. High-throughput screening of chromatographic separations. IV. Ion-exchange. Biotechnol Bioeng 2008;100 (5):950–963.
154. Mehta A, Tse ML, Fogle J, Len A, Shrestha R, Fontes N, Lebreton B, Wolk B, van Reis R. Purifying therapeutic monoclonal antibodies. Chem Eng Prog 2008;104(5):S14–S20.
155. Kelley B. Very large scale monoclonal antibody purification: the case for conventional unit operations. Biotechnol Prog 2007;23(5):995–1008.
156. Kelley BD, Tobler SA, Brown P, Coffman JL, Godavarti R, Iskra T, Switzer M, Vunnum S. Weak partitioning chromatography for anion exchange purification of monoclonal antibodies. Biotechnol Bioeng 2008;101(3):553–566.
157. Shultz J. Large-scale pool-less purification: applications towards reducing overhead cost and production time. AIChE National Meeting; 2006.
158. Shultz J, Case study: How to develop a 1-ton microbial protein process, when the world is not ready to manufacture it. Presented at Recovery of Biological Products XIII, Quebec, Canada, Jun 22–26 2008.
159. Datar RV, Cartwright T, Rosen CG. Process economics of animal cell and bacterial fermentations: a case study analysis of tissue plasminogen activator. Biotechnology 1993; 11(3):349–357.
160. Low D, O'Leary R, Pujar NS. Future of antibody purification. J Chromatogr B 2007;848 (1):48–63.
161. Vunnum S, Vedantham G, Hubbard B. Protein A based affinity chromatography. In Gottschalk U, editor. Process Scale Purification of Antibodies. John Wiley & Sons, Inc., 2009. p 79–102.
162. Imamoglu S. Simulated moving bed chromatography (SMB) for application in bioseparation. Adv Biochem Eng Biotechnol 2002;76: 211–231.
163. Boi C. Membrane adsorbers as purification tools for monoclonal antibody purification. J Chromatogr B 2007;848(1):19–27.
164. Gottschalk U. Bioseparation in antibody manufacturing: the good, the bad and the ugly. Biotechnol Prog 2008;24:496–503.
165. Etzel MR, Riordan WT. Membrane chromatography: breakthrough curves and viral clearance. In: Shukla AA, Etzel MR, Gadam S, editors. Process Scale Bioseparations for the Biopharmaceutical Industry. Boca Raton: CRC Press; 2007. p 277–365.
166. Zhou JX, Tressel T. Basic concepts in Q membrane chromatography for large-scale antibody production. Biotechnol Prog 2006;22 (2):341–349.
167. Zhou JX, Solamo F, Hong T, Shearer M, Tressel T. Viral clearance using disposable systems in monoclonal antibody commercial downstream processing. Biotechnol Bioeng 2008; 100(3):488–496.
168. Przybycien TM, Pujar NS, Steele LM. Alternative bioseparation operations: life beyond packed-bed chromatography. Curr Opin Biotechnol 2004;15(5):469–478.
169. Chang BS, Hershenson S. Practical approaches to protein formulation development. Pharm Biotechnol 2002;13:1–25.
170. Daugherty AL, Mrsny RJ. Formulation and delivery issues for monoclonal antibody therapeutics. Adv Drug Del Rev 2006;58(5–6): 686–706.

171. Krishnan S, Pallito MM, Ricci MS. Development of formulations for therapeutic monoclonal antibodies and Fc fusion proteins. Formulation and Process Development Strategies for Manufacturing of a Biopharmaceutical. John Wiley & Sons, Inc., 2010. P. 381–425.

172. Wang W, Singh S, Zeng DL, King K, Nema S. Antibody structure, instability, and formulation. J Pharm Sci 2007;96(1):1–26.

173. Manning MC, Patel K, Borchardt RT. Stability of protein pharmaceuticals. Pharm Res 1989;6(11):903–918.

174. Cleland JL, Powell MF, Shire SJ. The development of stable protein formulations: a close look at protein aggregation, deamidation, and oxidation. Crit Rev Ther Drug Carrier Syst 1993;10(4):307–377.

175. Volkin DB, Mach H, Middaugh CR. Degradative covalent reactions important to protein stability. Methods Mol Biol 1995;40:35–63.

176. Chu GC, Chelius D, Xiao G, Khor HK, Coulibaly S, Bondarenko PV. Accumulation of succinimide in a recombinant monoclonal antibody in mildly acidic buffers under elevated temperatures. Pharm Res 2007;24(6):1145–1156.

177. Dehart MP, Anderson BD. The role of the cyclic imide in alternate degradation pathways for asparagine-containing peptides and proteins. J Pharm Sci 2007;96(10):2667–2685.

178. Wakankar AA, Borchardt RT. Formulation considerations for proteins susceptible to asparagine deamidation and aspartate isomerization. J Pharma Sci 2006;95(11):2321–2336.

179. Patel K, Borchardt RT. Deamidation of asparaginyl residues in proteins: a potential pathway for chemical degradation of proteins in lyophilized dosage forms. J Parenter Sci Technol 1990;44(6):300–301.

180. Wakankar AA, Borchardt RT, Eigenbrot C, Shia S, Wang YJ, Shire SJ, Liu JL. Aspartate isomerization in the complementarity-determining regions of two closely related monoclonal antibodies. Biochemistry 2007;46(6):1534–1544.

181. Li S, Nguyen TH, Schoneich C, Borchardt RT. Aggregation and precipitation of human relaxin induced by metal-catalyzed oxidation. Biochemistry 1995;34(17):5762–5772.

182. Khossravi M, Shire SJ, Borchardt RT. Evidence for the involvement of histidine A(12) in the aggregation and precipitation of human relaxin induced by metal-catalyzed oxidation. Biochemistry 2000;39(19):5876–5885.

183. Thirumangalathu R, Krishnan S, Bondarenko P, Speed-Ricci M, Randolph TW, Carpenter JF, Brems DN. Oxidation of methionine residues in recombinant human interleukin-1 receptor antagonist: implications of conformational stability on protein oxidation kinetics. Biochemistry 2007;46(21):6213–6224.

184. Liu D, Ren D, Huang H, Dankberg J, Rosenfeld R, Cocco MJ, Li L, Brems DN, Remmele RL Jr. Structure and stability changes of human IgG1 Fc as a consequence of methionine oxidation. Biochemistry 2008;47(18):5088–5100.

185. Lam XM, Yang JY, Cleland JL. Antioxidants for prevention of methionine oxidation in recombinant monoclonal antibody HER2. J Pharm Sci 1997;86(11):1250–1255.

186. Yu L, Vizel A, Huff MB, Young M, Remmele RL Jr, He B. Investigation of N-terminal glutamate cyclization of recombinant monoclonal antibody in formulation development. J Pharm Biomed Anal 2006;42(4):455–463.

187. Kerwin BA, Remmele RL Jr. Protect from light: photodegradation and protein biologics. J Pharm Sci 2007;96(6):1468–1479.

188. Gadgil HS, Bondarenko PV, Pipes GD, Dillon TM, Banks D, Abel J, Kleemann GR, Treuheit MJ. Identification of cysteinylation of a free cysteine in the Fab region of a recombinant monoclonal IgG1 antibody using Lys-C limited proteolysis coupled with LC/MS analysis. Anal Biochemistry 2006;355(2):165–174.

189. Banks DD, Gadgil HS, Pipes GD, Bondarenko PV, Hobbs V, Scavezze JL, Kim J, Jiang XR, Mukku V, Dillon TM. Removal of cysteinylation from an unpaired sulfhydryl in the variable region of a recombinant monoclonal IgG1 antibody improves homogeneity, stability, and biological activity. J Pharm Sci 2008;97(2):775–790.

190. Andya JD, Maa YF, Costantino HR, Nguyen PA, Dasovich N, Sweeney TD, Hsu CC, Shire SJ. The effect of formulation excipients on protein stability and aerosol performance of spray-dried powders of a recombinant humanized anti-IgE monoclonal antibody. Pharm Res 1999;16(3):350–358.

191. Gadgil HS, Bondarenko PV, Pipes G, Rehder D, McAuley A, Perico N, Dillon T, Ricci M, Treuheit M. The LC/MS analysis of glycation of IgG molecules in sucrose containing formulations. J Pharm Sci 2007;96(10):2607–2621.

192. Quan C, Alcala E, Petkovska I, Matthews D, Canova-Davis E, Taticek R, Ma S. A study in glycation of a therapeutic recombinant humanized monoclonal antibody: where it is, how it

got there, and how it affects charge-based behavior. Anal Biochem 2008;373(2): 179–191.
193. Tzannis ST, Hrushesky WJM, Wood PA, Przybycien TM. Adsorption of a formulated protein on a drug delivery device surface. J Colloid Interface Sci 1997;189(2):216–228.
194. Roberts CJ. Non-native protein aggregation kinetics. Biotechnol Bioeng 2007;98(5): 927–938.
195. Chi EY, Krishnan S, Randolph TW, Carpenter JF. Physical stability of proteins in aqueous solution: mechanism and driving forces in nonnative protein aggregation. Pharm Res 2003;20 (9):1325–1336.
196. Wang W. Protein aggregation and its inhibition in biopharmaceutics. International J Pharm 2005;289(1–2):1–30.
197. Cromwell ME, Hilario E, Jacobson F. Protein aggregation and bioprocessing. AAPS J 2006;8 (3):E572–E579.
198. Carpenter JF, Kendrick BS, Chang BS, Manning MC, Randolph TW. Inhibition of stress-induced aggregation of protein therapeutics. Methods Enzymol 1999;309:236–255.
199. Carpenter JF, Randolph TW, Jiskoot W, Crommelin DJ, Middaugh CR, Winter G, Fan YX, Kirshner S, Verthelyi D, Kozlowski S, Clouse KA, Swann PG, Rosenberg A, Cherney B. Overlooking subvisible particles in therapeutic protein products: gaps that may compromise product quality. J Pharm Sci 2009;98 (4):1201–1205.
200. Patten PA, Schellekens H. The immunogenicity of biopharmaceuticals. Lessons learned and consequences for protein drug development. Dev Biol 2003;112:81–97.
201. Demeule B, Gurny R, Arvinte T. Where disease pathogenesis meets protein formulation: renal deposition of immunoglobulin aggregates. Eur J Pharm Biopharm 2006;62(2):121–130.
202. Gokarn YR, Kosky A, Drad E, McAuley A, Remmele RL. Excipients for protein drugs. In: Ashok Katdare MVC, editor Excipient Development for Pharmaceutical, Biotechnology, and Drug Delivery Systems. New York: CRC Press; 2006. p 291–332.
203. Arakawa T, Prestrelski SJ, Kenney WC, Carpenter JF. Factors affecting short-term and long-term stabilities of proteins. Adv Drug Del Rev 2001;46(1–3):307–326.
204. Laursen T, Hansen B, Fisker S. Pain perception after subcutaneous injections of media containing different buffers. Basic Clin Pharmacol Toxicol 2006;98(2):218–221.
205. Liu J, Nguyen MD, Andya JD, Shire SJ. Reversible self-association increases the viscosity of a concentrated monoclonal antibody in aqueous solution. J Pharm Sci 2005;94 (9):1928–1940. Erratum: J Pharm Sci 2006;95 (1): 234–235.
206. Wang W, Wang YJ, Wang DQ. Dual effects of Tween 80 on protein stability. Int J Pharm 2008;347(1–2):31–38.
207. Roy S, Jung R, Kerwin BA, Randolph TW, Carpenter JF. Effects of benzyl alcohol on aggregation of recombinant human interleukin-1-receptor antagonist in reconstituted lyophilized formulations. J Pharm Sci 2005;94 (2):382–396.
208. Ha E, Wang W, Wang YJ. Peroxide formation in polysorbate 80 and protein stability. J Pharm Sci 2002;91(10):2252–2264.
209. Stoner MR, Fischer N, Nixon L, Buckel S, Benke M, Austin F, Randolph TW, Kendrick BS. Protein-solute interactions affect the outcome of ultrafiltration/diafiltration operations. J Pharm Sci 2004;93(9):2332–2342.
210. Saluja A, Kalonia DS. Nature and consequences of protein–protein interactions in high protein concentration solutions. Int J Pharm 2008;358(1–2):1–15.
211. Kanai S, Liu J, Patapoff TW, Shire SJ. Reversible self-association of a concentrated monoclonal antibody solution mediated by Fab–Fab interaction that impacts solution viscosity. J Pharm Sci 2008;97(10):4219–4227.
212. Gokarn YR, Kras E, Nodgaard C, Dharmavaram V, Fesinmeyer RM, Hultgen H, Brych S, Remmele RL Jr, Brems DN, Hershenson S. Self-buffering antibody formulations. J Pharm Sci 2008;97(8):3051–3066.
213. Chen B, Bautista R, Yu K, Zapata GA, Mulkerrin MG, Chamow SM. Influence of histidine on the stability and physical properties of a fully human antibody in aqueous and solid forms. Pharm Res 2003;20 (12):1952–1960.
214. Shire SJ, Shahrokh Z, Liu J. Challenges in the development of high protein concentration formulations. J Pharm Sci 2004;93 (6):1390–1402.
215. Bhatnagar BS, Bogner RH, Pikal MJ. Protein stability during freezing: separation of stresses and mechanisms of protein stabilization. Pharm Dev Technol 2007;12(5):505–523.
216. Carpenter JF, Pikal MJ, Chang BS, Randolph TW. Rational design of stable lyophilized protein formulations: some practical advice. Pharm Res 1997;14(8):969–975.
217. Nail SL, Jiang S, Chongprasert S, Knopp SA. Fundamentals of freeze-drying. Pharm Biotechnol 2002;14:281–360.

218. Wang W. Lyophilization and development of solid protein pharmaceuticals. Int J Pharm 2000;203(1–2):1–60.
219. Breen ED, Curley JG, Overcashier DE, Hsu CC, Shire SJ. Effect of moisture on the stability of a lyophilized humanized monoclonal antibody formulation. Pharm Res 2001;18(9):1345–1353.
220. Iacocca RG, Allgeier M. Corrosive attack of glass by a pharmaceutical compound. J Mat Sci 2007;42:801–811.
221. Sluzky V, Tamada JA, Klibanov AM, Langer R. Kinetics of insulin aggregation in aqueous solutions upon agitation in the presence of hydrophobic surfaces. Proc Natl Acad Sci USA 1991;88(21):9377–9381.
222. Schwarzenbach MS, Reimann P, Thommen V, Hegner M, Mumenthaler M, Schwob J, Guntherodt HJ. Interferon alpha-2a interactions on glass vial surfaces measured by atomic force microscopy. PDA J Pharm Sci Technol 2002;56(2):78–89.
223. Qadry SS, Roshdy TH, Char H, Del Terzo S, Tarantino R, Moschera J. Evaluation of CZ-resin vials for packaging protein-based parenteral formulations. Int J Pharm 2003;252(1–2):207–212.
224. Smith EJ. Siliconization of parenteral packaging components. J Parenter Sci Technol 1988;42(4S):S1–S13.
225. Baldwin RN. Contamination of insulin by silicone oil: a potential hazard of plastic insulin syringes. Diabet Med 1988;5(8):789–790.
226. Jones LS, Kaufmann A, Middaugh CR. Silicone oil induced aggregation of proteins. J Pharm Sci 2005;94(4):918–927.
227. Goldberg M, Gomez-Orellana I. Challenges for the oral delivery of macromolecules. Nat Rev Drug Discov 2003;2(4):289–295.
228. Gomez-Orellana I, Paton DR. Advances in the oral delivery of proteins. Expert Opinion on Therapeutic Patents 1998;8(3):223–234.
229. Witschi C, Mrsny RJ. In vitro evaluation of microparticles and polymer gels for use as nasal platforms for protein delivery. Pharm Res 1999;16(3):382–390.
230. Smart JD. Recent developments in the use of bioadhesive systems for delivery of drugs to the oral cavity. Criti Rev Ther Drug Carrier Syst 2004;21(4):319–344.
231. Brown MB, Traynor MJ, Martin GP, Akomeah FK. Transdermal drug delivery systems: skin perturbation devices. Methods Mol Biol 2008;437:119–139.
232. Fineberg SE. Diabetes therapy trials with inhaled insulin. Expert Opin Investig Drugs 2006;15(7):743–762.
233. Kumar TR, Soppimath K, Nachaegari SK. Novel delivery technologies for protein and peptide therapeutics. Curr Pharm Biotechnol 2006;7(4):261–276.
234. Grainger DW. Controlled-release and local delivery of therapeutic antibodies. Expert Opin Biol Ther 2004;4(7):1029–1044.
235. Lee KY, Yuk SH. Polymeric protein delivery systems. Prog Polym Sci 2007;32(7):669–697.
236. Salmaso S, Bersani S, Semenzato A, Caliceti P. Nanotechnologies in protein delivery. J Nanosc Nanotechnol 2006;6(9–10):2736–2753.
237. Jorgensen L, Moeller EH, van de Weert M, Nielsen HM, Frokjaer S. Preparing and evaluating delivery systems for proteins. Eur J Pharm Sci 2006;29(3–4):174–182.
238. Bilati U, Allemann E, Doelker E. Protein encapsulation: protein drugs entrapped within micro- and nanoparticles: an overview of therapeutic challenges and scientific issues. Drug Deliv Technol 2005;5(8):41–47.
239. Cleland JL. Protein delivery from biodegradable microspheres. Pharm Biotechnol 1997;10:1–43.
240. Pettit DK, Lawter JR, Huang WJ, Pankey SC, Nightlinger NS, Lynch DH, Schuh JA, Morrissey PJ, Gombotz WR. Characterization of poly(glycolide-co-D,L-lactide)/poly(D,L-lactide) microspheres for controlled release of GM-CSF. Pharm Res 1997;14(10):1422–1430.
241. Veronese FM, Mero A. The impact of PEGylation on biological therapies. BioDrugs 2008;22(5):315–329.
242. Farr S, Boyd B, Bridges P, Linn LS. Using needle-free injectors for parenteral delivery of proteins. In: Eugene JEH, McNally J, Editors. Protein Formulation and Delivery. Boca Raton: CRC Press; 2007. p 255–283.
243. Prausnitz MR. Microneedles for transdermal drug delivery. Adv Drug Del Rev 2004;56(5):581–587.
244. Nayar R, Manning MC. High throughput formulation: strategies for rapid development of stable protein products. Pharm Biotechnol 2002;13:177–198.
245. Capelle MA, Gurny R, Arvinte T. High throughput screening of protein formulation stability: practical considerations. Eur J Pharm Biopharm 2007;65(2):131–148.
246. Jen A, Merkle HP. Diamonds in the rough: protein crystals from a formulation perspective. Pharm Res 2001;18(11):1483–1488.

247. Basu SK, Govardhan CP, Jung CW, Margolin AL. Protein crystals for the delivery of biopharmaceuticals. Expert Opin Biol Ther 2004;4(3):301–317.

248. Govardhan C, Khalaf N, Jung CW, Simeone B, Higbie A, Qu S, Chemmalil L, Pechenov S, Basu SK, Margolin AL. Novel long-acting crystal formulation of human growth hormone. Pharm Res 2005;22(9):1461–1470.

249. Flatman S, Alam I, Gerard J, Mussa N. Process analytics for purification of monoclonal antibodies. J Chromatogr B 2007;848(1):79–87.

250. Apostol I, Kelner DN. Managing the analytical lifecycle for biotechnology products: a journey from method development to validation. Part 1. Bioprocess Int 2008;6(8):12–19.

251. Apostol I, Schofield T, Koeller G, Powers S, Stawicki M, Wolfe RA. A rational approach for setting and maintaining specifications for biological and biotechnology-derived products. Part 1. BioPharm Int 2008;21(6):42–55.

252. Wehr T, Rodriguez-Diaz R. Use of size exclusion chromatography in biopharmaceutical development. In: Rodriguez-Diaz R, Wehr T, Tuck S, Editors. Analytical Techniques for Biopharmaceutical Development. New York: Marcel Dekker; 2005. p 95–112.

253. Ribarska JT, Jolevska ST, Panovska AP, Dimitrovska A. Studying the formation of aggregates in recombinant human granulocyte-colony stimulating factor (rHuG-CSF), lenograstim, using size-exclusion chromatography and SDS-PAGE. Acta Pharm 2008;58(2):199–206.

254. Gunturi SR, Ghobrial I, Sharma B. Development of a sensitive size exclusion HPLC method with fluorescence detection for the quantitation of recombinant human erythropoietin (r-HuEPO) aggregates. J Pharm Biomed Anal 2007;43(1):213–221.

255. Rosenberg AS. Effects of protein aggregates: an immunologic perspective. AAPS J 2006;8(3):E501–507.

256. Gagnon P. Practical strategies for protein contaminant detection by high-performance ion-exchange chromatography. In: Rodriguez-Diaz R, Wehr T, Tuck S, Editors. Analytical Techniques for Biopharmaceutical Development. New York: Marcel Dekker; 2005. p 67–79.

257. Weitzhandler M, Farnan D, Horvath J, Rohrer JS, Slingsby RW, Avdalovic N, Pohl C. Protein variant separations by cation-exchange chromatography on tentacle-type polymeric stationary phases. J Chromatgr A. 1998;828(1–2):365–372.

258. Weitzhandler M, Farnan D, Rohrer JS, Avdalovic N. Protein variant separations using cation exchange chromatography on grafted, polymeric stationary phases. Proteomics 2001;1(2):179–185.

259. Harris RJ, Kabakoff B, Macchi FD, Shen FJ, Kwong M, Andya JD, Shire SJ, Bjork N, Totpal K, Chen AB. Identification of multiple sources of charge heterogeneity in a recombinant antibody. J Chromatogr B 2001;752(2):233–245.

260. Dick LW Jr, Qiu D, Mahon D, Adamo M, Cheng KC. C-terminal lysine variants in fully human monoclonal antibodies: investigation of test methods and possible causes. Biotechnol Bioeng 2008;100(6):1132–1143.

261. Ahrer K, Jungbauer A. Chromatographic and electrophoretic characterization of protein variants. J Chromatog B 2006;841(1–2):110–122.

262. Kundu A, Barnthouse KA, Cramer SM. Selective displacement chromatography of proteins. Biotechnol Bioeng 1997;56(2):119–129.

263. Wypych J, Li M, Guo A, Zhang Z, Martinez T, Allen MJ, Fodor S, Kelner DN, Flynn GC, Liu YD, Bondarenko PV, Ricci MS, Dillon TM, Balland A. Human IgG2 antibodies display disulfide-mediated structural isoforms. J Biol Chem 2008;283(23):16194–16205.

264. Kannan K, Mulkerrin MG, Zhang M, Gray R, Steinharter T, Sewerin K, Baffi R, Harris R, Karunatilake C. Rapid analytical tryptic mapping of a recombinant chimeric monoclonal antibody and method validation challenges. J Pharma Biomed Anal 1997;16(4):631–640.

265. Bongers J, Cummings JJ, Ebert MB, Federici MM, Gledhill L, Gulati D, Hilliard GM, Jones BH, Lee KR, Mozdzanowski J, Naimoli M, Burman S. Validation of a peptide mapping method for a therapeutic monoclonal antibody: what could we possibly learn about a method we have run 100 times?. J Pharm Biomed Anal 2000;21(6):1099–1128.

266. Kroon DJ, Baldwin-Ferro A, Lalan P. Identification of sites of degradation in a therapeutic monoclonal antibody by peptide mapping. Pharma Res 1992;9(11):1386–1393.

267. Dillon TM, Bondarenko PV, Rehder DS, Pipes GD, Kleemann GR, Ricci MS. Optimization of a reversed-phase high-performance liquid chromatography/mass spectrometry method for characterizing recombinant antibody heterogeneity and stability. J Chromatogr A 2006;1120(1–2):112–120.

268. Dalmora SL, Masiero SMK, De Oliveira PR, Da Silva Sangoi M, Brum L Jr. Validation of an RP-LC method and assessment of rhG-CSF in

pharmaceutical formulations by liquid chromatography and biological assay. J Liq Chromatogr Relat Technol 2006;29:1753–1767.
269. Yang J, Wang S, Liu J, Raghani A. Determination of tryptophan oxidation of monoclonal antibody by reversed phase high performance liquid chromatography. J Chromatogr A 2007; 1156(1–2):174–182.
270. Scheich C, Leitner D, Sievert V, Leidert M, Schlegel B, Simon B, Letunic I, Bussow K, Diehl A. Fast identification of folded human protein domains expressed in E. coli suitable for structural analysis. BMC Struct Biol 2004;4:4.
271. Aizawa P, Winge S, Karlsson G. Large-scale preparation of thrombin from human plasma. Thromb Res 2008;122(4):560–567.
272. Battersby JE, Snedecor B, Chen C, Champion KM, Riddle L, Vanderlaan M. Affinity-reversed-phase liquid chromatography assay to quantitate recombinant antibodies and antibody fragments in fermentation broth. J Chromatogr A 2001;927(1–2):61–76.
273. Ma S. Analysis of protein therapeutics by capillary electrophoresis: applications and challenges. Dev Biol 2005;122:49–68.
274. Han M, Phan D, Nightlinger N, Taylor L, Jankhah S, Woodruff B, Yates Z, Freeman S, Guo A, Balland A, Pettit DK, Optimization of CE-SDS method for antibody separation based on multi-users experimental practices chromatographia 2006;64(5–6): 1–8.
275. Nunnally B, Park S, Patel K, Hong M, Zhang X, Wang S, Rener B, Reed-Bogan A, Salas-Solano O, Lau W, Girard M, Carnegie H, Garcia-Cañas V, Cheng KC, Zeng M, Ruesch M, Frazier R, Jochheim C, Natarajan K, Jessop K, Saeed M, Moffatt F, Madren S, Thiam SA, Altria K. Series of collaborations between various pharmaceutical companies and regulatory authorities concerning the analysis of biomolecules using capillary electrophoresis. Chromatographia 2006;64(5–6):359–368.
276. Salas-Solano O, Tomlinson B, Du S, Parker M, Strahan A, Ma S. Optimization and validation of a quantitative capillary electrophoresis sodium dodecyl sulfate method for quality control and stability monitoring of monoclonal antibodies. Anal Chem 2006;78(18):6583–6594.
277. Ma S. Analysis of protein therapeutics by capillary electrophoresis. Chromatographia 2001;53(Suppl 1):S75–S89.
278. Guo A, Han M, Martinez T, Ketcham RR, Novick S, Jochheim C, Balland A. Electrophoretic evidence for the presence of structural isoforms specific for the IgG2 isotype. Electrophoresis 2008;29(12):2550–2556.
279. Amon S, Zamfir AD, Rizzi A. Glycosylation analysis of glycoproteins and proteoglycans using capillary electrophoresis-mass spectrometry strategies. Electrophoresis 2008;29(12): 2485–2507.
280. Han M, Guo A, Jochheim C, Zhang Y, Martinez T, Kodama P, Pettit DK, Balland A. Analysis of glycosylated type II interleukin-1 receptor (IL-1R) by imaged capillary isoelectric focusing (i-cIEF). Chromatographia 2007;66(11–12): 969–976.
281. U.S. Food and Drug Administration Center for Biologics Evaluation and Research. Points to consider in the manufacture and testing of monoclonal antibody products for human use (1997). J Immun 1997;20(3):214–243.
282. Xu Y, Brorson K. An overview of quantitative PCR assays for biologicals: quality and safety evaluation. Dev Biol 2003;113:89–98.
283. Nissom PM. Specific detection of residual CHO host cell DNA by real-time PCR. Biologicals 2007;35(3):211–215.
284. Lovatt A. Applications of quantitative PCR in the biosafety and genetic stability assessment of biotechnology products. J Biotechnol 2002; 82(3):279–300.
285. Kubista M, Andrade JM, Bengtsson M, Forootan A, Jonak J, Lind K, Sindelka R, Sjoback R, Sjogreen B, Strombom L, Stahlberg A, Zoric N. The real-time polymerase chain reaction. Mol Asp Med 2006;27(2–3):95–125.
286. Eaton LC. Host cell contaminant protein assay development for recombinant biopharmaceuticals. J Chromatogr A 1995;705(1):105–114.
287. Champion K, Madden H, Dougherty J, Shacter E. Defining your product profile and maintaining control over it, Part 2. Bioprocess Int 2005; 52–57.
288. Krawitz DC, Forrest W, Moreno GT, Kittleson J, Champion KM. Proteomic studies support the use of multi-product immunoassays to monitor host cell protein impurities. Proteomics 2006;6(1):94–110.
289. Ehrnstrom R. Miniaturization and integration: challenges and breakthroughs in microfluidics. Lab Chip 2002;2(2):26N–30N.
290. Poulsen F, Jensen KB. A luminescent oxygen channeling immunoassay for the determination of insulin in human plasma. J Biomol Screen 2007;12(2):240–247.
291. Kange R, Selditz U, Granberg M, Lindberg U, Ekstrand G, Ek B, Gustafsson M. Comparison of different IMAC techniques used for enrich-

ment of phosphorylated peptides. J Biomol Tech 2005;16(2):91–103.
292. Meager A. Measurement of cytokines by bioassays: theory and application. Methods 2006;38 (4):237–252.
293. Jason-Moller L, Murphy M, Bruno J. Overview of Biacore systems and their applications In: Coligan J (Ed). Current Protocols in Protein Science. New York: John Wiley & Sons 2006. Chapter 19, Unit 19.13.1-19.13.14.
294. Mire-Sluis A. Expression of potency: why units of biological activity, not mass? Pharm Sci 1997;3:1–4.
295. Mire-Sluis AR. Progress in the use of biological assays during the development of biotechnology products. Pharm Res 2001;18:1239–1246.
296. Mire-Sluis AR. Setting specifications for potency assays: basic principles. Dev Biol 2002; 107:107–115.
297. Desjarlais JR, Lazar GA, Zhukovsky EA, Chu SY. Optimizing engagement of the immune system by anti-tumor antibodies: an engineer's perspective. Drug Discov Today 2007;12(21–22):898–910.
298. Cacia J, Keck R, Presta LG, Frenz J. Isomerization of an aspartic acid residue in the complementarity-determining regions of a recombinant antibody to human IgE: identification and effect on binding affinity. Biochemistry 1996;35(6):1897–1903.
299. Remmele RL Jr, Callahan WJ, Krishnan S, Zhou L, Bondarenko PV, Nichols AC, Kleemann GR, Pipes GD, Park S, Fodor S, Kras E, Brems DN. Active dimer of Epratuzumab provides insight into the complex nature of an antibody aggregate. J Pharm Sci 2006;95(1):126–145.
300. Kotarsky K, Antonsson L, Owman C, Olde B. Optimized reporter gene assays based on a synthetic multifunctional promoter and a secreted luciferase. Anal Biochem 2003;316(2):208–215.
301. Srebalus Barnes CA, Lim A. Applications of mass spectrometry for the structural characterization of recombinant protein pharmaceuticals. Mass Spectrom Rev 2007;26(3):370–388.
302. Balland A, Jochheim C. Mass spectrometry for biopharmaceutical development. Anal Tech Biopharm Dev 2005; 227–277.
303. Cruz HJ, Conradt HS, Dunker R, Peixoto CM, Cunha AE, Thomaz M, Burger C, Dias EM, Clemente J, Moreira JL, Rieke E, Carrondo MJ. Process development of a recombinant antibody/interleukin-2 fusion protein expressed in protein-free medium by BHK cells. J Biotechnol 2002;96(2):169–183.
304. Field M, Papac D, Jones A. The use of high-performance anion-exchange chromatography and matrix-assisted laser desorption/ionization time-of-flight mass spectrometry to monitor and identify oligosaccharide degradation. Anal Biochem 1996;239(1):92–98.
305. Yuk IH, Wang DI. Changes in the overall extent of protein glycosylation by Chinese hamster ovary cells over the course of batch culture. Biotechnol Appl Biochem 2002;36(Pt 2):133–140.
306. Dillon TM, Ricci MS, Vezina C, Flynn GC, Liu YD, Rehder DS, Plant M, Henkle B, Li Y, Deechongkit S, Varnum B, Wypych J, Balland A, Bondarenko PV. Structural and functional characterization of disulfide isoforms of the human IgG2 subclass. J Biol Chem 2008;283 (23):16206–16215.
307. Martinez T, Guo A, Allen M, Han M, Pace D, Jones D, Gillespie R, Ketchem RR, Zhang Y, Balland A. Disulfide connectivity of human immunoglobulin G2 structural isoforms. Biochemistry 2008;47(28):7496–7508.
308. Senderoff RI, Kontor KM, Kreilgaard L, Chang JJ, Patel S, Krakover J, Heffernan JK, Snell LB, Rosenberg GB. Consideration of conformational transitions and racemization during process development of recombinant glucagon-like peptide-1. J Pharm Sci 1998;87(2): 183–189.
309. Weinberger SR, Boschetti E, Santambien P, Brenac V. Surface-enhanced laser desorption-ionization retentate chromatography mass spectrometry (SELDI-RC-MS): a new method for rapid development of process chromatography conditions. J Chromatogr B 2002;782 (1–2):307–316.
310. Yen TY, Joshi RK, Yan H, Seto NO, Palcic MM, Macher BA. Characterization of cysteine residues and disulfide bonds in proteins by liquid chromatography/electrospray ionization tandem mass spectrometry. J Mass Spectrom 2000;35(8):990–1002.
311. Eng M, Ling V, Briggs JA, Souza K, Canova-Davis E, Powell MF, DeYoung LR. Formulation development and primary degradation pathways for recombinant human nerve growth factor. Anal Chem 1997;69(20): 4184–4190.
312. Wang L, Amphlett G, Lambert JM, Blattler W, Zhang W. Structural characterization of a recombinant monoclonal antibody by electrospray time-of-flight mass spectrometry. Pharm Res 2005;22(8):1338–1349.
313. Wen J, Arakawa T, Philo JS. Size-exclusion chromatography with on-line light-scattering,

absorbance, and refractive index detectors for studying proteins and their interactions. Anal Biochem 1996;240(2):155–166.
314. Philo JS. Is any measurement method optimal for all aggregate sizes and types?. AAPS J 2006;8(3):E564–571.
315. Kendrick BS, Kerwin BA, Chang BS, Philo JS. Online size-exclusion high-performance liquid chromatography light scattering and differential refractometry methods to determine degree of polymer conjugation to proteins and protein–protein or protein–ligand association states. Anal Biochem 2001;299(2):136–146.
316. Nobbmann U, Connah M, Fish B, Varley P, Gee C, Mulot S, Chen J, Zhou L, Lu Y, Shen F, Yi J, Harding SE. Dynamic light scattering as a relative tool for assessing the molecular integrity and stability of monoclonal antibodies. Biotechnol Genet Engin Rev 2007;24:117–128.
317. Fraunhofer W, Winter G. The use of asymmetrical flow field-flow fractionation in pharmaceutics and biopharmaceutics. Eur J Pharma Biopharm 2004;58(2):369–383.
318. Litzen A, Walter JK, Krischollek H, Wahlund KG. Separation and quantitation of monoclonal antibody aggregates by asymmetrical flow field-flow fractionation and comparison to gel permeation chromatography. Anal Biochem 1993;212(2):469–480.
319. Liu J, Andya JD, Shire SJ. A critical review of analytical ultracentrifugation and field flow fractionation methods for measuring protein aggregation. AAPS J 2006;8(3):E580–589.
320. Schuck P. A model for sedimentation in inhomogeneous media. I. Dynamic density gradients from sedimenting co-solutes. Biophys Chem 2004;108(1–3):187–200.
321. Cole JL, Lary JW, Moody TP, Laue TM. Analytical ultracentrifugation: sedimentation velocity and sedimentation equilibrium. Methods Cell Biol 2008;84:143–179.
322. Gabrielson JP, Randolph TW, Kendrick BS, Stoner MR. Sedimentation velocity analytical ultracentrifugation and SEDFIT/c(s): limits of quantitation for a monoclonal antibody system. Anal Biochem 2007;361(1):24–30.
323. Berkowitz SA. Role of analytical ultracentrifugation in assessing the aggregation of protein biopharmaceuticals. AAPS J 2006;8(3): E590–605.
324. Pekar A, Sukumar M. Quantitation of aggregates in therapeutic proteins using sedimentation velocity analytical ultracentrifugation: practical considerations that affect precision and accuracy. Anal Biochem 2007;367(2): 225–237.
325. Gabrielson JP, Brader ML, Pekar AH, Mathis KB, Winter G, Carpenter JF, Randolph TW. Quantitation of aggregate levels in a recombinant humanized monoclonal antibody formulation by size-exclusion chromatography, asymmetrical flow field flow fractionation, and sedimentation velocity. J Pharm Sci 2007;96 (2):268–279.
326. Doglia SM, Ami D, Natalello A, Gatti-Lafranconi P, Lotti M. Fourier transform infrared spectroscopy analysis of the conformational quality of recombinant proteins within inclusion bodies. Biotechnol J 2008;3(2):193–201.
327. Matheus S, Mahler HC, Friess W. A critical evaluation of T_m (FTIR) measurements of high-concentration IgG1 antibody formulations as a formulation development tool. Pharm Res 2006;23(7):1617–1627.
328. Matheus S, Friess W, Mahler HC. FTIR and nDSC as analytical tools for high-concentration protein formulations. Pharm Res 2006;23 (6):1350–1363.
329. Harn N, Allan C, Oliver C, Middaugh CR. Highly concentrated monoclonal antibody solutions: direct analysis of physical structure and thermal stability. J Pharm Sci 2007;96 (3):532–546.
330. Whitmore L, Wallace BA. Protein secondary structure analyses from circular dichroism spectroscopy: methods and reference databases. Biopolymers 2008;89(5):392–400.
331. Greenfield NJ. Using circular dichroism collected as a function of temperature to determine the thermodynamics of protein unfolding and binding interactions. Nat Protoc 2006;1 (6):2527–2535.
332. Ejima D, Tsumoto K, Fukada H, Yumioka R, Nagase K, Arakawa T, Philo JS. Effects of acid exposure on the conformation, stability, and aggregation of monoclonal antibodies. Proteins 2007;66(4):954–962.
333. Kats M, Richberg PC, Hughes DE. Conformational diversity and conformational transitions of a monoclonal antibody monitored by circular dichroism and capillary electrophoresis. Anal Chem 1995;67(17):2943–2948.
334. Vermeer AW, Norde W, van Amerongen A. The unfolding/denaturation of immunogammaglobulin of isotype 2b and its F(ab) and F(c) fragments. Biophys J 2000;79(4):2150–2154.
335. Walker KN, Bottomley SP, Popplewell AG, Sutton BJ, Gore MG. Equilibrium and pre-equilibrium fluorescence spectroscopic studies

of the binding of a single-immunoglobulin-binding domain derived from protein G to the Fc fragment from human IgG1. Biochem J 1995;310(Pt 1):177–184.
336. Hawe A, Friess W, Sutter M, Jiskoot W. Online fluorescent dye detection method for the characterization of immunoglobulin G aggregation by size exclusion chromatography and asymmetrical flow field flow fractionation. Anal Biochem 2008;378(2):115–122.
337. Hawe A, Sutter M, Jiskoot W. Extrinsic fluorescent dyes as tools for protein characterization. Pharm Res 2008;25(7):1487–1499.
338. Shah D, Johnston TP, Heitz S, Mitra AK. The effects of various excipients on the unfolding of basic fibroblast growth factor. PDA J Pharma Sci Technol 1998;52(5):209–214.
339. Krell T. Microcalorimetry: a response to challenges in modern biotechnology. Microb Biotechnol 2008;1(2):126–136.
340. Hill JJ, Shalaev EY, Zografi G. Thermodynamic and dynamic factors involved in the stability of native protein structure in amorphous solids in relation to levels of hydration. J Pharm Sci 2005;94(8):1636–1667.
341. Kett V, McMahon D, Ward K. Thermoanalytical techniques for the investigation of the freeze drying process and freeze-dried products. Curr Pharm Biotechnol 2005;6(3): 239–250.
342. Remmele RL Jr, Zhang-van Enk J, Dharmavaram V, Balaban D, Durst M, Shoshitaishvili A, Rand H. Scan-rate-dependent melting transitions of interleukin-1 receptor (type II): elucidation of meaningful thermodynamic and kinetic parameters of aggregation acquired from DSC simulations. J Am Chem Soc 2005;127(23):8328–8339.
343. Ahrer K, Buchacher A, Iberer G, Josic D, Jungbauer A. Thermodynamic stability and formation of aggregates of human immunoglobulin G characterised by differential scanning calorimetry and dynamic light scattering. J Biochem Biophys Methods 2006;66(1–3): 73–86.
344. Garber E, Demarest SJ. A broad range of Fab stabilities within a host of therapeutic IgGs. Biochem Biophys Res Commun 2007;355(3): 751–757.
345. Meyer BK, Ni A, Hu B, Shi L. Antimicrobial preservative use in parenteral products: past and present. J Pharm Sci 2007;96(12):3155–3167.
346. Sharma B. Immunogenicity of therapeutic proteins. Part 3. Impact of manufacturing changes. Biotechnol Adv 2007;25(3):325–331.
347. Schellekens H. How to predict and prevent the immunogenicity of therapeutic proteins. Biotechnol Annu Rev 2008;14:191–202.
348. Schenerman MA, Phillips K. Comparability (bioequivalence) testing of monoclonal antibodies. BioPharm 1997;10:20–27.
349. Chirino AJ, Mire-Sluis A. Characterizing biological products and assessing comparability following manufacturing changes. Nat Biotechnol 2004;22(11):1383–1391.
350. Chirino AJ, Mire-Sluis AR. State of the art analytical comparability: a review. Dev Biol 2005;122:3–26.
351. Yang JD, Lu C, Stasny B, Henley J, Guinto W, Gonzalez C, Gleason J, Fung M, Collopy B, Benjamino M, Gangi J, Hanson M, Ille E. Fed-batch bioreactor process scale-up from 3-L to 2,500-L scale for monoclonal antibody production from cell culture. Biotechnol Bioeng 2007;98(1):141–154.
352. Meuwly F, Weber U, Ziegler T, Gervais A, Mastrangeli R, Crisci C, Rossi M, Bernard A, von Stockar U, Kadouri A. Conversion of a CHO cell culture process from perfusion to fed-batch technology without altering product quality. J Biotechnol 2006;123(1):106–116.
353. Schenerman MA, Hope JN, Kletke C, Singh JK, Kimura R, Tsao EI, Folena-Wasserman G. Comparability testing of a humanized monoclonal antibody (Synagis) to support cell line stability, process validation, and scale-up for manufacturing. Biologicals 1999;27(3):203–215.
354. Defelippis MR, Larimore FS. The role of formulation in insulin comparability assessments. Biologicals 2006;34(1):49–54.
355. Combe C, Tredree RL, Schellekens H. Biosimilar epoetins: an analysis based on recently implemented European medicines evaluation agency guidelines on comparability of biopharmaceutical proteins. Pharmacotherapy 2005; 25(7):954–962.
356. Covic A, Kuhlmann MK. Biosimilars: recent developments. Int Urol Nephrol 2007; 39 (1):261–266.
357. Simmerman H, Donnelly RP. Defining your product profile and maintaining control over it. Part 1. Bioprocess Int 2005;3(6):32–40.
358. U.S. Food and Drug Administration, HHS. Substances generally recognized as safe. Code of Federal Regulations. Part 182. Title 21. 2004.
359. Huang MM, Penn L, Bongers J, Gurman S. Validation of a shielded-hydrophobic-phase high-performance liquid chromatography

method for the determination of residual methotrexate in recombinant protein biopharmaceuticals. J Chromatogr 1998;828(1–2):303–309.

360. Kennan JJ, Breen LL, Lane TH, Taylor RB. Methods for detecting silicones in biological matrixes. Anal Chem 1999;71(15):3054–3060.

361. Qiu J, Lee H, Zhou C. Analysis of guanidine in high salt and protein matrices by cation-exchange chromatography and UV detection. J Chromatogr A 2005;1073(1–2):263–267.

362. Food US, Drug Administration HHS. International Conference on Harmonisation; Guidance on Q8 Pharmaceutical Development; availability. Notice. Federal Register. 2006; 71(98):1–9.

363. Lindemann CJ, Rosolia A. Gas chromatographic assay for benzyl alcohol and phenylethyl alcohol in pharmaceutical formulations. J Pharma Sci 1969;58(1):118–119.

364. Yu LX. Pharmaceutical quality by design: product and process development, understanding, and control. Pharm Res 2008;25(4):781–791.

365. Wren SA, Tchelitcheff P. Use of ultra-performance liquid chromatography in pharmaceutical development. J Chromatogr A 2006;1119(1–2):140–146.

366. Zhang B, Foret F, Karger BL. High-throughput microfabricated CE/ESI-MS: automated sampling from a microwell plate. Anal Chem 2001;73(11):2675–2681.

367. Kang SH, Gong X, Yeung ES. High-throughput comprehensive peptide mapping of proteins by multiplexed capillary electrophoresis. Anal Chem 2000;72(14):3014–3021.

368. Londo T, Lynch P, Kehoe T, Meys M, Gordon N. Accelerated recombinant protein purification process development automated, robotics-based integration of chromatographic purification and analysis. J Chromatogr A 1998;798(1–2):73–82.

369. Haeberle S, Zengerle R. Microfluidic platforms for lab-on-a-chip applications. Lab Chip 2007;7(9):1094–1110.

370. Johlman CL, White RL, Wilkins CL. Applications of Fourier transform mass spectrometry. Mass Spectrom Rev 2005;2(3):389–415.

371. Bereman MS, Nyadong L, Fernandez FM, Muddiman DC. Direct high-resolution peptide and protein analysis by desorption electrospray ionization Fourier transform ion cyclotron resonance mass spectrometry. Rapid Commun Mass Spectrom 2006;20(22):3409–3411.

372. Morstadt L, Bohm A, Yuksel D, Kumar K, Stollar BD, Baleja JD. Engineering and characterization of a single chain surrogate light chain variable domain. Protein Sci 2008;17(3):458–465.

COST-EFFECTIVENESS ANALYSES THROUGHOUT THE DRUG DEVELOPMENT LIFE CYCLE

RENÉE J.G. ARNOLD[1,2,3]
[1]Arnold Consultancy & Technology LLC, New York, NY
[2]Master of Public Health Program, Mount Sinai School of Medicine, New York, NY
[3]Arnold and Marie Schwartz College of Pharmacy, Brooklyn, NY

1. INTRODUCTION

Cost-effectiveness analysis (CEA) is a systematic, quantitative method for summarizing health benefits and health resources of various treatment options into single numbers or ratios so that policy makers can choose among them. CEA compares the costs and consequences of treatment alternatives or programs where cost is measured in monetary terms and consequences are measured in natural units (e.g., millimeters of mercury) [1]. Typically, the preferable result is considered to be the option with the least cost per unit of measure gained; the results of the CEA are represented by the ratio of cost, a summation of costs of all resources surrounding a therapeutic option to effectiveness, the health benefit. With this type of analysis, various disease end points that are affected by therapy (risk markers, disease severity, and death) can be assessed by corresponding indices of therapeutic outcome (millimeters of mercury blood pressure reduction, hospitalizations averted, and life years saved or quality-adjusted life year (QALY), respectively]. There are two ways of performing or representing CEAs—average cost-effectiveness and incremental cost-effectiveness. "Average" cost-effectiveness is the result of dividing mean total costs by mean outcomes and is typically represented as per patient value. It is calculated, as follows, for each therapeutic option:

$$\frac{\bar{X} \text{ Cost}}{\bar{X} \text{ Effectiveness}}$$

The average cost-effectiveness of each therapy is then compared and the one with the lowest cost per unit of effectiveness would be preferred. Although this type of analysis allows one to view the actual numbers involved in the calculation, *average* cost-effectiveness does not illustrate differences between alternative strategies [2,3]. Thus, many researchers prefer to use or further explain the results of a CEA in terms of an "incremental" cost-effectiveness ratio, that is, additional cost for additional benefit, which may be calculated as follows:

$$\frac{\Delta C}{\Delta E} = \frac{\text{Cost}_1 - \text{Cost}_2}{\text{Effectiveness}_1 - \text{Effectiveness}_2},$$

where $\text{Effectiveness}_1 > \text{Effectiveness}_2$

The term "incremental" is commonly used interchangeably with the term "marginal" to denote the additional cost and outcome of one intervention in comparison with another [2].

Currently, in most countries, safety and efficacy are the only criteria required for drug approval. Cost is considered primarily in terms of rationing, such as the Australian Pharmaceutical Benefits Advisory Committee's or PBAC's role in listing a drug for coverage (e.g., who should be allowed to receive these new drugs?) and reimbursement (such as in Canada, where the price of the new drug is set by the government, primarily based on CEAs. CEAs are used to assess the benefit of drugs and other therapeutic strategies. CEA is applied in situations where trade-offs exist—typically, greater benefit for an increased cost over an existing therapy or strategic option or versus usual care. An effective CEA answers certain questions such as Is the treatment effective? What will it cost? and How do the gains compare with the costs? By combining answers to all of these questions, CEA helps decision makers to weigh the factors, compare alternative treatments, and decide which treatments are most appropriate for specific situations.

2. REGULATORY APPROVAL

In the United Kingdom, the National Institute for Health and Clinical Excellence (NICE), established in 1999, carries out economic analyses of health technologies, which are used to inform reimbursement policies of the National Health Service (NHS). As Drummond mentions, although "NICE does not control the reimbursement or pricing of drugs ... a negative recommendation from NICE usually results in highly restricted use of a given product" [4]. For example, NICE recently suggested that the effectiveness of four drugs used to treat advanced kidney cancer—Avastin (Roche AG), Nexavar (Bayer AG), Sutent (Pfizer, Inc.), and Torisel (Wyeth)—did not justify their high cost and should not be covered by the NHS [5]. In its preliminary findings, the economic model developed by NICE using clinical trial data demonstrated that the drugs cost between £71,500 and £171,300 (approximately US$139,800–335,000) for each QALY or year of healthy life they gave patients. The usual threshold for a cost-effective agent according to NICE is approximately £30,000 per QALY gained.

The debate of value in healthcare has recently begun in earnest in the United States. Former Senator Tom Daschle, then President-elect Barack Obama's pick for Secretary of the Department of Health and Human Services (DHHS), spoke of a federal health-care board charged with "recommending coverage of those drugs and procedures backed by solid evidence. It would exert influence by ranking services and therapies by their health and cost impacts" [6]. Daschle cited Britain's NICE, "the U.K.'s cost-effectiveness watchdog", in these discussions.

3. DECISION-ANALYTIC APPROACHES

Given the likelihood that cost-effectiveness is expected to play a greater role in the future, perhaps CEAs should be undertaken at the beginning of drug development, using computational methods to weigh probabilities and costs associated with all steps of the drug discovery and development process. As mentioned by Ekins and Arnold, the decisions suggested in such an approach could be based on one of a number of algorithms, such as a decision tree approach that has been widely used in health economic analysis [7] and drug innovation assessment algorithm analysis [8].

Indeed, continuous risk and uncertain timing of events may depend on when events occur.

3.1. Decision-Analytic Models

One of the most useful types of computational methodologies to take these probabilities and costs into consideration is decision-analytic or decision tree modeling. Special types of decision-analytic models, such as Markov models, [9] account for issues of time sensitivity. For example, Lewis and colleagues [10] employed a Markov model to discern the relative cost-effectiveness of Sandimmune® (an older formulation of cyclosporine) versus Neoral® (a newer formulation of cyclosporine) in the first 3 months following renal transplantation. Using the results from one of the multiple sources that informed the model, Neoral was shown to be both more effective and less costly than Sandimmune for both effectiveness criteria—non-functioning graft and rejection-free clinical course; thus, Neoral was the dominant strategy, a result that the pharmaceutical manufacturer would embrace. The practical application of these data for the healthcare providers would be that with a $10 million budget, it would be possible to transplant 115 patients on Sandimmune or 124 patients on Neoral; 49/115 (43%) patients on Sandimmune versus 84/124 (68%) patients on Neoral would have a rejection-free clinical course. This market evaluation bodes well for future sales of the more cost-effective agent. Risk assessment data from premarketing and postmarketing studies can also be linked using statistical analyses to determine the nature of the side effects that are observed and whether they represent sentinels for more serious events that may only occur in very large trials [11].

The influence of mathematical modeling has also reached pharmaceutical pricing and go/no-go decision making for in licensing [12]. A decision-analytic model was used to estimate the average cost per patient with heparin-induced thrombocytopenia (HIT) with or without thrombosis [13] (Fig. 1) Clinical probability data used to populate the model were obtained from clinical trials and from published clinical literature. Resource utilization data and cost data were also obtained from

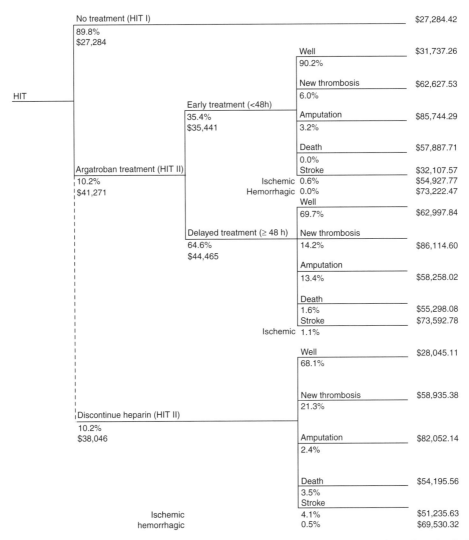

Figure 1. A decision tree for patients with heparin-induced thrombocytopenia without thrombosis [7].

available literature, the 2003 Physician's Fee Reference, the Healthcare Cost and Utilization Project 2000, the 2003 Drug Topics Red-Book, and a modified Delphi panel. The total per-patient cost included: hospital days, diagnostic tests, drug costs, major hemorrhagic events, and patient outcomes (i.e., amputation, new thrombosis, stroke, or death), multiplied by the probability of each event. The incremental cost-effectiveness ratio (ICER) was calculated by dividing the incremental cost between patients with and without treatment by the incremental effectiveness, or the cost per new thrombosis event avoided.

Another example of the use of CEA during drug development and reimbursement decision making is that of the cost-effectiveness of a vaccine for human papillomavirus (HPV) [13]. Over a lifetime (see Table 1 [13]), providing HPV vaccination for a patient costs, on average, US$310, that is, $Cost_1 - Cost_2$, more than screening for HPV infection and results in an average gain of 0.0092 QALYs, that is, $Effectiveness_1 - Effectiveness_2$, a marginal cost-effectiveness ratio of US$310/0.0092 or US$33,700/QALY. A marginal or incremental CEA is useful in the following two instances— (1) where the new strategy is more costly, but

Table 1. Health and Economic Outcomes of HPV Vaccination [14]

Outcome	Cost (US$) Per Patient	Incremental Cost (US$) Per Patient	Life Expectancy (QALY)	Incremental Life Expectancy (QALY)	Cost-Effectiveness (Cost/QALY)
No vaccination (current screening program)	1111		25.9815		
HPV vaccination at 70% efficacy	1421	310^a (1421 − 1111)	25.9907	0.0092^b (25.9907 − 25.9815)	33,700 (rounded) 310/0.0092

^{a} Total cost was derived by aggregating the costs of cervical cancer screening (i.e., cytology, HPV DNA test, office visit, and patient time), diagnosis and treatment and vaccination (vaccination series and patient time); incremental cost is the difference between "no vaccination" and "HPV Vaccination at 70% Efficacy" to determine the life expectancy of a hypothetical cohort.
^{b} Total Effectiveness was derived by aggregating the probabilities of the incidence and clearance of HPV infection, natural history of CIN, and natural history of invasive cervical cancer (probability of progression, developing symptoms, survival at 5 years) to determine the life expectancy of a hypothetical cohort; the results were calibrated to the lifetime risk of cervical cancer reported in the National Cancer Institute's Surveillance, Epidemiology, and End Results Program.
QALY: quality-adjusted life year.

expected to be more effective or (2) where the new strategy is less costly, but less effective [14,15]. To understand the multiple trade-offs inherent in the development of a HPV vaccine, it is first necessary to review some facts about cervical cancer and coverage of HPV strains by the vaccine that is currently available in the United States. Persistent infection with cancer-associated HPV (termed oncogenic or high-risk HPV) causes the majority of squamous cell cervical cancer, the most common type of cervical cancer, and its histologic precursor lesions, the low-grade cervical dysplasia cervical intraepithelial neoplasia-1 (CIN1) and the moderate-to-high-grade dysplasia CIN 2/3. Multiple HPV strains cause varying degrees of invasive cervical cancer (ICC) and its CIN precursors. HPV strains 16 and 18 cause approximately 70% of all cervical cancers [16,18] and CIN3, specifically, and 50% of CIN2 cases. In addition, HPV 16 and 18 cause approximately 35–50% of all CIN1. Low-oncogenic HPV risk types 6 and 11 account for 90% of genital wart cases [18]. Unfortunately, cytological and histological examinations cannot reliably distinguish between those patients who will progress from cervical dysplasia to ICC from those whose dysplasias will regress spontaneously, the latter being the vast majority of cases [19]. This inability to definitely ascertain the natural history of HPV infection is one of the primary reasons for the dilemma with HPV vaccination.

To argue for vaccination is the fact that although cervical cancer screening programs have substantially reduced the incidence and mortality of ICC in developed countries over the past 50 years [18,20], there has been a slowing of these declines in recent years due to poor sensitivity of cervical cytology, anxiety and morbidity of screening investigations, poor access to and attendance of screening programs, falling screening coverage, and poor predictive value for adenocarcinoma, an increasingly common cause of ICC [20]. Other factors to be considered on the "pro" side of the vaccination dilemma include the fact that HPV is the most common sexually transmitted disease in the United States, that virtually 100% of cervical cancer is due to HPV, that treatment of cervical cancer is very expensive (US$1.7 billion/year using U.S. Medicare dollars) and that HPV is also linked to head and neck cancer in men. On the "con" side is the presence of >100 HPV strains (thereby potentially reducing vaccine efficacy for oncogenic strains not covered by the vaccine) and the facts that HPV infection is often self-limited, routine screening is still necessary, and the vaccine is expensive (in the United States, it is priced at US$120/dose and given as 3 doses over 6 months). A mitigating factor for the argument against using the vaccine is the fact that the cost-effectiveness of screening with Pap smears is reduced (improves) from US$1 million/QALY if the patients continue to be screened annually, as is the common current recommendation, to US$150,000/QALY if the patients are screened every 3 years, the latter a likely scenario if the vaccine is used [13,16,21,22].

Worldwide, the incidence of cervical cancer is 470,000 new cases and 233,000 deaths per year; it is the second-leading cause of cancer deaths [23], with 80% of these cases observed in developing countries [24]. Women in developing countries are especially vulnerable as they lack access to both cervical cancer screening and treatment. The demographics of cervical cancer in the United States show that 9710 new cases of ICC were expected to be diagnosed in 2006 and approximately 3700 deaths in women were expected from ICC. Quadrivalent human papillomavirus (HPV) vaccine recombinant (Gardasil®), the vaccine recently approved for use in the United States and Europe, covers the two major oncogenic HPV strains (16 and 18) for cervical cancer. In addition, it covers HPV strains 6 and 11, the primary causes of genital warts. Therefore, the vaccine does not offer full protection against cervical cancer, since it does not protect against HPV strains 31 and 45, which are also implicated in ICC and cervical dysplasia. In order to significantly reduce the rate of cervical cancer in the population as a whole, 70% of girls need to be vaccinated to achieve what is called "herd immunity"—when the vaccine's impact goes beyond just people who are inoculated. So far, it is unknown if HPV strains will mutate as the vaccine is introduced, although this is not very likely seeing that HPV is a DNA-based virus [19].

3.2. HPV Vaccine Model CEAs

Multiple CEAs have been performed [13,16,21,22,25,26] to evaluate the cost-effectiveness of HPV vaccine in comparison to various screening policies and using various diagnostic and treatment scenarios. The models were used to inform the United States Advisory Committee on Immunization Practices (ACIP), which advises the US Centers for Disease Control and Prevention (CDC) on vaccination matters. Based largely on the model results, ACIP unanimously recommended that 11- and 12-year-old girls receive a new vaccine designed to protect against cervical cancer (Gardasil, Merck) and that Gardasil should be added to the routine vaccination schedule for children and adolescents [27].

4. PHARMACOGENOMICS AND PERSONALIZATION

Stallings and colleagues [29] developed a decision-analytic model to test the likely cost impact of a hypothetical pharmacogenomic test to determine a preferred initial therapy in patients with asthma. The authors compared annualized per patient cost distributions using a "test all" strategy for a nonresponse genotype prior to treating versus "test none." They found that the cost savings per patient of the testing strategy simulation ranged from US$200 to US$767 (95% confidence interval) and concluded that upfront testing costs were likely to be offset by avoided nonresponse costs. However, incentives will have to be created so that individualization of therapy using pharmacogenomic testing is not only cost-effective for the payers but also revenue generating for the developers. Some have suggested creation of a two-tiered system not dissimilar from that of Medicare drug reimbursement whereby initial clinical use of a test would be at a lower reimbursement rate on the basis of biomarkers and intermediate outcomes (see Volume 4, Chapter 26); long-term outcomes testing to show cost-effectiveness of the pharmacogenomic test would then result in a higher reimbursement rate [29,30].

5. POSTLAUNCH

Since practice, including available treatments and procedures, changes over time, it is essential to use retrospective data to continuously inform health policy decisions [28]. An example of use of data from a pharmacy benefits management claims database to evaluate two decision-analytic models regarding the cost-effectiveness of therapeutic regimens to eradicate *Helicobacter pylori* in ulcer patients is a case in point [32]. Indeed, the authors found that model results overstated the cost-effectiveness of the previously more cost-effective regimen and underestimated the cost-effectiveness of the other regimen such that the model assumptions and, ultimately, the outcomes, were not supported by the data.

6. APPLYING THE MODELS THROUGHOUT DRUG DEVELOPMENT

Therefore, it is apparent that computer modeling may play a significant role in informing the pharmaceutical decision-making process throughout a product's life cycle. Cost-effectiveness should be considered early in the drug development process to increase the probability that regulatory authorities will recommend reimbursement coverage and that payers will make the agent available to the broadest and most appropriate group possible.

REFERENCES

1. Drummond MF, Sculpher MJ, Torrance GW, O'Brien BJ, Stoddart GL. Methods for the Economic Evaluation of Health care Programmes. New York: Oxford University Press, Inc.; 2005.
2. Detsky A, Naglie I. A clinician's guide to cost-effectiveness analysis. Ann Intern Med 1990; 113:147–154.
3. Weinstein M, Fineberg H, Elstein A, et al. Clinical decisions and limited resources. In: Weinstein M, editor. Clinical Decision Analysis. Philadelphia: Saunders; 1980. p 228–265.
4. Drummond MF, Sorenson C. Use of pharmacoeconomics in drug reimbursement in Australia. Canada and the United Kingdom. In: Arnold R, editor. Introduction to Pharmacoeconomics: from Theory to Practice. Boca Raton: Taylor and Francis Group, LLC 2009; 175–196.
5. Whalen J. British Agency impugns value of four costly cancer drugs. Wall Street Journal. New York: 2008.
6. Gratzer D. American cancer care beats the rest. Wall Street Journal. New York: Dow Jones; 2008.
7. Arnold RJ, Kim R, Tang B. The cost-effectiveness of argatroban treatment in heparin-induced thrombocytopenia: the effect of early versus delayed treatment. Cardiol Rev 2006; 14:7–13.
8. Caprino L, Russo P. Developing a paradigm of drug innovation: an evaluation algorithm. Drug Discov Today 2006;11:999–1006.
9. Beck J, Pauker S. The Markov process in medical prognosis. Med Decis Making 1983;3: 419–458.
10. Lewis R, Canafax D, Pettit K, DiCesare J, Kaniecki D, Arnold R, et al. Use of Markov model for evaluating the cost-effectiveness of immunosuppressive therapies in renal transplant recipients. Transplant Proc 1996;28:2214–2217.
11. O'Neill RT. Biostatistical considerations in pharmacovigilance and pharmacoepidemiology: linking quantitative risk assessment in pre-market licensure application safety data, post-market alert reports and formal epidemiological studies. Stat Med 1998;17:1851–1858; discussion 9-62.
12. Vernon JA, Hughen WK, Johnson SJ. Mathematical modeling and pharmaceutical pricing: analyses used to inform in-licensing and developmental go/no-go decisions. Health Care Manag Sci 2005;8:167–179.
13. Goldie SJ, Kohli M, Grima D, Weinstein MC, Wright TC, Bosch FX, et al. Projected clinical benefits and cost-effectiveness of a human papillomavirus 16/18 vaccine. J Natl Cancer Inst 2004;96:604–15.
14. Arnold R. Use of interactive software in medical decision making. In: Ekins S, editor. Computer Applications in Pharmaceutical Research and Development. Hoboken: John Wiley & Sons, Inc.; 2006.
15. Goldberg Arnold R. Health economic considerations in cardiovascular drug utilization. In: Frishman W, Sonnenblick E, editors. Cardiovascular Pharmacotherapeutics. New York: McGraw Hill, Inc.; 2003. p 43.
16. Kulasingam SL, Myers ER. Potential health and economic impact of adding a human papillomavirus vaccine to screening programs. JAMA 2003;290:781–789.
17. Van de Velde N, Brisson M, Boily MC. Modeling human papillomavirus vaccine effectiveness: quantifying the impact of parameter uncertainty. Am J Epidemiol 2007;165:762–775.
18. Brisson M, Van de Velde N, De Wals P, Boily MC. The potential cost-effectiveness of prophylactic human papillomavirus vaccines in Canada. Vaccine 2007;25:5399–5408.
19. Woodman CB, Collins SI, Young LS. The natural history of cervical HPV infection: unresolved issues. Nat Rev Cancer 2007;7:11–22.
20. Adams M, Jasani B, Fiander A. Human papilloma virus (HPV) prophylactic vaccina-

20. tion: challenges for public health and implications for screening. Vaccine 2007;25: 3007–3013.
21. Sanders GD, Taira AV. Cost-effectiveness of a potential vaccine for human papillomavirus. Emerg Infect Dis 2003;9:37–48.
22. Taira AV, Neukermans CP, Sanders GD. Evaluating human papillomavirus vaccination programs. Emerg Infect Dis 2004;10: 1915–1923.
23. U.S. Food and Drug Administration. GARDASIL® Questions and Answers. 2006. FDA press release about Gardasil approval, includes HPV stats.
24. Cox T, Cuzick J. HPV DNA testing in cervical cancer screening: from evidence to policies. Gynecol Oncol 2006;103:8–11.
25. Dasbach EJ, Elbasha EH, Insinga RP. Mathematical models for predicting the epidemiologic and economic impact of vaccination against human papillomavirus infection and disease. Epidemiol Rev 2006;28:88–100.
26. Kim JJ, Goldie SJ. Health and economic implications of HPV vaccination in the United States. N Engl J Med 2008;359:821–832.
27. Arnold RJ. Cost-effectiveness analysis: should it be required for drug registration and beyond? Drug discovery today 2007;12:960–965.
28. Stallings SC, Huse D, Finkelstein SN, Crown WH, Witt WP, Maguire J, et al. A framework to evaluate the economic impact of pharmacogenomics. Pharmacogenomics 2006;7:853–862.
29. Califf RM. Defining the balance of risk and benefit in the era of genomics and proteomics. Health Aff (Millwood) 2004;23:77–87.
30. Deverka PA, McLeod HL. Harnessing economic drivers for successful clinical implementation of pharmacogenetic testing. Clin Pharmacol Ther 2008;84:191–193.
31. Fairman KA, Motheral BR. Do decision-analytic models identify cost-effective treatments? A retrospective look at *Helicobacter pylori* eradication. J Manag Care Pharm 2003;9:430–440.

PROVISIONAL BCS CLASSIFICATION OF THE LEADING ORAL DRUGS ON THE GLOBAL MARKET

Arik Dahan[1]
Gordon L. Amidon[2]

[1]Department of Clinical Pharmacology, School of Pharmacy, Ben-Gurion University of the Negev, Beer-Sheva, Israel.
[2]Department of Pharmaceutical Sciences, University of Michigan College of Pharmacy, Ann Arbor, MI

1. INTRODUCTION

In the early 1990s, a collaborative research effort between academia and the U.S. Food and Drug Administration (FDA) was initiated to establish a database of human jejunal permeabilities based on a Biopharmaceutics Classification System (BCS) [1,2]. The BCS broadly allowed the prediction of human absorption, fraction absorbed (F_{abs}) and the rate-limiting step in the intestinal absorption process of drugs following oral administration. The BCS classified compounds into one of four categories according to their water solubility and membrane permeability characteristics. Today, the BCS has generated remarkable impact on the global pharmaceutical sciences arena, in drug discovery, development, and regulation, and an extensive validation/discussion of the BCS is continuously published in the literature. The BCS has been effectively implanted by drug regulatory agencies around the world, and is widely practiced by the pharmaceutical industry. The aim of this chapter is to present the BCS and its scientific basis, to describe its impact on regulatory practice of oral drug products, and to access the BCS classification of the top drugs on the global market. Finally, current and future directions in BCS related extensions, and their impact on the pharmaceutical industry, will be discussed.

2. ORAL DRUG ABSORPTION AND THE BCS

The absorption and systemic availability of drugs following oral administration is a cascade of complex events. Separating absorption and systemic availability was the first step in advancing the science of absorption prediction. The fraction drug absorbed is the upper limit of systemic availability and thus of primary interest in development of a new oral therapeutic agents. We take drug absorption to be that first step of drug crossing a biological membrane and entering the "body." Subsequently, the drug may be metabolized or even exported back across this defining membrane (thus lowering the effective membrane permeability, see below). Luminal, brush border metabolism, and chemical instability are not considered in the review.

The rate and extent of the drug absorption, F_{abs}, are affected by many factors. These include physicochemical factors (e.g., pK_a, solubility, stability, diffusivity, lipophilicity, polar–nonpolar surface area, presence of hydrogen-bonding functionalities, particle size, and crystal form), physiological factors (e.g., GI pH, GI blood flow, and gastric emptying), factors related to the dosage form (e.g., tablet, capsule, solution, suspension, emulsion, and gel), and most importantly transit time throughout the different gastrointestinal (GI) segments and the absorption mechanism differences resulting in different permeabilities [3–5].

When Fick's First Law is applied to a membrane, the absorption, flux, of a drug across the GI mucosal surface under sink conditions can be written as

$$J_W = P_W \times C_W \quad (1)$$

where J_W is the mass transport across the GI wall (mass/area/time), P_W (or P_{eff}) is defined as the effective permeability, and C_W (C_{eff}) is the concentration of the drug at the membrane. This is a local law, pertaining to each point along the intestinal membrane. As developed by Amidon et al. [2], the drug absorption rate, that is, the rate of loss of drug from the intestinal lumen (assuming no luminal reactions) at any time is

$$\text{Absorption rate} = \frac{dm}{dt} = \iint_A P_W C_W dA \quad (2)$$

where the double integral is over the entire GI surface. The total mass, M, of drug absorbed at any time t is then given by

$$M(t) = \int_0^t \int\int_A P_W C_W \, dA \, dt \qquad (3)$$

Key to limited oral absorption of drugs is that this integration (Eq. 3) is limited to the residence time of the gastrointestinal tract ($t_{res} \sim 3\,h$ in humans [6]). This analysis revealed that the fundamental events controlling oral drug absorption are the permeability of the drug through the GI membrane, the dissolution of the drug in the GI milieu (because this determines C_W in a well-mixed fluid phase), and the dose [2,7,8]. These key parameters are characterized in the BCS by the following three dimensionless numbers: (1) absorption number (A_n), (2) dissolution number (D_n), and (3) dose number (D_0). These numbers take into account both physicochemical and physiological parameters and are the most fundamental descriptors controlling the oral absorption process.

The absorption number (A_n) is the ratio of permeability (P_{eff}) and the intestinal radius (R) multiplied by the residence time (t_{res}), which can be interpreted as the effective absorption rate constant (t_{abs}^{-1}) times the residence time:

$$A_n = \text{Absorption number} = \frac{P_{eff}}{R} t_{res} = t_{abs}^{-1} t_{res} \qquad (4)$$

The dissolution number (D_n) is the ratio of the residence time and the dissolution time (t_{Diss}), which comprises the equilibrium solubility (C_S), diffusivity (D), density (ρ), the initial particle radius (r_0), and the intestinal residence time (t_{res}):

$$D_n = \text{Dissolution number}$$
$$= \frac{DC_S}{r_0} \times \frac{4\pi r_0^2}{(4/3)\pi r_0^3 \rho} t_{res} = \frac{t_{res} 3 D C_S}{\rho r_0^2} = \frac{t_{res}}{t_{Diss}} \qquad (5)$$

Finally, the dose number (D_0) is the ratio of dose to dissolved drug:

$$D_0 = \text{Dose number} = \frac{M_0/V_0}{C_S} \qquad (6)$$

where C_S is the equilibrium solubility, M_0 is the dose, and V_0 is the volume of water taken with the dose, which is generally set to be 250 mL. This volume was selected based on a typical bioequivalence study protocol that administers an 8 oz (240 mL) glass of water with the oral dosage form. Thus 250 mL, allowing a small GI residual volume, represents the initial gastrointestinal volume to which an oral dosage form is exposed in the fasting state. This number may be viewed as the number of glasses of water required to dissolve the drug dose. When $A_n > 1$, $D_n > 1$, and $D_0 < 1$, a drug will be well absorbed [9].

Based on their solubility and intestinal permeability characteristics, drug substances have been classified into one of the following four categories according to the BCS proposed by Amidon et al. (Fig. 1) [2]:

BCS Class I: High solubility, high permeability drugs. BCS Class I drugs are generally very well absorbed ($F_{abs} \geq 90\%$). For this class of drugs (where $D_n > 1$) the fraction of drug absorbed (F_{abs}) can be expressed as

$$F_{abs} = 1 - \exp(-2A_n) \qquad (7)$$

As A_n increases, F_{abs} increases, with 90% absorption occurring when $A_n = 1.15$. Hence, an immediate release (IR) product of this class is expected to yield 100% intestinal absorption if at least 85% of the drug is dissolved within 30 min across the physiological pH.

BCS Class II: Low-solubility, high-permeability drugs. For this class of drugs, the dissolution number $D_n < 1$, and the relationship between D_0 and D_n is critical in determining the fraction of drug absorbed. In general, BCS Class II drug products are likely to be dissolution/solubility rate limited. Accordingly, factors that increase the rate and extant of *in vivo* dissolution/solubilization will also increase the oral bioavailability of this compound.

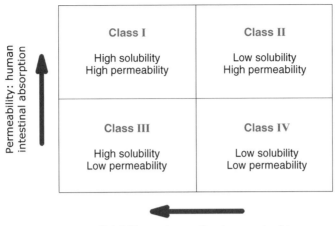

Figure 1. The BCS as defined by Amidon et al. [2]. The BCS is a classification of drug substances according to their solubility and permeability properties, in order to stand for the most fundamental view of the drug intestinal absorption process following oral administration. (This figure is available in full color at http://mrw.interscience.wiley.com/emrw/9780471266945/home.)

BCS Class III: High-solubility, low-permeability drugs. This class of drugs characterized by dose number, D_0, equals or smaller than 1, indicates that the dissolution of the drug is likely to occur rapidly, and the intestinal absorption of these drugs will be permeability rate limited.

BCS Class IV: Low-solubility, low-permeability drugs. As a low-solubility compound, these drugs have a high dose number (D_0), and in cases where both A_n and D_n are low, the compound will be classified as a Class IV drug [10,11]. These drugs are characterized by poor oral bioavailability, and tend to exhibit very large inter- and intrasubject variability. Hence, unless the dose is very low, they are generally poor oral drug candidates [12].

This BCS is one of the most significant prognostic tools created to facilitate product development in recent years, and has been adopted by the FDA, the European Medicines Agency (EMEA) and the World Health Organization (WHO) for setting bioavailability/bioequivalence (BA/BE) standards for oral drug product approval.

3. THE BCS IN REGULATORY PRACTICE

In the past decade, the BCS has become an increasingly important tool in drug product regulation worldwide, by presenting a new paradigm in bioequivalence. Bioequivalence (BE) is the critical step that connects the physical drug product with the clinical properties claimed on its label, ensuring continuing quality of the innovative products and the generic products. Before the presentation of the BCS, the BE standard was solely empirical, depending on bioavailability (BA) studies, that is, plasma levels. By revealing the fundamental parameters dictating the *in vivo* oral drug absorption process, the BCS is able to ensure BE by mechanistic tools, rather than empirical observation; if two drug products that contain the same active pharmaceutical ingredient (API) have a similar GI concentration-time profile, then a similar rate and extant of absorption is ensured for these products, that is, they are bioequivalent. Thus, BE can be guarantied based on *in vitro* dissolution tests that provide the mechanistic proof for similar bioavailability, rather than empirical *in vivo* human studies. This is the regulatory waiver of *in vivo* BE, based on the scientific and mechanistic rationale provided by the BCS. Initially, waivers of *in vivo* BE were accepted only for Scale Up and Post-Approval Changes (SUPAC), but later, the biowaiver principle was extended to the approval of new generic drug products, thus avoiding unnecessary human experiments, and reducing cost and time of developing generic drug products.

Currently, the solubility classification of a given drug is based on the highest dose strength in an IR product. A drug substance is considered highly soluble if the highest strength is soluble in 250 mL or less of aqueous media throughout the pH range of 1.0–7.5. Otherwise, the drug substance is considered poorly soluble [13]. A drug substance is considered highly permeable if the extent of intestinal absorption is determined to be 90% or higher. Otherwise, the drug substance is considered poorly permeable. The permeability classification is based either directly on the extent of intestinal absorption of a drug substance in humans or indirectly on the measurements of the rate of mass transfer across the human intestinal membrane. Alternatively, animal or *in vitro* models that predict human intestinal absorption, for example, intestinal rat perfusion models or epithelial cell culture models, can be used. An IR product is characterized as rapidly dissolved if not less than 85% of the labeled drug amount is dissolved within 30 min using USP Apparatus I at 100 rpm or USP Apparatus II at 50 rpm in a volume of 900 mL or less of each of the following media: (1) acidic media, such as USP simulated gastric fluid without enzymes; (2) pH 4.5 buffer; and (3) pH 6.8 buffer or USP simulated intestinal fluid without enzymes. Otherwise, the drug product is considered to be slow dissolving.

Up to now, The FDA has implemented the BCS system to allow waiver of *in vivo* BA/BE testing of IR solid dosage forms for Class I, high-solubility, high-permeability drugs. As for Class III (high-solubility, low-permeability) drugs, as long as the drug product does not contain permeability modifying agents, *in vitro* dissolution test can ensure BE. Hence, biowaivers for BCS Class III drugs are scientifically justified and have been recommended [14–18].

4. PROVISIONAL BCS CLASSIFICATION OF THE TOP DRUGS ON THE MARKET

Since its presentation in 1995, the validity and broad applicability of the BCS has been the subject of extensive research and discussion, including an effort to draw a BCS classification of many drug products. In order to determine the broad applicability and significance of BCS, we developed a provisional classification of first the WHO Essential Medicines List [19] and then extended this analysis to the top 200 drugs on the United States, Great Britain, Spain, and Japan lists [20]. This provisional classification was based on secondary solubility references (Merck Index, USP and JP) and a permeability estimation method based on *in silico* partition coefficient estimation relative to that of metoprolol. As such the classification is provisional and can be revised based on experimental data. However, as is shown by comparison with other methods the results are very similar. One noticeable short coming of the *in silico* method is that drugs whose intestinal absorption is carrier mediated, either in the absorptive direction or in the exorptive direction, will have their permeabilities underestimated or over estimated, respectively [20]. In this section, we will review the information gathered on the BCS classification of the top oral drug products on the global market.

4.1. BCS Classification Based on Literature Data

Since 1977, the World Health Organization (WHO) has published a list of essential medicines required for basic health care, based on public health relevance, efficacy, safety, and cost-effectiveness. There are 260 drugs on the 12th edition WHO list including 123 orally administered drugs [21]. In order to classify these drugs according to the BCS, the maximal dose strength was obtained from the WHO Essential Medicines Core List, values for the drugs solubility were obtained from Merck Index [22] and the United States Pharmacopeia [23], and the dose number (D_0) was calculated (Eq. 6). A dose number equal or lower than 1 indicated high-solubility, and $D_0 > 1$ signified a low-solubility compound.

Since information about the actual human jejunal permeability is available only for a small fraction of drugs on the list, the permeability classification was based on correlation of the estimated n-octanol/water partition coefficient, using both $\log P$ and $C \log P$, of the uncharged form of the drug molecule [24,25]. The correlation was based on a set of 29 reference drugs for which the actual human jejunal membrane permeability data is available. Drugs exhibiting n-octanol/water partition coefficient value greater than metoprolol ($\log P = 1.72$) were categorized as high permeability, since metoprolol is known to be 95% absorbed from the GI and hence may be used as a reference standard for the low/high class boundary. This list classification was subsequently compared with the classification of the top 200 prescribed drugs in the United States that include 141 orally administered drugs [26]. Only 43 oral drugs appear in both the WHO list and the top 200 prescribed US drugs, probably due to differences in treatment priorities, social acceptance, and awareness between the United States and the developing countries [19].

Solubility classification of the drugs on the WHO list and the top 200 US list revealed that 67% and 68%, respectively, are categorized as high solubility. This finding was obtained even though a conservative approach was applied for the dose number calculations. A total of 43 and 49 drugs on the WHO list and the top 200 US list, respectively, exhibited solubility lower than 0.1 mg/mL, however, some of these drugs were classified as high solubility based on the dose number (low-dose compounds). This reflects the recent trend toward development of highly-lipophilic, but high-potency drugs, leading to low dose that compensate for the poor water solubility [16].

The permeability classification revealed that 43% and 50% of the drugs on the WHO list and the top 200 US list, respectively, exhibited $\log P$ and $C \log P$ greater than the reference drug metoprolol, and are provisionally categorized as high permeability. For carrier-mediated absorbed drugs, for example, glucose, L-leucine, phenylalanine and L-dopa, permeability classification based on partition coefficient was false negative (as expected).

The percentage of drugs on the WHO list that were classified as BCS Class I and Class III was 28.5% and 35%, respectively (Fig. 2; $C \log P$). Thus, BE testing of 63% of the WHO drug products may be based on a suitable in vitro dissolution test procedure. On the basis of solubility alone, 67% of the drugs were high solubility, representing the potential number of drugs that may be eligible for a biowaiver. Hence, the majority of the drug products on the WHO List of Essential Drugs are candidates for waiver of in vivo BE testing based on in vitro dissolution test. The impact

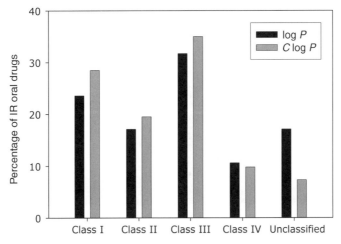

Figure 2. Provisional BCS classification of the 123 oral drugs in immediate release solid dosage forms on the WHO Essential Medicines List [19].

of waiver of expensive *in vivo* BE testing, and its replacement by rapid and affordable *in vitro* dissolution standards in developing countries is expected to be profoundly significant. The implementation of simpler, routinely monitored, and more reliable dissolution test would ensure clinical performance of marketed oral drug products worldwide [19].

A subsequent BCS classification of WHO list of Essential Medicines, utilizing primarily fraction of drug absorbed literature data for permeability classification, has produced similar results [27]. In this report, out of 61 drugs that could be classified with certainty, 34% were BCS Class I drugs, 17% were Class II drugs, 39% were classified as Class III drugs, and 10% were Class IV compounds. Hence, in this analysis, more than 70% of the classified drugs proved to be candidates for waiver of *in vivo* BE testing based on *in vitro* dissolution test [27].

Since many of the WHO drugs are not on the top 200 drug lists of the developed countries, a provisional BCS classification of the orally administered IR solid dosage forms in the top 200 drug products from the United States (US), Great Britain (GB), Spain (ES), and Japan (JP) was carried out [20]. More than 50% of the top 200 drug products on all four lists were oral IR drug products, ranging from 102 to 113. The classification was based on D_0 and $C \log P$ or $\log P$ criteria. The average maximum and lowest dose strengths on the US, ES, and GB were similar, indicating commonality with respect to use and efficacy standards. Conversely, dramatically lower dose strengths were found on the JP list compared to the other countries, reflecting differences in therapeutic categories and higher emphasis on safety issues. According to the Japanese Guideline for BE studies, the value for V_0 in calculation of D_0 (Eq. 6) is 150 mL, hence this value was used for the classification of the JP list. A volume of 250 mL was used for the classification of the other three lists.

The distribution of the top selling drugs in the four countries in terms of solubility values was very similar, despite of the fact that only 34–44 drugs on the JP list were in common with the US, GB, and ES lists. In the permeability classification, a slightly higher number of drugs were classified as high permeability using correlation to $C \log P$ than the $\log P$ criteria. This is a reflection of the greater number of drugs for which $C \log P$ values could be calculated compared to $\log P$.

The percentage of drugs that were classified as BCS Class I drugs were 31%, 30.4%, 30.2%, and 34.5% on the US, GB, ES, and JP, respectively (Fig. 3). Thus, according to the current FDA guidelines, BE testing of around 30% of the top selling drugs on these lists may be based on a suitable *in vitro* dissolution test procedure. On the basis of solubility criterion alone, that is, D_0, 55–60% of the

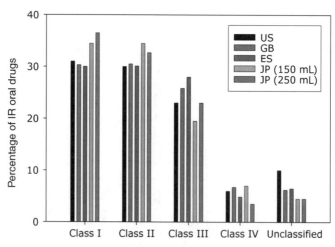

Figure 3. Provisional BCS classification of oral drugs in immediate release solid dosage forms on the top 200 US, GB, ES, and JP drugs lists, using dose number (D_0) for the solubility criterion and $C \log P$ for the permeability classification [20]. (This figure is available in full color at http://mrw.interscience.wiley.com/emrw/9780471266945/home.)

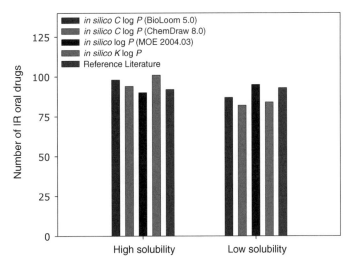

Figure 4. Comparison of the distribution of solubility classification of 185 drugs using estimated *in silico* approaches and reference literature solubility values [28]. (This figure is available in full color at http://mrw.interscience.wiley.com/emrw/9780471266945/home.)

drugs were classified as high-solubility drugs (Class I plus Class III) and may be eligible for biowaivers [20].

4.2. Provisional Classification Based on *In Silico* Calculations

It is well recognized that human permeability data is very expensive and difficult to obtain. In addition, at the very early stage of drug discovery and development, very little amount of the API is available for thorough evaluation of BCS classification. Hence, a reliable BCS classification based solely on an *in silico* approach can be highly valuable. Certainly, the underlying assumptions and methods used in any computational approach should be carefully evaluated; however, the continuous progress, convenience and feasibility of *in silico* methods attract increasing interest.

A set of 185 worldwide IR oral drug products was assigned with provisional BCS classification based on *in silico* solubility estimations and different *in silico* permeability approaches: $C\log P$ (BioLoom 5.0 and ChemDraw 8.0), $\log P$ (MOE Version 2004.03) and $K\log P$ using simplified approach based upon the Crippen fragmentation method that depend strictly on the element type in the molecule [28]. An excellent agreement was obtained between the solubility classification based on *in silico* methods and literature values (Fig. 4). The *in silico* permeability calculations demonstrated ∼75% accuracy in classifying 29 reference drugs with human permeability data, and ∼90% accuracy in classifying the 14 FDA reference drugs for permeability.

The provisional BCS classification of the 185 drugs using *in silico* or reference literature solubility classification and permeability classification with various *in silico* partition coefficients showed some interesting trends (Fig. 5). For a given solubility classification approach, the BCS classification was not significantly different when different *in silico* partition coefficient methods were used. The classification by the two solubility approaches for a given partition coefficient method, however, exhibited some systematic differences. The *in silico* solubility approach underestimated Class I and overestimated Class II drugs by an identical average of $4.3 \pm 1\%$, while it overestimated Class III and underestimated Class IV drugs by an identical average of $7.3 \pm 0.7\%$, compared to the classification using reference literature solubility [28]. This work suggests that when the *in silico* method is validated, it is convenient, efficient, and cost-effective in the early preclinical drug discovery setting. Further research should continuously improve the accuracy and reliability of *in silico* based BCS classification.

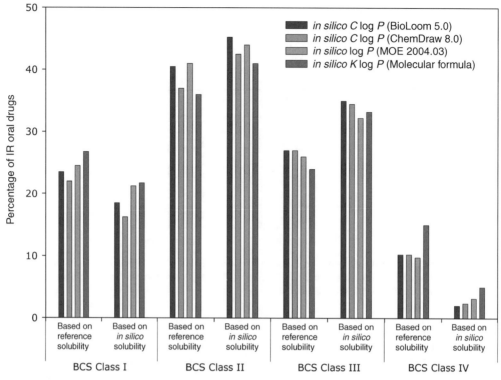

Figure 5. Comparison of the provisional BCS classification of 185 drugs using solubility classification based on *in silico* or reference literature solubility values and permeability classification using various *in silico* partition coefficient [28]. (This figure is available in full color at http://mrw.interscience.wiley.com/emrw/9780471266945/home.)

4.3. Biowaiver Monographs for IR Oral Dosage Forms

Starting in 2004, a series of monographs have been published in the *Journal of Pharmaceutical Sciences*, aiming to evaluate all pertinent data available from literature sources for a given API to assess the risk associated with a biowaiver. For this purpose, risk is defined as the probability of an incorrect biowaiver decision, as well as the consequences of an incorrect biowaiver decision in terms of public health and individual patient risks. On the basis of these considerations, a recommendation was made as to whether a biowaiver is advisable or not. The monographs have no formal regulatory status, but represent the best scientific opinions now available.

Class I assignment was certain for propranolol HCl [29], prednisolone [30], and chloroquine phosphate/sulfate/HCl [31], and according to the current guidelines, biowaiver for these APIs should be granted. Verapamil HCl could not be assigned as Class I due to solubility in pH 7.3–8, however, in the relevant pH, this is a Class I, high-solubility, high-permeability compound, and should be granted with a biowaiver [29]. Prednisone was assigned as Class I borderline, with a recommendation for biowaiver [32]. Amitriptyline HCl was classified as high-permeability compound with uncertainty in its solubility classification, and biowaiver for this API was recommended [33]. With some uncertainty regarding permeability, quinidine sulfate, and isoniazide were assigned as Class I drugs, and biowaivers were recommended [34,35]. Ethambutol dihydrochloride was classified as high solubility with borderline permeability and an *in vivo* BE waiver was recommended [36]. Ibuprofen and diclofenac sodium/potassium were classified as Class II drugs, however, a biowaiver for these APIs was scientifically justified and recommended [37,38]. The following APIs were assigned as Class III drugs: cimetidine [39], ranitidine HCl [40], atenolol [29], acetaminophen (paracetamol) [41], metoclopramide HCl [42], and pyrazinamide [43]. These Class III drugs were recommended as candidates for a biowaiver, as excipient interaction appears not to be critical with regard to the absorption, provided that the dosage form is formulated with well-known excipients, show rapid *in vitro* dissolution, and meet the dissolution profile comparison criteria as defined in the FDA Guidance. Acyclovir in strengths up to 400 mg was classified as Class III drug, and above that, that is, 800 mg, a Class IV compound, however, all strengths were recommended for a biowaiver under the usual restrictions [44]. No sufficient conclusive data were available for acetazolamide classification, and hence, a biowaiver for this API could not be recommended [45].

In conclusion, with the exception of one drug (acetazolamide), an *in vivo* BE waiver was scientifically justified and recommended for all APIs evaluated in these monographs. These monographs are part of an ongoing project, and the details and progress of this project are available at www.fip.org/bcs.

4.4. The Biopharmaceutics Drug Disposition Classification System

The major difficulty in assigning drugs to Class I is the determination of permeability. Solubility measurements are relatively easy to carry out, and in general most investigators agree when classifying drugs as either high or low solubility compounds. However, intestinal permeability is not as routinely measured, particularly using methods and laboratory practice that would allow granting a FDA *in vivo* biowaiver. To address this issue, Wu and Benet proposed a Biopharmaceutics drug Disposition Classification System (BDDCS), noting that BCS Class I and II compounds were eliminated by metabolism, while Class III (polar) compounds were eliminated by renal excretion, and suggested that if the major route of elimination of a given drug is metabolism, then the drug is high permeable [46]. On the other hand, if the major route of elimination is renal and biliary excretion of unchanged drug, then that drug should be classified as low permeability. Additional implications of the BDDCS, for example, food effect, were suggested as well [47]. The BDDCS has the advantage and disadvantage of being base on experimental (metabolism) results, and as such is not directly comparable to the *in silico* approach discussed above. However, the resulting classification is very similar to the classification based on the provisional BCS *in silico* method with the difference mainly residing in the choice of permeability (fraction absorbed) or percent metabolism dividing line for high/low classification (see below).

A total of 168 drugs were classified by the BDDCS based on solubility and metabolism [46]. Drugs with $\geq 50\%$ metabolism were defined as extensively metabolized and thus considered high permeability drugs (the 50% metabolism cutoff [48] has subsequently been revised). Takagi et al. compared this BDDCS classification of 164 drugs with the BCS approach using D_0 and $C \log P$ of different reference drugs for permeability classification (Fig. 6) [20]. The BDDCS classification indicated that 59 drugs are Class I, 51 Class II, 42 Class III, and 12 drugs out of the 164 are Class IV compounds. The conservative BCS classification based on metoprolol as the reference compound indicated a total of 42, 54, 57, and 11 drugs as Classes I, II, III, and IV, respectively. Hence, excellent agreement between BDDCS and BCS was obtained for the classification of Classes II and IV drugs, but not for Classes I and III. A total of 33 drugs were Class I compounds according to both BDDCS and BCS using metoprolol as the reference permeability drug, and 26 more drugs were Class I according to the BDDCS but not according to the BCS. The original choice for cutoff (i.e., metoprolol) was purposely conservative, and hence, the BCS classification was also carried out using cimetidine (65–70% fraction oral dose absorbed), ranitidine (30–70% fraction oral dose absorbed) and atenolol (~50% fraction oral dose absorbed) (Fig. 6). The agreement between the BCS and the BDDCS in assigning drugs to Class I increased from 56% (BCS/metoprolol) to 75% (BCS/cimetidine) to 86% (BCS/atenolol). Using cimetidine as the BCS permeability

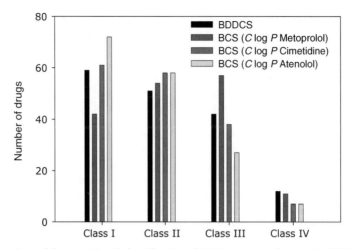

Figure 6. Comparison of the provisional classification of 164 drugs according to the BDDCS and the BCS using three different reference permeability drugs [20]. (This figure is available in full color at http://mrw.interscience.wiley.com/emrw/9780471266945/home.)

reference drug appeared to exhibit the best overall agreement with the BDDCS [20].

While permeability classification based solely upon metabolism might fail to correctly classify drugs that are highly absorbed but are excreted unchanged into urine and bile (e.g., amoxicillin, trimetoprim, lomefloxacin, zalcitabine, and chloroquine), lipophilicity considerations alone would not be able to predict active carrier-mediated transport of drugs. Despite these differences, the two approaches indicate that a waiver from *in vivo* BE studies should be granted to majority of drugs [48].

5. FUTURE DIRECTIONS

The FDA has implemented the BCS system to allow waiver of *in vivo* BA/BE testing of IR solid dosage forms only for Class I, high solubility, high-permeability drugs. It has been recommended at several workshops that biowaivers should be extended to Class III drugs with rapid dissolution properties. The absorption of a Class III drug is likely limited by its permeability, less dependent upon its formulation, and its bioavailability may be determined by its *in vivo* permeability pattern. If the dissolution of Class III drug product is rapid under all physiological pH conditions, its *in vivo* behavior will essentially be similar to oral solution (controlled by gastric emptying), and as long as the drug product does not contain permeability modifying agents (this potential effect is largely mitigated by the large gastric dilution), *in vitro* dissolution test can ensure BE. Hence, biowaivers for BCS Class III drugs are scientifically justified and have been recommended [14–18].

In addition to the extension of *in vivo* BE waivers to BCS classes other that I, the current FDA biowaiver guidance is generally considered highly conservative, especially with respect to the class boundaries of solubility, permeability and dissolution. Thus, the possibility of modifying these boundaries and criteria to allow biowaivers for additional drug products has received increasing attention [49–51]. For example, currently, drug substances are classified as high solubility compounds if the highest strength is soluble in 250 mL or less of aqueous media throughout the pH range of 1.0–7.5. Under fasting state, the GI pH vary from 1.4–2.1 in the stomach, 4.9–6.4 in the duodenum, 4.4–6.6 in the jejunum, and 6.5–7.4 in the ileum [52,53]. Hence, it seems reasonable to redefine the BCS class boundary pH range from 1.0–7.5 to 1.4–6.8. Moreover, if a drug product meets the dissolution criterion, that is, not less that 85% dissolved within 30 min, its dissolution process is probably completed during the jejunum, as it generally takes 85 min for a drug to reach the ileum [5,54]. Thus, it might be reasonable to narrow the pH range requirement even more.

Another example is the nature of the media used for solubility tests. While the solubility classification is based on the dissolution of the drug in aqueous buffers, the *in vivo* conditions for drug dissolution contain bile salts and phospholipids, even under fasting state [55]. These are powerful natural surfactants that aid in the dissolution/solubilization of the drug substances. A media which more adequately reflect physiological conditions may be more relevant in assessing *in vivo* solubility and dissolution, and potentially, drugs that are classified as Class II according to the current solubility definitions could be classified as Class I under these conditions [49,51,56].

With regard to drug discovery and development, BCS classification can proceed from early candidate screening based on *in silico* methods to *in vitro* (tissue culture) and *in situ* (jenunal perfusion in the rat) for candidate selection to full mass balance studies and permeability determination (if necessary) in humans. Early BCS Classification provides a basics for development decision making that focuses the development effort on the types of preclinical problems that will likely be faced and the studies that will be necessary to develop the candidate drug to NDA approval.

6. CONCLUSIONS

In conclusion, the BCS based new paradigm to approach BE provides a mechanistic rationale for drug development and can significantly reduced the regulatory burden. Consequently, the BCS contributes to the public health worldwide, by greatly enhancing the efficiency in drug development and regulatory approval processes. The information provided by the BCS classification of the top drugs on the global market shows that the clinical performance of majority of approved drugs essential for human health can be assured with an *in vitro* dissolution test. This information should help pharmaceutical manufacturers of both new medicines and generic drug products to avoid unnecessary human experiments, and reduce cost and time of the products development. This is of particular interest in countries with considerably limited health care budget. Further research into the BCS classification of drugs should be encouraged, in order to increase the classification certainty. This will enable a detailed risk analysis and will lead to confident recommendation as to whether a given drug product can be granted with *in vivo* biowaiver. Additionally, the current FDA guidelines on BCS classification are considered highly conservative, especially with respect to the class boundaries of solubility, permeability and dissolution. New regulatory policies, with criteria and class boundaries that will allow granting an *in vivo* biowaiver to larger number of drugs, should be constructively examined.

REFERENCES

1. Takamatsu N, Welage LS, Idkaidek NM, Liu DY, Lee PI, Hayashi Y, Rhie JK, Lennernas H, Barnett JL, Shah VP, Lesko L, Amidon GL. Human intestinal permeability of piroxicam, propranolol, phenylalanine, and PEG 400 determined by jejunal perfusion. Pharm Res 1997;14(9):1127–1132.
2. Amidon GL, Lennernas H, Shah VP, Crison JR. A theoretical basis for a biopharmaceutic drug classification: the correlation of *in vitro* drug product dissolution and *in vivo* bioavailability. Pharm Res 1995;12(3):413.
3. Dahan A, Amidon GL. Segmental dependent transport of low permeability compounds along the small intestine due to P-glycoprotein: the role of efflux transport in the oral absorption of BCS class III drugs. Mol Pharm 2009; 6(1):19–28.
4. Lennernas H. Human intestinal permeability. J Pharm Sci 1998;87(4):403–410.
5. Yu LX, Lipka E, Crison JR, Amidon GL. Transport approaches to the biopharmaceutical design of oral drug delivery systems: prediction of intestinal absorption. Adv Drug Deliv Rev 1996;19(3):359.
6. Yu LX, Amidon GL. A compartmental absorption and transit model for estimating oral drug absorption. Int J Pharm 1999;186(2):119–125.
7. Oh DM, Curl RL, Amidon GL. Estimating the fraction dose absorbed from suspensions of poorly soluble compounds in humans: a mathematical model. Pharm Res 1993;10(2):264.
8. Sinko PJ, Leesman GD, Amidon GL. Predicting fraction dose absorbed in humans using a

macroscopic mass balance approach. Pharm Res 1991;8(8):979.
9. Oh DM, Curl RL, Amidon GL. Estimating the fraction dose absorbed from suspensions of poorly soluble compounds in humans: a mathematical model. Pharm Res 1993;10(2):264–270.
10. Lobenberg R, Amidon GL. Modern bioavailability, bioequivalence and biopharmaceutics classification system. New scientific approaches to international regulatory standards. Eur J Pharm Biopharm 2000;50(1):3–12.
11. Martinez MN, Amidon GL. A mechanistic approach to understanding the factors affecting drug absorption: a review of fundamentals. J Clin Pharmacol 2002;42(6):620–643.
12. Dahan A, Hoffman A. In: Touitou E, Barry BW, editors. Enhancement in Drug Delivery. CRC press; 2006. p 111–127.
13. US, Food, Drug Administration. Waiver of *in vivo* bioavailability and bioequivalence studies for immediate release dosage forms containing certain active moieties/active ingredients based on a biopharmaceutical classification system. Center for Drug Evaluation and Research; 1999.
14. Blume HH, Schug BS. The biopharmaceutics classification system (BCS): Class III drugs— better candidates for BA/BE waiver? Eur J Pharm Sci 1999;9(2):117.
15. Cheng CL, Yu LX, Lee HL, Yang CY, Lue CS, Chou CH. Biowaiver extension potential to BCS Class III high solubility-low permeability drugs: bridging evidence for metformin immediate-release tablet. Eur J Pharm Sci 2004;22(4):297.
16. Dahan A, Amidon GL. In: Van de Waterbeemd H, Testa B, editors. Drug Bioavailability: Estimation of Solubility, Permeability, Absorption and Bioavailability. Wiley-VCH; 2008. p 33–51.
17. Jantratid E, Prakongpan S, Amidon GL, Dressman J. Feasibility of biowaiver extension to biopharmaceutics classification system class III drug products: cimetidine. Clin Pharmacokinet 2006;45(4):385–399.
18. Stavchansky S. Scientific perspectives on extending the provision for waivers of *in vivo* bioavailability and bioequivalence studies for drug products containing high solubility–low permeability drugs (BCS-Class 3). AAPS J 2008;10(2):300–305.
19. Kasim NA, Whitehouse M, Ramachandran C, Bermejo M, Lennernas H, Hussain AS, Junginger HE, Stavchansky SA, Midha KK, Shah VP, Amidon GL. Molecular properties of WHO Essential Drugs and provisional biopharmaceutical classification. Mol Pharm 2004;1(1):85–96.
20. Takagi T, Ramachandran C, Bermejo M, Yamashita S, Yu LX, Amidon GL. A provisional biopharmaceutical classification of the top 200 oral drug products in the United States, Great Britain, Spain, and Japan. Mol Pharm 2006;3(6):631–643.
21. WHO List of Essential Drugs, 12th ed. 2002. http://www.who.int/medicines/organization/par/edl/eml2002core.pdf.
22. The Merck Index, 13th ed. Rahway NJ: Merch Research Laboratories; 2001.
23. The United States Pharmacopeia. Philadelphia, PA: National Publishing; 2000.
24. Winiwarter S, Ax F, Lennernäs H, Hallberg A, Pettersson C, Karlén A. Hydrogen bonding descriptors in the prediction of human *in vivo* intestinal permeability. J Mol Graph Model 2003;21(4):273–287.
25. Winiwarter S, Bonham NM, Ax F, Hallberg A, Lennernas H, Karlen A. Correlation of human jejunal permeability (*in vivo*) of drugs with experimentally and theoretically derived parameters. A multivariate data analysis approach. J Med Chem 1998;41(25):4939–4949.
26. The top 200 prescriptions for 2002 by number of, US, prescriptions dispensed, 2002. http://www.rxlist.com/top200.htm.
27. Lindenberg M, Kopp S, Dressman JB. Classification of orally administered drugs on the World Health Organization Model list of Essential Medicines according to the biopharmaceutics classification system. Eur J Pharm Biopharm 2004;58(2):265–278.
28. Kim YH, Ramachandran C, Crippen GM, Takagi T, Bermejo M, Amidon GL. *In silico* approaches to prediction of permeability, solubility and BCS class: provisional classification of the top-selling IR oral drug products in the United States, Great Britain, Spain, Japan and South Korea, in preparation (2009).
29. Vogelpoel H, Welink J, Amidon GL, Junginger HE, Midha KK, Möller H, Olling M, Shah VP, Barends DM. Biowaiver monographs for immediate release solid oral dosage forms based on biopharmaceutics classification system (BCS) literature data: verapamil hydrochloride, propranolol hydrochloride, and atenolol. J Pharm Sci 2004;93(8):1945–1956.
30. Vogt M, Derendorf H, Krämer J, Junginger HE, Midha KK, Shah VP, Stavchansky S, Dressman JB, Barends DM. Biowaiver monographs for immediate release solid oral dosage forms: prednisolone. J Pharm Sci 2007;96(1):27–37.
31. Verbeeck RK, Junginger HE, Midha KK, Shah VP, Barends DM. Biowaiver monographs for

immediate release solid oral dosage forms based on biopharmaceutics classification system (BCS) literature data: chloroquine phosphate, chloroquine sulfate, and chloroquine hydrochloride. J Pharm Sci 2005;94(7):1389–1395.

32. Vogt M, Derendorf H, Krämer J, Junginger HE, Midha KK, Shah VP, Stavchansky S, Dressman JB, Barends DM. Biowaiver monographs for immediate release solid oral dosage forms: prednisone. J Pharm Sci 2007;96(6):1480–1489.

33. Manzo RH, Olivera ME, Amidon GL, Shah VP, Dressman JB, Barends DM. Biowaiver monographs for immediate release solid oral dosage forms: Amitriptyline hydrochloride. J. Pharm. Sci. 2006;95(5):966–973.

34. Becker C, Dressman JB, Amidon GL, Junginger HE, Kopp S, Midha KK, Shah VP, Stavchansky S, Barends DM. Biowaiver monographs for immediate release solid oral dosage forms: isoniazid. J Pharm Sci 2007;96(3):522–531.

35. Grube S, Langguth P, Junginger HE, Kopp S, Midha KK, Shah VP, Stavchansky S, Dressman JB, Barends DM. Biowaiver monographs for immediate release solid oral dosage forms: quinidine sulfate. J Pharm Sci 2009; 98(7):2238–2251.

36. Becker C, Dressman JB, Amidon GL, Junginger HE, Kopp S, Midha KK, Shah VP, Stavchansky S, Barends DM. Biowaiver monographs for immediate release solid oral dosage forms: ethambutol dihydrochloride. J Pharm Sci 2008;97(4):1350–1360.

37. Chuasuwan B, Binjesoh V, Polli JE, Zhang H, Amidon GL, Junginger HE, Midha KK, Shah VP, Stavchansky S, Dressman JB, Barends DM. Biowaiver monographs for immediate release solid oral dosage forms: diclofenac sodium and diclofenac potassium. J Pharm Sci 2009; 98(4):1206–1219.

38. Potthast H, Dressman JB, Junginger HE, Midha KK, Oeser H, Shah VP, Vogelpoel H, Barends DM. Biowaiver monographs for immediate release solid oral dosage forms: ibuprofen. J Pharm Sci 2005;94(10):2121–2131.

39. Jantratid E, Prakongpan S, Dressman JB, Amidon GL, Junginger HE, Midha KK, Barends DM. Biowaiver monographs for immediate release solid oral dosage forms: cimetidine. J Pharm Sci 2006;95(5):974–984.

40. Kortejärvi H, Yliperttula M, Dressman JB, Junginger HE, Midha KK, Shah VP, Barends DM. Biowaiver monographs for immediate release solid oral dosage forms: ranitidine hydrochloride. J Pharm Sci 2005;94(8):1617–1625.

41. Kalantzi L, Reppas C, Dressman JB, Amidon GL, Junginger HE, Midha KK, Shah VP, Stavchansky SA, Barends DM. Biowaiver monographs for immediate release solid oral dosage forms: acetaminophen (paracetamol). J Pharm Sci 2006;95(1):4–14.

42. Stosik AG, Junginger HE, Kopp S, Midha KK, Shah VP, Stavchansky S, Dressman JB, Barends DM. Biowaiver monographs for immediate release solid oral dosage forms: metoclopramide hydrochloride. J Pharm Sci 2008;97(9):3700–3708.

43. Becker C, Dressman JB, Amidon GL, Junginger HE, Kopp S, Midha KK, Shah VP, Stavchansky S, Barends DM. Biowaiver monographs for immediate release solid oral dosage forms: pyrazinamide. J Pharm Sci 2008;97(9):3709–3720.

44. Arnal J, Gonzalez-Alvarez I, Bermejo M, Amidon GL, Junginger HE, Midha KK, Shah VP, Stavchansky S, Dressman JB, Barends DM. Biowaiver monographs for immediate release solid oral dosage forms: aciclovir. J Pharm Sci 2008;97(12):5061–5073.

45. Granero GE, Longhi MR, Becker C, Junginger HE, Kopp S, Midha KK, Shah VP, Stavchansky S, Dressman JB, Barends DM. Biowaiver monographs for immediate release solid oral dosage forms: acetazolamide. J Pharm Sci 2008;97(9):3691–3699.

46. Wu C-Y, Benet LZ. Predicting drug disposition via application of BCS: transport/absorption/elimination interplay and development of a biopharmaceutics drug disposition classification system. Pharm Res 2005;22(1):11–23.

47. Custodio JM, Wu C-Y, Benet LZ. Predicting drug disposition, absorption/elimination/transporter interplay and the role of food on drug absorption. Adv Drug Deliv Rev 2008;60(6):717–733.

48. Benet L, Amidon GL, Barends D, Lennernäs H, Polli J, Shah V, Stavchansky S, Yu L. The use of BDDCS in classifying the permeability of marketed drugs. Pharm Res 2008;25(3):483–488.

49. Polli JE, Yu LX, Cook JA, Amidon GL, Borchardt RT, Burnside BA, Burton PS, Chen M-L, Conner DP, Faustino PJ, Hawi AA, Hussain AS, Joshi HN, Kwei G, Lee VHL, Lesko LJ, Lipper RA, Loper AE, Nerurkar SG, Polli JW, Sanvordeker DR, Taneja R, Uppoor RS, Vattikonda CS, Wilding IG, Zhang G. Summary workshop report: biopharmaceutics classification system—implementation challenges and extension opportunities. J Pharm Sci 2004;93(6):1375–1381.

50. Rinaki E, Dokoumetzidis A, Valsami G, Macheras P. Identification of biowaivers among class II drugs: theoretical justification and practical examples. Pharm Res 2004;21(9):1567.

51. Yu LX, Amidon GL, Polli JE, Zhao H, Mehta MU, Conner DP, Shah VP, Lesko LJ, Chen ML, Lee VHL, Hussain AS. Biopharmaceutics classification system: the scientific basis for biowaiver extensions. Pharm Res 2002;19(7):921.

52. Oberle R, Amidon GL. The influence of variable gastric emptying and intestinal transit rates on the plasma level curve of cimetidine; an explanation for the double peak phenomenon. J Pharmacokinet Biopharm 1987;15(5):529–544.

53. Youngberg C, Berardi R, Howatt W, Hyneck M, Amidon GL, Meyer J, Dressman J. Comparison of gastrointestinal pH in cystic fibrosis and healthy subjects. Dig Dis Sci 1987;32(5):472–480.

54. Kaus LC, Gillespie WR, Hussain AS, Amidon GL. The effect of *in vivo* dissolution, gastric emptying rate, and intestinal transit time on the peak concentration and area-under-the-curve of drugs with different gastrointestinal permeabilities. Pharm Res 1999;16(2):272.

55. Dahan A, Hoffman A. Rationalizing the selection of oral lipid based drug delivery systems by an *in vitro* dynamic lipolysis model for improved oral bioavailability of poorly water soluble drugs. J Control Release 2008;129(1):1–10.

56. Vertzoni M, Dressman J, Butler J, Hempenstall J, Reppas C. Simulation of fasting gastric conditions and its importance for the *in vivo* dissolution of lipophilic compounds. Eur J Pharm Biopharm 2005;60(3):413.

THE ROLE OF PERMEABILITY IN DRUG ADME/PK, INTERACTIONS AND TOXICITY, AND THE PERMEABILITY-BASED CLASSIFICATION SYSTEM (PCS)[1]

URBAN FAGERHOLM
Clinical Pharmacology & DMPK,
AstraZeneca R&D Södertälje,
Södertälje, Sweden

1. PERMEABILITY

Permeability (P_e) is one of the main determinants in drug and metabolite absorption, distribution, metabolism, excretion/pharmacokinetics (ADME/PK). Consideration, investigations, and understanding of the role of permeability in various organs are therefore crucial for predictions of ADME/PK, interactions, elimination routes, solubility/dissolution limitations, exposures (including organ/cellular levels), and toxicity, for selection of suitable candidate drugs (CDs), clinical doses and formulations, and for lead optimization recommendations.

Permeability is defined as the ability of molecules to permeate a cell, cell membrane, endothelium or epithelium, and also as the ability of these media to let molecules pass through. The P_e is usually expressed as the rate (commonly 10^{-6} cm/s) at which this occurs. In general, the passive transmembrane/cellular diffusion-driven P_e increases with decreased molecular weight and increased (up to a certain limit) lipophilicity of compounds. Paracellular (between cells) absorption and diffusion through unstirred fluid layers could potentially also be important, but studies have demonstrated that these are most likely of no significant importance for drugs and metabolites in humans *in vivo* [1–5]. Quite often active transport (uptake and/or efflux) by one or more of the many transport proteins (these will not be discussed in detail here) is also involved in permeation and makes a significant contribution to the ADME/PK profile.

The active component of permeation makes predictions of ADME/PK difficult and uncertain, especially when the passive P_e is low and there is polymorphism involved. The composition of cell membranes and expression, localization and activity of transport proteins usually differ among organs (and also within organs), and this together with differences in surface area (S) and transit time, makes the uptake and efflux capacities organ specific.

2. DRUG TRANSPORT INTERACTIONS AND POLYMORPHISM

Drug–drug, drug–metabolite, and metabolite–metabolite transport interactions (TIs) could imply safety risks and loss of pharmacological effects. It is therefore important that TIs can be well predicted during drug discovery and development (and the earlier they can be predicted the better).

Several factors determine the extent and significance of TIs, for example, the level of passive P_e, expression and activity of transport proteins, concentrations of drug, metabolite and inhibitor/inducer, time for uptake and induction/inhibition, and the relative importance of other ADME/PK factors, such as metabolism, glomerular filtration, and dissolution.

Many excellent presentations and reviews of transport proteins and clinically significant TIs and metabolism-related interactions (MIs) are available in the literature [6–14]. The most significant TIs have been due to inhibition processes in intestines, liver, brain, kidneys, and bile [6–11]. Sometimes, and as a consequence, unwanted serious adverse effects occur. Among these TIs, biliary TIs are the most common and pronounced. Examples (affected drugs) include pitavastatin, pravastatin, rosuvastatin, and digoxin, and possibly also cerivastatin. Inhibition of hepatic uptake and bile excretion of these comparably hydrophilic OATP1B1 substrates pitavastatin, pravastatin, and rosuvastatin in man (by cyclosporin) has led to 4.5-, 8- and 7-fold average increases in systemic area under the curve (AUC). Examples of drugs with affected renal secretion include digoxin, ACE inhibitors, antibiotics, cimetidine, procainamide, metformin, furosemide, dofetilide, fexofenadine, and

---

[1] This paper includes personal opinions of the author, which do not necessarily represent the views or policies of AstraZeneca.

methotrexate. Changes in renal clearance (CL_R) and AUC reach maximally 50–100%. The AUC for dofetilide and metformin increased by 50% in the presence of cimetidine, and probenecid caused a doubling of the AUC for cephalosporin. The drugs with biliary and/or renal TIs generally have low/moderate passive permeability, low metabolic hepatic CL (CL_H) and excretion as main elimination route.

The interplay between permeability and metabolism, and how that affects the type of interactions has been demonstrated for statins [7,8].

Compounds with high passive permeability (thus, high renal tubular and intestinal reabsorption potential) and/or poor metabolic stability are (and are anticipated to be) eliminated mainly via metabolism and demonstrate MIs (rather than TIs). Overall, the changed AUC in the presence of inhibitors or inducers is quantitatively less important for TIs than for MIs. Average MI-related AUC changes of 20-fold or more (for inhibition and induction) have been demonstrated for highly permeable CYP3A4 substrates (including lovastatin, midazolam, and nisoldipine) [8].

Intestinal TIs has been demonstrated for substrates of P-gp and BCRP with limited passive permeability [6,9–11]. Inhibited intestinal efflux is believed to be involved in the increased systemic exposure of fexofenadine, digoxin and topotecan, and maybe also talinolol. The AUCs of these compounds increased by maximally approximately 150%, 200% and 140% on average in the presence of inhibitors (such as ketoconazole, erythromycin, and elacridar), respectively. A systemic component (bile and renal TIs) is probably involved in such exposure changes. The P-gp inducer rifampicin caused approximately 30% and 60% reductions in the AUCs for digoxin and fexofenadine, respectively.

Among reasons for the modest intestinal TIs effects are a significant passive component, high intestinal concentrations and (thereby) saturation of transporters, different major absorption regions of substrates and inducers/inhibitors, and that the uptake occurs mainly in an intestinal region with comparably low transport activity. For verapamil, a drug with P-gp efflux (the efflux is more pronounced and permeability is lower at lower intestinal concentrations) and high passive permeability, the intestinal fraction absorbed (f_a) is unaffected by dose [15]. Its passive and total permeability is sufficiently high to provide complete gastrointestinal (GI) absorption.

There are (at least) two known clinical TIs at the human blood–brain barrier (BBB)—those with the highly (passive) permeable P-pg substrates loperamide and verapamil. The P-gp inhibitor quinidine was able to increase the brain uptake of loperamide (increased CNS levels of loperamide could not be explained by increased plasma exposure), and thereby, causing respiratory depression [16], and cyclosporin A caused almost a doubling of the brain-to-blood AUC ratio of verapamil (^{11}[C]-radioactivity measured) [17]. It is interesting to note that (in contrast to the intestines, kidneys, and bile) significant brain TIs occur for compounds with high passive permeability. One of possible reasons to this difference is the comparably short transit time (of blood and compounds) through the brain.

No apparent clincial TIs on metabolic CL_H has been demonstrated. For this to happen and to be of importance, it is required that the intrinsic CL_H (CL_{int}) exceeds the uptake CL (which generally requires that the liver has a low passive uptake capacity), and that CL_H is low in relation to the liver blood flow rate (Q_H). A few examples of such TIs have been demonstrated in rat liver perfusion experiments. Hepatic uptake transporter(s) has/have been shown to play a role in the CL_H of erythromycin [18], and the CL_H of enalaprilat seems to be influenced by uptake [19]. In addition, the CL_H of the P-pg- and CYP3A4-substrate tacrolimus increased in the presence of the P-gp inhibitor GG918 [20], and this was proposed to be a result of inhibited canalicular P-gp efflux and (thereby) increased exposure within hepatocytes. These compounds all have low passive permeability.

Transporter polymorphism could play a role for absorption and exposure. Average AUC differences between genotypes for OATP1B1 substrates pitavastatin, pravastatin, rosuvastatin, and atorvastatin, the OCT1 substrate metformin and the P-gp substrate digoxin were 2.9-, 2.1-, 0.65-, 1.4-, 0.6- (the CL_R of metformin did, however, not differ between OCT1 genotypes), and 2-fold, respectively [21–24].

Based on these clinical findings and the requirement to predict TIs and MIs well and early, establishment of P_e versus f_a relationships and interplay between permeability and metabolism in various organs is essential.

3. PERMEABILITY VERSUS FRACTION ABSORBED

The interplay between membrane P_e, S, pH, transit time and dispersion, blood component binding capacity, and the molecular characteristics, determine the extent and fraction of absorption during each passage through organs. Differences between (and also within) organs are expected to cause different shapes and shifts of P_e versus f_a relationships.

A (sigmoidal) relationship between P_e and f_a from the GI tract in man has been established for compounds with mainly passive transport [25]. The uptake of substances with a P_e at and near the steepest part of the passive P_e versus intestinal f_a relationships is potentially most sensitive to efflux than for compounds with high passive P_e (a small change in total P_e results in a considerable f_a change). When active efflux is involved such compounds often deviate significantly from the relationship between passive P_e and f_a [26]. The three intestinal TI compounds digoxin, fexofendine, and talinolol (see above) have a passive P_e in this critical region, and their GI f_a is approximately 0.70, 0.05–0.10, and $\geq$0.55, respectively. Active organ uptake is or is expected to be of greatest importance (for the f_a) for compounds with low passive P_e. Examples include levodopa, ACE inhibitors, and many antibiotics, which are taken up by nutrient transporters.

For many other important organs such relationships (including shapes and shifts) have not been available until recently. With the use of a passive P_e data set for 126 compounds (artificial membrane in vitro P_e obtained at pH 7.4; including a large amount of registered drugs; produced by Willmann et al. [26]), physiology data and clinical ADME/PK data, Fagerholm [27] was able to establish passive P_e versus f_a relationships for the intestines, liver, renal tubuli (f_{ra}), and brain. The fitted curves for these four organs are demonstrated in Fig. 1. Two curves were generated for renal tubular reabsorption—one for compounds with apparently no active transport involved and the upper limit for those with apparent active secretion. See this reference for a more detailed description of calculations and estimations, and establishment of relationships.

The P_e versus f_a relationships (f_a for unbound fraction (f_u) per each organ passage) were fitted based on Equation 1:

$$f_a = \frac{P_e^\lambda}{P_{e50}^\lambda + P_e^\lambda} \quad (1)$$

where P_{e50} is the P_e corresponding to a f_a (or f_{ra}) of 0.50 and λ is a shape factor.

Equation 1 assumes an infinite dispersion number (D_N) for the liver blood flow (well-stirred model). The data fit for the liver has now been updated using a recently proposed D_N-value of 0.5 and a distribution factor for correction of a longer hepatic transit time of unbound molecules (the basis of the newly developed modified dispersion model (MDM)) [5].

In this data set, P_e-values ranged between 0.01×10^{-6} cm/s (gentamicin) and 31.5×10^{-6} cm/s (testosterone). P_{e50} estimates for the liver, intestines, renal tubuli (passive and upper limit for secretion), and brain with this P_e data set are 0.33, 0.66, 1.4, 12, and 24×10^{-6} cm/s, respectively. These correspond to P_e estimates for nadolol/sulpiride, atenolol, sotalol, chloramphenicol, and propranolol/chlorpromazine, respectively. Corresponding λ-values are 1.0 (originally; higher slope at higher P_e with $D_N = 0.5$), 1.6, 3.3, 5.0, and 2.0 (0.7 at low P_e; two functions were applied for brain uptake), respectively. These shapes and shifts are in agreement with observed clinical TIs (see Section 2).

Based on these results, the liver, followed by the intestines, has the highest absorptive capacity (low P_{e50}), whereas the brain appears to have the lowest capacity (highest P_{e50}). The hepatic, brain, and intestinal f_a and renal tubular f_{ra} for the moderately permeable digoxin are approximately 0.75, 0.03, 0.66, and 0.25, respectively. After correction for differences in transit time through the organs it is,

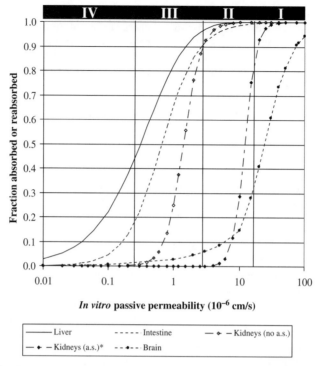

Figure 1. The PCS and its permeability classes (I–IV), and predicted average relationships between *in vitro* passive permeability and *in vivo* fraction absorbed or reabsorbed for the unbound fraction during each passage through the human intestines, liver, renal tubuli, and brain. *Approximated upper limit for compounds with active secretion (a.s.) in the kidneys.

however, apparent that the liver and brain have the highest uptake capacity. Interestingly, the brain has the highest f_a per S unit per time unit. Results are consistent with available *in vitro* P_e data obtained with intestinal, brain, liver, and kidney cells [28].

It should be noted that there is uncertainty in the data used in this evaluation. That includes uncertainties of clinical ADME/PK data, potential influences of active transport, and possible *in vitro* versus *in vivo* differences in passive permeability. Nevertheless, these approximate average curves are still very useful for understanding and predicting ADME/PK profiles (including elimination routes) and interaction potentials, and optimization of lead compounds.

Different shapes and shifts of the P_e versus f_a curves are expected when using P_e data obtained with other methods. See Ref. [29] for a comparison between different methods for assessment of P_e. Shapes and shifts might also change as new *in vivo* ADME/PK data become available.

4. PERMEABILITY CLASSIFICATION

Based on data and relationships presented in Section 3, a Permeability-Based Classification System (PCS) with four permeability classes (I–IV) was developed (Fig. 1):

I. Very high P_e; higher than for amlodipine, haloperidol, and nifedipine.
II. High P_e; lower limit similar to the P_e of cimetidine, caffeine, probenicide, and warfarin.
III. Intermediate P_e and incomplete f_a; lower limit similar to the P_e of nadolol and sulpiride.
IV. Low P_e and f_a.

Compounds of Class I are or are expected to be well reabsorbed, and therefore mainly eliminated from the blood circulation via metabolism. Thus, TIs could be avoided or reduced by aiming for increased/high passive P_e and/or moderate to high intrinsic metabolic CL (CL_{int}) of CDs, and also, finding/developing CDs that are not substrates for active transport. TIs of highly permeable compounds could however occur in the brain for drugs with efflux at the BBB (see above). Compounds of Class II without active secretion or without active secretion potential are also expected to be completely or near completely absorbed/reabsorbed by the intestines, liver and renal tubuli, and to be mainly metabolized. Class III compounds have intermediate passive uptake in the intestines, liver, and renal tubuli, and therefore, are potentially sensitive to involvement of active efflux. Class IV compounds have low/incomplete passive absorption in the intestines, liver, and renal tubuli, and highest excretion potential. For such compounds, in particular, active uptake could play a significant role for the f_a and f_{ra}.

Based on the Biopharmaceutics Classification System (BCS; see Section 7) Benet and colleagues have developed a Biopharmaceutics Drug Disposition Classification System (BDDCS) [13,30,31]. The BCS and BDDCS have two permeability classes (high/low), where 90% intestinal f_a, alternatively 90% metabolism of given oral dose, is the limit between high and low intestinal P_e and f_a. Evaluations have demonstrated that 90% metabolism is a better limit for high P_e than 90% GI f_a [32,33]. Khandelwal et al. [33] found that only 10 (18%) out of 56 drugs belonged to same BCS and BDDCS categories, and Fagerholm [32] showed that a considerable fraction of compounds with an intestinal $f_a < 0.9$ are extensively metabolized and that many drugs with an intestinal $f_a > 0.9$ are mainly excreted.

Advantages with the PCS (which actually was developed without the BDDCS in mind) compared to the BDDCS include a continuous P_e scale, consideration of P_e versus f_a relationships (including useful equations) of various important organs, four instead of two permeability classes, and the possibility to apply it for predictions (the BDDCS requires clinical data on fraction metabolized). It is also believed that it can give better predictions and understanding of the ADME/PK profile, polymorphism, potential TIs and MIs, and organ/cell trapping of drugs and their metabolites. The second component of BCS and BDDCS is solubility/dissolution, and as shown and discussed in Section 7, these systems appear to be very (too) strict on dissolution and for drugs with short half-life ($t_{1/2}$). A limitation with any classification system is the potential for incorrect classification of borderline compounds.-sec1-0

The PCS has now been updated to a matrix with several P_e and CL_{int} classes (preliminary P_e and CL_{int} categories and limits can be viewed in Ref. [5]). This has improved its applicability for prediction and overview of ADME/PK and interactions, selection and profiling of CDs, and lead optimization.

5. ROLE OF PERMEABILITY IN METABOLISM

5.1. Hepatic Metabolism

As demonstrated in Fig. 1, and in the perfused rat liver by Miyauchi et al. [34], the liver appears to have a high uptake capacity. Most compounds have an extraction ratio of the f_u ($E_{H,u}$) that is lower than the estimated f_a, and therefore, few compounds seem to have a metabolic CL_H that is uptake rate-limited. A recent evaluation of the determinants of CL_H showed that absorption, diffusion, and dissociation (from plasma proteins) are comparably rapid processes, and metabolism generally is rate-limiting [5]. Reference data and drugs were presented, and these are useful for predicting CL_H and the rate-limiting step. For example, enalaprilat, atenolol, and metoprolol were estimated to have a hepatic f_a (during each organ passage and for the f_u) of ~ 0.20, ~ 0.70, and ~ 0.99, respectively. Corresponding hepatic absorption rate constants (k_a) were estimated to ~ 0.007, ~ 0.04, and $\sim 0.15\,s^{-1}$, respectively. Tolbutamide, triazolam, and ketamine have $E_{H,u}$- and metabolism rate constant (k_{met})-values corresponding to these uptake estimates. The highly albumin-bound drug warfarin has an *in vitro* dissociation rate constant (k_{diss}) of $1.2\,s^{-1}$, predicted *in vivo* hepatic k_a of $\sim 0.1\,s^{-1}$ (predicted from passive

P_e), and measured *in vivo* k_{met} of ~0.006 s^{-1} [5]. Thus, its k_{diss} is comparably rapid (permissive binding) and metabolic CL_{int} is the rate-limiting step for the hepatic extraction. The highly permeable testosterone binds to albumin (50% bound) and globulin (44% bound), and its k_{diss} for these proteins have been estimated to >0.46 and 0.03 s^{-1}, respectively [35]. Its estimated hepatic k_a (estimated from passive P_e) is 0.42 s^{-1} [5], which demonstrates that dissociation is rate-limiting for its hepatic uptake.

Compounds with low P_e and no active hepatic uptake will demonstrate low CL_H and E_H. For example, drugs with a P_e below that of nadolol or sulpiride will have a maximum CL_H (CL_H for the f_u) of approximately 700 mL/min (45% of Q_H).

In vitro to *in vivo* extrapolation of metabolic CL_H (and also of metabolic CL and excretion CL in other organs) is associated with some difficulties, including the changed P_e, increased S, and potentially reduced transporter and metabolic activities during *in vitro* conditions [5]. Isolated hepatocytes are no longer surrounded by their normal compartments (resulting in three- to sevenfold increased S and changed P_e characteristics) and fail to reestablish normal canalicular networks [5]. The unphysiological direct contact between drug in an incubation medium and the canalicular membrane (with its specific transporters) could potentially lead to reduced uptake into hepatocytes. These changes could lead to erroneous predictions and understanding of the processes determining CL_H, especially for compounds with low passive P_e and high metabolic CL_{int}, and for substances with efficient canalicular efflux.

5.2. Gut-Wall Metabolism

Gut-wall metabolic extraction/elimination could be significant, especially in the absorptive direction (first-pass), and during small intestinal absorption for CYP3A4 substrates and small and large intestinal uptake for conjugated substances [36,37]. Because of the importance for CYP3A4 substrates, the low CYP3A4 levels in the colon, and difficulty to estimate the number and average activity CYP3A4 enzymes involved in the metabolism Fagerholm developed a P_e- and reference drug-based *in vitro* to *in vivo* method for prediction of the first-pass gut-wall extraction ratio (E_{GW}) for such substrates [36]. In this method, P_e data are used to predict the fraction of the intestinal f_a that takes place in the small intestine (where extraction is expected). *In vitro* CL_{int}, preferably using enterocytes (the fraction metabolized by CYP3A4 needs to be considered if hepatocytes are used), and human *in vivo* E_{GW} data of midazolam or verapamil (reference compounds and data) are used together with *in vitro* CL_{int} and predicted fraction of absorbed dose taken up by the small intestine of CDs to predict E_{GW}. This simple approach compensates for differences in absorption and metabolic activity along the intestine and villi, and does not require knowledge about the amount and fraction of cells involved in absorption and metabolism. The E_{GW} for a small number of CYP3A4 substances were reasonably well predicted [36]. A simple rule that can be used is that the E_{GW} will be maximally 50% if the CL_{int} of a CD is $\leq CL_{int}$ of midazolam or verapamil. Compounds with low/moderate P_e and efflux are challenging. This is because of the potential for entrapment in an enterocyte circulation, and thereby, higher E_{GW} than expected.

5.3. Oral Bioavailability

P_e is a determinant for both intestinal f_a, E_{GW} and CL_H, and therefore, it will play multiple roles in determining the oral F. The P_e versus intestinal f_a relationship will set an approximate upper limit for oral F. An approximate lower oral bioavailability limit (F_{min}) can also be predicted from the P_e uptake curves (including the E_{GW} prediction method for CYP3A4 substrates discussed in Section 5.2) [27]. A peak F_{min} of ~0.2 is obtained at a P_e similar to that of atenolol and digoxin. Thus, this simple P_e-based approach can be used for approximating the potential suitability of CDs for oral administration.

5.4. Renal Metabolism

The renal CL (CL_R) is difficult to predict, and reasons include the involvement of bidirectional passive and active tubular transport, differences in uptake capacity, pH and residence time on lumenal and blood sides of tubular cells, and limited knowledge about regional tubular residence time, P_e and metabolic capacity [38]. Predictions of metabolic CL_R are also limited by the lack of human *in vivo* data [38]. Renal concentrations and activities of CYP450s are, however, comparably low, which suggests that CYP450 substrates have negligible metabolic CL_R [38]. In contrast, kidneys have relatively high expression of UDP-glucuronyltransferases (UGTs) and high glucuronidation CL_{int}. Thus, it can be expected that metabolic CL_R of high P_e UGT substrates could contribute to the total CL to some (but yet unpredictable) extent [38].

6. ROLE OF PERMEABILITY IN EXCRETION

Metabolically stable drugs often undergo significant elimination in the bile, kidneys, and/or intestines. To be able to predict the CL of such compounds, we need to consider the processes included in the excretion and reabsorption.

6.1. Renal Excretion

As mentioned above, involvement of bidirectional passive and active tubular transport, differences in uptake capacity, pH and residence time on lumenal and blood sides of tubular cells, and limited knowledge about regional tubular residence time, P_e (passive and active) and metabolic capacity make prediction of the CL_R difficult. Available data clearly show that reabsorption and active transport are common and often significant in the kidneys.

Despite these difficulties, passive P_e data are useful for predicting the tubular reabsorption potential, CL_R and fraction excreted in urine (f_e). This is shown in Fig. 1 (PCS), where f_{ra} is complete or near complete for PCS Class II (only when there is no active secretion) or I drugs. Relationships between P_e and CL_R and P_e and f_e [27,38] demonstrated the impact on passive P_e on renal and total CL. Available renal data for Class I compounds showed ~0 CL_R and f_e. Zero or negligible values were also found for Class II compounds. In this class there were substances with CL_R- and f_e-values reaching ~400 mL/min and ~0.6, respectively. Values for compounds of Classes III and IV ranged between ~15 and ~500 mL/min and ~0.4 and ~1, respectively. These data could be useful to get an indication of renal excretion potential of CDs with different levels of P_e. It should be taken into account that the lower values for high P_e compounds could partly be results of the impact of lipophilicity on binding to blood components (reduced binding potential with increased lipophilicity) and metabolic CL_{int} (increased CL_{int} potential with increased lipophilicity).

6.2. Biliary Excretion

Data and predictions of biliary excretion have recently been reviewed [37]. Apparent limitations with such predictions are the lack of human *in vivo* bile CL (CL_{bile}) data where intestinal reabsorption and enterohepatic recirculation have been considered, and the neglect of blood component binding and intestinal reuptake in published *in vitro* to *in vivo* predictions [37]. Previously proposed distinct molecular weight-dependency and -thresholds (species-dependent) for bile excretion do not seem to hold [37].

A P_e-based mathematical method for prediction of the maximum bile excretion potential and CL_{bile} was developed by Fagerholm [27,37]. A simple rule was also presented—in addition to active transport, low metabolic CL_{int} (lower than that of diazepam, which is approximated to have an $E_{H,u}$ of 0.5 [5]) and low passive P_e (PCS III–IV; less than that of metoprolol) are required for substantial/significant CL_{bile} to occur.

6.3. Intestinal Excretion

Available data (for the low passive P_e P-gp substrates digoxin and talinolol and BCRP substrate rosuvastatin) show that intestinal secretion is generally not a major route of drug elimination [37]. These findings are in agreement with the absence of major transporter proteins such as P-gp and BCRP in basolateral enterocyte membranes, low Q in the intestinal mucosa compared to the liver and kidneys (~1/6 to ~1/5) and (anticipated) limited amount/fraction of enterocytes involved [37]. Although elimination via the intestinal route does not appear to be of great importance for these few compounds with available data, it could be equally important as bile excretion [37]. As for bile excretion, near complete to complete reabsorption (and therefore, no net secretion) are anticipated for compounds with an intestinal P_e greater than that of metoprolol (PCS Classes I and II).

7. PERMEABILITY VERSUS SOLUBILITY/DISSOLUTION IN GASTROINTESTINAL ABSORPTION

Prediction of the intestinal f_a for substances with low solubility and slow dissolution is difficult. *In vitro* to *in vivo* relationships for solubility/dissolution have not been fully investigated, the interplay between P_e and solubility/dissolution (P_e/absorption acts as a sink) is not fully understood, and available data and perceptions on the role of solubility/dissolution are not consistent.

Aqueous solubility is commonly used as a surrogate measurement of *in vivo* solubility, and Dose number (D_o) is a parameter that is commonly used as a measurement of solubility/dissolution potential. D_o is defined as the ratio of dose-concentration to solubility; $D_o =$ (highest dose strength/250 mL fluid)/solubility [39]. A D_o-value of 1 implies that the expected highest GI concentration is similar to the solubility, and a high D_o implies low dissolution potential. D_o versus f_a data for 73 high P_e compounds were collected from the literature, and results demonstrated a poor relationship between *in vitro* aqueous solubility and *in vivo* dissolution and absorption for these highly permeable compounds [32]. Only one drug product (danazol) has a human *in vivo* $f_a < 0.8$, and nine drug products have a $f_a < 0.9$. Drug products with very high aqueous D_o are completely or near completely absorbed (telmisartan $D_o = 660,000$; toremifene D_o 8700; oxatomide $D_o = 1500$). Based upon this finding it appears that many highly permeable drug products are likely to be sufficiently well absorbed regardless of the aqueous solubility (at least up to a D_o of 660,000) and that the P_e plays a more important role in GI absorption than expected. The good absorption of these low-solubility drugs could be signs of good *in vivo* sink conditions (P_e- and dispersion-driven), good *in vivo* dissolution potential, and/or successful formulation development strategies.

It must, however, be taken into account that there might exist more highly permeable compounds with low *in vivo* solubility (these could lack *in vivo* f_a data and/or P_e data to confirm high P_e), and that available results may include f_a data obtained in the fed-state or with fluid formulations.

Eight out of 10 compounds with D_o between 770 and 333,000 have demonstrated a more than 50% increase in the AUC when taken with food [40]. The greatest AUC increase (~fivefold) were observed for etrinate ($D_o = 333,000$) and albendazole ($D_o = 1600$). Halofantine with a D_o of 770 demonstrated a threefold increase, whereas no increases were observed for terfenadine ($D_o = 5000$) or oxaprozin ($D_o = 3000$). Incomplete uptake in the fasted state and increased dissolution in the fed state is one possible reason to the observed AUC increases. Changed first-pass metabolism due to changed Qs is another. The P_e of these compounds has not been measured (at least not to my knowledge), and therefore, the reason(s) to these food effects are not fully understood.

The BCS is a regulatory tool developed for enabling replacement of *in vivo* bioequivalence studies for immediate-release products by P_e and *in vitro* dissolution tests [41]. In the BCS, drug products are classified according to their intestinal P_e, and GI solubility in relation to dose: Class I, high P_e—high solubility; Class II, high P_e—low solubility; Class III, low P_e—high solubility; Class IV, low P_e—low solubility. A drug product is considered highly permeable and soluble when the P_e corresponds to a f_a following oral administration that is not less than 0.9, or when the measured f_a is ≥ 0.9, and the highest clinical dose is dissolved in 250 mL buffer or other aqueous media at pH 1–7.5. The BCS requires the test drug product to show *in vitro* dissolution profile similarity versus the comparator in three dissolution media. For IR Class I drug products, dissolution in HCl solution or simulated gastric fluid, pH 4.5 buffer, and pH 6.8 buffer or simulated intestinal fluid must be complete (>85%) within 30 min.

Evaluations have demonstrated that the BCS has a very strict solubility/dissolution limit, a generous P_e limit (≥ 14 times higher rate constant limit for dissolution than for permeation), and is stricter for drugs with long $t_{1/2}$ and for acids [32,42]. Other limitations include the difficulty to classify P_e for compounds with moderate to high P_e well (the approach by Benet et al. [31] is very helpful), and few and uncertain *in vitro* to *in vivo* correlations for solubility/dissolution rate. The results shown in [32] are inconsistent with the number and fraction of drug products in BCS Class II. Fagerholm [32] suggested some changes of the BCS that could give a better balance and increase the number of bioequivalence study biowaivers: (a) decrease the limit for high solubility/dissolution (to >40% and >95% dissolved within 30 min and 3 h, respectively), (b) increase the limit for high P_e (to >P_e of metoprolol), (c) use reliable P_e prediction/classification models (excluding intestinal perfusion methods), (d) perform solubility/dissolution tests using real or validated simulated gastrointestinal fluids, (e) establish *in vitro*/*in vivo* dissolution relationships, (f) consider the $t_{1/2}$, (g) determine the rate-limiting step for *in vivo* absorption, and (h) reduce the BCS into two classes: permeation-rate (Class I) or dissolution-rate (Class II) limited absorption.

8. ROLE OF PERMEABILITY IN DISTRIBUTION

8.1. Brain Distribution

It has been suggested that the BBB is nearly impermeable (that <2% of small drug molecules and essentially all large molecules do not cross the BBB) [43,44], and various cutoff values for zero or limited brain P_e have been suggested (polar surface area >70–90 or > 120 Å^2, log D above or below 1–4, and molecular weight >450 g/mol) [44,45]. On the basis of these suggestions/conclusions it appears that many CDs may not achieve sufficiently high brain exposure to elicit an effect and that brain drug development programs need to be adjusted so that compounds are designed for active brain uptake.

A recent review of the uptake capacity of the brain showed that this organ is (in contrast to previous perceptions) highly permeable [28]. The brain can absorb compounds with polar surface area >270 Å^2, molecular weight >1000 g/mol, log D <-3.5 and equilibrium brain-to-blood concentration ratio <0.01 well. It was proposed and generalized that sufficient intestinal uptake indicates good passive brain uptake potential, and that (unless there is significant efflux at the BBB) intestinal absorption is the rate-limiting step in the uptake into the brain. The high uptake capacity of the brain has been demonstrated *in vitro* (in general, considerably higher P_e across BBB than across intestinal mucosa), in rats *in vivo* (~70–100% brain uptake within 5–15 s for highly permeable (unbound) substances) [28], and also by the data in Fig. 1. In the PCS, the brain has the highest f_a per S unit per time unit [28]. The *in vivo* BBB passage time (not the same as time to equilibrium) for drug molecules, in general, is approximated to <1 s to ~1 min. For caffeine, morphine, and inulin (molecular weight ~5000 g/mol; considered impermeable) the passage time is approximated to 0.2 s, 7 s, and 50 min, respectively. The ~10–20 s that it takes from the start of smoking a cigarette to the experience of central nicotine effects can also be used as a reference for the rapid BBB passage.

Active influx and efflux takes place at the BBB, and due to the right-shifted P_e versus f_a curve (high P_{e50}; partly a result of short transit time) active transport is likely to influence the uptake and redistribution of compounds with both low and high passive P_e. Inhibition of BBB P-gp by quinidine resulted in an increased brain uptake and serious adverse event (respiratory depression) of loperamide [16]. Except for this finding BBB efflux effects in humans are small (doubled brain exposure of verapamil by cyclosporin A [17]) or have not been found/demonstrated.

The P_e versus f_a relationship in the PCS can be used to estimate/simulate the role of changed BBB efflux at different P_e levels. For example, for three hypothetical compounds with passive P_e of 0.2, 2, and 20×10^{-6} cm/s and efflux resulting in halved P_e (0.1, 1, and 10×10^{-6} cm/s), a total inhibition of the efflux would lead to ~60, ~60, and ~180% increases in f_a for each passage through the brain, respectively. These simulations demonstrate that the uptake of highly permeable drugs is (potentially) most sensitive to efflux. For a compound with passive and total P_e of 200 and 100×10^{-6} cm/s, respectively, complete efflux inhibition would result in a negligible change in f_a (~4%). However, P_e-values of this magnitude are considerable higher than for drugs in general. The highest passive P_e in the used data set was 31.5×10^{-6} cm/s (for testosterone).

8.2. Hepatic Distribution

For drugs efficiently absorbed into the liver by transporters, such as statins, concentrations within hepatocytes could exceed those in plasma by several folds. For example, liver:plasma ratios of atorvastatin and cerivastatin in the rat have been estimated to approximately 60 on average [46]. For such compounds, liver distribution could influence the overall volume of distribution (V_D) [13,47–49]. This has been demonstrated in humans for atorvastatin and glyburide [47,48]. Rifampin inhibited the hepatic uptake and reduced the V_D/F of these two compounds by 93% and 60% on average, respectively. Corresponding average CL/F decreased by 87% and 53%, respectively.

8.3. Red Blood Cell Distribution

Drug permeation of and binding to the membrane of red blood cells (RBCs) is common, and the extent and rate of this binding play a role in drug distribution and elimination. For many substances it may take as long as several hours to reach steady state between RBCs and the medium (blood or perfusion fluid) [50]. Rat liver *in situ* perfusion data have demonstrated that slow binding to and release from RBCs has an impact on the CL_H of doxorubicin, and RBC binding of propranolol does not influence its CL_H, but causes a significant doubling of the fraction escaping hepatic extraction (0.06 versus 0.03) [51].

Lipid solubility, degree of ionization, molecular size, and hydrogen-bonding ability have been identified as the main determinants for RBC uptake and binding [50,52]. Drugs generally enter RBCs more rapidly when the lipophilicity and degree of unionized form are high, and molecular size and ability to form hydrogen bonds are low [50,52]. Basic drugs, in particular, have been observed to bind nonspecifically to RBCs, and to α-adrenergic receptors on RBCs [53].

RBCs are bounded by a lipid-like membrane, which is perforated with positively charged aqueous pores of various diameters [50]. Small lipid-soluble molecules penetrate RBCs both by dissolving in the lipid phase and by diffusion through the aqueous channels, whereas larger lipophilic molecules enter mainly through the lipid phase [50]. Lipophilic drugs and anions mainly enter the cells if they are small enough to pass through the pores, whereas cations are largely excluded because of lipid insolubility and inability to pass through the positively charged pores [50]. Partially ionized lipid-soluble bases have been detected in the RBCs of various animals, and they usually penetrate RBCs so rapidly that the time for equilibration cannot be accurately determined [54]. Slow binding compounds have low oil:water partition coefficients (<0.001–0.011), which suggest that it takes or may take longer time for hydrophilic low P_e substances to equilibrate between plasma and RBCs.

RBCs appear comparably permeable. The P_e for water in human RBCs was nine times higher than in hepatocytes, and the P_e of urea was similar to that in hepatocytes [55]. Compounds that bind slowly to RBCs might potentially also have low GI P_e and f_a, low hepatocyte P_e and CL_{int}, and low CL_H.

RBCs express transporter proteins [56]. As an example, human RBCs have extremely high P_e to D-glucose (which is a hydrophilic compound known to utilize specific transporters in other cells). In comparison to another hydrophilic compound of similar size and hydrophilicity, mannitol, D-glucose had three orders of magnitudes higher RBC P_e [57]. Active transport across the RBC membrane makes estimation of the available f_u (and CL and unbound exposures) for substrates of these transporters difficult.

9. ROLE OF PERMEABILITY IN ORGAN ACCUMULATION AND TOXICITY

In general, drug and metabolite toxicity is exposure-related. Drug and/or metabolite exposures in plasma or blood are usually used as a measurement or surrogate measurement of exposures in organs where toxicities are found. This could be appropriate. In the GI tract, eliminating organs and within cells, drug and metabolite exposures might, however, not be related to or proportional to systemic exposures. This makes interspecies comparisons and predictions and evaluations of toxicity difficult.

Permeability plays a role in organ accumulation (as described in Chapter) of drugs and metabolites, and several liver diseases have been found to affect liver transporter expression and function [12]. It is therefore important that P_e is considered for predictions and evaluations of organ and systemic toxicity.

Significant hepatocyte accumulation is expected to occur for drugs with pronounced active uptake, especially when metabolic CL_{int} and passive P_e are comparably low. For example, high hepatocyte concentrations have been demonstrated for (actively absorbed) statins [46]. Coadministration of statins with specific transport inhibitors has led to decreased hepatic uptake, increased systemic exposure and toxic response [12,47,48].

Accumulation potential of metabolites within hepatocytes is highest when the metabolic CL_{int} is high and the metabolite P_e is low compared to that of the drug (especially when metabolite P_e is within the critical P_e versus f_a region for the liver and the P_e of the drug is comparably high).

High drug concentrations could be reached in renal tubular cells and fluid when active secretion is pronounced and elimination is mainly renal. High levels in tubular mucosal cells could also be reached when reabsorption is rapid and extensive.

In the upper GI mucosa, highest concentrations are expected for drugs with high P_e and dissolution rate and given at high oral doses (acids in particular; such as nonsteroidal anti-inflammatory drugs), and for rapidly and extensively formed metabolites of such drugs (substrates of CYP3A4 and conjugating enzymes in particular).

Pang et al. [58] recently published a useful review about active and toxic metabolites, and where models for prediction and understanding of metabolite formation and metabolite ADME/PK and accumulation are presented and discussed. It is also believed that the updated PCS (which includes a metabolic CL_{int} dimension) could be very useful for a deeper understanding and better predictions of organ exposures and toxicity.

REFERENCES

1. Fagerholm U, Nilsson D, Knutson L, Lennernäs H. Jejunal permeability in humans *in vivo* and rats *in situ*: investigation of molecular size selectivity and solvent drag. Acta Physiol Scand 1999;165:315–324.
2. Fagerholm U, In characteristics of intestinal permeability in humans *in vivo* and rats *in situ*. Doctoral thesis. Sweden: Uppsala University; 1997.
3. Fagerholm U, Lennernäs H. Experimental estimation of the effective unstirred water layer thickness in the human jejunum, and its importance in oral drug absorption. Eur J Pharm Sci 1995;3:247–253.
4. Shibata Y, Takahashi H, Chiba M, Ishii Y. Prediction of hepatic clearance and availability by cryopreserved human hepatocytes: an application of serum incubation method. Drug Metab Disp 2002;30:892–896.
5. Fagerholm U. Presentation of a Modified Dispersion Model (MDM) for hepatic drug extraction and new methodology for prediction of the rate-limiting step in hepatic metabolic clearance. Xenobiotica 2009;39:57–71.
6. Ho RH, Kim RB. Transporters and drug therapy: implications for drug disposition and disease. Clin Pharmacol Ther 2005;78:260–277.
7. Shitara Y, Sato H, Sugiyama Y. Evaluation of drug-drug interaction in the hepatobiliary and renal transport of drugs. Annu Rev Pharmacol Toxicol 2005;45:689–723.
8. Shitara Y, Sugiyama Y. Pharmacokinetic and pharmacodynamic alterations of 3-hydroxy-3-methylglutaryl coenzyme A (HMG-CoA) reductase inhibitors: drug–drug interactions and interindividual differences in transporter and metabolic enzyme functions. Pharmacol Ther 2006;112:71–105.

9. Endres JC, Hsiao P, Chung FS, Unadkat JD. The role of transporters in drug interactions. Eur J Pharm Sci 2006;27:501–517.
10. Lin JH. Transporter-mediated drug interactions: clinical implications and *in vitro* assessment. Expert Opin Drug Metab Toxicol 2007;3: 81–92.
11. Lin JH, Yamazaki M. Role of P-glycoprotein in pharmacokinetics: clinical implications. Clin Pharmacokinet 2003;42:59–89.
12. Li P, Wang G-J, Robertsson TA, Roberts MS. Liver transporters in hepatic drug disposition. Curr Drug Metab 2009;10:482–498.
13. Shugarts S, Benet LZ. The role of transporters in the pharmacokinetics of orally administered drugs. Pharm Res 2009;26:2039–2054.
14. Kindla J, Fromm MF, König J. *In vitro* evidence for the role of OATP and OCT uptake transporters in drug–drug interactions. Expert Opin Drug Metab Toxicol 2009;5:489–500.
15. Sandström R, Karlsson A, Knutson L, Lennernäs H. Jejunal absorption and metabolism of R/S-verapamil in humans. Pharm Res 1998;15:856–862.
16. Sadeque AJM, Wandel C, He H, Shah S, Wood AJ. Increased drug delivery to the brain by P-glycoprotein inhibition. Clin Pharmacol Ther 2000;68:231–237.
17. Sasongko L, Link JM, Muzi M, Mankoff DA, Yang X, Collier AC, Shoner SC, Unadkat JD. Imaging P-glycoprotein transport activity at the human blood–brain barrier with positron emission tomography. Clin Pharmacol Ther 2005;77: 503–514.
18. Lam JL, Okochi H, Huang Y, Benet LZ. *In vitro* and *in vivo* correlation of hepatic transporter effects on erythromycin metabolism: characterizing the importance of transporter-enzyme interplay. Drug Metab Disp 2006;34: 1336–1344.
19. Schwab AJ, Barker F III Goresky CA, Pang KS. Transfer of enalaprilat across rat liver cell membranes is barrier limited. Am J Physiol 1990;258:G461–G475.
20. Wu C-Y, Benet LZ. Disposition of tacrolimus in isolated perfused rat liver: influence of troleandomycin, cyclosporine, and GG918. Drug Metab Disp 2003;31:1292–1295.
21. Deng JW, Song IS, Shin HJ, Yeo CW, Cho DY, Shon JH, Shin JG. The effect of SLCO1B1*15 on the disposition of pravastatin and pitavastatin is substrate dependent: the contribution of transporting activity changes by SLCO1B1*15. Pharmacogen Genom 2008;18:424–433.
22. Pasanen MK, Fredrikson H, Neuvonen PJ, Niemi M. Different effects of *SLCO1B1* polymorphism on the pharmacokinetics of atorvastatin and rosuvastatin. Clin Pharmacol Ther 2007; 82:726–733.
23. Shu Y, Brown C, Castro RA, Shi RJ, Lin ET, Owen RP, Sheardown SA, Yue L, Burchard EG, Brett CM, Giacomini KM. Effect of genetic variation in the organic cation transporter 1, OCT1, on metformin pharmacokinetics. Clin Pharmacol Ther 2008;83:273–280.
24. Hoffmeyer S, Burk O, vonRichter O, Arnold HP, Brockmöller J, Johne A, Cascorbi I, Gerloff T, Roots I, Eichelbaum M, Brinkmann U. Functional polymorphisms of the human multidrug-resistance gene: multiple sequence variations and correlation of one allele with P-glycoprotein expression and activity *in vivo*. PNAS 2000;97: 3473–3478.
25. Karlsson J, Artursson P. A new diffusion chamber system for the determination of drug permeability coefficients across the human intestinal epithelium that are independent on the unstirred water layer. Biochim Biophys Acta 1992;1111:204–210.
26. Willmann S, Schmitt W, Keldenich J, Lippert J, Dressman JB. A physiological model for the estimation of the fraction dose absorbed in humans. J Med Chem 2004;47:4022–4031.
27. Fagerholm U. The role of permeability in drug ADME/PK, interactions and toxicity—presentation of a permeability-based classification system (PCS) for prediction of ADME/PK in humans. Pharm Res 2008;25:625–638.
28. Fagerholm U. The highly permeable blood–brain barrier: an evaluation of current opinions about brain uptake capacity. Drug Discovery Today 2007;12:1076–1082.
29. Fagerholm U. Prediction of human pharmacokinetics—gastrointestinal absorption. J Pharm Pharmacol 2007;59:905–916.
30. Wu CY, Benet L. Predicting drug disposition via application of BCS: transport/absorption/elimination interplay and development of a biopharmaceutics drug disposition classification system. Pharm Res 2005;22:11–23.
31. Benet LZ, Amidon GL, Barends DM, Lennernäs H, Polli JE, Shah VP, Stavchansky SA, Yu LX. The use of BDCCS in classifying the permeability of marketed drugs. Pharm Res 2007;52: 483–488.
32. Fagerholm U. Evaluation and suggested improvements of the Biopharmaceutics Classification System (BCS). J Pharm Pharmacol 2007;59:751–757.

33. Khandelwal A, Bahadduri PM, Chang C, Polli JE, Swaan PW, Ekins S. Computational models to assign biopharmaceutics drug disposition classification from molecular structure. Pharm Res 2007;24:2249–2262.
34. Miyauchi S, Sawada Y, Iga T, Hanano M, Sugiyama Y. Comparison of the hepatic uptake clearances of fifteen drugs with a wide range of membrane permeabilities in isolated rat hepatocytes and perfused rat livers. Pharm Res 1993;10:434–440.
35. Mendel CM. The free hormone hypothesis: a physiologically based mathematical model. Endocr Rev 1989;10:232–274.
36. Fagerholm U. Prediction of human pharmacokinetics—gut-wall metabolism. J Pharm Pharmacol 2007;59:1335–1343.
37. Fagerholm U. Prediction of human pharmacokinetics—biliary and intestinal clearance and enterohepatic circulation. J Pharm Pharmacol 2008;60:535–542.
38. Fagerholm U. Prediction of human pharmacokinetics—renal metabolic and excretion clearance. J Pharm Pharmacol 2007;59:1463–1471.
39. Amidon GL, Lennernäs H, Shah V, Crison J. A theoretical basis for a biopharmaceutical drug classification: the correlation of *in vitro* drug product dissolution and *in vivo* bioavailability. Pharm Res 1995;12:413–420.
40. Yazdanian M, Briggs K, Jankovsky C, Hawi A. The "high solubility" definition of the current FDA Guidance on Biopharmaceutical Classification System may be too strict for acidic drugs. Pharm Res 2004;2:293–299.
41. Dressmann JB, Amidon GL, Fleisher D. Absorption potential: estimating the fraction absorbed for orally administered drugs. J Pharm Sci 1985;74:588–589.
42. Singh BN. A quantitative approach to probe the dependence and correlation of food-effect with aqueous solubility, dose/solubility ration, and partition coefficient (log P) for orally active drugs administered as immediate-release formulations. Drug Dev Res 2005;65:55–75.
43. Su Y, Sinko PJ. Drug delivery across the blood–brain barrier: why is it difficult? How to measure and improve it?. Expert Opin Drug Del. 2006;3:419–435.
44. Pardridge WM. Blood–brain barrier delivery. Drug Disc Today 2007;1–2:54–61.
45. Kelder J, et al. Polar molecular surface as a dominating determinant for oral absorption and brain penetration of drugs. Pharm Res 1999;16:1514–1519.
46. Paine SW, Parker AJ, Gardiner P, Webborn PJH, Riley RJ. Prediction of the pharmacokinetics of atorvastatin, cerivastatin, and indomethacin using kinetic models applied to isolated rat hepatocytes. Drug Metab Disp 2008;36:1365–1374.
47. Lau YY, Huang Y, Frassetto L, Benet LZ. Effect of OATP1B1 transporter inhibition on the pharmacokinetics of atorvastatin in healthy volunteers. Clin Pharmacol Ther 2007;81:194–204.
48. Zheng HX, Huang Y, Frassetto L, Benet LZ. Elucidating rifampin's inducing and inhibiting effects on glyburide pharmacokinetics and blood glucose in healthy volunteers: unmasking the differential effects of enzyme induction and transporter inhibition for a drug and its primary metabolite. Clin Pharmacol Ther 2009;85:78–85.
49. Grover A, Benet LZ. Effects of drug transporters on volume of distribution. AAPS J 2009; 11:250–261.
50. Schanker LS, Nafpliotis PA, Johnson JM. Passage of organic bases into human red cells. J Pharmacol Ther 1961;133:325–331.
51. Lee HL, Chiou WL. Erythrocytes as barriers for drug elimination in the isolated liver: II. Propranolol Pharm Res 1989;6:840–843.
52. Naccache P, Sha'afi RI. Patterns of nonelectrolyte permeability in human red blood cell membrane. J Gen Physiol 1973;62:714–736.
53. Tillement J-P, Houin G, Zini R, Urien S, Albengres E, Barré J, Lecomte M, D'Athis P, Sebille B. The binding of drugs to blood plasma macromolecules: recent advances and therapeutic significance. Adv Drug Res 1984;13:59–94.
54. Sahin S, Rowland M. Effect of erythrocytes on the hepatic distribution kinetics of antipyrine. Eur J Drug Metab Pharmacokinet 2004;29: 37–41.
55. Alpini G, Garrick RA, Jones MJT, Nunes R, Tavoloni N. Water and nonelectrolyte permeability of isolated rat hepatocytes. Am J Physiol 1986;251:C872–C882.
56. Kock K, Grube M, Jedlitschky G, et al. Expression of adenosine triphosphate-binding cassette (ABC) transporters in peripheral blood cells: relevance for physiology and pharmacotherapy. Clin Pharmacokinet 2007;46:449–470.
57. Jung CY, Carlson LM, Whaley DA. Glucose transport carrier activities in extensively washed human red cell ghosts. Biochim Biophys Acta 1971;241:613–627.
58. Pang KS, Morris ME, Sun H. Formed and preformed metabolites: facts and comparisons. J Pharm Pharmacol 2008;60:1247–1275.

SALT SCREENING AND SELECTION

DAVID J. SEMIN[1]
JANAN JONA[1]
MATTHEW L. PETERSON[2]
ROGER ZANON[1]

[1] Small Molecule Process and Product Development, Amgen, Inc., Thousand Oaks, CA

[2] Small Molecule Process and Product Development, Amgen, Inc., Cambridge, MA

1. INTRODUCTION

Salt formation is frequently performed on acidic and basic drugs because it is relatively simple chemical manipulation. The salt formation may alter the physicochemical and mechanical properties of the compound and may improve the formulation, biopharmaceutical properties (this may also lead to improve therapeutic properties), and scale-up of the compound without irreversibly modifying the basic chemical structure.

Theoretically, every compound possessing, acidic or basic functional group, can participate in salt formation.

What is presented in this chapter is not intended to be another one of the many reviews on salt formation and selection already available in the literature [1,2] nor intended to be a comprehensive review of available crystallization screening tools currently available; it is the goal of the authors to provide a simple guide for a pragmatic screening strategy, which is based on salt formation theories and commercially available technology platforms, of how crystallization platforms can be utilized to help ensure the selection of the right salt at the most appropriate time, given the resources at hand.

1.1. Desired Properties of the Form

A developable active pharmaceutical ingredient (API) will give the required exposure for efficacy through the desired route of administration and will demonstrate chemical stability sufficient to support a useful product shelf life, which is typically 2 years at room temperature. In addition to these foundational requirements for a drug candidate, other physical properties are also important. Those that are most often considered are as follows:

1. *Crystalline*: The crystallinity of the form is the most important and the first property to consider. Amorphous forms are less attractive to develop due to inherent stability issues with amorphous materials. Crystallinity can be assessed easily using polarized light microscope (PLM) or powder X-ray diffraction (PXRD).

2. Acceptable thermal behavior that includes a high temperature ($>120°C$) single endothermic transition, preferably due to melting. This property is more important if there is a need to mill, or micronize, as milling will produce heat that may melt the material. The thermal behavior of the form is easily evaluated using differential scanning calorimetry (DSC) and thermogravimetric analysis (TGA).

3. Nonhygroscopic (less than 1% water uptake from 30% to 90% relative humidity at room temperature). The hygroscopicity, or moisture uptake of a material, is an important consideration because this parameter can significantly affect mechanical properties, stability, and excipient compatibility of the API. A quick way to evaluate this property is to conduct a dynamic moisture sorption experiment. It is preferred that the selected form does not undergo significant weight gain or weight loss when exposed to relative humidity between 10% and 90% (less than 3%), while simultaneously maintaining a single physical form.

4. *Good Water Solubility*: This property is important for the parenteral administration as well as for conventional oral dosage form. The dose should be taken into consideration when determining acceptable solubility.

5. Exists as a single physical form. If multiple physical forms exist, the relationship between the forms should be well understood.
6. No hydrate forms with undesirable properties (stability, hygroscopicity, aqueous solubility of the hydrate decreases dramatically compared to the anhydrous) and does not form multiple hydrates.
7. *Good Flow Properties*: Plates flow better than large needles. Selecting a desirable form for good flow properties facilitates the development of a robust drug product manufacturing process. Flow properties impact the compaction and compression aspects during the tabletting process [3].
8. Easy to scale up and purify.
9. Does not cause an undesirable side effect, for example, irritation on IV administration.

It is unlikely that any one physical form will have all these desired properties. If the free form does not have these properties, the salt approach should be undertaken to identify a form that has the best compliment of these properties. In terms of importance, the first two properties are critical for the developability of the form.

The salt-formation approach is usually undertaken to improve the bioavailability of the compound. Therefore, if most of the properties mentioned above were suitable for the free form but the exposure, and it is related to the solubility or dissolution of the selected free form, the ability of a salt to improve exposure should be evaluated. This could be done easily by conducting a PK study at a reasonable dose with a simple suspension of the free form with a known particle size, for example, in a 2% hydroxypropyl methylcellulose (HPMC) and a solution (if possible) of the *in situ* salt or any other solution in a canine and nonhuman primate to evaluate the exposure. The selected dose(s) for this study should take into consideration the expected human dose and not the doses that will be used in the toxicology studies. If the exposure from the suspension is >70% of the solution, from the solubility/dissolution perspective, the salt will not be expected to improve the exposure when the exposure is solubility/dissolution limited.

The toxicology study exposure could simply be addressed by forming an *in situ* salt without actually having crystalline salt isolated and scaled up or manipulating the formulation by other means such as adding surfactants, complexing agents, and cosolvents.

Although the free form may have all the desired properties, the increased emphasis on developing generics due to the low cost of entry into the market (no R&D costs) and low cost to consumers (due to competition), it has driven generic companies to look for salt forms of a marketed drug that is not covered by current patents from the original manufacturer. The preceding could be a driver for screening and identifying different salts of the compound for the purpose of strengthening intellectual property (IP).

1.2. Potential Disadvantages of the Salt

The first disadvantage of having a salt is the additional molecular weight associated with the counterion, especially for relatively low-potency compounds that may require high doses to achieve efficacy. It is not uncommon for some salts to realize 20–50% of their weight from the inactive counterion [4].

Second, high-dose, strong acids (e.g., dihydrochloride salts) may cause high gastrointestinal (GI) acidity or process issues such as equipment corrosion. The metal surfaces of tabletting equipment are constantly in contact with powder blends during manufacturing and are therefore most at risk of corrosion. The potential for corrosion is exacerbated by the presence of water, even in trace amounts that are often absorbed from the atmosphere. As a result, it is generally recommended that salts with a pH of unbuffered saturated aqueous solution below 4 be tested for corrosiveness [5].

Third, for low pK_a compounds the potential dissociation of the salt in aqueous media or in the GI fluid may lead to the precipitation of the neutral form that has undesirable properties.

Fourth, salts may create or exacerbate the tendency for hydrate or polymorph formation, as has been observed for phosphate salts in general [4,6,7].

Depending on the properties of a molecule, the free form (not ionized) of a weak acid or base may be acceptable for selection as the preferred form for development. To arrive at this disposition, a scientist may take one of the two approaches. Some will fully characterize the properties of the free form, prior to evaluating the existence and subsequent assets of any potential salts, while others may conduct some types of salt screen, regardless of the acceptability of the free form, in hope of generating promising salts with still superior properties. It is the opinion of these authors that each approach is acceptable and is accompanied by its own set of attributes and drawbacks.

1.3. Salt Improving Physicochemical Properties

Now that the fundamentals and approaches for salt screening and selection have been discussed, the reader is referred to the chapter by Badawy and coworkers [8] in the preformulation in solid dosage form development that will provide examples of how a salt can improve physicochemical properties (e.g., mechanical properties, crystallinity, hygroscopicity, solubility/dissolution rate, and physical and chemical stability) of the API [4].

One simple example of how a salt could improve the chemical stability of a compound is when the free base form of a primary amine-containing molecule would be susceptible to a typical Maillard-type reaction. This reaction would likely be suppressed when the amine is protonated as would be the case with the selection of an acid salt. In this case, an evaluation of the relative stability of the API alone (as a free form versus a salt) would not be sufficient to enable optimal formulation development moving forward. Consideration also must be given to the possibility of adduct formation such as the one that has been reported between secondary amines and fumaric acid [9].

1.4. Crystallization

The nature of chemical crystallography is still such that when asked if a particular compound will crystallize, the answer, in the absence of previous experience with the compound, must be, "Maybe." Odds are not better when the question is whether a particular salt of a compound will crystallize. While it is true that crystalline salts are used in many of today's drugs, that many drug-like small organic molecules do crystallize, and that for many of these compounds several crystalline salts are known, there is no way of predicting with certainty that a particular salt will crystallize until the event has occurred. Even then the reproduction of the result is not guaranteed. Significant difficulty may be encountered when trying to reproduce a particular crystallization, or alternative crystalline forms; a more stable polymorph or solvate may form and great effort may be needed to obtain the initial results again [10]. In this regard, salt screening is in the realm of experimentalism.

Experts can be called upon for advice in designing experiments that may yield crystalline salts, but none are able to state *a priori* that a particular salt will crystallize or how that end may be easily achieved. This is not meant to discredit the value of crystallization experience. In fact, in the pursuit of crystalline salts, few if any things are as importance as experience. With experience, an experimentalist will learn how to coax a compound to crystallize, modifying experiments in progress to guide their outcome, appreciating the subtleties that result from each manipulation.

Each compound poses its own unique challenges and only with a great deal of practice will the experimentalist be able to quickly render a developable crystalline material at the necessary scale. The recognition that experimentation is necessary in order to find crystalline forms of any new compound, or to find new crystalline forms of a well studied compound, has led to the development of many tools to assist the experimentalist. These tools are meant to compliment experience gained through hands on practice in identifying new crystalline forms and have also been extended to help with down stream crystallization activities.

Even with a great deal of practice, there will be compounds that evade crystallization for long periods of time or prove difficult to scale up appropriately. In the quest to identify the right crystalline form, time can be another ally of the experimentalist [11], and organizations need to find ways to advance important compounds while uncertainties surrounding the development form are being settled. Stories of new, often less soluble, crystalline forms being discovered during scale-up or manufacturing runs are well known in the pharmaceutical industry. Unfortunately, most do not make it into the primary scientific literature. The most highly publicized of these reports is that of Ritonavir, where a more stable crystalline form was discovered during a manufacturing campaign and resulted in the need for rapid reformulation of a marketed product [12].

To understand salt screening strategies, the reader must first understand the basics of acid/base ionization and dissociation concepts. Building upon that background will aid the reader in understanding the approach for selecting the free form of the active pharmaceutical ingredient or making the decision for evaluating and selecting a salt form.

2. ACID–BASE EQUILIBRIUM

2.1. Basic Equations

2.1.1. Weak Acids The ionization of a weak acid in water may be written in the Bronsted–Lowry manner [13] as

$$AH + H_2O \underset{K_2}{\overset{K_1}{\rightleftharpoons}} A^- + H_3O^+$$

Where AH represents the unionized species of the acid, A^- represents the ionized species of the acid, commonly called the conjugate base, K_1 represents the forward rate constant and K_2 represents the reverse rate constant.

Water ionizes slightly to produce hydronium ion and hydroxide ion and thus water may be thought of as a weak acid or a weak base and will be used as such in the subsequent discussion.

The forward reaction, R_f, and the reverse reaction, R_r, can be represented as follows:

$$R_f = K_1(AH + H_2O) \tag{1}$$

$$R_r = K_2(A^- + H_3O^+) \tag{2}$$

At equilibrium,

$$R_f = R_r \tag{3}$$

therefore

$$k = \frac{K_1}{K_2} = \frac{[A^-][H_3O^+]}{[AH][H_2O]} \tag{4}$$

Given the concentration of the reactants tend to be very low compared to the water concentration, as in diluted solution of the acid in water, the water concentration is always in excess and does not change significantly. The water concentration could be regarded as a constant at 55.3 mol/L.

$$K_a = k(55.3) = \frac{[A^-][H_3O^+]}{[AH]} \tag{5}$$

where K_a is the dissociation constant of the acid.

An important term that is used when discussing the solubility of an acid or a base is the dissociation constant, pK_a, where

$$pK_a = -\log K_a \tag{6}$$

As will be discussed later, pK_a values are often used as guides during salt selection and screening. When screening for salts, it is common practice to limit the acidic or basic counterions to those with pK_a values that differ from that of the API by two or more units.

2.1.2. Weak Bases For the basic compound, the same equations can be derived as

$$B + H_2O \underset{K_2}{\overset{K_1}{\rightleftharpoons}} BH^+ + HO^-$$

$$R_f = K_1(B + H_2O) \tag{7}$$

$$R_r = K_2(BH^+ + HO^-) \tag{8}$$

$$\frac{K_1}{K_2} = \frac{[BH^+][HO^-]}{[B][H_2O]} \tag{9}$$

$$K_b = \frac{K_1}{K_2}(55.3) = \frac{[BH^+][HO^-]}{[B]} \tag{10}$$

Or, we can write it as the conjugated acid

$$BH^+ + H_2O \underset{K_2}{\overset{K_1}{\rightleftharpoons}} B + H_3O^+$$

$$R_f = K_1(BH^+ + H_2O) \qquad (11)$$

$$R_r = K_2(B + H_3O^+) \qquad (12)$$

$$\frac{K_1}{K_2} = \frac{[B][H_3O^+]}{[BH^+][H_2O]} \qquad (13)$$

$$K_a = \frac{K_1}{K_2}(55.3) = \frac{[B][H_3O^+]}{[BH^+]} \qquad (14)$$

2.1.3. Zwitterions For a molecule that has both a basic and an acidic ionizable groups, the following equations could be derived:

$$H_3\overset{+}{B}CH_3COOH \underset{H_2O}{\overset{K_{a1}}{\rightleftharpoons}} H_3\overset{+}{B}CH_3CHOO^- + H_3\overset{+}{O}$$

$$H_3\overset{+}{B}CH_3COO^- \underset{F_2O}{\overset{K_{a2}}{\rightleftharpoons}} H_2BCH_3COO^- + H_3\overset{+}{O}$$

Or more generic equations as follows:

$$AHBH^+ + H_2O \overset{K_{a1}}{\rightleftharpoons} {}^-AHB^+ + H_3O^+$$

$${}^-AHB^+ + H_2O \overset{K_{a2}}{\rightleftharpoons} {}^-AB + H_3O^+$$

where $AHBH^+$ represents the positively charged species, the charged ionizable basic group, $^-AHB^+$ represents the positively and the negatively charged species, the charged ionizable basic group, and the negatively acidic charged group

^-AB represents the negatively charged species, the charged ionizable acidic group

$$K_{a1} = \frac{[^-AHB^+][H_3O^+]}{[AHBH^+]} \qquad (15)$$

$$K_{a1} = \frac{[^-AB][H_3O^+]}{[^-AHB^+]} \qquad (16)$$

2.2. pH-Solubility Profile

2.2.1. Acidic Compounds For any acidic compound and at any given pH, the total solubility is expressed as

$$S_t = S_0 + S_{A^-} \qquad (17)$$

Where S_t is the total solubility, S_0 is the solubility of the free acid (intrinsic solubility), and S_{A^-} is the concentration of the salt form.

Rearranging the equations above and substituting for A^-, we get,

$$S_t = S_0\left(\frac{K_a}{H^+} + 1\right) \qquad (18)$$

This implies that

(1) at $pH \ll pK_a$ (more than two units), the solubility of the acid will be constant and it will be equal to the intrinsic solubility of the acid and

(2) at $pH < pK_a$ and higher, the solubility of an acid will increase exponentially as the pH increases, but, in general, that is not the case. At certain pH, the solubility will be constant.

Therefore, the pH solubility profile of an acid could be divided into three areas. First, where the solubility of the free acid will increase as the pH increases as long as the excess solid at that pH is the free acid. Second, above a certain pH (pH_{max}), where the excess solid is converted to the salt, the solubility will not increase as a function of pH, but it will be constant and will represent the saturated solubility of the salt and this is expressed as the solubility product K_{sp}. Third, at pH_{max}, in the presence of excess solid, the solution is saturated with the free acid equals to its intrinsic solubility and the salt form [14].

The pH solubility profile of an acidic compound with a pK_a of 4 and an intrinsic solubility of 35 µg/mL, is shown in Fig. 1, assuming no pH_{max}.

But above certain pH, pH_{max}, the solubility of the salt will be constant. If we assume the pH_{max} is 7 and the solubility of the salt is 35 mg/mL, Fig. 2 now represents the pH solubility profile of the acid. The solubility of the salt could be expressed as a solubility product represented by

$$K_{sp} = [A^-][Na^+] \qquad (19)$$

where $[Na^+]$ is the concentration of the counterion of the salt formed. Equation 19 refers to sodium but it could be any other counterion

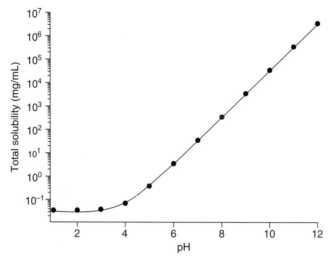

Figure 1. pH solubility of an acidic compound.

such as magnesium, potassium, etc. Therefore, the solubility of the salt will be $\sqrt{K_{sp}}$.

If the concentration of the counterion increases to keep the K_{sp} constant, the solubility of the salt will decrease, this phenomena is a called a common ion effect. Of course, it is critical that the solid form in equilibrium with the solution phase be characterized during the determination of pH solubility profiles. Usually, the changes in crystalline form in equilibrium are limited to changes between salt form and free form, but in some cases it has been shown that different hydration states are present at different pH values.

2.2.2. Basic Compounds The same approach discussed in Section 2.1.1 for the acidic compounds could be utilized for the basic compounds. For example,

$$S_t = S_0 + S_{BH+} \quad (20)$$

where S_t is the total solubility, S_0 is the solubility of the free acid (intrinsic solubility), and S_{BH+} is the concentration of the salt form.

Substituting for S_{BH+} will result

$$S_t = S_0\left(\frac{H^+}{K_a} + 1\right) \quad (21)$$

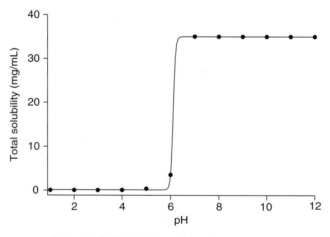

Figure 2. pH solubility of an acidic compound.

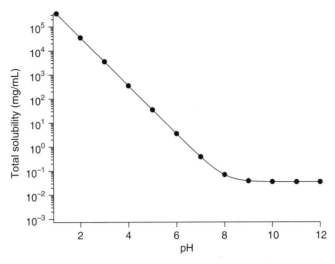

Figure 3. pH solubility of a basic compound.

If we have a basic compound with a pK_a of 7 and an intrinsic solubility of the free base of 35 μg/mL, the pH solubility profile is represented by Fig. 3 assuming no pH$_{max}$.

But above certain pH, pH$_{max}$, the solubility of the salt will be constant. If we assume the pH max is 5 and the solubility of the salt is 35 mg/mL, Fig. 4 now represents the pH solubility profile of the base.

The solubility of the salt could be expressed as a solubility product represented by

$$K_{sp} = [BH^+][Cl^-] \qquad (22)$$

where [Cl$^-$] is the concentration of the counterion of the salt formed. Equation 22 refers to chloride but it could be any other counterion such as acetate, sulfate, etc. Again, the solubility of the salt will be $\sqrt{K_{sp}}$.

2.2.3. Zwitterionic Compounds We can combine the equations used to describe the solubility behavior of acidic and basic compounds to describe the solubility behavior of zwitterionic compounds. For any zwitterionic compound at any given pH, the total solubility is expressed as

$$S_t = S_0 + S_{BH^+} + S_{A^-} \qquad (23)$$

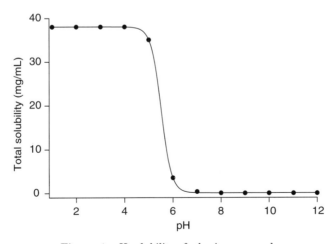

Figure 4. pH solubility of a basic compound.

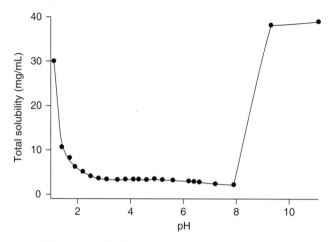

Figure 5. pH solubility of a zwitterionic compound.

where S_t is the total solubility, S_0 is the solubility of the zwitterion, and S_{BH+} is the concentration of the salt form of the base, and S_A^- is the concentration of the salt form of the acid.

At $pH \ll pK_{a2}$ and $\gg pK_{a1}$, then

$$S_t = S_0 + S_{BH}{}^+ \tag{24}$$

$$S_t = S_0 \left(\frac{H^+}{K_a} + 1 \right) \tag{25}$$

At $pH \gg pK_{a2}$,

$$S_t = S_0 + S_{A^-} \tag{26}$$

$$S_t = S_0 \left(\frac{K_a}{H^+} + 1 \right) \tag{27}$$

the pH solubility profile for such a molecule will could be represented in Fig. 5.

2.2.4. pH$_{max}$ As has been indicated by the preceding discussion, the location of pH$_{max}$ plays an important role in the pH solubility profile. If a salt of the free base is dissolved in water and the resulting pH is higher than the pH$_{max}$, the excess solid, which is originally the salt will convert to the free base. On the other hand, if the free base is dissolved in water, the excess solid will not convert to the salt unless the pH is lower than the pH$_{max}$. This is very important in evaluating the pH solubility profile [15].

According to Kramer and Flynn [16], the pH$_{max}$ could be obtained from the intersection of two curves based on the total solubility.

At $pH > pH_{max}$,

$$S_t = S_0 + S_{BH^+} \tag{28}$$

where S_t is the total solubility, S_0 is the solubility of the free acid (intrinsic solubility), and S_{BH^+} is the concentration of the salt form.

Substituting for S_{BH^+} will result

$$S_t = S_0 \left(\frac{H^+}{K_a} + 1 \right) \tag{29}$$

On the other hand, at $pH < pH_{max}$,

$$S_t = S_{BH^+} \left(\frac{K_a}{H^+} + 1 \right) \tag{30}$$

where S_{BH^+} is the solubility of the salt form.

The intersection of the plot S_t versus pH from Equations 29 and 30 represent the pH$_{max}$. The same relationship holds true for an acidic compound.

3. SCREENING FOR A SALT FORM OF THE API

Before we discuss in detail the salt screening processes, we need to first discuss the approach to the selection of the counterions and solvents.

3.1. Counterion Selection

Most often scientists utilize pK_a differences when selecting the acids or bases to use in these screens. However, with the recent demonstrations of pharmacokinetic improvements resulting from pharmaceutical cocrystals [17–22], the argument can be made for decreasing the emphasis on the pK_a guidelines during experimental design. If only salts are desired, it is generally accepted that a difference of at least two pK_a units between the acid and the base is needed, as apposed to cocrystals, which require no separation of pK_a values. Appropriate acids or bases should be chosen accordingly. Primary consideration must be given to the toxicological safety profile of the salt former; therefore, any counterion sources considered inherently unsafe should be excluded from screening. A most convenient approach to avoid potentially concerning counterions is to select from those that have established precedence on the market. Tables 1 and 2 list the currently acceptable counterion forming acids and bases, respectively. For convenience, the salt formers have been listed alphabetically [23].

Table 3 shows a two-tiered approach for counterion selection for salt screening. Tier 1 salts are preferred based on the following factors:

(1) Extensively used salts commonly found in marketed products.
(2) Salts that exhibit good tolerability and low toxicity.

A quick look through the salt formers listed in Tables 1 and 2 shows that for extremely weak bases ($pK_a < 3$) rather few options exist, and selecting the salt formers for screening is relatively straightforward. However, when dealing with stronger bases or typical weak acids, the options for counterions increase significantly. When this is encountered, a judicious approach is required to limit the scope of early screening activities. To help in this area, the development scientist should consider what primary problems exist with the molecule and how different salt options are likely to best overcome its liabilities.

3.2. Solvent Selection

When selecting solvents, the desired outcome of the experiments should be considered. If the goal is to identify new lead salts or to do an extensive form screen on a particular salt form the solvent selection need not be limited to any specific type. It is generally the case that solvents that span the spaces of functionality, dielectric constant, polarity, or other properties should be used to probe the most diverse set of solution crystallization conditions possible. Because the desire of these screens is form diversity and identification of the stable crystalline form of a given chemical composition solvent selection need not be limited. Useful solvents for these types of screens can be found in Tables 4–6. These tables provide a reasonable, but far from comprehensive, list of chemically diverse solvents.

If the goal of the experiment is to identify solvent systems that might be useful for scale-up of a specific form of a particular salt the selection is usually limited to solvents acceptable by the U.S. Food and Drug Administration (FDA) [25]. Recommendations based upon the International Conference on Harmonization (ICH) of Technical Requirements for Registration of Pharmaceuticals for Human Use guidance for industry (notably Q3C Impurities: Guideline for Residual Solvents) may be considered [26]. In an attempt to identify systems that are most likely to be process friendly in terms of scale-up, class 3 solvents (see Table 6), should be given priority, and it is recommended that a standard library be established although at early stages this is less important. This library should be used to first screen solubility (and stability if possible) of the free form, the results serving as a tool for selecting the solvents that will be incorporated into the salt screen. Solvents demonstrating the highest free form solubility, while still maintaining polar diversity should be prioritized. These solvents will allow process chemistry the greatest likelihood of smooth scale up, as crystallization of a salt form from a solution is considered ideal, both from control and purification perspectives. This does not mean that to get a crystalline salt in the early stage of the screening, class 3 solvents should not be used if class 1 and class 2 solvents fail to

Table 1. List of Acidic Counterion Sources

Acid	Molar Mass
Acetic acid	60.05
Acetic acid, 2,2-dichloro-	128.95
Adipic acid	146.14
Alginic acid	240.0
L-Arginine	174.2
L-Ascorbic acid	176.13
L-Aspartic acid	133.11
Benzenesulfonic acid	158.18
Benzoic acid	122.12
Benzoic acid, 4-acetamido-	179.18
(+)-Camphoric acid	200.24
(+)-Camphor-10-sulfonic acid	232.29
Capric acid	172.27
Caprioc acid	116.16
Caprylic acid	144.22
Carbonic acid	44.01
Cinnamic acid	148.06
Citric acid	192.13
Cyclamic acid	179.24
Dodecylsulfuric acid	266.40
Ethane-1,2-disulfonic acid	190.20
Ethanesulfonic acid	110.13
Ethanesulfonic acid, 2-hydroxy-	126.13
Formic acid	46.02
Fumaric acid	116.08
Galactaric acid	210.14
Gentisic acid	154.12
D-Glucoheptonic acid	226.18
D-Gluconic acid	196.16
D-Glucuronic acid	194.14
Glatamic acid	147.13
Glutaric acid	132.12
Glutaric acid, 2-oxo-	146.10
Glycerophosphoric acid	172.08
Glycolic acid	76.05
Hippuric acid	179.17
Hydrobromic acid	80.92

Table 1. (Continued)

Acid	Molar Mass
Hydrochloric acid	36.46
Isobutyric acid	88.11
DL-Lactic acid	90.08
Lactobionic acid	358.30
Lauric acid	200.32
Lysine	146.19
Maleic acid	116.08
(-)-L-Malic acid	134.09
Malonic acid	104.06
DL-Mandelic acid	152.15
Methanesulfonic acid	96.10
Naphthalene-1,5-disulfonic acid	332.26
Naphthalene-2-sulfonic acid	208.24
2-Naphthoic acid, 1-hydroxy-	188.17
Nicotinic acid	123.11
Nitric acid	63.02
Oleic acid	282.45
Orotic acid	156.10
Oxalic acid	90.04
Palmitic acid	256.42
Pamoic acid	388.38
Phosphoric acid	98.00
Propionic acid	74.08
(-)-L-Pyroglutamic acid	129.11
Salicylic acid	138.12
Salicylic acid, 4-amino-	153.14
Sebacic acid	202.25
Stearic acid	284.49
Succinic acid	118.09
Sulfuric acid	98.08
(+)-L-Tartaric acid	150.09
Thiocyanic acid	59.09
p-Toluenesulfonic acid	172.21
Undecylenic acid	184.27

A more comprehensive data set can be found in Ref. [23].

produce crystalline materials. The trick is to use class 3 solvents to generate seed crystals that could be used in class 2 or class 3 solvents to crystallize the salt.

If a search for hydrates is desired, mixtures of organic solvents in water to cover a good range of water activities can be employed. Such screens are sometimes best conducted on samples that are slurries. Solution processes for crystallization, including slurry conversions, sometimes yield solvates or hydrates that are only stable when in equilibrium with the crystallization media. These cases may be difficult to identify, especially considering the materials are almost always isolated from the solution before characterization. The final step of isolation can destroy the crystalline solvate or hydrate and depending on how rigorous the procedures and how weakly the solvent is held in the crystal lattice, completely erase all evidence of its existence. It is important that the scientist pay close attention to the shape and clarity of the crystals as they form in solution and after isolation from the solvent. If necessary, the solids should be characterized by PXRD from the slurries as the wet cake.

Table 2. List of Basic Counterion Sources

Base	Molar Mass
Ammonia	17.03
L-Arginine	174.20
L-Aspartate	133.11
Benethamine	197.28
Benzathine	240.35
Betaine	117.15
Calcium hydroxide	74.10
Choline	121.18
Deanol	89.14
Diethylamine	73.14
Diethanolamine (2-2′-iminobis(ethanol))	105.14
Ethanolamine (2-aminoethanol)	61.08
Ethanol, 2-(diethylamino)-	117.19
Ethylenediamine	60.10
Hydrabamine	596.99
Glucamine, N-methyl-	195.22
Glutamate	147.13
1H-Imidazole	68.08
Lysine	146.19
Magnesium hydroxide	58.33
Morpholine, 4-(2-hydroxyethyl)-	131.18
Piperazine	86.14
Potassium hydroxide	56.11
Pyrrolidine, 1-(2-hydroxyethyl)-	115.18
Sodium hydroxide	40.00
Triethanolamine (2,2′,2″-nitrilotris(ethanol))	149.19
Tromethamine	121.14
Zinc hydroxide	99.38

A more comprehensive data set can be found in Ref. [23].

Table 3. Commonly Used Salt Formers, Listed Alphabetically Within Category of Acids and Bases

	Acids	Bases
Tier 1	• Acetic • Benzenesulfonic • Citric • Ethanesulfonic • Fumaric • Hydrochloric • Methanesulfonic • Phosphoric • Sulfuric • p-Toluenesulfonic	• Calcium • Magnesium • Potassium • Sodium
Tier 2	• Benzoic • Gluconic • Hydrobromic • Lactic • Maleic • Salicyclic • Stearic • Tannic • Tartaric	• Aluminum • Arginine • Benzathine • Chloroprocaine • Choline • Diethanolamine • Ethylenediamine • Lithium • Meglumine • Procaine • TRIS • Zinc

3.3. Process

3.3.1. Manual One of the main advantages of manual salt screening is the involvement of scientists. This advantage comes in the form of observations made during the recrystallization process. Observations of dissolution rate, material wetting, color changes, dispersion of solids, precipitation, and conglomeration are made real time and should be noted. For the most part, these types of observations are only made and reacted to by the scientist during manual salt screening or salt recrystallization studies and help guide subsequent crystallization experiments. The techniques most often applied are based on solution recrystallizations and the subsequent manipulations of the materials obtained.

After selecting counterions and solvents, the next consideration is focused on crystallization conditions. Solution crystallization of salts can be done in several ways. In salt screening if the free form is soluble in the selected solvent at room temperature, the simple addition of the counterion may lead to salt formation and precipitation. If spontaneous nucleation and crystallization does not happen, adding an antisolvent or cooling the solution in a refrigerator or freezer may lead to the precipitation of the salt. There are several variations of both antisolvent addition and cooling that can be applied; most attempt to change the rate of antisolvent addition or rate of cooling applied. In some cases, evaporating the solvent is necessary to obtain solid material. Here again, the rate of evaporation can be modified or controlled in order to obtain more crystalline material.

Table 4. Class 1 Solvents Listed in ICH Guideline Q3C(R3) as Solvents to Be Avoided in The Manufacture of Drug Substances, Excipients, and Drug Products Because of Unacceptable Toxicity or Effects on the Environment [26]

Solvent	Reason for Concern	Concentration Limit (ppm)
Benzene	Carcinogenicity	2
Carbon Tetrachloride	Toxicity and environment	4
1,2-Dichloroethane	Toxicity	5
1,1-Dichloroethene	Toxicity	8
1,1,1-Trichloroethane	Environment	1500

It is not necessary to bring all of the solid materials fully into solution together for salt formation to occur. Salts can form by slurrying the free acid or base of a compound with the appropriate salt former in a solvent that only partially dissolves the free form and/or salt former. Solvent assisted grinding or milling, a technique recently demonstrated in the formation of cocrystals, may be thought of as a higher energy variation on slurrying and may also yield salts.

Solution recrystallization of a salt, as taught in basic organic chemistry laboratory classes, is exemplified in the synthesis of sodium toluene-p-sulfonate [27]. Here, the acidic reaction mixture containing crude toluene-p-

Table 5. Class 2 Solvents Listed in ICH Q3C(R3) as to Be Limited in Pharmaceutical Products [26]

Solvent	Type	ε	t_b (°C)
Acetonitrile	Nonprotic	36.64	81.65
Chlorobenzene	Nonprotic	5.6895	131.72
Chloroform	Nonprotic	4.8069	61.17
Cyclohexane	Nonprotic	2.0243	80.73
1,2-Dichloroethane	Nonprotic	10.42	8305
Dichloromethane	Nonprotic	8.93	40
1,2-Dimethoxyethane	Nonprotic	7.30	84.5
N,N-Dimethylacetamide	Nonprotic	38.85	165
N,N-Dimethylformamide	Nonprotic	38.25	153
1,4-Dioxane	Nonprotic	2.2189	101.5
2-Ethoxyethanol	Protic	13.38	135
Ethyleneglycol	Protic	41.4	197.3
Formamide	Nonprotic	111.0	220
Hexane	Nonprotic	1.8865	68.73
Methanol	Protic	33.0	64.6
2-Methoxyethanol	Protic	17.2	124.7
Methylbutyl ketone	Nonprotic	12.4	126
Methylcyclohexane	Nonprotic	2.024	100.93
N-Methylpyrrolidone	Nonprotic	32.55	202
Nitromethane	Nonprotic	37.27	101.19
Pyridine	Nonprotic	13.260	115.23
Sulfolane	Nonprotic	43.26	287.3
Tetralin	Nonprotic	2.771	207.6
Toluene	Nonprotic	2.379	110.63
1,1,2-Trichloroethene	Nonprotic	7.1937	113.8
Xylenes[a]	Nonprotic	2.275–2.562	138.37–144.5

The dielectric constant (ε) and boiling point (t_b) were obtained from Ref. [24].
[a] Typically ~60% m-xylene, ~14% p-xylene, and ~9% o-xylene along with ~17% ethyl benzene.

Table 6. Class 3 Solvents Listed in ICH Q3C(R3) as to Be Limited by GMP or Other Quality-Based Requirements [26]

Solvent	Type	ε	t_b (°C)
Acetic acid	Protic	6.20	117.9
Acetone	Nonprotic	21.01	56.05
Anisole	Nonprotic	4.3	153.7
1-Butanol	Protic	17.84	117.73
2-Butanol	Protic	17.26	99.51
Butyl acetate	Nonprotic	5.07	126.1
tert-Butylmethyl ether	Nonprotic	4.5	55.2
Cumene (isopropyl benzene)	Nonprotic	2.381	152.41
DMSO	Nonprotic	47.24	189
Ethanol	Protic	25.3	78.29
Ethyl acetate	Nonprotic	6.0814	77.11
Ethyl ether	Nonprotic	4.266	34.5
Ethyl formate	Nonprotic	8.57	54.4
Formic acid	Protic	51.1	101
Heptane	Nonprotic	1.9209	98.4
Isobutyl acetate	Nonprotic	5.068	116.5
Isopropyl acetate	Nonprotic	6.3	88.6
Methyl acetate	Nonprotic	7.07	56.87
2-Methyl-1-butanol	Protic	15.8	130
Methyl ethyl ketone	Nonprotic	18.5	79.6
Methyl isobutyl ketone	Nonprotic	13.1	117
2-Methyl-1-propanol	Protic	17.93	107.89
Pentane	Nonprotic	1.8371	36.06
1-Pentanol	Protic	15.13	137.98
1-Propanol	Protic	20.8	97.2
Propyl acetate	Nonprotic	5.62	101.54
Tetrahydrofuran	Nonprotic	7.52	65
Water	Protic	80.1	100.0

The dielectric constant (ε) and boiling point (t_b) were obtained from Ref. [24].

sulfonic acid is partially neutralized using sodium carbonate prior to saturation of the aqueous solution with sodium chloride forming the sodium salt in solution. The solution is then heated to boiling, hot filtered, and the crystallization vessel is cooled by stirring in ice. The crystalline material is isolated by filtration and recrystallized by again dissolving the solids, now the sodium salt, in hot solvent, cooling appropriately, isolating the solids by filtration and finally the resulting crystalline material is rinsed with ethanol. This example takes advantage of not only salt formation as part of a purification procedure but also the common ion effect during the first crystallization, where high concentrations of the sodium ion, from NaCl, are used to drive precipitation of the sodium salt of toluene-p-sulfonic acid. It also exemplifies the use of temperature change to cause crystallization.

An approach that can be used as an early assessment of the form diversity to expect with a new salt form can also be used to help gauge the solubility of the salt. Slurrying small amounts of the new salt in small amounts of solvents at multiple temperatures and characterizing the solids obtained in the end often leads to the identification of new forms of the salt. For example, 1 mg of a particular salt form can be dispensed by hand into small vials and a small amount of solvents (20–100 µL) can be added. The samples can be incubated at room temperature for 24 h and an aliquot of the solution removed and the concentration of the API in the solution determined by HPLC. These samples can go on directly to PXRD or Raman spectroscopy of the form or the remaining sample can be incubated for an additional 24 h at a higher temperature (~50°C) and another aliquot of the solution removed and solubility determined. Finally, the remaining solids can be isolated and analyzed by PXRD or Raman spectroscopy to determine if the form of the salt changed during the incubation. The form changes tend to manifest in three ways. First, the crystalline form of the salt can change, usually to a more stable crystalline phase of the salt or to a solvate or hydrate of the salt. The salt may disproportionate in the solvent leaving one of the forms of the free base or free acid of the API. Finally, the salt can collapse to an amorphous material. This procedure can constitute the first form screen on any given compound whether it is a salt or free form and often yields an initial snapshot of the polymorphism or solvatomorphism that may be expected for a given compound.

A general workflow for manual screening is shown in Fig. 6. The modern laboratory will be fitted to undertake many different routes of crystallization, including those described above. These should be augmented with equipment for conducting sublimations, hot stages for microscopic investigations, various solvents and additives, seeding materials, and a full assortment of analytical equipment. As the scale of the experiments increase, more sophisticated experimental equipment will be applied as the crystallization processes are formalized. By using ibuprofen as a model Lee and Wang [28] outlined a systematic initial salt screening procedure for use in early development for manufacturing applicability.

The greatest advantage of manual salt screening, the intimate involvement of the scientist, is also one of its limitations. While conducting a few experiments with a dozen or sometimes more salt formers is manageable by a single scientist much more than this becomes unwieldy without the assistance of automated dispensing and some level of data tracking. Even at this scale of experimentation, the overall process can be time consuming. The time consuming nature of these experiments comes in at least three forms. First, sample preparation can be tedious. Second, data acquisition can be time consuming. Preparing samples for PXRD analysis when starting with less than 1 mg of material can be challenging, but automated sample changers available on modern diffractometers make it possible to efficiently record useful PXRD patterns on dozens of samples. Third, data ana-

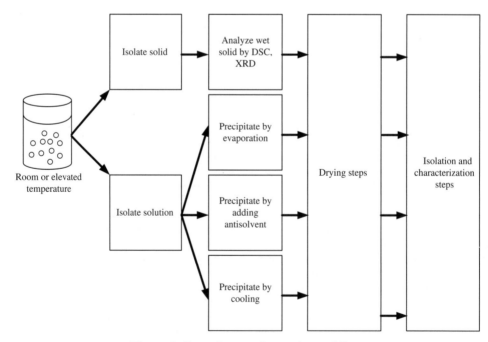

Figure 6. Example manual screening workflow.

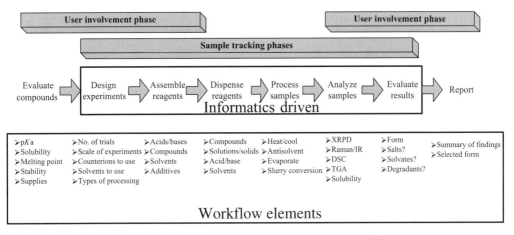

Figure 7. Example automated assisted screening workflow.

lysis can be time consuming, especially if a file naming convention is not applied. Analysis of PXRD patterns, for the purpose of bucketing patterns into groups based on similarity can be time consuming. Standard software packages available with modern diffractometers can aid in addressing those challenges. Because of these limitations and the high visibility of certain failures to obtain the stable crystalline form early enough in development many companies have developed automated or partially automated crystallization systems that can be used to conduct salt screening experiments [29–36].

By taking advantage of some automated data collection features available with modern instrumentation the experimentalist is able balance the line between manual and automated experimentation. While collecting data on a select few salts and evaluating them for developability, the staging of automation assisted for screening should be evaluated.

3.3.2. Automated Assisted Screening High-throughput, or automated assisted, screening can be conducted in several ways and with different desired outcomes. The automation is usually designed to aid the experimentalist with tasks that are time consuming and tedious to do by hand. Typical automated steps are heavily focused on the dispensing of solids and liquids into reaction containers, usually vials of some sort, and the collection of data on processed samples. Manipulation of the samples, in terms of moving plates between processing stations and in some cases harvesting individual samples has also been automated.

A general workflow for automated assisted screening is shown in Fig. 7. As with most experimentation automated assisted salt screening begins with an experimental design. The parameters to consider during the design stage include which solvents and counterions to use, how the compounds will be dispensed, how the experiments will be processed, isolated, and analyzed. Consideration to the number of experiments that can be efficiently dispensed and, often more importantly, efficiently harvested and analyzed is given. Because compound supply is usually relatively limited when these experiments are conducted, limiting the scale of each experiment can be a significant advantage.

Upon establishing a library from which to select counterions, one must next consider the strength of acid or base, which is reflected in the pK_a value. To help ensure proton transfer from the acid to base species, it is recommended that a difference of two units exists between the pK_a of interest on the drug candidate and the pK_a of the potential salt former [37,38]. For example, if one wishes to form an ionic salt with a weak base that demonstrates a pK_a of 6, acids with pK_a values below pH 4 would therefore be of interest. It is understood that the converse of this be true for salts of weak acids conjugated to stronger bases.

During these screening experiments, data acquisition is often limited to what is sometimes referred to as fingerprinting. It is advantageous for the technique(s) used for initial characterization to be nondestructive. Two techniques that have been used extensively are Raman spectroscopy and PXRD. With the proper equipment either of these techniques can allow for rapid and accurate characterization of materials recovered from screening experiments. The amount of recovered material necessary for the initial characterization is less than 1 mg.

Data, in the form of spectra or diffractograms, are compared against data of known forms and against each other. This stage of data analysis is gauged toward identifying new or different forms of the material so that subsequent follow-up work can begin. These comparisons can be done by hand, simply by overlaying spectra or powder patterns on a computer screen and making visual comparison, or by using more sophisticated data analysis packages available from vendors or developed internally. Because the number of samples requiring analysis is still limited during these screens it is usually reasonable for the scientist conducting the experiment to evaluate each individual piece of data and determine the number of unique crystalline salts identified during the screen. Attempts should be made to scale up, usually to 10–20 mg, as many of the salt forms as material and time allow.

Because both Raman spectroscopy and PXRD are nondestructive analytical techniques, a careful experimentalist can sometimes use the samples for additional analytical experiments, such as DSC or TGA. These experiments can give additional characteristics of the solids, melting point in the case of DSC and hydrations/solvation state in the case of TGA, or confirm that the compound is still chemically intact.

4. FROM SCREENING TO SELECTION

To avoid characterizing free form polymorphs as potential crystalline salts, negative controls should be run with free form material being exposed to the same crystallization conditions as those of the salt experiments, simply in the absence of any potential salt-forming counterion.

Whether the initial screening is done manually or with the help of automation, in individual vials or specialized well plates, a minimal amount of characterization data are required to evaluate any resulting solids. The first tier of analyses should begin with microscopy to assess birefringence of the sample, with birefringent material being assumed to be crystalline. PXRD data should be gathered on any samples suspected of being crystalline. Based on PXRD patterns, samples should be grouped into categories representing like forms.

Once sufficient first-tier data are generated and potential crystalline salt forms are grouped according to each salt, a thorough evaluation of the resulting data should be performed. At this stage, it is recommended that highest priority be given to the crystalline salts that demonstrate the lowest potential for polymorphism and a high temperature first endothermic transition. However, the caveat should be understood that this offers no guarantee that additional, problematic forms would not manifest in the future. Drug product processability is the obvious driver behind the recommended prioritization of forms demonstrating high first endothermic transition points. Based on these limited acceptance criteria, it is recommended that a lead salt and backup be selected for advancement.

If sufficient solid material is available, a DSC thermogram should be generated on a representative sample from each form. This will help to confirm crystallinity, provide early information on potential thermodynamic stability and help with the identification of potential hydrates and solvates. In addition, a suspected amorphous sample should be characterized by PXRD to confirm amorphicity and by DSC to evaluate the potential of the amorphous material to crystallize into a form that is different from those already identified in the screen. The greatest concern would be that this amorphous material can crystallize into a highest melting form that would potentially be thermodynamically stable.

Upon identification of a lead and backup salt through screening, the process of selecting a final salt form for development can be made as described in the literature [39–42].

5. SUMMARY

With the benefits outlined for selecting a salt, it is no surprise that greater than half of all marketed pharmaceuticals are a salt form and within that population the majority are acidic [40,43]. Commonly occurring acidic salts are hydrochloride, sulfate, and hydrobromide. There are many aspects to consider with salt screening and whether it is conducted through a manual or automated assisted approach the role of the scientist in assessing the data and selecting the form is pivotal. Automated assisted screening approaches enable larger experimental design spaces to be evaluated in a shorter period of time, and with less material needed and lower labor costs. Having this information early in the lifecycle mitigates risk throughout the lifecycle and can aid in protecting the intellectual property of the product. The discovery of possible salts and forms is not necessary pure science, but with guiding principles and practices coupled with close involvement and expertise of scientists it has and always will be central to the development of therapeutic products.

REFERENCES

1. Black SN, Collier EA, Davey RJ, Roberts RJ. Structure, solubility, screening, and synthesis of molecular salts. J Pharm Sci 2007;96(5): 1053–1068.
2. Gross TD, Schaab K, Ouellette M, Zook S, Reddy JP, Shurtleff A, Sacaan AI, Alebic-Kolbah T, Bozigian H. An approach to early-phase salt selection: application to NBI-75043. Org Process Res Dev 2007;11(3):365–377.
3. Faqih AMN, Mehrotra A, Hammond SV, Muzzio FJ. Effect of moisture and magnesium stearate concentration on flow properties of cohesive granular materials. Int J Pharm 2006;336(2):338–345.
4. Bastin RJ, Bowker MJ, Slater BJ. Salt selection and optimisation procedures for pharmaceutical new chemical entities. Org Proc Res Dev 2000;4(5):427–435.
5. Stahl PH, Nakano M. Pharmaceutical aspects of the drug salt form. In: Stahl PH, Camille GW, editors. Handbook of Pharmaceutical Salts. Properties, Selection, and Use. Zurich: Verlag Helvetica Chimica Acta; Weinheim, Germany: Wiley-VCH; 2002. p 97–99.
6. Stahl PH, Wermuth CG. Monographs on acids and bases. In: Stahl PH, Camille GW, editors. Handbook of Pharmaceutical Salts. Properties, Selection, and Use. Zurich: Verlag Helvetica Chimica Acta; Weinheim, Germany: Wiley-VCH; 2002. p 302.
7. Gwak H, Choi J, Choi H. Enhanced bioavailability of piroxicam via salt formation with ethanolamines. Int J Pharm 2005;297(1–2):156–161.
8. Badawy SI, Franchini MK, Hussain MA. Salt selection for pharmaceutical compounds. In: Brittain, Harry, Moji Adeyeye, editors. Preformulation in Solid Dosage Form Development. New York: Informa Healthcare USA; 2008. p 63–80.
9. Stahl PH, Nakano M. Pharmaceutical aspects of the drug salt form. In: Stahl PH, Camille GW, editors. Handbook of Pharmaceutical Salts. Properties, Selection, and Use. Zurich: Verlag Helvetica Chimica Acta; Weinheim, Germany: Wiley-VCH; 2002. p 100–102.
10. Dunitz JD, Bernstein J. Disappearing polymorphs. Acc Chem Res 1995;28(4):193–200.
11. Peresypkin A, Variankaval N, Ferlita R, Wenslow R, Smitrovich J, Thompson K, Murry J, Crocker L, Mathre D, Wang J, Harmon P, Ellison M, Song S, Makarov A, Helmy R. Discovery of a stable molecular complex of an API with HCl: a long journey to a conventional salt. J Pharm Sci 2008;97(9):3721–3726.
12. Miller JM, Collman BM, Greene LR, Grant DJW, Blackburn AC. Identifying the stable polymorph early in the drug discovery–development process. Pharm Dev Tech 2005;10(2):291–297.
13. Martin A, Bustamante P, Chun AHC. Ionic equilibria. In: Martin A, Pilar B, Chun AHC, editors. Physical pharmacy. Physical Chemical Principles in the Pharmaceutical Sciences. 4th ed. Philadelphia, PA: Lea & Febiger; 1993. p 143–168.
14. Serajuddin ATM. Salt formation to improve drug solubility. Adv Drug Deliv Rev 2007;59(7):603–616.
15. Serajuddin ATM, Pudipeddi M. Salt-selection strategies. In: Stahl PH, Camille GW, editors. Handbook of Pharmaceutical Salts. Properties, Selection, and Use. Zurich: Verlag Helvetica Chimica Acta; Weinheim, Germany: Wiley-VCH; 2002. p 138.
16. Kramer SF, Flynn GL. Solubility of organic hydrochlorides. J Pharm Sci 1972;61(12): 1896–904.
17. Remenar JF, Morissette SL, Peterson ML, Moulton B, MacPhee JM, Guzman HR, Almars-

son O. Crystal engineering of novel cocrystals of a triazole drug with 1,4-dicarboxylic acids. J Am Chem Soc 2003;125(28):8456–8457.
18. Childs SL, Chyall LJ, Dunlap JT, Smolenskaya VN, Stahly BC, Stahly GP. Crystal engineering approach to forming cocrystals of amine hydrochlorides with organic acids. Molecular complexes of fluoxetine hydrochloride with benzoic, succinic, and fumaric acids. J Am Chem Soc 2004;126(41):13335–13342.
19. Variankaval N, Wenslow R, Murry J, Hartman R, Helmy R, Kwong E, Clas SD, Dalton C, Santos I. Preparation and solid-state characterization of nonstoichiometric cocrystals of a phosphodiesterase-IV inhibitor and L-tartaric acid. Cryst Growth Des 2006;6(3):690–700.
20. Hickey MB, Peterson ML, Scoppettuolo LA, Morisette SL, Vetter A, Guzman H, Remenar JF, Zhang Z, Tawa MD, Haley S, Zaworotko MJ, Almarsson O. Performance comparison of a co-crystal of carbamazepine with marketed product. Eur J Pharm Biopharm 2007;67(1):112–119.
21. Bak A, Gore A, Yanez E, Stanton M, Tufekcic S, Syed R, Akrami A, Rose M, Surapaneni S, Bostick T, King A, Neervannan S, Ostovic D, Koparker A. The co-crystal approach to improve the exposure of a water-insoluble compound: AMG 517 sorbic acid co-crystal characterization and pharmacokinetics. J Pharm Sci 2008;97(9):3942–3956.
22. Stanton MK, Bak A. Physicochemical properties of pharmaceutical co-crystals: a case study of ten AMG 517 co-crystals. Cryst Growth Des 2008;8(10):3856–3862.
23. Stahl PH. Appendix. In: Stahl PH, Camille GW, editors. Handbook of pharmaceutical salts. properties, selection, and use. Zurich: Verlag Helvetica Chimica Acta; Weinheim, Germany: Wiley-VCH; 2002. p 334–345.
24. Practical laboratory data, laboratory solvents and other liquid reagents David RL, Editor-in-Chief. CRC Handbook of Chemistry and Physics, 87th ed. Boca Raton, FL: CRC Press, Taylor and Francis Group. p 15–13, 15–22.
25. Available at www.fda.gov.
26. Available at http://www.ich.org/cache/compo/276-254-1.html.
27. Aromatic sulphonic acids and their derivatives. In: Furnis BS, Hannaford AJ, Smith PWG, Tatchell AR, editors. Vogel's Textbook of Practical Organic Chemistry. 5th ed. Essex, UK: Pearson Education Limited; 1989. p 872–887.
28. Lee T, Wang YW. Initial salt screening procedures for manufacturing ibuprofen. Drug Dev Ind Pharm 2009;35(5):555–567.
29. Peterson ML, Morissette SL, McNulty C, Goldsweig A, Shaw P, LeQuesne M, Monagle J, Encina N, Marchionna J, Johnson A, Gonzalez-Zugasti J, Lemmo AV, Ellis SJ, Cima MJ, Almarsson O. Iterative high-throughput polymorphism studies on acetaminophen and an experimentally derived structure for Form III. J Am Chem Soc 2002;124(37):10958–10959.
30. Carlson ED, Cong P, Chandler WH, Chau HK, Crevier T, Desrosiers PJ, Doolen RD, Freitag C, Hall LA, Kudla T, Luo R, Masui C, Rogers J, Song L, Tangkilisan A, Ung KQ, Wu L. An integrated high throughput workflow for preformulations: polymorphs and salt selection studies. PharmaChem 2003;2(7–8):10–15.
31. Gardner CR, Almarsson O, Hongming C, Morissette S, Peterson M, Zhong Z, Szu W, Lemmo A, Gonzalez-Zugasti J, Monagle J, Marchionna J, Ellis S, McNulty C, Johnson A, Levinson D, Cima M. Application of high throughput technologies to drug substance and drug product development. Comput Chem Eng 2004;28(6–7):943–953.
32. Hammond RB, Hashim RS, Caiyun M, Roberts KJ. Grid-based molecular modeling for pharmaceutical salt screening: case example of 3,4,6,7,8,9-hexahydro-2H-pyrimido (1,2-a) pyrimidinium acetate. J Pharm Sci 2006;95(11):2361–2372.
33. Kojima T, Onoue S, Murase N, Katoh F, Mano T, Matsuda Y. Crystalline form Information from multiwell plate salt screening by use of Raman microscopy. Pharm Res 2006;23(4):806–812.
34. Trask AV, Haynes DA. Samuel Motherwell WD, Jones W. Screening for crystalline salts via mechanochemistry. Chem Commun 2006;1:51–53.
35. Kumar L, Amin A, Bansal AK. An overview of automated systems relevant in pharmaceutical salt screening. Drug Discov Today 2007;12(23–24):1046–1053.
36. Childs SL, Rodriguez-Hornedo N, Reddy LS, Jayasankar A, Maheshwari C, McCausland L, Shipplett R, Stahly BC. Screening strategies based on solubility and solution composition generate pharmaceutically acceptable cocrystals of carbamazepine. CrystEngComm 2008;10(7):856–864.
37. Anderson BD, Flora KP. Preparation of water soluble compounds through salt formation. In: Camille GW, editors. The Practice of Medicinal Chemistry. London: Academic Press; 1996. p 739–754.
38. Tong WQ, Whitesell G. *In situ* salt screening: a useful technique for discovery support and pre-

formulation studies. Pharm Dev Tech 1998;3(2):215–233.

39. Morris KR, Fakes MG, Thakur AB, Newman AW, Singh AK, Venit JJ, Spagnuolo CJ, Serajuddin ATM. An integrated approach to the selection of optimal salt form for a new drug candidate. Int J Pharm 1994;105(3):209–217.

40. Berge SM, Bighley LD, Monkhouse DC. Pharmaceutical salts. J Pharm Sci 1977;66(1):1–19.

41. Giron D. Characterization of salts of drug substances. J Therm Anal Cal 2003;73(2):441–457.

42. Gould PL. Salt selection for basic drugs. Int J Pharm 1986;33(1–3):201–217.

43. Bighley LD, Berge SM, Monkhouse DC. Encyclopedia of pharmaceutical technology. In: Swarbrick J, Boylan JC, editors. Vol. 13. New York: Marcel Dekker, Inc; 1995. p 453–499.

ENZYMATIC ASSAYS FOR HIGH-THROUGHPUT SCREENING

[1]DANIELE CARETTONI
[2]PHILIPPE VERWAERDE
[1] Axxam, Milano, Italy
[2] iNovacia, Stockholm, Sweden

1. INTRODUCTION

To date, high-throughput screening (HTS) represents the leading approach within the target-based drug discovery process for the identification of small-molecule hits to be progressed as validated lead compounds. Established in large pharmaceutical companies in 1980s and 1990s, HTS is currently disseminated in small–medium pharmaceutical and biotech companies, and in life science industries [1]. More recently, HTS has been adopted as the cornerstone method in prominent initiatives coordinated by academic institutions for the identification of molecular probes aimed at promoting the knowledge of basic biological processes and disease mechanisms [2]. The expansion of HTS has been paralleled with the growth of the compound collections: at present, the average size of a chemical library in a large pharmaceutical company is estimated to be almost two million of compounds, and it is expected to steadily increase [3]. Rapid diffusion of HTS and compound collections allowed different sites to implement their own screening strategies in terms of readouts, formats, type of assays, and architecture of the compound library. In this context, enzymatic assays represent a major route for drug discovery programs, accounting for more than 40% of all screened assay in HTS [3]. Rather than comprehensively covering the entire field of enzymatic assay for HTS, this chapter will focus on the overview of the main strategies adopted to develop enzymatic assays, and provide insights into its different steps, driving principles, and quality criteria that make enzyme targets accessible to HTS.

It is important to note that enzymatic assays for HTS are generally referred as biochemical assays. However, this definition is potentially misleading. Indeed, biochemical assays cover all the cell-free tests based on isolated target proteins and comprise, besides enzymatic assays, also binding assays, ELISA tests, protein–protein interaction analysis, and others. Hence, enzymatic assays for HTS consist exclusively of detection systems based on the catalytic activity of functional enzymes isolated from their cellular context. Thus, by definition, assays in which the functional enzyme is detected in living cells pertain to cell-based, rather than biochemical assay configuration. Consistently, this attractive option for drug discovery that relies on dedicated and independent strategies will not be treated in this chapter.

2. ENZYMES AS DRUG TARGETS

Preclinical drug discovery programs within the pharmaceutical companies and academic institutions have focused their efforts on five major target classes: G-protein coupled receptors, ion channels, nuclear hormone receptors, transporters, and enzymes [4,5]. Recent surveys have revealed that enzyme inhibitors account for more than 25% of the drugs with therapeutic application [6,7]. In particular, 74 enzymes are currently targeted by small-molecule inhibitors approved by the U.S. Food and Drug Administration (FDA). These targets are predominantly human enzymes (70%), although also bacterial (17%), viral (7%), fungal (5%), and protozoan (1%) enzymes are represented [6,7]. Interestingly, these targets belong to all six enzyme classes, with a strong prevalence of oxidoreductases, transferases, and hydrolases (accounting in total to 88% of the FDA-validated targets), and with limited, but significant cases of lyases, isomerases, and ligases [6,7]. However, a unique feature confers distinctive properties to enzymes. While all target classes accomplish their physiopathological role by a process mediated by ligand binding, only enzymes are capable to covalently modify their ligands (substrates), acting as optimized and selective catalysts to promote chemical transformations. This exclusive property is widely exploited for the development of functional assays based on the enzymatic activity of the target, and for the identification of specific inhibitors designed to interfere with its catalytic mechanism. Consequently, the vast majority of currently marketed inhibitors are structural analogs of enzyme substrates, or they chemically react with residues and prosthetic groups of the active site. Competitive inhibitors represent the prevalent group within the currently FDA-approved drugs [6,7]. Nevertheless, peculiar mode of actions have been described, including tight binding of the active site, covalent modification of the catalytic residues, inhibition through transition-state analogs, stabilization of reaction intermediates, and allosteric inhibition [6,7].

Another aspect makes enzymes a distinctive class of drug targets. While both positive and negative modulators have been developed as therapeutic drugs for all the other target classes, drug discovery efforts on enzymes have pursued almost exclusively enzyme inhibitors. However, an underexplored option of therapeutic intervention may be represented by the configuration of specific assays designed to identify positive modulators of highly regulated enzymes [6,8,9].

3. EXPRESSION AND PURIFICATION OF ENZYMES FOR HTS

A central part of the HTS assay development for enzymes is the isolation of the target. Enzyme production for HTS relies exclusively on recombinant expression because the requirements in terms of protein amount and reproducibility practically exclude the native sources. The design of the recombinant versions should take into consideration the structure of the enzyme, its subcellular localization under physiological conditions, and whether posttranslational modifications (site-directed proteolysis, phosphorylation, acylation, etc.) and possible noncatalytic subunits may be essential to recover the enzyme in a functional form. When this information is scarce or missing, data on orthologous and homologous enzymes have to be analyzed.

In the HTS field, the expression strategy should always consider the possibility to express the full-length protein, to try to keep unaltered the biochemical structure and the functional properties of the target (Fig. 1a). Since it is not possible to predict which protein version will fulfill the screening criteria, it is advisable to explore a panel of chimeric versions, including truncations, site-directed mutants, and gene fusions. Furthermore, full-length enzymes may turn out to be either produced in inadequate yield to support HTS or obtained in a poorly active or inactive form.

In almost all cases, recombinant constructs will include in-frame fusion tags that may be exploited for protein detection and affinity purification (Fig. 1b) [10]. To avoid any possible interference with the catalytic properties of the target, long tags (e.g., GST, MBP, Fc, etc.) are proteolytically removed. Instead, short tags (e.g., polyhistidine tag, Strep tag, etc.) may be tolerated, when it is demonstrated that enzyme regulation and catalytic parameters are not significantly altered in the uncleaved form [10].

The choice of the expression host is inherently dependent on the enzyme. *Escherichia coli* is mostly used to express prokaryotic targets investigated in antibacterial drug discovery programs [11]. When clear literature evidence demonstrates the possibility to express properly folded functional enzymes, a bacterial host can also be used to produce eukaryotic protein targets [12]. In all the other cases, the recombinant expression has to be performed in a eukaryotic host (Fig. 2). In this regard, expression systems predominantly used are yeast *Pichia pastoris* [13], the baculovirus system with insect cell lines [14], and mammalian cells, namely, modified CHO or HEK-293 cell lines [15]. The eukaryotic expression hosts are more time consuming and resource intensive than *E. coli*, but they have higher chances to support the production of posttransationally modified, active enzymes [12]. The choice of the most suitable eukaryotic system is determined by a trial-and-error approach, and often depends on the internal expertise of each research group as well as on the unique biochemical and structural properties of the target.

Recombinant expression requires an extensive optimization in terms of cell lines, culturing conditions, and expression time, with the aim of obtaining maximally active enzyme

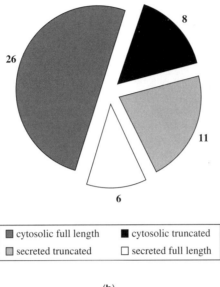

Figure 1. Enzymes as drug targets for HTS: chimeric versions and cellular localization. (a) A collection of 51 enzymes configured as HTS functional assays was analyzed. The enzyme list comprises 8 oxidoreductases (EC 1), 12 transferases (EC 2), 29 hydrolases (EC 3), and 2 ligases (EC 6) (Source: Axxam Assays Database). In total, 26 enzymes (dark gray) out of the 34 cytosolic enzymes (76%), and 6 enzymes (white) out of the 17 secreted enzymes (35%) were recombinantly expressed as full-length proteins. Overall, 63% of the enzymes configured as HTS assays were expressed as full-length proteins. (b) Only one enzyme of the analyzed collection was recombinantly expressed as untagged protein, while the remaining 50 enzymes were expressed as chimeric proteins with fusion tags, with a strong prevalence of the polyhistidine tag. Nt: amino terminal; Ct: carboxyl terminal; HIS: polyhistidine tag; GST: glutathione S-transferase; gp64: leader peptide of the major envelope glycoprotein of *Orgyia pseudotsugata* multicapsid polyhedrosis virus; Strep: Strep-Tag II peptide.

Enzymes for HTS and expression hosts

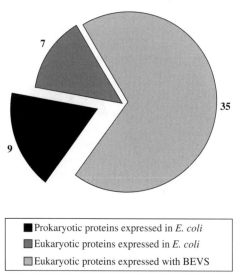

■ Prokaryotic proteins expressed in *E. coli*
■ Eukaryotic proteins expressed in *E. coli*
□ Eukaryotic proteins expressed with BEVS

Figure 2. Enzymes as drug targets for HTS: expression hosts. The enzyme collection described in Fig. 1 legend is composed of 9 prokaryotic enzymes and 42 eukaryotic enzymes. Prokaryotic enzymes were recombinantly expressed exclusively in *E. coli* (black). Seven eukaryotic enzymes (dark gray) were expressed in *E. coli* (17%), and the remaining 35 enzymes (83%) were expressed in insect cells using the baculovirus expression vector system (BEVS).

preparations and increased production yield. Recently, miniaturization and parallelization have been successfully applied to test arrays of different expression conditions, both for prokaryotic and eukaryotic expression hosts [16–18].

Purification of the target requires that the maximal homogeneity is achieved preserving the catalytic activity of the enzyme. This implies that the ideal protein preparation for HTS has to be not only homogeneously pure but also, and more importantly, enzymatically pure (i.e., no contaminant catalytic activity should be detectable using the HTS readout of the target). Protein purifications rely on chromatographic separations of crude cell extracts and culture media. A general purification scheme is represented by a first separation step based on affinity chromatography (in most cases, exploiting the fusion tag), and, when needed, by additional purifications usually performed through ion exchange and size-exclusion chromatography [19].

Following purification, both biochemical and functional identities of the enzyme have to be verified. Thus, the protein should migrate at the expected molecular weight when resolved by SDS-PAGE, and it should be recognized by specific antibodies in western blot analysis. Moreover, the signal produced using the HTS assay should be unequivocally ascribed to the purified enzyme. The latter evidence is generally achieved following two different approaches. A site-directed mutant corresponding to a catalytic-defective version of the target is expressed and purified following the same experimental strategy used for the functional enzyme. Alternatively, a mock control sample is obtained using a parental vector devoid of any insert and following the same expression–purification procedure applied for the recombinant target. When assayed in parallel with the functional enzyme, the catalytic-defective mutant and/or the mock control should not generate any detectable signal under the same experimental conditions. These

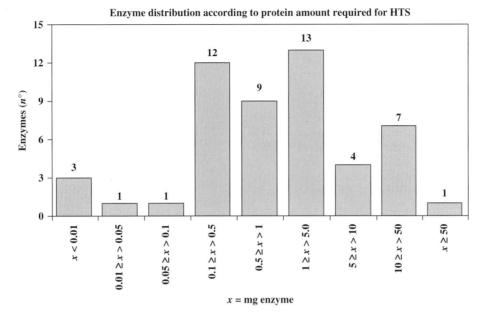

Figure 3. Enzymes as drug targets for HTS: protein amount. Graphical representation of the distribution of the 51 enzymes described in Fig. 1 legend according to the final amount of protein required to support a HTS campaign. For homogeneity, the following parameters were applied: size of the compound collection, 100,000 compounds; plate format, 384 well/plate; volume per test point, 40 µL; safety margin for liquid handling, 1.5×. More than 0.5 mg of protein had to be produced for 34 enzymes (67%), and more than 1 mg of protein for 25 enzymes (49%), to support the specified screening campaign. (This figure is available in full color at http://mrw.interscience.wiley.com/emrw/9780471266945/home.)

controls play a crucial role when integral transmembrane enzymes are recombinantly expressed. In fact, the peculiar structural features of transmembrane proteins impose that homogeneous purification is often incompatible with their recovery in a functional form. In this case, it is essential to ensure that the enzymatic activity derives unambiguously from the target, since additional membrane proteins will likely be copurified.

In general, high-throughput screening is a very demanding process in terms of protein amount, requiring on an average 0.5–5 mg of pure enzyme per 100,000 test points in 384 well/plate format (Fig. 3). Thus, the purification strategy needs to be scalable and reproducible because reiterated large-scale productions are often required to support the entire screening campaign. Distinct protein preparations will be independently assayed and pooled, to carry out the screening with a homogeneous protein batch. The enzyme pool will be retested to establish its specific activity, and different biophysical techniques will be applied to analyze its aggregation state (oligomerization) and its stability (folding and degradation) [20].

The final amount of enzyme, which may evidently represent a potential bottleneck in assay development, is determined by five critical factors. The first two of them, number of compounds to be tested and format of the plates supported by the available instrumentation (usually 384 well/plate and 1536 well/plate) can be regarded as structural elements, with minor, if any, possibility of optimization. On the contrary, the three remaining factors influencing the enzyme amount for HTS can be all addressed during assay development: expression–purification yield, catalytic efficiency of the target, and final enzyme concentration. These three variables may all significantly contribute, when adequately optimized, in reducing the time required for assay configuration and to create the permissive conditions for difficult-to-express targets to enter in the drug discovery process [19].

4. THE "IDEAL" ENZYMATIC ASSAY FOR HTS

The ideal enzymatic assay for HTS must be relevant, predictive, and as close as possible to the biological activity to be interfered. On the other hand, it should comply with the constraints imposed by the size of the compound collection and with the following basic requirements of HTS:

- *Homogeneity*: The assay should be configured with addition-only steps. Therefore, washings, separations, and centrifugations should be avoided and are usually considered incompatible with high-density microplate-based assay configuration. However, depending on internal expertise of the screening group, on the properties of the target, and on the available assay format, non-homogeneous assays (e.g., DELFIA, ELISA, etc.) are sometimes accepted [21].
- *Miniaturization*: The assay should be assembled and detected in a microplate. Under these conditions, the final volume of each test point, including compounds, enzyme, substrate, and detection reagents, is 20–50 µL in 384 well/plate format, 8–15 µL in low-volume 384 well/plate format, and 4–6 µL in 1536 well/plate format, respectively.
- *Sensitivity*: The miniaturized assay format imposes that the highest degree of sensitivity has to be achieved. Indeed, fluorescence-based and luminescence-based readouts cover more than 70% of the main detection modes currently used in HTS, followed by radiometric readouts (13%) and absorbance (8%) [3]. The wide use and the ranking order of these readout systems are intrinsically linked to their superior sensitivity compared to other detection signals.
- *Automation*: Addition steps, plate handling, and detection should be amenable to be integrated in a single screening unit and autonomously managed by a robotic system, with minimal manual intervention. The automation ensures precise and reproducible dispensing and timing along the screening, minimizing artifacts and system-dependent variability.

In addition, the assay must be as simple as possible, with a minimal number of plate manipulations and addition steps (usually from three to five). Furthermore, the assay should be fast, with a workflow from reaction start to signal acquisition comprised between 30 min and 2 h, and with a plate detection time below 5 min. In the ideal situation, the throughput will be high, with the aim of reducing the whole period of the screening campaign. Depending on the assay type and available instrumentation, throughput usually ranges between 10,000 and 50,000 test points per day in 384 well/plate, and 70,000–350,000 test points per day in 1536 well/plate format, respectively [1].

However, when the target, its substrate or the assay itself, does not allow such an "ideal" assay configuration, the strict quality criteria of HTS can be violated, to make valuable, yet challenging targets accessible to the drug discovery process. A remarkable example of a "breaking-rule" assay is represented by prostaglandin-E synthase (PTGES; EC 5.3.99.3), a potential therapeutic target for the treatment of inflammation [22]. PTGES substrate, prostaglandin-H2, displays an extraordinary instability, being non-enzymatically degraded with a half-life of approximately 10 min at room temperature. For this reason, high-throughput screening was run under exceptional conditions, consisting of a 30 s PTGES enzymatic reaction, followed by enzyme immunoassay performed on 3 instruments requiring 14 different steps, divided in 2 distinct phases. In spite of such extreme conditions, HTS was apparently able to identify 2600 potential hits for PTGES divided in 12 different chemical classes from a library of 315,000 compounds [22]. In practice, multiple infringement of the main quality criteria of HTS might have represented the only way for PTGES, as possibly for other targets, to be screened in HTS.

5. TYPE OF ASSAYS

The impressive effort of the drug discovery toward the development of detection systems optimized for HTS has ultimately made available multiple alternatives for the assay configuration of the most extensively explored enzyme classes. As a matter of fact, a wide variety of assay types have been described and made commercially available for protein kinases and proteases that jointly represent a predominant sector of the currently investigated enzyme targets (see Sections 8.2 and 8.3) [4,5]. Therefore, it is essential to evaluate at an early stage potentialities and drawbacks of the different alternatives because type and quality of hits retrieved during a screening campaign will strictly depend on the type of assay that is selected for assay development. Moreover, enzymes represent a structurally and functionally heterogeneous family of targets that require, in most cases, dedicated approaches for the design of robust assays suitable for HTS. Therefore, key factors such as detection of substrate/products, type of HTS readout, number of pipetting steps, sensitivity, and dynamic range of the assay, possibility to use the assay for a detailed kinetic characterization of the reaction, and the estimated cost of the assay should play a key role in the selection of the type of assay to be used for a screening campaign.

5.1. Continuous and Discontinuous Assays

Enzymatic assays can be divided according to their ability to detect over time substrate consumption or product formation as continuous or discontinuous. In continuous assays, the enzyme-generated signal can be uninterruptedly measured, allowing the detection in real time of the progress of the catalytic reaction [23]. Typical continuous readout systems applied in HTS are absorbance, fluorescence intensity, and fluorescence resonance energy transfer (FRET) for hydrolases, in which chemically modified surrogate substrates release chromogenic or fluorogenic groups [24]. On the contrary, discontinuous assays require that the reaction is sampled at certain time points, stopped and further processed to obtain a detectable signal [23]. Representative examples of discontinuous assays are scintillation proximity assay (SPA), time resolved-fluorescence energy transfer (TR-FRET), luciferase-based ATP consumption assay, and fluorescence polarization [25].

Provided that the sensitivity is adequate for the scope of the assay and the HTS assay adaptation does not impose insuperable technical limitations, continuous assays present remarkable advantages over discontinuous assays. This preference is not directly related to HTS. Rather, the advantages greatly impact steps upstream and downstream to the primary screening. In fact, regardless of whether the assay is continuous or discontinuous, primary screening is usually performed under endpoint measurement conditions [19]. In practice, this type of setting is mandatory because it would be almost impossible to keep the reading unit of the screening station fully engaged with a single plate all along the time of the enzymatic reaction, typically lasting from 30 min up to 2 h. Furthermore, data analysis of primary screening would be exceedingly complicated and time consuming, when continuous kinetic measurements were applied to the entire compound collection. Nevertheless, continuous assays can be advantageous during assay development and characterization of primary hits, since the ability to directly detect the progress curve of the reaction enables a more extensive and accurate optimization of the assay, a rapid identification of false-positive hits (i.e., fluorescent or quencher compounds), and a straightforward investigation of the mechanism of action of confirmed hits [19,23].

Continuous configuration is preferable, when possible, also for a practical reason. As the kinetic characterization of the enzymatic reaction is essential during assay development and hit characterization, it can be estimated that roughly 10 times less reagents are used in continuous assays compared to discontinuous assays, being the latter dependent on multiple sampling of the same reaction. This implication should be carefully considered for drug discovery programs based on difficult-to-express enzymes and/or expensive substrates and reagents.

5.2. Direct, Indirect, and Coupled Assays

In high-throughput screening, the ideal situation is represented by a target enzyme generating a product (or consuming a substrate) readily detectable by a HTS-compatible readout. This type of assays, named direct assays, comprises almost exclusively reactions catalyzed by hydrolases on surrogate chromogenic or fluorogenic substrates, and by oxidoreductases acting on NAD(P)H (see Sections 8.1 and 8.3) [23]. Thus, all the remaining enzymatic reactions that do not produce measurable changes in the spectral properties of substrates and products require that additional components are added to the target-mediated reaction, to transform the primary, nonmeasurable product into a HTS-detectable signal. These assays are referred as indirect assays [23]. Intuitively, all direct assays are continuous assays, while almost all indirect assays are discontinuous assay, although it is possible under particular circumstances to configure some indirect systems as continuous assays for HTS.

In HTS applications, indirect assays rely on reactive probes, labeled antibodies/affinity binders, and coupled enzymatic reactions. Reactive probes are chemical compounds that are able to specifically react with one of the products of the primary reaction, leading to a chromogenic or fluorogenic signal. A representative example of this class is the detection of free coenzyme A, a product of a large number of metabolic enzymes, by 7-diethylamino-3-(4'-maleimidylphenyl)-4-methylcoumarin (CPM), a molecular probe that becomes fluorescent upon reaction with the free thiol group of coenzyme A [26] (see also Section 8.1). On the other hand, an increasingly important class of indirect assays utilizes specific antibodies and affinity binders for the detection of the primary product. Techniques such as SPA, TR-FRET, and fluorescence polarization belong to this highly sensitive class of HTS assays (see also Section 7.1) [27–29]. In particular, scintillation proximity assay is a method that does not require a separation of free and bound radioligand and therefore is amenable to screening applications [30,31]. This technology relies either on microscopic beads that are filled with a scintillant liquid or on microtiter plates that are coated with a scintillant. Upon proximity to the bead or plate surface, a radioactive molecule will excite the scintillant, which will then emit photons. SPA has been applied to multiple assay formats in HTS, being an extremely versatile detection system. Hence, a variety of plates (type, surface and tags), isotopes (^{3}H, ^{125}I, ^{35}S, ^{33}P, ^{14}C), and beads with distinct spectral properties are available. In addition, beads can be functionalized with different tags for the binding of the radiolabeled substrate, further expanding the potentialities of the system [32].

To end, a large number of indirect assays employ auxiliary enzymes, called coupled enzymes, which use one product of the target-mediated reaction as substrate to generate a HTS-detectable signal [19,23,33]. Coupled systems have found a very broad application in HTS, due to their versatility and potentiality. Examples of primary substrates and products detected by coupled systems are, among the others, ATP (coupled to firefly luciferase to generate a luminescent signal) [33,34], ADP (coupled to pyruvate kinase and lactate dehydrogenase, detected as NADH consumption with a fluorescence-based readout) [33], free fatty acids (coupled to acyl-CoA synthetase, acyl-CoA oxidase, and horseradish peroxidase to generate fluorescent resorufin) [35], and deacetylated lysine residues in peptides labeled with 4-methylcoumarine-7-amide (primary reaction mediated by histone deacetylases, coupled to trypsin to generate fluorescent 7-amino-4-methylcoumarin) [36].

Coupled systems require that specific conditions are fulfilled to be applied in HTS. A prerequisite is that the rate-limiting step of the coupled system is the target-mediated reaction. However, in the initial part of the primary reaction, only a minimal amount of product used by the coupled enzyme will be available, and so a lag phase is expected to occur [33]. It is essential that in the assay configuration the lag phase is minimized, and the coupled reaction is detected only when the steady state is achieved (i.e., equilibrium between product formation by primary reaction and its consumption by the coupled system). Strategies to minimize the lag phase comprise the use of excess amount of coupled enzymes (generally 10-fold excess), use of saturating amount of any other substrate required to the coupled system (10-fold their K_m values), and selection of coupled enzymes with low K_m and high k_{cat} [23,33,39,40].

A drawback of the use of coupled assays in HTS is that primary hits will include a fraction of false positives targeting the auxiliary enzymes. Nevertheless, the elimination of false-positive hits can be achieved by including the appropriate controls. In practice, after the detection of the target reaction in the presence of compounds, plates are reinjected with substrate(s) of the coupled enzymes. Upon injection, those wells containing true-positive hits targeting the primary reaction will generate a detectable signal, while wells containing false-positive hits directed against the coupled system will remain unchanged. This strategy, based on differential rescue of the signal, can be almost universally applied to coupled assays used in HTS, allowing an early selection of true-positive hits.

To end, it is important to note that the higher is the number of enzymes coupled to the primary reaction, the more is difficult to investigate the mechanism of action of inhibitors identified during the screening campaign [23,33]. Therefore, the design of specific secondary assays may be valuable to confirm and deepen the analysis of the results obtained in HTS with coupled assays.

6. OPTIMIZATION OF THE ASSAY CONDITIONS

The enzymatic reactions are profoundly affected by the experimental conditions. Besides substrate and enzyme concentration, the most relevant factors that influence the catalytic activity of enzymes are the buffer composition and temperature. Since these two factors greatly affect the kinetic parameters of the enzymatic reaction (K_m, k_{cat}, K_i) [23,37,38], they are expected to directly impact the outcome of the screening campaign in terms of hit rate and quality. Thus, the optimization of the reaction conditions has to be regarded as a key step of the HTS assay configuration.

A driving principle for assay optimization is the reconstitution of the native conditions in which the target accomplishes its physiopathological role. This would imply the knowledge of the chemical microenvironment of the enzyme, including not only ions and solutes but also its macromolecular context (e.g., organelle membrane lipids, interacting proteins, nucleic acids, glycosamminoglycan high-order structures, etc.). However, it is practically impossible to reassemble—and in most cases, it is impossible even to know—the cell or tissue complexity of the target. So, a realistic and widely accepted compromise is the configuration of the enzymatic assay under experimental conditions that preserve the main kinetic parameters of the native reaction. Moreover, the assay conditions should enable the detection of the physiological regulation of the enzymatic activity, and should take into consideration the most likely composition of the solution (pH, salts, cofactors) in which the catalysis physiologically occurs.

The second driving principle for the optimization is the setup of the assay under experimental conditions that support the maximal catalytic efficiency (k_{cat}/K_m) of the target enzyme. Hence, optimized assays display a higher robustness, which in turn are expected to produce more reliable and reproducible results during HTS [1]. Moreover, the configuration of the assay under conditions maximally supporting the catalytic activity of the target implies a reduction of enzyme concentration, of substrate concentration, or both. In fact, any experimental condition that improves the turnover number (k_{cat}) will inversely decrease the enzyme working concentration ($k_{cat} = V_{max}/[Enzyme]$), assuming that the linear range of the reaction is maintained [23,39,40]. In addition, the assay configuration at minimal enzyme concentration has also the advantage to increase the sensitivity of the system to noncompetitive inhibitors. To end, since the expression and purification of the recombinant enzyme is an integral part of the assay development, a reduction of the enzyme working concentration results in a direct decrease of the time required for the production of the target (see Section 3). Conversely, optimized conditions decreasing the substrate concentration that half-saturates the active site of the enzyme allow the assay to be configured at a decreased substrate concentration, assuming that the substrate is used at a K_m-equivalent concentration [39,40]. Also in this case, the reduction of the working substrate concentration will favorably impact the cost of the HTS, in particular when expensive substrates are used (e.g., fluorogenic surrogate substrates for hydrolases, difficult-to-synthesize physiological substrates, etc.).

For all the above-mentioned reasons, extensive optimization of the assay is considered a prerequisite to obtain high-quality primary hits with affordable HTS assays. A basic source of information to guide the optimization of the assay conditions is represented by literature data. However, it is advisable to confirm and complement these evidences with an in-house performed characterization of the enzyme because public available data are very rarely obtained under experimental conditions that comply with the quality criteria of HTS. To investigate the effects of multiple alternative conditions on the reaction catalyzed by the target, a matrix of different buffers, pH, and salts can be screened with the aim of selecting the conditions that maximize the catalytic efficiency of the target. Integration of the data generated from these unbiased screenings with the knowledge of the physiological microenvironment of the target may effectively define the most suitable conditions to enter in a screening campaign.

To end, it is important to note that also the detection of the enzyme-generated signal may be deeply affected by the experimental conditions. In fact, changes in buffer composition and temperature may strongly influence the sensitivity and the dynamic range of the signal detected in HTS [23,24,38,41]. Therefore, to avoid misinterpretation and artifactual results, it is mandatory to include the proper controls (e.g., blank reactions with substrates and/or products without enzyme) to distinguish between on-target (catalytic efficiency) and off-target (signal detection, coupled system) effects.

6.1. Buffers and pH

The effect of proton concentration on enzyme activity can be considered similar to the effect produced by a positive or negative modulator of the reaction. In essence, the influence of pH on the catalytic activity can be explained for most of the enzymes in terms of protonation and deprotonation of their active site [23,37]. Thus, proton concentration may interfere with the formation of the enzyme–substrate complex, and/or with the catalytic rate. For this reason, both K_m and k_{cat} may be profoundly altered by pH variations [23,37]. In addition, pH may strongly change the ionization status of the substrate, hampering or promoting its recognition by the enzyme [23,37]. Consistently, also the interaction with potential enzyme inhibitors can display a strong pH-dependent behavior [23,37,39,40]. Moreover, attention must be paid when pH-sensitive fluorescence tracers, such as fluorescein and NAD(P)H, are used [19,41]. Hence, pH should be considered a key parameter for the assay configuration for HTS. Choice of the pH to be used in HTS should take the following criteria into consideration: physiological pH at which the target enzyme is most likely expected to accomplish its physiopathological role; optimal pH that promotes the maximal catalytic activity and the minimal K_m value for the substrate; compatibility of the compound collection with the pH value selected for screening. The latter criterion is particularly relevant when the target enzyme is functionally expressed in a tissue or subcellular context characterized by extreme pH values (e.g., gastric secretions, lysosomes, etc.). In this case, assay setup will compromise between the optimal pH of the enzyme and the confidence that the identity and stability of the chemical library is on average preserved under screening conditions [1].

As buffer systems have remarkably different effects on enzyme activity, buffer type should be considered as important as pH. Thus, chemical groups of the buffer may directly interact with the enzyme, substrate and cofactors [38,42]. Consequently, some buffers may negatively interfere with the catalytic mechanism of the target, by chelating essential divalent cations (e.g., citrate buffers), by

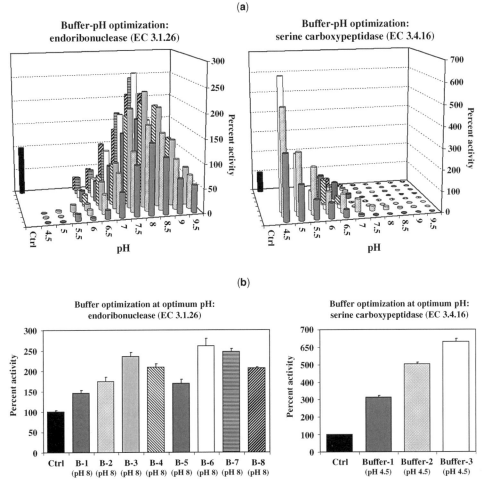

Figure 4. Optimization of the buffer/pH reaction conditions for two representative enzymes. (a) Screening of a matrix of 73 buffer/pH conditions, ranging from pH 4.5 to 9.5, including 16 different buffers, each tested at 25 mM within the interval $(pK_a - 1) \leq pH \leq (pK_a + 1)$ in 384 well/plate format for an endoribonuclease (EC 3.1.26) and for a serine carboxypeptidase (EC 3.4.16). The endoribonuclease and the serine carboxypeptidase displayed different optimal pH, corresponding to 8.0 and 4.5, respectively. (b) The endoribonuclease and the serine carboxypeptidase exhibited remarkably different activities when assayed in the presence of different buffers (B-1/B-8 and B-1/B-3, respectively) at their optimal pH. Selection of the optimal buffer type (at their optimal pH) resulted in 1.7-fold and 2-fold increase of the specific activity, respectively. Overall, the replacement of a standard buffer (Ctrl; black bars) with the optimal buffer/pH (white bars) resulted in a 2.5-fold and 6.3-fold increase of their specific activity, respectively (*Source*: Axxam Assays Database).

competing with the enzymatic reaction (e.g., phosphate buffers for phosphotransferases; glycylglycine for certain proteases), and by precipitating some divalent metal ions and polyvalent cations (e.g., phosphate buffers) [38,42].

As it is difficult to anticipate the requests of any enzymatic reaction in terms of buffer and pH, it is advisable to screen a panel of different buffers tested at ±1 pH unit from their pK_a, and covering a wide range of pH values. Figure 4 shows the results obtained from a screening of two enzymes—an endoribonuclease (EC 3.1.26) and a serine carboxypeptidase (EC 3.4.16)—on a matrix of 73 buffer/pH conditions. The replacement of a standard buffer commonly used in literature with the optimal buffer/pH often determines a remarkable increase of the enzyme specific activity (see Fig. 2 for details).

6.2. Monovalent and Divalent Salts

Similarly to pH, monovalent and divalent ions may strongly influence the catalytic activity of the target. While assaying the effect of salt concentration, it should be considered that the buffer system contributes to the ionic strength of the solution, and this contribution should be regarded as irrelevant only at very high salt concentration [42]. Therefore, for those enzymes whose activity is negatively affected by salts, it is important to select buffering systems with an effective pH control at low buffer concentration.

As anticipated by their physiological concentration, monovalent and divalent salts generally need to be tested at different concentration ranges, the former up to several hundred millimolar concentration, the latter at a concentration one order of magnitude lower. It may be particularly informative to test a few series of biologically relevant monovalent cations and anions, to establish whether any ion-selective modulation of the enzyme activity may occur. In this regard, minimal series are represented by Na^+, K^+, NH_4^+, and Cl^-, CH_3COO^-, respectively.

Divalent cations are well-recognized cofactors of enzymatic reactions, playing in many cases the role of essential prosthetic groups [43]. They either are tightly bound to enzymes and virtually indissociable using mild purification conditions or need to be added to the reaction to induce the activation of the otherwise inactive enzyme [43]. Enzymes activated by divalent cations represent a wide and extremely heterogeneous group, comprising extensively investigated pharmacological targets (e.g., protein kinases, polymerases, calcium-dependent proteases, etc.) [43]. Divalent cation-mediated activation sometimes occurs in a narrow range of salt concentrations because above a defined, enzyme-specific threshold divalent ions can be poisonous to the reaction. A minimal set of biologically relevant divalent cations includes Ca^{2+} and Mg^{2+}, but it can be usefully complemented with Mn^{2+}, Fe^{2+} (and Fe^{3+}), Cu^{2+}, and Zn^{2+}. Furthermore, it may be useful during assay development to assess the sensitivity of the target to heavy metals (e.g., Cd^{2+}, Ni^{2+}, Hg^{2+}, Pb^{2+}), since these toxic compounds may be present in traces in the buffer reagents and in the compound collection.

6.3. Additives, Stabilizers, and Other Cofactors

The configuration of enzymatic assays for high-throughput screening poses practical problems in terms of liquid handling, stability of the reagents, and adsorption of solutes to vessels and plates. In the attempt to circumvent these issues, it may be beneficial to test the effect of different type of chemical additives. These compounds may directly stimulate the enzymatic reaction, or they can improve the performance of the assay under screening conditions. A non-exhaustive list of reagent types comprises detergents, reducing agents, chelators, stabilizers, and other cofactors.

Detergents may greatly affect enzymatic assays—they can bind to the enzyme modulating its catalytic efficiency, and/or improving its stability [44]. Binding of detergents to hydrophobic substrates and products may increase the catalytic rate and/or their affinity for the enzyme [44]. Conversely, detergents may decrease the stickiness of certain enzymes and substrates to the plastic plates and vessels. In addition, they may improve the reliability of microdispensing devices used for liquid handling during screening. Most importantly, detergents can avoid nonspecific enzyme inhibition due to compound aggregation [44]. In practice, the investigation of the effect of zwitterionic and nonionic detergents has to be privileged over ionic detergents, as the latter have often a denaturant activity on enzymes [44]. Detergents should be tested in a concentration range spanning their respective critical micelle concentration (CMC) because physicochemical properties of detergents are remarkably different above and below their CMC [44].

Reducing agents are often included in enzymatic reactions for their ability to prevent the oxidation of thiol groups of the enzymes.

Oxidation of active-site cysteine residues is detrimental for the catalytic activity, while oxidation of other free thiol groups of the enzyme may compromise its structure and folding [40]. The requirement of certain enzymes for a reducing environment may be sensibly amplified in HTS assays, since enzyme mixture has generally to sustain a long-term storage before the reaction start (sometimes up to 15 h), especially in fully automated HTS campaigns running day-and-night screenings. In general, it is advisable to analyze the effect of reducing agents with different redox potential, including also physiological reducing agents, such as ascorbic acid and glutathione.

As already mentioned, a different, and somehow more frequent cause of potential instability of the enzyme and substrate mixtures is represented by the non-specific binding of the reagents to the plastic supports. In addition, poor solubility of the isolated enzyme and substrate in the assay buffer may potentially cause gradual aggregation, precipitation, and loss of activity. To circumvent these problems, reaction buffers for HTS usually include exogenous proteins as stabilizing agents, such as bovine serum albumin. Non-mutually exclusive alternatives are represented by small molecules known to preserve enzymes in solution and to prevent their absorption. These compounds include sulfobetaines, cyclodextrins, short-chain polyethylene glycols, gelatin, and glycerol [45–47]. However, care must be taken with compounds increasing the density and viscosity of the solutions because these conditions may be incompatible with the robotic liquid handling. Nevertheless, in some cases, viscous solutions can also be advantageous for the dispensing of heavy assay components, such as SPA beads [33–35]. Moreover, it should be taken into consideration that some of these components might alter the potency of potential inhibitors during screening [48].

Divalent cation chelators can be included in HTS assay buffers, when they are beneficial in shielding highly sensitive enzymes from the inhibitory effects of possible traces of heavy metals present as contaminants in protein preparations, buffer reagents, substrates, plastics, and compounds to be screened (see Section 6.2).

Together with the above-mentioned reagents, an almost infinite array of enzyme-specific cofactors can be tested to reconstitute the physiological environment and regulation of the target. Macromolecules such as triglycerides, fatty acids, glycosamminoglycans, polysaccharides, and nucleic acids have been reported to promote a remarkable stimulation of specific enzymatic activities [39,40]. On the other hand, physiological high-energy carriers (e.g., nucleoside triphosphates, NAD(P)H, FMN, FAD, coenzyme A) are worth to be tested even when they are not predicted to be directly involved in the catalytic mechanism, due to their universal role in biological systems [39,40]. Moreover, enzyme-specific regulatory subunits and effectors can allow the configuration of functional assays designed to identify allosteric modulators of the catalytic activity, which may represent an attractive opportunity to complement the extensively explored orthosteric route.

6.4. Temperature

The effect of temperature on enzyme kinetics has been comprehensively studied and treated elsewhere [38], and it is above the scope of the present account to give a detailed description of this topic. However, a few considerations have to be highlighted, as they have implications in assay development for enzymatic targets.

For practical reasons, the ideal temperature at which enzymatic assays should be performed is the room temperature of the screening station. In fact, the use of a homogeneous temperature to store the enzyme and substrate mix, to incubate the reaction with the compound collection, and to detect the signal minimizes artifacts that may arise from temperature transitions. Nevertheless, many enzymatic assays require specific temperature conditions that deviate from this ideal situation. The main reason is enzyme, and sometimes substrate, instability [22]. Screening campaigns of large compound collections require that enzyme and substrate mixes are prepared once or twice a day, and consequently stored up to several hours before the reaction start. To fulfill the quality criteria of HTS, assay plates run immediately after the preparations of the enzyme and substrate mix should have a performance comparable to the plates run after several hours. However, long-term storage of enzyme preparation may be detrimental to the enzyme activity, mostly due to progressive inactivation [38]. Hence, enzyme and substrate mix is routinely kept at low temperature (4°C or on ice), and let equilibrate at the desired temperature before the reaction start. For this reason, it is important to assess the stability of enzyme and substrate mix during assay configuration, and to elaborate a proper scheme for liquid handling during screening in compliance with the stability requirements of the reagents.

A second variable that should be taken into account is the strong temperature dependency of the catalytic activity of some enzymes [38]. When catalytic efficiency at room temperature is remarkably lower than that at the optimum temperature (i.e., the temperature at which the enzymatic rate is maximal), and unsatisfactory to achieve a valid signal-to-background ratio, the reaction has to be carried out at a higher temperature. However, assays are very rarely configured at the optimum temperature for the enzyme, because this may pose practical problems, including time-dependent thermal inactivation of the enzyme in long-lasting reactions, issues on the stability of the compound collection used for screening, and potential artifacts arising from evaporation of the samples in miniaturized HTS campaigns. In practice, for enzymes with a marked temperature dependency, assays are configured at a temperature value suitable for generating an unambiguous signal, without interfering with the performance of the assay. In this case, HTS assays are preferentially configured either at 30°C or 37°C, depending on the kinetic properties of the enzyme and on the experience of each screening group.

7. ASSAY CONFIGURATION

After having established an optimized reaction buffer, HTS enzymatic assays should undergo an extensive kinetic characterization to define the key parameters of the assay/substrate and enzyme concentration. Furthermore, additional factors will be addressed, including tolerance to organic vehicle of the compound collection, stability of the reagents, and response to standard pharmacological compounds. To end, the assay will be transferred to the screening station to test its actual performance under automated conditions. Most importantly, the assay system should be selected in the perspective of the type of hits that are intended to be identified during the screening campaign. Hence, enzymological characterization of the target will contribute to set up the reciprocal substrate and enzyme concentration, to favor the selection of competitive or non-competitive inhibitors, and of orthosteric or allosteric modulators.

7.1. Substrate and Enzyme Concentration

The configuration of the assay under steady-state velocity condition is of prime importance in HTS for achieving a reliable measurement of enzyme activity and inhibition. The catalytic rate under steady state is essentially constant in time, a condition that occurs only when a limited amount of the initial substrate, conventionally below 10%, is converted into product [23,39,40]. Under these conditions, the modulation of the signal produced by changes in substrate and enzyme concentration—or by the presence of potential inhibitors—is detected at the highest possible sensitivity, ensuring that the assay has the ability to identify also weak enzyme inhibitors. However, a potential challenge is represented by the constraints imposed by the assay automation. In practice, the linear range of enzymatic reactions should be maintained for at least 30 min and up to 2 h in high-throughput screening, to allow the proper dispensing of the reaction mixtures, handling of the plates, and reliable detection of the target-generated signal [1]. To achieve this prolonged time window of linear kinetic, enzyme and substrate concentrations should be reciprocally balanced. As far as substrate is concerned, the ideal condition for HTS assay configuration is represented by a concentration equivalent to its K_m value. In fact, below K_m value, the substrate become rapidly limiting, causing hyperbolic bending of the signal, loss of the steady state and flattening of the assay sensitivity [23,39,40]. On the other hand, the higher substrate concentration is used above its K_m, the more insensitive the assay gets to competitive inhibitors [23,39,40]. Hence, assay configuration at K_m-equivalent concentration represents in almost all cases the optimal compromise between assay sensitivity and reliable signal detection. However, in case sensitivity and robustness of the assay are inadequate for HTS at K_m-equivalent concentration, a different substrate concentration can be used. Potentiality and implications of these alternative assay formats will be presented in Section 7.2.

To achieve a long-lasting steady-state kinetic, the enzyme has to be generally used at three-order magnitude lower concentration than substrate [23,39,40]. This indication may be applied as a reference to experimentally select the enzyme concentration range that allows the maintenance of the linear kinetic for the requested time, ensuring that a HTS-suitable signal-to-background ratio and robustness are achieved. Having fulfilled the latter two conditions, the enzyme concentration will be set up at the minimal compatible value, to increase the sensitivity of the assay toward non-competitive inhibitors.

Indications of the reciprocal enzyme and substrate concentrations may be rapidly collected by testing a matrix of increasing enzyme and substrate concentrations, as described in Fig. 5a. This experiment is expected to provide valuable information concerning the kinetic properties of the assay. First, K_m value for the substrate can be calculated at different enzyme concentrations. By definition, K_m value should be demonstrated to be independent on enzyme concentration, unless artifactual effects are present [23,39,40]. Second, the relationship between substrate and initial velocity can be established, leading to the conclusion whether the reaction obeys to a hyperbolic, Michaelis–Menten equation, or a cooperative relationship exists between substrate concentration and initial velocity (Fig. 5b) [23,39,40]. Third, when an adequate standard curve of product is detected in parallel, the experiment allows the calculation of the turnover number (k_{cat}), which can be used to determine the catalytic efficiency (k_{cat}/K_m) of the reaction [23,39,40]. To end, if the reaction is detected in the linear range, the halving of the enzyme concentration should result in a twofold decrease of the initial velocity (Fig. 6a) [23,39,40]. This parameter is crucial for assay configuration, as it mimics 50% inhibition during screening. Other important parameters can be extrapolated from this experiment, as, for instance, the signal-to-background ratio as a function of enzyme and substrate concentrations (Fig. 6b) (see also Section 7.5).

In case the enzyme has more than one substrate, the experimental determination of the kinetic parameters will be carried out by varying one substrate at a time, and using saturating concentrations of the others [23,39,40]. Then, configuration for HTS will take into consideration whether it is advantageous to set up the assay at K_m value for all the substrates, or it is more appropriate to privilege the sensitivity of the assay toward only one of them. For instance, kinase assays can be configured at both ATP and peptide K_m-equivalent concentrations, or at saturating ATP concentrations, with the aim of imposing in the latter case a higher selectivity against ATP-competitive inhibitors [25].

Critical issues that may be revealed during the kinetic characterization of the reaction are represented, for instance, by the presence of a lag–phase, that is, a deviation from the linear range in the initial part of the reaction [23]. Lag-phase can be ascribed not only to inadequate temperature control of the reaction, but also to non-homogeneous enzyme and substrate solutions. This may be particularly relevant for integral transmembrane enzymes, or for substrates with a high partition coefficient, such as lipid-like molecules [23]. Alternatively, lag–phase may be caused by slow conformational changes of the enzyme induced by the substrate (e.g., oligomerization) [23]. When these issues cannot be experimentally addressed, it is essential to set up the detection of the assay far downstream from the lag–phase, to avoid the collection of misleading data.

Although time-dependent hyperbolic bending of the signal is very often caused by substrate depletion, other reasons may be also

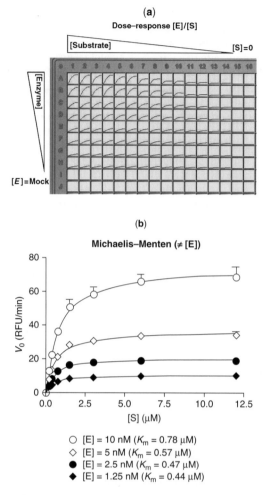

Figure 5. Assay setup and reciprocal substrate and enzyme concentration. (a) Minigraphs of a reciprocal enzyme and substrate dose–response experiment in 384 well/plate format for a monoxigenase (EC 1.14.14) (*Source*: Axxam Assays Database). The fluorogenic substrate was tested at seven concentrations (0.1875, 0.375, 0.75, 1.5, 3.0, 6.0, and 12.0 µM) in quadruplicate, while the monoxigenase was tested at four concentrations (1.25, 2.5, 5.0, and 10.0 nM) in quadruplicate. The mock control (a catalytic defective enzyme) was tested at 10 nM. (b) Data of the experiment reported in A were elaborated to assess the relationship between initial velocity (V_0, expressed as relative fluorescence units (RFU) per minute) and substrate concentration [S]. The monoxigenase reaction followed a hyperbolic Michaelis–Menten equation at all the enzyme concentrations tested. K_m value of the substrate calculated at different monoxigenase concentrations was independent on enzyme concentration.

taken into consideration and possibly addressed during assay development. Possible concurrent causes include product inhibition, occurrence of a reverse reaction, instability of substrate or enzyme, catalysis-dependent enzyme inactivation, and artifacts arising from an improper use of the detection systems (e.g., saturation effect) [23,39,40].

During assay configuration it is essential to kinetically characterize the influence of possible allosteric modulators of the reaction. Then, if therapeutically relevant, the assay will be set up at a concentration of the allosteric modulator equivalent to its K_a (i.e., the concentration of effector for half-maximal stimulation), to increase the sensitivity of the

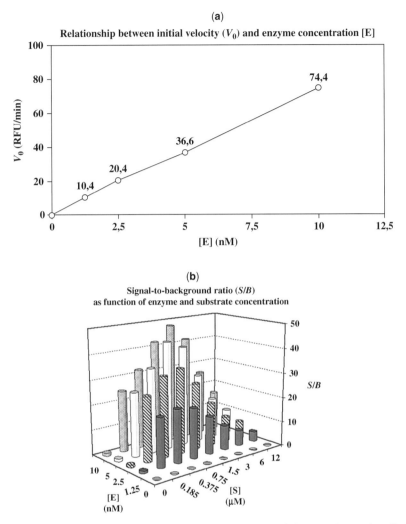

Figure 6. Assay setup, linerity and signal-to-background. (a) Data of the experiment described in Fig. 5 legend were elaborated to assess the relationship between initial velocity (V_0) and enzyme concentration. A linear relationship existed between monoxigenase concentration and initial velocity of the reaction, and doubling the enzyme concentration resulted in a twofold increase of the initial velocity. (b) Data of the experiment described in Fig. 5 legend were elaborated to determine the signal-to-background ratio (S/B) as a function of the enzyme (E) and substrate (S) concentrations. In the monoxigenase reaction, the increase in enzyme concentration resulted in an increase of the S/B, with a plateau effect at substrate concentrations below 0.75 µM. On the contrary, the increase of substrate concentration above 1.5 µM caused a drop of the S/B, determined primarily by the increase of the background signal of the substrate.

assay toward small-molecule compounds targeting not only the active site but also the allosteric site [23,39,40].

Although the above-mentioned kinetic criteria are widely applied for HTS assay configuration of enzymatic targets, they cannot be entirely extended to a distinct and increasingly important type of assays, the antibody-based and affinity binder-based assays for HTS. An example of antibody-based assay is represented by TR-FRET assay for protein kinases [28]. In this assay, a peptide modified with an acceptor dye contains a consensus sequence phosphorylated by the target kinase. Following kinase reaction, the phosphorylated amino acid is recognized by a sequence-specific antibody labeled with an europium chelate. The reaction is excited at a defined wavelength and the energy transferred by the europium chelate to the acceptor dye is measured [28]. Conversely, an example of affinity binder-based assay is immobilized metal affinity for phosphochemicals (IMAP) fluorescence polarization [50]. In IMAP fluorescence polarization assay for protein kinases, a fluorescent peptide is specifically phosphorylated by the target kinase. The reaction is injected with nanoparticles bearing a trivalent metal that anchors the phosphorylated peptide. Complex formation reduces the rotational speed of the peptide that is detected as an increase in fluorescence polarization [50]. The distinctive property of antibody and affinity binder assay systems is their exquisite sensitivity that allows the detection of the signal in the low picomolar range [28,49]. The advantage of these assays is clearly that a minimal amount of enzyme is required to elicit the production of the detectable signal. These assays require that the peptide substrates are used at a concentration several times lower than their K_m values, to avoid signal saturation [28,49]. On the one hand, this condition intensifies the sensitivity of the system toward competitive inhibitors. On the other hand, it makes impossible to perform a detailed enzymatic characterization, neither in terms of essential kinetic parameters (K_m-equivalent concentration of substrates, k_{cat} of the reaction), nor in terms of mode of action of the inhibitors identified during screening [23,39,40]. Consistently, under these conditions the reaction does not fulfill the criteria of the steady state, and doublings in enzyme concentration do not necessarily correspond to twofold increments in the detected signal [23,39,40]. However, these systems represent powerful first-line assays in primary screening that need to be complemented by secondary assays for the detailed characterization and confirmation of the retrieved hits.

7.2. Unconventional Assay Setup

The leading principle of the conventional enzymatic assay configuration is that the consumption of substrate has to be kept at less than 10% of its initial concentration and the detection of the inhibition has to be performed within this kinetic window (see Section 7.1). However, under these conditions several assay systems do not produce a signal-to-noise ratio sufficient to achieve an unequivocal identification of potential inhibitors in HTS. In these cases, screening is performed and measured at a conversion level higher than 10% of initial substrate concentration, to increase the sensitivity and the robustness of the assay. This alternative assay configuration poses potential issues according to the conventional kinetic paradigms in terms of the results that can be obtained from a screening campaign. Nevertheless, the theoretical analysis of this unconventional assay setup for HTS has been recently undertaken by independent groups [50,51]. The main criterion for the analysis is based on the definition of how much substrate should be converted into product when signals of the reaction with or without inhibitors are compared. Based on the observation that the largest difference between inhibited and non-inhibited reactions does not occur necessarily in the initial velocity region, it was proposed that the optimal point for measuring is in the maximal divergence region of the kinetic (Fig. 7a). Moreover, it was observed that the strongest signal occurs at substrate concentration higher than its K_m value. As expected, under these conditions the IC_{50} value for a reference inhibitor is higher than the value estimated under steady state (Fig. 7b). However, assay sensitivity in HTS should be privileged over the accuracy required to estimate IC_{50} value, and inhibition constants can be correctly determined afterward using secondary assays. Conversely, the advantages of a higher degree of inhibition together with the benefit due to optimized assay robustness and sensitivity are evident. The analysis takes into consideration the sensitivity of progress curves to variations in enzyme activity at different substrate conversion levels and the

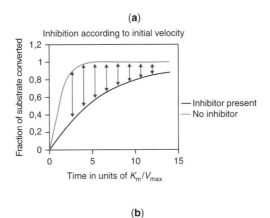

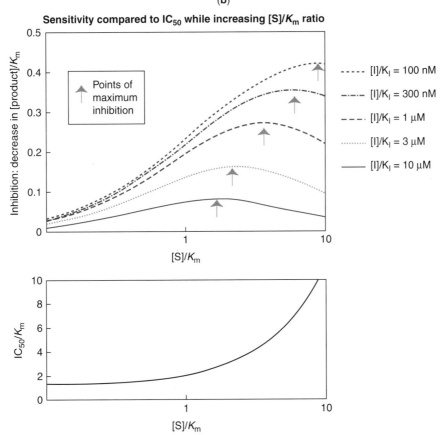

Figure 7. Unconventional assay setup: effect of high substrate concentration on inhibition assay sensitivity. (a) Initial enzyme velocity in presence (black line) or absence (gray line) of a competitive inhibitor (with no product induced inhibition). The predicted decrease in sensitivity does not occur past the point of non-linearity (*Source:* with permission from John C. Owicki, 2002). (b) Assaying different competitive inhibitor conditions at a fixed time ([I] = 10 µM). Increasing [S] above K_m, the amount of inhibition increases. Increased IC_{50} is largely compensated by a higher sensitivity (*Source:* with permission from John C. Owicki, 2002).

competitive effect of substrate/product on the inhibitors [52]. Thus, the limit of detection of competitive inhibitors for a typical endpoint enzymatic assay is modeled, and a novel parameter, termed IDL (Inhibition Detection Limit), was defined [52]:

$$\text{IDL} = \frac{\ln(1-P^c+3\text{SD}_p^c)}{(3K(K_{ps}-1)\text{SD}_p^c+(1+KK_{ps})(\ln(1-P^c)-\ln(1-P^c+3\text{SD}_p^c)))}$$

where P^c is the substrate conversion level of the enzyme activity in the absence of inhibitor, SD_p^c is the standard deviation of P^c, K is initial substrate concentration relative to the Michaelis–Menten constant, and K_{ps} is the ratio of the Michaelis–Menten constants for substrate and product.

When IDL is plotted as a function of the substrate conversion level in the absence of inhibitor (P^c), its curve has an inverse U-shape, with a single maximum (IDL_{max}), whose value represents the substrate conversion level where the ability to detect competitive inhibition is maximized (P^c_{max}). According to this analysis, the substrate conversion rate that maximizes the possibility to detect weak inhibitors in most cases significantly diverges from the steady-state initial measurement, and implies that more than 10% and up to 70% of the substrate is converted into product [52].

7.3. Type of Substrate

Although most enzymes are specialized catalysts capable of converting a single substrate to a single product, many are catalytically promiscuous, and they can recognize related, but structurally distinct substrates [53]. This property has been widely exploited in assay configuration for HTS to design surrogate substrates with favorable features for their detection with HTS-compatible readouts. Furthermore, surrogate molecules have been extensively applied as affordable substitutes of expensive and difficult-to-produce natural substrates. The use of non-physiological substrates in screening may pose some concerns for a possible lack of validation of the retrieved inhibitors when they are retested in the native context. However, if the catalytic mechanism is identical, surrogate molecules reliably mimic natural substrates, and potential inhibitors are expected to confirm their relative potency and selectivity [1,19,21].

Surrogate substrates have been used in HTS for all classes of enzymatic targets. In particular, luminogenic and fluorogenic substrates have been applied for oxidoreductases (e.g., cytochromes P450, monoamino oxidase) [54–56] and hydrolases (proteases, lipases, esterases, phosphatases and histone deacetylases) [36,57–59]. Moreover, short peptides have been used as substituents of substrate proteins in kinase assays [25], while modified oligonucleotides have been successfully used as surrogate substrates for DNA/RNA modifying enzymes [106,129,136].

Surrogate substrates can be designed on the basis of the structural features of their natural counterparts, or they can be identified from focused libraries. Using the same strategy adopted in HTS, it is possible to screen collections of molecules designed to be potentially recognized by a specific class of targets (Fig. 7). Representative examples are sets of peptides enriched in serine, threonine, and tyrosine that are screened with protein kinases [25] and libraries of peptides chemically modified with fluorogenic groups that are used to identify effective substrates for proteases [60]. Moreover, surrogate phospholipids and triglycerides are screened with lipases [21], while arrays of fluorogenic ester bond-containing molecules are screened with esterases [21]. This approach has the advantage of identifying artificial substrates on the basis of their kinetic performance and ability to generate sensitive signal under HTS conditions. In addition, a side advantage is that the selectivity of the target is rapidly profiled, providing valuable information on its catalytic mechanism.

7.4. Additional Parameters for HTS Assay Configuration

A screening campaign is an impressive investment of resources, time, and competences. For this reason, several quality criteria have been established to consider an assay eligible for

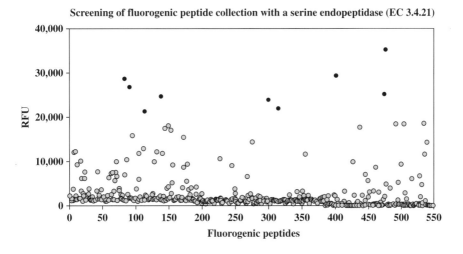

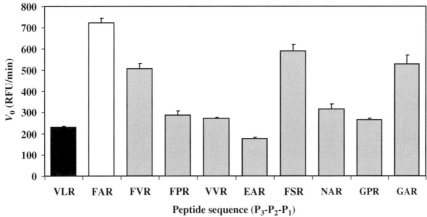

Figure 8. Identification of a surrogate substrate for a serine endopeptidase by screening a fluorogenic peptide library. (a) A library of 550 fluorogenic peptides was screened at 50 μM with a serine endopeptidase (EC 3.4.21) at 10 nM in 384 well/plate format. Nine primary hits (black) were identified above the arbitrary threshold of 20,000 RFU. (b) The fluorogenic peptides identified in primary screening as putative substrates of the serine endopeptidase were retested in the hit confirmation phase, in comparison with the reference peptide VLR, a generic substrate for this class of proteases (black bar). All the retrieved peptides displayed an obligate R residue in P_1, while hydrophobic residues were apparently preferred in P_2 and P_3 positions. The fluorogenic peptide FAR (white bar) was hydrolyzed 3.1-fold more efficiently than the generic substrate and was selected for HTS assay configuration (*Source:* Axxam Assays Database).

HTS. Although these criteria may vary considerably among different screening groups, a few of them are broadly agreed to distinguish between HTS-suitable and unsuitable assays [1,19,21].

Since the compounds of the chemical libraries are routinely dissolved in DMSO, the sensitivity of the assay to this organic solvent has to be assessed. In practice, assays should be demonstrated to be tolerant (i.e., recovery of at least 80% of the enzyme activity assayed in the absence of organic solvent) to a DMSO concentration ranging from 0.5% to 1.5% that corresponds to the concentration range contributed by the compound collection under screening conditions [1,19,21].

In addition, the assay should demonstrate a reproducible, dose-dependent response to a small panel of reference inhibitors active against the target [1,19,21]. Their respective IC_{50} values should comply with those reported in literature, although a certain degree of discrepancy may be tolerated if different types of assays have been used. Furthermore, the robustness of the assay can be assessed by screening a predefined array of pharmacological agents directed against known molecular targets [1,19,21]. This experiment indicates whether the assay is exceedingly responsive to off-target and nonspecific inhibition, a feature that may dramatically increase the rate of false-positive hits during primary screening.

To end, enzyme and substrate should be stable when diluted in the assay buffer at their working concentrations. The storage time for enzyme and substrate compatible with the detection of the signal (i.e., recovery of at least 80% of the initial activity) have to be experimentally determined by performing time–course incubation of either enzyme mix or substrate mix up to 24 h, and testing different storage temperature (typically, on ice, 4°C and room temperature).

Failure in achieving one of the above-mentioned criteria may make the high-throughput screening practically unfeasible. In this case, the assay has to be reoptimized to try to address its critical issues.

7.5. Assay Adaptation to HTS

Only a limited number of experiments are performed under automated, multiplate testing conditions throughout assay development. Rather, manual handling of limited number of replicates is normally enough during assay configuration. Therefore, the adaptation of the assay to the robotic unit requires dedicated efforts to ensure that the reliability and robustness of the assay is confirmed also under screening conditions.

A minimal set of experiments performed during assay adaptation comprises a multiplate test with maximum and minimum controls, to extrapolate the main statistical parameters of the assay, and a pilot screening with reference compounds or random compound plates, to estimate a tentative hit rate of primary screening [61].

In the multiplate test, a minimum of 10–20 plates is assembled with a predefined layout (in general, alternate quadrants), comprising maximum controls (MAX; enzymatic reactions without inhibitors), and minimum controls (MIN; enzymatic reactions incubated with a concentration of a reference compound causing complete inhibition). From this experiment, the main qualitative parameters of the assay can be calculated [61], including the signal-to-background ratio (S/B), an estimation of the dynamic range of the assay, as

$$\frac{S}{B} = \frac{\text{MAX}}{\text{MIN}}$$

and the signal-to-noise ratio (S/N), a measure of signal height relative to background noise [61] as

$$\frac{S}{N} = \frac{(\text{MAX} - \text{MIN})}{\text{SD}_{\text{MIN}}}$$

where SD_{MIN} is the standard deviation of MIN wells.

The third parameter is the coefficient of variation (CV), a measure of relative dispersion of the data [61], which is given as

$$\text{CV} = \frac{\text{SD}}{x_{\text{MEAN}}}$$

where x_{MEAN} is the mean of the entire data set.

Intuitively, the quality of a HTS assay increases by increasing the assay sensitivity (S/B), and by decreasing the data variability (SD, and consequently CV). Consistently, a unique "assay window coefficient", named Z', has been established to evaluate the quality of the assay on the basis of its MAX and MIN control values [61,62]:

$$Z' = \frac{1 - (3\text{SD}_{\text{MAX}} + 3\text{SD}_{\text{MIN}})}{\text{MAX} - \text{MIN}}$$

A $Z' = 1$ represents the upper limit of excellence, achievable only in the theoretical case that SD_{MAX} and SD_{MIN} are 0. Assays with

$1 > Z' \geq 0.5$ are considered suitable for generating high-quality data in HTS. On the contrary, assays with Z' values below 0.5 are generally unsuitable for HTS, especially when values approach 0 or become negative [61,62]. In the latter cases, a deep reoptimization of the assay is strictly required.

The statistical analysis of the assay adaptation is generally completed by calculating the intraplate, interplate, and day-to-day variability of the assay estimated with a multiplate test performed on at least three different days.

Pilot screening is an informative test to anticipate the performance of the assay during HTS. When an unbiased compound collection is used, the expected hit rate tentatively ranges between 0.1% and 1%. Hence, at least 10–20 plates in 384 well/plate format have to be tested, to derive a statistically significant analysis. To evaluate the quality of the assay in the presence of compounds, a "screening window index," termed Z, is applied [61,62], which is given as

$$Z = \frac{1-(3\mathrm{SD}_X + 3\mathrm{SD}_{\mathrm{MIN}})}{X-\mathrm{MIN}}$$

where SD_X is the standard deviation of samples incubated with compounds, and X is the mean sample response. Again, a $Z=1$ represents an ideal screening assay, $1 > Z \geq 0.5$ corresponds to an excellent assay, while $0.5 > Z > -1$ (being -1 the lowest practically achievable limit) indicates that different assay conditions, including possibly also compound concentration, have to be identified to obtain consistent data during screening [61,62].

It is important to note that, although Z and Z' factors are commonly used to indicate HTS data quality, they may not be adequate to ensure a sensitive enzyme inhibition assay during development (see Section 7.2). Indeed, these factors represent static parameters that may not reflect the capacity of an assay to detect inhibition [52].

8. ENZYME FAMILIES AND ASSAY TECHNOLOGIES

Enzyme targets currently investigated in drug discovery programs are dominated by protein kinases and proteases [4–7]. Indeed, multiple formats and technologies are available for the HTS assay configuration of these two subclasses [25,63]. For all the other enzymes, on the contrary, options are limited, and in some cases HTS requires that target-specific readout systems have to be designed and developed *de novo*, by integrating basic enzymology with knowledge of available reagents potentially exploitable for innovative purposes.

The general strategies adopted for the configuration of functional HTS assays for all enzyme classes—oxidoreductases, transferases, hydrolases, lyases, isomerases, and ligases—will be described in next sections. However, it is beyond the scope of this chapter to provide a detailed picture of all the possible readout systems suitable for HTS. Therefore, descriptive examples will be presented to illustrate the multiple approaches applied in HTS to detect the catalytic activity of several representative enzyme targets.

8.1. Oxidoreductases

Oxidoreductases (EC 1) catalyze the transfer of hydrogen atoms and electrons from a donor to an acceptor molecule [64]. This class accounts for more than 30% of the enzymes currently targeted by FDA-approved small-molecule inhibitors [6,7]. Among them, it is worth mentioning HMG-CoA reductase (EC 1.1.1.34), a key target in atherosclerosis [65]; ribonucleoside-diphosphate reductase (EC 1.17.4.1) and dihydrofolate reductase (EC 1.5.1.3), for cancer therapy [66,67]; monoamino oxidase (EC 1.4.3.4), a central target for depression [68]; cyclooxygenase-1, cyclooxygenase-2 (EC 1.14.99.1), and arachidonate 5-lipoxygenase (EC 1.13.11.34), targeted in inflammation [69,70].

For HTS assay configuration, NAD(P)H-dependent oxidoreductases (EC 1.6) have to be regarded as a distinct subclass with unique features. In fact, their catalytic activity can be detected with fluorescence-based assays by exploiting the spectral properties of NAD(P)H that emits at 460 nm when excited at 340 nm [19]. Despite this favorable feature, an accurate assay setup is required for NAD(P)H-dependent oxidoreductases, because of spontaneous weak oxidation of NAD(P)H in solution, its low quantum yield, and its near-UV spectral properties prone to be interfered by visible and UV-absorbing compounds [19]. For these reasons, direct, continuous assays based on NAD(P)H detection are frequently converted into indirect, discontinuous assays coupled to auxiliary enzymes to generate a more intense and stable signal (e.g., bacterial luciferase, resorufin derivatives) [19,34].

For all the other oxidoreductases, primary reactions should be coupled to reactive probes or, more frequently, to auxiliary enzymes to generate a HTS-compatible signal. Examples of reactive probes coupled to oxidoreductase-mediated reactions comprise fluorogenic lanthanide complexes for the detection of H_2O_2 [71], and 7-diethylamino-3-(4'-maleimidylphenyl)-4-methylcoumarin for measurement of free coenzyme A [26]. Conversely, coupled systems exploit auxiliary oxidoreductases that generate either a luminescent signal (e.g., luciferase) [19] or a fluorescent signal (e.g., horseradish peroxidase with resorufin-derivative substrates) [19]. In addition, some oxidoreductase reactions can be configured for HTS with SPA [27].

Besides NAD(P)H-dependent oxidoreductases, the subfamily that has been most intensively profiled with HTS-suitable assays is likely represented by cytochromes P450 [54,55]. This class of oxidoreductases contains highly promiscuous members that play a key role in xenobiotic and drug metabolism. For this reasons, alternative assay systems have been developed for cytochrome P450 activities, including direct, continuous assays with fluorogenic surrogate substrates [54], and discontinuous, coupled systems configured as luminescence-based assays [55].

A potential issue in oxidoreductase screening is the presence of compounds able to interfere with the enzyme-mediated reaction, due to their redox potential. Reducing or oxidizing agents, electron scavengers, and compounds with the ability to nonenzymatically transfer electrons in the presence of reducing agents can artifactually contribute to primary hits in HTS [72]. Although the identification of most of these compounds requires a dedicated characterization in secondary testing, rapid assays have been recently developed to profile entire compound collections and counterselect potential false-positive hits [72].

8.2. Transferases

Transferases (EC 2) catalyze the transfer of a functional group from a donor compound to an acceptor compound [64]. When the FDA-approved drugs targeting enzymes are considered, 23% of them are directed against transferases [6,7]. Besides protein kinases that likely represent the most abundant enzyme subgroup currently investigated in HTS, the FDA-validated targets comprise, among the others, farnesyl-diphosphate farnesyltransferase (EC 2.5.1.21), targeted in osteoporosis [73]; UDP-N-acetylglucosamine 1-carboxyvinyltransferase (EC 2.5.1.7), an antibacterial target [74]; and a complete panel of polymerases, including HIV reverse transcriptase (EC 2.7.7.49) [75], viral DNA-directed DNA polymerase (EC 2.7.7.7) [76], bacterial RNA-directed RNA polymerase (EC 2.7.7.48) [77], and bacterial DNA-directed RNA polymerase (EC 2.7.7.6) [78].

Available screening methods are unevenly distributed among transferases. While protein kinases can be configured in more than 15 alternative assay types [25], all the remaining transferases that do not transfer a phosphate group from ATP can be detected almost exclusively with either dedicated coupled systems or SPA assays [27,33]. For instance, fatty acid synthase (EC 2.3.1.85) was assayed using lipid-coated FlashPlates to capture the lipophylic radiolabeled product of the reaction [79]. On the contrary, polymerases can be configured as HTS assays using nucleoside triphosphate analogs that release a fluorescent signal upon their incorporation in RNA or DNA molecules [80].

Protein kinases (EC 2.7.10 and EC 2.7.11) catalyze the transfer of a phosphoryl group from ATP to hydroxyl groups of the amino acid side chains of serine, threonine, or tyrosine [64]. Kinase assays for HTS can be classified as ATP/ADP-detection assays and phosphopeptide-detection assays, according to the substrate/products pair that is revealed with the HTS-suitable readout. ATP/ADP-detection assays include coupled systems that use the residual ATP (through firefly luciferase) [34] or the produced ADP (through pyruvate kinase/lactate dehydrogenase) to generate either a luminescent signal or an NADH-dependent fluorescent signal, respectively [19,25]. Interestingly, while ATP consumption assay is a canonical indirect, discontinuous assay, ADP-production assay is an indirect, yet continuous assay [19,25,34]. Another approach in ATP/ADP-detection relies on an ADP-specific antibody to generate a fluorescence polarization competition assay in the presence of labeled ADP [81]. Being independent on the phosphorylated peptide, these detection systems have to be considered as generic assays applicable to any enzymatic reaction involving either ATP hydrolysis or ADP formation. The second type of assays reveals the kinase-mediated phosphopeptide formation and is dependent on detection systems that are almost exclusively antibody based or affinity binder based. A non-comprehensive list of antibody-based systems includes TR-FRET and HTRF [24,28,82], fluorescence polarization [29], Alphascreen [83], and electrochemiluminescence [84], while affinity-binder assays comprise SPA [27,30–32], IMAP [49], and fluorescence quenching [85]. In most cases, these indirect, discontinuous detection systems are extremely sensitive, with the ability to reveal a phosphorylation event in the low picomolar range [24,28]. An atypical HTS system is a mobility shift assay, in which substrates and phosphopeptide are electrophoretically separated in a dedicated microfluidic device [86]. A phosphopeptide-detection system that does not rely on antibodies or affinity binders is a coupled assay in which phosphorylation of a double-labeled peptide inhibits a protease-mediated cleavage, revealed as FRET signal [87]. To end, a peculiar detection system is represented by enzyme fragment complementation [88], in which a nonselective kinase inhibitor, staurosporine, is covalently linked to a defective portion of β-galactosidase. The kinase of interest, due to its affinity for staurosporine, is tethered to the defective portion of β-galactosidase, preventing the reconstitution of β-galactosidase enzymatic activity with a paired portion of the enzyme. In the presence of a competing inhibitor, the kinase dissociates from the β-galactosidase-staurosporine complex that in turn can associate with its counterpart and trigger the hydrolysis of fluorogenic or luminogenic β-galactosidase substrates.

8.3. Hydrolases

Hydrolases (EC 3) catalyze the hydrolysis of various chemical bonds and account for 35% of the enzyme targeted with a FDA-approved drug [6,7,64]. Among hydrolases, a leading role is played by proteases, although other target classes are significantly represented. Archetypal examples of non-protease hydrolases targeted by FDA-approved drug are pancreatic lipase (EC 3.1.1.3) for obesity [89]; histone deacetylase (EC 3.5.1.98), for cancer therapy [90]; β-lactamase (EC 3.5.2.6), an antibacterial target [91]; and two different phosphodiesterases (EC 3.1.4.17; EC 3.1.4.35), targeted for the treatment of asthma and erectile dysfunction, respectively [92,93]. Despite the relevance of these examples, the broadest opportunity for therapeutic intervention among all enzyme classes has been provided by proteases [6,7,57]. Hence, proteases are pharmacologically validated targets in different disease areas, such as bacterial infections (serine-type D-Ala-D-Ala carboxypeptidase [94], EC 3.4.16.4; folate hydrolase [95], EC 3.4.13.19), cancer (proteasome [96], EC 3.4.25.1), diabetes (dipeptidyl peptidase IV [97], EC 3.4.14.5), hypertension (angiotensin converting enzyme [98], EC 3.4.15.1; renin [99], EC 3.4.23.15), thrombosis (plasma kallikrein [100], EC 3.4.21.34; thrombin [101], EC 3.4.21.5; coagulation factor Xa [102], EC 3.4.21.6; plasminogen [103], EC 3.4.21.7), and AIDS (HIV-1 protease [104], EC 3.4.23.16).

Readout systems for hydrolase screening are dominated by the use of surrogate substrates, whose hydrolysis is revealed as a fluorescent signal [63]. In practice, physiological substrates are reduced to the minimal requirements for the catalytic activity of the target and labeled with fluorescence-emitting groups. For instance, in protease assays a short peptide containing the target-specific cleavage site is labeled either with a fluorogenic derivative or with a fluorescent donor, whose emission is quenched by an acceptor group (i.e., a FRET system) [24,28,63]. Upon protease cleavage, FRET is interrupted and the signal can be detected as an increase in fluorescence. These continuous, direct assays can be easily customized on the basis of the specificity of the different proteases, and usually have an excellent performance under HTS conditions [19,21]. Thus, this strategy has been widely applied also for other hydrolase members, including esterases, glycosidases, helicases, phospholipases, phosphatases, and triacylglycerol lipases [63,105–107]. However, the latter enzymes have also been configured with HTS assays alternative to FRET, with the aim of performing the screening campaign with the native substrates and to circumvent possible artifacts arising from compounds interfering with fluorescence detection. A representative example of this approach is endothelial lipase (EL; EC 3.1.1.3). Although fluorogenic substrates are available to reveal the EL activity [108], an alternative HTS enzymatic assay was established to detect the degradation of EL physiological substrate, HDL [35]. Hydrolysis of HDL by EL releases free fatty acids that are used as substrates by acyl-CoA synthetase in combination with coenzyme A to produce acyl-CoA. Acyl-CoA oxidase then oxidized acyl-CoA with the production of H_2O_2, which is revealed by horseradish peroxidase as a fluorescent signal [35]. Consistently, HTS assays alternative to FRET have also been developed for proteases. These assays include discontinuous, coupled systems, such as a luminescent assay based on protease-dependent maturation of proluminescent luciferase substrates [109], and enzyme fragment complementation assay, where a cyclic β-galactosidase conjugate unable to complement the paired portion of the enzyme is proteolytically converted into a complementation-proficient β-galactosidase fragment [88]. In addition, protease activity can be detected by indirect, discontinuous assays, such as SPA, fluorescence polarization and TR-FRET [27–29]. Usually all the HTS assays for proteases are developed with short peptide substrates that might inadequately mimic the structural complexity of the physiological protein substrate. However, FRET technology can be also applied to large protein domains. In this regard, an extreme example is represented by ADAMTS-13 (EC 3.4.24), which cleaves exclusively the unusually large forms of von Willebrand factor (ULF-VWF), a multimeric protein complex of 20,000 kDa. Nevertheless, also for ADAMTS-13 it was possible to design a FRET substrate, using the minimum cleavage region of the enzyme, a 73-amino acid sequence of VWF [110].

Besides proteases, another highly investigated hydrolase subgroup in drug discovery is phosphodiesterases (PDE; EC 3.1.4) [92,93]. HTS assays for PDE include, among the others, SPA, TR-FRET, IMAP, Alphascreen, and a dedicated coupled system generating a luminescent signal [27,28,111,112]. The latter assay detects the residual cAMP or cGMP after hydrolysis by phosphodiesterases. Before the reaction start (or in the presence of inhibitors), cAMP/cGMP binds and activates protein kinase A, which in turn phosphorylates a specific substrate, consuming ATP. Residual ATP concentration is detected by luciferase as a low luminescent signal. In noninhibited control samples, extensive hydrolysis of cAMP/cGMP induce the negative modulation of protein kinase A. Poorly active protein kinase A in turn causes the accumulation of high level of ATP, which is converted by luciferase in a strong luminescent signal. Thus, in this assay phosphodiesterase activity is directly proportional to luminescence emission [112].

8.4. Lyases

Lyases (EC 4) cleave chemical bonds with mechanisms that differ from oxidation and hydrolysis. In lyase reactions, two substrates are involved in the forward reaction, and only one in the backward reaction (or *vice versa*) [64]. Four lyases have been reported to be currently targeted by FDA-approved small-molecule inhibitors [6,7]. Three of them are decarboxylases: aromatic L-amino acid decarboxylase (EC 4.1.1.28), a target in Parkinson's disease [113]; ornithine decarboxylase (EC 4.1.1.17), for protozoan infection [114]; and orotidine-5'-phosphate decarboxylase (EC 4.1.1.23), for the treatment of gout [115]. The last validated lyase target is a dehydratase, α-carbonic anhydrase II (EC 4.2.1.1), for the treatment of glaucoma [116]. α-Carbonic anhydrase II catalyzes the reversible interconversion of carbon dioxide and bicarbonate [64]. HTS assays for α-carbonic anhydrase II involves the use of a synthetic surrogate substrate, 4-nitrophenyl acetate [117] that upon enzyme-mediated cleavage is detectable in absorbance.

Other remarkable examples of physiologically relevant lyases are adenylate cyclase (EC 4.6.1.1) and guanylate cyclase (EC 4.6.1.2) [64]. Assays for the detection of the catalytic activity of adenylate and guanylate cyclase are predominantly discontinuous antibody-based assays, including SPA, fluorescence polarization, Alphascreen, TR-FRET [27–29]. However, alternative assays have also been designed. Similarly to protein kinases and proteases, the activity of adenylate cyclase and guanylate cyclase can be detected by enzyme fragment complementation assay [88]. Moreover, coupled systems were also developed, in which the formation of the either product of the reaction, pyrophosphate, was coupled to nicotinamide mononucleotide adenylyltransferase and luciferase to generate a luminescent signal [118]. In addition, a direct, continuous assay has been described based on a GTP derivative covalently linked to a self-quenched fluorophore, whose fluorescence is released after cleavage by guanylate cyclase or by a mutant form of adenylate cyclase [119]. To end, reactive probes such as lanthanide complexes have been applied in direct, continuous assays to detect adenylate cyclase-dependent ATP conversion into cAMP [120].

8.5. Isomerases

Isomerases (EC 5) catalyze intramolecular rearrangements that convert the substrate in its isomer [64]. Four isomerases are targeted by FDA-approved drugs [6,7]: alanine racemase (EC 5.1.1.1), an antinfective target [121]; peptidylprolyl isomerase activity of cyclophilin A (EC 5.2.1.8), targeted by the immunosuppressant drug cyclosporin A [122]; and two topoisomerases, type I DNA topoisomerase (EC 5.99.1.2), for cancer therapy [123], and type II DNA topoisomerase (DNA gyrase; EC 5.99.1.3), a key antibacterial target [124].

There is no standard screening method for isomerases. Rather, different strategies have been devised based on the enzyme-specific substrate and product. Coupled systems have been successfully applied for cyclophilin A and for a potential antibacterial target, glutamate racemase. The forward reaction of glutamate racemase (EC 5.1.1.3), which catalyzes the conversion of D-glutamate into its L-enantiomer, is measured by coupling L-glutamate dehydrogenase, which in turn reduces NAD^+ to NADH [125]. On the contrary, glutamate racemase backward reaction from L- to D-glutamate can be detected by coupling UDP-N-acetylmuramoyl-L-alanine-D-glutamate ligase and purine nucleoside phosphorylase in the presence of 2-amino-6-mercapto-7-methylpurine ribonucleoside, to generate a signal detectable in absorbance [125]. Conversely, cyclophilin A catalyzes the isomerization of a noncleavable fluorogenic tetrapeptide containing a proline residue into its conformer that is recognized by chymotrypsin, whose activity is detected as an increase in fluorescence [126]. An alternative HTS-compatible detection system is represented by a direct, continuous assay based on cyclophilin A-dependent conversion of difluoroanilide-labeled tetrapeptide into a chromogenic molecule [127].

DNA topoisomerases are validated drug targets that are usually configured with cell-based assays for HTS [123,124]. However, also cell-free systems have been developed, to better investigate the mode of action of potential inhibitors. For instance, type I DNA topoisomerase activity has been configured either with SPA [128] or with a direct, continuous assay using modified oligonucleotides that mimic a reaction intermediate [129]. DNA ligation catalyzed by type I DNA topoisomerase on 3′-phospho-(para-nitrophenyl) oligonucleotides releases para-nitrophenol, a readily detectable chromogenic tracer.

To end, it is worth mentioning a subgroup of isomerases that are potential therapeutic targets for cardiovascular diseases and inflammatory response: thromboxane-A synthase (EC 5.3.99.5) and several prostaglandin synthases, namely, prostaglandin-E synthase (EC 5.3.99.3), prostaglandin-D synthase (EC 5.3.99.2), prostaglandin-A isomerase (EC 5.3.3.9), and prostaglandin-I synthase (EC 5.3.99.4) [64]. Most of these enzymes have been configured for HTS with highly sensitive antibody-based assays. However, also alternative assay systems have been developed. For instance, the fluorogenic reactive probe thiobarbituric acid has been applied to detect the formation of malondialdehyde, a reaction product of thromboxane-A synthase [130]. Moreover, as already described in Section 4, a noncanonical multistep screening based on enzyme immunoassay has been performed to identify potential inhibitors of prostaglandin-E synthase [22].

8.6. Ligases

Ligases (EC 6) catalyze the formation of covalent bonds between two molecules with the concomitant hydrolysis of a triphosphate compound [64]. Two ligases are validated targets with known FDA-approved inhibitors: D-alanine-D-alanine ligase (EC 6.3.2.4) [131] and isoleucine tRNA ligase (EC 6.1.1.5) [132], both playing an essential role in bacterial proliferation. As clearly anticipated by their mechanism of action, it is possible to develop generic systems for the HTS configuration of ligases, based on either ATP hydrolysis or pyrophosphate/ADP formation. These generic assays rely in most cases on coupled enzymes that generate a fluorescent or luminescent signal (see also Section 8.2) [19,25,34]. However, also customized HTS assay have been developed for ligases, to overcome possible limitations in terms of ATP concentration during screening and to emphasize the identification of competitive inhibitors of the enzyme-specific substrate. Examples of therapeutically relevant ligases configured with specific HTS assays comprise ubiquitin-protein ligase, DNA ligase, tRNA synthetase, and acetyl-coenzyme A carboxylase.

Ubiquitin-protein ligase (EC 6.3.2.19) is a potential target in cancer therapy and other degenerative diseases [133]. Ubiquitin-protein ligase assays have been developed based on electrochemiluminescence detection of both intramolecular (self-ubiquitination) and intermolecular ubiquitinations. In these assays, protein substrates are captured on the surface of electrode-equipped plates and ubiquitination events are detected as luminescent signals directly, by using electrochemiluminescent-labeled ubiquitin, or indirectly through labeled anti-polyubiquitin antibodies [134]. An alternative option for ubiquitin-protein ligase is represented by sensitive, antibody-based assays such as SPA or TR-FRET [27,135]. In the latter assay, the incorporation of ubiquitin labeled with europium chelate into a biotinylated protein substrate is detected as a fluorescent TR-FRET signal through a streptavidin-acceptor dye complex [135].

DNA ligase (EC 6.5.1.2), an essential enzyme for DNA replication, is an attractive target for new antibacterial agents. It catalyzes the formation of a phosphodiester at the site of a single-strand break in duplex DNA, through a ligase-AMP covalent intermediate unique to the bacterial DNA enzyme [136]. In this case, a SPA-based HTS assay was developed to detect the ligase-AMP formation, using biotinylated DNA ligase, radiolabeled NAD^+, and streptavidin-coated scintillant plates [136]. Also the activity of tRNA synthetases (tRNA ligases; EC 6.1.1) was configured as SPA assays, exploiting the properties of Yttrium-silicate beads, whose negatively charged surface binds tRNAs with highaffinity [137].

Inhibitors of acetyl-coenzyme A carboxylase (EC 6.4.1.2) are actively pursued as therapeutics for diabetes, obesity, and metabolic syndrome [138–140]. Acetyl-coenzyme A carboxylase produces malonyl-CoA, ADP, and inorganic phosphate from acetyl-CoA, ATP and HCO_3^-, offering multiple opportunities for assay configuration. Indeed, at least five different HTS-compatible coupled systems have been developed to detect the acetyl-coenzyme A carboxylase activity, based on either substrate consumption or product formation [138–140]. The multiple detection systems for acetyl-coenzyme A carboxylase highlight the remarkable potentialities—and the relative complexity—of certain enzymatic reactions to accommodate alternative HTS assay configurations, which may be applied not only in primary screening but also as valuable opportunities in secondary testing.

9. CONCLUSIONS

In drug discovery programs based on high-throughput screening, three factors are inherently correlated to downstream success: the choice of validated and druggable targets, the quality and size of the compound collection, and the assay setup. The main purpose of this chapter was to provide an overview of the strategic decisions and quality assessments that can be applied for the development of HTS enzymatic assays. To contribute to the successful outcome of a screening campaign, quality and perspective view should guide the decisions throughout the project. The choice of recombinant versions of the target and the expression–purification strategy are as important as the selection of the HTS-suitable readout and the setup of the kinetic and operating parameters for the assay configuration and adaptation. Basically, high-throughput screening and the assay development for HTS are synergistically interconnected, and the current commitment of pharmaceutical companies and academic institutions toward high-throughput screening as cardinal source of novel lead compounds will impose a constant evolution of the techniques and approaches applied in assay development that in turn are expected to increase potentialities and performance of HTS.

REFERENCES

1. Hüser J, Lohrmann E, Kalthof B, Burkhardt N, Brüeggemeier U, Bechem M. High-throughput screening for targeted lead discovery. In: Hüser J, Vol. editor. High-Throughput Screening in Drug Discovery. Mannhold R, Kubinyi H, Folker G (Ser. editors), Series: *Methods and Principles in Medicinal Chemistry*. Weinheim: Wiley-VCH Verlag Gmbh & Co. KGaA; 2006. p 15–34.
2. Verkman AS. Drug discovery in academia. Am J Physiol Cell Physiol 2004;286(3):C465–C474.
3. Comley J. Tools and technologies that facilitate automated screening. In: Hüser J, Vol. editor. High-Throughput Screening in Drug Discovery. Mannhold R, Kubinyi H, Folker G (Ser. editors), Series: *Methods and Principles in Medicinal Chemistry*. Weinheim: Wiley-VCH Verlag Gmbh & Co. KGaA; 2006. p 37–70.
4. Hopkins AL, Groom CR. The druggable genome. Nat Rev Drug Discov 2002;1:727–730.
5. Russ AP, Lampel S. The druggable genome: an update. Drug Discov Today 2005;10(23/24):1607–1610.
6. Robertson JG. Mechanistic basis of enzyme-targeted drugs. Biochemistry 2005;44:5561–5571.
7. Robertson JG. Enzymes as a special class of therapeutic target: clinical drugs and modes of action. Curr Opin Struct Biol 2007;17:1–6.
8. Grimsby J, Sarabu R, Corbett WL, Haynes NE, Bizzarro FT, Coffey JW, Guertin KR, Hilliard DW, Kester RF, Mahaney PE, Marcus L, Qi L, Spence CL, Tengi J, Magnuson MA, Chu CA, Dvorozniak MT, Matschinsky FM, Grippo JF. Allosteric activators of glucokinase: potential role in diabetes therapy. Science 2003;301:370–373.
9. Cool B, Zinker B, Chiou W, Kifle L, Cao N, Perham M, Dickinson R, Adler A, Gagne G, Iyengar R, Zhao G, Marsh K, Kym P, Jung P, Camp HS, Frevert E. Identification and characterization of a small molecule AMPK activator that treats key components of type 2 diabetes and the metabolic syndrome. Cell Metab 2006;3(6):403–416.
10. Arnau J, Lauritzen C, Petersen GE, Pedersen J. Current strategies for the use of affinity tags and tag removal for the purification of recombinant proteins. Protein Expr Purif 2006;48(1):1–13.
11. Payne DJ, Gwynn MN, Holmes DJ, Pompliano DL. Drugs for bad bugs: confronting the challenges of antibacterial discovery. Nat Rev Drug Discov 2007;6(1):29–40.
12. Leder L, Freuler F, Forstner M, Mayr LM. New methods for efficient protein production in drug discovery. Curr Opin Drug Discov Devel 2007;10(2):193–202.
13. Li P, Anumanthan A, Gao XG, Ilangovan K, Suzara VV, Düzgüneş N, Renugopalakrishnan V. Expression of recombinant proteins in *Pichia pastoris*. Appl Biochem Biotechnol 2007;142(2):105–124.
14. Murphy CI, Piwnica-Worms H. Overview of the baculovirus expression system. Curr Protoc Neurosci 2001;4:4–18.
15. Baldi L, Hacker DL, Adam M, Wurm FM. Recombinant protein production by large-scale transient gene expression in mammalian cells: state of the art and future perspectives. Biotechnol Lett 2007;29(5):677–684.
16. Hunt I. From gene to protein: a review of new and enabling technologies for multi-parallel protein expression. Protein Expr Purif 2005;40(1):1–22.

17. Redaelli L, Zolezzi F, Nardese V, Bellanti B, Wanke V, Carettoni D. A platform for high-throughput expression of recombinant human enzymes secreted by insect cells. J Biotechnol 2005;120(1):59–71.
18. Chapple SD, Crofts AM, Shadbolt SP, McCafferty J, Dyson MR. Multiplexed expression and screening for recombinant protein production in mammalian cells. BMC Biotechnol 2006; 6(49):1–15.
19. Mallender WD, Bembenek M, Dick LR, Kuranda M, Li P, Menon S, Pardo E, Parsons T. Biochemical assays for high-throughput screening. In: Hüser J, Vol. editor. High-Throughput Screening in Drug Discovery. Mannhold R, Kubinyi H, Folker G (Ser. editors), Series: *Methods and Principles in Medicinal Chemistry*. Weinheim: Wiley-VCH Verlag Gmbh & Co. KGaA; 2006. p 93–125.
20. Capelle MA, Gurny R, Arvinte T. High throughput screening of protein formulation stability: practical considerations. Eur J Pharm Biopharm 2007;65(2):131–148.
21. Goddard JP, Reymond JL. Recent advances in enzyme assays. Trends Biotechnol 2004;22 (7):363–370.
22. Massé F, Guiral S, Fortin LJ, Cauchon E, Ethier D, Guay J, Brideau C. An automated multistep high-throughput screening assay for the identification of lead inhibitors of the inducible enzyme mPGES-1. J Biomol Screen 2005;10(6):599–605.
23. Tipton KF. Principles of enzyme assay and kinetic studies. Eisenthal R, Danson M, editors. Enzyme Assays: A Practical Approach. New York: Oxford University Press; 2002. pp 1–53.
24. Bremer C. Optical methods. Semmler W, Schweiger M, editors. Molecular Imaging II, Handbook of Experimental Pharmacology. Berlin: Springer; 2008. pp 3–12.
25. Ishida A, Kameshita I, Sueyoshi N, Taniguchi T, Shigeri Y. Recent advances in technologies for analyzing protein kinases. J Pharmacol Sci 2007;103(1):5–11.
26. Chung CC, Ohwaki K, Schneeweis JE, Stec E, Varnerin JP, Goudreau PN, Chang A, Cassaday J, Yang L, Yamakawa T, Kornienko O, Hodder P, Inglese J, Ferrer M, Strulovici B, Kusunoki J, Tota MR, Takagi T. A fluorescence-based thiol quantification assay for ultra-high-throughput screening for inhibitors of coenzyme A production. Assay Drug Dev Technol 2008;6(3):361–374.
27. Wu S, Liu B. Application of scintillation proximity assay in drug discovery. BioDrugs 2005;19(6):383–392.
28. Wu P, Brand L. Resonance energy transfer: methods and applications. Anal Biochem 1994;218(1):1–13.
29. Burke TJ, Loniello KR, Beebe JA, Ervin KM. Development and application of fluorescence polarization assays in drug discovery. Comb Chem High Throughput Screen 2003;6 (3):183–194.
30. Hart HE. Scintillation proximity assay. US Patent 4,271,139. 1981; June 2.
31. Cook ND, Jessop RA, Robinson PS. Scintillation proximity enzyme assay. A rapid and novel assay technique applied to HIV proteinase. Adv Exp Med Biol 1991;306:525–528.
32. Glickman JF, Schmid A, Ferrand S. Scintillation proximity assays in high-throughput screening. Assay Drug Dev Technol 2008;6 (3):433–455.
33. Rudolph FB, Baugher BW, Beissner RS. Techniques in coupled enzyme assays. Methods Enzymol 1979;63:22–42.
34. Ford SR, Leach FR. Improvements in the application of firefly luciferase assays. LaRossa RA, editor. Bioluminescence Methods and Protocols. Totowa, NJ: Humana Press; 1998; pp 3–20.
35. Keller PM, Rust T, Murphy DJ, Matico R, Trill JJ, Krawiec JA, Jurewicz A, Jaye M, Harpel M, Thrall S, Schwartz B. A high-throughput screen for endothelial lipase using HDL as substrate. J Biomol Screen 2008;13 (6):468–475.
36. Wegener D, Hildmann C, Riester D, Schwienhorst A. Improved fluorogenic histone deacetylase assay for high-throughput-screening applications. Anal Biochem 2003;321 (2):202–208.
37. Tipton KF, Dixon HB. Effect of pH on enzymes. Methods Enzymol 1979;63:183–233.
38. Laidler KJ, Peterman BF. Temperature effects in enzyme kinetics. Methods Enzymol 1979;63:234–256.
39. Segel IH. Enzyme Kinetics: Behavior and Analysis of Rapid Equilibrium and Steady-State Enzyme Systems. New York: Wiley-Interscience; 1993.
40. Cook PF, Cleland WW. Enzyme Kinetics and Mechanism. New York: Garland Science. 2007; pp 1–416.
41. Ohkuma S, Poole B. Fluorescence probe measurement of the intralysosomal pH in living

cells and the perturbation of pH by various agents. Proc Natl Acad Sci USA 1978;75(7):3327–3331.

42. Stevens L. Buffers and the determination of protein concentrations. Eisenthal R, Danson M, editors. Enzyme Assays: A Practical Approach. New York: Oxford University Press; 2002; pp 317–336.

43. Morrison JF. Approaches to kinetic studies on metal-activated enzymes. Methods Enzymol 1979;63:257–293.

44. Helenius A, McCaslin DR, Fries E, Tanford C. Properties of detergents. Methods Enzymol 1979;56:734–749.

45. Yamamoto E, Yamaguchi S, Nagamune T. Effect of β-cyclodextrin on the renaturation of enzymes after sodium dodecyl sulfate–polyacrylamide gel electrophoresis. Anal Biochem 2008;381(2):273–275.

46. Chong Y, Chen H. Preparation of functional recombinant proteins from *E. coli* using a nondetergent sulfobetaine. Biotechniques 2000;29(6):1166–1167.

47. Parveen S, Sahoo SK. Nanomedicine: clinical applications of polyethylene glycol conjugated proteins and drugs. Clin Pharmacokinet 2006;45(10):965–988.

48. Koch-Weser J, Sellers EM. Binding of drugs to serum albumin. N Engl J Med 1976;294(6):311–316.

49. Sharlow ER, Leimgruber S, Yellow-Duke A, Barrett R, Wang QJ, Lazo JS. Development, validation and implementation of immobilized metal affinity for phosphochemicals (IMAP)-based high-throughput screening assays for low-molecular-weight compound libraries. Nat Protoc 2008;3(8):1350–1363.

50. Owicki JC. 8th Annual Conference of Society Biomolecular Screening, The Hague, The Netherlands, 2002.

51. Gutierrez OA, Chavez M, Lissi EA. Theoretical approach to some analytical properties of heterogeneous enzymatic assays. Anal Chem 2004;76:2664–2668.

52. Gutierrez OA, Danielson UH. Detection of competitive enzyme inhibition with end point progress curve data. Anal Biochem 2006;358(1):11–19.

53. Copley SD. Enzymes with extra talents: moonlighting functions and catalytic promiscuity. Curr Opin Chem Biol 2003;7:265–272.

54. Trubetskoy OV, Gibson JR, Marks BD. Highly miniaturized formats for *in vitro* drug metabolism assays using Vivid® fluorescent substrates and recombinant human cytochrome P450 enzymes. J Biomol Screen 2005; 10:56–66.

55. Cali JJ, Ma D, Sobol M, Simpson DJ, Frackman S, Good TD, Daily WJ, Liu D. Luminogenic cytochrome P450 assays. Expert Opin Drug Metab Toxicol 2006;2(4):629–645.

56. Tenne M, Finberg JP, Youdim MB, Ulitzur S. A new rapid and sensitive bioluminescence assay for monoamine oxidase activity. J Neurochem 1985;44(5):1378–1384.

57. Barrett AJ, Rawlings ND, Woessner J. Handbook of Proteolytic Enzymes. London: Academic Press; 1998.

58. Schmidt M, Bornscheuer UT. High-throughput assays for lipases and esterases. Biomol Eng 2005;22:51–56.

59. Montalibet J, Skorey KI, Kennedy BP. Protein tyrosine phosphatase: enzymatic assays. Methods 2005;35(1):2–8.

60. Thomas DA, Francis P, Smith C, Ratcliffe S, Ede NJ, Kay C, Wayne G, Martin SL, Moore K, Amour A, Hooper NM. A broad-spectrum fluorescence-based peptide library for the rapid identification of protease substrates. Proteomics 2006;6(7):2112–2120.

61. Gubler H. Methods for statistical analysis, quality assurance and management of primary high-throughput screening data. In: Hüser J, Vol. editor. High-Throughput Screening in Drug Discovery. Mannhold R, Kubinyi H, Folker G (Ser. editors), Series: *Methods and Principles in Medicinal Chemistry*. Weinheim: Wiley-VCH Verlag Gmbh & Co. KGaA; 2006. p 151–201.

62. Zhang JH, Chung TD, Oldenburg KR. A simple statistical parameter for use in evaluation and validation of high throughput screening assays. J Biomol Screen 1999;4(2):67–73.

63. Mugherli L, Burchak ON, Chatelain F, Balakirev MY. Fluorogenic ester substrates to assess proteolytic activity. Bioorg Med Chem Lett 2006;16(17):4488–4491.

64. Fleischmann A, Darsow M, Degtyarenko K, Fleischmann W, Boyce S, Axelsen KB, Bairoch A, Schomburg D, Tipton KF, Apweiler R. IntEnz, the integrated relational enzyme database. Nucleic Acids Res 2004;32:D434–D437.

65. Istvan ES. Structural mechanism for statin inhibition of 3-hydroxy-3-methylglutaryl coenzyme A reductase. Am Heart J 2002;144:S27–S32.

66. Cerqueira NM, Fernandes PA, Ramos MJ. Ribonucleotide reductase: a critical enzyme for

cancer chemotherapy and antiviral agents. Recent Pat Anticancer Drug Discov 2007;2(1):11–29.
67. Williams JW, Morrison JF, Duggleby RG. Methotrexate, a high-affinity pseudosubstrate of dihydrofolate reductase. Biochemistry 1979;18:2567–2573.
68. Mallinger AG, Smith E. Pharmacokinetics of monoamine oxidase inhibitors. Psychopharmacol Bull 1991;27:493–502.
69. Gierse JK, McDonald JJ, Hauser SD, Rangwala SH, Koboldt CM, Seibert K. A single amino acid difference between cyclooxygenase-1 (COX-1) and -2 (COX-2) reverses the selectivity of COX-2 specific inhibitors. J Biol Chem 1996;271:15810–15814.
70. Bell RL, Young PR, Albert D, Lanni C, Summers JB, Brooks DW, Rubin P, Carter GW. The discovery and development of zileuton: an orally active 5-lipoxygenase inhibitor. Int J Immunopharmacol 1992;14:505–510.
71. Wu M, Lin Z, Schäferling M, Dürkop A, Wolfbeis OS. Fluorescence imaging of the activity of glucose oxidase using a hydrogen-peroxide-sensitive europium probe. Anal Biochem 2005;340(1):66–73.
72. Johnston PA, Soares KM, Shinde SN, Foster CA, Shun TY, Takyi HK, Wipf P, Lazo JS. Development of a 384-well colorimetric assay to quantify hydrogen peroxide generated by the redox cycling of compounds in the presence of reducing agents. Assay Drug Dev Technol 2008;6(4):505–518.
73. Bergstrom JD, Bostedor RG, Masarachia PJ, Reszka AA, Rodan G. Alendronate is a specific, nanomolar inhibitor of farnesyl diphosphate synthase. Arch Biochem Biophys 2000;373:231–241.
74. Skarzynski T, Mistry A, Wonacott A, Hutchinson SE, Kelly VA, Duncan K. Structure of UDP-N-acetylglucosamine enolpyruvyl transferase, an enzyme essential for the synthesis of bacterial peptidoglycan, complexed with substrate UDP-N-acetylglucosamine and the drug fosfomycin. Structure 1996;4:1465–1474.
75. Sluis-Cremer N, Temiz NA, Bahar I. Conformational changes in HIV-1 reverse transcriptase induced by nonnucleoside reverse transcriptase inhibitor binding. Curr HIV Res 2004;2:323–332.
76. Villarreal EC. Current and potential therapies for the treatment of herpesvirus infections. Prog Drug Res 2001; Spec. No: 185-228.
77. Maag D, Castro C, Hong Z, Cameron CE. Hepatitis C virus RNA-dependent RNA polymerase (NS5B) as a mediator of the antiviral activity of ribavirin. J Biol Chem 2001;276:46094–46098.
78. Campbell EA, Korzheva N, Mustaev A, Murakami K, Nair S, Goldfarb A, Darst SA. Structural mechanism for rifampicin inhibition of bacterial RNA polymerase. Cell 2001;104:901–912.
79. Weiss DR, Glickman JF. Characterization of fatty acid synthase activity using scintillation proximity. Assay Drug Dev Technol 2003;1:161–166.
80. Summerer D. DNA polymerase profiling. Methods Mol Biol 2008;429:225–235.
81. Kleman-Leyer KM, Klink TA, Kopp AL, Westermeyer TA, Koeff MD, Larson BR, Worzella TJ, Pinchard CA, van de Kar SA, Zaman GJ, Hornberg JJ, Lowery RG. Characterization and Optimization of a Red-Shifted Fluorescence Polarization ADP Detection Assay. Assay Drug Dev Technol 2009; Feb 2.
82. Jia Y, Quinn CM, Gagnon AI, Talanian R. Homogeneous time-resolved fluorescence and its applications for kinase assays in drug discovery. Anal Biochem 2006;356(2):273–281.
83. Von Leoprechting A, Kumpf R, Menzel S, Reulle D, Griebel R, Valler MJ, Büttner FH. Miniaturization and validation of a high-throughput serine kinase assay using the AlphaScreen® platform. J Biomol Screen 2004;9(8):719–725.
84. Xiao SH, Farrelly E, Anzola J, Crawford D, Jiao X, Liu J, Ayres M, Li S, Huang L, Sharma R, Kayser F, Wesche H, Young SW. An ultra-sensitive high-throughput electrochemiluminescence immunoassay for the Cdc42-associated protein tyrosine kinase ACK1. Anal Biochem 2007;367(2):179–189.
85. Morgan AG, McCauley TJ, Stanaitis ML, Mathrubutham M, Millis SZ. Development and validation of a fluorescence technology for both primary and secondary screening of kinases that facilitates compound selectivity and site-specific inhibitor determination. Assay Drug Dev Technol 2004;2(2):171–181.
86. Perrin D, Martin T, Cambet Y, Frémaux C, Scheer A. Overcoming the hurdle of fluorescent compounds in kinase screening: a case study. Assay Drug Dev Technol 2006;4(2):185–196.
87. Rodems SM, Hamman BD, Lin C, Zhao J, Shah S, Heidary D, Makings L, Stack JH, Pollok BA. A FRET-based assay platform for ultra-high density drug screening of protein kinases and phosphatases. Assay Drug Dev Technol 2002;1 (1 Pt 1):9–19.

88. Eglen RM. Enzyme fragment complementation: a flexible high throughput screening assay technology. Assay Drug Dev Technol 2002;1(1 Pt 1):97–104.
89. Nelson RH, Miles JM. The use of orlistat in the treatment of obesity, dyslipidaemia and type 2 diabetes. Expert Opin Pharmacother 2005;6(14):2483–2491.
90. Liu T, Kuljaca Sa, Tee A, Marshall GM. Histone deacetylase inhibitors: multifunctional anticancer agents. Cancer Treat Rev 2006;32:157–165.
91. Livermore DM. Determinants of the activity of β-lactamase inhibitor combinations. J Antimicrob Chemother 1993;31(Suppl A):9–21.
92. Kodimuthali A, Jabaris SS, Pal M. Recent advances on phosphodiesterase 4 inhibitors for the treatment of asthma and chronic obstructive pulmonary disease. J Med Chem 2008;51(18):5471–5489.
93. Corbin JD, Francis SH. Cyclic GMP phosphodiesterase-5: target of sildenafil. J Biol Chem 1999;274:13729–13732.
94. Lee W, McDonough MA, Kotra L, Li ZH, Silvaggi NR, Takeda Y, Kelly JA, Mobashery S. A 1.2-Å snapshot of the final step of bacterial cell wall biosynthesis. Proc Natl Acad Sci USA 2001;98:1427–1431.
95. Campbell BJ, Forrester LJ, Zahler WL, Burks M. β-Lactamase activity of purified and partially characterized human renal dipeptidase. J Biol Chem 1984;259:14586–14590.
96. Adams J, Palombella VJ, Sausville EA, Johnson J, Destree A, Lazarus DD, Maas J, Pien CS, Prakash S, Elliott PJ. Proteasome inhibitors: a novel class of potent and effective antitumor agents. Cancer Res 1999;59:2615–2622.
97. Peters JU. 11 Years of cyanopyrrolidines as DPP-IV inhibitors. Curr Top Med Chem 2007;7:579–595.
98. Cushman DW, Cheung HS, Sabo EF, Ondetti MA. Development and design of specific inhibitors of angiotensin-converting enzyme. Am J Cardiol 1982;49:1390–1394.
99. Gradman AH, Schmieder RE, Lins RL, Nussberger J, Chiang Y, Bedigian MP. Aliskiren, a novel orally effective renin inhibitor, provides dose-dependent antihypertensive efficacy and placebo-like tolerability in hypertensive patients. Circulation 2005;111:1012–1018.
100. Scott CF, Wenzel HR, Tschesche HR, Colman RW. Kinetics of inhibition of human plasma kallikrein by a site-specific modified inhibitor Arg15-aprotinin: evaluation using a microplate system and comparison with other proteases. Blood 1987;69:1431–1436.
101. Markus G, DePasquale JL, Wissler FC. Quantitative determination of the binding of ε-aminocaproic acid to native plasminogen. J Biol Chem 1978;253:727–732.
102. Keam SJ, Goa KL. Fondaparinux sodium. Drugs 2002;62:1673–1685.
103. Rydel TJ, Ravichandran KG, Tulinsky A, Bode W, Huber R, Roitsch C, Fenton JWII. The structure of a complex of recombinant hirudin and human R-thrombin. Science 1990;249:277–280.
104. Rodriguez-Barrios F, Gago F. HIV protease inhibition: limited recent progress and advances in understanding current pitfalls. Curr Top Med Chem 2004;4:991–1007.
105. Shulman ML, Kulshin VA, Khorlin AY. A continuous fluorimetric assay for glycosidase activity: human N-acetyl-beta-D-hexosaminidase. Anal Biochem 1980;101(2):342–348.
106. Boguszewska-Chachulska AM, Krawczyk M, Stankiewicz A, Gozdek A, Haenni AL, Strokovskaya L. Direct fluorometric measurement of hepatitis C virus helicase activity. FEBS Lett 2004;567(2–3):253–258.
107. Hong SB, Lubben TH, Dolliver CM, Petrolonis AJ, Roy RA, Li Z, Parsons TF, Li P, Xu H, Reilly RM, Trevillyan JM, Nichols AJ, Tummino PJ, Gant TG. Expression, purification, and enzymatic characterization of the dual specificity mitogen-activated protein kinase phosphatase, MKP-4. Bioorg Chem 2005;33(1):34–44.
108. Mitnaul LJ, Tian J, Burton C, Lam MH, Zhu Y, Olson SH, Schneeweis JE, Zuck P, Pandit S, Anderson M, Maletic MM, Waddell ST, Wright SD, Sparrow CP, Lund EG. Fluorogenic substrates for high-throughput measurements of endothelial lipase activity. J Lipid Res 2007;48(2):472–482.
109. Liu JJ, Wang W, Dicker DT, El-Deiry WS. Bioluminescent imaging of TRAIL-induced apoptosis through detection of caspase activation following cleavage of DEVD-aminoluciferin. Cancer Biol Ther 2005;4(8):885–892.
110. Kokame K, Nobe Y, Kokubo Y, Okayama A, Miyata T. FRETS-VWF73, a first fluorogenic substrate for ADAMTS13 assay. Br J Haematol 2005;129(1):93–100.
111. Marchand C, Lea WA, Jadhav A, Dexheimer TS, Austin CP, Inglese J, Pommier Y, Simeonov A. Identification of phosphotyrosine mimetic inhibitors of human tyrosyl-DNA phosphodiesterase I by a novel AlphaScreen®

high-throughput assay. Mol Cancer Ther 2009;8(1):240–248.
112. Worzella T, Hsiao K, Goueli1 S, Grevelis H, Ahrweiler P. An automated bioluminescent assay for measuring phosphodiesterase activity. 2007; http://www.promega.com/scientific_posters/ps007/ps007.pdf.
113. Burkhard P, Dominici P, Borri-Voltattorni C, Jansonius JN, Malashkevich VN. Structural insight into Parkinson's disease treatment from drug-inhibited DOPA decarboxylase. Nat Struct Biol 2001;8:963–967.
114. Bacchi CJ, Nathan HC, Hutner SH, McCann PP, Sjoerdsma A. Polyamine metabolism: a potential therapeutic target in trypanosomes. Science 1980;210:332–334.
115. Fyfe JA, Miller RL, Krenitsky TA. Kinetic properties and inhibition of orotidine 5′-phosphate decarboxylase. Effects of some allopurinol metabolites on the enzyme. J Biol Chem 1973;248:3801–3809.
116. Wells JW, Kandel SI, Kandel M, Gornall AG. The esterase activity of bovine carbonic anhydrase B above pH 9. Reversible and covalent inhibition by acetozolamide. J Biol Chem 1975;250:3522–3530.
117. Iyer R, Barrese AA, Parakh S, Parker CN, Tripp BC. Inhibition profiling of human carbonic anhydrase II by high-throughput screening of structurally diverse, biologically active compounds. J Biomol Screen 2006; 11(7):782–791.
118. Graeff R, Lee HC. A novel cycling assay for nicotinic acid-adenine dinucleotide phosphate with nanomolar sensitivity. Biochem J2002;367(Pt 1):163–168.
119. Vadakkadathmeethal K, Cunliffe JM, Swift J, Kennedy RT, Neubig RR, Sunahara RK. Fluorescence-based adenylyl cyclase assay adaptable to high throughput screening. Comb Chem High Throughput Screen 2007; 10(4):289–298.
120. Spangler CM, Spangler C, Göttle M, Shen Y, Tang WJ, Seifert R, Schäferling M. A fluorimetric assay for real-time monitoring of adenylyl cyclase activity based on terbium norfloxacin. Anal Biochem 2008;381(1):86–93.
121. Mustata GI, Soares TA, Briggs JM. Molecular dynamics studies of alanine racemase: a structural model for drug design. Biopolymers 2003;70:186–200.
122. Ke H, Mayrose D, Belshaw PJ, Alberg DG, Schreiber SL, Chang ZY, Etzkorn FA, Ho S, Walsh CT. Crystal structures of cyclophilin A complexed with cyclosporin A and N-methyl-4-[(E)-2-butenyl]-4,4-dimethylthreonine cyclosporin A. Structure 1994;2(1):33–44.
123. Arun B, Frenkel EP. Topoisomerase I inhibition with topotecan: pharmacologic and clinical issues. Expert Opin Pharmacother 2001;2:491–505.
124. Anderson VE, Osheroff N. Type II topoisomerases as targets for quinolone antibacterials: turning Dr. Jekyll into Mr. Hyde. Curr Pharm Des 2001;7:337–353.
125. Lundqvist T, Fisher SL, Kern G, Folmer RH, Xue Y, Newton DT, Keating TA, Alm RA, de Jonge BL. Exploitation of structural and regulatory diversity in glutamate racemases. Nature 2007;447(7146):817–822.
126. Janowski B, Wöllner S, Schutkowski M, Fischer G. A protease-free assay for peptidyl prolyl cis/trans isomerases using standard peptide substrates. Anal Biochem 1997;252: 299–307.
127. Kofron JL, Kuzmic P, Kishore V, Colón-Bonilla E, Rich DH. Determination of kinetic constants for peptidyl prolyl cis–trans isomerases by an improved spectrophotometric assay. Biochemistry 1991;30:6127–6134.
128. Lerner CG, Chiang Saiki AY, Mackinnon AC. High throughput screening for inhibitors of bacterial DNA topoisomerase 1 using the scintillation proximity assay. J Biomol Screening 1996;1(3):135–143.
129. Woodfield G, Cheng C, Shuman S, Burgin AB. Vaccinia topoisomerase and Cre recombinase catalyze direct ligation of activated DNA substrates containing a 3′-para-nitrophenyl phosphate ester. Nucleic Acids Res 2000;28 (17):3323–3331.
130. Ledergerber D, Hartmann RW. Development of a screening assay for the in vitro evaluation of thromboxane A2 synthase inhibitors. Enzyme Inhib 1995;9(4):253–261.
131. Fan C, Park IS, Walsh CT, Knox JR. D-Alanine: D-alanine ligase: phosphonate and phosphinate intermediates with wild type and the Y216F mutant. Biochemistry 1997; 36: 2531–2538.
132. Brown MJ, Mensah LM, Doyle ML, Broom NJ, Osbourne N, Forrest AK, Richardson CM, O'Hanlon PJ, Pope AJ. Rational design of femtomolar inhibitors of isoleucyl tRNA synthetase from a binding model for pseudomonic acid-A. Biochemistry 2000;39:6003–6011.
133. Ostrowska H. The ubiquitin-proteasome system: a novel target for anticancer and anti-inflammatory drug research. Cell Mol Biol Lett 2008;13(3):353–365.

134. Kenten JH, Davydov IV, Safiran YJ, Stewart DH, Oberoi P, Biebuyck HA. Assays for high-throughput screening of E2 AND E3 ubiquitin ligases. Methods Enzymol 2005; 399:682–701.
135. Murray MF, Jurewicz AJ, Martin JD, Ho TF, Zhang H, Johanson KO, Kirkpatrick RB, Ma J, Lor LA, Thrall SH, Schwartz B. A high-throughput screen measuring ubiquitination of p53 by human mdm2. J Biomol Screen 2007;12(8):1050–1058.
136. Miesel L, Kravec C, Xin AT, McMonagle P, Ma S, Pichardo J, Feld B, Barrabee E, Palermo R. A high-throughput assay for the adenylation reaction of bacterial DNA ligase. Anal Biochem 2007;366(1):9–17.
137. Macarrón R, Mensah L, Cid C, Carranza C, Benson N, Pope AJ, Díez E. A homogeneous method to measure aminoacyl-tRNA synthetase aminoacylation activity using scintillation proximity assay technology. Anal Biochem 2000;284(2):183–190.
138. Liu Y, Zalameda L, Kim KW, Wang M, McCarter JD. Discovery of acetyl-coenzyme A carboxylase 2 inhibitors: comparison of a fluorescence intensity-based phosphate assay and a fluorescence polarization-based ADP Assay for high-throughput screening. Assay Drug Dev Technol 2007;5(2):225–235.
139. Santoro N, Brtva T, Roest SV, Siegel K, Waldrop GL. A high-throughput screening assay for the carboxyltransferase subunit of acetyl-CoA carboxylase. Anal Biochem 2006;354 (1):70–77.
140. Webb MR. A continuous spectrophotometric assay for inorganic phosphate and for measuring phosphate release kinetics in biological systems. Proc Natl Acad Sci USA 1992; 89:4884–4887.

CRYSTALLOGRAPHIC SURVEY OF ALBUMIN DRUG INTERACTION AND PRELIMINARY APPLICATIONS IN CANCER CHEMOTHERAPY

Daniel C. Carter
New Century Pharmaceuticals, Inc.,
Huntsville, AL

1. INTRODUCTION

Several proteins in the circulatory system have been determined to be important effectors of drug disposition. These proteins include, serum albumin, alpha-1-acid glycoprotein and to a lesser extent, various lipoproteins. Chief among the plasma proteins is albumin, which as the most abundant protein of the circulatory and lymphatic system, is well known as the principal determinant of drug disposition and delivery. Synthesized by the liver and secreted into the circulatory system, albumin supports the livers role by sequestering and off loading an immense variety of endogenous and exogenous ligands for metabolism. Endogenous ligands include fatty acids, vitamins, hormones and metabolic products, such as bilirubin. Generally, albumin has a preferential affinity for small molecular weight anionic and hydrophobic compounds. This specialized role in binding and transport of small molecules has a significant impact on drug pharmacokinetics. Drug affinity and binding location to albumin can significantly alter the half-life, distribution, and metabolism of drugs, thereby playing a major role in the ADME (absorption, distribution, metabolism, and excretion) of many important pharmaceuticals. At high plasma concentrations of 30–50 mg/mL, albumin contributes 80% to colloidal osmotic blood pressure and to maintaining the pH of the blood. In addition to its principal role in the circulatory system, more than 40% of the total albumin is extravascular, where it is found in every organ and bodily secretion. The structure of albumin and the chemical basis for the molecules of unusual ligand-binding properties were revealed by the successful crystallographic determination at low resolution in 1989 [1,2] that was completed at atomic resolution in 1992 [3]. Because of albumin's low cost, availability in purified form and propensity to bind reversibly an immense variety of endogenous and exogenous ligands, it has historically been one of most studied and applied proteins in biochemistry. Still, as it will become apparent in this chapter, much continues to be learned about this fascinating and important protein. Structurally, it is a large protein of 585 amino acids (66,500 Da,) comprised of three structurally homologous repeating domains, denoted I, II, and III, each of which in turn is comprised of two subdomains (IA, IB; IIA, IIB; and IIIA, IIIB) (Fig. 1) [3,4]. Albumin is unusual among the plasma proteins in its absence of glycosylation and possessing a highly activated sulphydryl (Cys 34) that reacts with cysteine, glutathione, and a variety of metals and other substances. The mystery surrounding albumin's unusually long half-life in circulation (18 days), a property alone that has been the subject of major and successful therapeutic development programs, has been elucidated in recent years by the important work of Anderson and colleagues who determined that plasma retention is through the MHC (FcRn) receptor mechanism, with the albumin-binding interaction similar to, but distinct from, IgG [5]. For an excellent overview of albumin biochemistry and applications, the reader is referred to the book, *All about Albumin: Biochemistry, Genetics, and Medical Applications*, by Theodore Peters [6].

Albumin's abundance in the circulatory system and its affinity for anionic and hydrophobic compounds, often (95–98% bound), often leads to increased dosing requirements to achieve the desired pharmacological effect, making it a significant factor in the safety and efficacy for many pharmaceuticals. This is particularly problematic for drugs that are highly cytotoxic (or neurologically active) and where the concentrations required for efficacy create narrow therapeutic indices. Higher dosings of cytotoxic drugs, particularly antineoplastic agents, contribute to dose-limiting secondary toxicities, including life-threatening cardio toxicity in some cases, diarrhea, and the longer term issues of potential secondary cancer. These inherent properties of albumin make it an attractive target for improving

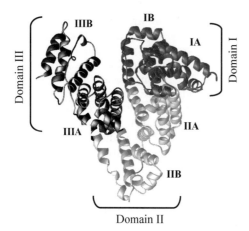

Figure 1. A ribbon diagram illustrating the overall topology of human serum albumin and its domain and subdomain structure.

the drug development process through the detailed understanding of the chemistry of the drug interaction at atomic resolution. This information has immense application throughout the current pharmacopoeia and can guide the synthesis of new analogs with improved ADME profiles ultimately resulting in safer more efficacious drugs.

The purpose of this chapter is to summarize a comprehensive multiyear, multimillion-dollar commercial program conducted during 2000–2004 to map the endogenous drug transport system at atomic resolution for purposes which included the development of safer more efficacious drugs [7]. Here, we describe an overview of the results of this unprecedented albumin structural survey and present some of the immediate and direct applications of this information in improving the therapeutic performance for several important families of oncology drugs.

2. RESULTS FROM THE SERUM ALBUMIN STRUCTURAL SURVEY

2.1. Experimental Approach

A systematic approach was developed to address a number of variables that limited the reproducible growth and quality of human albumin crystals suitable for X-ray structural studies. Crystal quality issues have plagued albumin X-ray structural studies since the first serious efforts in the early 1970s. The development of reproducible systems to grow new or improved high-resolution crystal forms was thus a high priority. We addressed this issue by securing and analyzing protein from a variety of sources, including our own internal program in recombinant production of human albumin, ultimately yielding standardized purification protocols and reproducible systems for manifold crystal forms. Secondly, and perhaps equally critical were the methods developed to cryogenically preserve and archive albumin crystals which were not reproducibly amenable to standard protocols. Successful cryogenics allowed in-house prescreening and the accumulation and banking of hundreds of crystals for later evaluation at synchrotron facilities.

Additionally, chemical libraries were established of current pharmaceuticals, drug-like molecules, and natural ligands. Analysis of over 1000 drug or drug-like molecules based on literature reports of high plasma binding and/or high affinity to serum albumin, resulted in the acquisition of 350 targeted compounds which included representatives of every major therapeutic indication. In order to improve on the success ratio of cocrystallization of the targeted ligands with albumin, albumin-binding properties were completely reassessed on a common platform. This uniformity in methodology allowed for the preselection of ligands with the requisite albumin affinities necessary to visualize the drug interaction through crystallographic methods. The prequalified compounds were combined with a series of well established albumin preparations and placed through a series of proprietary crystallization screens that routinely produced multiple crystal forms of each potential drug/ligand target. It was not always possible to identify which crystals would ultimately yield a drug complex, consequently, a high percentage of the crystals produced through the screening process represented the native unligated structure. All crystals were prescreened for quality and other characteristics using an in-house Rigaku RU200 equipped with an Osmic Blue mirror system and a Rigaku R-Axis II area detector. Prequalified crystals were then archived on liquid

Table 1. Representative Therapeutic Categories of Structures Solved

Analgesics	Anesthetics	Antiamyotrophic
Antiarrhythmic	Antiasthma	Antibacterial
Anticancer	Anticoagulant	Anticonvulsants
Antidepressant	Antidiabetic	Antihistamine
Antihypertensive	Anti-infective	Anti-inflammatory
Antiporphyria	Antipsychotic	Antiulcerative
Anxiolitic	Calcium blockers	Cholesterol lowering
Diagnostics	Hormone replacement	Muscle relaxant
NSAIDs		

nitrogen for later evaluation at one of several national synchrotron facilities. The quality of these crystal forms was routinely acceptable for crystallographic studies with diffraction resolution ranging from 2.8 to 1.9 Å. The highest resolution complex from the study to date was the previously reported human serum albumin complex with hemin at 1.9 Å [8] which is comparable to the highest resolution native unligated human albumin structure (1.9 Å) collected from a high-quality specimen grown in microgravity [9]. Synchrotron data were collected at the National Synchrotron Light Source (NSLS) at Brookhaven National Laboratory, the Advanced Photon Source (APS) at Argonne National Laboratory, and the Advanced Light Source (ALS) at Berkeley Laboratories.

The coordinated proprietary process described above was referred to internally as CADEX™. At this time, CADEX has resulted in more than 230 resolved structures at atomic resolution, representing examples from virtually every therapeutic indication and providing an unprecedented insight into human albumin's diverse drug and ligand-binding chemistry (Table 1).

2.2. Survey Results

The characterization of albumin drug interaction was greatly influenced by the early work of Sudlow et al. [10] who identified two dominant drug-binding locations by equilibrium dialysis methods, which were denoted as Sites I and II. Our initial crystallographic studies based on the small number of compounds studied, identified two major drug-binding regions within subdomains IIA and IIIA that were consistent with Sudlow's observations [1–4]. As a result of the unprecedented scope of this survey, 14 independent binding regions have now been determined, requiring the development of a more descriptive nomenclature (Fig. 2). Although several of these sites have been previously identified crystallographically, many are novel. Consequently, we have deviated from our original earlier expansion of Sudlow's nomenclature [4] to a more unambiguous subdomain location outlined in Fig. 1 and Table 2.

Evaluation of the first 142 complexes identified the 14 binding sites, three of which clearly dominate, accounting for ~90% of drug-binding locations currently determined (Fig. 2, Table 2). More than 70% of these represent drugs bound to discrete single-occupancy sites, 23 drugs show two sites and fifteen have more than two binding locations. Three drugs exhibited preferential binding of two molecules within a single site; a notable example of this phenomenon is presented in Section 2.2.4. Of the single-occupancy sites, 39% are located within Site IB, 23.5% within Site IIA and 20.5% within Site IIIA (Fig. 2, Table 2). Unexpectedly, 49% of all compounds have at least one binding site at subdomain IB. While subdomain IB was previously identified with the binding of long-chain fatty acids [4,11] and hematin [8] at the time of this survey, the identification of IB as the principal dominant drug-binding region was entirely unexpected. More recently the antiviral drug AZT has been independently determined crystallographically as a Site IB binder [12]. The chemical basis for the sites versatile binding properties is described in detail in the following sections.

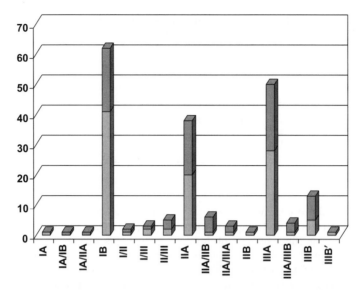

Figure 2. Histogram of drug-binding location and frequency. Green: number of total observations; blue: single-site binders. (This figure is available in full color at http://mrw.interscience.wiley.com/emrw/9780471266945/home.)

Table 2. Location, Frequency, and Description of Human Serum Albumin Drug-Binding Sites

Drug-Binding Site	Frequency (%)	Single-Site Frequency (%)	Residues Surrounding the Sites
IA	1 (0.52%)	1 (0.95%)	V007, F019, V023, F027, E045, V046, F049, A050, E060, N061, K064, L066, L069, F070, G071, D072, K073, C075, T076, C091, R098, L251
IA/IB	1 (0.52%)	0 (0.00%)	E017, N018, A021, E132, L135, L139, L155, A158, K159
IA/IIA	1 (0.52%)	0 (0.00%)	V007, F019, V023, A026, V046, F049, L066, H067, L069, F070, G248, D249, L250, L251, E252
IB	62 (32.46%)	41 (39.05%)	F036, F037, D108, P110, N111, L112, P113, R114, L115, V116, R117, P118, V122, M123, A126, N130, T133, F134, L135, K137, Y138, L139, Y140, E141, I142, A143, R145, H146, P147, Y148, F149, Y150, L154, F157, A158, Y161, F165, L182, D183, L185, R186, D187, G189, K190, K190, S192, S193, A194, Q196, R197, L219, R222, F223, L234, L238, R257, L260, A261, I264, S287, I290, A291, E425, Q459
I/II	2 (1.05%)	1 (0.95%)	E100, L103, Q104, D108, R145, H146, P147, Y148, F149, S193, Q196, R197, C200, A201, Q204, K205, C245, C246, H247, G248, N458, V462
I/III	3 (1.57%)	2 (1.90%)	D108, N109, R145, H146, R186, D187, E188, K190, A191, S193, A194, R197, P421, T422, E425, R428, N429, K432, V433, K436, Y452, V455, V456, Q459, V462, L463, K519, I523

Table 2. (*Continued*)

Drug-Binding Site	Frequency (%)	Single-Site Frequency (%)	Residues Surrounding the Sites
II/III	5 (0.62%)	2 (1.90%)	A194, K195, R197, L198, K199, A201, S202, L203, K205, F206, G207, A210, F211, A213, W214, H242, V343, V344, L347, E450, L453, S454, L457, N458, C461, V462, E465, C477, T478, E479, S480, L481, V482, R484, R485
IIA	38 (19.90%)	20 (19.05%)	F149, Y150, E153, A191, S192, K195, Q196, L198, K199, C200, S202, F211, W214, A215, R218, L219, R222, F223, L234, L238, V241, H242, C245, C246, C253, D256, R257, L260, A261, I264, K286, S287, H288, I290, A291, E292, V293, V343, P447, D451, Y452, V455
IIA/IIB	6 (3.14%)	1 (0.95%)	L198, K199, S202, F206, R209, A210, F211, K212, A213, W214, V216, F228, V231, S232, D324, V325, L327, G328, L331, V343, V344, L347, A350, K351, E354, D451, S454, E479, S480, L481, V482, N483
IIA/IIIA	3 (1.57%)	1 (0.95%)	A194, K195, L198, K199, C200, S202, L203, F206, G207, A210, F211, W214, H242, C246, V344, L347, E450, D451, S454, V455, L481, V482
IIB	1 (0.52%)	1 (0.95%)	D308, F309, N318, E321, A322, V325, F326, M329
IIIA	50 (26.18%)	28 (26.67%)	E383, P384, K387, L387, I388, Q390, A391, N391, C392, F395, F403, L407, L408, R410, Y411, K414, V415, V418, T422, L423, V424, V426, S427, L430, G431, V433, G434, S435, C437, C438, R445, M446, A449, E450, W450, Y452, L453, V456, L457, L460, V473, R484, R485, F488, S489, A490, L491, W492
IIIA/IIIB	4 (2.09%)	1 (0.95%)	L398, Y401, K402, N405, A406, L407, V409, R410, K413, K525, L529, A539, T540, K541, E542, L544, K545, M548, D549, A552
IIIB	13 (6.81%)	5 (4.76%)	Y401, K402, N405, F502, F507, F509, I513, L516, R521, K524, K525, Q526, A528, L529, E531, L532, V533, H535, K536, A539, T540, Q543, L544, K545, V547, M548, D549, D550, F551, A552, F554, V555, E556, G572, L575, V576, A577, A578, S579
IIIB'	1 (0.52%)	1 (0.95%)	C514, E518, R521, V555, E556, C559, K560

"/" indicates binding at the interface between two domains or subdomains. The numbers in the table were derived from the first 142 complex structures determined. There are 105 single-site complexes. The remaining 37 show multiple binding locations. The indicated residues in each binding region represent the combination of all observed contacts for all ligands, i.e., There would be fewer contacts per individual ligand.

2.2.1. Chemistry of the IB Site For the purposes of describing the drug-binding location and the chemical nature of the IB site, it would be helpful to examine the anatomy of the site in its unligated form. As in the case for heme [8], drug binding occurs in the hydrophobic groove created by helices h7, h8, h9, and h10 [3], which is covered in a bridge-like fashion by the flexible polypeptide loop, L1 (Fig. 3). This binding region is distinctive among the three main binding sites on albumin in having a capacity for larger heterocyclic compounds. The tremendous binding versatility observed can be explained by the conform-

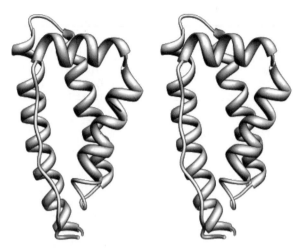

Figure 3. Ribbon stereo view illustrating the secondary structure and topology of the IB subdomain of human albumin as viewed from the proximal side. The topology from the n-terminal extended polypeptide L1 (forefront), followed in sequence by helices h7, h8, h9, and h10. The molecular graphics for this and many of the following figures were created with the program Chimera [13].

ability demonstrated by the IB site. A selected compilation of IB complexes identified by the survey has been presented in Table 3. These molecules range in MW from 160 for hydralazine HCl to 995 for the idarubicin dimer complex, with an average MW of 362 for the compounds listed in Table 3. The compounds include representative drugs for a variety for therapeutic indications, including several important classes of oncology drugs, as well as many endogenous ligands, such as bilirubin. Access to the groove is through two cavernous openings, one located between helices h7 and h8, that will be referred to as the "proximal" site (Fig. 4a), and the other between L1 and h10, the "distal" site (Fig. 4b). Anterior to the distal opening (Fig. 4b) is a small unprotected bowl-shaped hydrophobic depression, also observed as a minor drug-binding region, see below. Ligand access to the distal site is guarded by three closely grouped arginines: 114, 117, and 186. In addition to salt bridge formation, depending on the ligand involved, these three residues were found to make important binding contributions through hydrophobic and/or hydrogen bonding interactions. The proximal site opening is less guarded and lined primarily by hydrophobic residues.

2.2.2. Bilirubin The important endogenous ligand, bilirubin has been chosen to further illustrate the chemistry of IB. Bilirubin, owing to its medical importance, is considered among the most extensively studied of the ligand interactions with albumin. Since bilirubin is a neurotoxin, displacement interactions by drugs are important in neonates where accumulation of bilirubin in the brain can lead to akernicterus. Moreover, altered bilirubin binding is the only known example involving a potentially fatal albumin mutation, characterized among members of the indigenous Yanomama tribe as Yanomama-2 [14]. The site of substitution in this albumin variant is Arg 114 ⇒ Gly. The atomic structure of bilirubin albumin complex produced from the survey is well determined at 2.4 Å resolution and clearly places the site in IB and in direct chemical contact with the Arg 114 [15]. The bilirubin/HSA complex produced from the survey is shown in Fig. 5.

Unlike, many of the binding interactions to be discussed, bilirubin is bound by a full range of ligand/protein interactions that explains the molecules unusually high affinity to HSA ($\sim 10^7 \, M^{-1}$) [16]. An affinity no doubt imparted to protect the body from bilirubin toxicity. Specific binding interactions include a hydro-

Table 3. Therapeutic Catagories of Human Serum Albumin IB Binding Compounds

	Compound Name	IUPAC	Binding Site	Molecular Weight
Analgesic	Indomethacin	2-{1-[(4-Chlorophenyl)carbonyl]-5-methoxy-2-methyl-1H-indol-3-yl}acetic acid	IB	357.8
	Tolmetin sodium dihydrate	[1-Methyl-5-(4-methylbenzoyl)-1H-pyrrol-2-yl]acetic acid	IB	257.29
Anesthetic	Buplvacaine HCL	1-Butyl-N-(2,6-dimethylphenyl)piperidine-2-carboxamide	IB	288.43
Anti-ALS	Riluzole	6-(Trifluoromethoxy)benzothiazol-2-amine	IB, II/III	234.2
Antiarrhythmic	Quinidine gluconate	(9S)-6′-Methoxycinchonan-9-ol	IB	324.4
	Lidocaine	2-(Diethylamino)-N-(2,6-dimethylphenyl) acetamide	IB	234.34
Antiarthritic	Teriflunomide (A77 1726)	(2Z)-2-Cyano-3-hydroxy-N-[4-(trifluoromethyl)phenyl]but-2-enamide	IB	270.2
Antibacterial	Ampicillin	(2S,5R,6R)-6-{[(2R)-2-Amino-2 phenylacetyl] amino}-3,3-dimethyl-7-oxo-4-thia-1-azabicyclo [3.2.0] heptane-2-carboxylic acid	IB	349.4
	Cefamandole Nafate	5-Thia-1-azabicyclo [4.2.0]oct-2-ene-2-carboxylic acid, 7-[[(formyloxy)phenylacetyl]amino]-3-[[(1-methyl-1H-tetrazol-5-yl)thio]methyl]-8-oxo-, monosodium salt, [6R-[6(alpha),7(beta)(R*)]]	IB	512.49
	Metampicillin	3,3-Dimethyl-6-[[2-(methylideneamino)-2-phenylacetyl]amino]-7-oxo-4-thia-1-azabicyclo[3.2.0]heptane-2-carboxylic acid	IB	361.4

(continued)

Table 3. (*Continued*)

	Compound Name	IUPAC	Binding Site	Molecular Weight
	Penicillin G	4-Thia-1-azabicyclo(3.2.0)heptane-2-carboxylic acid, 3,3-dimethyl-7-oxo-6-((phenylacetyl)amino)-(2S-(2α,5α,6β))-	IB	334.4
	Sulfisoxazole	4-Amino-*N*-(3,4-dimethyl-1,2-oxazol-5-yl)benzenesulfonamide	IB	267.3
Anticancer	Bicalutamide	*N*-[4-Cyano-3-(trifluoromethyl)phenyl]-3-[(4-fluorophenyl)sulfonyl]-2-hydroxy-2-methylpropanamide	IB	430.4
	Camptothecin, (S)(+)	4-Ethyl-4-hydroxy-1*H*-pyrano[3',4':6,7]indolizino[1,2-b]quinoline-3,14-(4*H*,12*H*)-dione	IB	348.36
	Camptothecin, 9-nitro	(4*S*)-4-Ethyl-4-hydroxy-11-nitro-1*H*-pyrano[3',4':6,7]indolizino[1,2-b]quinoline-3,14(4*H*,12*H*)-dione	IB	393.35
	Idarubicin	(1*S*,3*S*)-3-Acetyl-3,5,12-trihydroxy-6,11-dioxo-1,2,3,4,6,11-hexahydrotetracen-1-yl 3-amino-2,3,6-trideoxo-α-L-*lyxo*-hexopyranoside	IB	497.5
	Teniposide	(5*S*,5a*R*,8a*R*,9*R*)-9-(4-Hydroxy-3,5-dimeth oxyphenyl)-8-oxo-5,5a,6,8,8a,9-hexahydrofuro[3',4':6,7]naphtho[2,3-d][1,3] dioxol-5-yl 4,6-*O*-(2-thienylmethylene)-β-D-glucopyranoside	IB	656.7
Anticoagulant	Dicumarol	3,3'-Methylenebis(4-hydroxy-2*H*-chromen-2-one)	IB; IIA	336.3
Anticonvulsant	Methsuximide	1,3-Dimethyl-3-phenyl-pyrrolidine-2,5-dione	IB	203.2
Antidepressant	Trazodone	2-(3-[4-(3-Chlorophenyl)piperazin-1-yl]propyl)-[1,2,4]triazolo[4,3-*a*]pyridin-3(2*H*)-one	IB; IIIA	371.9

Antidiabetic	Glimepiride	3-Ethyl-4-methyl-*N*-(4-[*N*-((1*r*,4*r*)-4-methylcyclohexylcarbamoyl)sulfamoyl]phenethyl)-2-oxo-2,5-dihydro-1*H*-pyrrole-1-carboxamide	IB, IIA*	490.6
	Glipizide	*N*-(4-[*N*-(Cyclohexylcarbamoyl)sulfamoyl]phenethyl)-5-methylpyrazine-2-carboxamide	IB	445.5
	Glyburide	5-Chloro-*N*-(4-[*N*-(cyclohexylcarbamoyl)sulfamoyl]phenethyl)-2-methoxybenzamide	IB	494.01
	Tolbutamide	*N*-[(Butylamino)carbonyl]-4-methylbenzenesulfonamide	IB; IIA	270.4
Antihistimine	Fexofenadine	2-[4-[1-Hydroxy-4-[4-(hydroxy-diphenyl-methyl)-1-piperidyl]butyl]phenyl]-2-methyl-propanoic acid	IB	501.66
Antihypertensive	Alprenolol	{2-Hydroxy-3-[2-(prop-2-en-1-yl)phenoxy]propyl}(propan-2-yl)amine	IB	249.34
	Doxazosin mesylate	2-{4-[(2,3-Dihydro-1,4-benzodioxin-2-yl)carbonyl]piperazin-1-yl}-6,7-dimethoxyquinazolin-4-amine	IB; IIA/IIB	451.5
	Hydralazine HCl	1-Hydrazinylphthalazine	IB	160.18
	Irbesartan	2-Butyl-3-({4-[2-(2*H*-1,2,3,4-tetrazol-5-yl)phenyl]phenyl}methyl)-1,3-diazaspiro[4.4]non-1-en-4-one	IB	428.5
	Methyldopa	(2*S*)-2-Amino-3-(3,4-dihydroxyphenyl)-2-methyl-propanoic acid	IB	211.2
	Prazosin HCl	2-[4-(2-Furoyl)piperazin-1-yl]-6,7-dimethoxyquinazolin-4-amine	IB	383.41
	Quinapril	(3*S*)-2-[(2*S*)-2-{[(2*S*)-1-Ethoxy-1-oxo-4-phenylbutan-2-yl]amino}propanoyl]-1,2,3,4-tetrahydroisoquinoline-3-carboxylic acid	IB	438.5

(*continued*)

Table 3. (Continued)

	Compound Name	IUPAC	Binding Site	Molecular Weight
	Ramipril	(2S,3aS,6aS)-1-[(2S)-2-{[(2S)-1-Ethoxy-1-oxo-4-phenylbutan-2-yl]amino}propanoyl]-octahydrocyclopenta[b]pyrrole-2-carboxylic acid	IB	416.5
	Telmisartan	2-(4-{[4-Methyl-6-(1-methyl-1H-1,3-benzodiazol-2-yl)-2-propyl-1H-1,3-benzodiazol-1-yl]methyl}phenyl)benzoic acid	IB	514.63
	Terazosin	6,7-Dimethoxy-2-[4-(tetrahydrofuran-2-ylcarbonyl)piperazin-1-yl]quinazolin-4-amine	IB	387.4
	Valsartan	(2S)-3-Methyl-2-[N-({4-[2-(2H-1,2,3,4-tetrazol-5-yl) phenyl]phenyl}methyl)pentanamido]butanoic acid	IB	435.53
Anti-infective	Ceftiaxone Sodium	(6R,7R,Z)-7-(2-(2-Aminothiazol-4-yl)-2-(methoxyimino)acetamido)-3-((6-hydroxy-2-methyl-5-oxo-2,5-dihydro-1,2,4-triazin-3-ylthio)methyl)-8-oxo-5-thia-1-aza-bicyclo[4.2.0]oct-2-ene-2-carboxylic acid	IB	554.57
Anti-inflammatory	Budesonide	16,17-(Butylidenebis(oxy))-11,21-dihydroxy-, (11-β,16-α)-pregna-1,4-diene-3,20-dione	IB	430.5
Antilipemic	Fenofibric acid		IB	318.8
Antiporphyria	Hemin, bovine		IB	651.9
Antipsychotic	Ziprasidone	5-[2-[4-(1,2-Benzisothiazol-3-yl)-1-piperazinyl]ethyl]-6-chloro-1,3-dihydro-2H-indol-2-one	IB	412.9

Category	Compound	IUPAC name	Class	MW
Bile pigment	Bilirubin	3-[2-[(3-(2-Carboxyethyl)-5-[(3-ethenyl-4-methyl-5-oxo-pyrrol-2-ylidene)methyl]-4-methyl-1H-pyrrol-2-yl]methyl)-5-[(4-ethenyl-3-methyl-5-oxo-pyrrol-2-ylidene)methyl]-4-methyl-1H-pyrrol-3-yl]propanoic acid	IB	584.65
	Biliverdin dihydrochloride	3-[2-[(E)-[(5E)-3-(2-Carboxyethyl)-5-[(4-ethenyl-3-methyl-5-oxo-pyrrol-2-yl) methylidene]-4-methylpyrrol-2-ylidene]methyl]-5-[(E)-(3-ethenyl-4-methyl-5-oxo-pyrrol-2-ylidene)methyl]-4-methyl-1H-pyrrol-3-yl]propanoic acid	IB	582.65
Cholesterol lowering	Cerivastatin	(3R,5S,6E)-7-[4-(4-Fluorophenyl)-5-(methoxymethyl)-2,6-bis(propan-2-yl)pyridin-3-yl]-3,5-dihydroxyhept-6-enoic acid	IB	459.56
	Clofibric acid	2-(4-Chlorophenoxy)-2-methylpropanoic acid	IB, IIIA	214.65
	Gemfibrozil	5-(2,5-Dimethylphenoxy)-2,2-dimethyl-pentanoic acid	IB, IIIA	250.3
Contraceptive	Norethindrone	(17β)-17-Ethynyl-17-hydroxyestr-4-en-3-one	IB	298.4
Diuretic (antihypertensive)	Chlorothiazide	6-Chloro-1,1-dioxo-2H-1,2,4-benzothiadiazine7-sulfonamide	IB	295.73
Hormone replacement	Estadiol, beta	(17β)-Estra-1,3,5(10)-triene-3,17-diol	IB, IA	272.4
	Ethinyl estradiol	17-Ethynyl-13-methyl 7,8,9,11,12,13,14,15,16,17-decahydro-6H-cyclopenta[a]phenanthrene-3,17-diol	IB	296.4
Muscle relaxant	Chlorzoxazone	5-Chloro-3H-benzooxazol-2-one	IB	169.6
	Cyclobenzaprine	3-(5H-Dibenzo[a,d]cyclohepten-5-ylidene)-N,N-dimethyl-1-propanamine	IB	275.4

(continued)

Table 3. (Continued)

	Compound Name	IUPAC	Binding Site	Molecular Weight
NSAID	6-MNA		IB, IIIA, IIA/IIB*	216.2
	Diflunisal	5-(2,4-Difluorophenyl)-2-hydroxybenzoic acid	IB; IIA/IIB; IIIA; IIIA/IIIB	250.2
	Etodolac	2-(1,8-Diethyl-4,9-dihydro-3H-pyrano[3,4-b]indol-1-yl)acetic acid	IB, IIIA	287.4
	Ibuprofen, S(+)	2-[4-(2-Methylpropyl)phenyl]propanoic acid	IB, IIIA	206.3
	Ketoprofen	2-(3-Benzoylphenyl)propanoic acid	IB	254.29
	Ketorolac	(±)-5-Benzoyl-2,3-dihydro-1H-pyrrolizine-1-carboxylic acid,2-amino-2-(hydroxymethyl)-1,3-propanediol	IB, IIIA, minor IIA/IIB	255.27
	Fenoprofen calcium hydrate	2-[3-(Phenoxy)phenyl]propanoic acid	IB, IIIA*	242.27
	Naproxen sodium	(+)-(S)-2-(6-Methoxynaphthalen-2-yl) propanoic acid	IB, IIIA, IIA/IIB, IIA	230.26
Stimulant	Caffeine	1,3,7-Trimethyl-1H-purine-2,6(3H,7H)-dione	IB, II/III	194.19

"/" indicates binding at the interface between two domains or subdomains.
"*" indicates minor site

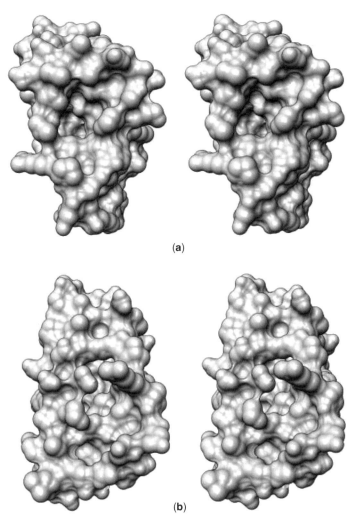

Figure 4. Stereo views of the surface of IB in the native uncomplexed state (a) view of the opening to the proximal site; (b) the distal site opening guarded by Arginines 114, 117, and 186. Note the anterior hydrophobic bowl-like depression that has also been observed to participate in ligand-binding of anticancer drugs.

phobic and electrostatic role by Arg 186, where the hydrocarbon chain is fully extend to maximize the hydrophobic packing with the pyrridole rings A and B, while simultaneously contributing to an electrostatic interaction with the ring B, C9 proprionic carboxyl (Fig. 5b). The C9 proprionic carboxyl also forms a key salt bridge with Arg 114. The C12 proprionic acid of ring C forms a salt bridge with Arg 117 creating a conformation where rings A and B remain in the original heme plane (Z conformation), with ring C–D pair of pyrridole rings rotating approximately 180° to optimize salt bridge with Arg 117. The ring C–D pair, however, is not coplanar as expected, but shows a significant rotation of approximately 120° about the normally aromatic C4–C15 bond. Bilirubin is known to undergo photoisomerization when exposed to light to the (ZZ to ZE) form that involves photoisomerization of double bond at C4–C15, a phenomenon used to advantage in

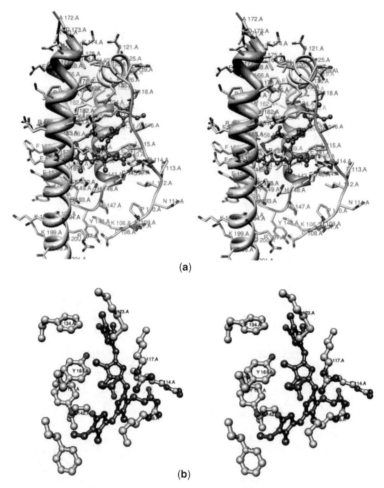

Figure 5. Stereo diagram illustrating the topology of subdomain IB and the placement of bilirubin within the subdomain IB-binding crevice. Bilirubin is bound in extended conformation stabilized by salt bridges from arginines 114, 117, and 184 to the proprionic carboxyls of bilirubin. The elimination of key salt bridge from Arg 114 by the substitution of Gly, accounts for the reduced affinity of Yanomama-2 to bilirubin and the fatal consequence in neonates as noted by Putnam and colleagues [14]. (See color insert.)

the treatment of hyperbilirubinemic neonates ("bilibabies") to promote bilirubin elimination (the product has greater solubility in plasma). The structure, determined under cryogenic conditions, reveals ring D frozen in a conformation virtually midstream between the two isoforms. It is interesting to note that Tyr 161 phenolic hydroxyl appears to be accepting a hydrogen bond from the pyrrole nitrogen of ring D, suggesting a potential role in the photoisomerization transition state. The structure of bilirubin IXα reported here, therefore appears transitory between the 4Z, 15Z and the 4Z, 15E conformations. The reader is referred to a recent independent structure determination of bilirubin IXα human albumin complex in the photoisomerized state (4Z, 15E), which includes a more in-depth analysis [17].

The protein surface surrounding the bilirubin-binding interaction is shown in Fig. 6, where it occupies a central position in the binding groove accessible through either the proximal or the distal opening.

While bilirubin is a complex heterocyclic ligand, whose binding with albumin has

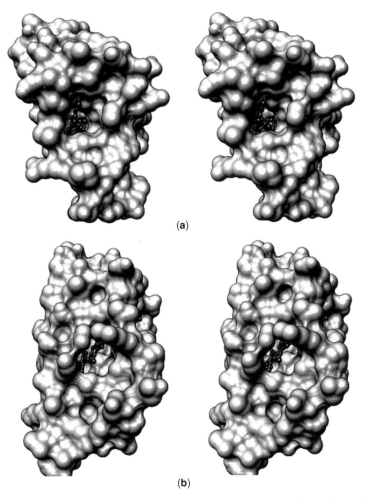

Figure 6. Stereo views of the surface illustrating the proximal (a) and distal (b) openings to the IB bilirubin complex. Gray: carbon; red: oxygen; blue: nitrogen. (See color insert.)

evolved in vertebrates [18], the interaction with many other ligands at the IB site, as shown in the following sections, is generally less specific.

2.2.3. The Camptothecins Today, there are several important anticancer drugs, such as irinotecan, topotecan, SN38, and others that owe their origin to the discovery of camptothecin. Camptothecin was originally isolated in the 1960s as an alkaloid compound from plants and found to have anticancer activity effective against more than 13 human cancer xenograft lines carried by immunodeficient (nude) mice [19,20]. As a result of its potent activity against cancer cells, it was quickly ushered into clinical trials shortly after its discovery. These early trials, however, were soon terminated due to very disappointing results which included severe toxicity issues in humans in spite of excellent activity against tumor cells in xenografts [21–24]. Interest in camptothecin analogs were rekindled after it was identified as an inhibitor of topoisomerase I by forming a covalent complex with DNA and topoisomerase. This category of compounds is attractive because of the selective toxicity against cells undergoing DNA replication, a process cancer cells are going through much more frequently than normal cells.

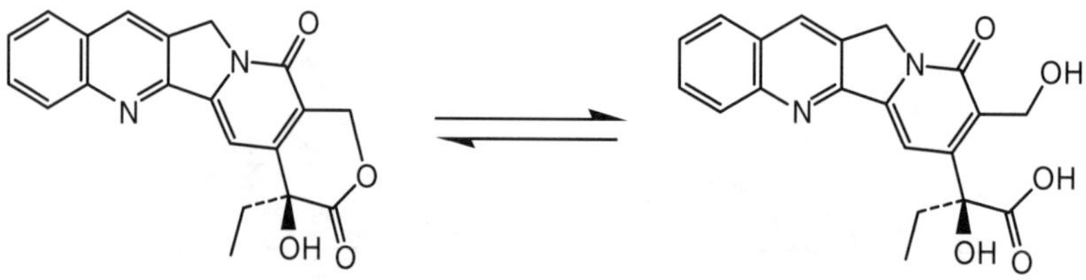

Figure 7. Reversible hydrolysis of the camptothecin lactone.

There were at least two important issues in the path to the optimum development of camptothecin-based therapeutics: (1) secondary toxicity and (2) inactivation in humans by high affinity of the carboxylate form to albumin [25,26]. Generally, camptothecin and its derivatives have two equilibrium forms in solution, the open carboxylate (inactive) and the closed lactone (active) form (Fig. 7). In aqueous solution above pH 7.0, the two forms essentially exist as 50:50 mixtures. In whole blood, human albumin binds preferentially to the carboxylate form with a binding affinity of $\sim 10^6 \, M^{-1}$, rapidly diminishing the available lactone form from the blood stream [25,26]. On the other hand, mouse albumin has a differential affinity to these compounds resulting in more proportionate active concentrations [26]. The differences in lactone-free concentration between the mouse and the human correlate with the observation that camptothecin and many of its derivatives have shown powerful antineoplastic activity against a broad spectrum of human cancer cells introduced in nude mice.

The survey produced X-ray structures of human serum albumin complexed with camptothecin and several of its important therapeutic derivatives. Camptothecin is bound in the lower section of the proximal IB site (Fig. 8a). The unbiased difference density produced from the camptothecin complex clearly indicated the bound moiety is the open carboxylate form consistent with expectations from the published literature. Hydrophobic interactions stabilize and completely enclose the heterocyclic structure with the carboxylate located in proximity to Arg 117, which, while in position to form a key salt bridge, appears to be maintaining its normal coordination with Asp 183. Additionally, a significant hydrogen bond is suggested by the close proximity of a guanidinium N from Arg 186 (Fig. 8b) to the gamma hydroxyl of camptothecin. The remaining heterocyclic ring structure is enclosed by predominantly hydrophobic interactions with Tyr 138, Leu 142, His 146, Phe 149, Leu 154, Phe 157, Tyr 161, and Leu 182 (Fig. 8b). Figure 9 shows the IB molecular surface and the views of the bound ligand from the proximal and distal sites.

The structure of the complex also provides insight into the reason for the marked differences in camptothecin blood lactone levels between mouse and man. Two of the three key interactive residues associated with the carboxylate or gamma hydroxyl in the human albumin complex are substituted in the mouse: Arg 114 $\Rightarrow$ Pro 114 and Arg 186 $\Rightarrow$ Lys 186. Arg 114 contributes a key hydrogen bonding interaction with the gamma hydroxyl, an interaction not available to a bound camptothecin in the closed ring form. Further the substitution with Pro (mouse) would place two prolines in sequence, creating a conformational change to L1 polypeptide strand and disruption of this important region of the camptothecin binding site. These two differences provide a compelling explanation for the experimentally observed decreased preferential affinity to the carboxylate form and more favorable fractional distribution of camptothecin lactone in mouse plasma.

2.2.4. The Anthracyclines Doxorubicin and Idarubicin are members of the powerful anthracycline family of antitumor antibiotics with efficacy across a broad spectrum of malignancies. The anthracyclines owe there origin to pigmented antibiotic isolates from the soil microbe, *Streptomyces peucetius*, first discovered during the 1950s [27]. The anticancer activity has been attributed to intercalation of double helical DNA. As with other neoplastic agents discussed in this chapter, the signifi-

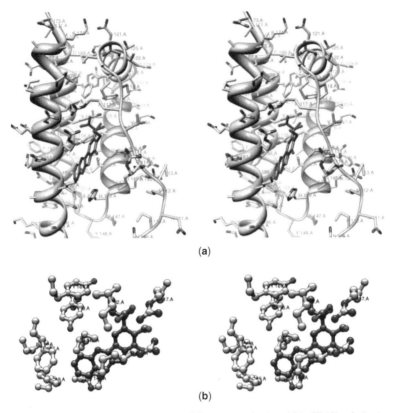

Figure 8. (a) Stereo view illustrating the position of the camptothecin within IB (distal view) and (b) details of its binding interaction with Arg 117 and Arg 186. (See color insert.)

cant and potentially life-threatening toxicities of the anthracyclines limit the more widespread application. In the case of doxorubicin, cardiotoxicity is a serious limitation, where overdosing can lead to drug induced congestive heart failure [28]. In addition, toxicity is accumulative, often excluding its continued application for follow-on treatments involving the recurrence of cancer.

The binding of the idarubicin by albumin illustrates another example of the exceptional conformability of the IB site. Here, the interaction with HSA is almost entirely hydrophobic, showing a striking dinucleotide/DNA-like stacking of two molecules within the IB pocket at the proximal site (Fig. 10a). This complex represents one of only three compounds from the survey exhibiting the property of paired binding. The tetrahydrotetracene ring is stabilized in the pocket base by interactions with Tyr 161 and Tyr 138. In this case, Tyr 161 has rotated from its usual π-stacking interaction with Tyr 138 to contribute a direct hydrophobic contact to the lower base of the tetracene ring. Phe 134 together with Pro 118, and Leu 115 combine to provide further hydrophobic stabilization of the complex around the exterior of the opening. Lys 137 is extended along the groove between the two rings with the lysine ε nitrogen in close proximity and equally spaced between the two acetyl groups at C9. One of the few discernable nonhydrophobic influences involves a potential hydrogen bond with the Glu 141 carboxyl to the C9 hydroxyl of the lower idarubicin molecule (Fig. 10a). Although it appears that the hexopyranoside rings may be engaged in cooperative hydrogen bonding between the two molecules, the potential hydrogen bonding interactions were outside the accepted distances. This could perhaps be a limitation in the resolution of the crystallographic complex, but

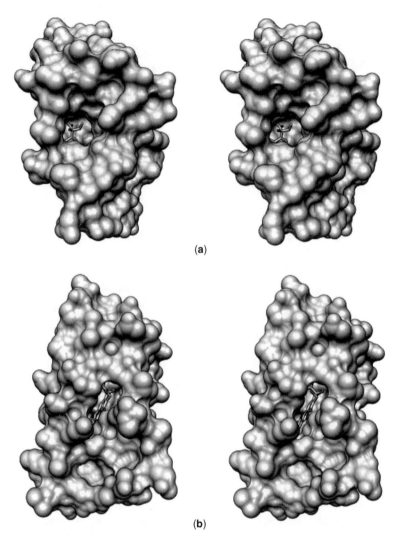

Figure 9. Stereo views of the surface illustrating the proximal (a) and distal (b) openings to the IB camptothecin complex. Gray: carbon; red: oxygen; blue: nitrogen; stereo gray: carbon; red: oxygen; blue: nitrogen. (See color insert.)

does not, however, rule out bridging water molecules widely observed in carbohydrates in the crystalline state [29]. The occurrence of the dimer as the albumin-binding form and the clear indications of intermolecular stabilization of the complex, strongly suggested that this drug may be present as a dimer in solution, an observation which was indeed supported in the literature [30]. The enlargement of the proximal site to accommodate the idarubicin dimer (MW 995) is further illustrated in Fig. 10b.

2.2.5. Podophyllotoxin Derivatives Etoposide and teniposide are important cancer drugs derived from podophyllotoxin, a lignan isolated from the resin produced from the *Podophyllum* species [31]. They are used in the treatments of testicular, small-cell lung cancers, lymphoma, leukemias, Kaposi's sarcoma and other neoplasms. In addition to anticancer activity, this family of lignans has also proved useful in several other therapeutic applications, for example, antiviral. The mode of action has been attributed to inhibiting

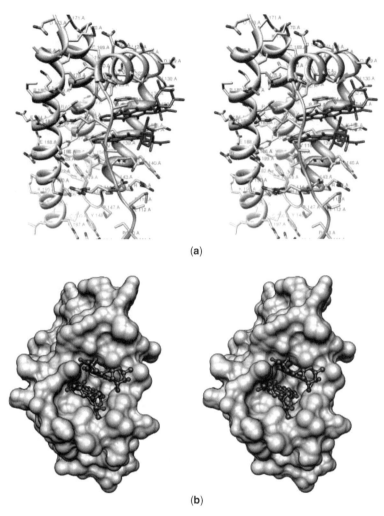

Figure 10. Stereo views of (a) the IB idarubicin dimer complex showing the binding site location at the proximal site; and the packing interactions with residues: Tyr 138, Tyr 161, Phe 134, Pro 118, and Leu 115. Note the close association of the pyranosides that suggests the potential for intermolecular hydrogen bonding interaction, possibly through bridging water molecules not identified by the present structure resolution. (b) Stereo view of the IB molecular surface illustrating the enlargement of the proximal site to accommodate the idarubicin dimer. (See color insert.)

topoisomerase II, thereby arresting the cell cycle in the metaphase. Again, as with the other neoplastic agents outlined in this chapter, toxicity to normal cells remains an important limitation.

Structurally, etoposide and teniposide represent the largest and most complex pharmaceutical compounds determined by the survey. The large podophyllotoxin derivative occupies the IB site entirely, from the proximal opening through to the distal. Despite the molecules complexity in chemical structure, it is a neutral molecule tending to maximize the hydrophobic interactions without major disturbance to the native structure (Fig. 11a). For example, the Tyr 138–Tyr 161 interhelical π-stacking interaction remains undisturbed, despite the demonstrated conformational adaptability of both tyrosines in the binding of other ligands. Most notable among the structural changes is the large exposed opening at the distal site created by the significant repositioning of Arg

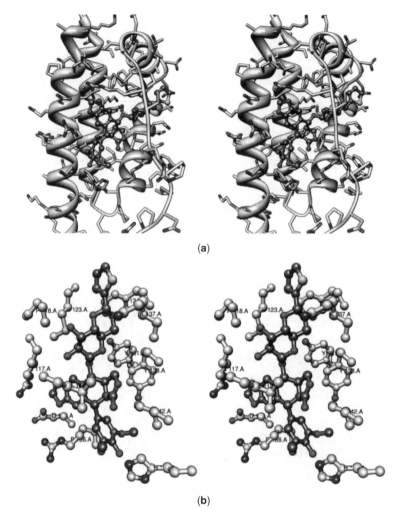

Figure 11. Stereo views of (a) the teniposide complex showing the binding site location within the IB-binding site and (b) a selection of key residues surrounding the teniposide molecule. (See color insert.)

114 away from the ligand (Fig. 11b). Other interactive residues include Leu 115, Arg117, Pro 118, Met 123, Phe 134, Lys 137, Tyr 138, Leu 142, His 146, Tyr 161, Leu 182, Asp 183, Arg 186. A surface view of the bound ligand is shown in Fig. 12.

3. THERAPEUTIC APPLICATIONS OF THE ALBUMIN STRUCTURE SURVEY

The design of new and potentially improved cancer chemotherapeutics can be supported by examination of the structures of the crystalline complexes of albumin previously described in this chapter. Design modifications must, however, be made with an understanding of the SAR data and where possible, knowledge of the drug complex structure with its therapeutic target. Our initial applications of albumin-based structure guided drug design were focused on the development of novel warfarin analogs for the purpose of modifying drug plasma stability (pharmacokinetics) to improve the challenging patient management and safety issues of this important anticoagulant. Another approach involves modifying the albumin-based drug pharmacokinetics with the coadmi-

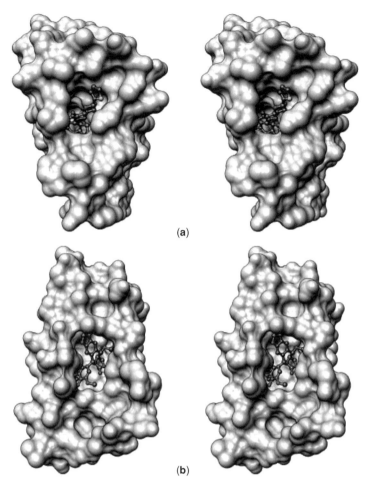

Figure 12. (a) Stereo view of the IB molecular surface with views of teniposide within the proximal site and (b) from the distal site. Note the unusually large access to the distal opening created by the rotation of Arg 114 side chain away from the ligand toward L1. (See color insert.)

nistration a blocking agent or inhibitor using a second drug or ligand. These albumin-binding inhibitors or "pharmacokinetic modulators" must be specially selected for a variety of key properties, to be effective *in vivo*. This approach has several advantages that will be described in detail in the following sections.

3.1. Albumin Pharmacokinetic Modulation: Applications of Clofibrate

One of the early and most expedient applications of the drug survey was the opportunity to tune the pharmacokinetic properties of a target drug (active drug), with a second drug or ligand based on the knowledge of the albumin-binding chemistry in molecular detail [32,33]. The therapeutic approach of albumin pharmacokinetic modulation (APM) with a second compound in combination was targeted toward applications with highly cytotoxic drugs with narrow therapeutic indices where the benefits would be most pronounced. Attractive target drugs naturally fell into two major therapeutic categories, anticancer and antiinfective, where toxicity and/or serious side effects are problematic, and IB-binding chemistry is dominant.

Ideally, the APM drug or ligand IB candidates would have the following properties: (1)

high affinity and selectivity to Site IB; (2) safety, with the ability to achieve millimolar plasma levels with few side effects at established dosing regimens; (3) drug half-life and pharmacokinetics that do not negatively interfere with metabolic pathways of the target (active) drug; and (4) stable and continuous pharmacokinetic modulation throughout the chemotherapy treatment phase.

One such drug, clofibrate (ethyl-2-4-(4-chlorophenoxy)-2-methylpropanoate), an FDA approved drug used to lower blood cholesterol and triacylglyceride levels (now replaced by modern statin inhibitors in the United States) [34], clearly met these criteria for many applications. In addition to being an FDA approved generic, clofibrate offered the advantage of chemical simplicity and low manufacturing costs. Clofibrate is a prodrug that is rapidly hydrolyzed in the gut from the ethyl ester form to clofibric acid. As a prodrug, clofibrate is not highly bound to albumin. Clofibric acid, however, exhibits high affinity to albumin and reaches stable plasma levels within 48 h at standard clinical dosing.

Structurally, the binding of clofibric acid to human serum albumin occurs in both Sites IIIA and IB. For brevity, we shall limit the discussion of clofibric acid binding at Site IB, which occurs in the lower base of the distal opening. The carboxylic acid of clofibric acid forms a strong salt bridge with His 142 (Fig. 13a) with the aromatic ring stabilized by hydrophobic packing interactions with the hydrocarbon chain of Arg186, and residues Phe 157, Phe 149, Ile 142, and Leu 154 (Fig. 13b). The hydrophobic insertion of the chlorophenoxy group into the binding pocket causes the reorientation of the Ile 142 side chain, which together with the proximity of the chlorine atom, disrupts Tyr 138 from its normal interhelical π-stacking interaction with Tyr 161. Together, these disturbances cause the side chain of Tyr 138 to rotate about the $C\alpha$-$C\beta$ bond by $\sim 180°$ toward the proximal opening to optimize hydrophobic interactions with Phe 134, Met 123, and Pro 118. This clofibric acid "trigger" has the fortunate and important therapeutic consequence of closing the entrance to the proximal opening (Fig 14a). Closure of the proximal site may have important implications with the originally observed cholesterol lowering properties of the drug, since at least one class of steroid-based therapeutics were found by the survey to bind at the IB proximal site in a manner similar to idarubicin (Fig. 10). If this indeed represents the causal basis for the drugs original application, it would provide a potential explanation for its poor performance in reducing cardiovascular risk in a WHO trial in 1984 [35]. The initial rapid triglyceride lowering effects achieved in 48 h [34], are consistent with this suggestion and match the time required for stable clofribic acid plasma concentrations (maximum displacement) under typical clinical dosing. Other clofibric acid-based drugs and possibly selected analgesics, if present in the required millimolar concentration, would be predicted to have similar effects on the binding site. A further discussion of this proposed mode of action will be described in Section 4.

3.1.1. In Vitro Studies So, what is the performance of clofibrate in modulating the albumin-based pharmacokinetics of oncology drugs? To answer this question, we initially examined the ability of physiologically achievable levels of clofibric acid to favorably reduce albumin binding of selected target drugs, which were determined by the survey to be bound in Site IB. Figure 15 shows an example of the affect of clofibric acid in the pharmacokinetic modulation of teniposide in 30 mg/mL concentrations of human serum albumin. Under these plasma-like conditions in the absence of red blood cells, 1 mM concentrations of clofibric acid resulted in an approximate 2.5-fold increase in teniposide-free drug. In the case of another oncology drug, camptothecin, we examined free lactone levels of camptothecin versus the hydrolyzed lactone (free carboxylate) to evaluate improvements in blood chemistry of this family of drugs. Under normal physiological conditions in humans, the free lactone (up to 1000 times more active) of camptothecin is rapidly reduced to levels of 1% over the course of approximately 2 h by selective binding of the hydrolyzed free acid by human albumin as previously discussed in Section 2.2.3. In solutions containing 30 mg/mL human serum albumin and 1 mM clofibric

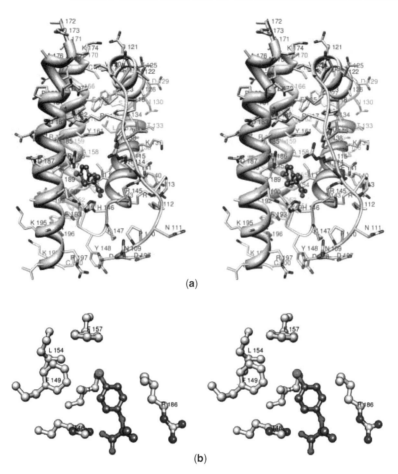

Figure 13. (a) Stereo views of the clofibric acid complex showing the binding site location within subdomain IB and (b) the detailed interactions of clofibric acid with His 146 and other key residues. (See color insert.)

acid, concentrations of free lactone increased to 20%, a 20-fold improvement (Fig. 16). Further examples of this approach were also successfully demonstrated with 9-nitro-camptothecin and 10-hydroxy-camptothecin, indicating that the APM agent, clofibrate, is generally applicable to the camptothecin family. In a separate study, the increase in free fraction of active drug directly correlated with increased inhibition of topoisomerase I using commercially available *in vitro* assays.

To further assess the therapeutic potential of this approach to the pharmacokinetic modulation of cancer drugs, we designed *in vitro* assays with the human breast cell cancer line MDA-MB-435S, supplemented with physiological levels of serum albumin to mimic blood plasma conditions. Under the conditions of these experiments, the target oncology drugs were examined using concentration ranges with midpoints approximating typical clinically observed peak plasma values (Fig. 17). Clofibric acid was added to model blood concentrations of 0.5 and 1.0 mM achieved under typical recommended dosing regimens. Human serum albumin concentrations approximated physiological concentrations of 35 mg/mL for each series, (excluding the required concentrations of alpha-fetoprotein (albumin's fetal counterpart) in the culture media). The *in vitro* results against human breast cell cancer line MDA-MB-

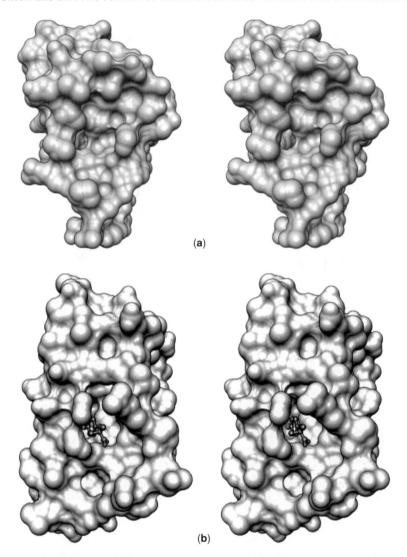

Figure 14. Stereo views of the surface charge and hydrophobicity illustrating the proximal (a) and (b) distal openings to the IB clofibric acid complex. Note that while the view from the proximal opening shows little indication of change, the view of the distal shows the conformation change created by Tyr 138 that resulted in the remarkable closing of the cavity entrance. (This figure is available in full color at http://mrw.interscience.wiley.com/emrw/9780471266945/home.)

435S for these experiments are shown in Fig. 17 for three classes of important oncology drugs. As can be seen, the results are compelling, producing marked improvements in cancer cell death throughout the concentration range, including those values that would formerly be referred to as subtherapeutic. Controls with only clofibric acid show no decrease in cancer cell viability. The marked efficacy improvements can be easily related to the increased free drug concentrations (or repartitioned drug) produced through competitive displacement of the target therapeutic from albumin with physiologically relevant millimolar plasma levels of clofibric acid. These impressive results represent more than just simple competitive inhibition of the target ligand, since clofibric acid also induces subtle, but powerful chemical changes to the IB-binding site as previously described in Section 3.1.

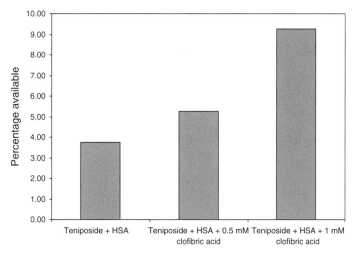

Figure 15. Illustration of the changes in free teniposide induced by the addition of 0.5 mM and 1.0 mM concentrations of clofibric acid as determined by high-performance size exclusion liquid chromatography. (This figure is available in full color at http://mrw.interscience.wiley.com/emrw/9780471266945/home.)

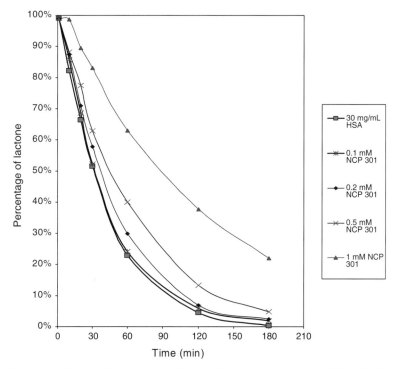

Figure 16. Percentage of active lactone form of Camptothecin in the presence of 30 mg/mL human serum albumin. After 3 h, the level of active camptothecin is 20% in the presence of 1 mM Salus agent (▲) versus essentially zero in the absence of the agent (■). (This figure is available in full color at http://mrw.interscience.wiley.com/emrw/9780471266945/home.)

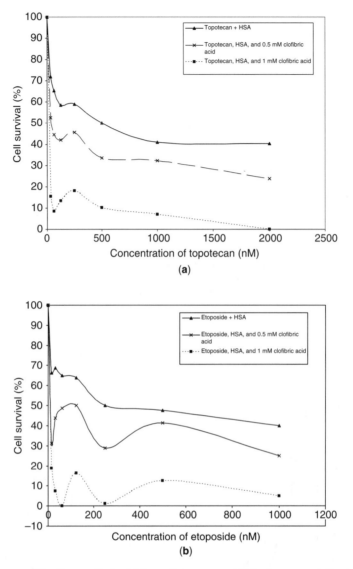

Figure 17. Summary of *in vitro* results for (a) the anticancer agent 10-hydroxycamptothecin, (b) teniposide, and (c) etoposide. Shown are (i) the control in a typical clinical dose, (ii) Salus with control at half clinical blood plasma levels, and (iii) Salus with control at typical plasma levels achieved at normal dosing. Human Breast Cancer cells (MDA-MB-435S) grown under standard ATCC conditons for 48 h. Cells were then incubated with HSA; drug, HSA, and 0.5 mM clofibric acid; drug, HSA, and 1 mM clofibric acid for 24 h. Cell survival was tested using an *In Vitro* Toxicology Assay Kit, XTT based. Controls with only clofibric acid show no decrease in cell viability.

3.1.2. Pharmacokinetic Studies Analysis of APM phamacokinetics were conducted using male Sprague Dawley rats and irinotecan, a camptothecin derived oncology drug used in the treatment of advanced colorectal cancer and nonsmall cell lymphoma. Irinotecan, unlike the oncology drugs described in previous sections, is a prodrug with low albumin affinity (~50%) whereas SN38, the principal active metabolite, is highly bound to human albumin (~98%).

To examine the pharmacokinetics of irinotecan with clofibric acid, animals were pre-

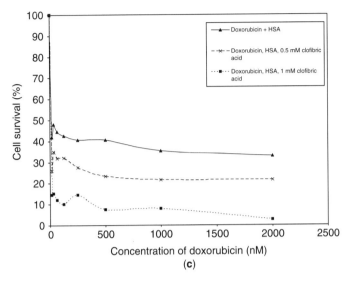

Figure 17. (*Continued*)

treated with clofibric acid for 72 h prior to administration of irinotecan. Twice daily dosing of clofibrate was maintained throughout the experimental period to ensure continued stable blood concentrations. CPT-11 was administered by intraperitoneal injection on day 3. The results illustrated in Fig. 18 show that the plasma concentrations of CPT-11 in the control group follow the expected drug levels and metabolic clearance of CPT-11 over the course of a 24 h period. The clofibrate APM drug combination, however, shows at the outset, a ∼50% lower level of CPT-11 in the plasma fraction which could not be explained by metabolic processes. After further analysis, it became apparent that the large fraction of CPT-11 shifted from albumin by clofibric acid, had been partitioned to the erythrocyte reservoir that is customarily removed and discarded prior to analysis [36]. Consequently,

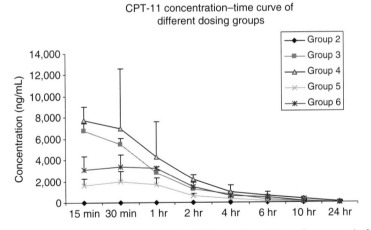

Figure 18. Group 2, control, no CPT11. Group 3 and 4, CPT11 at 60 and 30 mg/kg, respectively; Group 5 and 6, CPT11 60 and 30 mg/kg, respectively, plus Salus 30 mg/kg. (This figure is available in full color at http://mrw.interscience.wiley.com/emrw/9780471266945/home.)

traditional drug-binding analysis of the plasma fraction results in an erroneous interpretation of the APM pharmacokinetics. This dynamic interaction between the protein and the cell compartments of blood is supported by Combes et al. [37] who identified that a significant fraction (~30%) of radioisotope labeled CPT-11 in whole blood samples is carried on erythrocytes. Moreover, Kruszewski and Burke [38] using specialized fluorescence spectroscopy methods demonstrated a high affinity of a variety of camptothecin analogs to model membranes. They attributed this cell-associated fraction principally with only the hydrophobic lactone moiety that they maintain is stabilized from hydrolysis during this interaction.

The partitioning effect between albumin and the plasma blood cell component (PBC), illustrated by the *in vivo* clofibrate studies, sets the stage for a more complete understanding and more reliable modeling of drug pharmacokinetics in the future. The powerful albumin-based modulation of drugs by clofibrate illuminates the potential mechanism for this poorly understood family of lipid lowering drugs. Clofibrate, which was used as popular lipid lowering drug for more than 30 years before the introduction of modern statin inhibitors, such as Lipitor™ and Crestor™, produces a rapid drop in blood cholesterol within 48 h [34]. The 48 h period coincides with the minimum time required to reach stable plasma clofibric acid levels and the maximum (albeit hypothetical) displacement of cholesterol from IB. Theoretically, the cholesterol, shifted from its presumed albumin IB proximal site would be partitioned to maximize its hydrophobic interaction with membrane surfaces of the blood cell compartment. Since the blood cell fraction is routinely separated and discarded prior to determining blood cholesterol levels; this would lead to an incorrect interpretation that the patients cholesterol levels had actually been lowered, further leading to a false interpretation of therapeutic benefit as a lipid lowering agent.

While the pharmacokinetic implications of the APM approach on *in vivo* efficacy are still under evaluation, the substantial repartitioning of the drug to the plasma blood cell membrane surface (in the special case of campothecins, also a presumed enrichment of the camptothecin lactone fraction) and the observation of much lower secondary toxicity in preliminary animal model studies [36], shows promise of significant therapeutic benefit to cancer patients. Indeed, the dynamic shift in drug equilibrium between albumin and the blood cell compartment, could also result in enhanced drug partitioning to the cancer cell surface and may account for the unusually enhanced efficacy against cancer cells *in vitro* for all of the drug families examined in this study.

4. SUMMARY AND CONCLUSIONS

This chapter briefly describes the immense body of information generated by the albumin structural survey and focused on an early application of the information to improve cancer therapeutics. Highlights of the survey include the identification of the Site IB as a major drug-binding site on human albumin, the importance of which was not previously recognized. The results revealed at least 14 distinct binding locations on human albumin with the overall binding frequency dominated by three key binding regions representing sites IIA, IIIA and IB, with IB (based on the compounds studied), having the greatest capacity of the three. The broad chemical selectivity of Site IB can be attributed to the flexibility of extended loop L1 which bridges the binding groove, and that of several key residues: Tyr 161, Tyr 138, His 146, Arg 114, Arg 117, and Arg 186, all of which have shown versatility in accommodating and coordinating with complex ligands within the site.

Early applications of the information generated from the survey included research into drug combinatorial approaches to favorably alter the albumin-based drug pharmacokinetic properties. Since the pharmacokinetics of highly cytotoxic drugs could best be modified to advantage, the target indications naturally fell into two principal therapeutic categories, anticancer and anti-infectives. This APM approach combines an albumin "blocking agent" with known anticancer agents to improve efficacy by increasing drug availability while simultaneously decreasing the required therapeutic dose to lower toxicity. Increasing the free drug fraction can also be an important consideration for facilitating drug availability to tumors and across organ circulatory interfaces, such as the blood brain barrier. Clofibrate (clofibric acid), with its high affinity to site IB, stable blood plasma levels in the millimolar range and elegant conformational induced closure of the IB proximal site, was a drug of choice for further evaluation as an APM agent. Preclinical studies with clofibric acid began with *in vitro* demonstrations using three well characterized families of oncology drugs which were crystallographically determined as IB binders: the camptothecins: camptothecin, 10-hydoxy camptothecin, 9-nitro camptothecin, irinotecan, topotecan, and SN38; the anthracyclines: adriamycin (doxorubicin) and idarubicin; and the podophyllotoxins: etoposide and teniposide. APM studies produced unusually marked improvements in the *in vitro* efficacy profiles (Fig. 17), with the enhanced cancer cell death directly correlating with the displacement effects associated with the increases in clofibric acid concentrations. Significant improvements in therapeutic indices were also observed from *in vitro* studies using lung fibroblasts as surrogates for normal cells. In these studies, therapeutic indices increased from 8 to 138, for 10-hydroxy camptothecin, a 14-fold improvement, and for etoposide, from 68 to 1350, an approximate 20-fold improvement.

The *in vivo* pharmacokinetic and safety studies of APM drug combinations of clofibrate together with irinotecan (CPT-11), a camptothecin derivative used for the treatment of advanced colorectal cancer and non-small cell lymphoma, were evaluated in male Sprague Dawley rats. While the prodrug, CPT-11, is not highly bound to albumin, SN38, the principal active metabolic product is bound with high affinity. Clofibrate dosed at the standard mg/kg dosing regimen previously established in humans, resulted in displacement (Fig. 2) of about 50% of CPT-11 from rat albumin—shifting it to a more dynamic, nonspecific interaction with the erythrocyte cell surface rather than, necessarily, to free drug (although free concentrations of the drug were also increased). This is potentially very advantageous for camptothecin-based drugs, since according to Kruszewski and Burke [38], the cell-associated camptothecin is entirely active lactone, stabilized from hydrolysis during the cell surface interaction. We believe this repartitioning effect of drugs by clofibric acid is also the basis for the marked improvements in efficacy seen for all three families of oncology drugs studied to date and belies the drugs original application as a lipid lowering drug.

In conclusion, the more accurate view of albumin chemistry now revealed by this study, together with illumination of the principal albumin ligand-binding site, will bring further clarity to an immense body of work in the literature and promises to improve the utility and predictive outcome of applications in a variety of disciplines (e.g., *in silico* ADME). The high frequency in single-site drugs observed further strengthens the applicability of the albumin drug structural information to improve the safety and efficacy of many existing approved and developmental stage pharmaceuticals, whether through structure-guided design or other albumin-based pharmacokinetic modulation. In this chapter, we have briefly demonstrated several powerful applications of APM drug combination to improving the safety and efficacy of highly cytotoxic drugs. Clearly, the magnitude of the improvements are suggestive that modifying a drug's albumin-based pharmacokinetic properties are at least as important as optimizing the uM affinity of the drug to its target. In the case of clofibrate applications, the albumin-based performance advantages can be achieved with an FDA approved drug, without reengineering or forming adducts (e. g., polyethylene glycol derivatives) with the target drug of interest. The combinatorial use of FDA approved drugs may offer in some cases the advantage for an abbreviated FDA approval process. Finally, the potential for large reductions in dose while maintaining or exceeding the drugs original performance, provides new avenues to existing therapies, avenues that can increase the treatable patient population, reduce serious side effects, such as life-threatening secondary toxicities, and ultimately improve the clinical outcome. Further details of these studies and a more comprehensive discussion of the more than 230 albumin complexes determined by the study, will be reported elsewhere.

ACKNOWLEDGMENTS

The extensive and unprecedented work reported here was made possible by the dedication and support of the following individuals during employment or association with New Century Pharmaceuticals: Dr. Joseph X. Ho, Dr. Zhong-min Wang, Dr. John R. Ruble, Dr. Florian Ruker, Melanie Ellenburg, Robert Murphry, Dr. Justin Roberts, Dr. Mark Wardell, Pei Ye, Dr. Chester Li, James Click, Elizabeth Soistman, Leslie Wilkerson, Yolanda Kirksey, Simon McKenzie and Brenda Wright. We gratefully acknowledge the generous support of the beam line staff of the National Synchrotron Light Source at Brookhaven National Laboratory, the Advanced Photon Source at Argonne National Laboratory and the Advanced Light Source at Berkeley Laboratories. This work was made possible by the generous financial support of the shareholders of New Century Pharmaceuticals, Inc.

REFERENCES

1. Carter DC, He X-M, Munson SH, Twigg PD, Gernert KM, Broom MB, Miller TY. Three-dimensional structure of human serum albumin. Science 1989;244:1195–1198.
2. Carter DC, He X-M. Structure of human serum albumin. Science 1990;24:302–303.
3. He XM, Carter DC. Atomic structure and chemistry of human serum albumin. Nature 1992;358(6383):209–215.
4. Carter DC, Ho JX. Structure of serum albumin. Adv Protein Chem 1994;45:153–203.
5. Chaudhry C, Brooks CL, Carter DC, Robinson JM, Anderson CL. Albumin binding to FcRn: distinct from the FcRn-IgG interaction. Biochemistry 2006;45(15):4983–4990.
6. Peters T. All about Albumin: Biochemistry, Genetics, and Medical Applications. San Diego: Academic Press; 1996.
7. Carter DC, Ho JX, Wang ZH. Albumin binding sites for evaluating drug interaction and methods of evaluating or designing drugs based on their albumin binding properties, WO 2004/102151; WO 2005041895, WO 2007/130905.
8. Wardell M, Wang Z, Ho JX, Robert J, Ruker F, Ruble J, Carter DC. The atomic structure of human methemalbumin at 1.9 Å. Biochem Biophys Res Commun 2002;291(4):813–819.
9. Carter DC, Ho JX, Wang ZH, Ruble JR, Wright B. The native unligated structure of human serum albumin at 1.9 Å from crystals produced in microgravity, New Century Pharmaceuticals. To be deposited in the Protein Data Bank.
10. Sudlow G, Birkett DJ, Wade DN. Further characterization of specific drug binding sites on human serum albumin. Mol Pharmacol 1976;12(6):1052–1061.
11. Curry S, Mandelkow H, Brick P, Franks N. Crystal structure of human serum albumin complexed with fatty acid reveals an asymmetric distribution of binding sites. Nat Struct Biol 1998;5(9):827–835.
12. Zhu L, Yang F, Chen L, Meehan EJ, Huang M. A new drug binding subsite on human serum albumin and drug–drug interaction studied by X-ray crystallography. J Struct Biol 2008;162(1):40–49.
13. Pettersen EF, Goddard TD, Huang CC, Couch GS, Greenblatt DM, Meng EC, Ferrin TE. UCSF Chimera: a visualization system for exploratory research and analysis. J Comput Chem 2004;25(13):1605–1612.
14. Takahashi N, Takahashi Y, Isobe T, Putnam FW, Fujita M, Satoh C, Neel JV. Amino acid substitution in inherited albumin variants from Amerindian and Japanese populations. Proc Natl Acad Sci USA 1987;84:8001–8005.
15. Carter DC, Ho JX, Wang ZH, Ruble, JR. The structure of human serum albumin complexed with bilirubin at 2.4 Å. New Century Pharmaceuticals. To be deposited in the Protein Data Bank.
16. Brodersen R. Bilirubin: solubility and interaction with albumin and phospholipid. J Biol Chem 1979;254:2364–2369.
17. Zunsain PA, Ghuman J, McDonagh AF, Curry S. Crystallographic analysis of human serum albumin complexed with 4Z,15E-bilirubin-IX-alpha. J Mol Biol 2008;381(2):394–406.
18. Peters T. All about Albumin: Biochemistry, Genetics, and Medical Applications. San Diego: Academic Press; 1996. p 95–97.
19. Wall ME, Wani MC, Cook CE, Palmer KH, McPhail AT, Sim GA. Plant antitumor agents. I. The isolation and structure of camptothecin, a

novel alkaloidal leukemia and tumor inhibitor from camptotheca acuminata. J Am Chem Soc 1966;88(16):3888–3890.
20. Ewys WD, Humphreys SR, Goldin A. Studies on therapeutic effectiveness of drugs with tumor weight and survival time indices of Walker 256 carcinosarcoma. Cancer Chemother Rep 1968;52(2):229–242.
21. Gottlieb JA, Luce JK. Treatment of malignant melanoma with camptothecin (NSC-100880). Cancer Chemother Rep 1972;56(1):103–105.
22. Muggia FM, Creaven PJ, Hanson HH, Cohen MH, Selawry OS. Phase I clinical trial of weekly and daily treatment with camptothecin (NSC-100880): correlation with preclinical studies. Cancer Chemother Rep 1972;56(4):515–521.
23. Moertel CG, Schutt AJ, Reitemerer RC, Hahn RG. Effect of resection of the primary neoplasm on responsiveness to chemotherapy of patients with large bowel cancer. Cancer Chemother Rep 1972;56(4):551–552.
24. Giovanella BC, Stehlin JS, Wall ME, Wani MC, Nicholas AW, Liu LF, Silber R, Potmesil M. DNA topoisomerase I: targeted chemotherapy of human colon cancer in xenografts. Science 1989;246(4933):1046–1048.
25. Mi Z, Burke TG. Differential interactions of camptothecin lactone and carboxylate forms with human blood components. Biochemistry 1994;33(34):10325–10336.
26. Mi Z, Burke TG. Marked interspecies variations concerning the interactions of camptothecin with serum albumins: a frequency-domain fluorescence spectroscopic study. Biochemistry 1994;33(42):12540–12545.
27. Arcamone F. Properties of antitumor anthracyclines and new developments in their application: Cain memorial award lecture. Cancer Res 1985;45:5995–5999.
28. Chlebowsk RT. Adramycin (doxorubicin) cardiotoxicity: a review. West J Med 1979;131:364–368.
29. Jeffrey GA, Schienger W. Hydrogen Bonding in Biological Structures. Berlin: Springer-Verlag; 1994. p 169–214. ISBN 0-3875703-6.
30. Menozzi M, Valentini L, Vannini E, Acromone F. Self-association of doxorubicin and related compounds in aqueous solution. J Pharm Sci 1984;73:766–770.
31. Gordaliza M, Garcia PA, Miguel del Corral JM, Castro MA, Gomez-Zurita MA. Podophyllotoxin: distribution, sources, applications and new cytotoxic derivatives. Toxicon 2004;44(4):441–459.
32. Wang ZH, Ho JX, Carter DC. Methods and compositions for optimizing blood and tissue stability of camptothecin and circulatory availability of other albumin-binding therapeutic compounds 2006, March 2; EPO 06736461.2.
33. Burke T, Ho JX, Carter DC. Methods and compositions for optimizing blood and tissue stability of camptothecin and other albumin-binding therapeutic compounds 2002, March 20; WIPO PCT/US02/08301.
34. Fallon HJ, Adams LL, Lamb RG. A review of the mode of action of clofibrate and betabenzalbutyrate. Lipids 1972;7(2):106–109.
35. Committee of Principal Investigators WHO co-operative trial on primary prevention of ischaemic heart disease with clofibrate to lower serum cholesterol: final mortality follow-up. Lancet 1984;2(8403):600–604.
36. Kirksey Y, Ellenburg M, Tran H, Schreeder M, Carter DC, unpublished results.
37. Combes O, Barre J, Duche J-C, Vernillet C, Archimbaud Y, Marietta MP, Tillement J-P, Urien S. In vitro binding and partitioning of irinotecan (CPT-11) and its metabolite, SN38, in human blood. Invest New Drugs 2000;18:1–5.
38. Kruszewski S, Burke TG. Camptothecin affinity to HSA and membranes determined by fluorescence anisotropy measurements. Opt Appl 2002;32:721–730.

NANOTECHNOLOGY IN DRUG DELIVERY

IJEOMA F. UCHEGBU[1]
ANDREAS G. SCHATZLEIN[2]

[1] Department of Pharmaceutics, School of Pharmacy, University of London, London, UK
[2] Department of Pharmaceutical and Biological Chemistry, School of Pharmacy, University of London, London, UK

1. INTRODUCTION

Nanotechnology (technology at the nanoscale <1 μm) has been exploited to produce a new class of pharmaceutical in which the active is presented in the form of a nanoparticle (∼5–800 nm, e.g., Fig. 1). Such drug-loaded nanoparticles are sometimes termed nanomedicines. Essentially, presentation of the drug within a nanoparticle allows drug biodistribution to be controlled (Fig. 2) and in some cases the drug can be targeted to various areas of the anatomy such as the brain [1] or pathological sites such as solid tumors [2,3]. However, it must be made clear at the outset that, in essence, the achievement of significantly higher drug concentrations in target tissues with nanoparticles (when compared to target tissue drug concentrations achieved with the drug in solution) is what is meant by drug targeting and the use of the term targeting does not mean that other nontarget tissues are actually excluded from drug exposure. Nanomedicines are prepared from self-assembled low molecular weight amphiphiles (liposomes, niosomes, and micelles) [4–7], self-assembled amphiphilic polymers [8–13], polymer–drug conjugates [14,15], various water insoluble polymers [16,17], dendrimer-based electrostatic self-assemblies [3,18], carbon nanotubes [19], and even viruses [20].

Although there are a variety of chemistries that may be exploited to make various nanomedicines, to date only liposomal formulations (e.g., Doxil), micellar formulations (e.g., Fungizone), polymer–drug conjugates (e.g., Oncaspar), and a viral gene medicine—Gendicine® have been licensed for clinical use. Most of the nanomedicine research is in the preclinical phase of development, but as will be seen from the foregoing chapter, promising preclinical and premarket clinical data are beginning to emerge with some technologies and we will certainly see more nanomedicines being commercialized over the coming years.

This chapter will focus on synthetically derived nanoparticles that when loaded with drugs produce nanomedicines and the reader is directed to a number of reviews that exist on the use of viruses, the latter of which are mainly used to deliver genes [20–22].

2. SELF-ASSEMBLED LOW MOLECULAR WEIGHT AMPHIPHILES: LIPOSOMES, NIOSOMES, AND MICELLES

2.1. Chemical Structure and Liposome/Niosome/Micelle Preparation

Amphiphilic molecules with hydrophilic head groups and hydrophobic alkyl/acyl chains, such as phospholipids and nonionic surfactants (Fig. 3) self-assemble into closed bilayer vesicles and micelles (Fig. 4) in aqueous media. The nature of the self-assembly depends on the relative proportion of hydrophilic and hydrophobic portions of the molecule, as defined by Israelachvili's critical packing parameter (CPP) (Eq. 1) [23].

$$\text{CPP} = \frac{v}{a_0 l}$$

where v is the volume of hydrophobic moiety, a_0 is the hydrophilic head group area, and l is the length of hydrophobic chain. Spherical micelles are formed when $\text{CPP} < 1/3$, bilayer vesicles are formed when $\text{CPP} = 1/2 - 1$ and reverse micelles when $\text{CPP} > 1$.

Closed bilayer vesicles are termed liposomes [4,24] if derived from phospholipids or termed niosomes [5,6] if derived from nonionic surfactants (Fig. 4). Self-assembly into micelles and vesicles is driven by the entropy gain [25] associated with the liberation of water bounding the hydrophobic entities in the molecule, as the hydrophobic portion of the molecule is unable to hydrogen bond with water molecules. This entropy-driven hydrophobic association effectively shields the hydrophobic regions of the molecule from the

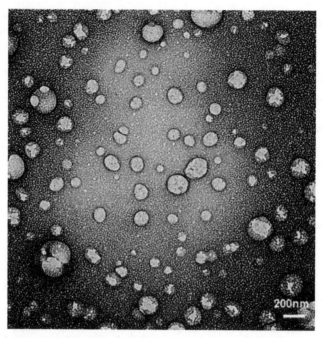

Figure 1. A nanomedicine comprising quaternary ammonium cetyl poly(ethylenimine) (10 mg/mL) and cyclosporine A (2 mg/mL). This formulation improves the oral absorption of cyclosporine A by threefold when compared to a suspension of cyclosporine A alone [79].

aqueous phase. Liposome and niosome formation require the input of energy in the form of kinetic energy, heat energy, or both kinetic and heat energies and as such these formulations are prepared using various methods: heating with agitation, microfluidization, high-pressure homogenization or probe sonication, for example, see Refs [4,5,26,27].

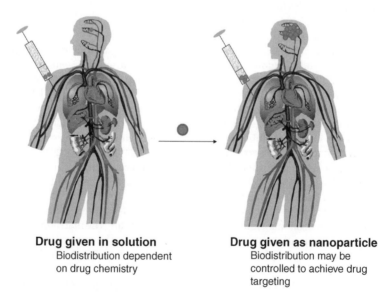

Drug given in solution
Biodistribution dependent on drug chemistry

Drug given as nanoparticle
Biodistribution may be controlled to achieve drug targeting

Figure 2. Schematic illustration of the use of nanotechnology to control biodistribution. (This figure is available in full color at http://mrw.interscience.wiley.com/emrw/9780471266945/home.)

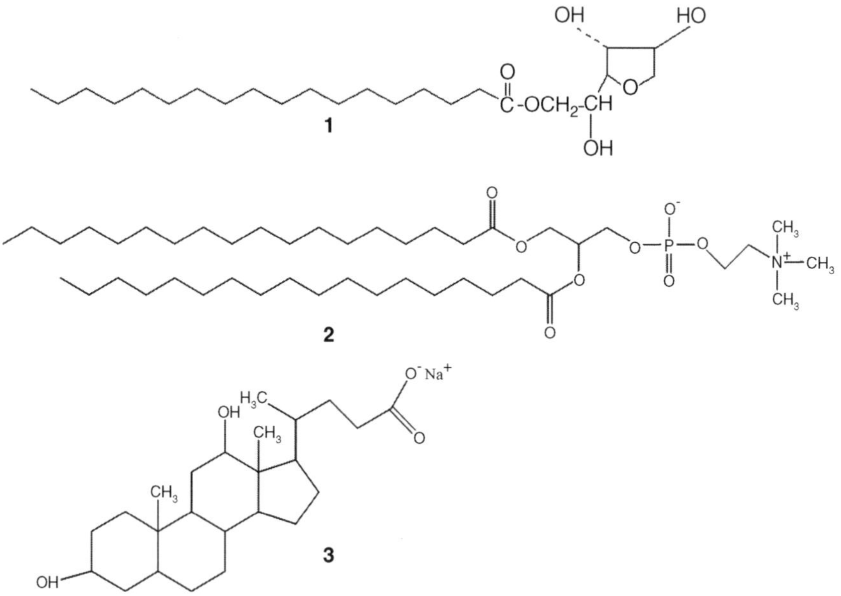

Figure 3. Low molecular weight amphiphiles. **1**: sorbitan monostearate that forms niosomes on self-assembly; **2**: 1,2-distearoyl-sn-glycerol-3-phosphocholine that forms liposomes on self-assembly; and **3**: sodium deoxycholate that forms micelles on self-assembly.

The resulting vesicles that possess a hydrophilic core (Fig. 4) may be loaded with hydrophilic drug solutions to give drug-loaded liposomes or niosomes [28–30]. It is these drug-loaded liposomes or niosomes that alter drug biodistribution such that more of the drug is targeted to pathological sites.

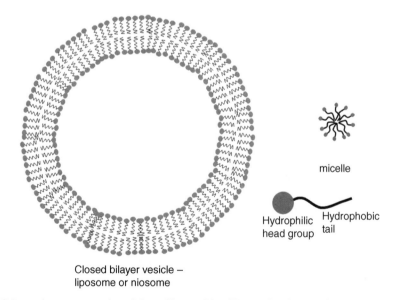

Figure 4. Schematic representation of the self-assembly of low molecular weight amphiphiles into closed bilayer vesicles (liposomes and niosomes) and micelles; liposomes and niosomes possess a hydrophilic core within which solutions of hydrophilic drugs may be encapsulated and micelles possess a hydrophobic core within which hydrophobic drugs may be encapsulated.

Micelles are formed on aggregation of comparatively hydrophilic surfactants and aggregation is seen at a critical concentration known as the critical micellar concentration (CMC), when the level of individual molecules in the aqueous phase reaches a critical level; determined largely by the molecule's chemistry and nature of the disperse phase [7,31]. Micelles are loaded with hydrophobic drugs within the micelle hydrophobic core and are usually used to improve the solubility of hydrophobic compounds, thus enabling intravenous administration into the aqueous blood compartment or to facilitate absorption from the gastrointestinal tract [32,33].

2.2. Drug Delivery with Liposomes

On intravenous injection, the surface chemistry of the liposomes is a crucial determinant of their distribution as the interactions of endogenous molecules or structures with the surface of the particle ultimately govern the path taken by these structures once administered [34,35]. A coating with a hydrophilic polymer increases the plasma half-life of intravenously injected liposomes by limiting uptake of the liposomes by the liver and spleen [34–36]. This reduced uptake of the liposomes by the liver and spleen has been exploited in the treatment of solid tumors as the prolonged circulation of the liposomes allows the liposomes to accumulate [37,38] and extravasate [39,40] within tumor tissue. Ultimately, this provides a clinical benefit to cancer patients [2]. Extravasation is achieved due to the disorganized tumor vasculature [41,42] (Fig. 5). This phenomenon has been termed the enhanced permeation and retention (EPR) effect and was first used to describe the tumor tropism shown by polymers [43].

Examples of liposomal drugs licensed for the treatment of various cancers include liposomal doxorubicin (Doxil or Caelyx®) indicated for the treatment of AIDS related Kaposi's sarcoma, refractory ovarian cancer, and, in combination with bortezomib, for the treatment of multiple myeloma [2,40,44–46] as well as liposomal daunorubicin (DaunoXome®) indicated for the treatment of AIDS related Kaposi's sarcoma, non-Hodgkin's lymphoma, and certain leukemias [47–50].

While a number of workers have exploited the fact that liposomes passively target tu-

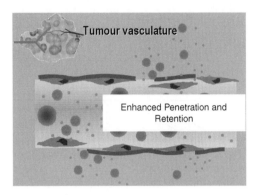

Figure 5. An illustration of the enhanced permeation and retention effect [43]—the hypothesized mechanism by which nanoparticles and polymers extravasate via the leaky vasculature of solid tumors and are retained within the tumor tissue, thus actually targeting tumors *in vivo*. (This figure is available in full color at http://mrw.interscience.wiley.com/emrw/9780471266945/home.)

mors, as detailed above, some active targeting strategies have also been employed in preclinical studies, although none of these endeavors have yet resulted in licensed products. Ligands [51–53] and antibodies [54,55] have been used to increase drug accumulation within the liver parenchyma [51] (since plain liposomes typically target the liver macrophages [56]) and solid tumors [54,55,57].

However, the use of liposomes is not limited to the treatment of neoplastic diseases and the systemic antifungal agent, amphotericin B (Ambisome®) was the first liposomal formulation to be licensed for clinical use [58]. This formulation reduces the incidence of nephrotoxic side effects by limiting distribution of the drug to the kidneys.

Transdermal formulations may also be enhanced with liposomes as specially engineered liposomes with "deformable" membranes—transfersomes, transport low and high molecular weight actives more efficiently across the skin [59–62]. A transfersome-based ketoprofen formulation (Diractin®) has undergone advanced clinical testing for pain relief in osteoarthritis and is able to localize the drug to the site of application without any side effects (side effects that usually originate from systemic exposure are eliminated) and confers pain relief that is similar to the market leader Celebrex® [63].

Most liposome formulations have their drugs encapsulated within the liposome interior, however, anticancer drugs may also be conjugated to the liposomal lipid which is subsequently used to make the liposomes [64], ensuring that the drug is covalently linked to the liposome. Such a strategy has resulted in improved chemotherapeutics in preclinical studies [64,65]. Liposomes may also be targeted to tumors with the use of a magnetic field originating from an intratumoral magnet and the administration of magnetic liposomes [66].

A promising area of research for which preclinical proof of concept data exist is the use of environmentally responsive liposomes: liposomes which release their contents in response to an external stimuli such as heat [67,68], ultrasound [69,70], or pH [71]. Thermosensitive formulations, for example, show improved tumoricidal activity in preclinical studies in the presence of an external heat stimulus at the tumor site [67].

2.3. Drug Delivery with Niosomes

Non-ionic surfactants also assemble into non-ionic surfactant vesicles or niosomes in a similar manner to phospholipids [6,30] (Figs 3 and 4). These species are able to target anticancer drugs such as doxorubicin to tumor tissue [30,72] and anti-infectives such as sodium stibogluconate to the liver and spleen for the treatment of leishmaniasis [73]. These formulations may offer an opportunity to use alternative materials to phospholipids, since niosomes have been prepared from materials such as sorbitan monostearate [30,74], a permitted food additive [75] that is cheap and widely available.

2.4. Drug Delivery with Micelles

Micellar formulations are largely used to increase the aqueous solubility of hydrophobic drugs and promote the gut permeation (and in turn oral absorption) of drug compounds [32,33]. Interestingly, these commonly used micelle forming surfactants such as polysorbate 80 are known to cause toxic effects in the submillimolar range, such as total inhibition of acetylcholine-induced endothelium-derived relaxing factor (EDRF) release arising from destruction of the blood vessel endothelial lining [76]. These toxic effects have been observed despite the fact that this surfactant is generally regarded as safe.

3. AMPHIPHILIC POLYMERS

Amphiphilic polymers are polymers that have separate hydrophobic and separate hydrophilic moieties. These amphiphiles may exist as block copolymers, in which there is a hydrophobic block of monomers linked to a hydrophilic block of monomers or may take the form of a hydrophobic or hydrophilic polymer backbone conjugated to either hydrophilic or hydrophobic pendant groups, respectively—pendant amphiphilic polymers (Fig. 6).

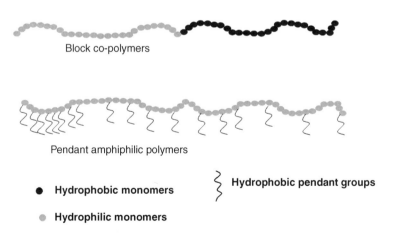

Figure 6. Schematic representation of amphiphilic block copolymers and amphiphilic pendant polymers.

3.1. Pendant Amphiphilic Polymers: Chemical Structure and Nanomedicine Synthesis

Examples of amphiphilic polymers comprising a water-soluble polymer backbone and hydrophobic pendant groups (pendant amphiphilic polymers) [77–79] are given in Fig. 7. Other water-soluble polymer backbones that have been used as starting materials in the synthesis of these pendant amphiphilic polymers include poly(L-lysine) [80–83] and poly(L-ornithine) [80]. These amphiphilic polymers either simply bear relatively low molecular weight hydrophobic pendant groups or, in some cases, possess additional hydrophilic substituents conjugated to the soluble polymer backbone (e.g., Fig. 7). The products are obtained in good yield (>70%) [77,79] and the level of hydrophobic modification is adequately controlled by the feed ratio of the long hydrophobic chain acyl or alkyl groups [77,84].

These polymers, by virtue of their chemistry, self-assemble in aqueous media [79,84] in a similar manner to the low molecular weight amphiphiles. Self-assembly yields extremely stable entities with critical micellar concentrations in the micromolar range [78,84], compared to the millimolar range seen for the pluronic block copolymers [85]. This exceptional stability stems from the ability of each amphiphilic molecule to engage in multiple hydrophobic associations; proof of this multiple hydrophobic contact hypothesis is provided by evidence that more stable micelles (with a lower CMC) are formed from molecules with more hydrophobic pendant groups [78,84] or from molecules with a higher molecular weight [84]. On a practical level, self-assembly is achieved by the introduction of energy and this could originate from simple handshaking such as with the relatively hydrophilic polymers shown in Fig. 7 or by the use of probe sonication [77,78,82].

The actual morphology of the self-assembly is largely controlled by polymer architecture with the hydrophilic polymers producing small 20–50 nm spherical micelles, polymers of intermediate hydrophobicity producing 100–600 nm spherical bilayer vesicle structures and the relatively hydrophobic materials forming 100–600 nm dense spherical and

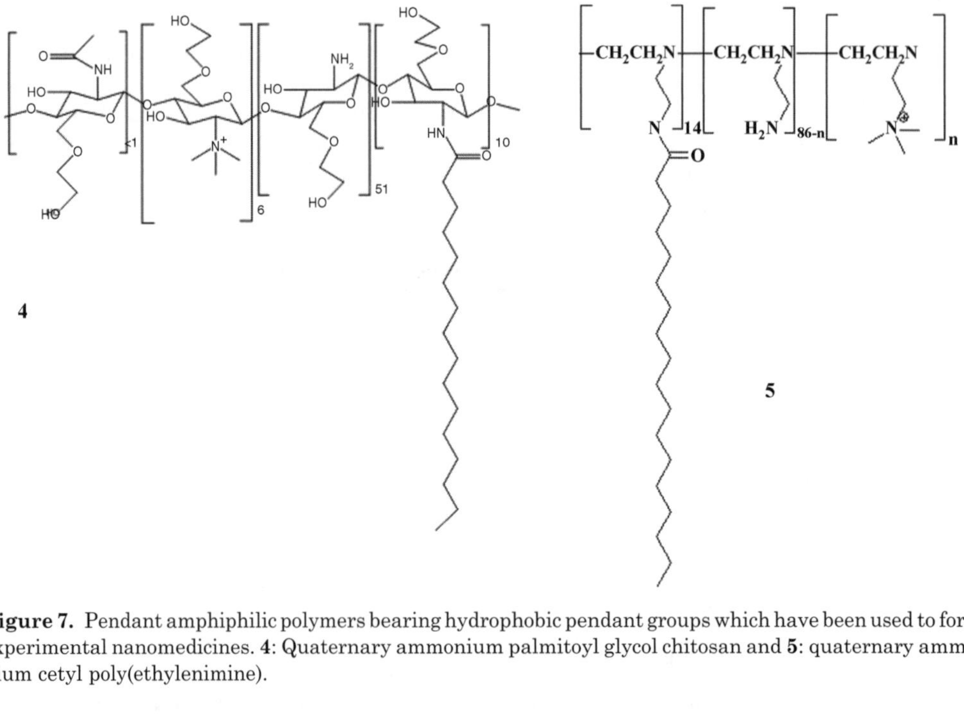

Figure 7. Pendant amphiphilic polymers bearing hydrophobic pendant groups which have been used to form experimental nanomedicines. **4**: Quaternary ammonium palmitoyl glycol chitosan and **5**: quaternary ammonium cetyl poly(ethylenimine).

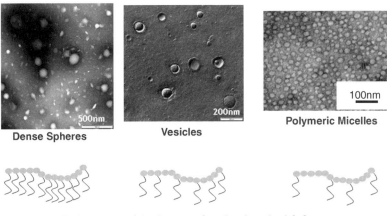

Figure 8. Nanosystems resulting from the self-assembly of cetyl poly(ethylenimine) amphiphiles [77]. The morphology of the resulting nanosystem depends on the level of hydrophobic substitution of the polymer backbone. With more hydrophilic polymers a high curvature micellar specie results, with intermediate hydrophobicity polymers, closed bilayer vesicles result and dense nanoparticles result when the level of hydrophobicity exceeds a critical value. The dense nanoparticles are in effect nanoprecipitates stabilized by a surface layer of the amphiphile. The size of the vesicular and dense nanoparticle assemblies are dependent on the level of cetylation of the amphiphiles [77]. (This figure is available in full color at http://mrw.interscience. wiley.com/emrw/9780471266945/home.)

oval nanoparticles [77,83] (Fig. 8). This finding builds upon Israelachvili's work [23], which taught us that the relative sizes of the hydrophobic and hydrophilic portions of low molecular weight amphiphiles governs the self-assembly as detailed above.

While flattened vesicles are sometimes observed with the more hydrophobic amphiphiles, a route to disc-shaped self-assemblies has recently been discovered (Fig. 9) branched polymers because of their lower hydrophilic head group area give rise to disc-shaped self-assemblies in the presence of cholesterol [78]. Shape variants are far from a scientific curiosity as particle shape appears to have a profound effect on particle cellular internalization [86]. The effect of nanomedicine shape on drug delivery *in vivo* is an area that awaits investigation.

A further property of these amphiphilic polymer-based nanosystems is the ability to precisely control their particle size. Particle size is linearly related to molecular weight [87] and also to the level of hydrophobic substitution [77]. Evidence is emerging that the self-assembly of these amphiphilic polymers in terms of particle size and morphology may be controlled to an unprecedented degree, due principally to the sheer number of variables that may be manipulated in a single molecule.

On interaction with hydrophobic drugs, the polymeric micelles, form nanomedicines that may take the form of drug swollen particles of >100 nm (e.g., Fig. 1) or drug filled micelles of <50 nm. In the laboratory, nanomedicine preparation is achieved by the introduction of energy and as such particles are either prepared by handshaking a mixture of the amphiphilic polymer, diluent, and drug or by probe sonication of this mixture [79]. These pendant amphiphilic polymer nanomedicines are colloidally stable; with unchanged quantities of drug present in the colloidal fraction after up to 270 days of storage in the liquid or freeze dried state [79].

3.2. Block copolymers: Chemical Structure and Nanomedicine Preparation

One of the most widely used amphiphilic block copolymers is poly(D,L-lactide-co-glycolide)-*block*-poly(ethylene oxide) (**6**, Fig. 10) [88]; this polymer has been used to form micelles [89–91], polymeric vesicles (termed

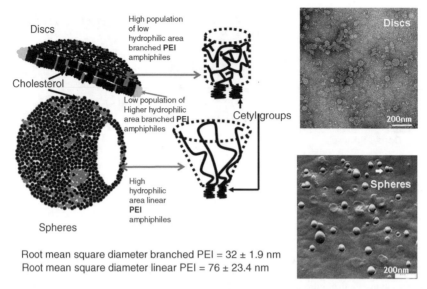

Figure 9. Disc-shaped nanoparticles are produced by the use of branched polymers and cholesterol while linear polymer amphiphiles give rise to mostly spherical self-assemblies [78]. This is exemplified using branched and linear poly(ethylenimine). The lower hydrodynamic area of the hydrophilic head group of the branched amphiphile enables the planar disc-shaped structures to evolve during self-assembly. The high curvature edges of the disc are thought to be composed of less hydrophobically substituted polymer amphiphiles. Discs shown in the transmission electron micrograph were prepared from branched cetyl poly(ethylenimine) containing 1.9 cetyl chains per molecule (MW = ~1050 Da) and the spheres shown in the transmission electron micrograph were prepared from the linear cetyl poly(ethylenimine) containing 1.4 cetyl chains per molecule (MW = ~700 Da), arrow indicates unusual fused particles [78].

polymerosomes when vesicles are prepared from a block copolymer) [91], and dense core nanoparticles [88]. Polymersomes may be converted to polymeric micelles on hydrolysis of the hydrophobic block [91]; indicative of the nature in which subtleties in polymer chemistry fundamentally govern self-assembly. These nanosystems are generally prepared by a water in oil in water (w/o/w) double emulsion technique [92] or by rehydrating films of the polymer deposited from an organic solvent [93]. In the w/o/w double emulsion method, the polymer in an organic solvent and drug in aqueous media are homogenized to produce a water in oil emulsion; this emulsion is then dispersed in aqueous media containing surfactant stabilizers such as polyvinyl pyrollidone to form a w/o/w emulsion, which is then homogenized; the procedure is complete on evaporation of the organic solvent to produce the nanoparticles with the nanoparticles being collected by filtration [92]. In the film rehydration method, which is used to prepare polymerosomes, the block copolymer is dissolved in an organic solvent, the solvent evaporated and the resulting film hydrated with a solution of the drug [93].

Polymersomes were first reported in the 1990s [94] and have been prepared from a variety of block copolymers (e.g., Fig. 10). Other polymersome forming block copolymers include poly(ethylene oxide)-*block*-poly(ethylethylene) [12], poly(ethylene oxide)-*block*-poly(caprolactone) [91], poly(ethylene oxide)-*block*-poly([3-(trimethyoxy-silyl)propyl methacrylate] [95,96], poly(ethylene oxide)-block-poly(butadiene), [97–99], poly(ethylene oxide)-*block*-poly(propylene sulphide) [100], and poly(ethylene oxide)-*block*-polystyrene [94].

With block copolymers (Fig. 2), it has been established that the critical packing parameter [23] should approach unity for vesicular self-assemblies to prevail [101]. However, once again polymer-specific factors also govern the self-assembly of block copolymers; the flexibility of the hydrophobic block in block copoly-

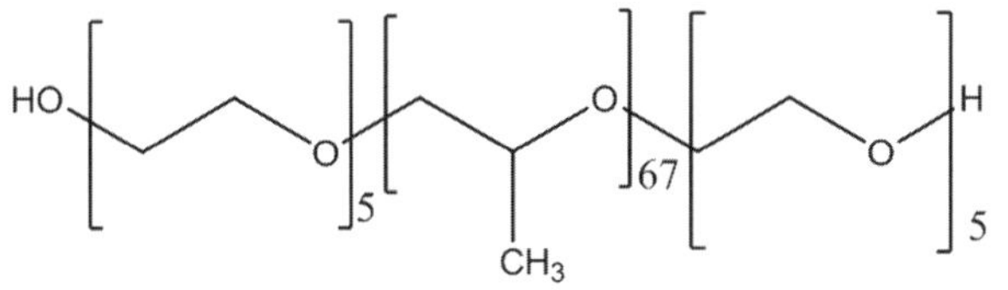

Figure 10. Block copolymers used to prepare nanomedicines. **6**: A typical poly(ethylene oxide)-*block*-poly(propylene oxide)-*block*-poly(ethylene oxide) triblock copolymer and **7**: poly(D,L-lactide-co-glycolide)-*block*-poly(ethylene oxidel) diblock copolymer.

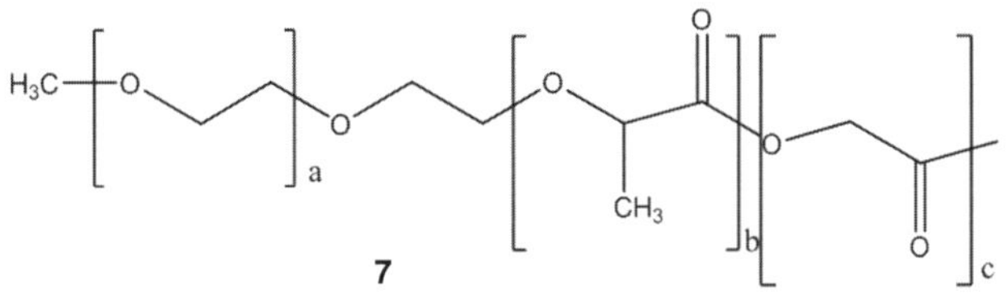

mers determines which self-assemblies will be formed; with more flexible hydrophobic portions of the polymer able to form vesicles and the more rigid polymers unable to self-assemble into three dimensional structures [101].

In a similar manner to the pendant type polymer vesicle membranes [77], polymersome membranes are also thicker than conventional vesicle membranes, ranging from 8–21 nm in thickness, two to five times thicker than the 4 nm hydrophobic cores displayed by conventional low molecular weight amphiphile membranes [12,23,91,102,103]. Polymersome membrane thickness is determined by the degree of polymerization in the hydrophobic block [102]. These extra thick membranes confer exceptional stability to organic solvents and water-soluble surfactants [101] and superior mechanical stability [12,101,104,105]. Polymersomes are thus less likely to rupture on perturbation [12] and their mechanical stability increases with polymer molecular weight [102].

3.3. Polymerization After Self-Assembly

An example of a polymerizable vesicle forming monomer is shown in Fig. 11. A polymerization step follows self-assembly with these vesicles. This method of polymeric vesicle production was one of the earliest methods of producing polymeric vesicles [106,107]. However, it must be stated that vesicles produced from polymerized self-assembling monomers are essentially polymer shells and it is unclear how much of the bilayer assembly actually survives the polymerization step.

The main advantage of this technology is the ability to produce extremely stable carriers that resist degradation from detergents [108,109] and organic solvents [108,109], are also less leaky [110], thermostable [111], and because the vesicle forming components are kinetically trapped (in essence actually fundamentally altered) by the polymerization process have improved colloidal stability [112]. The resulting nanosystems may be isolated as dry powders which are readily dispersible in water to give 50–100 nm particles [109]; thus, potentially enabling the formulation of nanomedicine-based solid dosage forms, in which the nanomedicine is more likely to survive pharmaceutical processing. However, as polymerization involves fairly reactive species, this technology is best applied prior to drug loading and thus has some limitations.

3.4. Drug Delivery with Block Copolymers

Block copolymer particles have been shown to circulate for prolonged periods on intravenous administration [88,113], a property that is known to favor tumor targeting as the particles then have enough time to extravasate the leaky tumor vasculature (Fig. 5). Polymersomes composed of poly(ethylene oxide)-*block*-polybutadiene or poly(ethylene

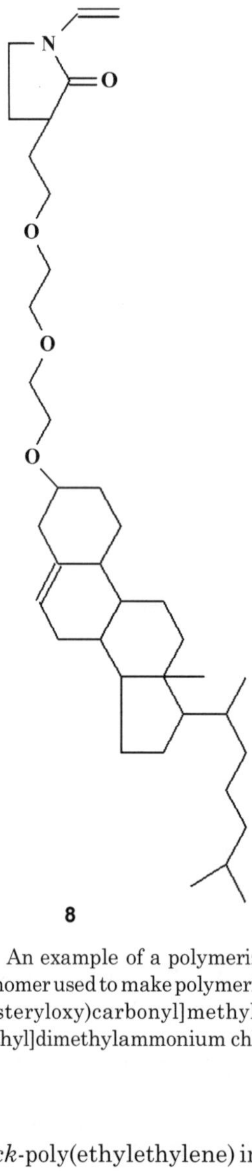

Figure 11. An example of a polymerizable vesicle forming monomer used to make polymerized vesicles. 8: [[(Cholesteryloxy)carbonyl]methyl][2-(methacryoyloxy)ethyl]dimethylammonium chloride [109].

oxide)-*block*-poly(ethylethylene) in which the entire vesicle surface is covered with a poly(ethylene oxide) coat have blood circulation half lives of up to 28 h in rats [113]. The circulation time of poly(ethylene oxide)-coated polymersomes is directly dependent on the length of the poly(ethylene oxide) block and half lives of up to 28 h are obtained with a poly(ethylene oxide) degree of polymerization of 50 [113]. This half-life compares favorably with a half-life of 14 h recorded for poly(ethylene oxide)-coated liposomes [114]. It is assumed that the 100% surface coverage of the polymeric vesicles is responsible for the reduced clearance of these polymersomes from the blood. [105]. The long half-life of these polymersomes makes them possible candidates for the development of antitumor medicines.

Furthermore, drug release may be controlled in the polymersomes by controlling the hydrolysis rate of the hydrophobic blocks [91]. This has been demonstrated with poly(L-lactic acid)-*block*-poly(ethylene oxide) and poly(caprolactone)-*block*-poly(ethylene oxide) vesicles [91]. Polymer hydrolysis of the hydrophobic block causes the polymer to move from a vesicular to a micellar assembly as the overall level of hydrophobic content diminishes and this in turn leads to drug release [91]. Hydrolysis rates and in turn release rates may be controlled by varying the relative level of the biodegradable hydrophobic blocks. It is conceivable that an optimum combination of prolonged circulation and programmed release rates may yield nanomedicines that are able to deliver high numbers of drug molecules to a target site at precise times after dosing; thus, yielding extreme levels of combined temporal and locoregional control of drug activity. Recent evidence that these block copolymer species are taken up by cells [89] also makes these materials attractive for the construction of nanomedicines.

The ultimate goal of all drug delivery efforts is the simple fabrication of responsive systems that are capable of delivering precise quantities of their pay load in response to pathological stimuli. Preprogrammable and intelligently responsive pills, implants, and injectables are so far merely the unobtainable ideal, however, polymeric nanomedicines have been fabricated with responsive capability.

Diblock polypeptides in which the hydrophilic block consists of ethylene oxide derivatized amino acids (L-lysine) and the hydrophobic block consists of poly(L-leucine) form pH responsive vesicles which disaggregate at low pH (pH = 3.0) [115]. It is possible that such L-lysine systems, if suitably developed, may be applied to facilitate release in intracellular acidic compartments such as the endosome, since a pH gradient of ~2 pH units exist between the cytosol and the endosome.

A number of reports have recently emerged that demonstrate that these block copolymer nanoparticles result in the delivery of greater proportions of the administered dose to tumors: both paclitaxel and cisplatin are targeted to tumors using these nanotechnologies [116,117] and these block copolymer-based nanomedicines produce significant tumor growth delay when compared to the pristine drug.

3.5. Drug Delivery with Pendant Amphiphilic Polymers

While still at an experimental stage, pendant amphiphilic polymer nanomedicines have proved in preclinical studies to be able to significantly improve the bioavailability of medicines via the oral [79], parenteral [84,118], and topical ocular [84] routes.

For molecules to be delivered in clinically relevant amounts via the oral route, said molecules have to have appropriate dissolution kinetics within the gastrointestinal tract, so as to be able to dissolve within a finite time window, and also be permeable to the gut epithelium of the absorptive villi [119]. Permeability through the gastrointestinal tract may occur via the transcellular route—the route taken by most hydrophobic compounds or the paracellular route—the route taken by most hydrophilic compounds [120,121]. Compounds must additionally evade the efflux transporters located on the luminal side of the gastrointestinal epithelial cells, such as the P-glycoprotein efflux pump, in order to be absorbed [122].

Quaternary ammonium cetyl poly(ethylenimine) (4, Fig. 7) nanoparticles enhance the oral absorption of cyclosporine A by threefold [79]. These polymers act by increasing the dissolution of the drug within the gastrointestinal tract [79] and promoting transcellular transport. The polymer did not inhibit the P-glycoprotein efflux pump or significantly promote transport of cyclosporine A via the paracellular route [79]. However, these polymers were not progressed into clinical development because of their relatively poor biocompatibility [79].

For molecules to be delivered to the brain parenchyma in the treatment of CNS diseases, such molecules have to cross the blood–brain barrier. The blood–brain barrier is a formidable transport barrier for most compounds as capillaries, characterized by an absence of fenestrae, are surrounded by astrocyte foot processes and capillary endothelial cells are characterized by tight intercellular junctions, low pinocytotic activity, and efflux transporters at their luminal surface; all of which limit the passage of 95% molecules into the brain [123–126].

A quaternary ammonium palmitoyl glycol chitosan (4 in Fig. 7) nanomedicine with a particle size of 150 nm, on intravenous delivery, is able to enhance the brain activity of the anesthetic propofol by 10-fold [84]. Although a reduced particle size is integral to the activity of this nanomedicine, as a particle size in excess of 400 nm does not yield the bioavailability gains seen with the smaller particle size, its mechanism of action is not entirely clear [84].

This chitosan amphiphile nanomedicine also promotes the transport of drugs across the cornea on topical administration [84].

Environmentally responsive experimental nanomedicines have also been prepared from pendant polymer amphiphiles. Vesicles that release their contents in the presence of an enzyme may be formed by loading polymeric pendant amphiphile vesicles with an enzyme activated prodrug [8]. The particulate nature of the drug delivery system should allow the drug to accumulate in tumors as shown in Fig. 5, where the particulates may be activated by an externally applied enzyme or an enzyme specific to the tumor cell. Alternatively, a membrane-bound enzyme may be used to control and ultimately prolong the activity of either an entrapped hydrophilic drug (entrapped in the vesicle aqueous core) or an entrapped hydrophobic drug (entrapped in the vesicle membrane) [8].

3.6. Drug Delivery with Polymerized Vesicles

Magnetically responsive polymerized liposomes composed of 1,2-di-(2,4-octadecadienoyl)-sn-glycerol-3-phosphorylcholine, loaded with ferric oxide and subsequently polymerized may be localized by an external magnetic field to the small intestine and specifically the Peyer's patches [108]. These polymerized vesicles are stable to the degradative influence of solubilizing surfactants such as triton-X

100 [108] and hence should not suffer excessive bile salt mediated degradation during gut transit. These magnetically responsive polymeric vesicles may be used to improve the absorption of drugs via the oral route and the authors of this report found higher levels of a radioactive marker in the liver of magnetically treated mice when compared to control animals which had not been subjected to magnetic treatment [108].

4. POLYMER–DRUG CONJUGATES

4.1. Chemical Structure and Nanomedicine Preparation

A schematic representation of a polymer–drug conjugate is given in Fig. 12. The polymer backbone is usually a water-soluble material and may be a poly(ethylene oxide), poly(glutamic acid), or hydroxypropylmethacrylamide (HPMA) [14]. These polymer–drug conjugates are 10 nm in size and usually have a molecular mass of <40,000 Da.

4.2. Drug Delivery with Polymer–Drug Conjugates

The first polymer–drug conjugates to be licensed for clinical use were the protein poly (ethylene oxide) conjugates [127]. These act by improving the half-life of the protein and maintaining therapeutic blood levels of the protein for longer periods. A large variety of drug delivery technologies have been developed for the area of oncology, since the high potency and narrow therapeutic index of most antitumor compounds means that they benefit excellently from drug targeting strategies. Cytotoxic polymer–drug conjugates [128] have thus joined liposomes as a means of targeting these drugs. A HPMA polymer–drug conjugate in which the cytotoxic drug is conjugated to the polymer backbone via enzymatically cleavable linkers was shown to accumulate within tumors in preclinical models [128] and also to reduce the incidence of clinical side effects in patients with some responses also being observed in patients refractory to other treatments [15]. While HPMA anticancer polymeric prodrugs were the first cytotoxic polymers to enter clinical trials [129], a number of other polymeric prodrugs such as poly(glutamic acid)-based polymers have also undergone clinical testing. A poly(glutamic acid)-paclitaxel polymer–drug conjugate (Xyotax® or Opaxio®) has undergone clinical testing for the treatment of women with advanced small lung cell cancer; this

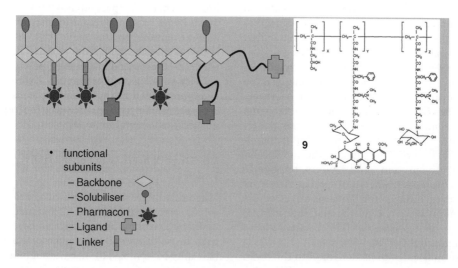

Figure 12. A schematic representation of a polymer–drug conjugate (left) and an example of a polymer–drug conjugate that has been clinically tested (right). 9: PK2 [14] bearing the drug doxorubicin and the galactose targeting agent bound via a peptide linker to 2-(hydroxypropyl)methacryamide. (This figure is available in full color at http://mrw.interscience.wiley.com/emrw/9780471266945/home.)

polymer is cleaved by Cathepsin B and its cleavage is linked to oestrogen status, making it active in women [130,131]. Xyotax is currently under evaluation for a product license.

Further refinements to the polymer–drug conjugate technology have involved a number of different variations on the polymer–drug conjugate theme, such as the use of thermally responsive polymer backbones that in combination with localized hyperthermia (42°C), produce high tumor levels of the polymer [132,133].

Combining the polymer prodrug and particulate encapsulation technologies has also yielded interesting results [134,135]. The encapsulation of polymer–drug conjugates within niosomes [134,135] improves the accumulation of the drug within the liver [135], while the encapsulation of polymer–drug conjugates within nanoparticles improves the tumoricidal activity [136] when compared to the unencapsulated polymer–drug conjugate.

In recognition of the fact that ligand targeted systems may offer even more tumor specificity, polymer–drug conjugates bearing galactose ligands targeted to hepatocytes have been developed [137]. Delivery of siRNA (small interfering ribonucleic acid) and DNA is challenging because of the anionic and large nature of these molecules and as such a polymer conjugate with targeting ligands has been used to deliver siRNA. An siRNA delivery system bearing a targeting ligand specific for hepatocytes produced apolipoprotein A silencing, resultant reduced blood cholesterol levels and the accumulation of fat in the liver [138]. These studies provide proof of concept data on *in vivo* gene silencing using siRNA and pave the way for siRNA treatments to be developed.

5. WATER INSOLUBLE POLYMERS AND CROSS-LINKED POLYMERIC NANOPARTICLES

5.1. Chemical Structure and Nanomedicine Preparation

Water insoluble polymers such as poly(lactic acid), poly(D,L-lactide-co-glycolide) (PLGA) (Fig. 13) and poly(alkyl cyanoacrylates) are regularly used to prepare nanoparticles using the w/o/w double emulsion technique outlined above [16]. Furthermore, water-

Figure 13. 10: Poly(D,L-lactide-co-glycolide) a nanoparticle forming water insoluble polymer.

soluble polymers have been used to prepare nanoparticles by chemical cross-linking, for example, cross-linked albumin [1] and cross-linked chitosan [139].

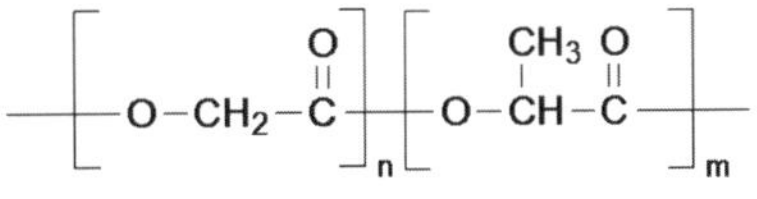

5.2. Drug Delivery with Water Insoluble Polymers and Cross-Linked Polymeric Nanoparticles

Surfactant-coated poly(butylcyanoacrylate) nanoparticles [17] have been used to deliver drugs across the blood–brain barrier as exemplified by the production of a pharmacodynamic response from a neuropeptide—dalargin; dalargin does not usually cross the blood–brain barrier [17,140]. While the mechanism of action is not clear the activity of these surfactant-coated nanoparticles has been linked to a perturbation of the brain endothelial barrier, the result of a possible toxic effect at the blood–brain barrier [141]. The use of nanoparticles bearing ligands for transport receptors at the blood–brain barrier is also a strategy used to deliver drugs across the blood–brain barrier and as such the intravenous administration of cross-linked albumin particles bearing apolipoprotein E (ApoE) ligands results in receptor mediated transcytosis via ApoE brain endothelial cell receptors [1] and improved brain drug delivery [142].

Nanomedicines based on water insoluble polymers have also been employed in preclinical cancer chemotherapy studies: PLGA nanoparticles in preclinical studies have been shown to deliver oligonucleotides and cause radiosensitization in head and neck squamous cell carcinomas in mice [143].

Finally, nanoparticles may also be used simply to improve the oral bioavailability of a drug with a low aqueous solubility [144–146] and have also been found to improve the delivery of ibuprofen to the aqueous humor on ocular topical application [147].

6. DENDRIMERS

6.1. Chemical Structure and Nanomedicine Preparation

Dendrimers (from the Greek "dendron": tree, and "meros": part) are highly ordered, branched monodisperse macromolecules [148] (Fig. 14). Such dendritic structures first emerged in a new class of polymers named "cascade molecules," initially reported by Vögtle and his group at the end of the 1970s [149]. Further development by Tomalia's group [150] and Newkome's group [151,152] gave rise to larger dendritic structures. These hyperbranched molecules are termed "dendrimers" or "arborols" (from the Latin "arbor" for tree). Their unique molecular architecture means that dendrimers have a number of distinctive properties that differentiate them from other polymers; specifically they are not the result of statistical polymerization events but are built up in a stepwise fashion from a core group (a convergent method of synthesis) or by the addition of dendrimer arms to a core group (a divergent method of synthesis). The controlled synthetic approach means that dendrimers tend to be monodisperse with a well-defined size and structure.

Dendrimers used for gene delivery contain amine functional groups and are usually protonated at physiological pH (cationic); such protonated amine groups may be electrostatically bound to DNA phosphate groups to give dendriplexes [153–158]. This electrostatic interaction protects DNA from degradation *in vivo* [155,159].

6.2. Gene Delivery with Dendrimers

As stated earlier, gene delivery is a challenge due to the large and anionic nature of DNA. Poly(amidoamine) and poly(propylenimine) dendrimers efficiently deliver nucleic acids into cells; ranging from small oligonucleotides to plasmids and artificial chromosomes [3,157,160–167]. The higher generation poly(amidoamine) dendrimers (e.g., generation 5 to generation 10) efficiently transfect cells and are more efficient gene carriers than the lower generation materials [168,169]. However, cell toxicity has been shown to be directly related to molecular weight [170,171], making the lower generation dendrimers, such as the lower generation poly(propylenimine) dendrimers more attractive for the formulation of gene nanomedicines [3].

When higher generation poly(amidoamine) dendriplexes (200 µg DNA complexed with 650 µg poly(amidoamine)$_{G9}$) are injected intravenously into healthy mice, gene expression

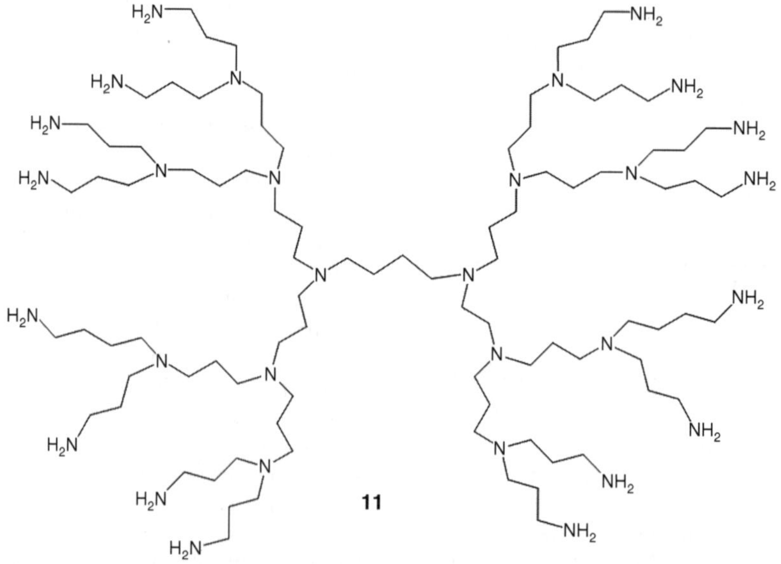

Figure 14. 11: Poly(propylenimine) generation 3 dendrimer (DAB 16).

occurs mainly in the lung parenchyma but not in other organs [172]. Lower generation dendrimer conjugate (poly(amidoamine)$_{G3}$-α-cyclodextrin) dendriplexes, however, transfect primarily the spleen [173], while the lower generation poly(propylenimine) (generation 2 and generation 3) dendriplexes (50 μg DNA complexed with 250 μg poly(propylenimine)$_{G3}$) transfect primarily the liver [158].

Cancer gene therapy is currently limited by the difficulty of efficiently delivering therapeutic genes to remote tumors and metastases [174,175]. However, poly(propylenimine)$_{G3}$ dendriplexes on intravenous administration localize within the tumor [176] and induce tumor gene expression [3]. This localization within the tumor is likely to be linked to the fact that the poly(propylenimine)$_{G3}$ dendriplexes avoid the lung endothelial cells and thus have a chance to extravasate to tumor tissue, unlike the poly(amidoamine)$_{G9}$ dendriplexes [172], poly(ethylenimine)–DNA complexes [158], and liposome–DNA complexes [177,178], which all localize gene expression to the lung. This lung avoidance is hypothesized to be linked to the low molecular weight of the poly(propylenimine)$_{G3}$ dendrimers, which may enable them to avoid excessive aggregation within the blood and thus evade being trapped in the lung capillaries. Others have observed low molecular weight (MW = ~4 kDa) oligoethylenimine pseudo-dendrimers also direct gene expression away from the lung tissue and to tumor tissue on intravenous administration, when compared to linear (MW = ~25 kDa) poly(ethylenimine) [179].

When murine xenografts were treated by intravenous injection of poly(propylenimine)$_{G3}$ dendriplexes containing a tumor necrosis factor alpha (TNFα) expression plasmid under control of a tumor-specific promoter, regression of established tumors was observed in 100% of the animals. The antitumor activity is the result of a combination of the effects of the tumor-specific expression of TNFα and an intrinsic antiproliferative effect of the dendrimer [3]. This novel antiproliferative effect was also observed with other cationic polymers. The lack of apparent toxicity and significant weight loss compared to untreated controls suggests that the treatment is relatively well tolerated [3]. Furthermore, the administration of poly(propylenimine)$_{G3}$ dendriplexes containing genes expressing either a p53 or a minimal p53-derived apoptotic peptide, or alternatively a small hairpin RNA (shRNA) sequence that targets the p73 apoptosis inhibitor—iASPP, results in tumor regression independent of p53 status [166]. The gene expressing the shRNA sequence leads to downregulation of iASPP and consequently activation of the p73 apoptosis pathway [166]. These studies conducted by Bell and coworkers were the first to validate p73 as a target in cancer therapy.

7. CARBON NANOTUBES

7.1. Chemical Structure and Nanomedicine Preparation

Carbon nanotubes comprise a wall of graphene wrapped into a seamless tube (Fig. 15) [180]. Graphene consists of carbon atoms covalently linked in a hexagonal aromatic arrangement of carbon atoms. Carbon nanotubes are largely insoluble in organic and aqueous solvents and require derivatization [180] prior to use in biomedical applications [19]. Carbon nanotubes have thus been derivatized on their surface with amino acids, polymers, proteins such as bovine serum albumin, and enzymes such as peroxidize [180–182].

7.2. Drug Delivery with Carbon Nanotubes

Functionalized carbon nanotubes have been used experimentally to deliver various drugs.

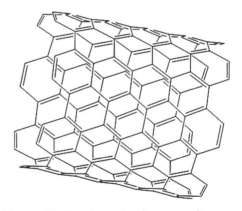

Figure 15. Structure of carbon nanotubes—prepared from a layer of graphene wound into a seamless tube.

For example, carbon nanotubes grafted with poly(acrylic acid) and decorated with magnetic particles and gemcitabine by physical adsorption [183] were used to deliver gemcitabine to lymph nodes on subcutaneous injection. The carbon nanotubes were absorbed by lymph vessels and transferred to lymph nodes under the direction of a magnetic field and these studies point to a possible treatment for lymphatic metastasis. Carbon nanotubes have also been derivatized with both drugs and targeting ligands [184]. When such materials are decorated with epidermal growth factor and with cisplatin, good tumor regression results and such tumor regression is superior to that seen with the nontargeted nanotubes [184]. Carbon nanotubes derivatized with 1,6-diaminohexane adsorbed siRNA and were then used to produce gene silencing and tumor regression [185].

Although evidence of these materials being metabolized is not apparent, carbon nanotubes are reported to be excreted in the urine on intravenous administration [186,187]. Carbon nanotube-based therapeutics are still highly experimental and a long way from clinical development and routine clinical use, however, the fact that carbon nanotubes appear to be eliminated in the urine in animal models does make further study of these nanosystems warranted.

8. CONCLUSIONS

Nanotechnology has been exploited to yield a variety of experimental, and a few commercial, drug delivery systems. These pharmaceutical nanoparticles may be prepared from a variety of chemical compounds and, when combined with drugs, produce nanomedicines. Nanoparticles arise from the aqueous self-assembly of polymer and low molecular weight amphiphiles, from the nanoprecipitation of water insoluble polymers, from water insoluble materials such as carbon nanotubes and from elaborate water-soluble polymer constructs. Although the commercial use of these nanomedicines is currently limited there is much experimental evidence to show that these nanomedicines do offer benefit over conventional medicines in that once a drug is presented to the body in nanoparticulate form, the distribution of the drug may be controlled and drug targeting achieved. As such formulations have emerged that are able to target drugs and genes to tumor sites, thus limiting the exposure of healthy tissue to narrow therapeutic index drugs in particular and also producing efficient gene therapeutics. Nanomedicine formulations have also resulted in delivery of drugs across the blood–brain barrier; technologies that are badly needed to treat diseases of the central nervous system. Finally, nanomedicines are able to improve the oral absorption of hydrophobic compounds; thus, reducing the likelihood of therapeutic failure.

The preclinical testing arena is crowded with these technologies and it is predicted that a number of new technologies should emerge into the patient arena within the coming years. It is pertinent to note that only formulations that offer real clinical benefit will survive this busy race to the clinic and ultimately the market place.

REFERENCES

1. Zensi A, Begley D, Pontikis C, Legros C, Mihoreanu L, Wagner S, Buchel C, von Briesen H, Kreuter J. Albumin nanoparticles targeted with Apo E enter the CNS by transcytosis and are delivered to neurones. J Control Rel 2009;137:78–86.
2. Gabizon AA. Pegylated liposomal doxorubic: Metamorphosis of an old drug into a new form of chemotherapy. Cancer Investig 2001;19: 424–436.
3. Dufes C, Keith WN, Bilsland A, Proutski I, Uchegbu JF, Schatzlein AG. Synthetic anticancer gene medicine exploits intrinsic antitumor activity of cationic vector to cure established tumors. Cancer Res 2005;65:8079–8084.
4. Gregoriadis G, editor. Liposome Technology. Vols I–III. Boca Raton: CRC Press; 2006.
5. Uchegbu IF, Vyas SP. Non-ionic surfactant based vesicles (niosomes) in drug delivery. Int J Pharm 1998;172:33–70.
6. Uchegbu IF. Synthetic Surfactant Vesicles. Amsterdam: Harwood Academic Publishers; 2000.
7. Florence AT, Attwood D. Physicochemical Principles of Pharmacy. 4th ed. London: McMillan Press; 2006. p 564.

8. Uchegbu IF. Pharmaceutical nanotechnology: polymeric vesicles for drug and gene delivery. Expert Opin Drug Deliv 2006;3:629–640.
9. Discher DE, Ahmed F. Polymersomes. Ann Rev Biomed Engineer 2006;8:323–341.
10. Kita-Tokarczyk K, Grumelard J, Haefele T, Meier W. Block copolymer vesicles: using concepts from polymer chemistry to mimic biomembranes. Polymer 2005;46:3540–3563.
11. Uzun O, Xu H, Jeoung E, Thibault RJ, Rotello VM. Recognition-induced polymersomes: structure and mechanism of formation. Chem Eur J 2005;11:6916–6920.
12. Discher B, Won YY, Ege JCM, Bates FS, Discher D, Hammer DA. Polymersomes: tough vesicles made from diblock copolymers. Science 1999;284:1143–1146.
13. Discher DE, Ortiz V, Srinivas G, Klein ML, Kim Y, David CA, Cai SS, Photos P, Ahmed F. Emerging applications of polymersomes in delivery: From molecular dynamics to shrinkage of tumors. Prog Polym Sci 2007;32:838–857.
14. Duncan R. The dawning era of polymer therapeutics. Nat Rev Drug Discov 2003;2:347–360.
15. Seymour LW, Ferry DR, Kerr DJ, Rea D, Whitlock M, Poyner R, Boivin C, Hesslewood S, Twelves C, Blackie R, Schatzlein A, Jodrell D, Bissett D, Calvert H, Lind M, Robbins A, Burtles S, Duncan R, Cassidy J. Phase II studies of polymer-doxorubicin (PK1, FCE28068) in the treatment of breast, lung and colorectal cancer. Int J Oncol 2009;34:1629–1636.
16. Mundargi RC, Babu VR, Rangaswamy V, Patel P, Aminabhavi TM. Nano/micro technologies for delivering macromolecular therapeutics using poly(D,L-lactide-co-glycolide) and its derivatives. J Control Rel 2008;125:193–209.
17. Kreuter J, Ramge P, Petrov V, Hamm S, Gelperina SE, Engelhardt B, Alyautdin R, von Briesen H, Begley DJ. Direct evidence that polysorbate-80-coated poly(butylcyanoacrylate) nanoparticles deliver drugs to the CNS via specific mechanisms requiring prior binding of drug to the nanoparticles. Pharm Res 2003;20:409–416.
18. Dufes C, Uchegbu IF, Schatzlein AG. Dendrimers in gene delivery. Adv Drug Del Rev 2005;57:2177–2202.
19. Bianco A, Prato M. Can carbon nanotubes be considered useful tools for biological applications? Adv Mater 2003;15:1765–1768.
20. Peng Z. Current status of Gendicine in China: recombinant Ad-p53 agent for treatment of cancers. Hum Gene Ther 2005;16:1016–1027.
21. Mah C, Byrne BJ, Flotte TR. Virus-based gene delivery systems. Clin Pharm 2002;41: 901–911.
22. Gaspar HB, Thrasher AJ. Gene therapy for severe combined immunodeficiencies. Expert Opin Biol Ther 2005;5:1175–1182.
23. Israelachvili J. Intermolecular and Surface Forces. 2nd ed. London: Academic Press; 1991.
24. New RRC. Liposomes: a Practical Approach. Practical Approach Series. Oxford: Oxford University Press; 1990.
25. Tanford C. The Hydrophobic Effect: Formation of Micelles and Biological Membranes. New York: John Wiley & Sons; 1980.
26. Gregoriadis G, editor. Liposome Technology. Liposome Preparation and Related Techniques. 2nd ed. Vol. 1. Boca Raton: CRC Press; 1993.
27. Gregoriadis G, Saffie R, Hart SL. High yield incorporation of plasmid DNA within liposomes: effect on DNA integrity and transfection efficiency. J Drug Target 1996;3:469–475.
28. Haran G, Cohen R, Bar LK, Barenholz Y. Transmembrane ammonium-sulfate gradients in liposomes produce efficient and stable entrapment of amphipathic weak bases. Biochim Biophys Acta 1993;1151:201–215.
29. Gregoriadis G, editor. Liposome Technology. Entrapment of Drugs and Other Materials. 2nd ed. Vol. 2. Boca Raton: CRC Press; 1993.
30. Uchegbu IF, Double JA, Turton JA, Florence AT. Distribution, metabolism and tumoricidal activity of doxorubicin administered in sorbitan monosterate (Span 60) niosomes in the mouse. Pharm Res 1995;12:1019–1024.
31. Hildebrand A, Garidel P, Neubert RAB. Thermodynamics of demicellisation of mixed micelles composed of sodium oleate and bile salts. Langmuir 2004;20:320–328.
32. Strickley RG. Solubilizing excipients in oral and injectable formulations. Pharm Res 2004;21:201–230.
33. van Zuylen L, Verweij J, Sparreboom A. Role of formulation vehicles in taxane pharmacology. Invest New Drugs 2001;19:125–141.
34. Blume G, Cevc G. Liposomes for the sustained drug release *in vivo*. Biochim Biophys Acta 1990;1029:91–97.
35. Torchilin VP, Omelyanenko VG, Papisov MI, Bogdanov AA, Trubetskoy VS, Herron JN, Gentry CA. Poly(ethylene glycol) on the liposome surface: on the mechanism of polymer-coated liposome longevity. Biochim Biophys Acta-Biomembr 1994;1195:11–20.

36. Whiteman KR, Subr V, Ulbrich K, Torchilin VP. Poly(HPMA)-coated liposomes demonstrate prolonged circulation in mice. J Liposome Res 2001;11:153–164.
37. Gabizon A, Catane R, Uziely B, Kaufman B, Safra T, Cohen R, Martin F, Huang A, Barenholz Y. Prolonged circulation time and enhanced accumulation in malignant exudates of doxorubicin encapsulated in polyethylene-glycol coated liposomes. Cancer Res 1994;54:987–992.
38. Papahadjopoulos D, Allen TM, Gabizon A, Mayhew E, Matthay K, Huang SK, Lee KD, Woodle MC, Lasic DD, Redemann C, et al. Sterically stabilized liposomes: improvements in pharmacokinetics and antitumor therapeutic efficacy. Proc Natl Acad Sci USA 1991;88:11460–11464.
39. Huang SK, Lee K-D, Hong K, Friend DS, Papahadjopoulos D. Microscopic localisation of sterically stabilised liposomes in colon carcinoma bearing mice. Cancer Res 1992;52:5135–5142.
40. Gabizon A, Shmeeda H, Barenholz Y. Pharmacokinetics of pegylated liposomal doxorubicin: review of animal and human studies. Clin Pharm 2003;42:419–436.
41. Eatock MM, Schatzlein AG, Kaye SB. Tumour vasculature as a target for anti-cancer therapy. Cancer Treat Rev 2000;26:191–204.
42. Baish JW, Gazit Y, Berk DA, Nozue M, Baxter LT, Jain RK. Role of tumour vascular architecture in nutrient and drug delivery. Microvasc Res 1996;51:327–346.
43. Maeda H. The tumor blood vessel as an ideal target for macromolecular anticancer agents. J Control Rel 1992;19:315–324.
44. Cattel L, Ceruti M, Dosio F. From conventional to stealth liposomes a new frontier in cancer chemotherapy. Tumori 2003;89:237–249.
45. Rose PG. Pegylated liposomal doxorubicin: optimizing the dosing schedule in ovarian cancer. Oncologist 2005;10:205–214.
46. Moreau P. Combination regimens using doxorubicin and pegylated liposomal doxorubicin prior to autologous transplantation in multiple myeloma. Expert Rev Anticancer Ther 2009;9:885–890.
47. Cervetti G, Caracciolo F, Cecconi N, Azzara A, Petrini M. Efficacy and toxicity of liposomal daunorubicin included in PVABEC regimen for aggressive NHL of the elderly. Leuk Lymphoma 2003;44:465–469.
48. Presant CA, Scolaro M, Kennedy P, Blayney DW, Flanagan B, Lisak J, Presant J. Liposomal daunorubicin treatment of HIV-associated Kaposis-sarcoma. Lancet 1993;341:1242–1243.
49. Fassas A, Anagnostopoulos A. The use of liposomal daunorubicin (DaunoXome) in acute myeloid leukemia. Leuk Lymphoma 2005;46:795–802.
50. Latagliata R, Breccia M, Iacobelli S, Fazi P, Martinelli G, Di Raimondo F, Sborgia M, Fabbiano F, Lauria F, Zuffa E, Venditti A, Crugnola M, Peta A, Candoni A, Cilloni D, Mattei E, Morselli M, Pastore D, Adamo F, Centra A, Falzetti F, Vignetti M, Alimena G, Mandelli F. Long term follow-up of the gimema GSI 103 AMLE randomized trial: Daunoxome seems to improve disease-free survival (DFS) of elderly patients with acute myelogenous leukemia (AML). 48th Annual Meeting of the American Society of Hematology 2006: 1979.
51. Kawakami S, Fumoto S, Nishikawa M, Yamashita F, Hashida M. *In vivo* gene delivery to the liver using novel galactosylated cationic liposomes. Pharm Res 2000;17:306–313.
52. Kawakami S, Wong J, Sato A, Hattori Y, Yamashita F, Hashida M. Biodistribution characteristics of mannosylated, fucosylated, and galactosylated liposomes in mice. Biochim Biophys Acta 2000;1524:258–265.
53. Hattori Y, Kawakami S, Yamashita F, Hashida M. Controlled biodistribution of galactosylated liposomes and incorporated probucol in hepatocyte-selective drug targeting. J Control Rel 2000;69:369–377.
54. Nam SM, Kim HS, Ahn WS, Park YS. Sterically stabilized anti-G(M3), anti-Le(x) immunoliposomes: targeting to B16BL6, HRT-18 cancer cells. Oncol Res 1999;11:9–16.
55. Cao Y, Suresh MR. Bispecific MAb aided liposomal drug delivery. J Drug Target 2000;8:257–266.
56. Gregoriadis G, Ryman B. Fate of protein containing liposomes injected into rats. Eur J Biochem 1972;24:484–491.
57. Marty C, Odermatt B, Schott H, Neri D, Ballmer-Hofer K, Klemenz R, Schwendener RA. Cytotoxic targeting of F9 teratocarcinoma tumours with anti-ED-B fibronectin scFv antibody modified liposomes. Br J Cancer 2002;87:106–112.
58. Boswell GW, Buell D, Bekersky I. AmBisome (liposomal amphotericin B): a comparative review. J Clin Pharmacol 1998;38:583–592.
59. Schatzlein A, Cevc G. Non-uniform cellular packing of the stratum corneum and permeability barrier function of intact skin: a high-resolution confocal laser scanning microscopy

study using highly deformable vesicles (transfersomes). Br J Dermatol 1998;138:583–592.
60. Cevc G, Blume G. New, highly efficient formulation of diclofenac for the topical, transdermal administration in ultradeformable drug carriers, Transfersomes. Biochim Biophys Acta 2001;1514:191–205.
61. Cevc G, Schatzlein A, Richardsen H. Ultradeformable lipid vesicles can penetrate the skin and other semi-permeable barriers unfragmented. Evidence from double label CLSM experiments and direct size measurements. Biochim Biophys Acta 2002;1564:21–30.
62. Cevc G, Schatzlein AG, Richardsen H, Vierl U. Overcoming semipermeable barriers, such as the skin, with ultradeformable mixed lipid vesicles, transfersomes, liposomes, or mixed lipid micelles. Langmuir 2003;19:10753–10763.
63. Cevc G. A small company's approach to developing innovative pharmaceuticals for treating peripheral conditions: from bench to registration. J Pharm Pharmacol (Suppl 1): 2009; A143.
64. Asai T, Shuto S, Matsuda A, Kakiuchi T, Ohba H, Tsukada H, Oku N. Targeting and antitumor efficacy of liposomal 5′-O- dipalmitoylphosphatidyl 2′-C-cyano-2′-deoxy-1-beta-D-arabino-pentofuranosylcytosine in mice lung bearing B16BL6 melanoma. Cancer Lett 2001;162:49–56.
65. Vodovozova EL, Moiseeva EV, Grechko GK, Gayenko GP, Nifant'ev NE, Bovin NV, Molotkovsky JG. Antitumour activity of cytotoxic liposomes equipped with selectin ligand SiaLe(X), in a mouse mammary adenocarcinoma model. Eur J Cancer 2000;36:942–949.
66. Kubo T, Sugita T, Shimose S, Nitta Y, Ikuta Y, Murakami T. Targeted delivery of anticancer drugs with intravenously administered magnetic liposomes in osteosarcoma-bearing hamsters. Int J Oncol 2000;17:309–315.
67. Yuyama Y, Tsujimoto M, Fujimoto Y, Oku N. Potential usage of thermosensitive liposomes for site-specific delivery of cytokines. Cancer Lett 2000;155:71–77.
68. Lindner LH, Eichhorn ME, Eibl H, Teichert N, Schmitt-Sody M, Issels RD, Dellian M. Novel temperature-sensitive liposomes with prolonged circulation time. Clin Cancer Res 2004;10:2168–2178.
69. Myhr G. Multimodal cancer treatment: real time monitoring, optimization, and synergistic effects. Technol Cancer Res Treat 2008;7: 409–414.
70. Schroeder A, Honen R, Turjeman K, Gabizon A, Kost J, Barenholz Y. Ultrasound triggered release of cisplatin from liposomes in murine tumors. J Control Rel 2009;137:63–68.
71. Ishida T, Kirchmeier MJ, Moase EH, Zalipsky S, Allen TM. Targeted delivery and triggered release of liposomal doxorubicin enhances cytotoxicity against human B lymphoma cells. Biochim Biophys Acta 2001;1515:144–158.
72. Rogerson A, Cummings J, Willmott N, Florence AT. The distribution of doxorubicin in mice following administration in niosomes. J Pharm Pharmacol 1988;40:337–342.
73. Baillie AJ, Coombs GH, Dolan TF, Laurie J. Nonionic surfactant vesicles, niosomes, as a delivery system for the antileishmanial drug, sodium stibogluconate. J Pharm Pharmacol 1986;38:502–505.
74. Uchegbu IF, Vyas SP. Non-ionic surfactant based vesicles (niosomes) in drug delivery. Int J Pharm 1998;172:33–70.
75. UK, Government. Statutory Instrument 1989 No. 876, The Emulsifiers and Stabilisers in Food Regulations 1989. London: The Stationery Office; 1989.
76. Uluoglu C, Korkusuz P, Uluoglu O, Zengil H. Tween 80 and endothelium: functional reduction due to tissue damage. Res Commun Mol Pathol Pharmacol 1996;91:173–183.
77. Wang W, Qu XZ, Gray AI, Tetley L, Uchegbu IF. Self-assembly of cetyl linear polyethylenimine to give micelles, vesicles, and dense nanoparticles. Macromolecules 2004;37:9114–9122.
78. Qu X, Omar L, Le TBH, Tetley L, Bolton K, Chooi KW, Wang W, Uchegbu IF. Polymeric amphiphile branching leads to rare nano-disc shaped planar self assemblies. Langmuir 2008;24:9997–10004.
79. Cheng WP, Gray AI, Tetley L, Hang TLB, Schatzlein AG, Uchegbu IF. Polyelectrolyte nanoparticles with high drug loading enhance the oral uptake of hydrophobic compounds. Biomacromolecules 2006;7:1509–1520.
80. Brown MD, Schatzlein A, Brownlie A, Jack V, Wang W, Tetley L, Gray AI, Uchegbu IF. Preliminary characterization of novel amino acid based polymeric vesicles as gene and drug delivery agents. Bioconj Chem 2000;11:880–891.
81. Brown MD, Gray AI, Tetley L, Santovena A, Rene J, Schatzlein AG, Uchegbu IF. In vitro and in vivo gene transfer with poly(amino acid) vesicles. J Control Rel 2003;93:193–211.
82. Wang W, Tetley L, Uchegbu IF. A new class of amphiphilic poly-L-lysine based polymers forms nanoparticles on probe sonication in aqueous media. Langmuir 2000;16:7859–7866.

83. Wang W, Tetley L, Uchegbu IF. The level of hydrophobic substitution and the molecular weight of amphiphilic poly-L-lysine-based polymers strongly affects their assembly into polymeric bilayer vesicles. J Coll Interf Sci 2001;237:200–207.
84. Qu XZ, Khutoryanskiy VV, Stewart A, Rahman S, Papahadjopoulos-Sternberg B, Dufes C, McCarthy D, Wilson CG, Lyons R, Carter KC, Schatzlein A, Uchegbu IF. Carbohydrate-based micelle clusters which enhance hydrophobic drug bioavailability by up to 1 order of magnitude. Biomacromolecules. 2006;7: 3452–3459.
85. Alexandridis P, Holzwarth JF, Hatton TA. Micellization of poly(ethylene oxide)-poly(propylene oxide)-poly(ethylene oxide) triblock copolymers in aqueous solutions: thermodynamics of copolymer association. Macromolecules 1994;27:2414–2425.
86. Gratton SEA, Ropp PA, Pohlhaus PD, Luft JC, Madden VJ, Napier ME, DeSimone JM. The effect of particle design on cellular internalization pathways. Proc Natl Acad Sci USA 2008;105:11613–11618.
87. Wang W, McConaghy AM, Tetley L, Uchegbu IF. Controls on polymer molecular weight may be used to control the size of palmitoyl glycol chitosan polymeric vesicles. Langmuir 2001; 17:631–636.
88. Gref R, Minamitake Y, Peracchia MT, Trubetskoy VS, Torchilin VP, Langer R. Biodegradable long circulating polymeric nanospheres. Science 1994;263:1600–1603.
89. Savic R, Luo L, Eisenberg A, Maysinger D. Micellar nanocontainers distribute to defined cytoplasmic organelles. Science 2003;300: 615–618.
90. Savic R, Eisenberg A, Maysinger D. Block copolymer micelles as delivery vehicles of hydrophobic drugs: micelle-cell interactions. J Drug Target 2006;14:343–355.
91. Ahmed F, Discher DE. Self-porating polymersomes of PEG-PLA and PEG-PCL: hydrolysis-triggered controlled release vesicles. J Control Rel 2004;96:37–53.
92. Huh KM, Cho YW, Park K.PLGA-PEG copolymers. Drug Deliv Technol 2003; 3. Available at http://www.drugdeliverytech.com/ME2/dirmod.asp?sid = 4306B1E9C3CC4E07A4-D64E23FBDB232C&nm=Back + Issues&type =Publishing&mod=Publications%3A%3AArticle &mid= 8F3A7027421841978F18BE895 F87F791 & tier =4&id= BB85E8579021481 EACBC7C 3F0674348F.
93. Discher DE, Discher BM, Won YY, Lee JC-M, Bates FS, Hammer DA.Polymersomes and related encapsulating membranes. US patent 6,835,394.
94. Yu KH, Eisenberg A. Bilayer morphologies of self-assembled crew-cut aggregagtes of amphiphilic PS-b-PEO copolymers in solution. Macromolecules 1998;31:3509–3518.
95. Du JZ, Chen YM. Preparation of organic/inorganic hybrid hollow particles based on gelation of polymer vesicles. Macromolecules 2004;37: 5710–5716.
96. Du JZ, Chen YC, Zhang Y, Han CC, Fishcer K, Schmidt M. Organic/inorganic hybrid vesicles based on a reactive block copolymer. J Am Chem Soc 2003;125:14710–14711.
97. Lin JJ, Silas JA, Bermudez H, Milam VT, Bates FS, Hammer DA. The effect of polymer chain length and surface density on the adhesiveness of functionalized polymersomes. Langmuir 2004;20:5493–5500.
98. Ahmed F, Hategan A, Discher DE, Discher BM. Block copolymer assemblies with cross-link stabilization: from single-component monolayers to bilayer blends with PEO-PLA. Langmuir 2003;19:6505–6511.
99. Bermudez H, Aranda-Espinoza H, Hammer DA, Discher DE. Pore stability and dynamics in polymer membranes. Europhys Lett 2003; 64:550–556.
100. Napoli A, Valentini M, Tirelli N, Muller M, Hubbell JA. Oxidation-responsive polymeric vesicles. Nat Mater 2004;3:183–189.
101. Antonietti M, Forster S. Vesicles and liposomes: a self-assembly principle beyond lipids. Adv Mater 2003;15:1323–1333.
102. Bermudez H, Brannan AK, Hammer DA, Bates FS, Discher DE. Molecular weight dependence of polymerosome membrane structure, elasticity, and stability. Macromolecules 2002;35:8203–8208.
103. Discher BM, Hammer DA, Bates FS, Discher DE. Polymer vesicles in various media. Curr Opin Coll Interf Sci 2000;5:125–131.
104. Dimova R, Seifert U, Pouligny B, Forster S, Dobereiner HG, Hyperviscous diblock copolymer vesicles. Eur Phys J E 2002;7:241–250.
105. Lee JCM, Bermudez H, Discher BM, Sheehan MA, Won YY, Bates FS, Discher DE. Preparation, stability, and in vitro performance of vesicles made with diblock copolymers. Biotechnol Bioengineer 2001;73:135–145.
106. Hub HH, Hupfer B, Koch H, Ringsdorf H. Polymerizable phospholipid analogues: new

stable biomembrane cell models. Angew Chem Int Ed Engl 1980;19:938–940.
107. Eaton PE, Jobe PG, Kayson N. Polymerised vesicles. J Am Chem Soc 1980;102: 6638–6640.
108. Chen HM, Langer R. Magnetically-responsive polymerized liposomes as potential oral delivery vehicles. Pharm Res 1997;14:537–540.
109. Cho I, Chung KC. Cholesterol-containing polymeric vesicles: syntheses, characterization, and separation as a solid powder. Macromolecules 1988;21:565–571.
110. Cho I, Kim YD. Synthesis and properties of tocopherol-containing polymeric vesicle systems. Macromol Symp 1997;118:631–640.
111. Cho IH, Kim YD. Formation of stable polymeric vesicles by tocopherol-containing amphiphiles. Macromol Rapid Commun 1998;19:27–30.
112. Cho I, Dong S, Jeong SW. Vesicle formation by nonionic polymerizable cholesterol-based amphiphiles. Polymer 1995;36:1513–1515.
113. Photos PJ, Bacakova L, Discher B, Bates FS, Discher DE. Polymer vesicles *in vivo*: correlations with PEG molecular weight. J Control Rel 2003;90:323–334.
114. Blume G, Cevc G. Molecular mechanism of the lipid vesicle longevity *in vivo*. Biochim Biophys Acta 1993;1146:157–168.
115. Bellomo EG, Wyrsta MD, Pakstis L, Pochan DJ, Deming TJ. Stimuli-responsive polypeptide vesicles by conformation-specific assembly. Nat Mater 2004;3:244–248.
116. Mattheolabakis G, Taoufik E, Haralambous S, Roberts ML, Avgoustakis K. *In vivo* investigation of tolerance and antitumor activity of cisplatin-loaded PLGA-mPEG nanoparticles. Eur J Pharm Biopharm 2009;71:190–195.
117. Danhier F, Lecouturier N, Vroman B, Jerome C, Marchand-Brynaert J, Feron O, Preat V. Paclitaxel-loaded PEGylated PLGA-based nanoparticles: *in vitro* and *in vivo* evaluation. J Control Rel 2009;133:11–17.
118. Park JH, Kwon S, Lee M, Chung H, Kim JH, Kim YS, Park RW, Kim IS, Seo SB, Kwon IC, Jeong SY. Self-assembled nanoparticles based on glycol chitosan bearing hydrophobic moieties as carriers for doxorubicin: *in vivo* biodistribution and anti-tumor activity. Biomaterials 2006;27:119–126.
119. Martinez MN, Amidon GL. A mechanistic approach to understanding the factors affecting drug absorption: a review of fundamentals. J Clin Pharmacol 2002;42:620–643.
120. Daugherty AL, Mrsny RJ. Transcellular uptake mechanisms of the intestinal epithelial barrier. Part I. Pharm Sci Technol Today 1999;2:144–151.
121. Salama NN, Eddington ND, Fasano A. Tight junction modulation and its relationship to drug delivery. Adv Drug Del Rev 2006;58: 15–28.
122. Hunter J, Hirst BH. Intestinal secretion of drugs. The role of P-glycoprotein and related drug efflux systems in limiting oral drug absorption. Adv Drug Del Rev 1997;25:129–157.
123. Deeken JF, Loscher W. The blood–brain barrier and cancer: transporters, treatment, and Trojan horses. Clin Cancer Res 2007;13: 1663–1674.
124. Pardridge WM. The Blood–brain barrier: bottleneck in brain drug development. J Am Soc Exp NeuroTherap 2005;2:3–14.
125. Begley DJ. The blood–brain barrier: principles for targeting peptides and drugs to the central nervous system. J Pharm Pharmacol 1996;48: 136–146.
126. Begley DJ. ABC transporters and the blood–brain barrier. Curr Pharm Des 2004;10: 1295–1312.
127. Harris JM, Martin NE, Modi M. Pegylation: a novel process for modifying pharmacokinetics. Clin Pharm 2001;40:539–551.
128. Seymour L, Ulbrich K, Steyger P, Brereton M, Subr V, Strohalm J, Duncan R. Tumour tropism and anti-cancer efficacy of polymer-based doxorubicin prodrugs in the treatment of subcutaneous murine B16F10 melanoma. Br J Cancer 1994;70:636–641.
129. Vasey PA, Kaye SB, Morrison R, Twelves C, Wilson P, Duncan R, Thomson AH, Murray LS, Hilditch TE, Murray T, Burtles S, Fraier D, Frigerio E, Cassidy J. Phase I clinical and pharmacokinetic study of PK1 [*N*-(2-hydroxypropyl)methacrylamide copolymer doxorubicin]: first member of a new class of chemotherapeutic agents–drug–polymer conjugates. Cancer Research Campaign Phase I/II Committee. Clin Cancer Res 1999;5:83–94.
130. Edelman MJ. Novel taxane formulations and microtubule-binding agents in non-small-cell lung cancer. Clin Lung Cancer 2009;10:S30–S34.
131. Mita M, Mita A, Sarantopoulos J, Takimoto CH, Rowinsky EK, Romero O, Angiuli P, Allievi C, Eisenfeld A, Verschraegen CF. Phase I study of paclitaxel poliglumex administered weekly for patients with advanced solid malignancies. Cancer Chemother Pharmacol 2009; 64:287–295.

132. Meyer DE, Shin BC, Kong GA, Dewhirst MW, Chilkoti A. Drug targeting using thermally responsive polymers and local hyperthermia. J Control Rel 2001;74:213–224.

133. Meyer DE, Kong GA, Dewhirst MW, Zalutsky MR, Chilkoti A. Targeting a genetically engineered elastin-like polypeptide to solid tumors by local hyperthermia. Cancer Res 2001;61: 1548–1554.

134. Gianasi E, Cociancich F, Uchegbu I, Florence A, Duncan R. Pharmaceutical and biological characterisation of a doxorubicin-polymer conjugate (PK1) entrapped in sorbitan monostearate Span 60 niosomes. Int J Pharm 1997;148:139–148.

135. Uchegbu I, Duncan R. Niosomes containing N-(2-hydroxypropyl)methacrylamide copolymer-doxorubicin (PK1): effect of method of preparation and choice of surfactant on niosome characteristics and a preliminary study of body distribution. Int J Pharm 1997; 155:7–17.

136. Mitra S, Gaur U, Ghosh PC, Maitra AN. Tumour targeted delivery of encapsulated dextran-doxorubicin conjugate using chitosan nanoparticles as carrier. J Control Rel 2001;74:317–323.

137. Seymour LW, Ulbrich K, Wedge SR, Hume IC, Strohalm J, Duncan R. N-(2-hydroxypropyl) methacrylamide copolymers targeted to the hepatocyte galactose-receptor: pharmacokinetics in DBA2 mice. Br J Cancer 1991; 63:859–866.

138. Rozema DB, Lewis DL, Wakefield DH, Wong SC, Klein JJ, Roesch PL, Bertin SL, Reppen TW, Chu Q, Blokhin AV, Hagstrom JE, Wolff JA. Dynamic polyconjugates for targeted in vivo delivery of siRNA to hepatocytes. Proc Natl Acad Sci USA 2007;104:12982–12987.

139. Banerjee T, Mitra S, Singh AK, Sharma RK, Maitra A. Preparation, characterization and biodistribution of ultrafine chitosan nanoparticles. Int J Pharm 2002;243:93–105.

140. Schroder U, Sabel BA. Nanoparticles, a drug carrier system to pass the blood–brain barrier, permit central analgesic effects of iv dalargin injections. Brain Res 1996;710:121–124.

141. Olivier JC, Fenart L, Chauvet R, Pariat C, Cecchelli R, Couet W. Indirect evidence that drug brain targeting using polysorbate 80-coated polybutylcyanoacrylate nanoparticles is related to toxicity. Pharm Res 1999;16:1836–1842.

142. Kreuter J, Hekmatara T, Dreis S, Vogel T, Gelperina S, Langer K. Covalent attachment of apolipoprotein A-I and apolipoprotein B-100 to albumin nanoparticles enables drug transport into the brain. J Control Rel 2007;118:54–58.

143. Zou J, Qiao XM, Ye HP, Zhang Y, Xian JM, Zhao HY, Liu SX. Inhibition of ataxia-telangiectasia mutated by antisense oligonucleotide nanoparticles induces radiosensitization of head and neck squamous-cell carcinoma in mice. Cancer Biother Radiopharm 2009;24: 339–346.

144. Demirel M, Yazan Y, Muller RH, Kilic F, Bozan B. Formulation and in vitro-in vivo evaluation of piribedil solid lipid micro- and nanoparticles. J Microencap 2001;18:359–371.

145. Sahana DK, Mittal G, Bhardwaj V, Kumar M. PLGA nanoparticles for oral delivery of hydrophobic drugs: influence of organic solvent on nanoparticle formation and release behavior in vitro and in vivo using estradiol as a model drug. J Pharm Sci 2008;97:1530–1542.

146. Shaikh J, Ankola DD, Beniwal V, Singh D, Kumar M. Nanoparticle encapsulation improves oral bioavailability of curcumin by at least 9-fold when compared to curcumin administered with piperine as absorption enhancer. Eur J Pharm Sci 2009;37:223–230.

147. Pignatello R, Bucolo C, Ferrara P, Maltese A, Puleo A, Puglisi G. Eudragit RS100 (R) nanosuspensions for the ophthalmic controlled delivery of ibuprofen. Eur J Pharm Sci 2002;16: 53–61.

148. Klajnert B, Bryszewska M. Dendrimers: properties and applications. Acta Biochim Pol 2001;48:199–208.

149. Buhleier E, Wehner W, Vögtle F. Cascadechain-like and nonskid-chain-like syntheses of molecular cavity topologies. Synthesis 1978: 155–158.

150. Tomalia DA, Baker H, Dewald J, Hall M, Kallos G, Martin S, Roeck J, Ryder J, Smith P. A new class of polymers: starburst-dendritic macromolecules. Polym J 1985;17:117–132.

151. Newkome GR, Yao ZQ, Baker GR, Gupta VK, Russo PS, Saunders MJ. Cascade Molecules. 2. Synthesis and characterization of a benzene[9] 3-arborol. J Am Chem Soc 1986;108:849–850.

152. Newkome GR, Yao ZQ, Baker GR, Gupta VK. Micelles. 1. Cascade molecules: a new approach to micelles. A [27]-arborol. J Org Chem 1985;50:2003–2004.

153. Bielinska AU, Chen CL, Johnson J, Baker JR. DNA complexing with polyamidoamine dendrimers: implications for transfection. Bioconjug Chem 1999;10:843–850.

154. Tang MX, Szoka FC. The influence of polymer structure on the interactions of cationic polymers with DNA and morphology of the resulting complexes. Gene Ther 1997;4:823–832.

155. Bielinska AU, Kukowska-Latallo JF, Baker JR. The interaction of plasmid DNA with polyamidoamine dendrimers: mechanism of complex formation and analysis of alterations induced in nuclease sensitivity and transcriptional activity of the complexed DNA. Biochim Biophys Acta 1997;1353:180–190.

156. Chen W, Turro NJ, Tomalia DA. Using ethidium bromide to probe the interactions between DNA and dendrimers. Langmuir 2000;16:15–19.

157. Zinselmeyer BH, Mackay SP, Schatzlein AG, Uchegbu IF. The lower-generation polypropylenimine dendrimers are effective gene-transfer agents. Pharm Res 2002;19:960–967.

158. Schatzlein AG, Zinselmeyer BH, Elouzi A, Dufes C, Chim YTA, Roberts CJ, Davies MC, Munro A, Gray AI, Uchegbu IF. Preferential liver gene expression with polypropylenimine dendrimers. J Control Rel 2005;101:247–258.

159. Abdelhady HG, Allen S, Davies MC, Roberts CJ, Tendler SJB, Williams PM. Direct real-time molecular scale visualisation of the degradation of condensed DNA complexes exposed to DNase I. Nucleic Acids Res 2003;31:4001–4005.

160. Bielinska A, Kukowska-Latallo JF, Johnson J, Tomalia DA, Baker JR Jr. Regulation of in vitro gene expression using antisense oligonucleotides or antisense expression plasmids transfected using starburst PAMAM dendrimers. Nucleic Acids Res 1996;24:2176–2182.

161. Delong R, Stephenson K, Loftus T, Fisher M, Alahari S, Nolting A, Juliano RL. Characterization of complexes of oligonucleotides with polyamidoamine starburst dendrimers and effects on intracellular delivery. J Pharm Sci 1997;86:762–764.

162. Yoo H, Sazani P, Juliano RL. PAMAM dendrimers as delivery agents for antisense oligonucleotides. Pharm Res 1999;16:1799–1804.

163. Yoo H, Juliano RL. Enhanced delivery of antisense oligonucleotides with fluorophore-conjugated PAMAM dendrimers. Nucleic Acids Res 2000;28:4225–4231.

164. Hollins AJ, Benboutera M, Omidi Y, Zinselmeyer B, Schatzlein AG, Uchegbu IF, Akhtar S. Evaluation of generation 2 and 3 poly(propylenimine) dendrimers for the potential cellular delivery of antisense oligonucleotides targeting epidermal growth factor receptor. Pharm Res 2004;21:458–466.

165. Santhakumran LM, Thomas T, Thomas TJ. Enhanced cellular uptake of a triplex-forming oligonucleotide by nanoparticle formation in the presence of polypropylenimine dendrimers. Nucl Acid Res 2004;32:2102–2112.

166. Bell HS, Dufes C, O'Prey J, Crighton D, Bergamaschi D, Lu X, Schatzlein AG, Vousden KH, Ryan KM. A p53-derived apoptotic peptide derepresses p73 to cause tumor regression in vivo. J Clin Invest 2007;117:1008–1018.

167. de Jong G, Telenius A, Vanderbyl S, Meitz A, Drayer J. Efficient in vitro transfer of a 60-Mb mammalian artificial chromosome into murine and hamster cells using cationic lipids and dendrimers. Chromosome Res 2001;9:475–485.

168. Kukowska-Latallo JF, Bielinska AU, Johnson J, Spindler R, Tomalia DA, Baker JR. Efficient transfer of genetic material into mammalian cells using starburst polyamidoamine dendrimers. Proc Natl Acad Sci USA 1996;93:4897–4902.

169. Tang MX, Redemann CT, Szoka FC. In vitro gene delivery by degraded polyamidoamine dendrimers. Bioconjug Chem 1996;7:703–714.

170. Haensler J, Szoka FC. Polyamidoamine cascade polymers mediate efficient transfection of cells in culture. Bioconjug Chem 1993;4:372–379.

171. Malik N, Wiwattanapatapee R, Klopsch R, Lorenz K, Frey H, Weener JW, Meijer EW, Paulus W, Duncan R. Dendrimers: relationship between structure and biocompatibility in vitro, and preliminary studies on the biodistribution of I-125-labelled polyamidoamine dendrimers in vivo. J Control Rel 2000;65:133–148.

172. Kukowska-Latallo JF, Raczka E, Quintana A, Chen CL, Rymaszewski M, Baker JR. Intravascular and endobronchial DNA delivery to murine lung tissue using a novel, nonviral vector. Hum Gene Ther 2000;11:1385–1395.

173. Kihara F, Arima H, Tsutsumi T, Hirayama F, Uekama K. In vitro and in vivo gene transfer by an optimised a-cyclodextrin conjugate with polyamidoamine dendrimer. Bioconjug Chem 2003;14:342–350.

174. Schatzlein AG. Non-viral vectors in cancer gene therapy: principles and progress. Anticancer Drugs 2001;12:275–304.

175. Schatzlein AG. Targeting of synthetic gene delivery systems. J Biomed Biotechnol 2003;2003:149–158.

176. Chisholm EJ, Vassaux G, Martin-Duque P, Chevre R, Lambert O, Pitard B, Merron A, Weeks M, Burnet J, Peerlinck I, Dai M-S, Alusi G, Mather SJ, Bolton K, Uchegbu IF, Schatzlein AG, Baril P. Cancer-specific transgene

176. expression mediated by systemic injection of nanoparticles. Cancer Res 2009;69:2955–2962.
177. Song YK, Liu F, Chu S, Liu D. Characterization of cationic liposome-mediated gene transfer *in vivo* by intravenous administration. Hum Gene Ther 1997;8:1585–1594.
178. Song YK, Liu F, Liu D. Enhanced gene expression in mouse lung by prolonging the retention time of intravenously injected plasmid DNA. Gene Ther 1998;5:1531–1537.
179. Russ V, Elfberg H, Thoma C, Kloeckner J, Ogris M, Wagner E. Novel biodegradeable oligoethylenimine acrylate ester-based pseudodendrimers for *in vitro* and *in vivo* gene transfer. Gene Ther 2008; 15:18–29.
180. Tasis D, Tagmatarchis N, Bianco A, Prato M. Chemistry of carbon nanotubes. Chem Rev 2006;106:1105–1136.
181. Pulikkathara MX, Khabashesku VN. Covalent sidewall functionalization of single-walled carbon nanotubes by amino acids. Russ Chem Bull 2008;57:1054–1062.
182. Prencipe G, Tabakman SM, Welsher K, Liu Z, Goodwin AP, Zhang L, Henry J, Dai HJ. PEG branched polymer for functionalization of nanomaterials with ultralong blood circulation. J Am Chem Soc 2009;131:4783–4787.
183. Yang D, Yang F, Hu JH, Long J, Wang CC, Fu DL, Ni QX. Hydrophilic multi-walled carbon nanotubes decorated with magnetite nanoparticles as lymphatic targeted drug delivery vehicles. Chem Commun 2009: 4447–4449.
184. Bhirde AA, Patel V, Gavard J, Zhang GF, Sousa AA, Masedunskas A, Leapman RD, Weigert R, Gutkind JS, Rusling JF. Targeted killing of cancer cells *in vivo* and *in vitro* with EGF-directed carbon nanotube-based drug delivery. ACS Nano 2009;3:307–316.
185. Yang R, Yang X, Zhang Z, Zhang Y, Wang S, Cai Z, Jia Y, Ma Y, Zheng C, Lu Y, Roden R, Chen Y. Single-walled carbon nanotubes-mediated *in vivo* and *in vitro* delivery of siRNA into antigen-presenting cells. Gene Ther 2006;13:1714–1723.
186. Wang HF, Wang J, Deng XY, Sun HF, Shi ZJ, Gu ZN, Liu YF, Zhao YL. Biodistribution of carbon single-wall carbon nanotubes in mice. J Nanosci Nanotechnol 2004;4:1019–1024.
187. Singh R, Pantarotto D, Lacerda L, Pastorin G, Klumpp C, Prato M, Bianco A, Kostarelos K. Tissue biodistribution and blood clearance rates of intravenously administered carbon nanotube radiotracers. Proc Natl Acad Sci USA 2006;103:3357–3362.

INDEX

Abbreviated new drug application (ANDA):
  FDA guidelines:
    approval, 92
    delays and exclusivities, 86
    generic drug review and approval, 87–91
    historical background, 67
    labeling, 91
    preapproval inspection, 91–92
    withdrawal of approval, 92
  patent infringement and, 151–152
  submission requirements, 85, 88
Ab initio calculations, oral drug melting point, 30–31
Absolute novelty principle:
  patent protection, 110–111
  U.S.C.§102 provision, 134
Absorption:
  oral drug delivery:
    Biopharmaceutics Classification System, 353–355
    class III drugs, 50–52
    class V drugs, barriers, 52–54
    prodrug development, 223–226
    CK-2130 prodrug case study, 269–276
Abuse of discretion standard, infringement of patent proceedings, 164
ABYFG C-glycoside derivative, cocrystal engineering, 211–212
Acid-base equilibrium, salt compounds, 384–388

pH solubility profile, 385–388
  weak acids, 384–385
  weak bases, 384–385
  zwitterions, 385, 387
Acids, salt compounds, acid-base equilibrium, 384–385
Acquired immunodeficiency syndrome (AIDS), FDA reforms and, 68–69
Active pharmaceutical ingredients (APIs):
  commercial-scale operations, continuous microfluidic reactors, 13–14
  crystal engineering, research background, 187–190
  large-scale synthesis, 1
  polymorph crystallization, 190–192
    case studies, 196–198
  salt compounds:
    basic properties, 381–382
    screening process, 388–396
  scale-up operations, 1–14
Addition order, bench-scale experiments, 6–7
Additives, high-throughput screening, assay optimization, 412–413
ADMET (absorption, distribution, metabolism, excretion, and toxicity):
  FDA preclinical testing guidelines, 71–72
  permeability:
    basic principles, 367

classification, 370–371
distribution, 375–376
excretion, 373–374
fraction absorption, 369–370
gastrointestinal absorption, solubility *vs.* permeability, 374–375
metabolism, 371–373
organ accumulation and toxicity, 377
transport interactions and polymorphism, 367–369
prodrug properties, 220–226
  CK-2130 prodrug case study, 269–276
classical drug discovery paradime, 226–227
current drug discovery paradigm, 226, 228–229
design principles, 230
metabolic issues, 231–233
Administration protocols, prodrug development, 221–226
Affinity chromatography, protein therapeutics:
  analytic development, 320
  downstream processing, 304
  protein A, 308–309
A4F separation technique, protein therapeutics, 324
Aggregation, protein therapeutics formulation and delivery, 313–314
Agitation functions, pilot plant development, 8–9

INDEX

Alternative dispute resolution, infringement of patent proceedings, 164–165
AMG 517 antagonists, cocrystal engineering, 210–211
Amino gorups, prodrug moieties, 251–252
Amorphous solids, oral dosage systems, 29
Amphiphilic molecules, nanoscale drug delivery:
  polymers, 473–480
  self-assembled low molecular weight structures, 469–473
Analog design, prodrug developmenet, esmolol case study, 259–268
Analytical development:
  defined, 289–290
  protein therapeutics, 317–327
    bioactivity assays, 321–322
    biophysical characterization, 323–325
    electrophoresis, 320
    future trends, 326–327
    immunoassays, 321
    in-process/product release testing, 318–320
    mass spectrometry, 322–323
    nucleic acid testing, 320–321
Angiotensin-converting enzyme (ACE) inhibitors, prodrug development, 242
Animal drugs, regulatory guidelines, 97
Anthracyclines, serum albumin binding, 452–454
Antibiotic/antibacterial drugs, prodrug development, 242–246
Antibody-directed enzyme prodrug therapy (ADEPT), prodrug design, 229
Anticancer agents, prodrug development, 246–248
Appeals process, infringement of patent proceedings, 163–164
Aqueous solubility:
  gastrointestinal absorption, permeability vs., 374–375
  oral drug physicochemistry, 32–33

salt compounds, 381–382
Area under the curve (AUC) measurements:
  permeability, ADMET interactions, transport interactions, 368–369
  protein therapeutics, 324
Aspirin:
  polymorphism, 192–193
  prodrug development, metabolic issues, 232–233
Assay techniques:
  carboxylesterases, preclinical properties, 239
  high-throughput screening, enzymatic assays:
    adaptation mechanisms, 422–423
    additives, stabilizers, and other cofactors, 412–413
    basic principles, 401, 406
    buffers and pH, 410–412
    configuration, 414–423
      substrate/enzyme concentration, 414–418
      substrate selection, 420
      unconventional setup, 418–420
    continuous/discontinuous assays, 406–407
    direct, indirect, and coupled assays, 407–409
    drug targets, 401–402
    expression and purification, 402–406
    hydrolases, 425–426
    isomerases, 427–428
    ligases, 428
    lyases, 426–427
    monovalent/divalent salts, 412
    optimization conditions, 409–414
    oxidoreductases, 423–424
    temperature, 413–414
    transferases, 424–425
Australia, patent requirements, 122
Automated assisted screening, salt compound selection, 395–396
Automation, high-throughput screening, enzymatic assays, 406

Baby hamster kidney (BHK) cells, protein therapeutics, 293–294
Bases, salt compounds, acid-base equilibrium, 384–385
Batch processing:
  commercial-scale operations, continuous microfluidic reactors, 13–14
  protein therapeutics, upstream process development, 298–300
Bench-scale experiments, scale-up operations, 4–7
  pilot plant scale-up, 7–8
  reaction solvent selection, 5–6
  reaction temperature, 6
  reaction times, 6
  solid-state requirements, 7
  stoichiometry and addition order, 6–7
Best mode principle, patent specifications, 126, 130–131
Beyond our Borders Initiative (FDA), 69
Biliary excretion, permeability, ADMET interactions, 373
Bilirubin, serum albumin binding, 442–451
Binding sites, serum albumin structural survey, 440–442
  anthracyclines, 452–454
  bilirubin binding, 442–451
  podophyllotoxin derivatives, 454–456
Bioactivity assays, protein therapeutics, analytic development, 321–322
Bioadhesives, oral drug systems, class III bioadhesives, 52
Bioavailability:
  Biopharmaceutics Classification System, regulatory practices, 355–356
  permeability, ADMET interactions, oral bioavailability, 372
  pharmaceutical cocrystals, 198–213
  polymorph cocrystals, 196–198

Bioconversion mechanisms, prodrug design, 252–254
Bioequivalence (BE):
  Biopharmaceutics Classification System, regulatory practices, 355–356
  generic drug guidelines, 85, 91
Biologic Control Act, 93
Biologics license applications (BLAs), FDA guidelines, 94–95
Biopharmaceutics Classification System (BCS):
  crystal engineering, 187–190
  drug ADMET/PK interactions, Permeability-Based Classification System, 370–371
  drug disposition classification system, 361–362
  future trends and issues, 362–363
  gastrointestinal absorption, permeability vs. solubility, 374–375
  oral drug delivery, 41–55
    absorption, 353–355
    class I drugs, 42, 354
    class II drugs, 42–50, 354
    class III drugs, 50–52, 355
    class IV drugs, 52, 355
    class V drugs, 52–55
    global market, 353–363
  provisional classification process, 356–362
    biowaiver monographs, 360–361
    literature data, 356–359
    in silico calculations, 359–360
  regulatory practices, 355–356
Biopharmaceutics Drug Disposition Classification System (BDDCS):
  drug ADMET/PK interactions, 371
  evolution of, 361–362
Biophysical techniques, protein therapeutics, 323–325
Bioseparation, protein therapeutics, downstream processing, 301–311

Biotechnology:
  drug development, FDA guidelines for, 93–96
  intellectual property:
    future research issues, 186
    miscellaneous protections, 185–186
  patents:
    enforcement, 148–165
      alternative dispute resolution, 162–163
      claim construction proceedings, 162
      defenses to infringement, 154–157
      discovery phase, 160–161
      geographic scope, 148
      infringement determination, 152–154
      parties to suit, 150–152
      proceedings commencement, 159–160
      remedies for infringement, 157–159
      summary judgment, 161
      trial and appeal processes, 148–150, 162–164
    global patent protection, 107–108, 165–171
      international agreements, 165–168
      international laws and regulations, 169–171
      PCT patent practice, 168–169
    protection and strategy, 105–118
      absolute novelty principle, 110–111
      application publication, 114–115
      first to invent vs. first to file, 108–110
      global strategy, 107–108, 165–171
      patent term, 111–114
      PTO prosecution strategy, 115–118
      trade secrets and, 182–183
    requirements, 118–148

    corrections, 146–148
    interference, 142–146
    international requirements, 121–123
    new and nonobvious inventions, 133–138
    provisional applications, 127–133
    PTO procedures, 138–142
    specifications, 123–127
    U.S. requirements, 118–121
  research background, 101–105
  trademarks, 171–179
    global rights, 177–178
    Lanham Act protections, 178–179
    as marketing tools, 172
    oppositions and cancellations, 176–177
    preservation of rights through proper use, 177
    registration process, 174–176
    selection criteria, 172–174
  trade secrets, 179–184
    defined, 180
    enforcement, 181–182
    Freedom of Information Acts, 183–184
    global protection, 184
    patent protection and, 182–183
    protection requirements, 180–181
Biowaiver monographs, Biopharmaceutics Classification System, 360–361
Block copolymers, nanoscale drug delivery, 475–479
Blood-brain barrier (BBB), permeability, ADMET interactions:
  distribution, 375–376
  transport interactions, 368–369
Budapest Treaty, drug patent requirements, 126–127
Buffers, high-throughput screening, assay optimization, 410–412

INDEX

CADEX process, serum albumin structural survey, 439–440
Calcium phosphate, protein therapeutics, transafection, 295
Calibration validation, commercial-scale operations, 12
Cambridge Structural Database (CSD):
  cocrystal design, 194–196
  crystal engineering, 189–190
Camptothecins, serum albumin binding, 451–453
  in vitro studies, 458–462
Canada, patent requirements, 122
Cancer therapy, serum albumin:
  chemical properties, 437–438
  therapeutic applications, 456–464
    clofibrate pharmacokinetic modulation, 457–464
    pharmacokinetic studies, 462–464
    in vitro studies, 458–462
  X-ray structural analysis, 438–456
    anthracyclines, 452–454
    bilirubin binding, 442–451
    CADEX categories, 439–440
    camptothecins, 451–453
    drug-binding sites, 440–442
    podophyllotoxin derivatives, 454–456
Carbamazepine, cocrystal engineering, 198, 208
Carboxylesterases:
  bioconversion mechanisms, 252–254
  prodrug development, 233–241
    classification, 233–234
    function and distribution, 234
    molecular biology, 234, 237–238
    preclinical assays, 239, 241
    substrates and inhibitors, 235–240
  structure-metabolism relationships, 254–256
Carboxylic acids, prodrug moieties, 250–252
Carboxylic dimers, cocrystal design, 195–196

cDNA, protein therapeutics, expression vectors, 294–295
Cell line development, protein therapeutics, 293–294
  attachment-dependent culture, 296–297
  banking systems, 296
  suspension culture, 298
Centrifugation, mammalian protein purification, harvest operations, 306–307
Ceramic hydroxyapatite chromatography, protein therapeutics, downstream processing, 304
Change control program, commercial-scale operations, 12–13
Chemical degradation, protein therapeutics formulation and delivery, 311–314
Chinese hamster ovary (CHO) cells, protein therapeutics, 293–294
Chirality, scale-up operations, 4
2-[4-(4-Chloro-2-fluorphenoxy)phenyl]pyrimidine-4-carboxamide (CFPPC), cocrystal engineering, 210
Chromatography, protein therapeutics, in-process and product release testing, 318–320
Circular dichroism (CD), protein therapeutics, biophysical assessment, 324–325
Citizens petition, generic drug approval, FDA guidelines, 86–87
CK-2130 prodrug case study, cardiotonicity, 268–276
Claims requirements:
  infringement of patents:
    alternative dispute resolution, 164–165
    appeal process, 163–164
    claim construction, 162
    commencement of proceedings, 159–160

    defenses to, 154–157
    determination of, 152–154
    discovery process, 160–161
    licensee rights, 150–152
    remedies for, 157–159
    summary judgment, 161
    trial, 162–163
    U.S. trial and appellate courts, 148–150
  patent specifications, 126–127
  composition of matter claims, obviousness standards and, 137–138
Class I drugs, oral drug physicochemistry, high solubility/high permeability, 41, 354
Class II drugs, oral drug physicochemistry:
  cocrystallization, 47–48
  complexation, 46–47
  lipid technologies, 48–50
  low solubility/high permeability, 42–50, 354
  metastable forms, 45–46
  particle size reduction, 44–45
  precipitation inhibition, 45
  salt compounds, 42, 44
  solid dispersion, 46
Class III drugs, oral drug physicochemistry, 50–52
Class IV drugs, oral drug physicochemistry, 52, 355
Class V drugs, oral drug physicochemistry, 52–55
Cleaning validation, commercial-scale operations, 12
Clinical trials, FDA guidelines, 76–78
  research subjects guidelines, 80
  site guidelines, 80, 82
Clofibrate, serum albumin modulation, 457–464
Cloning techniques, protein therapeutics, 296
Closure systems, protein therapeutics formulation, material interactions, 315
Cocrystallization:
  evolution of, 192–196
  oral drug physicochemistry, class II drugs, 47–48

pharmaceutical case studies, 198–213
  AMG 517 antagonist, 210–211
  carbamazepine, 198, 208
  C-glycoside derivative, 211–212
  fluoxetine hydrochloride, 208–209
  itraconazole, 209–210
  melamine and cyanuric acid, 213
  monophospate salt I, 212–213
  sildenafil, 211
  sodium channel blockers, 210
  polymorphs, solvates, and hydrates, 196
  structural properties, 188–190
Cofactor-based drug discovery, high-throughput screening, assay optimization, 412–413
Combinatorial chemistry, prodrug discovery, 226, 228–229
Commercial-scale operations:
  nevirapine case study, 21–23
  scale-up of, 10–13
    chemical safety, 13
    environmental controls, 13
    in-process controls, 11
    processing equipment requirements, 11
    validation, 12–13
    trade secrets limitations and, 180–181
Common technical document (CTD), generic drug approvals, 90–91
Comparability techniques, protein therapeutics process development, 325–326
Complexation, oral drug physicochemistry, class II drugs, 46–47
Composition of matter claims, patent obviousness standards and, 137–138
"Comprising" terminology, infringement of patents, 152–154

Concentration profiles, high-throughput screening, assay configuration, 414–418
Conformational polymorphism:
  crystal engineering, 191–192
  protein therapeutics, biophysical assessment, 324–325
Congestive heart failure (CHF), CK-2130 prodrug case study, 268–276
"Consisting essentially of" terminology, infringement of patents, 152–154
"Consisting of" terminology, infringement of patents, 152–154
Container material interactions, protein therapeutics formulation, 315
Contaminants, protein therapeutics, downstream processing, 301–303
Continuation-in-part (CIP) application, drug discovery and biotechnology patentsd:
  current practices, 104–105
  written description requirements, 123–124
Continuous assays, high-throughput screening, 407–408
Continuous microfluidic reactors, commercial-scale operations, 13–14
Cooling systems, pilot plant development, 8
Copolymers, nanoscale drug delivery, 475–479
Copyright protections, drug discovery and, 185
Cost-effectiveness analysis (CEA), drug development:
  basic principles, 345
  decision-analytic approaches, 346–349
  pharmacogenomics and personalization, 349
  postlaunch analysis, 349–350

regulatory approval, 345–346
Counterions, salt compound selection, 389–391
Coupled assays, high-throughput screening, 408–409
Court of Federal Claims (U.S.), patent infringement suits, 149–150
Cre/loxP system, protein therapeutics, 296
Critical micelle concentration (CMC):
  nanoscale drug delivery, self-assembled low molecular weight amphiphiles, 472–473
  oral drug physicochemistry, class II drugs, micelle solubilization, 48–49
Critical packing parameter, nanoscale drug delivery, self-assembled low molecular weight amphiphiles, 469–473
Cross-linked polymeric nanoparticles, nanoscale drug delivery, 481
Crystal engineering:
  cocrystals, 189–190, 192–196, 198–213
  future research issues, 213–214
  hydrates and solvates, 189
  polymorphs, 188, 190–192, 196–198
  protein therapeutics, 316–317
  research background, 187–190
  salts, 188–189
Crystalline solids:
  oral drug delivery, 25–26
  salt compounds, 381–383
Current Good Laboratory Practices (cGLP), FDA guidelines, 81
Current Good Manufacturing Practices (cGMP), FDA guidelines, 81–82
Cyanuric acid, cocrystal engineering, 213
Cyclodextrin complexes, oral drug physicochemistry, class II drugs, 47
Cysteinylation, protein therapeutics formulation and delivery, 313–314

Cytochrome P450 enzymes:
  oral dosage forms, class V drugs, 54
  prodrug development, CK-2130 prodrug case study, 274–276

Deamidation, protein therapeutics formulation and delivery, 312–314
Decision analysis, cost-effectiveness analysis of drug development, 346–349
Decision trees, cost-effectiveness analysis of drug development, 347–349
Degradation pathways, protein therapeutics formulation and delivery, 311–314
Denaturation, protein therapeutics formulation and delivery, 313–314
Dendrimers, nanoscale drug delivery, 482–483
Deposition process, infringement of patent proceedings, 161
Depth filtration, mammalian protein purification, harvest operations, 307
Design patents:
  drug discovery and, 185
  protection and strategy, 105–118
Diethylaminoethanol (DEAE), protein therapeutics, attachment-dependent cell culture, 296–297
Differential scanning calorimetry (DSC):
  oral drug melting point, 30–31
  protein therapeutics, 325
Dihydrofolate reductase (DHFR), protein therapeutics:
  dhfr- development, 293–294
  transfection, 295–296
Direct assays, high-throughput screening, 408–409
Discontinuous assays, high-throughput screening, 407–408

Discovery process, infringement of patents, 160–161
Dispersion methods, oral drug physicochemistry, class II drugs, solid dispersion, 46
Dissociation constant, acid-base equilibrium, salt compounds, 384–388
Dissolution:
  oral drug delivery:
    absorption, 354–355
    physicochemistry evaluation, 33
  permeability, ADMET interactions, 374–375
Distribution:
  permeability, ADMET interactions and, 375–376
  prodrug development, 223–226
Divalent salts, high-throughput screening, assay optimization, 412
Doctrine of equivalents, infringement of patents, 153–155
Dopamine, prodrug development, 223–226
  metabolic issues, 231–233
Dosage forms, oral drug delivery:
  biopharmaceutics classification system, 41–43
  class I drugs, 42
  class II drugs, 42–50
  class III drugs, 50–52
  class IV drugs, 52, 355
  class V drugs, 52–55
Dose number, gastrointestinal absorption, permeability *vs.* solubility, 374–375
Downstream processing:
  defined, 289
  protein therapeutics, 301–311
    classifications, 303–304
    design criteria, 303
    E. coli protein purification, 304–306
    future trends, 309–311
    impurities and contaminants, 301–303
    mammalian protein purification, 306–309

Doxorucibin:
  nanoscale drug delivery, liposomes, 472–473
  serum albumin binding, 452–454
Drug delivery systems:
  crystal engineering:
    cocrystals, 189–190, 192–196, 198–213
    future research issues, 213–214
    hydrates and solvates, 189
    polymorphs, 188, 190–192, 196–198
    research background, 187–190
    salts, 188–189
  nanotechnology:
    amphiphilic polymers, 473–480
    basic principles, 469
    carbon nanotubes, 483–484
    dendrimers, 482–483
    polymer-drug conjugates, 480–481
    self-assembled low molecular weight amphiphiles, 469–473
    water-insoluble and crosslinked polymeric nanoparticles, 481
  protein therapeutics, 311–317
    alternative modes, 315–316
    container materials interactions, 315
    degradation pathways, 311–314
    future trends, 316–317
    stability parameters, 314–315
Drug development and discovery:
  cost-effectiveness analysis:
    basic principles, 345
    decision-analytic approaches, 346–349
    pharmacogenomics and personalization, 349
    postlaunch analysis, 349–350
    regulatory approval, 345–346
  intellectual property law:
    future research issues, 186
    miscellaneous protections, 185–186
    patents:
      enforcement, 148–165
      alternative dispute resolution, 162–163

claim construction
proceedings, 162
defenses to
infringement,
154–157
discovery phase,
160–161
geographic scope, 148
infringement
determination,
152–154
parties to suit, 150–152
proceedings
commencement,
159–160
remedies for
infringement,
157–159
summary judgment, 161
trial and appeal
processes, 148–150,
162–164
global patent protection,
107–108, 165–171
international
agreements, 165–168
international laws and
regulations, 169–171
PCT patent practice,
168–169
protection and strategy,
105–118
absolute novelty
principle, 110–111
application publication,
114–115
first to invent *vs.* first to
file, 108–110
global strategy, 107–108,
165–171
patent term, 111–114
PTO prosecution
strategy, 115–118
trade secrets and,
182–183
requirements, 118–148
corrections, 146–148
interference, 142–146
new and nonbovious
inventions, 133–138
patentable subject
matter, U.S. and
international
requirements,
118–123
provisional applications,
127–133

PTO procedures,
138–142
specifications, 123–127
research background,
101–105
trademarks, 171–179
global rights, 177–178
Lanham Act protections,
178–179
as marketing tools, 172
oppositions and
cancellations, 176–177
preservation of rights
through proper use,
177
registration process,
174–176
selection criteria, 172–174
trade secrets, 179–184
defined, 180
enforcement, 181–182
Freedom of Information
Acts, 183–184
global protection, 184
patent protection and,
182–183
protection requirements,
180–181
Drugs, FDA defnition of, 53
Drying operations, pilot plant
development, 9–10
Dynamic light scattering (DLS),
protein therapeutics,
324
Dynamic moisture sorption, oral
drug
physicochemistry,
38–39

eBay doctrine, infringement of
patent claims,
158–159
Electrophoretic techniques,
protein therapeutics,
analytic development,
320
Electroporation, protein
therapeutics,
transfection, 295
Electrostatic potential, CK-2130
prodrug case study,
270–276
Elimination, oral dosage forms,
class V drugs, 53
Emulsions, oral drug
physicochemistry,
class II drugs, 48–49

Enablement requirement, patent
specifications,
124–126
Endogenous substances:
enzymes, prodrug design, 229
serum albumin drug transport
system:
basic principles, 437–438
structural survey, 438–456
therapeutic applications,
456–464
Environmental controls,
commercial-scale
operations, 13
Enzymatic assays, high-
throughput screening:
adaptation mechanisms,
422–423
additives, stabilizers, and other
cofactors, 412–413
basic principles, 401, 406
buffers and pH, 410–412
configuration, 414–423
substrate/enzyme
concentration,
414–418
substrate selection, 420
unconventional setup,
418–420
continuous/discontinuous
assays, 406–407
direct, indirect, and coupled
assays, 407–409
drug targets, 401–402
expression and purification,
402–406
hydrolases, 425–426
isomerases, 427–428
ligases, 428
lyases, 426–427
monovalent/divalent salts,
412
optimization conditions,
409–414
oxidoreductases, 423–424
temperature, 413–414
transferases, 424–425
Equipment/systems qualification,
commercial-scale
operations, 12
Escherichia coli, protein
purification, 304–306
Esmolol, prodrug development,
258–268
Esters, prodrug development,
242–248
esmolol case study, 260–268

Eukaryotic expression:
  high-throughput screening, enzymatic assays, 402–406
  protein therapeutics, upstream processing, 291–294
European Patent Convention (EPC), 166–168
Event reporting, FDA guidelines, 81
Excipients:
  oral dosage forms, class V drugs, 54–55
  protein therapeutics, 326
Exclusivity, generic drug approval, FDA guidelines, 86–88
Excretion:
  permeability, ADMET interactions, 373–374
  prodrug development, 224–226
Ex parte patent procedures:
  infringement of patent claims, defense based on, 155–156
  U.S. PTO requirements, 138–142
"Experimental use" defense, infringement of patent claims, 156
Expert testimony, infringement of patent proceedings, 162–163
Expressed sequence tags (ESTs), patent requirements, 120–123
Expression augmenting sequence element (EASE) sequence, protein therapeutics, expression vectors, 294–295
Expression systems:
  high-throughput screening, enzymatic assays, 402–406
  protein therapeutics, upstream processing, 291–294
Expression vectors, protein therapeutics, upstream processing, 294–295

Fed-batch systems, protein therapeutics:
  comparability analysis, 325–326
  upstream process development, 298–300
Federal Courts Improvement Act (FCIA), patent infringement suits, 150
Federal Declaratory Judgments Act, patent infringement and, 152
Festo I and *II* decisions, infringement of patents, 154
Fick's laws of passive diffusion, oral drug delivery:
  absorption, 353–355
  permeablity, 34–35
First-pass metabolism, oral dosage forms, class V drugs, 53
First to invent *vs.* first to file principle, patent protection, 108–110
502(b)(2) applications, FDA guidelines, 92
Flow properties, salt compounds, 382
Flp/FRT system, protein therapeutics, 296
Fluidized bed reactors, protein therapeutics, 297–298
Fluorescence spectroscopy, protein therapeutics, 325
Fluoxetine hydrochloride, cocrystal engineering, 208–209
Food and Drug Administration (FDA):
  Amendments Act (FDAAA), 69
  chemistry, manufacturing and control information, 74–75
  clinical site guidelines, 80
  event reportin guidelines, 81
  IND applications, 72–74
  inspections guide lines, 81
  Institutional Review Board, 80–81
  meetings guidelines, 80
  new drug application guidelines, 78
  new drug definitions, 82
  sponsor rights, 80
  pharmacology and toxicology information, 75
  problems and challenges of, 80
  prodrug approvals, 242
  protocols, 74
  regulatory role
    biotechnology-derived drugs, 93–96
    generic drug review and approval, 84–92
    historical background, 63–69
    new drug approval process, 71–82
    Orange Book, 92–93
    organizational structure, 70
    over the counter drug approval, 82–84
    postapproval process, 96
    preclinical testing, 71–72
    review process, 70–71
    statutes, 69–70
    United States Pharmacopoeia and National Formulary, 96–97
  reviewing process, 78–80
  trade secrets limitations and, 179–180
Food and Drug Administration Modernization Act (FDAMA), 67–68
Formulation development:
  defined, 289
  prodrugs, 221–226
  protein therapeutics, 311–317
    alternative modes, 315–316
    container materials interactions, 315
    degradation pathways, 311–314
    future trends, 316–317
    stability parameters, 314–315
Fraction absorbed (f_a), permeability, ADMET interactions, 369–370
France, patent requirements, 122
Fraud defense, infringement of patent claims, 155–156
Freedom of Information Act (FOIA), trade secrets, 183–184
Functional groups, CK-2130 prodrug case study, 270–276

Gene amplification, protein therapeutics, 295–296

Gene-directed enzyme prodrug
therapy (GDEPT),
prodrug design, 229
Generic drugs:
biologics, FDA guidelines, 96
FDA approval process, 84–92
ANDA approvals and
exclusivities, 86
ANDA review process, 87–91
bioequivalence, 85
505(b)(2) provision, 92
citizens petition, 86–87
Hatch-Waxman Amendment
provisions, 67, 85–86
historical background, 85
labeling, 91
patient term extensions, 87
pharmaceutically equivalent
compounds, 84–85
preapproval inspection,
91–92
withdrawal of approval, 92
Gene therapy, nanoscale
dendrimer delivery
systems, 482–483
Genome sequencing, patent
requirements,
120–123
Geographic jurisdiction, patent
enforcement, 148
Germany, patent requirements,
122
Globalization of drug discovery:
intellectual property law:
international agreements,
165–168
patententable subject matter
requirements,
121–123
patent strategies, 107–108,
165–171
U.S. patent applications fron
non-U.S. applicants,
133
trademark rights, international
regulations, 177–178
trade secrets protection, 184
Glutamine synthase, protein
therapeutics, cell line
expression, 294
GLUT5 transporter, protein
therapeutics,
upstream process
development, 300–301
Glycation, protein therapeutics
formulation and
delivery, 313–314

C-Glycoside derivative, cocrystal
engineering, 211–212
"Goldilocks" compounds, prodrug
design, esmolol case
study, 264–268
Good clinical practices (GCPs),
FDA Twenty-first
Century Initiative,
68–69
Graham v. John Deere Co. test,
patent obviousness
principle, 134–135
Graph sets, crystal engineering,
evolution of, 189–190
Gut-wall metabolism,
permeability, ADMET
interactions, 372

Half-life estimates, prodrug
design, esmolol case
study, 264–268
Harvest operations, mammalian
protein purification,
306–307
Hatch-Waxman Act, generic drug
approval, 85–88
Heating systems, pilot plant
development, 8
Heat-transfer correlation,
agitation functions,
pilot plant
development, 8–9
Hepatotoxicity, permeability,
ADMET interactions,
drug transport
mechanisms,
368–369
High-throughput formulation
(HTF), protein
therapeutics, 316, 327
High-throughput screening
(HTS):
enzymatic assays:
adaptation mechanisms,
422–423
additives, stabilizers, and
other cofactors,
412–413
basic principles, 401, 406
buffers and pH, 410–412
configuration, 414–423
substrate/enzyme
concentration,
414–418
substrate selection, 420
unconventional setup,
418–420

continuous/discontinuous
assays, 406–407
direct, indirect, and coupled
assays, 407–409
drug targets, 401–402
expression and purification,
402–406
hydrolases, 425–426
isomerases, 427–428
ligases, 428
lyases, 426–427
monovalent/divalent salts,
412
optimization conditions,
409–414
oxidoreductases, 423–424
temperature, 413–414
transferases, 424–425
prodrug discovery, 226,
228–229
Hollow fiber reactors, protein
therapeutics, 297
Homogeneity, high-throughput
screening, enzymatic
assays, 406
Human embryonic kidney
(HEK-293) cells,
protein therapeutics,
293–294
Human papilloma virus (HPV),
vaccine development
and outcomes, cost-
effectiveness analysis
of, 347–349
Humidity, oral drug
physicochemistry,
moisture uptake,
37–39
Hydrates:
cocrystal design, 196
crystal engineering, defined,
189
oral dosage systems, 27–28
salt compounds, 382
Hydrogenation reactions,
agitation functions,
pilot plant
development, 8–9
Hydrolases, high-throughput
screening, enzymatic
assays, 425–426
Hydrophobic interaction
chromatography
(HIC), protein
therapeutics:
analytic development, 320
downstream processing, 304

Hydrophobicity, oral drug physicochemistry, 37–39
Hydroxyapatite chromatography, protein therapeutics, downstream processing, 304

Idarubicin, serum albumin binding, 452–454
Ignition prevention, commercial-scale operations, 11
Immunoassays, protein therapeutics, analytic development, 321
Improper use principle, trademark preservationand, 177
Impurities, protein therapeutics, downstream processing, 301–303
Inclusion bodies (IBs), protein purification, *E. coli* systems, 304–306
Inclusion complexes, oral drug physicochemistry, class II drugs, 47
Indirect assays, high-throughput screening, 408–409
Inequitable conduct defense, infringement of patent claims, 155–156
Information Disclosure Statement (IDS), U.S. PTO procedures, 140–142
Information protection, trade secrets limitations and, 180–181
Infringement claims:
  patents:
    alternative dispute resolution, 164–165
    appeal process, 163–164
    claim construction, 162
    commencement of proceedings, 159–160
    defenses to, 154–157
    determination of, 152–154
    discovery process, 160–161
    licensee rights, 150–152
    remedies for, 157–159
    summary judgment, 161
    trial, 162–163
    U.S. trial and appellate courts, 148–150
  trademarks, 173–177

Inhibition Detection Limit (IDL), high-throughput screening, assay configuration, 420
Inhibitor binding (IB), serum albumin structural survey, 440–451
  anthracyclines, 452–454
  bilirubin binding, 442–451
  podophyllotoxin derivatives, 454–456
Inhibitors, carboxylesterases, 235–240
In-process controls (IPCs):
  bench-scale experiments, 5
  commercial-scale operations, 11
  protein therapeutics, 317–320
Insect cells, protein therapeutics, 292–294
In silico methods:
  Biopharmaceutics Classification System, provisional classification, 359–361
  prodrug design, 230
Inspections, FDA guidelines, 81
Institutional Review Board (IRB) (FDA), guidelines, 80–81
Intellectual property law. *See also* Patents
  drug discovery and biotechnology:
    future research issues, 186
    miscellaneous protections, 185–186
  patents:
    enforcement, 148–165
      alternative dispute resolution, 162–163
      claim construction proceedings, 162
      defenses to infringement, 154–157
      discovery phase, 160–161
      geographic scope, 148
      infringement determination, 152–154
      parties to suit, 150–152
      proceedings commencement, 159–160
      remedies for infringement, 157–159
      summary judgment, 161
      trial and appeal processes, 148–150, 162–164
    global patent protection, 107–108, 165–171
      international agreements, 165–168
      international laws and regulations, 169–171
      PCT patent practice, 168–169
    protection and strategy, 105–118
      absolute novelty principle, 110–111
      application publication, 114–115
      first to invent *vs.* first to file, 108–110
      global strategy, 107–108, 165–171
      patent term, 111–114
      PTO prosecution strategy, 115–118
      trade secrets and, 182–183
    requirements, 118–148
      corrections, 146–148
      interference, 142–146
      new and nonbovious inventions, 133–138
      patentable subject matter, international requirements, 121–123
      patentable subject matter, U.S. and international requirements, 118–123
      provisional applications, 127–133
      PTO procedures, 138–142
      specifications, 123–127
    research background, 101–105
  trademarks, 171–179
    global rights, 177–178
    Lanham Act protections, 178–179
    as marketing tools, 172

oppositions and cancellations, 176–177
preservation of rights through proper use, 177
registration process, 174–176
selection criteria, 172–174
trade secrets, 179–184
  defined, 180
  enforcement, 181–182
  Freedom of Information Acts, 183–184
  global protection, 184
  patent protection and, 182–183
  protection requirements, 180–181
Intent requirements, infringement of patent claims, 155–156
Intent-to-use provisions, trademark registration, 175–176
Interference in patent claims, 142–146
Internal ribosomal entry site (IRES), protein therapeutics, expression vectors, 294–295
International Conference on Harmonization (ICH), drug development guidelines, 97
International patent agreements, 165–168
International Trade Commission (ITC) (U.S.), patent infringement suits, 149–150
Intestinal absorption:
  oral dosage forms, class V drugs, 53
  permeability vs. solubility/dissolution, 374–375
Intestinal excretion, permeability, ADMET interactions, 373–374
Intravenous drug delivery, protein therapeutics, 316
Investigational new drugs (INDs), FDA guidelines, 72–74

additional and relevant information guidelines, 76
previous human experience, 75–76
In vitro studies:
  prodrug design, ADMET issues, 230
  serum albumin drug transport, 458–462
Ion exchange chromatography (IEC), protein therapeutics:
  analytic development, 319
  downstream processing, 304
Ionization constants, oral drug physicochemistry, 35–37
Ion pairing, oral drug delivery, class III drugs, 51–52
Isomerases, high-throughput screening, enzymatic assays, 427–428
Isomers and isomerization, protein therapeutics formulation and delivery, 312–314
Italy, patent requirements, 122
Itraconazole, cocrystal engineering, 209–210

Japan, patent requirements, 122

Kefauver-Harris Amendments, over the counter drug approval, 83–84
Ketoconazole, CK-2130 prodrug case study, 274–276

Labeling, FDA requirements, 91
β-Lactams, prodrug approvals, 242–246
Lanham Act (U.S.):
  false designation of trademarks and, 178–179
  trademark selection guidelines, 172–174
Large-scale synthesis:
  nevirapine case study, 14–23
  commercial production and process optimization, 20–23
  medicinal chemistry synthetic route, 15–16
  pilot-plant scale-up:
    chemical development, 16–17

process development, 17–20
research background, 14–15
research background, 1
scale-up operations, 1–14
  basic principles, 1
  bench-scale experiments, 4–7
    reaction solvent selection, 5–6
    reaction temperature, 6
    reaction times, 6
    solid-state requirements, 7
    stoichiometry and addition order, 6–7
  chiral requirements, 4
  commercial-scale operations, 10–13
    chemical safety, 13
    environmental controls, 13
    in-process controls, 11
    processing equipment requirements, 11
    validation, 12–13
  continuous microfluidic reactors, 13–14
  pilot plant scale-up, 7–10
    agitation, 8–9
    drying and solid handling, 9–10
    heating and cooling, 8
    liquid-solid separations, 9
    safety issues, 10
  process analytical technology, 14
  route selection, 3–4
  synthetic strategies, 1–4
Lethality dose (LD_{50}), FDA preclinical testing guidelines, 71–72
Levodopa (L-DOPA), prodrug development, 223–226
Licensing agreements, international patent laws, 169–171
Ligases, high-throughput screening, enzymatic assays, 428
Lipid solubility, oral drug physicochemistry, class II drugs, 48–50
Lipofection, protein therapeutics, transfection, 295
Lipophilicity, oral drug physicochemistry:
  partition coefficient, 31–32
  permeability, 34–35

Liposomes:
  nanoscale drug delivery, 469–473
  oral drug physicochemistry, class II drugs, 49
Liquid formulation, protein therapeutics stability, 314–315
Liquid-solid separation, pilot plant development, 9
Literature searches, Biopharmaceutics Classification System, 356–359
Liver, permeability, ADMET interactions:
  distribution, 376
  drug transport mechanisms, 368–369
  metabolism, 368–369, 371–372
Lost profits remedy provisions, infringement of patent claims, 157–159
Luminal degradation, oral dosage forms, class V drugs, 53
Lyases, high-throughput screening, enzymatic assays, 426–427
Lyophilization, protein therapeutics formulation, 315

Madrid Protocol, international trademark rights, 178
Mammalian cell lines, protein therapeutics:
  purification, 306–309
  upstream processing, 293–294
Marketed compounds:
  prodrugs, 241–248
    examples, 242, 244–248
    FDA approvals, 242–243
    prevalence, 241–242
  trademarks and, 172
Markman claim, infringement of patent proceedings, 159–162
Mass spectrometry (MS), protein therapeutics, 322–323
Master cell bank (MCB), protein therapeutics, 296
Material interactions, protein therapeutics formulation, container closure materials, 315
Materiality requirement, infringement of patent claims, 155–156
Mathematical modeling, cost-effectiveness analysis of drug development, 346–349
MAX/MIN values, high-throughput screening, assay configuration, 422–423
Medicinal chemistry, nevirapine synthesis, 15–16
Medium design and composition, protein therapeutics, upstream process development, 299–300
Melamine, cocrystal engineering, 213
Melting point, oral drug delivery, 30–31
Metabolism:
  oral dosage forms, class V drugs, 53
  permeability, ADMET interactions:
    gut-wall metabolism, 372
    hepatic metabolism, 371–372
    oral bioavailability, 372
    polymorphism and, 367–369
    renal metabolism, 372–373
  prodrug development, 224–226
    design issues, 230–233
    rule-of-one metabolism, 248–250
  protein therapeutics, upstream process development, 300
Metastable drug forms, oral drug physicochemistry, 45–46
Methods validation, commercial-scale operations, 12
Methotrexate, protein therapeutics:
  cell line development, 293–294
  transfection and gene amplification, 295–296
Micelle preparation:
  nanoscale drug delivery, 469–473
  oral drug physicochemistry, class II drugs, 48–49
Michaelis-Menten equation, high-throughput screening, assay configuration, 420
Microcarriers, protein therapeutics, attachment-dependent cell culture, 296–297
Microemulsions, oral drug physicochemistry, class II drugs, 49
Microfiltration, mammalian protein purification, harvest operations, 307
Microorganism depositories, drug patent requirements, 126–127
Miniaturization, high-throughput screening, enzymatic assays, 406
Modified dispersion model (MDM), permeability, ADMET interactions, fraction absorbed (f_a) vs., 369–370
Moieties, prodrug design, 250–252
Moisture uptake, oral drug physicochemistry, 37–39
Molecular biology, carboxylesterases, 234, 237–238
Monoamine oxidase inhibitors, prodrug development, 224–226
Monoclonal antibodies, mammalian protein purification, 306–309
Monophosphate salt I, cocrystal engineering, 212–213
Monovalent salts, high-throughput screening, assay optimization, 412
Mouse myeloma (NS0) cells, protein therapeutics, 293–294
mRNA, protein therapeutics, expression vectors, 294–295

Mucosal drug delivery, protein therapeutics, 315–316
Multiangle laser light scattering (MALLS), protein therapeutics, 323–324
Myocardial infarction, esmolol prodrug design, 260–268

Nanotechnology, drug delivery systems:
 amphiphilic polymers, 473–480
 basic principles, 469
 carbon nanotubes, 483–484
 dendrimers, 482–483
 polymer-drug conjugates, 480–481
 self-assembled low molecular weight amphiphiles, 469–473
 water-insoluble and crosslinked polymeric nanoparticles, 481
Nevirapine case study, large-scale synthesis, 14–23
 commercial production and process optimization, 20–23
 medicinal chemistry synthetic route, 15–16
 pilot-plant scale-up:
  chemical development, 16–17
  process development, 17–20
 research background, 14–15
New and nonobvious invention standard:
 patent requirements and, 133–138
 U.S.C. §103 provision, 134–138
New Animal Drug Application (NADA), regulatory guidelines, 97
New drug applications (NDAs):
 FDA guidelines, 78
 new drug definitions, 82
 over the counter drug approval, 83–84
 patent infringement and, 151–152
 submission requirements, 85, 88
New molecular entities (NMEs), prodrug development, 276–279
Noisomes, nanoscale drug delivery, 469–473

Nonhygroscopic properties, salt compounds, 381
Nonsolid drug forms, oral dosage systems, 29
North American Free Trade Agreement (NAFTA), patent interference limitations, 144–146
Nucleic acid testing, protein therapeutics, analytic development, 320–321

Obviousness standards, patent law and, 134–138
Octanol-water partition coefficients, oral drug delivery, physicochemistry evaluation, 34–35
Oil/water partition coefficient, oral drug physicochemistry, 31–32
Optimization in drug development:
 high-throughput screening, assay optimization, 409–414
 nevirapine case study, 21–23
Oral drug delivery:
 Biopharmaceutics Classification System, 41–43
 biowaiver monographs, 360–361
 global market, 353–363
 class I drugs, 42
 class II drugs, 42–50
 class III drugs, 50–52
 class IV drugs, 52, 355
 class V drugs, 52–55
 dosage form development strategies, 40–55
 nonsolid forms, 29
 permeability, ADMET interactions, oral bioavailability, 372
 physicochemical property evaluation, 29–40
  aqueous solubility, 32–33
  dissolution rate, 33
  hygroscopicity, 37–39
  ionization constant, 35–37
  melting point, 30–31
  partition coefficient, 31–32
  permeability, 33–35

rule of five, 34–35
stability, 39–40
prodrug properties, 220–226
protein therapeutics, 315–316
solid form selection criteria, 25–29
 amorphous solids, 29
 polymorphs, 26–27
 pseudopolymorphs, 27–29
 salts, 26
 solvates, 28–29
Orange Book (FDA), drug approval data, 92–93
Organ barriers, ADMET distribution properties, permeability and toxicity, 377
Orphan Drug Act, 67–68
Ostwald's step rule, polymorph crystallization, 190–192
Over the counter (OTC) drugs, FDA approval, 82–84
Oxidation, protein therapeutics formulation and delivery, 312–314
Oxidoreductases, high-throughput screening, enzymatic assays, 423–424

Packing polymorphism, crystal engineering, 191–192
Para-aminobenzoic acid (PABA), prodrug development, prontosil case study, 258
Paris Convention for the Protection of Industrial Property, 165–166
Particle size parameters, oral dosage systems, class II drugs, low solubility/high permeability, 44–45, 354
Particulate removal, protein therapeutics, downstream processing, 303–304
Partition coefficient, oral drug physicochemistry, 31–32
Passive diffusion, oral drug delivery, permeablity, 34–35

Patent and Trademark Office (PTO) (U.S.):
  copending applications, 109–110
  corrections of patents, 146–148
  intellectual property law, 103–105
  interference procedures, 142–146
  patent delay classifications, 113–114
  patent procedures at, 138–142
  patent prosecution strategy, 115 118
  patent protection, 106–107
  provisional applications to, 127–133
  trademark registration, 174–175
  trademark selection guidelines, 173–174
Patent Cooperation Treaty (PCT), 167–169
Patents, drug discovery and. See also Intellectual property law; specific types of patents, e.g., Utility patents
  applicants from non-U.S. countires, 133
  delays classifications, 113–114
  enforcement, 148–165
    alternative dispute resolution, 162–163
    claim construction proceedings, 162
    defenses to infringement, 154–157
    discovery phase, 160–161
    geographic scope, 148
    infringement determination, 152–154
    parties to suit, 150–152
    proceedings commencement, 159–160
    remedies for infringement, 157–159
    summary judgment, 161
    trial and appeal processes, 148–150, 162–164
  first to invent vs. first to file distinction, 103–104, 108–110
  generic drug approval, FDA term extension guidelines, 87

global patent protection, 107–108, 165–171
  international agreements, 165–168
  international laws and regulations, 169–171
  PCT patent practice, 168–169
historical background, 101–105
protection and strategy, 105–118
  absolute novelty principle, 110–111, 134
  application publication, 114–115
  first to invent vs. first to file, 108–110
  global strategy, 107–108, 165–171
  patent term, 111–114
  PTO prosecution strategy, 115–118
  trade secrets and, 182–183
  reexamination process, 107
requirements, 118–148
  best mode requirement, 126, 130–131
  claims, 126–127
  corrections, 146–148
  enablement requirement, 124–126
  interference, 142–146
  new and nonbovious inventions, 133–138
  patentable subject matter, international requirements, 121–123
  patentable subject matter, U. S. and international requirements, 118–123
  provisional applications, 127–133
  PTO procedures, 138–142
  specifications, 123–127
  statutory requirements, 129–130
  written description, 123–124
  time limits for, 106–107, 111–114
  trade secrets vs., 182–183
Pediatric Rule (FDA), 68
Pendant amphiphilic polymers, nanoscale drug delivery, 474–475, 479
Permeability:
  ADMET-PK drug interactions:

basic principles, 367
classification, 370–371
distribution, 375–376
excretion, 373–374
fraction absorption, 369–370
gastrointestinal absorption, solubility vs. permeability, 374–375
metabolism, 371–373
organ accumulation and toxicity, 377
transport interactions and polymorphism, 367–369
Biopharmaceutics Drug Disposition Classification System, 361–362
oral drug delivery:
  Biopharmaceutics Classification System, 41–42, 357–362
  class I drugs, 42, 354
  class II drugs, 42–50, 354
  class III drugs, 50–52, 355
  class IV drugs, 52, 355
  physicochemistry evaluation, 33–35
Permeability-Based Classification System (PCS), drug ADMET/PK interactions, 370–371
Permeation enhancers, oral drug delivery, class III drugs, 51
Personalized drug design, cost-effectiveness analysis, 349
Personnel training, commercial-scale operations, 12
P-glycoprotein (P-gp), permeability, ADMET interactions, transport interactions, 368–369
P-gp efflux mechanism, oral dosage forms, class V drugs, 54
Pharmaceutical equivalence, generic drug guidelines, 84–85
Pharmaceutical industry, cocrystal engineering applications, 193–196
Pharmacodynamics, CK-2130 prodrug case study, 272–273

Pharmacogenomics, cost-
effectiveness analysis
of drug development
and, 349
Pharmacokinetics:
permeability, ADMET
interactions:
basic principles, 367
classification, 370–371
distribution, 375–376
excretion, 373–374
fraction absorption, 369–370
gastrointestinal absorption,
solubility vs.
permeability, 374–375
metabolism, 371–373
organ accumulation and
toxicity, 377
transport interactions and
polymorphism,
367–369
serum albumin modulation,
457–464
clofibrates, 462–464
Pharmacology studies:
FDA guidelines, 75
prodrug design, esmolol case
study, 265–268
Phase 1 clinical trials, FDA
guidelines, 77
Phase 2 clinical trials, FDA
guidelines, 77
Phase 3 clinical trials, FDA
guidelines, 77
Phase 4 clinical trials, FDA
guidelines, 77–78
Phenolic hydroxyl groups,
prodrug moieties,
250–252
pH levels:
high-throughput screening,
assay optimization,
410–412
salt compounds, acid-base
equilibrium, 385–388
Phosphodiesterase inhibitors,
CK-2130 prodrug case
study, 268–276
Photodegradation, protein
therapeutics
formulation and
delivery, 313–314
Physicochemical property assays,
salt compounds, 382
Pilot plant scale-up, 7–10
agitation, 8–9
drying and solid handling, 9–10

heating and cooling, 8
liquid-solid separations, 9
nevirapine case study, 16–20
safety issues, 10
Pilot screening, high-throughput
screening, assay
configuration, 423
Piracetam, polymorphism, 192
Plant patents, protection and
strategy, 105–118
Podophyllotoxin derivatives,
serum albumin
binding, 454–456
Polarity, oral drug
physicochemistry,
34–35
Polishing steps, mammalian
protein purification,
307–309
Polymers, nanoscale drug
delivery:
amphiphilic molecules,
473–480
polymer-drug conjugates,
480–481
water-insoluble polymers and
cross-linked polymeric
nanoparticles, 481
Polymersomes, nanoscale drug
delivery, 476–477
Polymorphic crystals:
case studies, 196–198
cocrystal design, 196
crystal engineering:
defined, 188
evolution of, 190–192
oral dosage systems, 26–27
Postapproval process:
cost-effectiveness analysis,
349–350
FDA guidelines, 96
Potency assays, protein
therapeutics, analytic
development,
321–322
Power x-ray diffraction (PXRD),
salt compound
selection:
automated assisted screening,
396–397
crystallinity, 381
manual screening, 390–395
selection process, 396
Preapproval inspections, FDA
requirements, 91
Precipitation, oral drug
physicochemistry,

class II drugs,
inhibition, 45
Preclinical testing, FDA
regulations, 71–72
Pre-IND phase, FDA regulations,
71
Premafloxacin, polymorph
design, 196–198
Prescription Drug User Fee Act
(PDUFA), 67–68
President's Emergency Plan for
AIDS, 69
Pressure requirements,
commercial-scale
operations, 11
Presystemic elimination, oral
dosage forms, class V
drugs, 53
Prevalence, prodrug marketing,
241–242
Preventive maintenance,
commercial-scale
operations, 12
Prima facie obviousness, patent
specification and,
135–138
Privileged structures, prodrug
designesmolol case
study, 267–268
Process analytical technology
(PAT), commercial-
scale operations, 14
Process chromatography, protein
therapeutics,
downstream
processing, 303–304
Process claims, patent
obviousness standards
and, 138
Process flow diagramas (PFDs),
commercial-scale
operations, 11
Processing systems and
equipment:
commercial-scale operations,
11
nevirapine case study, 17–20
oral drug systems, class V drug
design, 54–55
protein therapeutics, 289–291
comparability, 325–326
reagent clearance and
excipient testing, 326
Process intensification, protein
therapeutics, current
and future
applications, 309–310

Process validation, commercial-scale operations, 12
Prodrugs:
  carboxylesterases, 233–241
    classification, 233–234
    function and distribution, 234
    molecular biology, 234, 237–238
    preclinical assays, 239, 241
    substrates and inhibitors, 235–240
  case studies, 257–276
    CK-2130, 268–276
    esmolol, 258–268
    prontosil, 257–258
  defined, 219–220
  design principles, 229–233
    ADMET issues, 230
    bioconversion mechanisms, 252–254
    drug targeting, 229
    metabolic issues, 230–233
    moieties, 250–252
    "rule-of-one" metabolism, 248–250
    structure-metabolism relationships, 265–256
  drug discovery paradigms, 226–229
  early examples and applications, 220–226
  future trends, 276
  marketed prodrugs, 241–248
    examples, 242, 244–248
    FDA approvals, 242–243
    prevalence, 241–242
  oral drug delivery, class III drugs, 50–51
  pharmacologically active products, 276–279
Product release testing, protein therapeutics, 318–320
Prokaryotic expression:
  high-throughput screening, enzymatic assays, 402–406
  protein therapeutics, upstream processing, 291–294
Prontosil, prodrug development, 257–258
"Prosecution history estoppel," infringement of patents, 154
Protein A affinity chromatography, protein therapeutics, 308–309
  mimetics research, 310
Protein folding, protein purification, *E. coli* systems, 305–306
Protein therapeutics:
  analytical methods, 317–327
    bioactivity assays, 321–322
    biophysical characterization, 323–325
    electrophoresis, 320
    future trends, 326–327
    immunoassays, 321
    in-process/product release testing, 318–320
    mass spectrometry, 322–323
    nucleic acid testing, 320–321
  downstream processing, 301–311
    classifications, 303–304
    design criteria, 303
    E. coli protein purification, 304–306
    future trends, 309–311
    impurities and contaminants, 301–303
    mammalian protein purification, 306–309
  formulation and delivery, 311–317
    alternative modes, 315–316
    container materials interactions, 315
    degradation pathways, 311–314
    future trends, 316–317
    stability parameters, 314–315
  process development, 289–291
    comparability, 325–326
    reagent clearance and excipient testing, 326
    research background, 289
    upstream processing, 291–301
      cell culture, 298–300
      expression plurality, 291–294
      formats, 296–298
      future trends, 300–301
      recombinant cell lines, 294–296
Provisional patent applications, requirements for, 127–133
Pseudopolymorphs, oral dosage systems, 27–29
Purification process:
  high-throughput screening, enzymatic assays, 402–406
  protein therapeutics:
    downstream processing, 301–311
    E. coli systems, 304–306
    mammalian cell lines, 306–309
Pyroglutamate formation, protein therapeutics formulation and delivery, 313–314

Quality-adjusted life years (QALY), cost-effectiveness analysis of drug development, 345
Quality by design (QbD), protein therapeutics, 327
Quantitative polymerase chain reaction (QPCR), protein therapeutics, analytic development, 321

Raman spectroscopy, salt compound selection, automated assisted screening, 396
"Random walk" profile:
  prodrug development, rule-of-one metabolism, 248–250
  prodrug properties, 220–226
Ranitidine, polymorph crystallization, 190–192
Reaction solvents, bench-scale experiments, 5–6
Reaction temperature, bench-scale experiments, 6
Reaction time, bench-scale experiments, 6
Reagent development, protein therapeutics, clearance process, 326
Recombinant DNA technology, protein therapeutics, upstream processing, 294–295
Red blood cells (RBCs), permeability, ADMET

interactions, distribution, 376
Reduction to practice principle, patent protection, 109–110
Refold concentrations, protein therapeutics, 310
Regulatory issues. *See also* specific drug legislation, e.g., Orphan Drug Act
  animal dugs, 97
  Biopharmaceutics Classification System, 355–356
  cost-effectiveness analysis of drug development, 345–346
  FDA role in:
    biotechnology-derived drugs, 93–96
    generic drug review and approval, 84–92
    historical background, 63–69
    new drug approval process, 71–82
    Orange Book, 92–93
    organizational structure, 70
    over the counter drug approval, 82–84
    postapproval process, 96
    review process, 70–71
    statutes, 69–70
    United States Pharmacopoeia and National Formulary, 96–97
  International Conference on Harmonization, 97
Remedies in infringement of patent claims, 157–159
Renal system, permeability, ADMET interactions:
  excretion, 373
  metabolism, 372–373
Request for continued examination (RCE), drug patent delays, 114
Residence times, oral dosage forms, class III bioadhesives, 52
Reverse phase high-performance liquid chromatography (RP-HPLC), protein therapeutics, analytic development, 319–320
Reversible hydrolysis, camptothecin lactone, 452–453
Roller bottle cell cultures, protein therapeutics, 297
Route selection criteria, scale-up operations, 3–4
ROY pharmaceutical intermediate, polymorphism, 191–192
Rule of five (Lipinski), oral drug physicochemistry, permeability, 34–35
Rule-of-one metabolism, prodrug design, 248–250

Safe harbor provisions, infringement of patent claims, 156–157
Safety issues:
  commercial-scale operations, 13
  pilot plant development, 10
Salt compounds:
  crystal engineering:
    defined, 188–189
    monophosphate salt I cocrystals, 212–213
  high-throughput screening, assay optimization, monovalent/divalent salts, 412
  oral dosage systems, 26, 28
  class II drugs, low solubility/high permeability, 42, 44, 354
  screening and selection:
    acid-base equilibrium, 384–388
    active pharmaceutical ingredients, 388–396
    automated assisted screening, 395–396
    basic properties, 381–382
    counterion selection, 389–390
    crystallization, 383–384
    limitations, 382–383
    manual process, 391–395
    physicochemical properties, 383
    selection technology, 396–397
    solvent selection, 389–391
Scaffold structures, prodrug design, esmolol case study, 263–268
Scale-up operations, large-scale synthesis, 1–14
  basic principles, 1
  bench-scale experiments, 4–7
    reaction solvent selection, 5–6
    reaction temperature, 6
    reaction times, 6
    solid-state requirements, 7
    stoichiometry and addition order, 6–7
  chiral requirements, 4
  commercial-scale operations, 10–13
    chemical safety, 13
    environmental controls, 13
    in-process controls, 11
    processing equipment requirements, 11
    validation, 12–13
  continuous microfluidic reactors, 13–14
  pilot plant scale-up, 7–10
    agitation, 8–9
    drying and solid handling, 9–10
    heating and cooling, 8
    liquid-solid separations, 9
    safety issues, 10
  process analytical technology, 14
  process trends and technologies, 13–14
  route selection, 3–4
  synthetic strategies, 1–4
Seagate doctrine, infringement of patent claims, 158–159
Secondary meaning principle, trademark selection guidelines, 173–174
Secondary protein structure, protein therapeutics, biophysical assessment, 324–325
Self-assembled low molecular weight amphiphiles, drug delivery systems, 469–473
  polymerization, 477
Self-emulsifying drug delivery systems (SEDDS), oral drug physicochemistry, class II drugs, 49–50

Self-micro-emulsifying drug delivery system (SMEDDS), oral drug physicochemistry, class II drugs, 49–50
Sensitivity, high-throughput screening, enzymatic assays, 406
Sequential multicolumn chromatography (SMCC), protein therapeutics, 310–311
Serum albumin:
  chemical properties, 437–438
  therapeutic applications, 456–464
    clofibrate pharmacokinetic modulation, 457–464
    pharmacokinetic studies, 462–464
    in vitro studies, 458–462
  X-ray structural analysis, 438–456
    anthracyclines, 452–454
    bilirubin binding, 442–451
    CADEX categories, 439–440
    camptothecins, 451–453
    drug-binding sites, 440–442
    podophyllotoxin derivatives, 454–456
Sildenafil, cocrystal engineering, 211
Simulated moving bed (SMB) techniques, protein therapeutics, 310–311
Single nucleotide polymorphisms (SNPs), patent requirements, 120–123
Size-based nanometer filtration, mammalian protein purification, 309
Size exclusion chromatography (SEC), protein therapeutics, 319
Small-molecule compounds, protein therapeutics and, 290–291
Sodium channel blockers, cocrystal engineering, 210
Soft drugs (SDs), prodrug developmenet, esmolol case study, 258–268
Solid drug forms:
  handling systems, pilot plant development, 9–10

oral dosage systems, 25–29
  amorphous solids, 29
  dispersion mechanisms, class II drugs, 46
  polymorphs, 26–27
  pseudopolymorphs, 27–29
  salts, 26
  solvates, 28–29
Solid-state requirements, bench-scale experiments, 7
Solid state stability, oral drug physicochemistry, 40
Solubility:
  Biopharmaceutics Classification System, oral drug delivery, 41–52, 355, 357–362
  oral drug delivery:
    aqueous solubility, 32–33
    class I drugs, 42, 354
    class II drugs, 42–50, 354
    class III drugs, 50–52, 355
    class IV drugs, 52, 355
  permeability, ADMET interactions, 374–375
  salt compounds, acid-base equilibrium, pH solubility, 384–388
Solution recrystallization, salt compound selection, 391–395
Solutions, oral drug physicochemistry:
  class II drugs, lipid solutions, 48
  stability, 40
Solvates:
  cocrystal design, 196
  crystal engineering, defined, 189
  oral dosage systems, 28–29
Solvent selection:
  bench-scale experiments, 5–6
  high-throughput screening, assay configuration, 421–422
  salt compound screening, 389–393
Sponsorship, FDA definition, 71
Stability:
  oral drug physicochemistry, 39–40
  protein therapeutics formulation and delivery, 314–315
Stabilizers, high-throughput screening, assay optimization, 412–413

Static light scattering (SLS), protein therapeutics, 323–324
Statutory invention registration (SIR), drug discovery and, 185–186
Stay provisions, generic drug approval, FDA guidelines, 86
Steric hindrance, prodrug design, 254–255
  esmolol case study, 265–268
Stirred-tank reactor (STRs), protein therapeutics, suspension cell cultures, 298
Stoichiometry, bench-scale experiments, 6–7
Structure-activity relationship (SAR):
  prodrug moieties, 251–252
  esmolol case study, 262–268
  serum albumin, 438–456
Structure-metabolism relationships (SMRs), prodrug design, 254–256
Substantial evidence standard, infringement of patent proceedings, 164
Substrates:
  carboxylesterases, 235–239
  high-throughput screening, assay configuration, 414–420
Sulfonilamide, prodrug development, prontosil case study, 258
Summary judgment, infringement of patents, 161
Supersaturation, class II drugs, precipitation inhibition, 45
Supramolecular heterosynthons:
  cocrystal design, 194–196
  evolution of, 189–190
Supramolecular homosynthons:
  cocrystal design, 195–196
  evolution of, 189–190
Supramolecular synthons:
  cocrystal design, 194–197
  crystal engineering, evolution of, 189–190
  polymorphism, 191–192

Surfactants:
  nanoscale drug delivery, noisomes, 473
  oral drug physicochemistry, class II drugs, micelle solubilization, 48–50
Suspensions:
  oral drug physicochemistry, class II drugs, 48
  protein therapeutics, cell cultures, 298
Sweden, patent requirements, 122

Tangential flow microfiltration, mammalian protein purification, harvest operations, 307
Targeted drug development:
  high-throughput screening, enzymatic assays, 401–402
  prodrug design, 229
    esmolol case study, 260–268
"Teaching, suggestion, or motivation" (TSM) test, patent obviousness principle and, 135–138
Temperature requirements:
  commercial-scale operations, 11
  high-throughput screening, assay optimization, 413–414
Tertiary protein structure, protein therapeutics, biophysical assessment, 324–325
Thermal behavior, salt compounds, 381
Toxicity:
  ADMET properties, permeability, ADMET/PK interactions, 377
  CK-2130 prodrug case study, 273–276
  FDA information guidelines, 75
  prodrug development, 225–226
Trademarks, drug discovery and biotechnology, 171–179
  global rights, 177–178
  Lanham Act protections, 178–179
  as marketing tools, 172
  oppositions and cancellations, 176–177
  preservation of rights through proper use, 177
  registration process, 174–176
  selection criteria, 172–174
Trade secrets, drug discovery and biotechnology, 179–184
  defined, 180
  enforcement, 181–182
  Freedom of Information Acts, 183–184
  global protection, 184
  patent protection and, 182–183
  protection requirements, 180–181
Transdermal drug delivery:
  liposomes, nanoscale technology, 472–473
  protein therapeutics, 316
Transfection, protein therapeutics, 295–296
Transferases, high-throughput screening, enzymatic assays, 424–425
Transgenic mouse models, protein therapeutics, upstream processing, 291–294
Translational medicine, prodrug development, prontosil case study, 258
Transport proteins, permeability, ADMET interactions, 367–369

United Kingdom, patent requirements, 122
United States Pharmacopoeia and National Formulary, 96–97
Upstream processing:
  defined, 289
  protein therapeutics, 291–301
    cell culture, 298–300
    expression plurality, 291–294
    formats, 296–298
    future trends, 300–301
    reagent clearance and excipient testing, 326
    recombinant cell lines, 294–296
U.S. Courts, patent trials in, 148–150
Utility patents:
  protection and strategy, 105–118
  requirements, 120–121

Validation:
  commercial-scale operations, 12–13
  infringement of patent claims, defense based on, 155–156
Vent treatment requirements, commercial-scale operations, 11
Vesicle polymerization, nanoscale drug delivery, 477, 479–480
Vinylogous cyclic carbonates, prodrug development, 246–247
Virus-directed enzyme prodrug therapy (VDEPT), prodrug design, 229
Viruses, mammalian protein purification, 308–309

Water-insoluble polymers, nanoscale drug delivery, 481
"Willful infringement" damages, infringement of patent claims, 158–159
Withdrawal of approval, FDA requirements, 92
Working agreements, international patent laws, 169–171
Working cell bank (WCB), protein therapeutics, 296
World Trade Organization (WTO), patent interference limitations, 144–146

Yalkowsky-Valvani equation, oral drug physicochemistry, aqueous solubility, 33
Yeast systems, protein therapeutics, 292–294

Zwitterions, salt compounds, acid-base equilibrium, 385, 387